T0215627

Lecture Notes in Computer Science 11307

Commenced Publication in 1973
Founding and Former Series Editors:
Gerhard Goos, Juris Hartmanis, and Jan van Leeuwen

More information about this series at http://www.springer.com/series/7407

Long Cheng · Andrew Chi Sing Leung
Seiichi Ozawa (Eds.)

Neural
Information Processing

25th International Conference, ICONIP 2018
Siem Reap, Cambodia, December 13–16, 2018
Proceedings, Part VII

 Springer

Editors
Long Cheng 🔟
The Chinese Academy of Sciences
Beijing, China

Seiichi Ozawa
Kobe University
Kobe, Japan

Andrew Chi Sing Leung
City University of Hong Kong
Kowloon, Hong Kong SAR, China

ISSN 0302-9743 ISSN 1611-3349 (electronic)
Lecture Notes in Computer Science
ISBN 978-3-030-04238-7 ISBN 978-3-030-04239-4 (eBook)
https://doi.org/10.1007/978-3-030-04239-4

Library of Congress Control Number: 2018960916

LNCS Sublibrary: SL1 – Theoretical Computer Science and General Issues

This Springer imprint is published by the registered company Springer Nature Switzerland AG
The registered company address is: Gewerbestrasse 11, 6330 Cham, Switzerland

Preface

The 25th International Conference on Neural Information Processing (ICONIP 2018), the annual conference of the Asia Pacific Neural Network Society (APNNS), was held in Siem Reap, Cambodia, during December 13–16, 2018. The ICONIP conference series started in 1994 in Seoul, which has now become a well-established and high-quality conference on neural networks around the world. Siem Reap is a gateway to Angkor Wat, which is one of the most important archaeological sites in Southeast Asia, the largest religious monument in the world. All participants of ICONIP 2018 had a technically rewarding experience as well as a memorable stay in this great city.

In recent years, the neural network has been significantly advanced with the great developments in neuroscience, computer science, cognitive science, and engineering. Many novel neural information processing techniques have been proposed as the solutions to complex, networked, and information-rich intelligent systems. To disseminate new findings, ICONIP 2018 provided a high-level international forum for scientists, engineers, and educators to present the state of the art of research and applications in all fields regarding neural networks.

With the growing popularity of neural networks in recent years, we have witnessed an increase in the number of submissions and in the quality of submissions. ICONIP 2018 received 575 submissions from 51 countries and regions across six continents. Based on a rigorous peer-review process, where each submission was reviewed by at least three experts, a total of 401 high-quality papers were selected for publication in the prestigious Springer series of *Lecture Notes in Computer Science*. The selected papers cover a wide range of subjects that address the emerging topics of theoretical research, empirical studies, and applications of neural information processing techniques across different domains.

In addition to the contributed papers, the ICONIP 2018 technical program also featured three plenary talks and two invited talks delivered by world-renowned scholars: Prof. Masashi Sugiyama (University of Tokyo and RIKEN Center for Advanced Intelligence Project), Prof. Marios M. Polycarpou (University of Cyprus), Prof. Qing-Long Han (Swinburne University of Technology), Prof. Cesare Alippi (Polytechnic of Milan), and Nikola K. Kasabov (Auckland University of Technology).

We would like to extend our sincere gratitude to all members of the ICONIP 2018 Advisory Committee for their support, the APNNS Governing Board for their guidance, the International Neural Network Society and Japanese Neural Network Society for their technical co-sponsorship, and all members of the Organizing Committee for all their great effort and time in organizing such an event. We would also like to take this opportunity to thank all the Technical Program Committee members and reviewers for their professional reviews that guaranteed the high quality of the conference proceedings. Furthermore, we would like to thank the publisher, Springer, for their sponsorship and cooperation in publishing the conference proceedings in seven volumes of *Lecture Notes in Computer Science*. Finally, we would like to thank all the speakers, authors, reviewers, volunteers, and participants for their contribution and support in making ICONIP 2018 a successful event.

October 2018

Jun Wang
Long Cheng
Andrew Chi Sing Leung
Seiichi Ozawa

ICONIP 2018 Organization

General Chair

Jun Wang   City University of Hong Kong,
Hong Kong SAR, China

Advisory Chairs

Akira Hirose   University of Tokyo, Tokyo, Japan
Soo-Young Lee   Korea Advanced Institute of Science and Technology,
South Korea
Derong Liu   Institute of Automation, Chinese Academy of Sciences,
China
Nikhil R. Pal   Indian Statistics Institute, India

Program Chairs

Long Cheng   Institute of Automation, Chinese Academy of Sciences,
China
Andrew C. S. Leung   City University of Hong Kong, Hong Kong SAR,
China
Seiichi Ozawa   Kobe University, Japan

Special Sessions Chairs

Shukai Duan   Southwest University, China
Kazushi Ikeda   Nara Institute of Science and Technology, Japan
Qinglai Wei   Institute of Automation, Chinese Academy of Sciences,
China
Hiroshi Yamakawa   Dwango Co. Ltd., Japan
Zhihui Zhan   South China University of Technology, China

Tutorial Chairs

Hiroaki Gomi   NTT Communication Science Laboratories, Japan
Takashi Morie   Kyushu Institute of Technology, Japan
Kay Chen Tan   City University of Hong Kong, Hong Kong SAR,
China
Dongbin Zhao   Institute of Automation, Chinese Academy of Sciences,
China

Publicity Chairs

Zeng-Guang Hou	Institute of Automation, Chinese Academy of Sciences, China
Tingwen Huang	Texas A&M University at Qatar, Qatar
Chia-Feng Juang	National Chung-Hsing University, Taiwan
Tomohiro Shibata	Kyushu Institute of Technology, Japan

Publication Chairs

Xinyi Le	Shanghai Jiao Tong University, China
Sitian Qin	Harbin Institute of Technology Weihai, China
Zheng Yan	University Technology Sydney, Australia
Shaofu Yang	Southeast University, China

Registration Chairs

Shenshen Gu	Shanghai University, China
Qingshan Liu	Southeast University, China
Ka Chun Wong	City University of Hong Kong, Hong Kong SAR, China

Conference Secretariat

Ying Qu	Dalian University of Technology, China

Program Committee

Hussein Abbass	University of New South Wales at Canberra, Australia
Choon Ki Ahn	Korea University, South Korea
Igor Aizenberg	Texas A&M University at Texarkana, USA
Shotaro Akaho	National Institute of Advanced Industrial Science and Technology, Japan
Abdulrazak Alhababi	UNIMAS, Malaysia
Cecilio Angulo	Universitat Politècnica de Catalunya, Spain
Sabri Arik	Istanbul University, Turkey
Mubasher Baig	National University of Computer and Emerging Sciences Lahore, India
Sang-Woo Ban	Dongguk University, South Korea
Tao Ban	National Institute of Information and Communications Technology, Japan
Boris Bačić	Auckland University of Technology, New Zealand
Xu Bin	Northwestern Polytechnical University, China
David Bong	Universiti Malaysia Sarawak, Malaysia
Salim Bouzerdoum	University of Wollongong, Australia
Ivo Bukovsky	Czech Technical University, Czech Republic

Ke-Cai Cao	Nanjing University of Posts and Telecommunications, China
Elisa Capecci	Auckland University of Technology, New Zealand
Rapeeporn Chamchong	Mahasarakham University, Thailand
Jonathan Chan	King Mongkut's University of Technology Thonburi, Thailand
Rosa Chan	City University of Hong Kong, Hong Kong SAR, China
Guoqing Chao	East China Normal University, China
He Chen	Nankai University, China
Mou Chen	Nanjing University of Aeronautics and Astronautics, China
Qiong Chen	South China University of Technology, China
Wei-Neng Chen	Sun Yat-Sen University, China
Xiaofeng Chen	Chongqing Jiaotong University, China
Ziran Chen	Bohai University, China
Jian Cheng	Chinese Academy of Sciences, China
Long Cheng	Chinese Academy of Sciences, China
Wu Chengwei	Bohai University, China
Zheru Chi	The Hong Kong Polytechnic University, SAR China
Sung-Bae Cho	Yonsei University, South Korea
Heeyoul Choi	Handong Global University, South Korea
Hyunsoek Choi	Kyungpook National University, South Korea
Supannada Chotipant	King Mongkut's Institute of Technology Ladkrabang, Thailand
Fengyu Cong	Dalian University of Technology, China
Jose Alfredo Ferreira Costa	Federal University of Rio Grande do Norte, Brazil
Ruxandra Liana Costea	Polytechnic University of Bucharest, Romania
Jean-Francois Couchot	University of Franche-Comté, France
Raphaël Couturier	University of Bourgogne Franche-Comté, France
Jisheng Dai	Jiangsu University, China
Justin Dauwels	Massachusetts Institute of Technology, USA
Dehua Zhang	Chinese Academy of Sciences, China
Mingcong Deng	Tokyo University of Agriculture and Technology, Japan
Zhaohong Deng	Jiangnan University, China
Jing Dong	Chinese Academy of Sciences, China
Qiulei Dong	Chinese Academy of Sciences, China
Kenji Doya	Okinawa Institute of Science and Technology, Japan
El-Sayed El-Alfy	King Fahd University of Petroleum and Minerals, Saudi Arabia
Mark Elshaw	Nottingham Trent International College, UK
Peter Erdi	Kalamazoo College, USA
Josafath Israel Espinosa Ramos	Auckland University of Technology, New Zealand
Issam Falih	Paris 13 University, France

Bo Fan	Zhejiang University, China
Yunsheng Fan	Dalian Maritime University, China
Hao Fang	Beijing Institute of Technology, China
Jinchao Feng	Beijing University of Technology, China
Francesco Ferracuti	Università Politecnica delle Marche, Italy
Chun Che Fung	Murdoch University, Australia
Wai-Keung Fung	Robert Gordon University, UK
Tetsuo Furukawa	Kyushu Institute of Technology, Japan
Hao Gao	Nanjing University of Posts and Telecommunications, China
Yabin Gao	Harbin Institute of Technology, China
Yongsheng Gao	Griffith University, Australia
Tom Gedeon	Australian National University, Australia
Ong Sing Goh	Universiti Teknikal Malaysia Melaka, Malaysia
Iqbal Gondal	Federation University Australia, Australia
Yue-Jiao Gong	Sun Yat-sen University, China
Shenshen Gu	Shanghai University, China
Chengan Guo	Dalian University of Technology, China
Ping Guo	Beijing Normal University, China
Shanqing Guo	Shandong University, China
Xiang-Gui Guo	University of Science and Technology Beijing, China
Zhishan Guo	University of Central Florida, USA
Christophe Guyeux	University of Franche-Comte, France
Masafumi Hagiwara	Keio University, Japan
Saman Halgamuge	The University of Melbourne, Australia
Tomoki Hamagami	Yokohama National University, Japan
Cheol Han	Korea University at Sejong, South Korea
Min Han	Dalian University of Technology, China
Takako Hashimoto	Chiba University of Commerce, Japan
Toshiharu Hatanaka	Osaka University, Japan
Wei He	University of Science and Technology Beijing, China
Xing He	Southwest University, China
Xiuyu He	University of Science and Technology Beijing, China
Akira Hirose	The University of Tokyo, Japan
Daniel Ho	City University of Hong Kong, Hong Kong SAR, China
Katsuhiro Honda	Osaka Prefecture University, Japan
Hongyi Li	Bohai University, China
Kazuhiro Hotta	Meijo University, Japan
Jin Hu	Chongqing Jiaotong University, China
Jinglu Hu	Waseda University, Japan
Xiaofang Hu	Southwest University, China
Xiaolin Hu	Tsinghua University, China
He Huang	Soochow University, China
Kaizhu Huang	Xi'an Jiaotong-Liverpool University, China
Long-Ting Huang	Wuhan University of Technology, China

Panfeng Huang	Northwestern Polytechnical University, China
Tingwen Huang	Texas A&M University, USA
Hitoshi Iima	Kyoto Institute of Technology, Japan
Kazushi Ikeda	Nara Institute of Science and Technology, Japan
Hayashi Isao	Kansai University, Japan
Teijiro Isokawa	University of Hyogo, Japan
Piyasak Jeatrakul	Mae Fah Luang University, Thailand
Jin-Tsong Jeng	National Formosa University, Taiwan
Sungmoon Jeong	Kyungpook National University Hospital, South Korea
Danchi Jiang	University of Tasmania, Australia
Min Jiang	Xiamen University, China
Yizhang Jiang	Jiangnan University, China
Xuguo Jiao	Zhejiang University, China
Keisuke Kameyama	University of Tsukuba, Japan
Shunshoku Kanae	Junshin Gakuen University, Japan
Hamid Reza Karimi	Politecnico di Milano, Italy
Nikola Kasabov	Auckland University of Technology, New Zealand
Abbas Khosravi	Deakin University, Australia
Rhee Man Kil	Sungkyunkwan University, South Korea
Daeeun Kim	Yonsei University, South Korea
Sangwook Kim	Kobe University, Japan
Lai Kin	Tunku Abdul Rahman University, Malaysia
Irwin King	The Chinese University of Hong Kong, Hong Kong SAR, China
Yasuharu Koike	Tokyo Institute of Technology, Japan
Ven Jyn Kok	National University of Malaysia, Malaysia
Ghosh Kuntal	Indian Statistical Institute, India
Shuichi Kurogi	Kyushu Institute of Technology, Japan
Susumu Kuroyanagi	Nagoya Institute of Technology, Japan
James Kwok	The Hong Kong University of Science and Technology, SAR China
Edmund Lai	Auckland University of Technology, New Zealand
Kittichai Lavangnananda	King Mongkut's University of Technology Thonburi, Thailand
Xinyi Le	Shanghai Jiao Tong University, China
Minho Lee	Kyungpook National University, South Korea
Nung Kion Lee	University Malaysia Sarawak, Malaysia
Andrew C. S. Leung	City University of Hong Kong, Hong Kong SAR, China
Baoquan Li	Tianjin Polytechnic University, China
Chengdong Li	Shandong Jianzhu University, China
Chuandong Li	Southwest University, China
Dazi Li	Beijing University of Chemical Technology, China
Li Li	Tsinghua University, China
Shengquan Li	Yangzhou University, China

Ya Li	Institute of Automation, Chinese Academy of Sciences, China
Yanan Li	University of Sussex, UK
Yongming Li	Liaoning University of Technology, China
Yuankai Li	University of Science and Technology of China, China
Jie Lian	Dalian University of Technology, China
Hualou Liang	Drexel University, USA
Jinling Liang	Southeast University, China
Xiao Liang	Nankai University, China
Alan Wee-Chung Liew	Griffith University, Australia
Honghai Liu	University of Portsmouth, UK
Huaping Liu	Tsinghua University, China
Huawen Liu	University of Texas at San Antonio, USA
Jing Liu	Chinese Academy of Sciences, China
Ju Liu	Shandong University, China
Qingshan Liu	Huazhong University of Science and Technology, China
Weifeng Liu	China University of Petroleum, China
Weiqiang Liu	Nanjing University of Aeronautics and Astronautics, China
Dome Lohpetch	King Mongkut's University of Technology North Bangoko, Thailand
Hongtao Lu	Shanghai Jiao Tong University, China
Wenlian Lu	Fudan University, China
Yao Lu	Beijing Institute of Technology, China
Jinwen Ma	Peking University, China
Qianli Ma	South China University of Technology, China
Sanparith Marukatat	Thailand's National Electronics and Computer Technology Center, Thailand
Tomasz Maszczyk	Nanyang Technological University, Singapore
Basarab Matei	LIPN Paris Nord University, France
Takashi Matsubara	Kobe University, Japan
Nobuyuki Matsui	University of Hyogo, Japan
P. Meesad	King Mongkut's University of Technology North Bangkok, Thailand
Gaofeng Meng	Chinese Academy of Sciences, China
Daisuke Miyamoto	University of Tokyo, Japan
Kazuteru Miyazaki	National Institution for Academic Degrees and Quality Enhancement of Higher Education, Japan
Seiji Miyoshi	Kansai University, Japan
J. Manuel Moreno	Universitat Politècnica de Catalunya, Spain
Naoki Mori	Osaka Prefecture University, Japan
Yoshitaka Morimura	Kyoto University, Japan
Chaoxu Mu	Tianjin University, China
Kazuyuki Murase	University of Fukui, Japan
Jun Nishii	Yamaguchi University, Japan

Haruhiko Nishimura	University of Hyogo, Japan
Grozavu Nistor	Paris 13 University, France
Yamaguchi Nobuhiko	Saga University, Japan
Stavros Ntalampiras	University of Milan, Italy
Takashi Omori	Tamagawa University, Japan
Toshiaki Omori	Kobe University, Japan
Seiichi Ozawa	Kobe University, Japan
Yingnan Pan	Northeastern University, China
Yunpeng Pan	JD Research Labs, China
Lie Meng Pang	Universiti Malaysia Sarawak, Malaysia
Shaoning Pang	Unitec Institute of Technology, New Zealand
Hyeyoung Park	Kyungpook National University, South Korea
Hyung-Min Park	Sogang University, South Korea
Seong-Bae Park	Kyungpook National University, South Korea
Kitsuchart Pasupa	King Mongkut's Institute of Technology Ladkrabang, Thailand
Yong Peng	Hangzhou Dianzi University, China
Somnuk Phon-Amnuaisuk	Universiti Teknologi Brunei, Brunei
Lukas Pichl	International Christian University, Japan
Geong Sen Poh	National University of Singapore, Singapore
Mahardhika Pratama	Nanyang Technological University, Singapore
Emanuele Principi	Università Politecnica elle Marche, Italy
Dianwei Qian	North China Electric Power University, China
Jiahu Qin	University of Science and Technology of China, China
Sitian Qin	Harbin Institute of Technology at Weihai, China
Mallipeddi Rammohan	Nanyang Technological University, Singapore
Yazhou Ren	University of Science and Technology of China, China
Ko Sakai	University of Tsukuba, Japan
Shunji Satoh	The University of Electro-Communications, Japan
Gerald Schaefer	Loughborough University, UK
Sachin Sen	Unitec Institute of Technology, New Zealand
Hamid Sharifzadeh	Unitec Institute of Technology, New Zealand
Nabin Sharma	University of Technology Sydney, Australia
Yin Sheng	Huazhong University of Science and Technology, China
Jin Shi	Nanjing University, China
Yuhui Shi	Southern University of Science and Technology, China
Hayaru Shouno	The University of Electro-Communications, Japan
Ferdous Sohel	Murdoch University, Australia
Jungsuk Song	Korea Institute of Science and Technology Information, South Korea
Andreas Stafylopatis	National Technical University of Athens, Greece
Jérémie Sublime	ISEP, France
Ponnuthurai Suganthan	Nanyang Technological University, Singapore
Fuchun Sun	Tsinghua University, China
Ning Sun	Nankai University, China

Norikazu Takahashi	Okayama University, Japan
Ken Takiyama	Tokyo University of Agriculture and Technology, Japan
Tomoya Tamei	Kobe University, Japan
Hakaru Tamukoh	Kyushu Institute of Technology, Japan
Choo Jun Tan	Wawasan Open University, Malaysia
Shing Chiang Tan	Multimedia University, Malaysia
Ying Tan	Peking University, China
Gouhei Tanaka	The University of Tokyo, Japan
Ke Tang	Southern University of Science and Technology, China
Xiao-Yu Tang	Zhejiang University, China
Yang Tang	East China University of Science and Technology, China
Qing Tao	Chinese Academy of Sciences, China
Katsumi Tateno	Kyushu Institute of Technology, Japan
Keiji Tatsumi	Osaka University, Japan
Kai Meng Tay	Universiti Malaysia Sarawak, Malaysia
Chee Siong Teh	Universiti Malaysia Sarawak, Malaysia
Andrew Teoh	Yonsei University, South Korea
Arit Thammano	King Mongkut's Institute of Technology Ladkrabang, Thailand
Christos Tjortjis	International Hellenic University, Greece
Shibata Tomohiro	Kyushu Institute of Technology, Japan
Seiki Ubukata	Osaka Prefecture University, Japan
Eiji Uchino	Yamaguchi University, Japan
Wataru Uemura	Ryukoku University, Japan
Michel Verleysen	Universite catholique de Louvain, Belgium
Brijesh Verma	Central Queensland University, Australia
Hiroaki Wagatsuma	Kyushu Institute of Technology, Japan
Nobuhiko Wagatsuma	Tokyo Denki University, Japan
Feng Wan	University of Macau, SAR China
Bin Wang	University of Jinan, China
Dianhui Wang	La Trobe University, Australia
Jing Wang	Beijing University of Chemical Technology, China
Jun-Wei Wang	University of Science and Technology Beijing, China
Junmin Wang	Beijing Institute of Technology, China
Lei Wang	Beihang University, China
Lidan Wang	Southwest University, China
Lipo Wang	Nanyang Technological University, Singapore
Qiu-Feng Wang	Xi'an Jiaotong-Liverpool University, China
Sheng Wang	Henan University, China
Bunthit Watanapa	King Mongkut's University of Technology, Thailand
Saowaluk Watanapa	Thammasat University, Thailand
Qinglai Wei	Chinese Academy of Sciences, China
Wei Wei	Beijing Technology and Business University, China
Yantao Wei	Central China Normal University, China

Guanghui Wen	Southeast University, China
Zhengqi Wen	Chinese Academy of Sciences, China
Hau San Wong	City University of Hong Kong, Hong Kong SAR, China
Kevin Wong	Murdoch University, Australia
P. K. Wong	University of Macau, SAR China
Kuntpong Woraratpanya	King Mongkut's Institute of Technology Chaokuntaharn Ladkrabang, Thailand
Dongrui Wu	Huazhong University of Science and Technology, China
Si Wu	Beijing Normal University, China
Si Wu	South China University of Technology, China
Zhengguang Wu	Zhejiang University, China
Tao Xiang	Chongqing University, China
Chao Xu	Zhejiang University, China
Zenglin Xu	University of Science and Technology of China, China
Zhaowen Xu	Zhejiang University, China
Tetsuya Yagi	Osaka University, Japan
Toshiyuki Yamane	IBM, Japan
Koichiro Yamauchi	Chubu University, Japan
Xiaohui Yan	Nanjing University of Aeronautics and Astronautics, China
Zheng Yan	University of Technology Sydney, Australia
Jinfu Yang	Beijing University of Technology, China
Jun Yang	Southeast University, China
Minghao Yang	Chinese Academy of Sciences, China
Qinmin Yang	Zhejiang University, China
Shaofu Yang	Southeast University, China
Xiong Yang	Tianjin University, China
Yang Yang	Nanjing University of Posts and Telecommunications, China
Yin Yang	Hamad Bin Khalifa University, Qatar
Yiyu Yao	University of Regina, Canada
Jianqiang Yi	Chinese Academy of Sciences, China
Chengpu Yu	Beijing Institute of Technology, China
Wen Yu	CINVESTAV, Mexico
Wenwu Yu	Southeast University, China
Zhaoyuan Yu	Nanjing Normal University, China
Xiaodong Yue	Shanghai University, China
Dan Zhang	Zhejiang University, China
Jie Zhang	Newcastle University, UK
Liqing Zhang	Shanghai Jiao Tong University, China
Nian Zhang	University of the District of Columbia, USA
Tengfei Zhang	Nanjing University of Posts and Telecommunications, China
Tianzhu Zhang	Chinese Academy of Sciences, China

Ying Zhang	Shandong University, China
Zhao Zhang	Soochow University, China
Zhaoxiang Zhang	Chinese Academy of Sciences, China
Dongbin Zhao	Chinese Academy of Sciences, China
Qiangfu Zhao	University of Aizu, Japan
Zhijia Zhao	Guangzhou University, China
Jinghui Zhong	South China University of Technology, China
Qi Zhou	University of Portsmouth, UK
Xiaojun Zhou	Central South University, China
Yingjiang Zhou	Nanjing University of Posts and Telecommunications, China
Haijiang Zhu	Beijing University of Chemical Technology, China
Hu Zhu	Nanjing University of Posts and Telecommunications, China
Lei Zhu	Unitec Institute of Technology, New Zealand
Pengefei Zhu	Tianjin University, China
Yue Zhu	Nanjing University, China
Zongyu Zuo	Beihang University, China

Contents – Part VII

Robotics and Control

Prescribed Performance Control of Double-Fed Induction Generator
with Uncertainties 3
   Yuqi Liu, Haojie Li, Wenjie Wu, Dan Wang, and Zhouhua Peng

Event-Triggered Adaptive Dynamic Programming for Continuous-Time
Nonlinear Two-Player Zero-Sum Game 15
   Shan Xue, Biao Luo, Derong Liu, and Yueheng Li

A Learning Based Recovery for Damaged Snake-Like Robots 26
   *Zhuoqun Guan, Jianping Huang, Zhiyong Jian, Linlin liu, Long Cheng,
   and Kai Huang*

Learning to Cooperate in Decentralized Multi-robot Exploration
of Dynamic Environments 40
   Mingyang Geng, Xing Zhou, Bo Ding, Huaimin Wang, and Lei Zhang

Q Value-Based Dynamic Programming with Boltzmann Distribution
by Using Neural Network 52
   *Wenxin Yu, Liang Yu, Gang He, Yibo Fan, Gang He, Jiu Xu, Zhuo Yang,
   and Zhiqiang Zhang*

Data-Driven and Collision-Free Hybrid Crowd Simulation Model
for Real Scenario 62
   Qingrong Cheng, Zhiping Duan, and Xiaodong Gu

Aligning Manifolds of Double Pendulum Dynamics Under the Influence
of Noise. 74
   *Fayeem Aziz, Aaron S. W. Wong, James S. Welsh,
   and Stephan K. Chalup*

Pinning Synchronization of Complex Networks with Switching Topology
and a Dynamic Target System 86
   Guanghui Wen, Xinghuo Yu, Peijun Wang, and Wenwu Yu

The Deep Input-Koopman Operator for Nonlinear Systems 97
   Rongrong Zhu, Yang Cao, Yu Kang, and Xuefeng Wang

Multi-UAV Collaborative Monocular SLAM Focusing on Data Sharing. 108
   *Zhuoyue Yang, Dianxi Shi, Yongjun Zhang, Shaowu Yang, Fu Li,
   and Ruoxiang Li*

Comparing Computing Platforms for Deep Learning
on a Humanoid Robot   120
  Alexander Biddulph, Trent Houliston, Alexandre Mendes,
  and Stephan K. Chalup

Min-Max Consensus Algorithm for Multi-agent Systems Subject
to Privacy-Preserving Problem   132
  Aijuan Wang, Nankun Mu, and Xiaofeng Liao

Robot Navigation on Slow Feature Gradients   143
  Muhammad Haris, Mathias Franzius, and Ute Bauer-Wersing

Neurodynamics-Based Distributed Receding Horizon Trajectory
Generation for Autonomous Surface Vehicles   155
  Jiasen Wang and Jun Wang

Adaptive Finite-Time Synchronization of Inertial Neural Networks
with Time-Varying Delays via Intermittent Control   168
  Lin Cheng, Yongqing Yang, Xianyun Xu, and Xin Sui

Adaptive Critic Designs of Optimal Control for Ice Storage Air
Conditioning Systems   180
  Zehua Liao and Qinglai Wei

Impulsive Constraint Control of Coupled Neural Network Model
with Actual Saturation   189
  Deqiang Ouyang, Tingwen Huang, Chuandong Li, Caiping Chen,
  and Hongfei Li

Value Iteration Algorithm for Optimal Consensus Control
of Multi-agent Systems   200
  Qichao Zhang and Dongbin Zhao

Potential and Sampling Based RRT Star for Real-Time Dynamic Motion
Planning Accounting for Momentum in Cost Function   209
  Saurabh Agarwal, Ashish Kumar Gaurav, Mehul Kumar Nirala,
  and Sayan Sinha

Dynamic Control of Storage Bandwidth Using Double Deep
Recurrent Q-Network   222
  Kumar Dheenadayalan, Gopalakrishnan Srinivasaraghavan,
  and V. N. Muralidhara

Adaptive Modeling and Control of an Upper-Limb Rehabilitation Robot
Using RBF Neural Networks   235
  Liang Peng, Chen Wang, Lincong Luo, Sheng Chen, Zeng-Guang Hou,
  and Weiqun Wang

Modelling Predictive Information of Stochastic Dynamics in the Retina   246
  Min Yan, Yiko Chen, C. K. Chan, and K. Y. Michael Wong

Local Tracking Control for Unknown Interconnected Systems
via Neuro-Dynamic Programming   258
  Bo Zhao, Derong Liu, Mingming Ha, Ding Wang, Yancai Xu,
  and Qinglai Wei

Optimal Control for Dynamic Positioning Vessel Based
on an Approximation Method   269
  Xiaoyang Gao, Tieshan Li, and Qihe Shan

Interactive Incremental Online Learning of Objects Onboard
of a Cooperative Autonomous Mobile Robot   279
  Stephan Hasler, Jennifer Kreger, and Ute Bauer-Wersing

Resilient Consensus for Multi-agent Networks with Mobile Detectors   291
  Haofeng Yan, Yiming Wu, Ming Xu, Ting Wu, Jian Xu, and Tong Qiao

Multi-feature Fusion for Deep Reinforcement Learning: Sequential Control
of Mobile Robots   303
  Haotian Wang, Wenjing Yang, Wanrong Huang, Zhipeng Lin,
  and Yuhua Tang

Dynamics Based Fuzzy Adaptive Impedance Control for Lower Limb
Rehabilitation Robot   316
  Xu Liang, Weiqun Wang, Zengguang Hou, Zihao Xu, Shixin Ren,
  Jiaxing Wang, and Liang Peng

A New Overvoltage Control Method Based on Active and Reactive
Power Coupling   327
  Guangbin Li, Yanhong Luo, and Dongsheng Yang

A Neural Network Compensation Technique for an Inertia Estimation Error
of a Time-Delayed Controller for a Robot Manipulator   339
  Seul Jung

Biomedical Applications

Estimating Criticality of Resting-State Phase Synchronization Network
Based on EEG Source Signals   349
  Li Zhang, Bo Shi, Mingna Cao, Sai Zhang, Yiming Dai, and Yanmei Zhu

A Spatio-Temporal Fully Convolutional Network for Breast Lesion
Segmentation in DCE-MRI   358
  Mingjian Chen, Hao Zheng, Changsheng Lu, Enmei Tu, Jie Yang,
  and Nikola Kasabov

Glomerulus Detection on Light Microscopic Images of Renal Pathology
with the Faster R-CNN   369
   Ying-Chih Lo, Chia-Feng Juang, I-Fang Chung, Shin-Ning Guo,
   Man-Ling Huang, Mei-Chin Wen, Cheng-Jian Lin, and Hsueh-Yi Lin

One-Bit DNA Compression Algorithm.   378
   Deloula Mansouri and Xiaohui Yuan

Robust Segmentation of Overlapping Cells in Cervical Cytology Using
Light Convolution Neural Network   387
   Shusong Xu, Chen Sang, Yulan Jin, and Tao Wan

Semantic Similarity Measures to Disambiguate Terms in Medical Text   398
   Kai Lei, Jiyue Huang, Shangchun Si, and Ying Shen

Age Estimation from MR Images via 3D Convolutional Neural Network
and Densely Connect.   410
   Qi Qi, Baolin Du, Mingyong Zhuang, Yue Huang, and Xinghao Ding

Low-Shot Multi-label Incremental Learning for Thoracic
Diseases Diagnosis   420
   Qingfeng Wang, Jie-Zhi Cheng, Ying Zhou, Hang Zhuang,
   Changlong Li, Bo Chen, Zhiqin Liu, Jun Huang, Chao Wang,
   and Xuehai Zhou

Continuous Convolutional Neural Network with 3D Input for EEG-Based
Emotion Recognition.   433
   Yilong Yang, Qingfeng Wu, Yazhen Fu, and Xiaowei Chen

3D Large Kernel Anisotropic Network for Brain Tumor Segmentation.   444
   Dongnan Liu, Donghao Zhang, Yang Song, Fan Zhang,
   Lauren J. O'Donnell, and Weidong Cai

Saliency Supervision: An Intuitive and Effective Approach for Pain
Intensity Regression   455
   Conghui Li, Zhaocheng Zhu, and Yuming Zhao

Identification of Causality Among Gene Mutations Through Local Causal
Association Rule Discovery   465
   Ruichu Cai, Qiqi Zhen, and Zhifeng Hao

EEG Sparse Representation Based Alertness States Identification Using
Gini Index   478
   Muna Tageldin, Talal Al-Mashaikki, Hamza Bali, and Mostefa Mesbah

Attention-Based Network for Cross-View Gait Recognition   489
   Yuanyuan Huang, Jianfu Zhang, Haohua Zhao, and Liqing Zhang

Experimental Validation of Minimum-Jerk Principle in Physical
Human-Robot Interaction   499
    Chen Wang, Liang Peng, Zeng-Guang Hou, Lincong Luo,
    Sheng Chen, and Weiqun Wang

Residual Semantic Segmentation of the Prostate from Magnetic
Resonance Images.   510
    Md Sazzad Hossain, Andrew P. Paplinski, and John M. Betts

The Relationship Between the Movement Difficulty and Brain Activity
Before Arm Movements.   522
    Tomoki Semoto, Isao Nambu, and Yasuhiro Wada

A Deep Learning Assisted Gene Expression Programming Framework
for Symbolic Regression Problems   530
    Jinghui Zhong, Yusen Lin, Chengyu Lu, and Zhixing Huang

Automated Tongue Segmentation in Chinese Medicine Based
on Deep Learning   542
    Yushan Xue, Xiaoqiang Li, Pin Wu, Jide Li, Lu Wang, and Weiqin Tong

Deep Feature Learning and Visualization for EEG Recording
Using Autoencoders   554
    Yue Yao, Jo Plested, and Tom Gedeon

A Feature Filter for EEG Using Cycle-GAN Structure.   567
    Yue Yao, Jo Plested, and Tom Gedeon

Influence of Difference of Spatial Information Obtained from a Moving
Virtual Sound Presentation on Auditory BCI   577
    Yuki Onodera, Isao Nambu, and Yasuhiro Wada

Association Study of Alzheimer's Disease with Tree-Guided Sparse
Canonical Correlation Analysis.   585
    Shangchen Zhou, Shuai Yuan, Zhizhuo Zhang, and Zenglin Xu

Relevance of Frequency of Heart-Rate Peaks as Indicator of 'Biological'
Stress Level   598
    Meena Santhanagopalan, Madhu Chetty, Cameron Foale, Sunil Aryal,
    and Britt Klein

Development of a Real-Time Motor-Imagery-Based EEG
Brain-Machine Interface.   610
    Gal Gorjup, Rok Vrabič, Stoyan Petrov Stoyanov,
    Morten Østergaard Andersen, and Poramate Manoonpong

Robust Eye Center Localization Based on an Improved SVR Method   623
    Zhiyong Wang, Haibin Cai, and Honghai Liu

Hardware

Hopfield Neural Network with Double-Layer Amorphous Metal-Oxide
Semiconductor Thin-Film Devices as Crosspoint-Type Synapse Elements
and Working Confirmation of Letter Recognition   637
   Mutsumi Kimura, Kenta Umeda, Keisuke Ikushima, Toshimasa Hori,
   Ryo Tanaka, Tokiyoshi Matsuda, Tomoya Kameda,
   and Yasuhiko Nakashima

FPGA Based Hardware Implementation of Simple Dynamic Binary
Neural Networks.   647
   Shunsuke Aoki, Seitaro Koyama, and Toshimichi Saito

Fast Depthwise Separable Convolution for Embedded Systems   656
   Byeongheon Yoo, Yongjun Choi, and Heeyoul Choi

An Analog Circuit Design for k-Winners-Take-All Operations   666
   Xiaoyang Liu and Jun Wang

NVM Weight Variation Impact on Analog Spiking Neural Network Chip . . .   676
   Akiyo Nomura, Megumi Ito, Atsuya Okazaki, Masatoshi Ishii,
   Sangbum Kim, Junka Okazawa, Kohji Hosokawa, and Wilfried Haensch

Author Index   687

Robotics and Control

Prescribed Performance Control
of Double-Fed Induction Generator
with Uncertainties

Yuqi Liu, Haojie Li, Wenjie Wu, Dan Wang, and Zhouhua Peng[(✉)]

School of Marine Electrical Engineering, Dalian Maritime University, Dalian, China
{dwang,zhpeng}@dlmu.edu.cn

Abstract. This paper considers the vector control of double-fed induction generator in the presence of uncertainties. An electromagnetic torque controller and a rotor current controller are proposed based on an error transformation technique and a reduced-order extended state observer. Specifically, the error transformation technique is used to achieve the prescribed transient and steady performance. The reduced-order extended state observer is utilized to estimate and compensate for system uncertainties in real time. By using the proposed controllers, the tracking performance of the system is improved. Compared with the full-order extended state observer, the reduced-order extended state observer reduces the adjustment parameters, which renders it easier to implement in practice. The effectiveness of proposed scheme is validated via theoretical analysis and simulations.

Keywords: Double-fed induction generator
Extended state observer · Prescribed performance · Vector control

1 Introduction

In recent years, double-fed induction generator has been widely used as the core component of variable speed constant frequency wind power generation system [1]. Compared with synchronous generator and fixed speed asynchronous generator, double-fed induction generator has many advantages, such as variable speed operation, four quadrant active power, reactive power regulation, low converter cost and power loss [2].

D. Wang—This work was supported in part by the National Natural Science Foundation of China under Grants 61673081, 51579023, and in part by the Innovative Talents in Universities of Liaoning Province under Grant LR2017014, and in part by High Level Talent Innovation and Entrepreneurship Program of Dalian under Grant 2016RQ036, and in part by the National Key Research and Development Program of China under Grant 2016YFC0301500, and in part by the Fundamental Research Funds for the Central Universities under Grants 3132016313, 3132018306.

L. Cheng et al. (Eds.): ICONIP 2018, LNCS 11307, pp. 3–14, 2018.
https://doi.org/10.1007/978-3-030-04239-4_1

Most traditional double-fed induction generators are controlled based on linearization technology. However, due to random wind speeds and some unavoidable disturbances, the work points of double-fed induction generators often change randomly. Therefore, the control effect of the linearization control method is not ideal, and the stability is not assured. Especially when the system parameters change, the linearization controller cannot guarantee the system performance. In [3], prescribed performance control combined with adaptive control is applied to the control of double-fed induction generator. In [4], robust differentiator techniques are utilized to eliminate the need for an acceleration of the wind and rotor. In [5], fuzzy logic system is applied to the angle control, which is well known for its capability in dealing with nonlinear systems. However, most control studies only focus on the steady-state performance of the system. There is little research and discussion on the transient performance of the system. In the controller design, analyzing the transient performance of the system has a very important role in improving system performance.

This paper focuses on the vector control of double-fed induction generator. A control strategy is used for guaranteeing the transient and steady control performance. A reduced-order extended state observer is used to estimate and compensate the uncertainty of the system in real time. Based on an error transformation technique and the reduced-order extended state observer, an electromagnetic torque controller and a rotor current controller are designed. Specifically, the error transformation technique is used to achieve the prescribed transient and steady performance. By using the developed controller, the dependence on system models and parameters are eliminated, the tracking performance of the system is improved and the adjustment parameters are reduced.

2   Design and Analysis

2.1   Turbine Model

As the prime mover of wind power generation system, wind turbine is the primary component to convert the wind energy to mechanical energy in the entire system. The performance of wind turbine determines the effective output power of the wind power system and the safe, stable and reliable operation of wind turbine. From the aerodynamic analysis, the mechanical power P_a and aerodynamic torque T_a captured by the wind turbine from the wind are:

$$P_a = \frac{1}{2}\rho\pi R^2 C_p(\lambda, \beta)v^3, T_a = \frac{P_a}{\omega_r} = \frac{1}{2}\rho\pi R^2 v^3 \frac{C_p(\lambda, \beta)}{\omega_r} \qquad (1)$$

where ρ is the air density, R is the radius of the rotor, $C_p(\lambda, \beta)$ is the wind energy utilization factor, v is the wind speed, and ω_r is the rotation speed of the wind wheel. The value of $C_p(\lambda, \beta)$ depends on the pitch angle β and tip speed ratio λ. According to Betz theory, the theoretical maximum value of $C_p(\lambda, \beta)$ is 0.593, which is also known as the Betz limit. For horizontal axis wind turbines, the value of C_p is generally less than 0.5. The tip speed ratio is expressed as:

$\lambda = (\omega_r R)/v$. In practical applications, the wind energy utilization factor $C_p(\lambda, \beta)$ is generally considered as a complex nonlinear function. Generally, the following empirical formula is used: $C_p(\lambda, \beta) = 0.5176(116/\bar{\lambda} - 0.4\beta - 5)e^{-21/(\bar{\lambda})} + 0.0068\lambda$, $1/\bar{\lambda} = 1/(\lambda + 0.08\beta) - 0.035/(\beta^3 + 1)$. According to [6], the following simplified model is utilized:

$$J_t \dot{\omega}_{mr} = T_a - K_t \omega_{mr} - T_g. \tag{2}$$

2.2   Generator Model

The DFIG is utilized in the wind turbine system. Compared with fully fed synchronous generator, the DFIG can reduce converter costs and power consumption. For the proposed control strategy, the dynamic model of the generator expressed in a synchronously rotating frame dq is given by

$$\begin{cases} u_{sd} = R_s i_{sd} + \dot{\phi}_{sd} - \omega_s \phi_{sq} \\ u_{sq} = R_s i_{sq} + \dot{\phi}_{sq} - \omega_s \phi_{sd} \\ u_{rd} = R_r i_{rd} + \dot{\phi}_{rd} - \omega_r \phi_{rq} \\ u_{rq} = R_r i_{rq} + \dot{\phi}_{rq} - \omega_r \phi_{rd} \\ \phi_{sd} = L_s i_{sd} + L_m i_{rd} \\ \phi_{sq} = L_s i_{sq} + L_m i_{rq} \\ \phi_{rd} = L_r i_{rd} + L_m i_{sd} \\ \phi_{rq} = L_r i_{rq} + L_m i_{sq} \\ T_{em} = n_p L_m (i_{rd} i_{sq} - i_{rq} i_{sd}) \end{cases} \tag{3}$$

where u_{rd}, u_{rq}, u_{sd} and u_{sq} are the rotor and the stator d-axis and q-axis voltage components, respectively. i_{rd}, i_{rq}, i_{sd} and i_{sq} are the rotor and the stator d-axis and q-axis current components, respectively. ϕ_{rd}, ϕ_{rq}, ϕ_{sd} and ϕ_{sq} are the rotor and the stator d-axis and q-axis flux components, respectively. R_r and R_s are the rotor and the stator winding resistances. L_r and L_s are the self-inductances of the rotor winding and the stator winding in a synchronous rotating coordinate system. n_p is the number of pole pairs. L_m is the equivalent mutual inductance in the stator windings in a synchronous rotating coordinate system.

According to the stator voltage orientation rule, the d-axis of the synchronous rotational coordinate system is aligned with the stator voltage vector U_s. Neglecting the influence of stator resistance, the voltage equation is given by

$$\begin{cases} u_{rd} = R_r i_{rd} + \sigma L_r \dot{i}_{rd} - \omega_{sl}(-\frac{L_m}{\omega_1 L_s} U_s + \sigma L_r i_{rq}) \\ u_{rq} = R_r i_{rq} + \sigma L_r \dot{i}_{rd} + \sigma L_r \omega_{sl} i_{rd} \end{cases} \tag{4}$$

where ω_{sl} is the rotational angular velocity, $\omega_{sl} = \omega_1 - \omega_r$. $\sigma = 1 - L_m^2/(L_s L_r)$ is the magnetic flux leakage parameter of DFIG. The electromagnetic torque equation and stator output reactive power are given by

$$\begin{cases} T_{em} = \frac{3}{2} n_p \frac{U_s L_m}{\omega_1 L_s} i_{rd} \\ Q_s = -\frac{3 U_s}{2 \omega_1 L_s}(U_s + \omega_1 L_m i_{rq}). \end{cases} \tag{5}$$

3   Problem Formulation

In different operating areas, the control objectives of wind power generators are different. The wind turbines running at low wind speed are focused in this paper. The control objective is to make the wind power generator run in the vicinity of the maximum power point. When the wind power generator runs at the maximum value of C_p, the reference value of the electromagnetic torque is [2]:

$$T_{ref} = k_{opt}\omega_g^2, k_{opt} = \frac{\pi \rho^5 C_{pmax}}{2n_g^3 \lambda_{opt}^3}. \tag{6}$$

The stator reactive power needs to be controlled to satisfy the grid requirements. When wind power plants do not require wind turbines to generate reactive power, the DFIG can run under the unit power factor, i.e., $Q = 0$. From (5), the reference value of the q-axis component of the rotor current can be calculated

$$i_{rq\_ref} = -\frac{U_s}{\omega_1 L_m}. \tag{7}$$

In actual wind power system, electrical subsystems respond more quickly than mechanical subsystems. Therefore, the controllers of generator and wind power generator are designed separately. A double closed loop control structure is used, in which the inner loop is concerned with the electrical subsystem and the outer loop is concerned with the mechanical subsystem. The outer loop provides reference torque input T_{ref} for the inner loop. When designing the outer loop controller, it is usually assumed that the inner loop of the electrical subsystem can track the reference torque T_{ref} in real time [7–12].

For the convenience of analysis, the electromagnetic torque tracking error $e_{T_{em}}$ and the reactive current tracking error $e_{i_{rq}}$ are defined as:

$$e_{T_{em}} = T_{em} - T_{ref}, e_{i_{rq}} = i_{rq} - i_{rq\_ref}. \tag{8}$$

In this paper, the transient and stability performance of the controller are considered. Therefore, the control target is described in detail as follows. By controlling the dq-axis components of the rotor voltage u_{rd}, u_{rq}, we can achieve:

(1) Electromagnetic torque T_{em} tracks its reference signal T_{ref} to achieve maximum power tracking;
(2) Reactive current i_{rq} tracks its reference signal i_{rq_ref} to achieve unit power factor operation;
(3) The system tracking error satisfies the prescribed transient and steady state control performance.

More specifically, tracking error must be satisfied as

$$\underline{e}_{T_{em}} < e_{T_{em}} < \bar{e}_{T_{em}}, \underline{e}_{i_{rq}} < e_{i_{rq}} < \bar{e}_{i_{rq}} \tag{9}$$

where $\bar{e}_{T_{em}}$ and $\underline{e}_{T_{em}}$ are the upper and lower bounds of torque tracking error $e_{T_{em}}$ respectively, and satisfy $\underline{e}_{T_{em}}(t) < 0$, $\bar{e}_{T_{em}}(t) > 0$. $\bar{e}_{i_{rq}}$ and $\underline{e}_{i_{rq}}$ are the upper and lower bounds of current tracking error $e_{i_{rq}}$ respectively, and satisfy $\underline{e}_{i_{rq}}(t) < 0$, $\bar{e}_{i_{rq}}(t) > 0$. By preselecting the upper and lower bound functions of the appropriate tracking error, the pre-defined transient and stability performance can be achieved. Taking the torque tracking error $e_{T_{em}}$ as an example, select the initial value $\underline{e}_{T_{em}}(0)$ and $\bar{e}_{T_{em}}(0)$. The overshoot must be between $\underline{e}_{T_{em}}(0)$ and $\bar{e}_{T_{em}}(0)$. The shapes of $\underline{e}_{T_{em}}(t)$ and $\bar{e}_{T_{em}}(t)$ determine the convergence rate of the tracking error in the transient process. The steady-state error signal $e_{T_{em}}(\infty)$ is maintained in a prescribed error domain, i.e., $\underline{e}_{T_{em}}(\infty) < e_{T_{em}}(\infty) < \bar{e}_{T_{em}}(\infty)$.

For the design of the controller, the following assumption is needed.

Assumption 1. The upper and lower bounds of tracking error $\bar{e}_{T_{em}}$, $\underline{e}_{T_{em}}$, $\bar{e}_{i_{rq}}$, $\underline{e}_{i_{rq}}$ and their derivatives are known and can be used directly in the controller design.

3.1 Reduced-Order Extended State Observer

The extended state observer (ESO) has a good performance. It does not depend on the accurate mathematical model. The ESO can be used not only to recover the unmeasured system state, but also to estimate system uncertainty in real time. By using the estimated information of uncertainty, a better control performance can be achieved. Since the output of the general control system can be measured directly, a reduced-order extended state observer (RESO) is used to estimate the system uncertainty. In the RESO, only one parameter is needed to adjust. The parameter selection is simpler, and it is easier to implement in practice [13].

From (4) and (5), the state equation of the electromagnetic torque T_{em} and the rotor current q-axis current i_{rq} can be obtained

$$\dot{T}_{em} = f(\cdot) + b_1 u_{rd}, \dot{i}_{rq} = g(\cdot) + b_2 u_{rq} \tag{10}$$

where $f(\cdot) = (3n_p U_s L_m)/(2\omega_1 L_s)(-R_r/(\sigma L_s)i_{rd} + \omega_{sl}(i_{rq} - (L_m U_s)/(\sigma L_r L_s \omega_1)) + 1/(\sigma L_r)u_{rd}) - b_1 u_{rd}$ and $g(\cdot) = -R_r/(\sigma L_r)i_{rq} - \omega_{sl}i_{rd} + 1/(\sigma L_r)u_{rq} - b_2 u_{rq}$ are the system uncertainties, including motor parameters and coupling items. b_1 and b_2 are control gains. Because the general motor parameters have a general value (L_r', L_s', L_m'), let $b_1 = 3n_p U_s L_m'/(2\sigma' L_r' \omega_1 L_s')$, $b_2 = 1/(\sigma' L_r')$. This can reduce the burden of ESO, and improve the estimation precision and estimation speed.

To estimate the unknown parts, two RESOs are proposed as follows

$$\begin{cases} \dot{p}_1 = -\beta_1 p_1 - \beta_1^2 T_{em} - \beta_1 b_1 u_{rd} \\ \hat{f}(\cdot) = p_1 \beta_1 T_{em} \\ \dot{p}_2 = -\beta_2 p_2 - \beta_2^2 i_{rq} - \beta_2 b_2 u_{rq} \\ \hat{g}(\cdot) = p_2 + \beta_2 i_{rq} \end{cases} \tag{11}$$

where p_1 and p_2 are the auxiliary state variables of the extended state observer. β_1 and β_2 are the observer gains. $\beta_1 > 0$, $\beta_2 > 0$. $\hat{f}(\cdot)$ and $\hat{g}(\cdot)$ are the estimated value for $f(\cdot)$ and $g(\cdot)$, respectively.

For the uncertainties $f(\cdot)$ and $g(\cdot)$, the following assumptions are made.

Assumption 2. There exist positive constants f^* and g^*, such that $|\dot{f}(\cdot)| \leq f^*, |\dot{g}(\cdot)| \leq g^*$ are satisfied.

The estimation error of RESO is defined

$$\tilde{f}(\cdot) = \hat{f}(\cdot) - f(\cdot), \tilde{g}(\cdot) = \hat{g}(\cdot) - g(\cdot). \tag{12}$$

Taking the time derivative of $\tilde{f}(\cdot)$ and $\tilde{g}(\cdot)$, and using (11), it follows that

$$\dot{\tilde{f}}(\cdot) = -\beta_1 \tilde{f}(\cdot) - \dot{f}(\cdot), \dot{\tilde{g}}(\cdot) = -\beta_2 \tilde{g}(\cdot) - \dot{g}(\cdot). \tag{13}$$

3.2   Error Conversion

In order to ensure the prescribed performance, an error conversion technique is proposed. It transforms the original constraint error into an unconstrained error,

$$e_i = \frac{\bar{e}_i(t) - \underline{e}_i(t)}{\pi} \arctan(z_i) + \frac{\bar{e}_i(t) + \underline{e}_i(t)}{2} \tag{14}$$

or

$$z_i(t) = \tan(\frac{\pi}{2} \times \frac{2e_i - \bar{e}_i(t) - \underline{e}_i(t)}{\bar{e}_i(t) - \underline{e}_i(t)}) \tag{15}$$

where $i = T_{em}, i_{rd}$. $\tan(\cdot)$ is a tangent function. $\arctan(\cdot)$ is an inverse tangent function. $z_i(t)$ is the unconstrained error. According to (14), we obtain $\lim_{z_i \to -\infty} e_i = \underline{e}_i(t), \lim_{z_i \to \infty} e_i = \bar{e}_i(t)$.

As $z_i(t)$ is bounded, it follows that e_i satisfies $\underline{e}_i(t) < e_i < \bar{e}_i(t)$. Therefore, the boundedness of the conversion error $z_i(t)$ can guarantee the prescribed performance of the original error. i.e., (9).

Taking the time derivative of (15), we have

$$\begin{aligned} \dot{z}_i &= \frac{\partial z_i}{\partial e_i} \dot{e}_i(t) + \frac{\partial z_i}{\partial \bar{e}_i} \dot{\bar{e}}_i(t) + \frac{\partial z_i}{\partial \underline{e}_i} \dot{\underline{e}}_i(t) \\ &= \gamma_i \dot{e}_i(t) + \sigma_i \end{aligned} \tag{16}$$

where $\gamma_i = \frac{\partial z_i}{\partial e_i}$ and $\sigma_i = \frac{\partial z_i}{\partial \bar{e}_i} \dot{\bar{e}}_i(t) + \frac{\partial z_i}{\partial \underline{e}_i} \dot{\underline{e}}_i(t)$. Since the signals $\sigma_i, e_i(t), \bar{e}_i(t), \dot{\bar{e}}_i(t)$, $\underline{e}_i(t)$ and $\dot{\underline{e}}_i(t)$ are known, the signals γ_i and σ_i can be calculated directly and applied to the controller design. In addition, from the nature of the conversion error, we obtain $\gamma_i > 0$.

Substituting (8) into (16), the equation of conversion error $z_{T_{em}}$ and $z_{i_{rq}}$ are obtained

$$\dot{z}_{T_{em}} = \gamma_{T_{em}}(\dot{T}_{em} - \dot{T}_{ref}) + \sigma_{T_{em}}, \dot{z}_{i_{rq}} = \gamma_{i_{rq}}(\dot{i}_{rq} - \dot{i}_{rq\_ref}) + \sigma_{i_{rq}}. \tag{17}$$

3.3 Controller Design

In order to capture as much energy as possible from the wind, it is necessary to design a corresponding controller. The controller drives the electromagnetic torque to track its reference torque T_{ref}. At the same time, in order to let the generator run in unit power factor, it is also necessary to design a controller, which drives the reactive current component i_{rq} to track its reference signal i_{rq_ref}. The controller design also needs to ensure the corresponding prescribed performance (9). Considering the transient and stability performance of the system, a controller is designed as follows

$$u_{rd} = (\frac{-k_1 z_{T_{em}} - \sigma_{T_{em}}}{\gamma_{T_{em}}} + \dot{T}_{ref} - \hat{f}(\cdot))/b_1, u_{rq} = (\frac{-k_2 z_{i_{rq}} - \sigma_{i_{rq}}}{\gamma_{i_{rq}}} - \hat{g}(\cdot))/b_2$$

$$(18)$$

where k_1 and k_2 are positive constants.

Using (10), (12) and (17), the following conversion error subsystem is obtained:

$$\dot{z}_{T_{em}} = -k_1 z_{T_{em}} - \gamma_{T_{em}} \tilde{f}(\cdot), \dot{z}_{i_{rq}} = -k_2 z_{i_{rq}} - \gamma_{i_{rq}} \tilde{g}(\cdot). \tag{19}$$

3.4 Stability Analysis

Considering the mathematical model of DFIG in two phase synchronous rotating coordinate system, the voltage equation is given by

$$\begin{cases} u_{sd} = R_s i_{sd} + \frac{d\psi_{sd}}{dt} - \omega_1 \psi_{sq} \\ u_{sq} = R_s i_{sq} + \frac{d\psi_{sq}}{dt} - \omega_1 \psi_{sd} \\ u_{rd} = R_r i_{rd} + \frac{d\psi_{rd}}{dt} - \omega_{sl} \psi_{rq} \\ u_{rq} = R_r i_{rq} + \frac{d\psi_{rq}}{dt} - \omega_{sl} \psi_{rd}, \end{cases} \tag{20}$$

the magnetic chain equation is given by

$$\begin{cases} \psi_{sd} = L_s i_{sd} + L_m i_{rd} \\ \psi_{sq} = L_s i_{sq} + L_m i_{rq} \\ \psi_{rd} = L_m i_{sd} + L_r i_{rd} \\ \psi_{rq} = L_m i_{sq} + L_r i_{rq}. \end{cases} \tag{21}$$

In the following, the cascade system theory is applied to prove the stability of the closed loop system. The system can be regarded as a cascade system, which is composed of the RESO estimation error subsystem (14) and the conversion error subsystem (19). Firstly, the stability of the RESO estimation error subsystem (14) is analyzed.

Lemma 1. The RESO estimation error subsystem (14) is input-to-state stable with the input being $(\dot{f}(\cdot), \dot{g}(\cdot))$ and the state being $(\tilde{f}(\cdot), \tilde{g}(\cdot))$.

Proof 1. A Lyapunov function is constructed as:

$$V_1 = \frac{1}{2}\tilde{f}(\cdot)^2 + \frac{1}{2}\tilde{g}(\cdot)^2. \tag{22}$$

Taking the time derivative of V_1 and considering (13), we have

$$
\begin{aligned}
\dot{V}_{k1} &= \tilde{f}(\cdot)\dot{\tilde{f}}(\cdot) + \tilde{g}(\cdot)\dot{\tilde{g}}(\cdot) \\
&= -\beta_1\tilde{f}(\cdot)^2 - \tilde{f}(\cdot)\dot{f}(\cdot) - \beta_2\tilde{g}(\cdot)^2 - \tilde{g}(\cdot)\dot{g}(\cdot) \\
&\leq -\lambda_{\min}(K_1)\|\xi_1\|^2 + \|h\|\|\xi_1\|
\end{aligned} \tag{23}
$$

where $\xi_1 = [\tilde{f}(\cdot), \tilde{g}(\cdot)]^T$, $K_1 = \mathrm{diag}\{\beta_1, \beta_2\}$, $h = [|\dot{f}(\cdot)|, |\dot{g}(\cdot)|]^T$.
From inequalities

$$\|\xi_1\| \geq \frac{|\dot{f}(\cdot)| + |\dot{g}(\cdot)|}{\theta_1\lambda_{\min}(K_1)} \geq \frac{\|h\|}{\theta_1\lambda_{\min}(K_1)}, \tag{24}$$

we have $\dot{V}_1 \leq -(1 - \theta_1)\lambda_{\min}(K_1)\|\xi_1\|^2$, where $0 < \theta_1 < 1$. Therefore, the RESO estimation error subsystem (14) is input-to-state stable, and $\|\xi_1(t)\| \leq \max\{\sigma_{a1}(\|\xi_1(0)\|, t), \sigma_{a2}(|\dot{f}(\cdot)|) + \sigma_{a3}(|\dot{g}(\cdot)|)\}$, where σ_{a1} is a class $\mathcal{KL}$ function, and $\sigma_{a2}(s) = \sigma_{a3}(s) = s/(\theta_1\lambda_{\min}(K_1))$.

Next, the stability of the conversion error subsystem (19) is analyzed.

Lemma 2. The conversion error subsystem (19) is input-to-state stable with the input being $(\tilde{f}(\cdot), \tilde{g}(\cdot))$ and the state being $(z_{T_{em}}, z_{i_{rq}})$.

Proof 2. A Lyapunov function is constructed as follows

$$V_2 = \frac{1}{2}z_{T_{em}}^2 + \frac{1}{2}z_{i_{rq}}^2. \tag{25}$$

Taking the time derivative of the above function and considering (19), we have

$$
\begin{aligned}
\dot{V}_2 &= z_{T_{em}}\dot{z}_{T_{em}} + z_{i_{rq}}\dot{z}_{i_{rq}} \\
&= -k_1 z_{T_{em}}^2 - \gamma_{T_{em}}\tilde{f}(\cdot)z_{T_{em}} - k_2 z_{i_{rq}}^2 - \gamma_{i_{rq}}\tilde{g}(\cdot)z_{i_{rq}} \\
&\leq -\lambda_{min}(K_2)\|\xi_2\|^2 + \|h_1\|\|\xi_2\|
\end{aligned} \tag{26}
$$

where $\xi_2 = [z_{T_{em}}, z_{i_{rq}}]^T$, $K_2 = \mathrm{diag}\{k_1, k_2\}$, $h_1 = [\gamma_{T_{em}}|\tilde{f}(\cdot)|, \gamma_{i_{rq}}|\tilde{g}(\cdot)|]^T$.
From inequalities

$$\|\xi_2\| \geq \frac{\gamma_{T_{em}}|\tilde{f}(\cdot)|}{\theta_2\lambda_{\min}(K_2)} + \frac{\gamma_{i_{rq}}|\tilde{g}(\cdot)|}{\theta_2\lambda_{\min}(K_2)} \geq \frac{\|h_1\|}{\theta_2\lambda_{\min}(K_2)}, \tag{27}$$

we have $\dot{V}_2 \leq -(1 - \theta_2)\lambda_{\min}(K_2)\|\xi_2\|^2$, where $0 < \theta_2 < 1$. The conversion error subsystem (19) is input-to-state stable, and $\|\xi_2(t)\| \leq \max\{\sigma_{b1}(\|\xi_3(0)\|, t), \sigma_{b2}(|\tilde{f}(\cdot)|) + \sigma_{b3}(|\tilde{g}(\cdot)|)\}$, where σ_{b1} is a class $\mathcal{KL}$ function, $\sigma_{b2}(s) = (\gamma_{T_{em}}s)/(\theta_2\lambda_{\min}(K_2))$, and $\sigma_{b3}(s) = (\gamma_{i_{rq}}s)/(\theta_2\lambda_{\min}(K_2))$.

Finally, the stability of cascaded systems is analyzed.

Theorem 1. When *Assumption 2* is valid, the cascade system composed of the RESO estimation error subsystem (14) and the conversion error subsystem (19) are input-to-state stable. All the error signals in the closed loop system are bounded.

Proof 3. *Lemmas* 1 and 2 separately illustrate the subsystem (14) is input-to-state stable with the input being $(\dot{f}(\cdot),\ \dot{g}(\cdot))$ and the state being $(\tilde{f}(\cdot),\ \tilde{g}(\cdot))$. Subsystem (19) is input-to-state stable with the input being $(\tilde{f}(\cdot),\ \tilde{g}(\cdot))$ and the state being $(z_{T_{em}}, z_{i_{rq}})$. It is known from the lemma C.4 in the literature [14] that the closed loop system is input-to-state stable with the input being $(\dot{f}(\cdot),\ \dot{g}(\cdot))$ and the state being $(z_{T_{em}},\ z_{i_{rq}},\ \tilde{f}(\cdot),\ \tilde{g}(\cdot))$.

Therefore, there exist a class $\mathcal{KL}$ function $\bar{\omega}_1$ and class $\mathcal{K}$ function ϕ_1, ϕ_2, such that $\|\xi_3(t)\| \leq \bar{\omega}(\|\xi_3(0)\|, t) + \phi_1(|\dot{f}(\cdot)|) + \phi_2(|\dot{g}(\cdot)|)$ are satisfied. where $\xi_3 = [z_{T_{em}}, z_{i_{rq}}, \dot{f}(\cdot), \dot{g}(\cdot)]$.

From the theory of the stability, signals $z_{T_{em}}$, $z_{i_{rq}}$, $\tilde{f}(\cdot)$ and $\tilde{g}(\cdot)$ are asymptotically converge to zero. From the property of error conversion, the boundedness of $z_{T_{em}}$ and $z_{i_{rq}}$ can guarantee the predetermined performance (9). Thus $e_{T_{em}}$ and $e_{i_{rq}}$ are also bounded. From $\hat{f}(\cdot) = \tilde{f}(\cdot) + f(\cdot)$, $\hat{g}(\cdot) = \tilde{g}(\cdot) + g(\cdot)$ and *Assumption* 2, $\hat{f}(\cdot)$ and $\hat{g}(\cdot)$ are bounded. From (18), the inputs u_{rd} and u_{rq} are also bounded. To sum up, all the signals in the closed loop system are bounded.

The signal $\dot{T}_{ref}$ is used during the controller design. From (6), we get $\dot{T}_{ref} = k_{opt}\omega_g\dot{\omega}_g$. If we can measure the values of ω_g and $\dot{\omega}_g$, the value of $\dot{T}_{ref}$ can be calculated. In practice, in order to reduce the operation cost and the calculation load, we can construct $\dot{\omega}_g$ and calculate the value of $\dot{T}_{ref}$ by the nonlinear tracking differentiator in the case of only ω_g is known.

Table 1. DFIG characteristics (converted to the stator side)

Parameter name	Number
Rated power	1.5 MW
Rated voltage	690 V
Stator resistance	2.05 MΩ
Rotor resistance	1.82 MΩ
Stator self-inductance	3.99 MH
Rotor self-inductance	3.977 MH
Mutual inductance	3.915 MH
Rated frequency	50 Hz
Polar logarithm	2

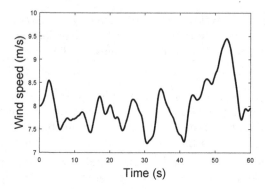

Fig. 1. Wind speed profile.

4   Simulation Results

In order to verify the effectiveness of the algorithm, a 1.5 MW DFIG was used. The parameters are listed in Table 1. The kaimal turbulence probability model is utilized in the simulation. The average wind speed is 10.8 m/s, and the turbulence intensity is 25%. The wind speed is shown in Fig. 1.

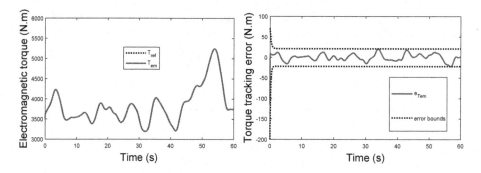

Fig. 2. Electromagnetic torque and its reference.

Fig. 3. Tracking error e_{em} along with its bounds.

The electromagnetic torque reference value T_{ref} and the reactive current reference value i_{rq_ref} are calculated by (6) and (7), and k_{opt} is chosen as 0.159. The upper bounds and lower bounds for $e_{T_{em}}$ and $e_{i_{rq}}$ are set as $\bar{e}_{T_{em}} = 50 \times e^{(-3t)+20}$, $\underline{e}_{T_{em}} = -200 \times e^{(-5t)-20}$, $\bar{e}_{i_{rq}} = 1 \times e^{(-3t)+0.2}$, $\underline{e}_{i_{rq}} = -2 \times e^{(-5t)-0.2}$. The parameters of the controller are set as $k_1 = 100$, $k_2 = 150$, and the parameters of RESO are set as $\beta_1 = 3000$, $\beta_2 = 1000$. In order to verify the performance of the controller and reduce the burden of the controller, 80% of the nominal value of the motor parameter is added to the controller simulation.

Electromagnetic torque T_{em} and its reference value T_{ref} are shown in Fig. 2, and the tracking error $e_{T_{em}}$ and its upper and lower bounds are shown in Fig. 3.

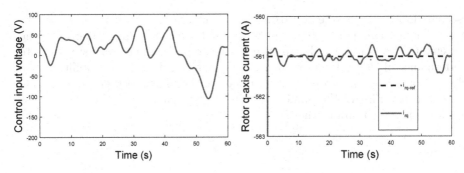

Fig. 4. Control input voltage u_{rd}.

Fig. 5. Reactive current i_{rq} and its reference i_{rq_ref}.

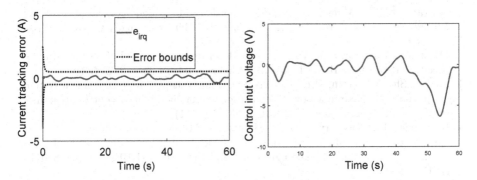

Fig. 6. Tracking error $e_{i_{rq}}$ along with its bounds.

Fig. 7. Control input voltage u_{rq}.

It can be seen that the electromagnetic torque T_{em} can track its reference signal T_{ref} well, and the tracking error $e_{T_{em}}$ remains between its upper and lower bounds. The transient and stability control performance is achieved. The system performance such as steady-state error, overshoot and convergence time can be ensured by reasonably setting the upper and lower bounds of the error. The control input voltage u_{rd} is shown in Fig. 4. The q-axis component of the rotor current i_{rq} and its reference value i_{rq_ref} are shown in Fig. 5, and the tracking error $e_{i_{rq}}$ and its upper and lower bounds are shown in Fig. 6. In addition, the control input voltage u_{rd} is shown in Fig. 7.

5   Conclusion

For the vector control of DFIG, a control strategy is proposed for double-fed induction generator with prescribed transient and steady performance, which is able to achieve maximum power tracking and unit power factor operation. The coupling relationship between the upper and lower bounds of the error is eliminated by the error conversion technique. The system uncertainties are estimated and compensated by RESO, which eliminates the dependence on the system model and parameters and improves the tracking performance of the system. In the controller design, the transient performance and steady-state performance of the system are quantitatively analyzed, and the stability of the closed-loop system is proved by the input-state stability theorem and cascade system theory. Finally, the simulation results are given to verify the effectiveness of the proposed control method.

References

1. Marques, G.D., Iacchetti, M.F.: Stator frequency regulation in a field-oriented controlled DFIG connected to a dc link. IEEE Trans. Ind. Electron. **61**(11), 5930–5939 (2014)
2. Beltran, M.B.B., Ahmed-Ali, T.: Second-order sliding mode control of a doubly fed induction generator driven wind turbine. IEEE Trans. Energy Convers. **27**(2), 261–269 (2012)
3. Meng, W., Yang, Q., Sun, Y.: Guaranteed performance control of DFIG variable-speed wind turbines. IEEE Trans. Ind. Electron. **24**(6), 2215–2223 (2016)
4. She, Y., She, X., Baran, M.E.: Universal tracking control of wind conversion system for purpose of maximum power acquisition under hierarchical control structure. IEEE Trans. Energy Convers. **26**(3), 766–775 (2011)
5. Roshandel, E., Gheasaryan, S.M., Mohamadi, M.: A control strategy for DFIG wind turbine using SLS-TLBO and fuzzy logic. In: IEEE 4th International Conference on Knowledge-Based Engineering and Innovation (KBEI), pp. 0108–0113 (2017)
6. Beltran, B., Ahmed-Ali, T., Benbouzid, M.E.H.: Sliding mode power control of variable-speed wind energy conversion systems. IEEE Trans. Energy Convers. **23**(2), 551–558 (2008)
7. Zhao, H., Wu, Q., Rasmussen, C.N., Blanke, M.: L_1 adaptive speed control of a small wind energy conversion system for maximum power point tracking. IEEE Trans. Energy Convers. **29**(3), 576–584 (2014)
8. Mansour, S., Reza, S., Nooshad, Y.: An optimal fuzzy PI controller to capture the maximum power for variable-speed wind turbines. Neural Comput. Appl. **23**(5), 359–1368 (2013)
9. Boukhezzar, B., Siguerdidjane, H., Hand, M.M.: Nonlinear control of variable-speed wind turbines for generator torque limiting and power optimization. J. Sol. Energy Eng. **128**(4), 516–530 (2006)
10. Boukhezzar, B., Siguerdidjane, H.: Nonlinear control of a variable-speed wind turbine using a two-mass model. IEEE Trans. Energy Convers. **26**(1), 149–162 (2011)
11. Beltran, B., AhmedAli, T., Benbouzid, M.E.H.: Sliding mode power control of variable-speed wind energy conversion systems. IEEE Trans. Energy Convers. **23**(2), 551–558 (2008)
12. Beltran, B., Ahmed-Ali, T., Benbouzid, M.E.H.: High-order sliding-mode control of variable-speed wind turbines. IEEE Trans. Ind. Electron. **56**(9), 3314–3321 (2009)
13. Su, X., Wang, H.: Back-stepping active disturbance rejection control design for integrated missile guidance and control system via reduced-order ESO. ISA Trans. **57**(4), 10–22 (2015)
14. Miroslav, K., Ioannis, K., Petar, K.: Nonlinear adaptive control design. Lect. Notes Control. Inf. Sci. **5**(2), 4475–4480 (1995)

Event-Triggered Adaptive Dynamic Programming for Continuous-Time Nonlinear Two-Player Zero-Sum Game

Shan Xue[1], Biao Luo[2(⊠)], Derong Liu[3], and Yueheng Li[1]

[1] School of Automation and Electrical Engineering,
University of Science and Technology Beijing, Beijing 100083, China
shan.xue0807@foxmail.com, yuehengli@hotmail.com
[2] The State Key Laboratory of Management and Control for Complex Systems,
Institute of Automation, Chinese Academy of Sciences, Beijing 100190, China
biao.luo@hotmail.com
[3] School of Automation, Guangdong University of Technology,
Guangzhou 510006, China
derongliu@foxmail.com

Abstract. In this paper, an event-triggered adaptive dynamic programming (ADP) algorithm is developed to solve the two-player zero-sum game problem of continuous-time nonlinear systems. First, a critic neural network is employed to approximate the optimal value function. Then, an event-triggered ADP method is proposed, which guarantees the stability of the closed-loop system. The developed method can save the amount of computation as the control law and disturbance law that update only when the pre-designed triggering condition is violated. Finally, its effectiveness is verified through simulation results.

Keywords: Event-triggering control
Adaptive dynamic programming · Two-player zero-sum game
Hamilton-Jacobi-Isaacs equation

1 Introduction

The development of cutting-edge technologies such as AlphaGo and driverless has made artificial intelligence a hot topic of research [1]. Machine learning, acts as a core technology of artificial intelligence, has been well studied. Therefore, adaptive dynamic programming (ADP) [2–23], an important branch of machine learning, is playing an increasingly important role in solving the optimal control problems of nonlinear systems. It is based on the principle of reinforcement learning, combined with neural network (NN) technology, through the construction of actor NN and critic NN to solve optimal control problems. However, most of these methods are time-triggered scheme, which is difficult to deal with the cases when network resources are limited.

© Springer Nature Switzerland AG 2018
L. Cheng et al. (Eds.): ICONIP 2018, LNCS 11307, pp. 15–25, 2018.
https://doi.org/10.1007/978-3-030-04239-4_2

Event-triggered control technology was firstly proposed in network control systems [24]. This communication mechanism is very effective in the case of limited resources. In recent years, this technology has been introduced into the field of ADP [6,9,12]. It is worth noticing that disturbance is unavoidable in real life. Therefore, the study of the system with disturbance is more practical.

Inspired by [15,18], this paper focus on the two-player zero-sum (ZS) game problem of nonlinear continuous-time systems. First, a novel triggering condition is designed to ensure the stability of the system. Second, a critic NN is used to implement the ZS game problem. Finally, the control law and disturbance law are updated only when the pre-designed triggering condition is violated, which reduces the amount of computation.

The remainder of the paper is organized as follows. Problem descriptions is introduced in Sect. 2. An event-triggered ADP method is described in Sect. 3. In Sect. 4, the effectiveness of the method is proved by simulation results. Some conclusions are presented in Sect. 5.

Notation: $\mathbb{R}$ represents the set of all real numbers. $\mathbb{R}^n$ denotes a set of all real vectors and $\mathbb{R}^{n \times m}$ denotes the set of all real matrices. $\mathbb{N}$ represents the non-negative integers. The superscript T and $\| \cdot \|$ represent the transpose and the 2-norm of a vector or matrix, respectively. I_m describes the identify matrix with dimension m and $\lambda_{\min}(A)$ is the minimal eigenvalue of the matrix A. $\nabla \triangleq \partial / \partial x$ is the gradient operator. $F(t^-)$ means the function $F(\cdot)$ is left continuous.

2   Problem Description

In this section, let us consider the following nonlinear systems:

$$\dot{x}(t) = f(x) + g(x)u(t) + k(x)v(t), x(0) = x_0, \tag{1}$$

where $x \in \mathbb{R}^n$ denotes the state vector, $f(x) \in \mathbb{R}^n$, $g(x) \in \mathbb{R}^{n \times m}$, $k(x) \in \mathbb{R}^{n \times l}$. $u(t) \in \mathbb{R}^m$ and $v(t) \in \mathbb{R}^l$ describe control input and disturbance, respectively. $x = 0$ is an equilibrium of (1). Consider the following form of the performance index,

$$J(x_0, u, v) = \int_0^\infty (x^\mathsf{T} Q x + u^\mathsf{T} R u - \gamma^2 v^\mathsf{T} v) dt,$$

where Q and R are positive definite matrices with appropriate dimensions. γ is a constant represents the disturbance attenuation level, which is given in advance in this paper. The control and disturbance law can be seen as two players is the ZS game problem. Then, the value function is given as

$$V(x(t), u, v) = \int_t^\infty \left(x^\mathsf{T}(\mu) Q x(\mu) + u^\mathsf{T}(\mu) R u(\mu) - \gamma^2 v^\mathsf{T}(\mu) v(\mu) \right) d\mu. \tag{2}$$

In this paper, we aim to design control input $u(x)$, which can not only stabilize the disturbance-free system (1), but also min-maximize the value function (2). For simplicity, use $V(x)$ to represent $V(x(t), u, v)$. Assume that the value function (2) is differentiable and continuous, we have

$$(\nabla V(x))^\mathsf{T}(f + gu + kv) + x^\mathsf{T} Q x + u^\mathsf{T} R u - \gamma^2 v^\mathsf{T} v = 0. \tag{3}$$

The corresponding Hamiltonian is denoted as follows

$$H(\nabla V(x), x, u, v) = (\nabla V(x))^{\mathsf{T}}(f + gu + kv) + x^{\mathsf{T}}Qx + u^{\mathsf{T}}Ru - \gamma^2 v^{\mathsf{T}}v.$$

According to Bellman's optimality principle, the Hamilton-Jacobi-Isaacs (HJI) equation is given by

$$\min_u \max_v H(\nabla V^*(x), x, u, v) = 0, \tag{4}$$

where $V^*(x)$ describes the optimal value function. According to the stationary conditions, we have

$$\frac{\partial H(\nabla V^*(x), x, u, v)}{\partial u} = 0, \tag{5}$$

$$\frac{\partial H(\nabla V^*(x), x, u, v)}{\partial v} = 0. \tag{6}$$

Then, based on (4)–(6), we get

$$u^*(x) = -\frac{1}{2}R^{-1}g^{\mathsf{T}}\nabla V^*(x), \tag{7}$$

$$v^*(x) = \frac{1}{2\gamma^2}k^{\mathsf{T}}\nabla V^*(x), \tag{8}$$

where $u^*(x)$ and $v^*(x)$ are the optimal control input and the worst disturbance, respectively. (u^*, v^*) is the saddle point solution. Then, the HJI Eq. (4) is redescribed as

$$(\nabla V(x))^{*\mathsf{T}}f + x^{\mathsf{T}}Qx - \frac{1}{4}(\nabla V(x))^{*\mathsf{T}}gR^{-1}g^{\mathsf{T}}\nabla V^*(x)$$

$$+ \frac{1}{4\gamma^2}(\nabla V(x))^{*\mathsf{T}}kk^{\mathsf{T}}\nabla V^*(x) = 0, V^*(0) = 0.$$

3   An Event-Triggered Adaptive Dynamic Programming Method

3.1   Event-Triggered Control

Suppose that there is no time delay in the sampled data system. All sampling moments of the system form a monotonically increasing sequence. The event-triggered error $e_j(t)$ is given as

$$e_j(t) = x(s_j) - x(t), \forall t \in [s_j, s_{j+1}),$$

where s_j denotes the j-th sampling moment, $j \in \mathbb{N}$. In the event-triggered control method, the controller is only updated when the pre-designed triggering condition is violated, i.e., when the event-triggered error $e_j(t)$ exceeds the threshold e_t. At each triggering moment s_j, the state is sampled and $e_j(t)$ is reset to zero.

In light of the event-triggered control, the control law and disturbance law are described as

$$u\big(x(s_j)\big) = u\big(e_j(t) + x(t)\big),$$
$$v\big(x(s_j)\big) = v\big(e_j(t) + x(t)\big),$$

respectively. The Eq. (1) becomes

$$\dot{x} = f + gu\big(x(s_j)\big) + kv\big(x(s_j)\big).$$

The associated Hamiltonian is denoted by

$$H\left(\nabla V(x), x, u(x(s_j)), v(x(s_j))\right) = (\nabla V(x))^{\mathsf{T}}(f + gu(x(s_j)) + kv(x(s_j))) + x^{\mathsf{T}}Qx$$
$$+ u^{\mathsf{T}}(x(s_j))Ru(x(s_j)) - \gamma^2 v^{\mathsf{T}}(x(s_j))v(x(s_j)).$$

Based on (7) and (8), the event-triggered control law and the disturbance law are derived as the following form:

$$u^*(x(s_j)) = -\frac{1}{2}R^{-1}g^{\mathsf{T}}(x(s_j))\nabla V^*(x(s_j)),$$
$$v^*(x(s_j)) = \frac{1}{2\gamma^2}k^{\mathsf{T}}(x(s_j))\nabla V^*(x(s_j)).$$

The event-triggered HJI equation can be obtained,

$$(\nabla V(x))^{*\mathsf{T}}f + x^{\mathsf{T}}Qx + \frac{1}{4}(\nabla V(x(s_j)))^{*\mathsf{T}}g(x(s_j))R^{-1}g^{\mathsf{T}}(x(s_j))\nabla V^*(x(s_j))$$
$$- \frac{1}{2}(\nabla V(x))^{*\mathsf{T}}gR^{-1}g^{\mathsf{T}}(x(s_j))\nabla V^*(x(s_j))$$
$$- \frac{1}{4\gamma^2}(\nabla V(x(s_j)))^{*\mathsf{T}}k(x(s_j))k^{\mathsf{T}}(x(s_j))\nabla V^*(x(s_j))$$
$$+ \frac{1}{2\gamma^2}(\nabla V(x))^{*\mathsf{T}}kk^{\mathsf{T}}(x(s_j))\nabla V^*(x(s_j)) = 0, V^*(0) = 0.$$

3.2   Implementation

To implement the event-triggered control, the value function $V(x)$ is denoted as

$$V(x) = C^{*\mathsf{T}}\Psi(x) + \Upsilon(x), \tag{9}$$

where $C^* \in \mathbb{R}^c$ is the ideal weight vector, c denotes the number of hidden neurons and $\Psi(x) \in \mathbb{R}^c$ denotes the activation function. $\Upsilon(x) \in \mathbb{R}$ represents the reconstruction error. Due to the ideal weights C^* are unknown, a critic NN is employed to approximated $V(x)$, i.e.,

$$\hat{V}(x) = \hat{C}^{\mathsf{T}}\Psi(x), \tag{10}$$

where $\hat{C} \in \mathbb{R}^c$ is an estimation of C^*. Then,

$$\hat{u}(x(s_j)) = -\frac{1}{2}R^{-1}g^{\mathsf{T}}(x(s_j))\nabla\Psi^{\mathsf{T}}(x(s_j))\hat{C}, \tag{11}$$

$$\hat{v}(x(s_j)) = \frac{1}{2\gamma^2}k^{\mathsf{T}}(x(s_j))\nabla\Psi^{\mathsf{T}}(x(s_j))\hat{C}. \tag{12}$$

Based on (10), the Hamiltonian is approximated as

$$\hat{H}(\hat{C}, x, \hat{u}(x(s_j)), \hat{v}(x(s_j))) = \hat{C}^{\mathsf{T}}\nabla\Psi(x)\dot{x} + x^{\mathsf{T}}Qx + \hat{u}^{\mathsf{T}}(x(s_j))R\hat{u}(x(s_j))$$
$$- \gamma^2\hat{v}^{\mathsf{T}}(x(s_j))\hat{v}(x(s_j))$$
$$\triangleq e_{\hat{c}}.$$

The weights $\hat{C}$ are tuned such that the objective function $E = \frac{1}{2}e_{\hat{c}}^{\mathsf{T}}e_{\hat{c}}$ is minimized. Based on [15,25], the normalized steepest descent algorithm is applied to updated the weight vector. Thus, we have

$$\dot{\hat{C}} = -\frac{\varrho}{(1+\Theta^{\mathsf{T}}\Theta)^2}\left(\frac{\partial E}{\partial\hat{C}}\right)^{\mathsf{T}} = -\frac{\varrho\Theta e_{\hat{c}}^{\mathsf{T}}}{(1+\Theta^{\mathsf{T}}\Theta)^2}, \tag{13}$$

where $\Theta = \nabla\Psi(x)\dot{x}$ and $\varrho > 0$ describes the learning rate. The weight estimation error is defined as

$$\tilde{C} = C^* - \hat{C}. \tag{14}$$

Then, we can obtain

$$\dot{\tilde{C}} = \frac{\varrho\Theta}{(1+\Theta^{\mathsf{T}}\Theta)^2}((C^{*\mathsf{T}} - \tilde{C}^{\mathsf{T}})\nabla\Psi(x)\dot{x} + x^{\mathsf{T}}Qx$$
$$+ \hat{u}^{\mathsf{T}}(x(s_j))R\hat{u}(x(s_j)) - \gamma^2\hat{v}^{\mathsf{T}}(x(s_j))\hat{v}(x(s_j)))$$
$$= \frac{\varrho\Theta}{(1+\Theta^{\mathsf{T}}\Theta)^2}(e_h - \Theta^{\mathsf{T}}\tilde{C}),$$

where $e_h = -(\nabla\Upsilon(x))^{\mathsf{T}}\dot{x}$ denotes the residual error [15,18,25,26].

3.3   Stability Analysis

Assumption 1. $g(x)$ and $\nabla\Psi(x)$ are Lipschitz continuous, with positive numbers L_g and L_Ψ. $\|\nabla\Psi(x)\|$, $\|\nabla\Upsilon(x)\|$ and $\|e_h\|$ have positive upper bound $\nabla\Psi_M$, $\nabla\Upsilon_M$ and e_M, respectively.

Theorem 1. *Consider system (1) with the critic network (10), the corresponding updating law (13), the event-triggered laws (11) and (12). Assume that Assumption 1 is satisfied. If the event-triggering threshold satisfies the following equality,*

$$\|e_t\|^2 = \frac{(1-\Lambda_0^2)x^{\mathsf{T}}Qx + \hat{u}^{\mathsf{T}}(x(s_j))R\hat{u}(x(s_j)) - \gamma^2\hat{v}^{\mathsf{T}}(x(s_j))\hat{v}(x(s_j))}{R^{-1}\Lambda_1^2\|\hat{C}\|^2}, \tag{15}$$

and the weight estimation error $\tilde{C}$ satisfies

$$\|\tilde{C}\|^2 \geq \frac{\Lambda_3 + \Lambda_4}{\Lambda_5 - \Lambda_2}, \tag{16}$$

where $\Lambda_1^2 = L_g^2 g_M^2 + L_\Psi^2 \nabla \Psi_M^2$, $\Lambda_2 = \frac{1}{2} R^{-1} g_M^2 \nabla \Psi_M^2$, $\Lambda_3 = \frac{1}{2} R^{-1} g_M^2 \nabla \Upsilon_M^2$, $\Lambda_0 \in (0,1)$, $\Lambda_4 = \frac{\varrho e_M^2}{2\lambda_{\min}(1+\Theta^\mathsf{T}\Theta)^\mathsf{T}(1+\Theta^\mathsf{T}\Theta)}$ and $\Lambda_5 = \frac{\varrho\lambda_{\min}(\Theta^\mathsf{T}\Theta)}{2\lambda_{\min}(1+\Theta^\mathsf{T}\Theta)^\mathsf{T}(1+\Theta^\mathsf{T}\Theta)}$ are positive constants. $\Lambda_5 > \Lambda_2$ can be satisfied by designing an appropriate learning rate. Then, we have x, $x(s_j)$, $\tilde{C}$ are uniformly ultimately bounded.

Proof. Select the following Lyapunov function,

$$L_c(t) = L_{c1}(t) + L_{c2}(t) + L_{c3}(t), \tag{17}$$

where $L_{c1}(t) = V^*(x)$, $L_{c2}(t) = V^*(x(s_j))$ and $L_{c3}(t) = \frac{1}{2}\tilde{C}^\mathsf{T}\tilde{C}$.
   The proof process is divided into two parts.

Part 1: Event is not triggered. Considering $\dot{L}_{c1}$, we obtain

$$\begin{aligned}
\dot{L}_{c1}(t) &= (\nabla V(x))^{*\mathsf{T}} f + (\nabla V(x))^{*\mathsf{T}} g\hat{u}(x(s_j)) + (\nabla V(x))^{*\mathsf{T}} k\hat{v}(x(s_j)) \\
&= -x^\mathsf{T} Q x + u^{*\mathsf{T}} R u^* - \gamma^2 v^{*\mathsf{T}} v^* - 2u^{*\mathsf{T}} R\hat{u}(x(s_j)) + 2\gamma^2 v^{*\mathsf{T}}\hat{v}(x(s_j)) \\
&= -x^\mathsf{T} Q x + R\|u^* - \hat{u}(x(s_j))\|^2 - \gamma^2\|v^* - \hat{v}(x(s_j))\|^2 \\
&\quad - \hat{u}^\mathsf{T}(x(s_j))R\hat{u}(x(s_j)) + \gamma^2\hat{v}^\mathsf{T}(x(s_j))\hat{v}(x(s_j)) \\
&\leq -x^\mathsf{T} Q x + R\|u^* - \hat{u}(x(s_j))\|^2 - \hat{u}^\mathsf{T}(x(s_j))R\hat{u}(x(s_j)) \\
&\quad + \gamma^2\hat{v}^\mathsf{T}(x(s_j))\hat{v}(x(s_j)).
\end{aligned}$$

Based on (9), (11), and (14), we have

$$\begin{aligned}
R\|u^* - \hat{u}(x(s_j))\|^2 = \frac{1}{4} R^{-1} \| &g^\mathsf{T}(x)(\nabla\Psi(x)^\mathsf{T}\tilde{C} + \nabla\Upsilon(x)) \\
&+ (g^\mathsf{T}(x)\nabla\Psi(x)^\mathsf{T} - g^\mathsf{T}(x(s_j))(\nabla\Psi(x(s_j)))^\mathsf{T})\hat{C}\|^2.
\end{aligned}$$

According to Assumption 1, we obtain

$$\begin{aligned}
&\|g^\mathsf{T}(x)\nabla\Psi^\mathsf{T}(x) - g^\mathsf{T}(x(s_j))\nabla^\mathsf{T}\Psi(x(s_j))\|^2 \\
&\leq 2(\|\nabla\Psi(x)(g(x) - g(x(s_j)))\|^2 + \|(\nabla\Psi(x) - \nabla\Psi(x(s_j)))g(x(s_j))\|^2) \\
&\leq 2\Lambda_1^2\|e_j(t)\|^2. \tag{18}
\end{aligned}$$

Based on (18), we have

$$R\|u^* - \hat{u}(x(s_j))\|^2 \leq \Lambda_1^2 R^{-1}\|e_j(t)\|^2\|\hat{C}\|^2 + \Lambda_2\|\tilde{C}\|^2 + \Lambda_3.$$

Then, we have

$$\begin{aligned}
\dot{L}_{c1}(t) \leq &-x^\mathsf{T} Q x - \hat{u}^\mathsf{T}(x(s_j))R\hat{u}(x(s_j)) + \gamma^2\hat{v}^\mathsf{T}(x(s_j))\hat{v}(x(s_j)) \\
&+ \Lambda_1^2 R^{-1}\|e_j(t)\|^2\|\hat{C}\|^2 + \Lambda_2\|\tilde{C}\|^2 + \Lambda_3.
\end{aligned}$$

Due to event is not triggered, $\dot{L}_{c2}(t) = 0$. Then, we can obtain

$$
\begin{aligned}
\dot{L}_{c3}(t) &= \frac{\tilde{C}^{\mathsf{T}}\varrho\Theta(e_h - \Theta^{\mathsf{T}}\tilde{C})}{(1+\Theta^{\mathsf{T}}\Theta)^2} \\
&\leq \frac{\varrho(\frac{1}{2}(\tilde{C}^{\mathsf{T}}\Theta)^{\mathsf{T}}(\tilde{C}^{\mathsf{T}}\Theta) + \frac{1}{2}e_h^2 - \tilde{C}^{\mathsf{T}}\Theta\Theta^{\mathsf{T}}\tilde{C})}{(1+\Theta^{\mathsf{T}}\Theta)^2} \\
&\leq \frac{\varrho(e_M^2 - \lambda_{\min}(\Theta\Theta^{\mathsf{T}})\|\tilde{C}\|^2)}{2\lambda_{\min}(1+\Theta^{\mathsf{T}}\Theta)^{\mathsf{T}}(1+\Theta^{\mathsf{T}}\Theta)} \\
&= \Lambda_4 - \Lambda_5\|\tilde{C}\|^2.
\end{aligned}
$$

Hence, we have

$$
\begin{aligned}
\dot{L}_c \leq &- \Lambda_0^2 x^{\mathsf{T}}Qx - (1-\Lambda_0^2)x^{\mathsf{T}}Qx - \hat{u}^{\mathsf{T}}(x(s_j))R\hat{u}(x(s_j)) + \gamma^2\hat{v}^{\mathsf{T}}(x(s_j))\hat{v}(x(s_j)) \\
&+ \Lambda_1^2 R^{-1}\|e_j(t)\|^2\|\hat{C}\|^2 + \Lambda_2\|\tilde{C}\|^2 + \Lambda_3 + \Lambda_4 - \Lambda_5\|\tilde{C}\|^2.
\end{aligned}
$$

According to (15) and (16), for all $x \neq 0$, $\dot{L}_c(t) < -\Lambda_0^2 x^{\mathsf{T}}Qx < 0$.

Part 2: Event is triggered. By using (17), we have

$$
\Delta L_c(t) = \Delta L_{c1}(t) + \Delta L_{c2}(t) + \Delta L_{c3}(t),
$$

where $\Delta L_{c1}(t) = V^*(x(s_{j+1})) - V^*(x(s_{j+1}^-))$, $\Delta L_{c2}(t) = V^*(x(s_{j+1})) - V^*(x(s_j))$ and $\Delta L_{c3}(t) = \frac{1}{2}\tilde{C}^{\mathsf{T}}(x(s_{j+1}))\tilde{C}(x(s_{j+1})) - \frac{1}{2}\tilde{C}^{\mathsf{T}}(x(s_{j+1}^-))\tilde{C}(x(s_{j+1}^-))$. Since the system state is continuous, we have $\Delta L_{c1} \leq 0$. Noting that $\tilde{C}$ is uniformly ultimately bounded for Part 1, we have $\Delta L_{c3}(t) \leq 0$. Then, $\Delta L_{c2}(t) \leq -v(\|\hat{e}_{j+1}(s_j)\|)$, where $\hat{e}_{j+1}(s_j) = x(s_{j+1}) - x(s_j)$ and $v(\cdot)$ is a class-κ function [27]. It can be conclude that (17) is still decreasing. $\square$

4   Simulation Results

We use the following form system for simulation results, i.e.,

$$
\dot{x} = \begin{bmatrix} x_2 \\ \frac{-x_1 + 0.2x_4^2\sin x_3}{1-0.04\cos^2 x_3} \\ x_4 \\ \frac{0.02\cos x_3(x_1 - 0.02x_4^2\sin x_3)}{1-0.04\cos^2 x_3} \end{bmatrix} + \begin{bmatrix} 0 \\ \frac{-0.2\cos x_3}{1-0.04\cos^2 x_3} \\ 0 \\ \frac{1}{1-0.04\cos^2 x_3} \end{bmatrix} u + \begin{bmatrix} 0 \\ \frac{1}{1-0.04\cos^2 x_3} \\ 0 \\ \frac{-0.2\cos x_3}{1-0.04\cos^2 x_3} \end{bmatrix} v
$$

where the state $x = [x_1, x_2, x_3, x_4]^{\mathsf{T}}$. u and v are control and disturbance laws, respectively. Let $x_0 = [1, -1, 1, -1]^{\mathsf{T}}$, $\gamma = 10$, $R = 1$, $Q = I_4$, $\Lambda_0 = 0.5$, $\Lambda_1 = 5$ and $\varrho = 0.5$. The sampling time is set to 0.01s. The initial weights of $\hat{C}$ are randomly selected between $[-1, 1]$. The activation function is set to the following form [8]

$$
\begin{aligned}
\Psi(x) = &[x_1^2, x_1 x_2, x_1 x_3, x_1 x_4, x_2^2, x_2 x_3, x_2 x_4, x_3^2, x_3 x_4, x_4^2, x_1^3 x_2, \\
&x_1^3 x_3, x_1^3 x_4, x_1^2 x_2^2, x_1^2 x_2 x_3, x_1^2 x_2 x_4, x_1^2 x_3^2, x_1^2 x_3 x_4, x_1^2 x_4^2, x_1 x_2^3].
\end{aligned}
$$

It can be seen from Fig. 1 that the event-triggered sampling state $x(s_j)_1$ is a segmented signal and the event-triggered and time-triggered state adjustments are similar. The adjustment process of other state variables is similar to $x(s_j)_1$. Figure 2 describes the adjustment of the event-triggered control input. Figure 3 denotes the adjustment process of the $||\hat{C}||$. In Fig. 4 states the error and threshold adjustment process. The time-triggered method obtained an optimal control used 15000 samples, while the event-triggered method used only 985 samples. This can be seen from Fig. 5. Figure 6 denotes the adjustment of the sampling period, the maximum period is 0.18 s. Therefore, simulation results show that the algorithm can effectively much reduce the amount of calculations and save resources.

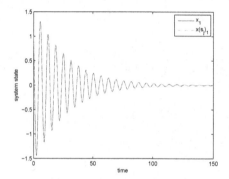

Fig. 1. The adjustment of x_1 and $x(s_j)_1$.

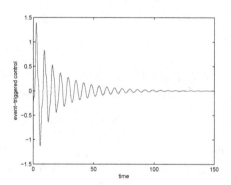

Fig. 2. The adjustment of u.

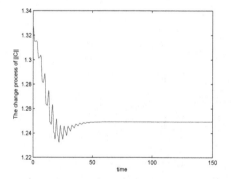

Fig. 3. The adjustment of $||\hat{C}||$.

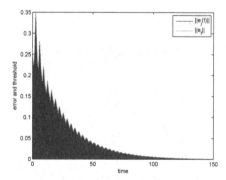

Fig. 4. The adjustment of error and threshold.

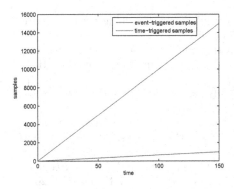

Fig. 5. The adjustment of samples.

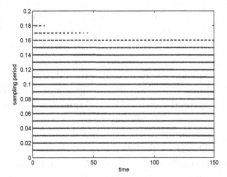

Fig. 6. The adjustment of sampling period.

5   Conclusions

In this paper, an event-triggered ADP algorithm is applied to solve the two-player ZS game problem. A critic NN is employed to approximate the value function. Then, the control law and the disturbance law are updated when the pre-designed triggering condition is violated. The stability of the system is guaranteed. The practically of the method is demonstrated by simulation results.

Acknowledgements. This work was supported in part by the National Natural Science Foundation of China under Grants 61873350, 61503377, 61533017 and U1501251.

References

1. Singh, S., Okun, A., Jackson, A.: Artificial intelligence: learning to play go from scratch. Nature **550**(7676), 336–337 (2017)
2. Liu, D., Wei, Q., Wang, D., Yang, X., Li, H.: Adaptive Dynamic Programming with Applications in Optimal Control. Springer, Cham (2017). https://doi.org/10.1007/978-3-319-50815-3
3. Wang, F.Y., Zhang, H., Liu, D.: Adaptive dynamic programming: an introduction. IEEE Comput. Intell. Mag. **4**(2), 39–47 (2009)
4. Zhao, Q., Xu, H., Jagannathan, S.: Near optimal output feedback control of nonlinear discrete-time systems based on reinforcement neural network learning. IEEE/CAA J. Autom. Sin. **1**(4), 372–384 (2014)
5. Luo, B., Wu, H.N., Huang, T.: Optimal output regulation for model-free quanser helicopter with multi-step Q-learning. IEEE Trans. Ind. Electron. **65**(6), 4953–4961 (2018)
6. Wang, D., Mu, C., He, H., Liu, D.: Event-driven adaptive robust control of nonlinear systems with uncertainties through NDP strategy. IEEE Trans. Syst. Man Cybern.: Syst. **47**(7), 1358–1370 (2017)
7. Luo, B., Wu, H.N., Huang, T., Liu, D.: Data-based approximate policy iteration for affine nonlinear continuous-time optimal control design. Automatica **50**(12), 3281–3290 (2014)

8. Luo, B., Wu, H.N., Huang, T.: Off-policy reinforcement learning for H_∞ control design. IEEE Trans. Cybern. **45**(1), 65–76 (2015)

9. Zhu, Y., Zhao, D., He, H., Ji, J.: Event-triggered optimal control for partially unknown constrained-input systems via adaptive dynamic programming. IEEE Trans. Ind. Electron. **64**(5), 4101–4109 (2017)

10. Luo, B., et al.: Policy gradient adaptive dynamic programming for data-based optimal control. IEEE Trans. Cybern. **47**(10), 3341–3354 (2017)

11. Luo, B., Yang, Y., Liu, D.: Adaptive Q-learning for data-based optimal output regulation with experience replay. IEEE Trans. Cybern. https://doi.org/10.1109/TCYB.2016.2623859 (2018)

12. Dong, L., Tang, Y., He, H., Sun, C.: An event-triggered approach for load frequency control with supplementary ADP. IEEE Trans. Power Syst. **32**(1), 581–589 (2017)

13. Luo, B., Wu, H.N., Li, H.X.: Adaptive optimal control of highly dissipative nonlinear spatially distributed processes with neuro-dynamic programming. IEEE Trans. Neural Netw. Learn. Syst. **26**(4), 684–696 (2015)

14. Luo, B., Huang, T., Wu, H.N., Yang, X.: Data-driven H_∞ control for nonlinear distributed parameter systems. IEEE Trans. Neural Netw. Learn. Syst. **26**(11), 2949–2961 (2015)

15. Wang, D., Mu, C., Liu, D., Ma, H.: On mixed data and event driven design for adaptive-critic-based nonlinear H_∞ control. IEEE Trans. Neural Netw. Learn. Syst. **29**(4), 993–1005 (2018)

16. Luo, B., Liu, D., Huang, T., Wang, D.: Model-free optimal tracking control via critic-only Q-learning. IEEE Trans. Neural Netw. Learn. Syst. **27**(10), 2134–2144 (2016)

17. Luo, B., Liu, D., Wu, H.N.: Adaptive constrained optimal control design for data-based nonlinear discrete-time systems with critic-only structure. IEEE Trans. Neural Netw. Learn. Syst. **29**(6), 2099–2111 (2018)

18. Zhang, Q., Zhao, D., Zhu, Y.: Event-triggered H_∞ control for continuous-time nonlinear system via concurrent learning. IEEE Trans. Syst. Man Cybern.: Syst. **47**(7), 1071–1081 (2017)

19. Luo, B., Wu, H.N.: Approximate optimal control design for nonlinear one-dimensional parabolic PDE systems using empirical eigenfunctions and neural network. IEEE Trans. Syst. Man Cybern. Part B (Cybernetics) **42**(6), 1538–1549 (2012)

20. Xue, S., Luo, B., Liu, D.: Event-triggered adaptive dynamic programming for zero-sum game of partially unknown continuous-time nonlinear systems. IEEE Trans. Syst. Man Cybern.: Syst. (2018). https://doi.org/10.1109/TSMC.2018.2852810

21. Luo, B., Liu, D., Huang, T., Liu, J.: Output tracking control based on adaptive dynamic programming with multistep policy evaluation. IEEE Trans. Syst Man Cybern.: Syst. (2017). https://doi.org/10.1109/TSMC.2017.2771516

22. Luo, B., Wu, H.N., Li, H.X.: Data-based suboptimal neuro-control design with reinforcement learning for dissipative spatially distributed processes. Industr. Eng. Chem. Res. **53**(19), 8106–8119 (2014)

23. Luo, B., Wu, H.N.: Online policy iteration algorithm for optimal control of linear hyperbolic PDE systems. J. Process Control **22**(7), 1161–1170 (2012)

24. Tabuada, P.: Event-triggered real-time scheduling of stabilizing control tasks. IEEE Trans. Autom. Control **52**(9), 1680–1685 (2007)

25. Zhong, X., He, H.: An event-triggered ADP control approach for continuous-time system with unknown internal states. IEEE Trans. Cybern. **47**(3), 683–694 (2017)

26. Vamvoudakis, K.G.: Event-triggered optimal adaptive control algorithm for continuous-time nonlinear systems. IEEE/CAA J. Autom. Sin. **1**(3), 282–293 (2014)
27. Khalil, H.K.: Noninear Systems, pp. 1–5. Prentice-Hall, New Jersey (1996)

A Learning Based Recovery for Damaged Snake-Like Robots

Zhuoqun Guan, Jianping Huang, Zhiyong Jian, Linlin liu, Long Cheng,
and Kai Huang[✉]

School of Data and Computer Science, Sun Yat-sen University, Guangzhou 511400,
Guangdong Province, China
huangk36@mail.sysu.edu.cn

Abstract. Snake-like robots have been widely studied and developed to exploit their flexible mobility and versatility. However, when encoutering powerful damages, how to recover the functionality is seldom investigated. This paper proposed a trial-and-error learning approach for damage recovery for 3-dimensional snake-like robots. The proposed method can guide snake-like robots to find compensation behavior in the absence of the pre-specified damage models. Our proposed method is evaluated by experiments in real world and various simulations.

Keywords: Snake-like robot · Damage recovery

1 Introduction

Snake-like robots [13] are a specific kind of robots that mimic the body structure as well as kinematics of snakes, in order to move in different complex terrains, such as deserts, woods, swamp lakes, and other amphibious environments. Although snake-like robots have been continuously developed for more than forty years, deploying such robots in real scenarios for practical uses is still limited. The complex body structure brings difficulties to design an appropriate dynamic model. Even for an appropriate dynamic model, designing a corresponding control is non-trivial. Therefore, most existing snake-like robots work in a predefined manner and lack the ability to adapt itself to the environment.

Utilizing predefined gaits for movement has one critical drawback. Once the robot is partially damaged, the model of the snake robot, which is the assumption for gait design, will change. In this case, the predefined gaits are not applicable anymore for the damaged robot. While animals can quickly adapt to a variety of injuries, current robots lack of the capability to find compensatory behavior by "thinking outside the box" when they are damaged.

The current damage recovery of a robot usually involves two stages: the self-diagnosis and the best pre-designed contingency plan [2]. It is impractical for snake-like robots. The reason is multi-folds. First, snake-like robots usually have multiple joints and multiple degrees of freedom. It is difficult to foresee all possible damage situations and pre-designed plans for all situations [1,3,9]. Second,

© Springer Nature Switzerland AG 2018
L. Cheng et al. (Eds.): ICONIP 2018, LNCS 11307, pp. 26–39, 2018.
https://doi.org/10.1007/978-3-030-04239-4_3

snake-like robots usually have contingent weight and space requirements, which do not allow full fledged self-monitoring sensors. In this case, a sophisticated self-diagnosis system is not possible even when recovery plan can be defined.

One possible way to conduct damage recovery for snake-like robots is to imitate animals through testing and error detection [6]. Recently, Mouret et al. introduced a new algorithm: Intelligent Trial and Error (IT&E) method [5] that was validated in a variety of ways on six-legged robots and robot arm. Most damage compensation can be found in 2 min by the six-legged robot, and it works better than most existing algorithms [5]. The motions of the snake-like robots are far more complex than the six-legged robots because the snake-like robot has many joints which are mutually independent and need different controllers. This controllers make up the serpenoid curve and all the controllers have different range which don't like the controllers of six-legged robot. So that this method is not directly applicable for snake-like robots.

This paper used a advanced trial-and-error learing approach for damage recovery for 3-dimentional snake-like robots. Our approach helps snake-like robot learn compensation behavior autonomously and then go out of the current environment without being artificially repaired. Robots can even continue to complete their tasks autonomously or return to safe spots. We complete damage recovery on the snake-like robot, not only in the simulation environment, but also on our real snake-like robot. In the simulation environment, the experimental results which use the damage recovery method can be 17 times better than the results which didn't use this method. The speed and motor pattern of snkae-like robot which didn't use this method is bad and we think it didn't find the compensation behavior. On the real snake-like robot platform, the experimental results which use the damage recovery method can be 15 times higher than the results which don't use this method. The contributions of our work are summarized as follows.

- In order to meet our experimental requirements, we extend the serpenoid curve model to the orthogonality structure snake-like robot. The new formula covers all the serpenoid curve motion modes of controlling snake-like robot through seven parameters. And by the formula and seven parameters, we define the snake-like robot's controller space.
- To cope with the structure of snake-like robot, we extend the IT&E algorithm with respect to our snake-like robot. We determine IT&E's main parameters through a certain amount of experimental data and support vector machine (SVM) to make IT&E more suitable for the platform of snake-like robots and make IT&E achieve better results.
- We firstly apply damage recovery to snake-like robots. We evaluate our approach on both simulation and real-life prototype.

2   Related Work

The first motion model of snakes can be dated bcak to 1946 [8]. Not until 1993, the first snake-like robot was developed by Hirose et al. [7]. In 2001, Hirose developed the snake-like robot ACM-R3, which used Orthogonal connection so that it can work on the 3d plane. Howie Choset [11], developed a series of snake-like robots. The most famous robot is "Uncle Sam". "Uncle Sam" firstly realized tree climbing motion. Some other scientists and research institutes also contributed to the development of snake-like robots, such as NASA, GMD lab in Germany [14], university of michigan [15], Kyriakopoulos [10], Nilsson [12] etc.

For damage recovery, most researches focus on legged-type robots. Vonásek et al. proposed a failure recovery for modular robot movements without reassembling modules that would allow the robot to move after some modulars failed. Mouret et al. propose a new improved method, Reset Free Trial and Error Learning: the researchers need to reset the robot when researchers proceed the trial and error process. The method allows robots with efficient data feedback to test multiple behaviors and select the optimal compensation behavior ultimately without resetting. They test the method on six-legged robots in a variety of ways, and without any intervention, six-legged robots could recover most of their exercise capacity in minutes [4].

Because different robots have different controller, all these aforementioned approaches cannot directly apply to 3-dimensioned snake-like robots.

3   Background

This section introduces the process and basis of the IT&E algorithm. IT&E algorithm has two main part. The one refers to the creation steps of the controller-performance map, which is created by the multi-dimensional archive of phenotypic elites (MAP-Elites) algorithm. The other refers to the adaptation step, which is completed by a map-based Bayesian optimization algorithm (M-BOA). The controller space is made up of multiple groups of controller parameters and the performance space comprises the same amount of indexes quantified by the performance of robots. In general, a part of controller parameters are selected as a group to add in the controller space for the robot of avoiding the dimension disaster. Under such background, MAP-Elites is used to solve how to select a part of controller parameters and recombine them to produce a new group.

At the beginning of the algorithm, it randomly generating a set of candidate controller points, each of which corresponds to a group of controller parameters. Once the initialization process is complete, MAP-Elites enters a loop, in which the algorithm make a mutation on the selected candidate point, obtaining a new one. As for each mutation point, it get involve in the process of simulation, which outputs a performance index. Finally, a new pair of controller-performance values are stuck in the map and next iteration begins.

The adaptation process is implemented through the M-BOA algorithm, a Bayesian optimization based on the controller-performance map. M-BOA is a process which starts from prediction of the controller-performance map and corrected by Gaussian process constantly. It includes the following subprocess: selecting the next controller which has the optimal expectation, building and updating the Gaussian process and judging whether the adaptation step should be stopped. The information acquisition function selects the next controller that will be evaluated on the physical robot. The information acquisition function used in here is Upper Confidence Bound (UCB).

There are two stop conditions. One is that the performance is less than 90% maximum performance predicted in the controller-performance map. The other is the iteration process excessing a threshold value. The process flow chart can be seen in the Fig. 1.

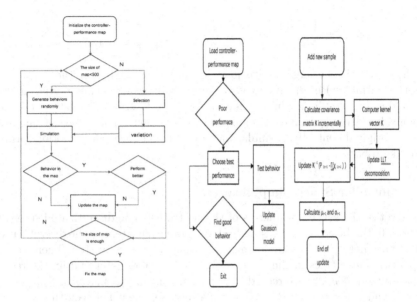

Fig. 1. The process flow chart of MAP-Elites, M-BOA and the update process of Guassion model.

4 Our Recovery Approach

In this section, we extend the controller formula and simplify the MAP-Elites step in order to make IT&E algorithm better perform on the damaged recovery of snake-like robots.

4.1 Controller Extension

The classical motion control formula of snake-like robots uses serpenoid curve with four parameters [7]: α, β, γ and ω. α controls the rotation amplitude of each joint angle; β controls the phase difference between two adjacent joints; γ controls the intermediate state when robot is moving; and ω controls the speed of movement. The formula is shown in Eq. 1.

$$\phi_i(t) = \alpha sin(\omega t + (i-1) * \beta) + \gamma \qquad i = 1, 2, ..., n-1 \qquad (1)$$

However, the above formula can only control serpentine motion. In general, snake-like robot's movement is symmetrical. So we consider γ is constant. In this paper, the structure of snake-like robot is morphometric. Thus, when the serpenoid curve is adopted to describe the gait of snake-like robot, the lateral and lengthwise joint should be considered. The phase difference between the lateral and lengthwise joints is 90°, so we can get a new formula to describe our snake-like robot (Eq. 2):

$$\phi_i(t) = \begin{cases} \alpha_1 sin(\omega t + \theta_s + i * \theta_1) & i = 1, 3, 5, ... \\ \alpha_2 sin(\omega t + \theta_s + \theta_p + i * \theta_2) & i = 2, 4, 6, ... \end{cases} \tag{2}$$

where $\phi_i(t)$ is the angle of ith joint at time t, α_1 and α_2 is the amplitude, θ_s is the initial phase, θ_1 and θ_2 is the phase difference between adjacent two joint and θ_p is the holistic phase difference between lateral and lengthwise joints. We consider later joints have same speed as lengthwise joints have. Compared with the original controller formula, the new one can cover more motion form of snake-like robots.

4.2   Controller Space Simplification

We discretize the seven parameters of the controller. The discretizated steps are more small, the best parameters we find are more accurate. For the discretization of each parameter, we first divide each parameter by a small step. Then we test the performance values in the case of each specific parameter and fix the rest of the parameters. Finally, we record each parameter and the performance value in the test, and calculate the Pearson correlation coefficient between them.

Because the controllers are continuous, we need to discretize them. Although we can refine the discrete operations, but too much refining operations can not improve the experiment result too much and will cost a lot of time when we realize the map creation step and the adaptation step.

We use the normalized Pearson correlation coefficient as the standard to determine the granularity of the discretization of parameters. First, a reference for the number of discrete values is selected as the standard granularity. Then the standard granularity is weighted by the Pearson correlation coefficient of each parameter to get the granularity of the discretization of parameters, so that the number of discrete values of each parameter is obtained.

By this way, we can balance the influence of discretization steps and the number of discrete values to reduce the accuracy loss of discretization and the time cost of sampling.

The value of controller are listed in Table 1, including step size (SS), initial value (IV) and the number of values (NV) and its reference (RNV). Moreover, if each joint uses independent control parameters, the whole machine needs to be controlled by $7*N$ (N means joint numbers) parameters. Since this paper focuses on damaged recovery rather than the control of snake-like robots, we adopt a classic scheme that all joints are controlled by a same set of parameters in order to reduce complexity of controlling and concentrate on damaged recovery.

Table 1. Controllers' value of the serpenoid curve. SS means step size. IV means initial value. NV means number of values

Parameter	Data range	SS, IV, NV
α_1	$0° - 90°$	$\frac{90°}{NV_{\alpha_1}}, 0°, RNV*\mathrm{W}_{\alpha_1}$
α_2	$0° - 90°$	$\frac{90°}{NV_{\alpha_2}}, 0°, RNV*\mathrm{W}_{\alpha_2}$
ω	$1 - \omega_{MAX}$	$\frac{\omega_{MAX}}{NV_\omega}, 1, RNV*\mathrm{W}_\omega$
θ_s	$0° - 360°$	$\frac{360°}{NV_{\theta_s}}, 0°, RNV*\mathrm{W}_{\theta_s}$
θ_1	$0° - 360°$	$\frac{360°}{NV_{\theta_1}}, 0°, RNV*\mathrm{W}_{\theta_1}$
θ_2	$0° - 360°$	$\frac{360°}{NV_{\theta_2}}, 0°, RNV*\mathrm{W}_{\theta_2}$
θ_p	$0° - 360°$	$\frac{360°}{NV_{\alpha_p}}, 0°, RNV*\mathrm{W}_{\alpha_p}$

4.3 Internal Parameter Determination

Since IT&E is an algorithmic framework, there are quite a number of variable parameters in which they can take different values to adjust the execution results of the algorithm. For the sake of simplicity, in the original IT&E algorithm, the value of these parameters are only determined according to the experience. For example, the parameter ν in the kernel function of Gaussian process takes a empirical value of 2.5.

We want to get a better selection of the variable parameters through regression algorithm. In this paper, we count all the variable parameters in the IT&E algorithm, and use their permutations to produce different models, in order to avoid the influence of human subjective factors. Then, the models and their test results constitute the training set, which is put into SVM model. We use the libsvm (A Library for Support Vector Machines[1]) We use cross-validation to find parameters which perform well. We use the SVR with the epsilon to build the regression model. Because SVR is fast, accurate and sophisticated. Our program mainly focuses on choosing the best values of gamma in kernel function, the parameter C (cost) of epsilon-SVR and the type of kernel function. At last, we put all the combinations into the regression model and get a parameter selection model. With this model, we can get the optimal combination of variable parameters so as to avoid empirical error.

Although this process is time-consuming, it can be carried out offline, and exactly executes one time to find the best parameter value. Therefore, it doesn't prolong the time needed for damage recovery.

4.4 Complexity Analysis

In the controller-performance map stage, each motion needs to enter the controller-performance map, which means it has to go through an equal-probability selection and an equal-probability mutation process. Then, through

---

[1] https://www.csie.ntu.edu.tw/~cjlin/libsvm/.

simulation test we will get the performance value and put the controller, performance to the map. These operations mentioned above are only related to themselves. So the time complexity and space complexity of establishing the controller-performance map are both $O(N_c)$. N_c is the size of controller-performance map.

In the adaptation step, trial and error process mainly comprises optimal controller searching, controller testing, model updating, end condition judging.

- Optimal controller searching. We use global traversal method in this paper. As for every controller data, we need covariance matrix to get correspond μ and σ. In this step, we need to perform inverse operation of covariance matrix. Thus, in this step, the time complexity is $O(N_c * N_t^3)$ where N_t is the number of iterations. So the complexity of the process is $O(N_c * N_t^4)$.
- Controller testing. These tests cost constant time, so the complexity is $O(1)$.
- Updated by GP. Gaussian process needs to compute the inverse of covariance matrix K and the product of covariance matrix K, kernel vector k and observed value vector $P_{1:t}$. The time complexity of the inverse operation of covariance matrix K is $O(t^3)$, where t is the number of observed values. So the time complexity of updating model is $O(N_t^4)$
- Judge end condition. This process will traverses the controller-performance map. So the time complexity of this process is $O(max(N_c, N_t))$ and the complexity of this process is $O(N_c * N_t)$.

From the above, the time complexity of damage recovery is $O(N_c * N_t^4)$ as $N_t > 1$ and $N_c > 1$.

5   Experiment Design and Realization

Intelligent trial and error learning algorithm has two main parts. First part is the realization of the controller-performance map, also known as data collecting process. Second part is the adaptation step.

Fig. 2. We use V-REP to simulate our snake-like robot. The model has 17 joints, which include 16 body-joints and 1 head-joint.

In different snake-like robots and even different robots, the parameters setting of the controller space in the data collecting process are different. In our experiment, after the discretization process, the values of the parameters are set as follows: $\omega_{MAX} = 8$, $NV_\omega = 1$, $NV_{\alpha1} = 20$, $NV_{\alpha2} = 20$, $NV_{\theta s} = 10$, $NV_{\theta1} = 5$, $NV_{\theta2} = 5$, $NV_{\theta p} = 10$. In data collecting process, we collect the controller and performance data when the snake-like robot is not damaged. As shown in Fig. 2, this step is completed in simulation tools. We use V-REP as simulation tools. V-REP is a good robot simulator, it can be used for fast algorithm development, factory automation simulations, fast prototyping and verification and so on.[2] We do the research mainly on the simulator tool due to following two reasons. On the one hand, our algorithm needs precise data of speed and gait, but in general, most physical robots fail to offer precise information. On the other hand, our algorithm model has to be trained many times. At the beginning of each training process, we need to reset the robot for a consistent state, which causes significant time expense and reduces efficiency if we test on physical robots.

Adaptation step has five threads, which are main thread, data transceiver thread, M-BOA process thread, simulation control thread and V-REP simulate thread.

Main thread is responsible for the creation and initialization of damage recovery environment, including load controller-performance map created by sample collecting process and launch other cooperative threads. The main work of main thread is to detect whether damage exists, which is judged by the decrease of performance. After finding the compensatory actions, the main thread use the compensatory actions to guide snake-like robot to move.

Data transceiver thread is one thread of the V-REP master client, the main work of the data transceiver thread is receiving the TCP connect request from V-REP and send the motion parameter which should to be simulated in the V-REP thread. After the simulation process, data transceiver thread receives the performance data.

The M-BOA process thread is responsible for adaptation step, including selecting the next motion which has the optimal expectation, building and updating the Gaussian process and judging whether the adaptation step should be stopped.

The main task of communication control thread is synchronizing with V-REP thread and controlling the simulation schedule V-REP thread to make V-REP thread cooperate with M-BOA thread. Simulation control thread controls the enablement of V-REP thread and when the action simulation test begins and ends.

V-REP simulate thread is the major carrier of the simulation. There are two main tasks for it. One is the simulation of snake-like robots. The other is receiving the controller-performance data from the master client.

6   Effect Validation and Result Analysis

In this section, we do the experimental verification and comparison the performance of our approach for different possible damage types. In Sects. 6.1 and 6.2,

---

[2] http://www.coppeliarobotics.com/.

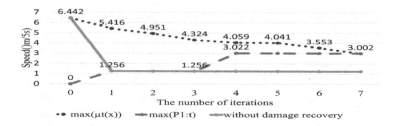

Fig. 3. Damage recovery process-6:90°.

we verify the damage recovery method on simulation environment. In Sect. 6.3, we apply the compensation behavior to the our real snake-like robot.

There are many kinds of damage for snake-like robots, such as non-rotation failure, joint-broken failure and non-control failure. In the experiment we found that the first failure type occupied the vast majority. Due to space limit, we only show the results of the first failure type.

6.1 Single Joint Damage Recovery Verification

In this subsection, we do the verification when the sixth joint is broken. In general, if the joint near the middle of the snake-like robot is broken, it will cause serious influence to the robot. Because our snake-like robot has 16 joints, we choose the sixth joint as the damaged joint. At the same time, we assume that its output axis remains at 90° after injured.

In Fig. 3, the lower line shows maximum observed values, that is, $max(P_{1:t})$ in each iteration. It can be seen that as the number of iterations increases, the observed performance value is increasing. The upper line shows the maximum performance prediction values, that is, $max(\mu_t(x))$ given in the controller-performance map after each iteration. It can be seen that the maximum predicted value decreases as the number of iterations increases because the observed performance decreases the value of other controllers around its corresponding controller through updating process of the Gaussian mode. The degree of reduction depends on the covariance matrix K and the kernel function $\kappa(x_1, x_2)$. It can be observed from Fig. 3 that the target recovery controller is found after seven iterations of this damage recovery process. When the damage recovery algorithm is not used, the speed of the snake-like robot travels decreases directly from the original 6.442 (m/5s) to 1.255 (m/5s). However, the best speed without damage recovery was found in the first three iterations of the damage recovery process. Finally, the compensation behavior is performed to achieve a mean speed of 3.022 (m/5s), which is 2 times faster than that of case without damage recovery. It shows the results of the exercise selected for the seven iterations. Figure 3 shows that the 4th iteration has achieved the best compensation effect.

Figure 4 shows the speed of joint six under different angles of damage recovery. This figure shows that when the damage angle is small, the non-rotation failure of the joint six caused by the damage has a small influence on the moving

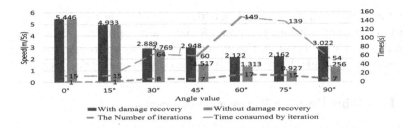

Fig. 4. The joint 6 under different angles of damage recovery.

speed of the snake-like robot; When the injury angle is larger than 30°, as shown in Fig. 4, the angle of the body injury can greatly affect the moving speed of the snake-like robot, decreasing the speed from 6.442 (m/5s) to 0.926–2.769 (m/5s). After looking for the compensatory behavior through the damage recovery algorithm, the velocity recover back to 2.122–3.021 (m/5s). The algorithm enhances the speed of motion up to 140.6%. The maximal number of iteration in damage recovery is seventeen. The iterative process spend 2.5 min to complete.

The results of case studies demonstrate that our approach is effective in regaining performance for damaged snake-like robots.

90° Single Joint Damage Recovery Verification. In this subsection, we verify the performance of damage recovery when the angle of the injury joint is 90°.

From the Fig. 5, it can be observed that the snake-like robot can keep a good moving speed if the damaged joints are 1, 2, 3, 5, 11, 15 and 16. The speed is about 44% of that before the robot is broken. The role of damage recovery algorithms on these joints is not apparent. In other joints, especially in joint 8, the damage recovery algorithm performs well. The more the motion speed decreases, the better the compensation behavior of the damage recovery algorithm performs. The speed of motion recovery after injury can be 10 times higher than that without damage recovery algorithm, and can let the injured robot move in a speed which is more than 38% of that before injuring. Moreover, the number of iterations is less than 12, and the consumed time is also within 2 min.

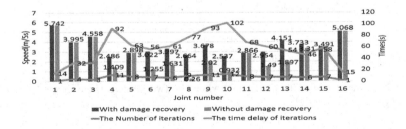

Fig. 5. Injury angle of 90° when a joint as the damaged joint

At the same time, we also find that the effect of using the damage recovery algorithm with and without the joints 8 and 9 are approximately symmetrical distributions. This is because, the snake-like robot structure and motion curve are similar to the snake body centroid as the center symmetrical distribution.

6.2   Two Joint Damage Recovery Verification

The damage recovery algorithm performs well on single joint damage recovery. This section verifies the injury of two joints. First, let us look at the recovery process of a bilateral joint injury. We choose two joints, namely joint 5, joint 11, and the damage angle is set to the maximum value of $90°$, $-90°$. Because in the experiment we find the joint 5 or joint 11 will affect the speed most if they are damaged.

From Fig. 6 it can be seen that after 19 iterations, the final compensation behavior was found. The final compensation motion speed is 2.337 (m/5s), which is 36.28% of the speed before damage and 3 times faster than that without using the damage recovery algorithm. In addition, the more iterations the algorithm takes, the better compensation it can find.

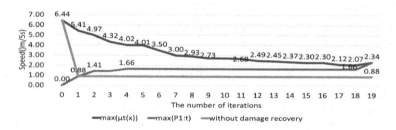

Fig. 6. Damage recovery process - 5:90°, 11:−90°

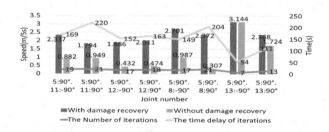

Fig. 7. Injury angle of 90° and −90° when the two joint as the damaged joints

Double Joint 90°, −90° Angle Damage Verification. There are many combinations of the joint failures, including the combination of the two joints, the combination of the angle values, etc. We first select several groups of representative cases to be verified in the case of the injury angle of $90°$ or $-90°$.

From Fig. 7, it can be observed that in some cases, the speed loss caused by the injury is serious. The compensation behavior found by the damage recovery algorithm can make a 41.9% movement speed recovery, which is 8 times faster than that without using the algorithm.

The number of iterations are significantly greater than that with single joint injuries, but the process also completes within 21 iterations, consuming less than 4 min.

There are other kinds of two joints damage recovery combinations, because the limit of space we don't introduce all of them. In our experiments, the compensation behavior found by the damage recovery algorithm can make the movement speed reach 64.7%, it is 17 times faster than that without using damage recovery. For the combination of multiple joints, this paper cannot cover all of them. Moreover, the probability that more than two joints fail at the same time is very small. Therefore, we just briefly list the experiment results of these cases. In our experiments, in the case of three, five, eight, and thirteen joints injury, the compensation behavior found by the damage recovery algorithm can make the movement speed reach 37.9% before injury, it is 12 times faster than that without using the algorithm. The number of iterations used is completed within 20 times.

6.3   Damage Recovery on Real Snake-Like Robot

In this subsection, we apply the damage recovery method on our real snake-like robot.

Figure 8 shows the real snake-like robot whose joint 7 is broken and the damage recovery method is not applied. We can see that, the behavior of the robot is irregular. the robot cannot move forward and the speed of the robot is very slow. So when some joints of the snake-like robot are broken, the real snake-like robot's motion become inefficient and the robot cannot work. The speed is about 20 times better than snake-like robot is in damaged situation and is about 0.79 times less than the snake-like robot is not damaged.

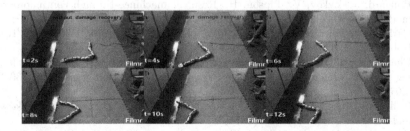

Fig. 8. Motion tracing without damage recovery.

Figure 9 shows how the snake-like robot moves when the damage recovery method is applied. Compared with Fig. 8, it can be seen that the snake-like robot moves toward the forward destination and the speed of the robot has an obvious promotion. For experimental videos, please refer this website.[3].

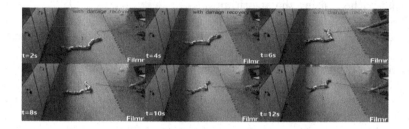

Fig. 9. Motion tracing after damage recovery.

7 Conclusion

In this paper, we introduce a novel Intelligent Trial and Error Learning algorithm for the orthomorphic snake-like robots, which can make continuous trial and error without learning the damage model to perform a compensation behavior. In the simulation environment, a variety of damage scenarios are verified for the damage recovery of the snake-like robot with orthogonal structure, and achieve good results. In the physical environment, the damaged snake-like robot to which the damaged recovery algorithm is applied can restore certain moving speed and keep forward to the destination. In the future, we will solve more kinds of damage situations and let the snake-like robot judge which damage is happened.

References

1. Blanke, M., Kinnaert, M., Staroswiecki, M.: Diagnosis and fault-tolerant control, pp. 1379–1384. Springer, Heidelberg (2006). https://doi.org/10.1007/978-3-662-47943-8. 49(4), 1379–1384
2. Bongard, J., Zykov, V., Lipson, H.: Resilient machines through continuous self-modeling. Science **314**(5802), 1118–1121 (2006)
3. Carlson, J., Murphy, R.R.: How UGVS physically fail in the field. IEEE Trans. Robot. **21**(3), 423–437 (2005)
4. Chatzilygeroudis, K., Vassiliades, V., Mouret, J.B.: Reset-free trial-and-error learning for data-efficient robot damage recovery. Robot. Auton. Syst. **100**, 236–250 (2016)
5. Cully, A., Clune, J., Tarapore, D., Mouret, J.B.: Robots that can adapt like animals. Nature **521**(7553), 503 (2014)
6. Fuchs, A., Goldner, B., Nolte, I., Schilling, N.: Ground reaction force adaptations to tripedal locomotion in dogs. Vet. J. **201**, 307–315 (2014)

---

[3] https://youtu.be/JCfBgXZbSb0.

7. Hirose, S., Cave, P., Goulden, C.: Biologically inspired robots: Serpentile locomotors and manipulators, pp. 351–363. Oxford University Press, Oxford (1993)
8. Gray, J.: The mechanism of locomotion in snakes. J. Exp. Biol. **23**(23), 101–120 (1946)
9. Katharine, S.: Mars rover spirit (2003–10). Nature **463**(7281), 600 (2010)
10. Kyriakopoulos, K., Migadis, G., Sarrigeorgidis, K.: The ntua snake: design, planar kinematics, and motion planning. J. Robot. Syst. **16**(1), 37–72 (1999)
11. Lipkin, K., Brown, I., Peck, A., Choset, H.: Differentiable and piecewise differentiable gaits for snake robots. In: IEEE/RSJ International Conference on Intelligent Robots & Systems, vol. 8, no. 1, pp. 1864–1869 (2007)
12. Nilsson, M.: Why snake robots need torsion-free joints and how to design them. In: IEEE International Conference on Robotics & Automation, vol. 1, pp. 412–417 (1998)
13. Ma, S.: Analysis of creeping locomotion of a snake-like robot. Auton. Robot. **15**(2), 205–224 (2006)
14. Paap, K.L., Dehlwisch, M., Klaassen, B.: GMD-Snake: a semi-autonomous snake-like robot. In: Asama, H., Fukuda, T., Arai, T., Endo, I. (eds.) Distributed Autonomous Robotic Systems, vol. 2. Springer, Tokyo (1996). https://doi.org/10.1007/978-4-431-66942-5_8
15. Shan, Y., Koren, Y.: Design and motion planning of a mechanical snake. IEEE Trans. Syst. Man Cybern. **23**(4), 1091–1100 (1993)

Learning to Cooperate in Decentralized Multi-robot Exploration of Dynamic Environments

Mingyang Geng[1(✉)], Xing Zhou[1], Bo Ding[1], Huaimin Wang[1], and Lei Zhang[2]

[1] National Key Laboratory of Parallel and Distributed Processing, College of Computer, National University of Defense Technology, Changsha, China
{gengmingyang13,zhouxing,dingbo,hmwang}@nudt.edu.cn
[2] National Key Laboratory of Integrated Automation of Process Industry, Northeastern University, Shenyang, China
zl999999990163.com

Abstract. This paper presents an approach to train a decentralized multi-robot system to learn cooperation strategy in the exploration of dynamic environments. The traditional approaches to multi-robot exploration problem are all based on the "pre-designed" cooperation strategy. However, many real-world settings are too complex for humans to "design" effective strategies. Besides, "pre-designed" strategy does not possess the ability to adapt to different task environment features, which also limits its application in real-world practices. Inspired by the superiority of deep reinforcement learning technique on complex individual behavior design, we apply the same technology to the cooperative learning process on the robot collective level. Our approach has been evaluated in a simulated multi-robot Disaster Exploration scenario and the results show that it could be applied in more complicated scenarios in contrast with two traditional "human-designed" methods.

Keywords: Distributed multi-robot exploration · Deep learning
Reinforcement learning · Dynamic environments · Cooperative behavior

1 Introduction

In this paper, we focus on the decentralized, dynamic cooperation problem in exploring a frequently-changing environment by a group of robots. Consider the following robotic search and rescue scenario: A group of unmanned aerial vehicles (UAVs) is sent to find the survivors in a group of high-rise buildings after an earthquake. These UAVs are supposed to act autonomously to suit themselves to the extreme environment. And because of the instability of the building structure and the possible aftershocks, the whole environment changes with time elapsing. Therefore, the task of each UAV could not be statically pointed in advance. In order to find the survivors as quickly as possible, these robots have to periodically exchange information with its neighbors and make decisions based both on its local view and the information from its neighbors.

© Springer Nature Switzerland AG 2018
L. Cheng et al. (Eds.): ICONIP 2018, LNCS 11307, pp. 40–51, 2018.
https://doi.org/10.1007/978-3-030-04239-4_4

The above-mentioned multi-robot cooperation problem has been thoroughly studied in the past decades in the robotic field. Various approaches were proposed, such as the frontier-based approach, the cost-utility approach, and the market-based approach [8]. All of them are based on the cooperation strategy (i.e., the communication and action rules in the collective) which is "pre-designed" by humans. However, the practice has proved that many real-world settings are too complex for humans to "design" effective strategies. Multi-robot task allocation with the ability to handle more complex real-world constraints is still an open challenge in the robotic society [10]. Besides, "pre-designed" strategy does not possess the ability to adapt to different task environment features, which also limits its application in real-world practices.

Today, neural network-based machine learning has shown its immense potential in tackling with real-life problems [7]. In particular, deep reinforcement learning has been proved to be an effective solution for enabling sophisticated and hard-to-design behaviors of robot individuals [16]. Naturally, on the robot collective level, could we apply the similar technology to enable them to actively learn cooperation strategies (instead of human designing) which is robust enough to handle complex and dynamic environments?

Our work is inspired by this simple idea. Although it sounds straight-forward, it is of great challenges to put it into practice. We have to enable the training of not only the action recipe of each individual robot but also the communication policies among these robots. To achieve this goal, our work presented in this paper abstracts the robots as agents and adopt an established neural network in the multi-agent learning field, CommNet (Communication Neural Network) [14]. And then, we adapt and enhance this network to the environment exploration problem. In concrete, firstly, we construct the modeling of dynamic environments on the basis of Occupancy Grid [1]. Secondly, we devise appropriate reward functions with the aim of encouraging the robots to learn the ability of cooperation. Finally, we design the Exploration Ratio-based Training Approach for the exploration task. Our approach has been evaluated in a simulated multi-robot Disaster Exploration scenario and the results show that it could be applied in more complicated scenarios in contrast with two traditional "human-designed" methods.

The remainder of this paper is organized as follows. In Sect. 2, the related work is presented emphasis on the novelty of our work. Section 3 gives a brief introduction to CommNet, the foundation of our work. Section 4 presents the details of our approach. Section 5 presents the experiments carried out on the simulation platform, as well as the analysis of the result and some discussions of our approach based on it.

2   Related Work

Our approach covers two research areas, including multi-robot exploration and learning to cooperate in multi-agent domains. In this section, we focus on the state of the art of these two fields, especially the difference between existing works and our work.

2.1   Multi-robot Exploration

Exploring an unknown environment by a team of autonomous robots are termed as the multi-robot exploration problem. Its goal is to cover the strange environment as soon as possible by the seamless cooperation of the teamed robots. It is the foundation of many real-life multi-robot applications, such as urban disaster search and rescue, reconnaissance and planetary exploration [8].

There has been a wide variety of approaches to multi-robot exploration. Several authors propose multi-robot exploration strategies based on market principles, in which robots place bids on subtasks of the exploration effort [12]. These bids are typically based on values such as expected information gain and travel cost to a particular location in the environment and may be assigned in a distributed fashion among team members, or by a central agent. Another common strategy for robotic exploration is to use frontiers [20], which can easily be extended for use by multiple robots [3].

For dynamic environments, a single auction based dynamic allocation scheme [13] was proposed to explore a set of targets with minimization of total cost objective. A method [4] which is based on difference evaluation functions is also proposed to develop successful control policies for dynamic and stochastic multi-robot exploration missions. However, the methods above are all based on the "pre-designed" cooperation strategy. In our method, we solve the dynamic cooperation by learning. That is to say, the final strategy learned by our approach could merge the advantages of the "pre-designed" strategies above, which will be too difficult to design in advance.

2.2   Learning to Cooperate in Multi-agent Domains

There are mainly three kinds of approaches to realize cooperation: assume full visibility of the environment, communicate with a predetermined protocol [21] and communicate using reinforcement learning approaches. Here, we only focus on the last kind of approaches. Kasai et al. [9] and Varshavskaya et al. [17], both use distributed tabular-RL approaches for simulated tasks. Guestrin et al. [6] use a single large MDP (Markov decision process) to control a collection of agents, via a factored message passing framework where the messages are learned. Foerster et al. [5] uses a deep reinforcement learning in multi- agent partially observable tasks, specifically two riddle problems which necessitate multi-agent communication. However, the cooperation strategies above have not been applied to multi-robot exploration tasks. This is the first paper that solves the multi-robot exploration problem through reinforcement learning technologies.

3   Preliminary

In this section, we introduce the communication model CommNet, which is the foundation of our approach. The function of CommNet is to make cooperating agents learn to communicate amongst themselves before taking actions. Each agent is controlled by a deep feed-forward network. In order to guarantee the communication process, each of the networks can get the summed transmissions of other agents by connecting with a communication channel carrying a continuous vector. The model is combined with policy gradient [19] algorithms and can be trained via back-propagation.

In order to understand the communication process, two questions need to be clarified: what does the agent transmit; how to deal with the summed transmissions of other agents. The transmissions of each agent are composed of two parts: the hidden state and the communication vector. For a single agent, the hidden state stands for its own observations (including its trajectory recorded by a list of history positions). The communication vector could be regarded as the information acquired from other agents. The communication process is divided into N steps. At each step i, the hidden state and communication vector are iterated as follows:

$$h_j^{i+1} = f^i(h_j^i, c_j^i) \tag{1}$$

$$c_j^{i+1} = \frac{1}{N(j)} \sum_{j' \in N(j)} h_{j'}^{i+1} \tag{2}$$

Here, f^i stands for the multi-layer neural networks, $i \in \{0, \ldots, N\}$, where N is the number of communication steps in the network. Each agent just deals with the communication message through the multi-layer neural networks. To take advantage of the recurrent neural network (RNN), the communication step i in Eqs. (1) and (2) is replaced by a time step t and the module f^t are used for all t. h_j^i stands for the hidden state of agent j at step i and c_j^i stands for the communication vector. $N(j)$ stands for the set of agents present within the communication range of agent j and will change over time. The main architecture of CommNet is shown in Fig. 1.

4 Approach

In this section, we will describe the details of our approach CommNet-Explore, which is based on CommNet. Our approach is composed of three aspects. Firstly, we construct the modeling of dynamic environments on the basis of Occupancy Grid. Secondly, we devise appropriate reward functions with the aim of encouraging the robots to learn the ability of cooperation. Finally, we design the Exploration Ratio-based Training Approach for the exploration task.

4.1 Modeling of Dynamic Environment

The exploration environment is represented by Occupancy Grids [1]. The basic idea behind it is to represent the map as a two-dimensional grid of binary random variables which stand for whether the locations are occupied.

We make some definitions first: z_t stands for the observation of the robot; c_t stands for the communication message from other robots within the certain range; a_t stands for the output actions given the corresponding inputs;

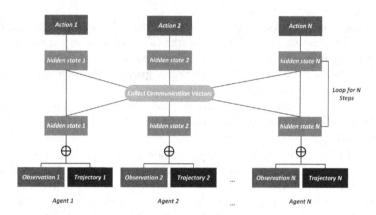

Fig. 1. The architecture of CommNet for n agents. The inputs are the observations and trajectories of the agents. The communication process is looped for N steps and finally gives out the actions.

$\pi(a_t|z_{1:t}, c_{1:t})$ stands for the policy for choosing controls based on the past observations and communication message.

For a better explanation, we make an assumption that there is an underlying grid map $m \in M = \{0,1\}^{N \times N}$ primarily unknown to the robots. Each robot wishes to calculate its belief over maps M at time t given all its previous observations and other robots' communication message leading up to that time step $b_t(m) = p(m|z_{1:t}, c_{1:t})$. To simplify the problem, we assume the individual map random variables, indexed as m_i, are independent:

$$b_t(m) = \prod_i p(m|z_{1:t}, c_{1:t}) = \prod_i b_t(m_i). \tag{3}$$

The robots learn to update the posteriors themselves. The information-theoretic [1] entropy of the belief state is used to quantify the uncertainty, which factories over the individual map random variables because they are assumed to be independent:

$$H(b_t(m)) = \sum_i H(b_t(m_i)) \tag{4}$$

$$= \sum_i b_t(m_i = 1) \log b_t(m_i = 1) + b_t(m_i = 0) \log b_t(m_i = 0) \tag{5}$$

4.2   Entropy-Oriented Reward Function

CommNet-Explore is designed for multi-agent learning to cooperate by sharing the observations and knowledge through communication. We choose CommNet as our communication model because it has successfully been applied in the traffic junction task and solved the problem of path conflicting in dynamic environments, which has some similarities to our application. In order to explain the feasibility from theory, for a single agent, the hidden state can be thought of as its previous observations memorized by the RNN (Recurrent Neural Network) network. The communication vector could be regarded as the action which it attempts to take on the next time-step. Based on the information above, the agent can learn how to express its state, understand the broadcasting information from other agents and finally learn to cooperate.

Now, we describe the approach to devising the reward functions. In the learning process, there is a central node that records the trajectory of each agent and gives the corresponding reward to each agent based on the performance. At each time-step t, the agent receives an observation z_t and the communication message c_t from the other agents within the specified range. It updates posterior and the belief over the map based on the experience. At this point, the agent is faced with the decision of what control input a_t to select given z_t and c_t.

The state can be defined as $s_t = [b_t, c_t]$. In our RL formulation, each agent seeks a policy $\pi(a|s)$ that reduces uncertainty for the whole map as quickly as possible. So at each step t, a agent gets the following reward:

$$R_t \doteq H(b_t) - H(b_{t+1}) + C^t r_{coll} + B^t r_{back} \tag{6}$$

where R_t is the combination of the reduction in the entropy of the agent's local view and the collision information with other agents at time step t. Here, C^t refers to the number of collisions (blocks or other agents) with the current agent. Two agents collide if their locations overlap. A collision with another agent or the blocks incurs a reward $r_{coll} = -10$ but does not affect the simulation in any other way. B^t refers to a Boolean variable to judge if the current robot has explored where the multi-robot system has already explored, and the value of r_{back} is -10. The simulation is terminated after the specified steps, and the standard of success is specified according to the difficulty of the task. We can then train a policy to maximize the expected reward in simulation and increase the success rate of the whole system.

4.3   Exploration Ratio-Based Training Approach

The task is trained for 50 epochs, each epoch being 100 weight updates on minibatch of 288 game episodes (distributed over multiple CPU cores). The training time takes about five days on our 24 GB-memory server.

To make the learned strategy robust to dynamic settings, we gradually add new blocks to the prime environment (n blocks every m time-steps). We use curriculum learning [2] and adjust the task difficulty (adding frequency of the blocks) to make the training easier. The value of n decreases during the training time, which means that the blocks are added more frequently so that the difficulty of the mission is increasing. However, the value of $\frac{n}{m}$ is constant all the time because we need to fix the value of el_{ratio} in the given time-steps, which is a crucial component to measure the success standard. The method of curriculum learning is expected to be essential for local minima and helps to learn superior models which can deal with the dynamic settings.

Each simulation is terminated after a specified number of time-steps and classified as a failure if collisions with blocks have occurred or the exploration ratio el_{ratio} is less than 90%. el_{ratio} here is calculated as follows:

$$el_{ratio} = \frac{count(S_{explored} \bigcup S_{finalblock})}{A} \tag{7}$$

being $S_{explored}$ the subset of explored cells in the map, $S_{ultimate_block}$ the subset of final blocks, A the number of cells in the map, $count(X)$ a function that counts the number of elements in the set X. Here, we take the operation $\bigcup$ because the positions of the newly generated blocks may overlap with the area that has been explored by the multi-robot system.

Additionally, the number of agents at a given time is also dynamic. Each agent has a life cycle and will be removed from the environment after the specified time-steps. This can be thought of as the situation that the energy of the robot has been used up. New agents can enter the environment with a fixed frequency to finish the remained task. This setting can strengthen the robustness towards dynamic settings because the newly entered agent needs to grasp the environment as quickly as possible and choose the best target area to complete the remaining task. Similarly, the removed agents also need to transfer the observations in a short time in case of information loss.

5   Evaluation and Experiments

In this section, we will introduce our experiments from the following three aspects: the experimental environments, the results, and the corresponding analyze.

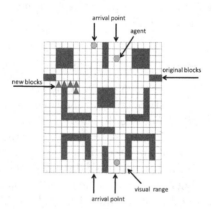

Fig. 2. The experimental environment which is dynamic in the number of agents and blocks.

5.1   Experimental Environments

We use MazeBase [15] to set up an environment to test our approach. The map represents an artificial environment with many bifurcations and loops on a 20×20 grid. This large and complex scenario was chosen because it shows more clearly the advantages of our method than a simple scenario. We discretize the locations of the agents so that each of the agents is in a single position in the grid. The agents can sense the information in the adjacent positions within their vision range (a surrounding $v \times v$ neighborhood). The agents can move to adjacent eight positions that do not contain blocks. Each agent can explore one cell in one step, and the goal for the whole system is to explore cells as many as possible in a fixed time. Each agent also has a communication range which can cover the whole environment.

Now, we describe the details in our environmental setting. There are four arrival points, situated at (1,9), (1,13), (20,9), (20,13). The survey region contains 70 newly introduced blocks (excluding the 81 blocks in the prime environment shown in Fig. 2) that are initialized according to a uniform random distribution across the search space. At initial execution, only the 81 static blocks are active in the entire survey region; as the mission progresses, more blocks become active at a rate of n blocks per m time-steps, such that all blocks are active after 90% of the mission time. For the number of agents, two new agents enter the grid randomly from the four arrival points every 4 time-steps. However, the total number of agents at a given time is limited to $N_{max} = 12$. Each agent's life cycle is 25 time-steps. The video of the trained policy is available at https:// youtu.be/9Ccklm3uJ10.

5.2   Results

After training, we evaluated our approach on 1000 random, unseen maps and explored how partial visibility within the environment affects the cooperative ability of the system. As the vision range of each agent increases, the average success rate increases as shown in Fig. 3(Left). The measurement of success standard here is calculated as above.

To examine the robustness to dynamic settings, we set up an initially obstacle-free testing environment and make comparisons with two traditional "pre-designed" methods to cope with the multi-robot exploration task. The two baseline approaches are:

Coordinated Frontier Based Approach. We use a simple agent-frontier assignment algorithm detailed in [18]; in short, every robot determines frontier utilities for itself and its nearby teammates, and iteratively calculates a robot to the frontier assignment that maximizes the joint utility. While this method is not necessarily optimal, it is fast, and in our experience entirely sufficient for distributed exploration.

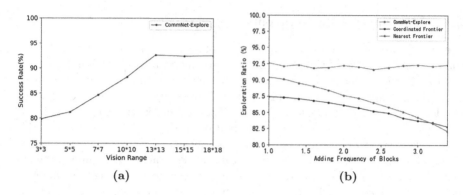

Fig. 3. Left: As visibility in the environment decreases, the importance of communication grows in the exploration task; Right: Robustness comparison amongst the three approaches in 70 s.

Nearest Frontier Approach. This approach is based on Yamauchi's technique [20], and it consists in selecting the shortest path to the nearest frontier. As it can be noticed, this method does not consider the utility of the frontiers and the coordination mechanism. Therefore, this method can save the time on frequently re-planning the target area.

We evaluate the three approaches on a 25 × 25 map which is initially free of blocks and spontaneously add the obstacles from a random distribution to the environment. In each test, the three approaches are faced with the same obstacle generating strategy. We repeat the process 1000 times and calculate the average performance in two methods: specified time-steps (65) and specified time (70 s), which is fixed by the average performance of the three approaches.

When fixed the time-steps, the planning time and overall performance of the three approaches are shown in Table 1. We can see that the planning time of CommNet-Explore has been largely decreased compared with the Coordinated Frontier based Approach method. This advantage is suitable for realistic surveillance missions in dynamic environments, which is more common for data from the team to be received in batches. For the limited time condition, as shown in Fig. 3(Right), the Coordinated Frontier Approach has shown obvious disadvantages because of the long planning time. As the adding frequency of blocks increases, the performance of CommNet-Explore is more stable than the others. Therefore, the superiority on the overall performance of CommNet-Explore suggests that it is robust enough to be adapted to more complicated environments.

5.3 Analysis of Communication

We now attempt to understand what the agents communicate themselves when performing the exploration task. We start by recording the hidden state h_j^i of each agent and the corresponding communication vectors $c_j^{i+1} = C_{i+1} h_j^i$ (the contribution agent j at step $i + 1$ makes to the hidden state of other agents).

Table 1. Average performance over 1000 out-of-sample episodes in 65 time-steps.

Approach	Planning time (ms)	Exploration ratio (%)
CommNet-Explore	**30 ± 10**	**92.6 ± 4.3**
Coordinated Frontier	230 ± 20	87.4 ± 6.7
Nearest Frontier	35 ± 5	90.4 ± 5.2

Figure 4(left) shows the t-SNE (t-Distributed Stochastic Neighbor Embedding) [11] of the communication vectors. Distinct clusters are clearly present, which indicates that the agents can coordinate themselves when necessary (the target area are conflicted). We also visualize the average norm of the communication vectors in Fig. 4(right) over the 20×20 grid which contains only the original blocks. We can clearly see that the agents have a stronger communication intention on the important positions such as the connection area between blocks, which indicates the effectiveness of CommNet-Explore.

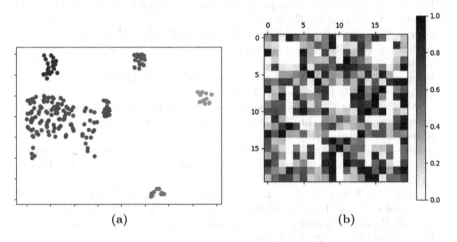

(a)   (b)

Fig. 4. Left: First two principal components of communication vectors c from multiple runs on the static exploration environment. Right: Average norm of communication vectors in the static environment.

6   Conclusion

In this work, we introduce an approach to train a decentralized multi-robot system to learn cooperation strategy in the exploration of dynamic environments via back-propagation. By using the multi-agent deep reinforcement learning technique and deliberately-designed reward functions, the robots can learn to communicate with each other, making decisions based on the communication messages and its own states, and thus seamlessly cooperate in multi-robot exploration process. The communication model learned by the multi-robot system is robust enough to adapt to the dynamic changing of environments. Experimental results show that the cooperation strategy gained by deep reinforcement learning achieves performance better than the traditional "human-designed" strategies.

This work is preliminary in that it only evaluated in a simulated scenario. In future work, we hope to apply this approach to problem setups with limited communication bandwidth. We would also like to integrate it with robotic localization process using recurrent network based on the speed information, leading to a fully trainable exploration system.

Acknowledgments. This work is partially supported by the National Natural Science Foundation of China (nos. 61751208)) and the special program for the applied basic research of the National University of Defense Technology (no. ZDYYJCYJ20140601).

References

1. Barratt, S.: Active robotic mapping through deep reinforcement learning. ArXiv Preprint ArXiv:1712.10069 (2017)
2. Bengio, Y., Louradour, J., Collobert, R., Weston, J.: Curriculum learning. In: Proceedings of the 26th Annual International Conference on Machine Learning, pp. 41–48. ACM (2009)
3. Burgard, W., Moors, M., Stachniss, C., Schneider, F.E.: Coordinated multi-robot exploration. IEEE Trans. Robot. **21**(3), 376–386 (2005)
4. Colby, M., Chung, J.J., Tumer, K.: Implicit adaptive multi-robot coordination in dynamic environments. In: 2015 IEEE/RSJ International Conference on Intelligent Robots and Systems (IROS), pp. 5168–5173. IEEE (2015)
5. Foerster, J.N., Assael, Y.M., de Freitas, N., Whiteson, S.: Learning to communicate to solve riddles with deep distributed recurrent q-networks. ArXiv Preprint ArXiv:1602.02672 (2016)
6. Guestrin, C., Koller, D., Parr, R.: Multiagent planning with factored MDPS. In: Advances in Neural Information Processing Systems, pp. 1523–1530 (2002)
7. Hadsell, R., et al.: Learning long-range vision for autonomous off-road driving. J. Field Robot. **26**(2), 120–144 (2009)
8. Juliá, M., Gil, A., Reinoso, O.: A comparison of path planning strategies for autonomous exploration and mapping of unknown environments. Auton. Robot. **33**(4), 427–444 (2012)
9. Kasai, T., Tenmoto, H., Kamiya, A.: Learning of communication codes in multi-agent reinforcement learning problem. In: 2008 IEEE Conference on Soft Computing in Industrial Applications, SMCia 2008, pp. 1–6. IEEE (2008)
10. Khamis, A., Hussein, A., Elmogy, A.: Multi-robot task allocation: a review of the state-of-the-art. In: Koubâa, A., Martínez-de Dios, J.R. (eds.) Cooperative Robots and Sensor Networks 2015. SCI, vol. 604, pp. 31–51. Springer, Cham (2015). https://doi.org/10.1007/978-3-319-18299-5_2
11. Maaten, L.V.D., Hinton, G.: Visualizing data using t-SNE. J. Mach. Learn. Res. **9**(2605), 2579–2605 (2008)
12. Nieto-Granda, C., Rogers III, J.G., Christensen, H.I.: Coordination strategies for multi-robot exploration and mapping. Int. J. Robot. Res. **33**(4), 519–533 (2014)

13. Sariel, S., Balch, T.: Real time auction based allocation of tasks for multi-robot exploration problem in dynamic environments. In: Proceedings of the AAAI-05 Workshop on Integrating Planning into Scheduling, pp. 27–33. AAAI Palo Alto, CA (2005)
14. Sukhbaatar, S., Fergus, R., et al.: Learning multiagent communication with back-propagation. In: Advances in Neural Information Processing Systems, pp. 2244–2252 (2016)
15. Sukhbaatar, S., Szlam, A., Synnaeve, G., Chintala, S., Fergus, R.: Mazebase: a sandbox for learning from games. ArXiv Preprint ArXiv:1511.07401 (2015)
16. Tai, L., Liu, M.: Deep-learning in mobile robotics-from perception to control systems: a survey on why and why not. ArXiv Preprint ArXiv:1612.07139 (2016)
17. Varshavskaya, P., Kaelbling, L.P., Rus, D.: Efficient distributed reinforcement learning through agreement. In: Asama, H., Kurokawa, H., Ota, J., Sekiyama, K. (eds.) Distributed Autonomous Robotic Systems, vol. 8. Springer, Heidelberg (2009). https://doi.org/10.1007/978-3-642-00644-9_33
18. Visser, A., Slamet, B.A.: Balancing the information gain against the movement cost for multi-robot frontier exploration. In: Bruyninckx, H., Pueučil, L., Kulich, M. (eds.) European Robotics Symposium 2008, pp. 43–52. Springer, Heidelberg (2008). https://doi.org/10.1007/978-3-540-78317-6_5
19. Williams, R.J.: Simple statistical gradient-following algorithms for connectionist reinforcement learning. Mach. Learn. **8**(3–4), 229–256 (1992)
20. Yamauchi, B.: Frontier-based exploration using multiple robots. In: Proceedings of the Second International Conference on Autonomous Agents, pp. 47–53. ACM (1998)
21. Zhang, C., Lesser, V.: Coordinating multi-agent reinforcement learning with limited communication. In: Proceedings of the 2013 International Conference on Autonomous Agents and Multi-agent Systems, pp. 1101–1108 (2013)

Q Value-Based Dynamic Programming with Boltzmann Distribution by Using Neural Network

Wenxin Yu[1,6(✉)], Liang Yu[1], Gang He[1(✉)], Yibo Fan[2], Gang He[3],
Jiu Xu[4], Zhuo Yang[5], and Zhiqiang Zhang[1]

[1] Southwest University of Science and Technology, Mianyang, Sichuan, China
star_yuwenxin27@163.com, yuwenxin@swust.edu.cn,
cosfrist@live.cn
[2] State Key Laboratory of ASIC & System, Shanghai, China
[3] Xidian University, Xi'an, China
[4] Rakuten Institute of Technology, Boston, USA
[5] Guangdong University of Technology, Guangzhou, China
[6] Sichuan Civil-Military Integration Institute, Mianyang, China

Abstract. In this paper, a feedback method using neural network is proposed with Q Value-based Dynamic Programming based on Boltzmann Distribution for static road network. The neural network can supply more distribute strategies and the feedback method chooses the best result from the strategies produced by neural network. The method distributes vehicles well on all the optimal routes from the origin to destination according to the gradual decreasing parameters, which are used in the neural network. This method can overcome local optimum problems to some extent by setting appropriate parameters at the beginning. The proposed method is evaluated by using the Kitakyushu city (Fukuoka, Japan) road network data. The simulation result shows that the better result can be obtained than conventional QDPBD method by training parameters.

Keywords: Q value · Boltzmann Distribution · Optimal routes
Feedback · Neural network · Vehicle distribution

1 Introduction

The Vehicle Navigation System [1, 2] could find the optimal route for drivers under current traffic conditions. With this optimal route, the driver could save travel time and petrol when driving from the origin to destination. So, finding out the optimal routes plays a significant role in the Vehicle Navigation System.

Many algorithms have been already proposed to find the optimal routes, such as Dijkstra algorithm [3, 4] and A* algorithm [5]. These algorithms could find only one shortest route, which may cause the traffic congestion if all the drivers choose the same route. On the other hand Q Value-based DP with Boltzmann Distribution [6, 7] could find many routes, and then it helps to avoid the traffic congestion because the drivers can choose different routes. Q Value-based DP with Boltzmann Distribution (QDPBD)

© Springer Nature Switzerland AG 2018
L. Cheng et al. (Eds.): ICONIP 2018, LNCS 11307, pp. 52–61, 2018.
https://doi.org/10.1007/978-3-030-04239-4_5

distributes the vehicles well on the sections, and it also could assure the efficiency for finding many routes whose traveling time is fairly short.

In the previous research [8], the conventional traffic assignment method distributes average amount of vehicles to each optimal route, which provides only one strategy to distribute vehicle, and cannot find better result. In this paper, a neural network is considered to find out more distribute strategies. And the better result can be elected by feedback method. By setting weights which vary greatly in the neural network [9] at the beginning, the method can overcome local optimum problem to some extent. A normalization method is considered in the hidden layer of the neural network which could adjust the weight value right. But, this improved method cost more calculate time than the conventional method.

The rest of the paper is organized as follows. In Sect. 2, the details of the basic Q Value-based DP with Boltzmann Distribution are introduced. In Sect. 3, the details of feedback method using neural network are introduced. Section 4 shows the simulation results. Section 5 concludes the paper.

2 Methodology

2.1 Conventional Q Value-Based Dynamic Programming with Boltzmann Distribution

The conventional method is based on the following iterative equation:

$$Q_d^{(n)}(i,j) \leftarrow t_{ij} + \sum_{k \in A(j)} P_d^{(n-1)}(j,k) Q_d^{(n-1)}(j,k) \tag{1}$$

$$P_d^{(n)}(i,j) \leftarrow \frac{e^{-\frac{Q_d^{(n)}(i,j)}{\tau}}}{\sum_{j \in A(i)} e^{-\frac{Q_d^{(n)}(i,j)}{\tau}}}, d \in D, i \in I - d - B(d), j \in A(i) \tag{2}$$

$$Q_d^{(n)}(d,j) = 0, d \in D, j \in A(d) \tag{3}$$

$$Q_d^{(n)}(i,d) = t_{id}, d \in D, i \in B(d) \tag{4}$$

$$P_d^{(n)}(d,j) = 0, d \in D, j \in A(d) - d \tag{5}$$

$$P_d^{(n)}(d,d) = 1.0, d \in D \tag{6}$$

where,

$i,j \in I$: Set of suffixes of intersections
$d \in D$: Set of suffixes of destinations
t_{ij}: Traveling time from intersection i to intersection j
A(i): Set of suffixes of intersections moving directly from intersection i
B(i):Set of suffixes of intersections moving directly to intersection i
τ: Temperature parameter

$Q_d^{(n)}(i,j)$: The expected traveling time to destination d when the car bound for destination d moves to intersection j at intersection i in the nth iteration.

$P_d^{(n)}(i,j)$: The probability that the vehicle bound for destination d moves to intersection j at intersection i in the nth iteration.

3 The Proposed Feedback Method

3.1 The Architecture of the Proposed Algorithm

Figure 1 shows the flow chart of improved QDPBD in static road networks with the proposed new feedback method in this paper and the parts with * are improved from the conventional method [8]. The proposed method can find more distribute strategies than the conventional method [8] in the Distribute vehicles part. The proposed method use the feedback method to choose the best result from the strategies by sending the

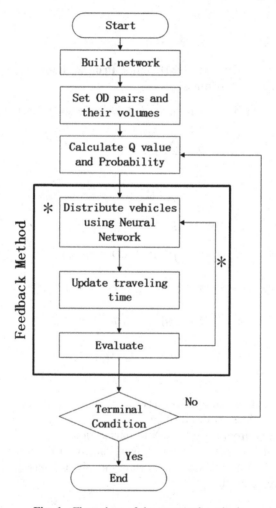

Fig. 1. Flow chart of the proposed method

result from Evaluate part to Distribute vehicles part showed in bold box of the feedback method in Fig. 1.

The basic neural network framework is like above. In the Fig. 2, the vehicle distributed on each route is used as an input node. The output is the evaluation value. A normalization method is used in the hidden layer in order to get appropriate weight value.

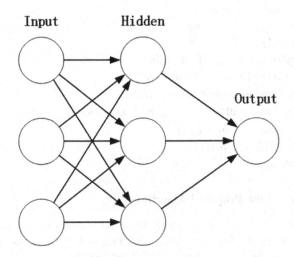

Fig. 2. Primary neural network

The mathematical definition of the proposed algorithm:

$$tv_r^{od} = tv^{od} * w_r^{od} \tag{7}$$

$$tv_{ij}^{od}(r) = tv_r^{od} \tag{8}$$

$$tv_{ij} \leftarrow \sum_{od \in OD} \sum_{r \in od} tv_{ij}^{od}(r) \tag{9}$$

$$t_{ij} \leftarrow t_{ij}^0 (1 + \alpha (\frac{tv_{ij}}{c_{ij}})^\beta) \tag{10}$$

$$t_{ij}^0 \leftarrow \frac{d_{ij}}{sp_{ij}} \tag{11}$$

$$w_r^{od} \leftarrow \frac{e^{-\frac{t_r^{od}}{\tau}}}{\sum_{r \in od} -\frac{t_r^{od}}{\tau}} \tag{12}$$

$$m \leftarrow m * n \tag{13}$$

where,

s_{ij} : section from intersection i to intersection j

t_{ij} : traveling time of section s_{ij}

t_{ij}^0 :free flow traveling time of section s_{ij}

d_{ij} : length of section s_{ij}

sp_{ij} : speed of section s_{ij}

c_{ij} : capacity of section s_{ij}

tv_{ij} :traffic volume of section s_{ij}

tv^{od} : traffic volume of $od \in OD$

r : candidate route

$od \in OD$: suffix and its set of OD pairs

w_r^{od} : weight value of route r of $od \in OD$

tv_r^{od} : traffic volume of route r of $od \in OD$

$tv_{ij}^{od}(r)$: traffic volume of section $s_{ij} \in r$ of $od \in OD$

t_r^{od} : traveling time of route $r \in od$

α, β : the congestion coefficients

m, n : parameters

3.2 The Details of the Proposed Algorithm

The traveling time which can be got easily from traffic states is used to calculate the weight value of route by Eq. (12). Then distribute the traffic value to each route according to this weight value by Eq. (7). Next, update the weight value randomly by adding the m parameter.

1) The brief steps of the algorithm:

1 > In one OD pair, each input note is defined as w_r^{od} which is initialized using Eq. (12)

2 > Add a random m to each w_r^{od}. $w_r^{od} = w_r^{od} + m$.

3 > Use normalization method to set sum of each w_r^{od} to 100%.

4 > Use Eq. (7) to distribute volume to each route.

5 > Evaluate E value and compare with best result BestE.

6 > Update m using Eq. (13). Go back to step 2.

2) The detail illustration of the steps:

In step 1, a basic OD pair showed in the Fig. 3 used to illustrate the algorithm. For example, three routes are chose out from the basic OD pair, which is OAED, OBFD and OCGD showed in Fig. 4. Assume that $t_{OAED}^{od} = 40.0$, $t_{OBFD}^{od} = 50.0$, and $t_{OCGD}^{od} = 60.0$, then we can get $w_{OAED}^{od} = 0.5$, $w_{OBFD}^{od} = 0.3$, and $w_{OCGD}^{od} = 0.2$ from Eq. (12).

In step 2, the Fig. 5 illustrates how to adjust the weight of each route. The parameter m is set as 0.1 at first, then use the Eq. (13) to update the m.

In step 3, the updated weight $w_{OAED}^{od} = 0.65$, $w_{OBFD}^{od} = 0.35$, and $w_{OCGD}^{od} = 0.45$ will be normalized to $w_{OAED}^{od} = 0.45$, $w_{OBFD}^{od} = 0.24$, and $w_{OCGD}^{od} = 0.31$.

In step 4, the Fig. 6 illustrates how to distribute volume on each route. It means the same volume will be distributed on each road section of each route.

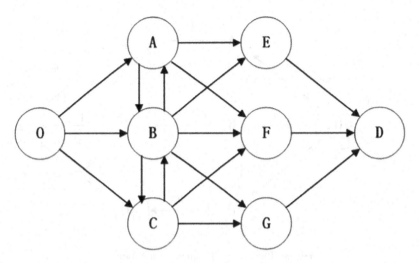

Fig. 3. Illustration of basic OD pair

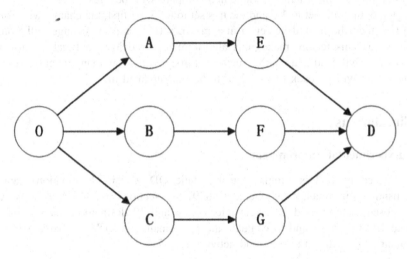

Fig. 4. Routes chose out from the basic OD pair

$$w^{od}_{OAED} = 0.5 \xrightarrow{\pm(-0.1\sim0.1)} w^{od}_{OAED} = 0.65 \xrightarrow{\text{Normalization}} w^{od}_{OAED} = 0.45$$

$$w^{od}_{OBFD} = 0.3 \xrightarrow{\pm(-0.1\sim0.1)} w^{od}_{OBFD} = 0.35 \qquad w^{od}_{OBFD} = 0.24$$

$$w^{od}_{OCGD} = 0.2 \xrightarrow{\pm(-0.1\sim0.1)} w^{od}_{OCGD} = 0.45 \qquad w^{od}_{OCGD} = 0.31$$

Fig. 5. Weight updating process

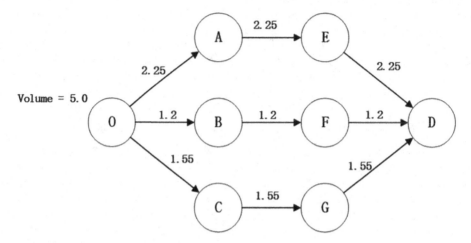

Fig. 6. Illustration of vehicle distribution

In step 5, we will evaluate E value and compare with best result BestE.

In step 6, the m is set to 1.0 and the n is set to 0.75. At first, the change will from - 0.1 to 0.1, then the m will be set to m = m * n = 0.75, and the change will from − 0.75 ~ 0.75. This feature makes this algorithm to avoid local optimal because the bigger change will lead to a random method at first and find out many potential results. The algorithm will go back to step 2 with the new updated m.

4   Simulations

4.1   Experimental Environment

The proposed method is simulated using static OD (Origin, Destination) data of Kitakyushu city in Japan, which includes 28397 intersections and 84456 sections. And the proposed method used 10 origin intersection and 10 destination intersections to generate 100 OD pairs and to evaluate the performance of Q Value-based DP with Boltzmann Distribution in static road network (Fig 7).

4.2   The Experimental Results

Parameters m and n are set as constant in this proposed paper. The performance of this proposed method is evaluated by Eq. (14). E of Eq. (14) means the average traveling time of all OD pairs.

$$E \leftarrow \frac{1}{|O|}\frac{1}{|D|}\sum_{o\in O}\sum_{d\in O}\sum_{(i,j)\in r_{od}} t_{ij} \qquad (14)$$

Table 1 shows the result of E of one iteration. The result also shows that the process of the algorithm. First, the m is set to 1.0, and the updated weight will be gotten from step 2 of algorithm steps. so the ten results could be gotten using different w, then

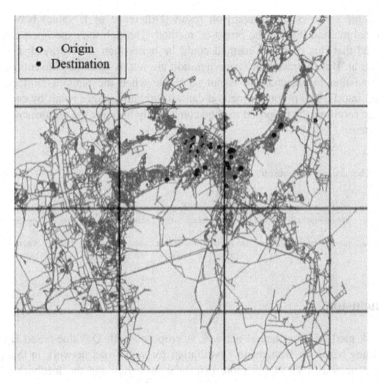

Fig. 7. Kitakyushu city's map

the best E = 608.68 will be chose out, secondly, the m is set to 0.75 by Eq. (13) where n = 0.75, and other ten results could be also gotten, and we can get better result 607.93 in m = 0.75, and the best E will be replaced by 607.93, also the distribute strategy will be recorded. In the results of m = 56, there are no result could better than best E = 607.93, so we keep the current distribute strategy which could get the best result. And in the iteration 0, the result will be 607.93.

Table 1. Result of iteration 0

m	1.0	0.75	0.56	0.42	0.32
w0	609.96	609.05	614.16	621.61	623.20
w1	608.68	615.61	628.56	613.86	622.78
w2	611.94	630.73	628.65	617.36	626.50
w3	617.08	617.43	629.07	614.55	605.77
w4	616.54	623.71	627.15	637.52	605.56
w5	616.54	623.65	624.93	627.69	605.40
w6	630.22	623.50	633.52	627.08	633.85
w7	629.82	618.09	623.85	633.03	615.57
w8	629.05	607.93	623.84	611.51	620.85
w9	635.84	612.26	632.10	617.57	622.42
Best	608.68	607.93	607.93	607.93	605.40

The Table 2 shows the comparison result (Difference of E value) between the conventional method [8] and the proposed method. Through the experimental results, we can find that this proposed method could be better than conventional method in most iteration. But the results in some iteration are not as good as conventional one, and we find that the mechanism is not so stable when the random numbers were introduced into it. The proposed method can also obtain a better result by comparing the average result, so the proposed method can actually improve the performance of the whole system.

Table 2. The comparison result, iteration i (i = 0, 1, 2..., 8, 9) where OD pairs = 400, cars = 120

Iteration	0	1	2	3	4	5	6	7	8	9	Average
Conventional method [8]	1099.33	1957.13	3974.96	1758.15	6203.89	3818.35	6666.3	3914.46	6318.37	4091.95	3980.29
Proposed method	1080.23	1863.73	4158.26	1665.43	6004.39	2816.25	7644.61	2808.16	4318.27	3087.45	3558.39
Differences	−19.1	−93.4	183.3	−92.72	−199.5	−1002.1	978.31	−1106.3	−2000.1	−1004.5	−421.9

5   Conclusion

A feedback method using neural network is proposed with Q Value-based Dynamic Programming based on Boltzmann Distribution for static road network in this paper. The neural network can supply more distribute strategies and the feedback method chooses the best result from the strategies produced by neural network. The method distributes vehicles well on all the optimal routes from the origin to destination according to the gradual decreasing parameters, which are used in the neural network. This method can overcome local optimum problems to some extent by setting appropriate parameters at the beginning. The proposed method is evaluated by using the Kitakyushu city (Fukuoka, Japan) road network data. Though the experimental results, the proposed method with novel idea of the feedback method by using neural network can provide better performance based on the more training times of neural network than the previous work with the same time consumption. By using this improved method, the better result can be obtained in most of the iteration without costing too much time. It means that the proposed method can provide better vehicle distribution strategy than the previous one.

Acknowledgments. This research was supported by A Project 【16ZA0131 】 which supported by Scientific Research Fund of Sichuan Provincial Education Department, 【2018GZ0517】 which supported by Sichuan Provincial Science and Technology Department, 【2018KF003】 Supported by State Key Laboratory of ASIC & System, Science and Technology Planning Project of Guangdong Province 【2017B010110007】 , the National Natural Science Foundation of China grants 【61672438】 .

References

1. Velaga, N.R., Quddus, M.A., Bristow, A.L., Zheng, Y.: Map-aided integrity monitoring of a land vehicle navigation system. IEEE Trans. Intell. Transp. Syst. **13**(2), 848–858 (2012)
2. Yang, J., Chou, L., Chang, Y.: Electric-vehicle navigation system based on power consumption. IEEE Trans. Veh. Technol. **65**(8), 5930–5943 (2015)
3. Dijkstra, E.: A note on two problems in connection with graphs. Numer. Math. **1**(1), 269–271 (1959)
4. Zhang, J., Feng, Y., Shi, F., Wang, G., Ma, B., Li, R., Jia, X.: Vehicle routing in urban areas based on the oil consumption weight –Dijkstra algorithm. Intell. Transp. Syst. **10**(7), 495–502 (2016)
5. Jagadeesh, G.R., Srikanthan, T., Quek, K.H.: Heuristic techniques for accelerating hierarchical routing on road networks. IEEE Trans. Intell. Transp. Syst. **3**(4), 301–309 (2002)
6. Mainali, M.K., Shimada, K., Mabu, S., Hirasawa, K.: Optimal route based on dynamic programming for road networks. J. Adv. Comput. Intell. Intell. Inf. **12**(6), 546–553 (2008)
7. Guggenheim, E.A., Green, M.S.: Boltzmann's distribution law. Phys. Today **9**(8), 34–36 (1955)
8. Yu, S., Xu, Y., Mabu, S., Mainali, M.K., Shimada, K., Hirasawa, K.: Q value-based dynamic programming with Boltzmann distribution in large scale road network. Sice JCMSI **4**(2), 129–136 (2011)
9. Halpin, S.M., Burch, R.F.: Applicability of neural networks to industrial and commercial power systems: a tutorial overview. IEEE Trans. Ind. Appl. **33**(5), 1355–1361 (1995)

Data-Driven and Collision-Free Hybrid Crowd Simulation Model for Real Scenario

Qingrong Cheng, Zhiping Duan, and Xiaodong Gu$^{(\boxtimes)}$

Department of Electronic Engineering, Fudan University,
Shanghai 200433, China
xdgu@fudan.edu.cn

Abstract. In order to take into account evading mechanism and make more realistic simulation results, we propose a data-driven and collision-free hybrid crowd simulation model in this paper. The first part of the model is a data-driven process in which we introduce an algorithm called MS-ISODATA (Main Streams Iterative Self-organizing Data Analysis) to learn motion patterns from real scenarios. The second part introduces an agent-based collision-free mechanism in which a steering approach is improved and this part uses the output from the first part to guide its agents. The hybrid simulation model we propose can reproduce simulated crowds with motion features of real scenarios, and it also enables agents in simulation evade from mutual collisions. The simulation results show that the hybrid crowd simulation model mimics the desired crowd dynamics well. According to a collectiveness measurement, the simulation results and real scenarios are very close. Meanwhile, it reduces the number of virtual crowd collisions and makes the movement of the crowd more effective.

Keywords: Crowd simulation · Data-driven · Main streams · MS-ISODATA
Collision-free · Collectiveness

1 Introduction

Crowd related research has been a significant theme for a long time in the realm of intelligent optimization, computer vision, robotics and computer graphics. People usually come across many circumstance with huge amount of people or other living creatures. It is meaningful to analyze the pattern of crowd motion. Consequently, we can extract a lot of useful information on crowd motion and put it into use. Related work usually involves counting [1], classification [2], tracking [3], abnormality detection [4] and understanding [5]. Pedestrian's movement are affected continuously by other pedestrian and their surrounding environment, so they should change their walking direction and speed frequently. Meanwhile, the state of emergency is different from the normal status. The multi-agent reinforcement learning model of pedestrian emergent behaviors are presented in [6]. Many crowd analysis methods related to crowd motion pattern and crowd simulation using computer vision techniques can be found in a survey [7].

Crowd simulation is an active research field that has drawn strong interest of researchers in industry, academia and government. Reynolds proposed the BOID

© Springer Nature Switzerland AG 2018
L. Cheng et al. (Eds.): ICONIP 2018, LNCS 11307, pp. 62–73, 2018.
https://doi.org/10.1007/978-3-030-04239-4_6

model [8] in 1987. He defined three behaviors, i.e. cohesion, alignment and separation for each agent in the group and the virtual group can mimic behaviors of bird group well. Helbing et al. constructed social force model [9] in 1995 and the model is especially helpful in simulating evacuation in public area [10]. In the past decades, many agents based or force based models were put forward and we saw crowd simulation flourished consequently such as [11, 12]. However, although the agent or force based model works well in small group of simulation, it has some drawbacks such as quantities of parameters tuning and complex rules defining, which makes it extremely difficult to fulfil simulation with large numbers of agents.

It is natural to think that we can directly learning some motion patterns from real crowds and simulating virtual crowds directly with these patterns, which is not only free the model from complex parameters and rules, but also enable the simulation results resemble to the real crowds. Accordingly, some approaches based on data-driven method emerged. Lerner presented an example-based crowd simulation technique [13] and examples were created from tracked video segments of real pedestrian crowds. Lee also presented a data-driven method [12] of simulating a crowd of virtual humans that exhibit behaviors imitating real human crowds.

Inspired by these works, we propose a data-driven and agent-based hybrid model for crowd simulation in this paper. The model mainly consists of two parts. The first part is a data-driven process in which motion patterns of specific scenarios are learned from videos of real crowds. The second part is an agent-based collision-free procedure, which is proposed by improving a velocity-based steering approach [14]. In the first part, global features we call Main Streams (MS) of crowds are extracted. While in the second part, the collision-free mechanism which guarantees the microscopic details are physically realistic by avoiding mutual collisions. With both macroscopic and microscopic characteristic we obtain, we can reproduce realistic simulating crowds just as real scenarios.

2   The Hybrid Model

2.1   Overview

The overall framework of the hybrid model is illustrated in Fig. 1.

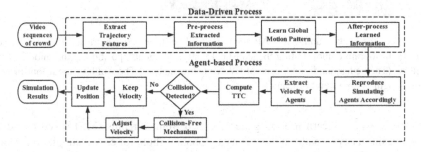

Fig. 1. The overall framework of our hybrid simulation model

2.2   The Data-Driven Process

Extract Trajectory Features. In our model, we would extract these common movement dynamics from observed video sequences and use them as global features of simulation. We know there exist some multi-target tracking algorithms and some of them work well, such as [15, 16]. Most of these algorithms are complex and high computational. In our model, we would extract point moving features from video sequences based on a method [17]. However, these trajectories are often highly fragmented and quite a lot of trajectories are missing because the tracker cannot keep track of all the moving points all the time. So the trajectories cannot be used directly and some pre-processing works are necessary.

Pre-process Extracted Information. The main purpose of the pre-processing procedure is to exclude outlets which result from tracking failures. So we would delete trajectories which are too short and wrongly recorded. As track captured might often shrink, the raw trajectories we get are usually slightly zigzag. So smoothing in a certain extent is also required. In our model, an arithmetic average filtering method is adopted, which is simple but effective.

Learn Global Motion Patterns. In a specific scenario, crowd behaviours actually share some common patterns based on adjacency of departure or destination positions, amplitude of velocities, similarity in motion paths or magnitude of collectiveness [18]. The paths of diverse pedestrians must be different from each other even under the circumstance of sharing the same departure and destination position. However, paths people chose should show some statistic features, which reflect the common preference of crowd flow. For example, in EWAP dataset [19] (Fig. 2), some statistic characteristics can be observed in the video, as is shown in Fig. 2 right. It implies that pedestrians tend to pass through the gallery in the middle.

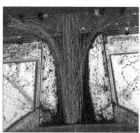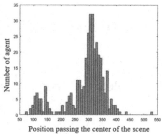

Fig. 2. Left, the background of EWAP Scenario. Middle, trajectories extracted from the video. Right, the histogram demonstrating the statistical feature of agents' position in the middle of its trajectory. (The pedestrian in the Scene of EWAP is low density. The width is about four meters and the length is about eight meters.)

In our work, we learn motion patterns called Main Streams (MS) based on adjacency and similarity of paths and use it as a global attribute for crowd simulation. Main Streams reflect the preference of crowds when they choose their paths. A trajectory of an agent is defined as the realistic path in the crowd. That is

$$\chi^p = \left(x_1^p, x_2^p, \cdots x_{np}^p \right), \tag{1}$$

where χ^p here denotes the l-th path and x_i^p denotes the i-th position recorded of the path. x_1^p and x_{np}^p here denote the departure and destination position of the path.

If there are k categories of patterns among χ, we define the set of paths sharing the same pattern as χ_i where $i = 1, 2, \cdots, k$. Then our next concern is to extract each χ_i from the overall set χ. Within χ_i, each path χ_i^p should be similar with each other. So if we regard a path χ_i^p as a sample in sample space χ_i, the distance between samples within sample space should be shorter than distance between samples outside the space. That means if there exists a centre C_i in space χ_i, the distance between any sample of the space and C_i should be the shortest one compared to other centres, which can be abstracted as the following equations. For all the centres defined in χ,

$$C = \{C_1, C_2, \cdots, C_k\}, \tag{2}$$

if the distance between a path χ^p and a centre C_i satisfies

$$D_i = \min\{\|\chi^p - C_i\|, i = 1, 2, \cdots k\}, \tag{3}$$

then we define $\chi^p \in \chi_i$. As a matter of fact, it is actually a process of clustering.

For a specific scene from real world, it is difficult to judge how many categories trajectories should be clustered. So some popular clustering method such as K-means and N-cut are not ideal choices. It is essential that the model is capable of automatically judging how many categories the set should be clustered. ISODATA [20] is an iterative self-organization data analysis or unsupervised clustering algorithm without any prior knowledge. We introduce ISODATA in clustering trajectories' set and extracting patterns Main Streams (MS-ISODATA). It is explained as follows.

Step 1, sample trajectories. Before clustering, we must acquire samples of the same length. Since the length of trajectories in the set χ is varied from each other, sampling trajectories to form sequences of the same length is necessary. Firstly, find out the shortest trajectory and denote its length as Ns. Secondly, linearly divide each path χ^p into Ns parts. It equals to constructing an arithmetic progression of Ns elements from 1 to np.

$$q^p = \left\{ 1, \frac{np}{Ns}, \frac{2np}{Ns}, \cdots, \frac{(Ns-1)np}{Ns}, np \right\}. \tag{4}$$

Thirdly, take one element from each parts of χ^p and form a sample of χ^p. It equals to form an index

$$index^l = floor(q^p), \tag{5}$$

by taking the nearest round number of each element of q^p in minus direction and extract point from χ^p according to this index. Then a sample of χ^p is acquired by

$$S^p = \chi^p(index^p). \tag{6}$$

Step 2, reshape each sample into one-dimension Eigenvector of the trajectory it represents.

Step 3, clustering. We apply ISODATA to cluster the trajectories set χ into categories.

Step 4, extract centre C_i of each clustered trajectories χ_i. These centres are the bases of our global motion patterns MS. They are reshaped into two-dimension position coordinate series.

$$C_i = \left(C_i^1, C_i^2, \cdots, C_i^{Ns}\right). \tag{7}$$

Step 5, obtain patterns of MS. The above centers cannot be used as MS directly or reflect complete global features, because that trajectories recorded might be highly fragmented. These ISODATA clustered centers are modified and the modified results are denoted as Main Streams of crowds. The modified process takes both trends of these original centers and positions of real departure or destination into account. At C_i^{Ns}, we denote direction vector to destination position as $\vec{v}_{des}$ and we compute

$$C_i^{Ns+1} = \frac{1}{\alpha}\vec{v}_{des} + \frac{1}{1-\alpha}\left(C_i^{Ns} - C_i^{Ns-1}\right) \tag{8}$$

iteratively until the modified center trajectory getting the edge of the scene. $\alpha \in (0,1)$ here and we use 0.5 in experiments following. For the departure direction the same process is repeated. If we denote this growing process as function $f(\cdot)$ then we get

In Fig. 3, key processes of extracting Main Streams from crowd scenario, EWAP dataset are demonstrated.

$$MS_i = f(C_i). \tag{9}$$

Fig. 3. Left, trajectories extracted from EWAP. Middle, clustering centers of paths. Right, MS extracted from EWAP.

2.3 Agent-Based Collision-Free Model

Paths Preference. In our hybrid model, Main Streams extracted from the real world are used as strong constraints for simulating agents when they choose appropriate paths. As illustrated in Fig. 2, agents prefer to pass through the scene in the middle and observation supports that agents' distribution among space is nearly a Gaussian distribution. Then we construct the model by defining the spatial distribution probability density function of agents as the following equation,

$$f_i(x, MS_i, \delta) = \frac{1}{\delta\sqrt{2\pi}}\left(-\frac{(x - MS_i)^2}{2\delta^2}\right). \tag{10}$$

Where δ is determined by the standard deviation in clustering. If no collision risks are detected around, agents would like to choose their paths according to Main Streams.

Collision-Free Steering Mechanism. Humans control their speed and direction based on their vision to avoid static and moving obstacles. Data-driven approaches cannot guarantee that agents are free of collision. It is necessary to add collision-free steering mechanism in crowd simulation. Collision avoidance research has drawn much attention in agent-based crowd simulation. Several approaches have been proposed to tackle the interaction and collision between agents. Helbing et al. [9] proposed the social forces model to avoid collision by repulse each other. However, Ondřej's vision-based approach (VISION) [21] performs more exactly and efficiently than RVO model and Helbing's model. Therefore, we introduce a collision-free mechanism by improving the VISION approach in our model. In this approach, velocities and directions of simulating agents can be controlled according to a threshold function

$$\tau_1(tti) = \left\{ \begin{array}{l} \tau_1(tti) = a - b \cdot tti^{-c}, \ if \dot{\alpha} < 0, \\ \tau_1(tti) = a + b \cdot tti^{-c}, \ otherwise. \end{array} \right\} \tag{11}$$

where τ is the threshold value controlling the angle magnitude of agents when making turns, tti is the time to interaction between two different agents, a, b and c are constant parameters, and $\dot{\alpha}$ is the relative angular acceleration between two different agents. In the original VISION, the constant parameter $a = 0$, $b = 0.6$ and $c = 1.5$. However, the author of VISION didn't mention the physical meanings of parameter a, b and c. We find that the better performance can be got during changing the value of a, b or c. In our model, we propose two different improvements with respect to the threshold function, as shown in Fig. 4.

Firstly, we reinforce the steering habit of agent by setting a suitable none-zero offset to parameter a. In experiments, we find that setting parameter a to 0 is not effective enough to fulfill completely collision free mechanism as the steering habit is not strong enough. By giving a suitable offset to a, the agent makes turns with larger magnitude.

Secondly, we restrict the range of reaction by adding cut-off effect to the threshold function. In the original approach of VISION, agents may behave unnecessary steering behaviors. Even agents without mutual collision risk would evade from each other

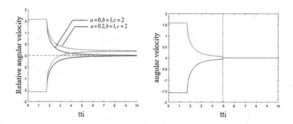

Fig. 4. Left, Reinforce the steering habit by adding an offset to the threshold function. Right, Bring cut-off effect into threshold function with cut-off threshold.

when they get closer if the reaction range is not restricted. The cut-off effect is brought in by refining (11) as follows:

$$
\tau_1(tti) = \left\{
\begin{array}{l}
\left.
\begin{array}{l}
\left\{
\begin{array}{l}
\tau_1(tti) = a - b \cdot tti^{-c},\ if\ \dot{\alpha} < 0, \\
\tau_1(tti) = a + b \cdot tti^{-c},\ otherwise.
\end{array}
\right\},\ if\ tti \leq \tau_0, \\
\end{array}
\right\} \\
0,\ otherwise.
\end{array}
\right\}
\tag{12}
$$

In (12), a threshold value of *tti* is brought in, which is denoted as τ_0. If the time of interaction *tti* between two agents is larger than the cut-off threshold τ_0, the steering behavior will not be activated. With the cut-off effect, unnecessary evading behaviors are avoided.

2.4 Centralized Post-process

Velocity Optimization. According to mechanism in 2.2 and 2.3, we can make sure that simulating agents move in the right ways and get rid of mutual collision. However, we still cannot guarantee simulation results to resemble the real scenario as we just define the global motion features according to real crowd for the simulation but the local velocity definition is still missing. It is necessary to optimize velocity choices of simulating agents after global paths and collision avoidance meet the requirements.

Let's assume that in example crowd video of the real scenario there are n agents and their velocities are denoted as $V_e = (v_1, v_2, \cdots, v_n)$. Their velocity are transformed to angle coordinate system respectively as $v_i = |v_i|\theta_i$. Then we get a vector of $\Theta_e = (\theta_1, \theta_2, \cdots, \theta_n)$. Accordingly, the vector of agents' direction in simulation can be computed as well, which is denoted as Θ_s. We compute the cross-correlation of agents' directions in each frame as

$$
\mathrm{cov}(\Theta_e, \Theta_s) = \sum_{i=1}^{n} (\Theta_e(i) - E(\Theta_e)) \cdot (\Theta_s(i) - E(\Theta_s))/n.
\tag{13}
$$

For each simulation epoch or frame, we make velocities magnitude of agents invariable in the basis of 2.2 and 2.3, and change directions of simulating agents within the range of $\theta_i = \pm\pi/4$ in the direction of gradient descent, using (13) as cost function. With this method of optimization, the distribution of simulating agents' velocities approximates the distribution of agents in real scenario, which makes simulation resemble real crowd locally in feature of motion.

3 Experimental Results

3.1 Hybrid Simulation Model Results

In the simulation, agents we reproduced succeed in passing through the scenario with the same motion pattern learned from real crowds in dataset Grand Central Hall (Grandhall) [22], in which 45,976 frames and 3,712 recorded agents are extracted. Using our model, we can automatically understand motion features of crowds in a specific real scenario and these features are successively used in crowd simulation. In the experimental results, we can see that our model successfully simulates a crowd of agents moving with global features like the real scenario, as demonstrated in Fig. 5. The simulation experiment comparison of three different approaches is demonstrated in Table 1. Approaches without collision-free mechanism, with VISION and with mechanism we propose are compared.

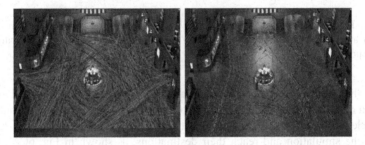

Fig. 5. Left, trajectories extracted from Grandhall. Right, one simulating frame of MS-based model.

Table 1. Comparison of experimental results based on different collision-free mechanism

Scenario	Evasion mechanism	Frames simulating	Number of agents	Times of collision	Agents reach goals
Grandhall	–	300	157	7,126	157
Grandhall	MS-VISION	300	157	0	59
Grandhall	MS-Improved VISION	300	157	0	121

The simulation results of the hybrid model can effectively mimic the behavior of the crowds on the global level, and the behavior of the simulating crowds is very similar to the real crowds' behavior. If the collision-free mechanism is not adopted, the simulation results based on the MS simulation method are developed smoothly along the MS direction. In the simulation, however, the spatial position is sometimes over-lapped by different agents. The simulation results are not real enough in details, as shown in Fig. 6B.

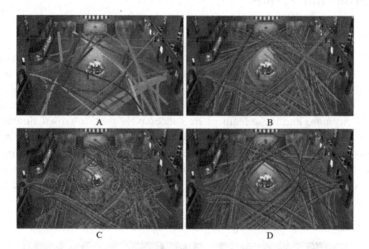

Fig. 6. A, MS extracted from grandhall dataset. B, simulation trajectories with MS. C, simulation trajectories with MS compounding VISION. D, simulation trajectories with hybrid model we proposed.

With the VISION collision-free mechanism, some agents are trapped in local oscillation due to the interference of neighboring agents. These individuals continue to make circles in some local areas and repeat searching for ideal paths. They failed to complete the simulation and reach their destinations, as shown in Fig. 6C. With our improved collision free mechanism, we can eliminate most local oscillation caused by mutual interference between individuals, so that individuals can finally reach their destinations while avoiding mutual collisions, as shown in Fig. 6D.

In this simulation experiment, we simulated 300 frames and the simulation contains 157 agents. We record times of collision occurred in the three approaches and agents finally reaching their goals within 300 simulation epochs. Fig. 6 and Table 1 show that the proposed hybrid simulation model can mimic the motion trajectories feature of real scenario and guide the agents reach their goals without collisions. In further experiments, we introduce collectiveness to measure the similarity between the simulation crowd and the real data in details.

3.2   Collectiveness Evaluation

The simulated crowd model needs a descriptor to evaluate whether its performances are similar to the real scenarios. In the past few years, some theoretic methods have been proposed to measure the similarity between the real-world data and the simulated results, such as [23, 24]. However, both of them are complex and heavy computing burden. In order to reduce the complexity of similarity measurement, we introduce a simple method to describe the similarity of real scenarios data to crowd simulators. Collectiveness is a common characteristic used to measure the behavior of a crowd motion from the individuals' behavior. Zhou et al. proposed a collectiveness descriptor based on path similarity and velocity correlation in 2013 [18].

Figures 7 and 8 show that the collectiveness of simulating crowds created by our simulation model is very close to the collectiveness of the real crowd scenes. The experiment shows that the maximum collectiveness error is about 0.3 for the first MS, and the error is below 0.1 for another. From the results of the collectiveness comparison, our simulation results are similar to real scenarios.

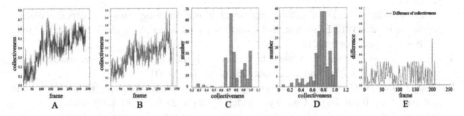

    A   B   C   D   E

Fig. 7. A, collectiveness of one streams of agents in 330 frames of real scenario. B, collectiveness of one streams of agents in 330 frames of simulation. C, histogram of simulation result's collectiveness. D, histogram of real scenario's collectiveness. E, the difference of collectiveness between simulation and real data.

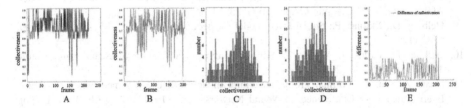

    A   B   C   D   E

Fig. 8. A, collectiveness of one streams of agents in 215 frames of real scenario. B, collectiveness of one streams of agents in 215 frames of simulation. C, histogram of simulation result's collectiveness. D, histogram of real scenario's collectiveness. E, the difference of collectiveness between simulation and real data.

4   Conclusion

In this paper, we have presented a data-driven and collision-free hybrid simulation model to achieve crowd simulation. As demonstrated above, our model could reproduce the crowd scenarios with real motion features by using Main Streams we introduce. Meanwhile, it combines improved collision-free mechanism, which makes the simulating agents evade from local mutual collisions. In the future work, we will explore more approaches (Deep Learning etc.) to extract global features of real crowds and develop advanced steering mechanisms in crowd simulation. Meanwhile, real-time crowd simulation or crowd analysis of large number of agents will be an important part of our next research.

Acknowledgments. This work is supported by national natural science foundation of China under grants 61771145 and 61371148.

References

1. Chan, A.B., Liang, Z.S.J., Vasconcelos, N.: Privacy preserving crowd monitoring: counting people without people models or tracking. In: 26th IEEE Conference on Computer Vision and Pattern Recognition, pp. 1–7. IEEE Press, Anchorage (2008)
2. Huang, X., Wang, W., Shen, G., Feng, X., Kong, X.: Crowd activity classification using category constrained correlated topic model. KSII Trans. Internet Inf. Syst. **10**, 5530–5546 (2016)
3. Bera, A., Manocha, D.: Realtime multilevel crowd tracking using reciprocal velocity obstacles. In: 22nd IEEE International Conference on Pattern Recognition, pp. 4164–4169. IEEE Press, Stockholm (2014)
4. Cui, J., Liu, W., Xing, W.: Crowd behaviors analysis and abnormal detection based on surveillance data. J. Vis. Lang. Comput. **25**(6), 628–636 (2014)
5. Shao, J., Loy, C.C., Wang, X.: Learning scene-independent group descriptors for crowd understanding. IEEE Trans. Circ. Syst. Video Technol. **27**(6), 1290–1303 (2017)
6. Martinez-Gil, F., Lozano, M., Fernández, F.: Emergent behaviors and scalability for multi-agent reinforcement learning-based pedestrian models. Simu. Model. Pract. Theory **74**, 117–133 (2017)
7. Junior, J.C.S.J., Musse, S.R., Jung, C.R.: Crowd analysis using computer vision techniques. IEEE Sig. Process. Mag. **27**(5), 66–77 (2010)
8. Reynolds, C.W.: Flocks, herds and schools: a distributed behavioral model. ACM SIGGRAPH Comput. Graph. **21**(4), 25–34 (1987)
9. Helbing, D., Molnar, P.: Social force model for pedestrian dynamics. Phys. Rev. E **51**(5), 4282 (1995)
10. Helbing, D., Farkas, I., Vicsek, T.: Simulating dynamical features of escape panic. Nature **407**(6803), 487–490 (2000)
11. Duan, Z., Gu, X.: Animal group behavioral model with evasion mechanism. In: IEEE International Joint Conference on Neural Networks, pp. 1167–1172. IEEE Press, Beijing (2014)
12. Lerner, A., Chrysanthou, Y., Lischinski, D.: Crowds by example. Comput. Graph. Forum **26**(3), 655–664 (2007)
13. Lee, K.H., Choi, M.G., Hong, Q., Lee, J.: Group behavior from video: a data-driven approach to crowd simulation. In: Proceedings of the 2007 ACM SIGGRAPH/Eurographics Symposium on Computer Animation, pp. 109–118. ACM Press, San Diego (2007)
14. Karamouzas, I., Skinner, B., Guy, S.J.: Universal power law governing pedestrian interactions. Phys. Rev. Lett. **113**(23), 238701 (2014)
15. Benfold, B., Reid, I.: Stable multi-target tracking in real-time surveillance video. In: 29th IEEE Conference on Computer Vision and Pattern Recognition, pp. 3457–3464. IEEE Press, Colorado (2011)
16. Berclaz, J., Fleuret, F., Fua, P.: Multiple object tracking using flow linear programming. In: 12th IEEE International Workshop on Performance Evaluation of Tracking and Surveillance (PETS-Winter), pp. 1–8. IEEE Press, Snowbird (2009)
17. Tomasi, C.: Detection and tracking of point features. Technical report 91(21), 9795–9802 (1991)

18. Zhou, B., Tang, X., Wang, X.: Measuring crowd collectiveness. In: 31st IEEE Conference on Computer Vision and Pattern Recognition, pp. 3049–3056. IEEE Press, Portland (2013)
19. Pellegrini, S., Ess, A., Schindler, K., Van Gool, L.: You'll never walk alone: Modeling social behavior for multi-target tracking. In: 12th IEEE International Conference on Computer Vision, pp. 261–268. Kyoto (2009)
20. Ball, G.H., Hall, D.J.: ISODATA, a novel method of data analysis and pattern classification. Stanford Research Institute, AD-699616 (1965)
21. Ondřej, J., Pettré, J., Olivier, A.H., Donikian, S.: A synthetic-vision based steering approach for crowd simulation. ACM Trans. Graph. (TOG) **29**(4), 123 (2010)
22. Zhou, B., Wang, X., Tang, X.: Understanding collective crowd behaviors: learning a mixture model of dynamic pedestrian-agents. In: 30th IEEE Conference on Computer Vision and Pattern Recognition, pp. 2871–2878. IEEE Press, Providence (2012)
23. Guy, S., Van Den Berg, J., Liu, W., Lau, R., Lin, M.C.: A statistical similarity measure for aggregate crowd dynamics. ACM Trans. Graph. (TOG) **31**(6), 190 (2012)
24. Wong, S.M., Yao, Y.Y.: A statistical similarity measure. In: Proceedings of the 10th Annual International ACM SIGIR Conference on Research and Development in Information Retrieval, pp. 3–12. ACM Press, New Orleans (1987)

Aligning Manifolds of Double Pendulum Dynamics Under the Influence of Noise

Fayeem Aziz[✉], Aaron S. W. Wong, James S. Welsh, and Stephan K. Chalup

School of Electrical Engineering and Computing, The University of Newcastle,
Callaghan, NSW 2308, Australia
MdFayeemBin.Aziz@uon.edu.au, Stephan.Chalup@newcastle.edu.au

Abstract. This study presents the results of a series of simulation experiments that evaluate and compare four different manifold alignment methods under the influence of noise. The data was created by simulating the dynamics of two slightly different double pendulums in three-dimensional space. The method of semi-supervised feature-level manifold alignment using global distance resulted in the most convincing visualisations. However, the semi-supervised feature-level local alignment methods resulted in smaller alignment errors. These local alignment methods were also more robust to noise and faster than the other methods.

Keywords: Manifold learning · Dimensionality reduction
Manifold alignment · Double pendulum · Robot motion

1 Introduction

Manifold alignment of two data sets assumes that they have similar underlying manifolds. The aim is to find a mapping between the two data sets so that the underlying manifold structure can be better recognised. Due to its generality manifold alignment has great potential to be useful in various domains. In the past it has been applied to facial expression analysis by image sequence alignment [6,10,17,24,26], graph matching [7], image classification [9,25], and bioinformatics [21]. However, due to its high computational demands and the complexity of manifold data, manifold alignment still requires substantial further research and development to increase its impact in practical applications.

Real-world data is often sampled from a high-dimensional space and can be modelled as a set of points that lie on a low-dimensional non-linear manifold [12]. This also applies to data collected from robot kinematics or human motion which could be represented in form of dynamical system trajectories on low-dimensional manifolds [4]. It has been shown that motion data can be transferred from a human to a robot or between robots by aligning the corresponding manifolds [3, 5]. In this context, the ability of manifold learning and manifold alignment to represent non-linear data is critical [18]. An example that can illustrate this is a double pendulum and its non-linear dynamics [13].

© Springer Nature Switzerland AG 2018
L. Cheng et al. (Eds.): ICONIP 2018, LNCS 11307, pp. 74–85, 2018.
https://doi.org/10.1007/978-3-030-04239-4_7

The contribution of this study is an experimental comparison of several existing manifold alignment algorithms using data sampled from the motions of two simulated double pendulums in three-dimensions. Previous restricted pilot experiments, had focused on a pendulum in two dimensions [1]. The present paper also it investigates the stability of the methods under the influence of noise. It introduces the concepts of normalised distances along with pairwise correspondence measures [8]. The experimental evaluation focuses on visualisations and measuring the proximity of alignments and the execution time of the algorithms.

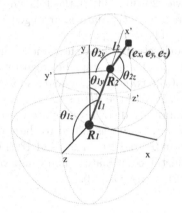

Fig. 1. Double pendulum motion in 3D. x, y and z are the local axes of limb l_1 and x', y' and z' are the local axes of limb l_2. (e_x, e_y, e_z) is the end-effector.

2 Double Pendulum Data

The simulated data used in our experiments represents the motion of a three-dimensional double pendulum similar to a two-limb robot arm as shown in Fig. 1. The two limbs are denoted l_1 and l_2 and the joints are denoted R_1 and R_2. One end of l_1 is fixed at R_1 and can rotate around this joint. The other end of l_1 is attached to l_2 at joint R_2. l_2 can rotate freely around joint R_2. The end point of the arm, that is, the end of l_2 not attached to R_2, is also known as the end-effector and has coordinates (e_x, e_y, e_z). The coordinate system follows the right-hand rule and has its origin at joint R_1. θ_{1y} and θ_{1z} are the angles of l_1 with the y and z axes, respectively, and $\theta_{2y'}$ and $\theta_{2z'}$ are the angles of l_2 with the y' and z' axes, respectively. The feature vector for each sample point was calculated from the kinematics at the joints and the end-effector coordinates were calculated using forward kinematics:

$$(e_x, e_y, e_z, \cos\theta_{1y}, \cos\theta_{1z}, \cos\theta_{2y'}, \cos\theta_{2z'}, \sin\theta_{1z}, \sin\theta_{1y}, \sin\theta_{2y'}, \sin\theta_{2z'}) \quad (1)$$

The data sets were generated from two similar double pendulums, which had different limb lengths and slightly different limb length ratios, where we restricted the experiments to the case $l_2 < l_1$ (Fig. 1):

Pendulum 1: $(l_2/l_1) = 0.75/1.25 = 0.60$
Pendulum 2: $(l_2/l_1) = 1.25/1.56 = 0.80$,

The data was acquired using increments of 30° on all four axes y, z, y' and z'. As a result, the number of instances was $(360/30)^4 = 20736$ and the size of each of these data sets was 20736×11. If the following sections refer to data sets X and Y it is assumed that the data is arranged in the form of matrices of dimension 20736×11. The correspondence subsets of X and Y comprise corresponding points in both sets that have the same or similar joint angles. They were selected by using 90° joint angle steps between two instances in the same data set. Therefore each of the two correspondence subsets had $(360/90)^4 = 256$ instances which was about 10% of the data set.

In the experiments, uniformly distributed white noise was added in two different ways to the data. First, noise was added to the joint angles and the noise range was incremented from 0° to ±10° in steps of 1°. The second type of noise was added to the end effector coordinates. The range of this coordinate noise was increased from 0 to ±1.0 in steps of 0.1.

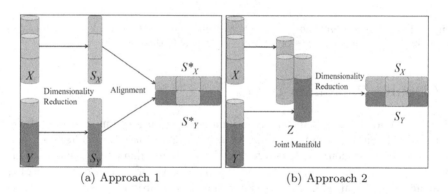

(a) Approach 1   (b) Approach 2

Fig. 2. (a) Approach 1: Dimensionality reduction is applied to X and Y, which yields low-dimensional manifolds S_X and S_Y, respectively. The manifolds are then mapped into a joint space as aligned manifolds S_X^* and S_Y^*. (b) Approach 2: A joint matrix Z is generated from X and Y by regularisation of correspondence information. Then, dimensionality reduction is applied to obtain the low-dimensional aligned manifolds S_X and S_Y. The green areas in both approaches indicate the correspondence subsets. (Color figure online)

3   Manifold Alignment Methods

We address two general approaches to align manifolds. Each of them has two stages (Fig. 2). In Approach 1 the data sets are first mapped into a low-dimensional space and then alignment is performed as described by Wang et al. [21]. In Approach 2, first a joint manifold is created to represent the union of the given manifolds and then the joint manifold is mapped to a lower dimensional latent space as described by Ham et al. [10]. Approach 2 was also described as a generalised semi-supervised manifold alignment framework by Wang and Mahadevan [22], where the joint manifolds can be constructed using different characteristics of the data based on the application.

Wang and Mahadevan [22] further distinguished between two levels of manifold alignment: instance-level alignment (Inst) and feature-level alignment (Feat). For instance-level alignment, their method computes a non-linear low-dimensional embedding based on an alignment of the instances in the data. The embedding results in a direct matching of corresponding instances in alignment space. In contrast, feature-level alignment builds on mapping functions of features which map any associated instance or set of instances into the newly aligned domain.

The present study applied the following four methods to the double pendulum data sets. Method 1 followed Approach 1 ([21] and Fig. 2(a).) and methods 2–3 followed the semi-supervised Approach 2 ([22] and Fig. 2(b)). Further, each of the four methods had an instance-level and a feature-level version that was addressed in separate experiments (see overview in Fig. 3):

- **Method 1:** Locality Preserving Projection (LPP) [11] is used for feature-level dimensionality reduction, and Laplacian eigenmaps (eigenmaps) [2] is used for instance-level dimensionality reduction, as the first stage, in two different experiments. In the second stage (Fig. 2(a)) we followed [21] and employed Procrustes analysis [14] to align the two data sets in low-dimensional space.
- **Method 2:** For semi-supervised manifold alignment preserving local geometry [22], first, the joint manifold Z was calculated using the graph Laplacians for X and Y. Then, eigenvalue decomposition of the joint manifold provided instance-level alignment and generalised eigenvalue decomposition provided feature-level alignment.
- **Method 3:** For semi-supervised manifold alignment using local weights [20] the intermediary joint manifold Z was formed by weights for X and Y, which were calculated for each set using a k-nearest neighbour graph and the heat kernel. Similar to Method 2 the low-dimensional manifolds were mapped by eigenvalue decomposition of joint matrices in two experimental components.
- **Method 4:** For semi-supervised manifold alignment preserving global geometry [23] the joint manifold Z was generated using the global distances of corresponding pairs in $X \cup Y$. The eigenvalue decomposition of Z provided the dimensionality reduction to obtain the aligned low-dimensional manifolds in the case of instance-level alignment. Generalised eigenvalue decomposition was used to reduce the dimensionality in the case of feature-level alignment.

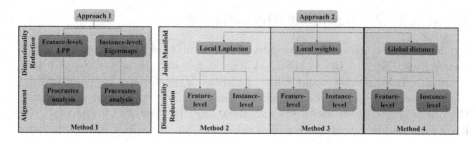

Fig. 3. Overview of methods used in the experiments where the general structure of Approach 1 and 2 is detailed in Fig. 2.

4   Performance Evaluation

To measure the performance of the alignment methods, we assumed that all instances of one data set can be mapped one-to-one to instances in the other data set. The widths of the resulting manifolds originating from the same data set could vary for the different methods. As a result of scaling, the distances between corresponding points could be larger between "wider" manifolds and smaller between "narrower" manifolds. To obtain comparable distance measurements between the two aligned low-dimensional manifolds, the distances were normalised by the maximum width of the two manifolds as follows:

$$D_i = \frac{\|S_X(i) - S_Y(i)\|}{\max_{\substack{j=1,\dots,n \\ k=1,\dots,n}} (\|S_X(j) - S_X(k)\|, \|S_Y(j) - S_Y(k)\|)} \tag{2}$$

where $S_X(i)$ and $S_Y(i)$ are corresponding points for $i = 1, \dots, n$ and n is the number of points that represent each of the manifolds. The matching errors of the alignments were measured using the average of the D_i, denoted by Δ. That is, Δ indicates the closeness of the aligned manifolds in low-dimensional space. The standard deviation of the D_i, denoted by σ, represents the consistency or smoothness of the alignment where a smaller σ indicates a smoother alignment. Δ and σ could be used to measure the quality of an alignment if it was successful.

5   Results

Both limbs of the double pendulum rotated freely in three dimensions. The resulting manifolds were too complex to be visualised in full in a three-dimensional graph. Therefore, we used snapshots of 90° steps for the motion of limb l_1 and of 10° steps for limb l_2. Figure 4 shows that feature-level manifold alignment using global distance resulted in six small spheres that were distributed regularly on a bigger sphere. The small spheres represent the motion of limb l_2 and the bigger sphere represents the motion of limb l_1. Ergo this figure can be interpreted as sections in time of the expected shape of the aligned manifolds of pendulum motion. The results from the other three methods for feature level alignment show complex shapes that include partially collapsed submanifolds. For the instance-level cases the manifolds collapsed completely into line segments.

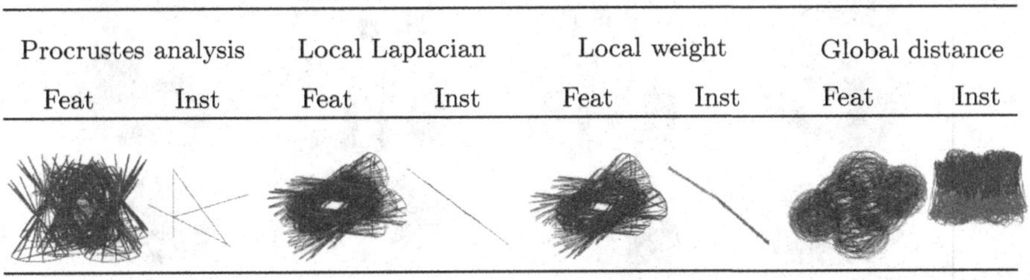

Procrustes analysis		Local Laplacian		Local weight		Global distance	
Feat	Inst	Feat	Inst	Feat	Inst	Feat	Inst

Fig. 4. 3D motion manifold alignments using snapshots in 90° steps for one limb and 10° steps for the other limb. The output of the feature-level manifold alignment using global distance (column 7) displays six small spheres representing the motion of l_2 which are distributed on a larger sphere representing the motion of l_1.

Table 1. Summary of the performance of manifold alignment methods: Feature-level manifold alignment using local weights achieved the lowest Δ and σ (3rd row).

	Method	Level	Δ	σ	Time (s)
1	Procrustes analysis	Feature	1.37×10^{-05}	6.20×10^{-06}	393
2	Local Laplacian	Feature	1.22×10^{-07}	5.08×10^{-08}	382
3	**Local weights**	**Feature**	$\mathbf{1.22 \times 10^{-07}}$	$\mathbf{4.92 \times 10^{-08}}$	**343**
4	Global distance	Feature	2.76×10^{-06}	7.38×10^{-07}	6.3×10^4

Table 1 shows results for the feature-level cases where Δ and σ were lowest for Method 3. The execution time of each method for same datasets is shown in the last column of 1. Due to the involvement of high dimensional matrix multiplication, the manifold alignment using global distance required a significantly higher execution time than the others. Figures 5 and 6 show the aligned manifolds in the feature level cases after application of joint angle and coordinate noise, respectively. Both limbs of the double pendulums were rotating in three-dimensional spherical motion. Ergo the overall motion manifold can be described as the cross product of two spheres (as indicated in column 7 of Fig. 4). The visualisations of the outcome of manifold alignment preserving global geometry are shown in column 4 of Figs. 5 and 6. The results seemed to be robust to noise and the visualisation were as expected. On the other hand, manifold alignments using the local Laplacian and local weights resulted in lower Δ and σ (Fig. 7) but visually more complex outcomes that also significantly varied under the influence of noise (Figs. 5 and 6).

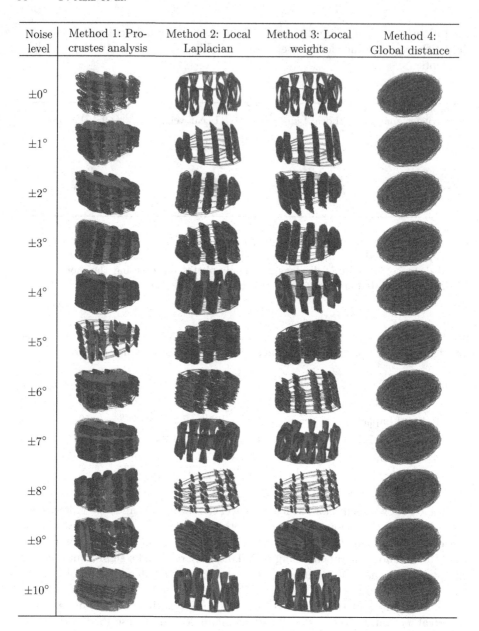

Fig. 5. Feature-level alignment of manifolds under the influence of different levels of joint angle noise: Each graph visualises a different way of aligning the two pendulum manifolds. Each row shows the results for a different level of joint angle noise.

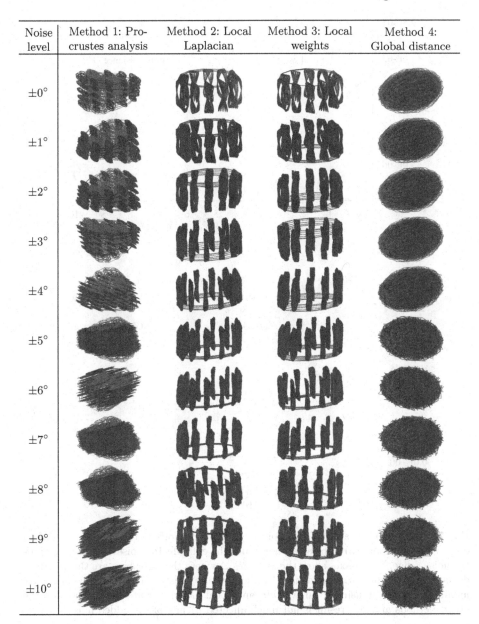

Fig. 6. Feature-level alignment of manifolds under the influence of different levels of coordinate noise: Each graph visualises a different way of aligning the two pendulum manifolds. Each row shows the results for a different level of coordinate noise.

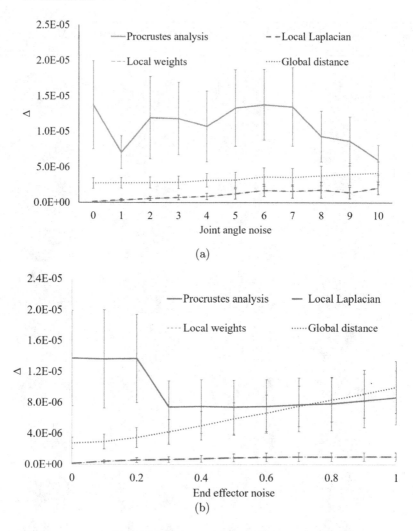

Fig. 7. Performance graphs of manifold alignment methods that align the pendulum data with additional (a) joint angle noise and (b) end effector noise. The x-axes show the noise range. The y-axes show the averages Δ where the error bars are the standard deviations σ of the correspondence distances of the aligned manifolds. Δ and σ for manifold alignment using local weights and local Laplacian are lower than for the other methods and this result is not much affected by any of the added noise.

6   Discussion and Conclusion

This study compared four different manifold alignment methods (Fig. 3). The data sets were generated by simulating the motion of two three-dimensional double pendulums that differed in the lengths of their arms. Where the joint angles of the two pendulums were equal, they were recorded as corresponding pairs, and this was exploited in the manifold alignment and performance evaluation.

We also investigated the effects of two different types of noise on the alignment of these data sets. First, random noise was added to the joint angles to observe the stability of the alignment with respect to actuator irregularities. Then, random noise was added to the end effector coordinates to observe the stability of the alignment methods with respect to jittery motions of the arm or noise in the data recordings. The performance of these methods was evaluated quantitatively by measuring the proximity of the corresponding points and qualitatively using visualisations. The experiments were conducted in Matlab 2016b on a high-performance computer cluster equipped with Xenon CPUs and 500GB RAM where all experiments for this paper could be executed in a total running time of about eighteen days. Individual running times are reported in Table 1.

The approach of alignment preserving global geometry produced the most convincing visualisations but was also much slower than the other methods. The methods using local weights resulted in the numerically smallest alignment errors (Table 1). One possible interpretation is that local methods had an advantage because the experiments were limited to data from two similarly configured pendulums. Another interpretation is that parts of the manifolds collapsed, i.e., they were projected onto line segments during the manifold learning process. This is supported by the visualisations in Figs. 4, 5 and 6. The observed instabilities of the local and instance level methods require further investigation.

Traditional robotic motion control engines often rely on inverse kinematics and can be difficult to adapt when the robot configuration changes [15]. Chalodhorn and Rao found that direct use of kinematics data from a human motion capture system to replicate human movement in a robot can result in dynamically unstable motion [5]. This instability is the result of differences in the degrees-of-freedom and inadequate physical parameters for a robot to match human motion. The outcomes of the present study support the hypothesis that manifold alignment could provide a mapping between motion trajectories of similar manifolds. Moreover, a feature-level alignment approach could align trajectories on manifolds that may include out-of-sample extensions that may be generated during changes of configurations.

One of the purposes of dimensionality reduction is to process data faster while requiring less memory than the calculations in high dimensions. Feature-level alignment provides a function that can map instances from high-dimensional feature space to low-dimensional alignment space. This function can map any new sample, whether it was included in the training data set or not, to the aligned manifolds. In contrast, the instance-level methods require a complete recalculation of the alignment mapping for each new sample. Therefore, feature-level alignment can perform better than instance-level alignment when out-of-sample extension processing is necessary [19]. The latter could be the case, for example, when the robot configuration changes.

In future research, the findings of this study could possibly become useful for the task of motion imitation [16]. Using manifold alignment transfer learning between robot arms could be achieved by copying previously trained stable motions from one robot to another where the robots' arms could have different limb length proportions.

Acknowledgements. FA was supported by a UNRSC50:50 PhD scholarship at the University of Newcastle, Australia. The authors are grateful to the UON ARCS team who facilitated access to the UON high performance computing system.

References

1. Aziz, F., Wong, A.S.W., Welsh, J., Chalup, S.K.: Performance comparison of manifold alignment methods applied to pendulum dynamics. In: Proceedings of the Applied Informatics and Technology Innovation Conference. Springer (2016, in press)
2. Belkin, M., Niyogi, P.: Laplacian eigenmaps for dimensionality reduction and data representation. Neural Comput. **15**(6), 1373–1396 (2003). https://doi.org/10.1162/089976603321780317
3. Bocsi, B., Csato, L., Peters, J.: Alignment-based transfer learning for robot models. In: The 2013 International Joint Conference on Neural Networks (IJCNN), pp. 1–7. IEEE (2013). https://doi.org/10.1109/IJCNN.2013.6706721
4. Chalodhorn, R., Grimes, D.B., Grochow, K., Rao, R.P.N.: Learning to walk through imitation. In: Proceedings of the 20th International Joint Conference on Artifical Intelligence, IJCAI 2007, pp. 2084–2090. Morgan Kaufmann Publishers Inc., San Francisco (2007)
5. Chalodhorn, R., Rao, R.N.: Learning to imitate human actions through eigenposes. In: Sigaud, O., Peters, J., (eds.) From Motor Learning to Interaction Learning in Robots. Studies in Computational Intelligence, vol. 264, pp. 357–381. Springer, Heidelberg (2010). https://doi.org/10.1007/978-3-642-05181-4-15
6. Cui, Z., Shan, S., Zhang, H., Lao, S., Chen, X.: Image sets alignment for video-based face recognition. In: 2012 IEEE Conference on Computer Vision and Pattern Recognition (CVPR), pp. 2626–2633, June 2012. https://doi.org/10.1109/CVPR.2012.6247982
7. Escolano, F., Hancock, E., Lozano, M.: Graph matching through entropic manifold alignment. In: 2011 IEEE Conference on Computer Vision and Pattern Recognition (CVPR), pp. 2417–2424 (2011). https://doi.org/10.1109/CVPR.2011.5995583
8. Fan, K., Mian, A., Liu, W., Li, L.: Unsupervised manifold alignment using soft-assign technique. Mach. Vis. Appl. **27**(6), 929–942 (2016)
9. Guerrero, R., Ledig, C., Rueckert, D.: Manifold alignment and transfer learning for classification of Alzheimer's disease. In: Wu, G., Zhang, D., Zhou, L. (eds.) MLMI 2014. LNCS, vol. 8679, pp. 77–84. Springer, Cham (2014). https://doi.org/10.1007/978-3-319-10581-9_10
10. Ham, J., Lee, D., Saul, L.: Semisupervised alignment of manifolds. In: Proceedings of the Annual Conference on Uncertainty in Artificial Intelligence, vol. 10, pp. 120–127. AISTATS (2005)
11. He, X., Niyogi, P.: Locality preserving projections. In: Thrun, S., Saul, L.K., Schölkopf, B. (eds.) Advances in Neural Information Processing Systems, vol. 16, pp. 153–160. MIT Press (2004)

12. Huang, D., Yi, Z., Pu, X.: Manifold-based learning and synthesis. IEEE Trans. Syst. Man Cybern. Part B (Cybernetics) **39**(3), 592–606 (2009). https://doi.org/10.1109/TSMCB.2008.2007499

13. Jensen, J.S.: Non-linear dynamics of the follower-loaded double pendulum with added support-excitation. J. Sound Vibr. **215**(1), 125–142 (1998). https://doi.org/10.1006/jsvi.1998.1620

14. Luo, B., Hancock, E.R.: Feature matching with Procrustes alignment and graph editing. In: Image Processing And Its Applications, 1999. Seventh International Conference on (Conf. Publ. No. 465), vol. 1, pp. 72–76, July 1999. https://doi.org/10.1049/cp:19990284

15. Mosavi, A., Varkonyi, A.: Learning in robotics. Int. J. Comput. Appl. **157**(1), 0975–8887 (2017)

16. Pan, S.J., Yang, Q.: A survey on transfer learning. IEEE Trans. Knowl. Data Eng. **22**(10), 1345–1359 (2010). https://doi.org/10.1109/TKDE.2009.191

17. Pei, Y., Huang, F., Shi, F., Zha, H.: Unsupervised image matching based on manifold alignment. IEEE Trans. Pattern Anal. Mach. Intell. **34**(8), 1658–1664 (2012). https://doi.org/10.1109/TPAMI.2011.229

18. Tenenbaum, J.B., de Silva, V., Langford, J.C.: A global geometric framework for nonlinear dimensionality reduction. Science **290**(5500), 2319–23 (2000). https://doi.org/10.1126/science.290.5500.2319

19. Wang, C.: A geometric framework for transfer learning using manifold alignment. Ph.d. thesis, Department of Computer Science, University of Massachusetts Amherst, UMass Amherst, September 2010

20. Wang, C., Krafft, P., Mahadevan, S.: Manifold alignment, Chap. Manifold alignment, pp. 95–120. CRC Press, December 2011. https://doi.org/10.1201/b11431-6

21. Wang, C., Mahadevan, S.: Manifold alignment using procrustes analysis. In: Proceedings of the 25th International Conference on Machine Learning, ICML 2008, pp. 1120–1127. ACM, New York (2008). https://doi.org/10.1145/1390156.1390297

22. Wang, C., Mahadevan, S.: A general framework for manifold alignment. In: AAAI Fall Symposium: Manifold Learning and Its Applications, pp. 79–86. AAAI Press (2009)

23. Wang, C., Mahadevan, S.: Manifold alignment preserving global geometry. In: Proceedings of the Twenty-Third International Joint Conference on Artificial Intelligence (IJCAI), pp. 1743–1749. AAAI Press (2013)

24. Wang, X., Yang, R.: Learning 3D shape from a single facial image via non-linear manifold embedding and alignment. IEEE Conference on Computer Vision and Pattern Recognition (CVPR) 2010, pp. 414–421 (2010). https://doi.org/10.1109/CVPR.2010.5540185

25. Yang, H.L., Crawford, M.M.: Manifold alignment for multitemporal hyperspectral image classification. In: 2011 IEEE International Geoscience and Remote Sensing Symposium (IGARSS), pp. 4332–4335, July 2011. https://doi.org/10.1109/IGARSS.2011.6050190

26. Zhai, D., Li, B., Chang, H., Shan, S., Chen, X., Gao, W.: Manifold alignment via corresponding projections. In: Proceedings of the British Machine Vision Conference, pp. 1–11. BMVA Press (2010). https://doi.org/10.5244/C.24.3

Pinning Synchronization of Complex Networks with Switching Topology and a Dynamic Target System

Guanghui Wen[1,2(✉)], Xinghuo Yu[2], Peijun Wang[1], and Wenwu Yu[1]

[1] Department of Systems Science, School of Mathematics,
Southeast University, Nanjing 211189, China
wenguanghui@gmail.com

[2] A School of Engineering, RMIT University, Melbourne, VIC 3001, Australia

Abstract. We address the global synchronization problem in this note for complex network with a target system whose control inputs are nonzero under directed switching communication topologies. To eliminate the impact of nonzero inputs, a discontinuous controller is designed. If the average dwell time is lower bounded by a positive constant, we show synchronization is achieved by selecting sufficiently large coupling strengths. In addition, some sufficient criteria are given for guaranteeing synchronization in the network under undirected switching communication topologies. Finally, we perform a numerical simulation to validate the theoretical results.

Keywords: Complex network · Pinning synchronization
Switching topology · Dynamical target system

1 Introduction

Over the last few decades, researchers have delved deeply into the complex networks as they relate many real-word networks in nature and our daily life, ecological web, aviation network, power network, urban public transport network, for example. In general, the complex networks are composed of a large amount of nodes (or individuals) which are interconnected. However, this large scale property has brought many difficulties for the network to reach cooperation functions based on local interactions among nodes. Among the various cooperation functions, synchronization behavior may be basic but really important. To date, synchronization of complex networks has attracted plenty of attentions

This work is supported by the National Nature Science Foundation of China through Grant Nos. 61722303 and 61673104, the Natural Science Foundation of Jiangsu Province of China through Grant No. BK20170079, the Australian Research Council through Grant No. DE180101268, and the Fundamental Research Funds for the Central Universities of China through Grant No. 2242018k1G004, Postgraduate Research & Practice Innovation Program of Jiangsu Province (No. KYCX17_0041).

L. Cheng et al. (Eds.): ICONIP 2018, LNCS 11307, pp. 86–96, 2018.
https://doi.org/10.1007/978-3-030-04239-4_8

from the fields of science and engineering, such as statistical physics [1], applied mathematics [2], system and control theory [3,4].

If the states of every node converge to the same constant vector or to the same trajectory finally, we call synchronization is reached. When the communication topology is fixed, the authors [5,6] proved that synchronization could be achieved by selecting sufficiently large coupling strength provided the communication graph is connected and the nonlinear dynamics satisfy Lipschitz condition. However, due to the limited sensing ability and complex communication environment, the communication topologies are time varying. More precisely speaking, it is more essential to study synchronization behaviors of networks with time varying communication topologies [7–11]. Such as in [11], Wen et. al. have settled the consensus tracking problem in multi-agent systems with switching communication topologies. More precisely, they showed that state consensus (synchronization) in multi-agent systems (complex networks) could be achieved by selecting suitable control parameters if each follower could sense the leader directly or indirectly frequently enough and the dwell time is lower bounded by a positive constant. It is worth noting that large scale network with switching communication topologies can characterize the real-world systems much better and thus has a wider application.

The target system in most existing works is modeled by an autonomous system. That is to say, the target system (leader) has zero control inputs. In practical applications, however, the target system (leader) may need some inputs for obstacle avoidance such that a security area would be formed for the followers to move along. Motivated by this observation, we [12] recently addressed the pinning synchronization problem for complex network with directed but fixed communication topologies, where the $\infty-$norm of the inputs on the target system is nonzero but upper bounded. By using non-smooth stability theory, we obtained few sufficient criteria to achieve synchronization. Inspired by the aforementioned works, in this paper, we further to show how to achieve pinning synchronization in a class of complex networks with directed but switching topologies. Unlike [12] where non-smooth stability theory was used for theoretical analysis, we use Lyapunov stability theory of switched systems. Firstly, we convert the synchronization problem into a stability problem of an error system. Then, a multiple Lyapunov function is established based on M-matrix theory. Finally, we show the error converges to a zero vector of appropriate dimension if the coupling strengths are sufficiently large and the average dwell time is lower bounded by a positive constant. In addition, the obtained theorem is validated by performing a simulation on Chua's circuit.

This paper mainly contains three parts. We present the preliminaries and model formulation in Sect. 2. And in Sect. 3, we give the theorems and their proofs. The numerical simulation is then performed in Sect. 4.

Notations. For an $N \times N$ dimensional matrix P with real eigenvalues, denotes the minimum one by $\lambda_{\min}(P)$. The Euclidean norm and $\infty-$norm are represented, respectively, by $\| \cdot \|$ and $\| \cdot \|_{\infty}$.

2   Preliminaries and Model Formulation

2.1   Preliminaries on Graph Theory

The interactions among neighboring nodes in the network are modeled by directed switching graph $\mathcal{G}(\mathcal{V}, \mathcal{E}^{\sigma(t)}, \mathcal{A}^{\sigma(t)})$ with $\mathcal{V} = \{1, 2, \cdots, N\}$, $\mathcal{E}^{\sigma(t)} \subset \mathcal{V} \times \mathcal{V}$, the adjacency matrix $\mathcal{A}^{\sigma(t)} = [a_{ij}^{\sigma(t)}] \in \mathbb{R}^{N \times N}$, where $a_{ij}^{\sigma(t)} > 0$ iff $(j, i) \in \mathcal{E}^{\sigma(t)}$ and $\sigma(t) : \mathbb{R}_{\geq 0} \to \mathbb{N}$ is a piecewise constant function that represents the switching signal. As N is finite, we assume $\sigma(t) \in \{1, 2, \cdots, \kappa\}$ with $\kappa \geq 2$ is a given positive integer. When no confusion will arise, denote $\mathcal{G}(\mathcal{V}, \mathcal{E}^{\sigma(t)}, \mathcal{A}^{\sigma(t)})$ by $\mathcal{G}^{\sigma(t)}$. If there is a node in the graph $\mathcal{G}^{\sigma(t)}$ that has at least one directed path to any other nodes, we call $\mathcal{G}^{\sigma(t)}$ has a directed spanning tree with common root. Denote the Laplacian matrix by $\mathcal{L}^{\sigma(t)} = [l_{ij}^{\sigma(t)}]$, where $l_{ij}^{\sigma(t)}$ is given by (1):

$$l_{ij}^{\sigma(t)} = \begin{cases} \sum_{j \in \mathcal{N}_i} a_{ij}^{\sigma(t)}, & \text{if } j = i, \\ -a_{ij}^{\sigma(t)}, & \text{if } j \neq i. \end{cases} \tag{1}$$

Associate with $\sigma(t)$, the definition of the average dwell time τ_a is given as follows.

Definition 1. For arbitrarily given $T_b > T_a \geq 0$, we call τ_a the average dwell time if it satisfies the following condition:

$$N_\sigma[T_a, T_b) \leq N_0 + \frac{T_b - T_a}{\tau_a}, \tag{2}$$

where $N_0 \in \mathbb{N}$ and $N_\sigma[T_a, T_b)$ is the number of switchings over $[T_a, T_b)$.

Definition 2. [13] Let $Q = [q_{ij}]$ be an $N \times N$ real matrix. If $q_{ij} \leq 0$ when $i \neq j$ and each eigenvalue has positive real part, we call Q a nonsingular $M-$matrix.

2.2   Formulation of the Problem

In the sequel, we assume the dynamical complex network consists of N coupled nodes and a target system, where the directed communication topologies among the $N + 1$ nodes are switching over time. For notation convenience, we label the N coupled nodes and the target system as node $1, 2, \cdots, N$ and node $N + 1$, respectively. And let $d_i^{\sigma(t)}$ be the pinning links between the target system and node i, $i = 1, 2, \cdots, N$, where $d_i^{\sigma(t)} = 1$ if node i is pinned at time instant t, $d_i^{\sigma(t)} = 0$ otherwise. The time evolution of node i is then given as

$$\dot{x}_i(t) = f(x_i(t), t) + \alpha \sum_{j=1}^{N} a_{ij}^{\sigma(t)} \Gamma(x_j(t) - x_i(t)) - \alpha d_i^{\sigma(t)} \Gamma(x_i(t) - x_{N+1}(t))$$

$$+ \beta \text{sgn} \left(\sum_{j=1}^{N+1} a_{ij}^{\sigma(t)} \Gamma(x_j(t) - x_i(t)) - d_i^{\sigma(t)} \Gamma(x_i(t) - x_{N+1}(t)) \right). \tag{3}$$

Here, $x_i(t) \in \mathbb{R}^n$ and $x_{N+1}(t)$ are, respectively, the states of node i and the target system, $f(\cdot, \cdot) : \mathbb{R}^n \times \mathbb{R}_{\geq 0} \rightarrow \mathbb{R}^n$ represents the inherent nonlinear dynamics which is continuous, $\beta > 0$, $\alpha > 0$ and $\Gamma = \text{diag}\{\gamma_1, \gamma_2, \cdots, \gamma_N\}$ are the control parameters to be designed, where the dynamics of $x_{N+1}(t)$ are given by (4):

$$\dot{x}_{N+1}(t) = f(x_{N+1}(t), t) + h(t)u_r(x_{N+1}(t), t), \tag{4}$$

where $h(\cdot) : \mathbb{R}_{\geq 0} \rightarrow \mathbb{R}^{n \times m}$ is the input function, and $u_r(\cdot, \cdot) : \mathbb{R}^n \times \mathbb{R}_{\geq 0} \rightarrow \mathbb{R}^m$ represents the inputs. For analytical convenience, we assume that there exists a positive constant u_0 such that $\|h(t)u_r(x_{N+1}(t), t)\|_\infty \leq u_0$.

Denote by the communication graph among the N coupled nodes $\mathcal{G}(\mathcal{V}, \mathcal{E}^{\sigma(t)}, \mathcal{A}^{\sigma(t)})$. Furthermore, let $\hat{\mathcal{G}}(\hat{\mathcal{V}}, \hat{\mathcal{E}}^{\sigma(t)} \hat{\mathcal{A}}^{\sigma(t)})$ be the communication graph among the $N+1$ nodes, where $\hat{\mathcal{V}} = \mathcal{V} + \{N+1\}$. For notational convenience, we write $\mathcal{G}(\mathcal{V}, \mathcal{E}^{\sigma(t)}, \mathcal{A}^{\sigma(t)})$ and $\hat{\mathcal{G}}(\hat{\mathcal{V}}, \hat{\mathcal{E}}^{\sigma(t)}, \hat{\mathcal{A}}^{\sigma(t)})$ as $\mathcal{G}^{\sigma(t)}$ and $\hat{\mathcal{G}}^{\sigma(t)}$, respectively. In the sequel, unless otherwise specified, we denote by a hat on all quantities about graph $\hat{\mathcal{G}}^{\sigma(t)}$, for instance, by $\hat{\mathcal{L}}^{\sigma(t)} = [\hat{l}_{ij}^{\sigma(t)}]$ the Laplacian matrix of graph $\hat{\mathcal{G}}^{\sigma(t)}$. According to the graph theory and matrix theory introduced in the preceding subsection, we have

$$\hat{\mathcal{L}}^{\sigma(t)} = \begin{bmatrix} \tilde{\mathcal{L}}^{\sigma(t)} & -\mathbf{d}^{\sigma(t)} \\ 0_N^T & 0 \end{bmatrix}, \tag{5}$$

here $\tilde{\mathcal{L}}^{\sigma(t)} = [\tilde{l}_{ij}^{\sigma(t)}] = \mathcal{L}^{\sigma(t)} + \text{diag}\{d_1^{\sigma(t)}, d_2^{\sigma(t)}, \cdots, d_N^{\sigma(t)}\}, \mathbf{d}^{\sigma(t)} = (d_1^{\sigma(t)}, d_2^{\sigma(t)}, \cdots, d_N^{\sigma(t)})^T$.

Let $e_i(t) = x_i(t) - x_{N+1}(t)$. Then, we obtain from (3) and (4) that

$$\dot{e}_i(t) = f(x_i(t), t) - f(x_{N+1}(t), t) + \alpha \sum_{j=1}^N a_{ij}^{\sigma(t)} \Gamma(e_j(t) - e_i(t)) - \alpha d_i^{\sigma(t)} \Gamma e_i(t)$$

$$+ \beta \text{sgn}\left(\sum_{j=1}^N a_{ij}^{\sigma(t)} \Gamma(e_j(t) - e_i(t)) - d_i^{\sigma(t)} \Gamma e_i(t)\right) - h(t)u_r(x_{N+1}(t), t).$$

This together with (5) gives

$$\dot{e}_i(t) = f(x_i(t), t) - f(x_{N+1}(t), t) - \alpha \sum_{j=1}^N \tilde{l}_{ij}^{\sigma(t)} \Gamma e_j$$

$$- \beta \text{sgn}\left(\sum_{j=1}^N \tilde{l}_{ij}^{\sigma(t)} \Gamma e_j\right) - h(t)u_r(x_{N+1}(t), t).$$

Let $e(t) = (e_1^T(t), e_2^T(t), \cdots, e_N^T(t))^T$. We can rewrite the preceding differential equations in the following compact form:

$$\dot{e}(t) = \tilde{f}(t) - \alpha(\tilde{\mathcal{L}}^{\sigma(t)} \otimes \Gamma)e(t) - \beta \text{sgn}\left((\tilde{\mathcal{L}}^{\sigma(t)} \otimes \Gamma)e(t)\right)$$

$$- (\mathbf{1}_N \otimes I_n)h(t)u_r(x_{N+1}(t), t), \tag{6}$$

of which $\tilde{f}(t) = \left(\left((f(x_1(t),t) - f(x_{N+1}(t),t))^T, \cdots, (f(x_N(t),t) - f(x_{N+1}(t), t))^T \right)^T \right)^T$.

Remark 1. The goal of this paper is to find appropriate coupling strengths α, β, such that under which pinning synchronization of network (3) and (4) is reached finally. That is, all the N coupled nodes will converge to the same trajectory determined by (4). From the above analysis, it is sufficient to show $e(t) \to \mathbf{0}_{Nn}$ when $t \to \infty$.

3   Main Results and Theoretical Analysis

We shall give two theorems to summarize the main theoretical results. Before presenting the two theorems, we make several assumptions.

Assumption 1. The average dwell time τ_a considered in this paper satisfies (2).

Assumption 2. The function $f(x,t)$ is Lipsctitz continuous. That is, for $\forall\, x, y \in \mathbb{R}^n$ and all $t \in \mathbb{R}_{\geq 0}$, there is a constant $\varrho > 0$ such that $\|f(x,t) - f(y,t)\| \leq \varrho\|x - y\|$.

Assumption 3. Each augmented communication graph $\hat{\mathcal{G}}^k$, $k = 1, 2, \cdots, \kappa$, contains a directed spanning tree and all the trees have a common root node $N + 1$.

If Assumption 3 holds, then the zero eigenvalue of $\hat{\mathcal{L}}^k$ is simple and the real parts of the other N eigenvalues are positive [8], where $\hat{\mathcal{L}}^k$, $k = 1, 2, \cdots, \kappa$, is defined in (5). Consequently, the real part of each eigenvalue of $\tilde{\mathcal{L}}^k$ is positive. This together with $\tilde{l}_{ij}^k \leq 0$ implies $\tilde{\mathcal{L}}^k$ is a nonsingular M-matrix (see Definition 2). Then, we can find some diagonal and positive definite matrices Ξ^k such that

$$\left(\tilde{\mathcal{L}}^k\right)^T \Xi^k + \Xi^k \tilde{\mathcal{L}}^k > 0. \tag{7}$$

It is worth noting that Ξ^k can be chosen as $\Xi^k = \mathrm{diag}\{\xi_1^k, \xi_2^k, \cdots, \xi_N^k\}$ with $(\xi_1^k, \xi_2^k, \cdots, \xi_N^k)^T = \left(\tilde{\mathcal{L}}^k\right)^{-T} \cdot \mathbf{1}_N$ (see Lemma 4 in [14]).

For notation brevity, denote by $\bar{\phi}^k$ and $\underline{\phi}^k$ the maximum and minimum singular value of $\tilde{\mathcal{L}}^k$, respectively. Let $\varepsilon^k = \bar{\phi}^k/\underline{\phi}^k$. Furthermore, let $\bar{\phi} = \max_{k=1,2,\cdots,\kappa} \bar{\phi}^k$, $\underline{\phi} = \min_{k=1,2,\cdots,\kappa} \underline{\phi}^k$, $\bar{\varepsilon} = \max_{k=1,2,\cdots,\kappa} \varepsilon^k$. Similarly, let $\bar{\xi}^k = \max_{i=1,2,\cdots,N} \xi_i^k$, $\underline{\xi}^k = \min_{i=1,2,\cdots,N} \xi_i^k$, $\bar{\xi} = \max_{k=1,2,\cdots,\kappa} \bar{\xi}^k$, $\underline{\xi} = \min_{k=1,2,\cdots,\kappa} \underline{\xi}^k$, $\bar{\gamma} = \max_{i=1,2,\cdots,N} \gamma_i$ and $\underline{\gamma} = \min_{i=1,2,\cdots,N} \gamma_i$.

Theorem 1. Suppose Assumptions 1, 2 and 3 hold. Then pinning synchronization of complex network (3) and (4) is reached if the coupling strengths $\beta \geq u_0$, $\alpha > (2\varrho\bar{\xi}\bar{\varepsilon})/(\underline{\gamma}\underline{\lambda})$, and the average dwell time satisfies

$$\tau_a > \frac{\ln \frac{\bar{\phi}^2 \bar{\xi}}{\underline{\phi}^2 \underline{\xi}}}{\left(\alpha\underline{\lambda} - 2\frac{\varrho\bar{\xi}\bar{\varepsilon}}{\underline{\gamma}}\right)} \cdot \frac{\bar{\xi}}{\underline{\gamma}},$$

where $\underline{\lambda} = \min_{k=1,2,\cdots,\kappa} \lambda_{\min}\left(\left(\tilde{\mathcal{L}}^k\right)^T \Xi^k + \Xi^k \tilde{\mathcal{L}}^k\right)$.

Proof. Construct the following multiple Lyapunov function:

$$V(t) = e(t)^T \left((\tilde{\mathcal{L}}^{\sigma(t)})^T \Xi^{\sigma(t)} \tilde{\mathcal{L}}^{\sigma(t)} \otimes \Gamma \right) e(t). \tag{8}$$

Taking the time derivative of $V(t)$ in (8) along the solution of (6) yields

$$\begin{aligned}
\dot{V}(t) =& 2e(t)^T \left((\tilde{\mathcal{L}}^{\sigma(t)})^T \Xi^{\sigma(t)} \tilde{\mathcal{L}}^{\sigma(t)} \otimes \Gamma \right) \tilde{f}(t) \\
& - \alpha e(t)^T \left(((\tilde{\mathcal{L}}^{\sigma(t)})^T)^2 \Xi^{\sigma(t)} \tilde{\mathcal{L}}^{\sigma(t)} \otimes \Gamma^2 \right) e(t) \\
& - \alpha e(t)^T \left((\tilde{\mathcal{L}}^{\sigma(t)})^T \Xi^{\sigma(t)} (\tilde{\mathcal{L}}^{\sigma(t)})^2 \otimes \Gamma^2 \right) e(t) \\
& - 2\beta e(t)^T \left((\tilde{\mathcal{L}}^k)^T \Xi^k \tilde{\mathcal{L}}^k \otimes \Gamma \right) \text{sgn} \left((\tilde{\mathcal{L}}^{\sigma(t)} \otimes \Gamma) e(t) \right) \\
& - 2e(t)^T \left((\tilde{\mathcal{L}}^{\sigma(t)})^T \Xi^{\sigma(t)} \tilde{\mathcal{L}}^{\sigma(t)} 1_N \otimes \Gamma \right) h(t) u_r(x_{N+1}(t), t).
\end{aligned} \tag{9}$$

Let $\delta(t) = \left(\tilde{\mathcal{L}}^{\sigma(t)} \otimes \Gamma \right) e(t)$, i.e., $\delta_i(t) = \sum_{j=1}^N \tilde{l}_{ij}^{\sigma(t)} \Gamma e_j(t)$. For any $t \in [t_m, t_{m+1})$, $m = 0, 1, \cdots$, there exist k such that $\sigma(t) = k$, $k \in \{1, 2, \cdots, \kappa\}$. Then, we may obtain that

$$\begin{aligned}
& e(t)^T \left((\tilde{\mathcal{L}}^{\sigma(t)})^T \Xi^{\sigma(t)} \tilde{\mathcal{L}}^{\sigma(t)} \otimes \Gamma \right) \tilde{f}(t) \\
=& \delta(t)^T \left(\Xi^k \tilde{\mathcal{L}}^k \otimes I_n \right) \tilde{f}(t) \\
\leq& \|\delta(t)\| \sqrt{ \tilde{f}(t)^T \left((\tilde{\mathcal{L}}^k)^T (\Xi^k)^2 \tilde{\mathcal{L}}^k \otimes I_n \right) \tilde{f}(t) } \\
\leq& \varrho \bar{\phi}^k \bar{\xi}^k \|\delta(t)\| \|e(t)\| = \varrho \bar{\phi}^k \bar{\xi}^k \|\delta(t)\| \left\| \left((\tilde{\mathcal{L}}^{\sigma(t)})^{-1} \otimes \Gamma^{-1} \right) \delta(t) \right\| \\
\leq& \frac{\varrho \bar{\xi}^k}{\underline{\gamma}} \frac{\bar{\phi}^k}{\underline{\phi}^k} \delta(t)^T \delta(t) = \frac{\varrho \bar{\xi}^k}{\underline{\gamma}} \epsilon^k \delta(t)^T \delta(t) \leq \frac{\varrho \bar{\xi} \bar{\epsilon}}{\underline{\gamma}} \delta(t)^T \delta(t),
\end{aligned} \tag{10}$$

where the second inequality follows from Assumption 2. Notice that $(\tilde{\mathcal{L}}^k)^T \Xi^k + \Xi^k \tilde{\mathcal{L}}^k > 0$, we have

$$\begin{aligned}
& - \alpha e(t)^T \left(((\tilde{\mathcal{L}}^k)^T)^2 \Xi^k \tilde{\mathcal{L}}^k \otimes \Gamma^2 \right) e(t) - \alpha e(t)^T \left((\tilde{\mathcal{L}}^k)^T \Xi^k (\tilde{\mathcal{L}}^k)^2 \otimes \Gamma^2 \right) e(t) \\
=& -\alpha \delta(t)^T \left(\left((\tilde{\mathcal{L}}^k)^T \Xi^k + \Xi^k \tilde{\mathcal{L}}^k \right) \otimes I_n \right) \delta(t) \\
\leq& -\alpha \underline{\lambda}^k \delta(t)^T \delta(t) \leq -\alpha \underline{\lambda} \delta(t)^T \delta(t),
\end{aligned} \tag{11}$$

where $\underline{\lambda}^k = \lambda_{\min} \left((\tilde{\mathcal{L}}^k)^T \Xi^k + \Xi^k \tilde{\mathcal{L}}^k \right)$, and $\underline{\lambda} = \min_{k=1,2,\cdots,\kappa} \underline{\lambda}^k$. From (5), we get

$$\begin{aligned}
& - 2\beta e(t)^T \left((\tilde{\mathcal{L}}^k)^T \Xi^k \tilde{\mathcal{L}}^k \otimes \Gamma \right) \text{sgn} \left((\tilde{\mathcal{L}}^{\sigma(t)} \otimes \Gamma) e(t) \right) \\
=& -2\beta \delta(t)^T \left(\Xi^k \tilde{\mathcal{L}}^k \otimes I_n \right) \text{sgn}(\delta(t)) \\
=& -2\beta \delta(t)^T \left(\Xi^k (\mathcal{L}^k + \text{diag}\{d_1^k, d_2^k, \cdots, d_N^k\}) \otimes I_n \right) \text{sgn}(\delta(t)).
\end{aligned} \tag{12}$$

Since Ξ^k is a diagonal matrix,

$$-\delta(t)^T \left(\Xi^k \cdot \text{diag}\{d_1^k, d_2^k, \cdots, d_N^k\} \otimes I_n\right) \text{sgn}(\delta(t)) = -\sum_{i=1}^{N} \xi_i^k d_i^k \|\delta_i(t)\|_1, \quad (13)$$

$$
\begin{aligned}
&- \delta(t)^T \left(\Xi^k \mathcal{L}^k \otimes I_n\right) \text{sgn}(\delta(t)) \\
&= \sum_{i=1}^{N} \xi_i^k l_{ij}^k \left(\|\delta_i(t)\|_1 - \delta_i^T(t)\text{sgn}(\delta_j(t))\right) \leq 0,
\end{aligned}
\quad (14)
$$

where the inequality follows from $l_{ij}^k \leq 0$ and $\delta_i^T(t)\text{sgn}(\delta_j(t)) \leq \|\delta_i(t)\|_1$ for $i \neq j$. Substituting (13) and (14) into (12) gives

$$
\begin{aligned}
&- 2\beta e(t)^T \left((\tilde{\mathcal{L}}^k)^T \Xi^k \tilde{\mathcal{L}}^k \otimes \Gamma\right) \text{sgn}\left((\tilde{\mathcal{L}}^{\sigma(t)} \otimes \Gamma)e(t)\right) \\
&\leq - 2\beta \sum_{i=1}^{N} \xi_i^k d_i^k \|\delta_i(t)\|_1.
\end{aligned}
\quad (15)
$$

Notice that $\tilde{\mathcal{L}}^{\sigma(t)} \mathbf{1}_N = (d_1^k, d_2^k, \cdots, d_N^k)^T$, we have

$$
\begin{aligned}
&- 2e(t)^T \left((\tilde{\mathcal{L}}^k)^T \Xi^k \tilde{\mathcal{L}}^k \mathbf{1}_N \otimes \Gamma\right) h(t)u_r(x_{N+1}(t), t) \\
&\leq 2 \sum_{i=1}^{N} \xi_i^k d_i^k \|\delta_i(t)\|_1 \|h(t)u_r(x_{N+1}(t), t)\|_\infty \leq 2u_0 \sum_{i=1}^{N} \xi_i^k d_i^k \|\delta_i(t)\|_1
\end{aligned}
\quad (16)
$$

where we use $\|h(t)u_r(x_{N+1}(t), t)\|_\infty \leq u_0$ to obtain the last inequality. Then, substituting (10), (11), (15) and (16) into (9) gives

$$
\begin{aligned}
\dot{V}(t) &\leq - (\alpha\underline{\lambda} - 2\frac{\varrho\bar{\xi}\bar{\varepsilon}}{\underline{\gamma}})\|\delta(t)\|^2 - 2(\beta - u_0) \sum_{i=1}^{N} \xi_i^k d_i^k \|\delta_i(t)\|_1 \\
&\leq - (\alpha\underline{\lambda} - 2\frac{\varrho\bar{\xi}\bar{\varepsilon}}{\underline{\gamma}})\frac{\underline{\gamma}}{\bar{\xi}}V(t),
\end{aligned}
\quad (17)
$$

where the last inequality follows from $\beta \geq u_0$ and $\|\delta(t)\|^2 = e(t)^T \left((\tilde{\mathcal{L}}^{\sigma(t)})^T \cdot \tilde{\mathcal{L}}^{\sigma(t)} \otimes \Gamma^2\right) e(t) \geq (\underline{\gamma}/\bar{\xi})V(t)$. Set $\nu = (\alpha\underline{\lambda} - 2\frac{\varrho\bar{\xi}\bar{\varepsilon}}{\underline{\gamma}})\cdot(\underline{\gamma}/\bar{\xi})$. Hence, for any $t \in [t_m, t_{m+1})$, $m = 0, 1, \cdots$,

$$V(t) \leq \exp(-\nu(t - t_m))V(t_m) \leq \frac{\bar{\phi}^2\bar{\xi}}{\underline{\phi}^2\underline{\xi}} \exp(-\nu(t - t_m))V(t_m^-) \quad (18)$$

holds, where $V(t_m^-) := \lim_{t \nearrow t_m} V(t)$. Let $\mu = \bar{\phi}^2\bar{\xi}/\underline{\phi}^2\underline{\xi}$. It is obviously that $\mu \geq 1$. By recursion,

$$V(t) \leq \mu^m \exp(-\nu(t - t_0))V(t_0), \quad \forall t \in [t_m, t_{m+1}). \quad (19)$$

For $\forall\, t \geq t_0$, we then obtain from (19) that

$$V(t) \leq \mu^{N_\sigma[t_0,t)} \exp(-\nu(t-t_0))V(t_0)$$
$$\leq \mu^{N_0} \exp\left(-\left(\nu - \frac{\ln\mu}{\tau_a}\right)(t-t_0)\right)V(t_0). \qquad (20)$$

Since $\nu > \ln\mu/\tau_a$, the proof is completed by letting $t \to \infty$.

When the communication graphs $\mathcal{G}^{\sigma(t)}$ among the N coupled nodes are undirected, $\mathcal{L}^{\sigma(t)}$ is symmetric. Hence, $\tilde{\mathcal{L}}^{\sigma(t)}$ is positive definite under Assumption 3. Then, we get Theorem 2.

Theorem 2. Suppose Assumptions 1, 2 and 3 hold. If the communication graphs $\mathcal{G}^{\sigma(t)}$ among the N coupled nodes are undirected, then pinning synchronization of complex network (3) and (4) is reached if the coupling strengths $\beta \geq u_0$, $\alpha > (2\varrho\bar{\rho}\bar{\gamma})/(\underline{\rho}^2\gamma^2)$, and the average dwell time satisfies

$$\tau_a > \bar{\rho}\bar{\gamma}\frac{\ln\frac{\bar{\varrho}}{\underline{\varrho}}}{\alpha\underline{\rho}^2\gamma^2 - 2\varrho\bar{\rho}\bar{\gamma}}$$

where $\bar{\rho} = \max_{k=1,2,\cdots,\kappa} \lambda_{\max}\left(\tilde{\mathcal{L}}^k\right)$ and $\underline{\rho} = \min_{k=1,2,\cdots,\kappa} \lambda_{\min}\left(\tilde{\mathcal{L}}^k\right)$.

Proof. Construct the multiple Lyaunov functions $V_2(t) = e(t)^T\left(\tilde{\mathcal{L}}^{\sigma(t)} \otimes \Gamma\right)e(t)$. Theorem 2 then can be proven by following the proof of Theorem 1.

4 Numerical Simulation

We perform a simulation example to validate Theorem 1.

Example: In this example, the complex network has 6 coupled nodes and a target system whose communication topology is plotted in Fig. 1. Clearly, Assumption 3 holds. From (5), we can get

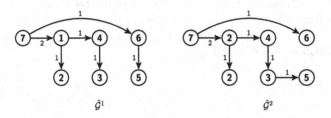

Fig. 1. Switching communication graphs, where the weight of each edge is represented by the number around it.

$$\tilde{\mathcal{L}}^1 = \begin{bmatrix} 2 & 0 & 0 & 0 & 0 & 0 \\ -1 & 1 & 0 & 0 & 0 & 0 \\ 0 & 0 & 1 & -1 & 0 & 0 \\ -1 & 0 & 0 & 1 & 0 & 0 \\ 0 & 0 & 0 & 0 & 1 & -1 \\ 0 & 0 & 0 & 0 & 0 & 1 \end{bmatrix} \quad \text{and} \quad \tilde{\mathcal{L}}^2 = \begin{bmatrix} 2 & 0 & 0 & 0 & 0 & 0 \\ -1 & 1 & 0 & 0 & 0 & 0 \\ 0 & 0 & 1 & -1 & 0 & 0 \\ -1 & 0 & 0 & 1 & 0 & 0 \\ 0 & 0 & -1 & 0 & 1 & 0 \\ 0 & 0 & 0 & 0 & 0 & 1 \end{bmatrix}.$$

Direct calculation gives $\bar{\phi} = 2.5344$, $\phi = 0.4146$, $\bar{\varepsilon} = 6.1125$, $\bar{\xi} = 3$, $\xi = 1$ and $\underline{\lambda} = 1.3317$. We assume the nonlinear dynamics $f(x_i(t), t)$ in (3) is given by (21):

$$f(x_i(t), t) = \begin{bmatrix} -\omega_1 & \omega_1 & 0 \\ 1 & -1 & 1 \\ 0 & -\omega_2 & -\omega_3 \end{bmatrix} \begin{bmatrix} x_{i1}(t) \\ x_{i2}(t) \\ x_{i3}(t) \end{bmatrix} + \Phi(x_{i1}(t)), \tag{21}$$

where $\Phi(x_{i1}(t)) = \omega_4 x_{i1}(t) + \frac{1}{2}(\omega_5 - \omega_4)\left(|x_{i1}(t) + 1| - |x_{i1}(t) - 1|\right)$, $\omega_1 = 2.7222$, $\omega_2 = 2.8700$, $\omega_3 = 0.01$, $\omega_4 = -0.714$, and $\omega_5 = -1.143$. And the target system is given by (4) with

$$f(x_{N+1}(t), t) = \begin{bmatrix} -\omega_1 & \omega_1 & 0 \\ 1 & -1 & 1 \\ 0 & -\omega_2 & -\omega_3 \end{bmatrix} \begin{bmatrix} x_{(N+1)1}(t) \\ x_{(N+1)2}(t) \\ x_{(N+1)3}(t) \end{bmatrix} + \Phi(x_{(N+1)1}(t)), \tag{22}$$

and $h(t)u_r(x_{N+1}(t), t) = [-0.01\sin(t), 0, 0.1\sin(t)]^T$,

where $\Phi(x_{(N+1)1}(t)) = \omega_4 x_{(N+1)1}(t) + \frac{1}{2}(\omega_5 - \omega_4)\left(|x_{(N+1)1}(t) + 1| - |x_{(N+1)1}(t) - 1|\right)$.

Direct calculation gives $u(0) = 0.1$ and $\varrho = 5.3328$. Choose $\Gamma = \text{diag}\{40, 40, 40\}$. According to Theorem 1, synchronization is reached if $\beta \geq 0.1$, $\alpha > 3.6715$ and $\tau_a > 0.0689$. Choose $\beta = 0.1$, $\alpha = 3.8$ and $\tau_a = 0.07$. Assume the communication topology switches from $\hat{\mathcal{G}}^1$ to $\hat{\mathcal{G}}^2$ at $t = 0.01s$. The evolutions of all the nodes are plotted in Fig. 2 which shows that synchronization of the considered complex network is indeed achieved. Let $Err(t) = \frac{1}{6}\sqrt{\sum_{i=1}^{6} \|x_i - x_7\|^2}$ be the synchronization error of the complex network under consideration. The evolution of $Err(t)$ with different values of α are plotted in Fig. 3. In addition, we discover from Fig. 3 that $Err(t)$ converges to zero faster with a larger α which can be got from the theoretical analysis directly. Hence, this example has validated Theorem 1 pretty well.

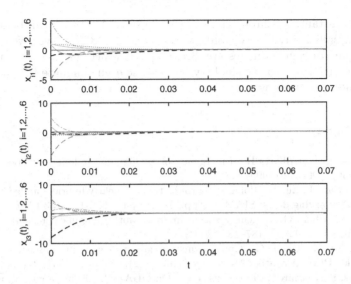

Fig. 2. State trajectories of target system and 6 coupling nodes.

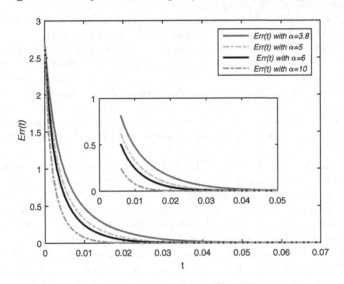

Fig. 3. The evolution of $Err(t)$ versus different values of α.

5   Conclusion

We have solved the pinning synchronization problem for a class of complex networks with directed switching communication topologies, where the inputs on the target system may be nonzero but the ∞-norm of the inputs is upper bounded. More precisely, we have proven that, if the average dwell time of the switching signal for switchings among different communication graphs is lower bounded by a positive constant, synchronization in the considered network can be achieved by selecting sufficiently large coupling strengths. From (3), however, we know the control inputs of coupled nodes are discontinuous which may result in undesirable chattering effect. So, we shall try to decrease even eliminate the chattering effect in the near future work.

References

1. Barabási, A.L., Albert, R.: Emergence of scaling in random networks. Science **286**(5439), 509–512 (1999)
2. Yu, W., Cao, J., Lü, J.: Global synchronization of linearly hybrid coupled networks with time-varying delay. SIAM J. Appl. Dyn. Syst. **7**(1), 108–133 (2008)
3. Wang, X., Chen, G.: Synchronization in small-world dynamical networks. Int. J. Bifurcat. Chaos **12**(1), 187–192 (2002)
4. Lü, J., Yu, X., Chen, G.: Chaos synchronization of general complex dynamical networks. Phys. A Statist. Mech. Appl. **334**(1–2), 281–302 (2004)
5. DeLellis, P., Bernardo, M.D., Russo, G.: On QUAD, Lipschitz, contracting vector fields for consensus and synchronization of networks. IEEE Trans. Circuits Syst. I Reg. Papers **58**(3), 576–583 (2011)
6. Yu, W., Chen, G., Cao, M.: Consensus in directed networks of agents with nonlinear dynamics. IEEE Trans. Autom. Control **56**(6), 1436–1441 (2011)
7. Zhao, J., Hill, D.J., Liu, T.: Synchronization of complex networks with switching topology: a switched system point. Automatica **45**(11), 2502–2511 (2009)
8. Ren, W., Beard, R.: Consensus seeking in multiagent systems under dynamically changing interaction topologies. IEEE Trans. Autom. Control **50**(5), 655–661 (2005)
9. Qin, J., Gao, H., Zheng, W.: Second-order consensus for multi-agent systems with switching topology and communication delay. Syst. Control Lett. **60**(6), 390–397 (2011)
10. Wen, G., Yu, W., Xia, Y., Yu, X., Hu, J.: Distributed tracking of nonlinear multiagent systems under directed switching topology: an observer-based protocol. IEEE Trans. Syst. Man Cybern. Syst. **47**(5), 869–881 (2017)
11. Wen, G., Duan, Z., Chen, G., Yu, W.: Consensus tracking of multi-agent systems with Lipschitz-type node dynamics and switching topologies. IEEE Trans. Circ. Syst. I Reg. Pap. **61**(2), 499–511 (2014)
12. Wen, G., Yu, W., Chen, M.Z.Q., Yu, X., Chen, G.: Pinning a complex network to follow a target system with predesigned control inputs. IEEE Trans. Syst. Man Cybern. Syst. (2018). https://doi.org/10.1109/TSMC.2018.2803147
13. Horn, R., Johnson, C.: Matrix Analysis. Cambridge Univ. Press, New York (1990)
14. Li, Z., Wen, G., Duan, Z., Ren, W.: Designing fully distributed consensus protocols for linear multi-agent systems with directed graphs. IEEE Trans. Autom. Control **60**(4), 1152–1157 (2015)

The Deep Input-Koopman Operator
for Nonlinear Systems

Rongrong Zhu, Yang Cao, Yu Kang$^{(\boxtimes)}$, and Xuefeng Wang

Department of Automation,
University of Science and Technology of China, Hefei, China
{Zhurr,mymcywxf}@mail.ustc.edu.cn, {forrest,kangduyu}@ustc.edu.cn

Abstract. In this paper, we propose a method that exploit the Koopman operator theory to make the strongly nonlinear dynamical systems approximately represented in the linear framework based on deep neural network (DNN) which is data-driven and equation-free. On account of the conventional Koopman operator is incapable for actuated systems, we introduce the notion of input-Koopman operator for the systems incorporated with the effects of inputs and controls. We construct the controllability gramian for nonlinear systems that are represented in the finite-dimensional input-Koopman operators. Moreover, we illustrate the several relationship between the space of full state observable functions and the original local controllability.

Keywords: Nonlinear dynamical system · Input-Koopman operator DNN · Controllability gramian

1 Introduction

Koopman operator theory is an idea of lifting nonlinear dynamical systems to a linear infinite-dimensional space thus we could yield a novel method for the analysis of nonlinear systems. The Koopman operator was firstly introduced in [1] describing how Hilbert space functions on the state of a dynamical system evolve in time. It can capture the nonlinear behavior of the system without losing any information and we can construct and analyze the operator only by experiment or simulation data without knowing the dynamic evolution equation of the system. These have made the study of Koopman operator an appealing work. A method for learning a finite-dimensional approximation Koopman operator from data before called extended dynamic mode decomposition(EDMD) [2,3]. However, people usually select observation functions manually by this method. The manual selection has the subjective limitation lead to a bad effect for the linear approximation of the nonlinear system. Likewise, the manual design needs a large number of attempts, the process is complicated and the efficiency is low.

The DNN [4] has gained much attention over the last decade because of their ability to efficiently represent complex functions from data. Moreover, the DNN only requires sufficient historical data to complete the training, so that the the the

© Springer Nature Switzerland AG 2018
L. Cheng et al. (Eds.): ICONIP 2018, LNCS 11307, pp. 97–107, 2018.
https://doi.org/10.1007/978-3-030-04239-4_9

complex mathematical derivation for observable functions can be avoided. Over the past few years, there some progress has been made by taking advantage of DNN to learn the Koopman operator which can be seen in [5,6]. Lusch et al. leverages DNN to get the representations of Koopman operator which provide intrinsic coordinates that globally linearize the dynamics for autonomous systems without considering actuated systems [7]. Therefore, in this paper we propose the DNN that have the potential to represent the Koopman operators with inputs called input-Koopman operator, which allows for the study of complex, input-output nonlinear systems.

Observability and controllability gramian were introduced in [8] and play a very important role in control theory. However, when it comes to nonlinear systems, it's hard to analysis the observability and controllability for the complex dynamics of the systems. The traditional method to analysis the controllability of nonlinear systems is the different geometric method [9] which needs a great deal of mathematical knowledge. In recent years, there are some works have been done for the extension of observability and controllability gramian from linear systems to nonlinear systems in [10,11]. Yeung et al. introduce the Koopman operator as a canonical representation of the system and apply a mathematical approach for the observable functions which may not suitable for complex nonlinear systems [12]. In this paper, we utilize input-Koopman operator to construct controllability gramian for nonlinear system with the DNN for observable functions. And through numerical experiments, we find the local controllability relationship between original systems and the linear framework.

Based on the above description, this paper proposes a Koopman operator using DNN. The contributions of this paper are summarized as:

(1) We use the DNN which has the capability of effectively representing for complex function and Koopman operator theory for globally linearize the actuated nonlinear dynamical systems. It can avoid complicated derivation and subjective factors of hand-crafting to get more accurately and naturally observable function and input-Koopman operator.
(2) We construct the Koopman controllability gramian with the observable function and input-Koopman operator obtained by DNN, so that we can get precisely analysis of the he local controllability relationship between original systems and the linear framework.

The rest of the paper is organized as follows. In Sect. 2, we introduce the formulation of input-Koopman operators. In Sect. 3, we present the analysis of Koopman controllability gramian. Our DNN for observable function and input-Koopman operators are described in Sect. 4. Numerical experiment is presented in Sect. 5.

2   The Formulation of the Input-Koopman Operator

The Koopman operator theory is a linear operator defined for any nonlinear system. In this section, we describe the conventional Koopman operator for the autonomous systems firstly. Then we introduce the input-Koopman operator for the actuated dynamical nonlinear systems. Consider the discrete nonlinear dynamical system:

$$x_{k+1} = f(x_k),\tag{1}$$

where $x_k \in \mathbb{R}^n$ stands for the state of the system and f is a map from $\mathbb{R}^n$ to itself. We could equivalently describe the Koopman operator for continuous-time systems in a discrete-time dynamics form that describe a continuous-time system which is sampled discretely in time. We also define a set of scalar valued observable functions g $(\mathcal{H} \to \mathbb{R})$. The Koopman operator acts on this set of observable functions is:

$$\mathcal{K}g(x_k) = g(x_{k+1})\tag{2}$$

As in (2) the Koopman operator is linear and infinite-dimensional. But the nonlinear dynamical system is often considered finite-dimensional. So representing the nonlinear dynamics in the linear framework by obtaining finite-dimensional approximations of infinite-dimensional Koopman operator has drawn many scholars' attention. The dynamical mode decomposition (DMD)[13] is a typical way to approximate the finite-dimensional representations of the Koopman operators.

Throughout this paper, we consider a discrete-time of nonlinear systems that allows for external inputs

$$x_{k+1} = f(x_k, u_k)\tag{3}$$

where $x \in \mathbb{R}^n$ and $u \in \mathbb{R}^u$. The Koopman operator with inputs acts on the space of observable function as following:

$$\mathcal{K}\psi(x, u) = \psi(f(x, u), *)\tag{4}$$

here * stands for different types of inputs. we don't apply control strategy on the system, so for the convenience of discussion, we define u as a step inputs in this paper. See Fig. 1 to get a better understand of Koopman operator with inputs.

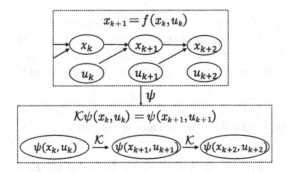

Fig. 1. Illustrate the nonlinear system evolving process by time according to the observable functions ψ and the input-Koopman operator $\mathcal{K}$

Next, we will demonstrate how to translate the nonlinear dynamics in linear frameworks numerically. Considering full-state access and full-input access by choosing ideal observable functions that are in the identity of $y_k = \psi_x(x_k)$ and $\gamma_k = \psi_u(u_k)$. In this case, the linear framework of nonlinear system can be rewritten in the following form:

$$\begin{bmatrix} y_{k+1} \\ \gamma_{k+1} \end{bmatrix} = \begin{bmatrix} \mathbf{A} & \mathbf{B} \\ \mathbf{C} & \mathbf{D} \end{bmatrix} \begin{bmatrix} y_k \\ \gamma_k \end{bmatrix} \tag{5}$$

where $A : \mathbb{R}^{n_D} \to \mathbb{R}^{n_D}, B : \mathbb{R}^{m_D} \to \mathbb{R}^{n_D}, C : \mathbb{R}^{n_D} \to \mathbb{R}^{m_D}$, and $D : \mathbb{R}^{m_D} \to \mathbb{R}^{m_D}$.

In this paper, we treat A and B as the input-Koopman operator which we approximated in the observable function space. In Sect. 4 we will discuss how to get A and B using DNN. As we previously defined u as step inputs, what should be only concerned are the evolution of the states, therefore we can suppose $C = \mathbf{0}$ and $D = \mathbf{I}$. Now, (5) can be the form as following:

$$\begin{bmatrix} y_{k+1} \\ \gamma_{k+1} \end{bmatrix} = \begin{bmatrix} \mathbf{A} & \mathbf{B} \\ \mathbf{0} & \mathbf{I} \end{bmatrix} \begin{bmatrix} y_k \\ \gamma_k \end{bmatrix} \tag{6}$$

This construction of input-Koopman operator forces the nonlinear dynamics systems in the canonical sate-space linear framework.

3   The Koopman Controllability Gramian

Consider the discrete-time nonlinear system described in (3). With inputs, the input-Koopman operator are computed on an input-state observable function $\psi(x_n, u_n) \in \mathbb{R}^{N_D}$ ($N_D = n_D + m_D$) space to satisfy the dynamical equation

$$\psi(x_{k+1}, u_{k+1}) = \mathcal{K}\psi(x_k, u_k), \tag{7}$$

Referring to paper [12], we can obtain (8) from (7) as following,

$$\psi_x(x_{k+1}) = \mathcal{K}_x\psi_x(x_k) + \mathcal{K}_u\psi_u(u_k) \tag{8}$$

where $\psi_x \in \mathbb{R}^{n_D}$ is a vector consisting of the elements of $\psi(x_k, u_k)$ that only depend on x_k. $\psi_u \in \mathbb{R}^{m_D}$ represents the elements of $\psi(x_k, u_k)$ that only depends on x_k. We don't have dependent on a mixture of x_k and u_k for u is a step inputs in this work. Now we can define the input to state operator as $\Phi_c^\psi : \mathbb{R}^{m_D} \to \mathbb{R}^{n_D}$

$$\Phi_c^\psi \equiv \mathcal{K}_x^j \mathcal{K}_u, \tag{9}$$

then the Koopman controllability gramian is defined as :

$$X_\psi^c = \sum_{j=0}^\infty \Phi_c^\psi \left(\Phi_c^\psi\right)^T = \sum_{j=0}^\infty \mathcal{K}_x^j \mathcal{K}_u \mathcal{K}_u^T \left(\mathcal{K}_x^j\right)^T \tag{10}$$

Here, we suppose that the Koopman observable function is state inclusive which means $\psi(x, u) = (x; N_{\theta_1}(x, u))$, where x stands for the original nonlinear system states, and $N_{\theta_1}(x, u) \in \mathbb{R}^{N_D - n}$ stands for the observable functions automatically obtained by training the DNN and the advantages of this method will be talked in next section. By this way, we could retain the local controllability of the underlying nonlinear system.

As we can see from (6) in Sect. 2, we could know that $\mathcal{K}_x^j = A^j$ and $\mathcal{K}_u = B$. In next section, we will illustrate the DNN we used in this work to obtain the matrix of A and B. And in Sect. 5, we will show the numerical results to prove that the Koopman controllability gramian maintain the local controllability of the nonlinear system to some extent.

4   The Input-Koopman Operator with DNN

Learning the Koopman observable functions for both the input and state can be computationally expensive. In addition, the number of observable functions required is not known a priori and often the number of terms requires multiple steps of manual refinement, even for the simple two or three state systems. Recently, it is popular to use DNN to generate Koopman observable function and input-Koopman operator that automatically update during the training process [14,15]. Unlike previous deep learning approaches to Koopman operator and observable function for autonomous systems [7], our network architecture is designed specifically to handle the system which are actuated and for the purpose of analysis of controllability of nonlinear system.

Our network architecture is shown in Fig. 2. The weights matrix are initialized by Xavier which can alleviate the problem of gradient explosion or disappearance. The activation function of all neurons in network is rectified linear unit (ReLU) which was superior to the traditional activation function (sigmoid, tanh, etc.) in terms of computation and convergence speed [16]. In the training process, we optimize over the weights and biases of the DNN using Adam for its better performance than other adaptive learning rate optimization method for training recent neural network models and is very simple to implement [17] meanwhile.

Owe to the universal function property of neural networks, N_{θ_1} is used to learn observable functions for approximating the input-Koopman operator. N_{θ_1} provide the fixed number of observable function after the DNN is trained. We learn the input-Koopman operator through evolving the observable function values one time step into future and minimize the loss function described in (11) in the meantime. The causation we build out N_{θ_2} is to make sure that the finite-dimensional input-Koopaman operator can be reconstructed to the original full states and inputs. In order to ensure that we can get the most accurate finite dimensional operators, we train the K as A and B.

According to the above, the loss function of the DNN is defined as following:

$$min_{K, \theta_1, \theta_2} \| N_{\theta_1}(x_{k+1}, u_{k+1}) - K N_{\theta_1}(x_k, u_k) \|_2 + \| N_{\theta_2}(y_k, \gamma_k) - (x_k, u_k) \|_2 \tag{11}$$

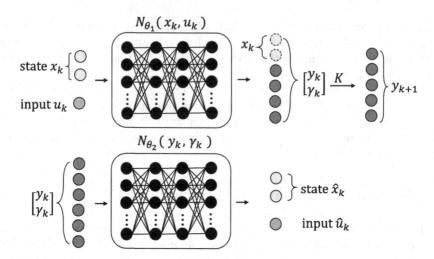

Fig. 2. Illustrate the suppose of finding the input-Koopman operators. N_{θ_1} and N_{θ_2} are two full connected networks with each network four hidden layers which are 84 neurons each. Dotted circle x_k is the original states which means state inclusive as talked in Sect. 4. K standing for input-Koopman operator is a hidden layer without activation functions and biases which means y_k and γ_k are in the linear framework.

By minimizing the loss function, we could get the proper K and N_{θ_1} as input-Koopman operator and observable function which is more accurate and natural than previous methods.

For the EDMD, the computational time is $\mathcal{O}\left(m^3\right)$, we could also evaluate that the computational time for the DNN that we proposed is $\mathcal{O}\left(m*n^2\right)$ which m stands for the amount of the training samples and n stands for the neurons of each hidden layers. For the DNN is fixed, n is constant. And as the training samples increases, we can find out that DNN is much more efficiency than EDMD.

In this section, we use the DNN to get the proper input-Koopman operator and observable function for nonlinear dynamical system which can be used to future study and analysis.

5   Numerical Experiment

In this section, we illustrate simulation results to show the effectiveness of the proposed method for dynamical systems. To show the correctness of this work, we firstly simulate on a well-studied linear system. We consider a simple linear system. Then we apply our proposed method on a nonlinear system.

Our DNN is implemented based on Google's TensorFlow framework. The inputs of the deep neural network are the time series data including states and inputs. We chose 50000 initial conditions for the data set, and we get these time series data from MATLAB which are divided into training (80%), validation

(10%) and testing (10%) sets (sampling time is 0.02 s). For both N_{θ_1} and N_{θ_2}, we set four hidden layers each. The layer width of each hidden layer is 84. In the training procedure, the learning rate is 0.001 initially and it's decreased during optimization. We set batch-size as 256.

5.1   Linear Dynamical System

Considering the linear dynamical system depicted in (12):

$$\dot{x}_1 = x_1$$
$$\dot{x}_2 = x_2 + u \tag{12}$$

As shown in Figs. 3 and 4, the predicted state trajectories is almost the same as the actual state trajectories and the error between these two items is almost zero. This indicate that we have obtained the proper finite-dimensional

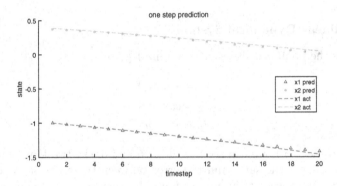

Fig. 3. Illustrate the one step prediction trajectories from the trained input-Koopman operator versus actual state trajectories for the system described in (12)

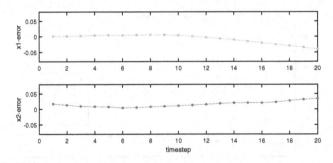

Fig. 4. Illustrate the errors between actual states and the one step prediction states of the system depicted in (12).

input-Koopman operator by the DNN. And the matrix we proposed before is as following:

$$A = \begin{bmatrix} 0.9929 & -0.1094 & 0.0667 & 0.1310 & 0.0816 \\ -0.1190 & 0.2375 & -0.3545 & 0.4016 & 0.3316 \\ 0.0999 & -0.2027 & 0.2497 & -0.2912 & -0.2384 \\ 0.1160 & 0.2322 & -0.3472 & 0.2293 & 0.3070 \\ 0.0717 & 0.1579 & -0.1970 & 0.2783 & 0.1578 \end{bmatrix} \tag{13}$$

$$B = \begin{bmatrix} 0.0008 & -0.0759 & 0.0479 & -0.0450 & -0.0762 \end{bmatrix}^{\mathrm{T}} \tag{14}$$

With deterministic A and B, we could get that the input to state operator Φ_c^{ψ} that mentioned in Sect. 3.

$$\Phi_c^{\psi} = \begin{bmatrix} 0.0002 & -0.0784 & 0.0487 & -0.0679 & -0.0459 \end{bmatrix}^{\mathrm{T}} \tag{15}$$

Note that the first column is essentially 0, indicating that the input gain from $\psi_u(u)$ to x_1 is negligible.

5.2 Nonlinear Dynamical System

Considering the nonlinear dynamical system depicted in (16):

$$\begin{aligned} \dot{x}_1 &= sin\,(x_1) \\ \dot{x}_2 &= cos\,(x_1) \\ \dot{x}_1 &= x_3 + u \end{aligned} \tag{16}$$

Similar to Sect. 5.1, we can see from Figs. 5 and 6, the two kind of state trajectories are almost the same and the error between these two items is almost zero which signifies we get the proper finite-dimensional input-Koopman operator for the nonlinear system as depicted in (16). And the matrix we proposed

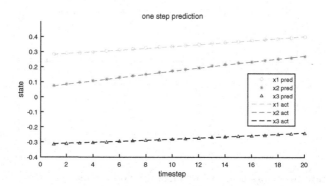

Fig. 5. Illustrate the one step prediction trajectories from the trained input-Koopman operator versus actual state trajectories for the system described in (16)

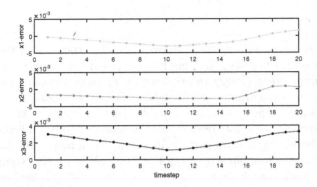

Fig. 6. Illustrate the errors between actual states and the one step prediction states of the system depicted in (16).

before are as following:

$$A = \begin{bmatrix} 0.7256 & -0.1184 & -0.0027 & 0.8506 & 0.2750 & 0.0733 & 0.0245 & -0.0020 \\ -0.1449 & 0.5222 & 0.2258 & 0.4461 & -0.2627 & 0.5992 & -0.6746 & -0.2407 \\ 0.0905 & 0.1716 & 0.8845 & -0.2897 & 0.0745 & -0.0168 & 0.4814 & -0.2718 \\ 0.2152 & 0.0726 & 0.0082 & 0.3853 & -0.1878 & -0.0393 & -0.0383 & -0.0715 \\ 0.1343 & 0.0212 & -0.0110 & -0.1909 & 0.2216 & 0.1895 & 0.1757 & 0.4512 \\ 0.0336 & 0.2895 & -0.0092 & 0.1909 & 0.0196 & 0.1432 & -0.0863 & 0.2549 \\ 0.0581 & -0.1827 & 0.2246 & -0.3899 & 0.1866 & -0.0972 & 0.0873 & 0.3534 \\ -0.0290 & 0.0223 & -0.0693 & -0.0459 & 0.3978 & 0.0776 & 0.2187 & 0.5171 \end{bmatrix} \tag{17}$$

$$B = \begin{bmatrix} 0.1596 & 0.1759 & 0.1025 & -0.1216 & 0.0376 & -0.0463 & 0.0419 & -0.0475 \end{bmatrix}^{\mathrm{T}} \tag{18}$$

$$\Phi_c^{\psi} = \begin{bmatrix} 0.0099 & 0.0053 & 0.2070 & -0.0084 & 0.0122 & -0.0106 & 0.057 & 0.0060 \end{bmatrix}^{\mathrm{T}} \tag{19}$$

Note that the first column and second column are essentially 0, indicating that the input gain from $\psi_u(u)$ to x_1 and x_2 are negligible.

In addition to these two system we exhibit here, we have done some experiments on many other nonlinear systems, we could get similar results to illustrate the effectiveness of the method proposed in this paper. Simulation results above prove that:

(1) The DNN we proposed this paper can lift a nonlinear system to a linear framework for future system analysis and control;
(2) With the theoretical derivation and experimental simulation, we can see from the Koopman controllability gramian that the local controllability of the original system is maintained by the input-Koopman operator in a way.

6 Conclusion

In this project, we introduce the Koopman operator theory and the DNN for nonlinear dynamical systems. The method proposed was used to find out the finite-dimensional input-Koopman operator to represent the nonlinear dynamical system in the linear framework. The DNN is used to obtain the more accurate input-Koopman operator and observable function, which is data-driven and avoids complex mathematical derivation. Simulation results indicate that the proposed method have effective performance, and it can figure out that we could remain the local controllability by using DNN to looking for the Koopman operators. Future work may include study of the control of the nonlinear system according to the Koopman operator theory with DNN.

Acknowledgments. This work was supported in part by the National Natural Science Foundation of China(61725304 and 61673361). Authors also gratefully acknowledge supports from the Youth Top-notch Talent Support Program, the 1000-talent Youth Program and the Youth Yangtze River Scholar.

References

1. Koopman, B.O.: Hamiltonian systems and transformation in hilbert space. Proc. Natl. Acad. Sci. **17**(5), 315–318 (1931)
2. Williams, M.O., Kevrekidis, I.G., Rowley, C.W.: A data-driven approximation of the koopman operator: extending dynamic mode decomposition. J. Nonlinear Sci. **25**(6), 1307–1346 (2015)
3. Li, Q., Dietrich, F., Bollt, E.M., et al.: Extended dynamic mode decomposition with dictionary learning: a data-driven adaptive spectral decomposition of the koopman operator. Chaos Interdisc. J. Nonlinear Sci. **27**(10), 103111 (2017)
4. Hinton, G.E., Salakhutdinov, R.R.: Reducing the dimensionality of data with neural networks. Science **313**(5786), 504–507 (2006)
5. Wehmeyer, C., Noé, F.: Time-lagged autoencoders: deep learning of slow collective variables for molecular kinetics. arXiv preprint arXiv:1710.11239 (2017)
6. Mardt, A., Pasquali, L., Wu, H., et al.: VAMPnets for deep learning of molecular kinetics. Nat. Commun. **9**(1), 5 (2018)
7. Lusch, B., Kutz, J.N., Brunton, S.L.: Deep learning for universal linear embeddings of nonlinear dynamics. arXiv preprint arXiv:1712.09707 (2017)
8. Moore, B.: Principal component analysis in linear systems: controllability, observability, and model reduction. IEEE Trans. Autom. Control **26**(1), 17–32 (1981)
9. Hermann, R., Krener, A.: Nonlinear controllability and observability. IEEE Trans. Autom. Control **22**(5), 728–740 (1977)
10. Lall, S., Marsden, J.E., Glavaški, S.: Empirical model reduction of controlled nonlinear systems. IFAC Proc. **32**(2), 2598–2603 (1999)
11. Fujimoto, K., Scherpen, J.M.A.: Nonlinear input-normal realizations based on the differential eigenstructure of hankel operators. IEEE Trans. Autom. Control **50**(1), 2–18 (2005)
12. Yeung, E., Liu, Z., Hodas, N.O.: A koopman operator approach for computing and balancing gramians for discrete time nonlinear systems. arXiv preprint arXiv:1709.08712 (2017)

13. Rowley, C.W., Mezić, I., Bagheri, S., et al.: Spectral analysis of nonlinear flows. J. Fluid Mech. **641**, 115–127 (2009)
14. Yeung, E., Kundu, S., Hodas, N.: Learning deep neural network representations for koopman operators of nonlinear dynamical systems. arXiv preprint arXiv:1708.06850 (2017)
15. Otto, S.E., Rowley, C.W.: Linearly-recurrent autoencoder networks for learning dynamics. arXiv preprint arXiv: 1712.01378 (2017)
16. Glorot, X., Bordes, A., Bengio, Y.: Deep sparse rectifier neural networks. In: Proceedings of the Fourteenth International Conference on Artificial Intelligence and Statistics, pp. 315–323. JMLR Press (2011)
17. Hinton, G., Deng, L., Yu, D., Dahl, G.E., Mohamed, A.R., Jaitly, N., Kingsbury, B.: Deep neural networks for acoustic modeling in speech recognition: the shared views of four research groups. IEEE Sig. Process. Mag. **29**(6), 82–97 (2012)

Multi-UAV Collaborative Monocular SLAM Focusing on Data Sharing

Zhuoyue Yang[1,3(✉)], Dianxi Shi[1,3,4(✉)], Yongjun Zhang[1,3,4(✉)],
Shaowu Yang[2,3], Fu Li[2,3], and Ruoxiang Li[1,3]

[1] Science and Technology on Parallel and Distributed Laboratory,
National University of Defense Technology, Changsha 410073, China
yangzhuoyue18@163.com, dxshi@nudt.edu.cn, zhang-vic@263.com
[2] State Key Laboratory of High Performance Computing,
National University of Defense Technology, Changsha 410073, China
[3] College of Computer, National University of Defense Technology,
Changsha 410073, China
[4] Artificial Intelligence Research Center (AIRC),
National Innovation Institute of Defense Technology (NIIDT), Beijing 100166, China

Abstract. Sharing data among Unmanned Aerial Vehicles (UAVs) is one of key issues in the field of multiple-robot SLAM. In this paper, aiming at problems of sharing data between UAVs during tracking lost and map fusion, we propose a robust, focusing on Date Sharing Multi-UAV visual SLAM (DSM-SLAM) based on centralized architecture. In addition, we present a two-step relocalization method based on sharing local maps, in order to support the UAV in using the data from other UAVs which have gone there before when the tracking is lost. Furthermore, we put forward a map fusion method based on hierarchical clustering to dynamically and adaptively select the order of map fusion that is more beneficial to data sharing between drones. Experimental results on popular public datasets demonstrate the feasibility and effectiveness of the system.

Keywords: Collaborative visual SLAM · Data sharing Workloads offload · Two-step relocalization · Map fusion

1 Introduction

Simultaneous localization and mapping (SLAM) refers to the problem of simultaneously estimating a model of the surroundings of a mobile robot and the robot's localization from a stream of sensor data [1]. It has been a fundamental technology for a robot to realize the autonomous mobility [2]. In the past decades, the basic characteristics of the single-robot SLAM have been intensively studied [3]. However, in more and more scenes such as disaster monitor and emergency anomaly detection, single robot can no longer meet the needs of applications. Therefore, multi-robot collaboration has become a hot issue in the

© Springer Nature Switzerland AG 2018
L. Cheng et al. (Eds.): ICONIP 2018, LNCS 11307, pp. 108–119, 2018.
https://doi.org/10.1007/978-3-030-04239-4_10

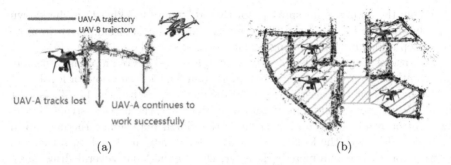

Fig. 1. (a) The result of using the two-step relocalization method. (b) The result of utilizing the map fusion method based on hierarchical clustering. The gray area isn't detected by drones. Different color areas are detected by different drones. (Color figure online)

robotics industry. In the robotics industry, Unmanned Aerial Vehicles (UAVs) play an increasingly important role. Multi-UAV collaborative SLAM can expand the range of detection, and shorten the time required to complete the task. Collaboration between UAVs is achieved through sharing the same type of information. At present, there are many related work to solve the problem based on visual SLAM. Main solutions have been fallen into two types. On the one hand, some systems are based on distributed architecture. The distributed systems mainly include D-RPGO [4], DDF-SAM [5] and DGS [6]. The disadvantage of distributed systems is that it is difficult to ensure data consistency and avoid duplicate calculation information [7]. On the other hand, other systems are based on centralized architectured, such as [7–13] and so on. The literature considers that a system using a centralized server can provide more advantages in the process of an algorithm that is expensive and has no real-time limitation. For example, Bundle Adjustment (BA) and Place Recognition can be processed on the server side [8]. This design allows the robot to make full use of its limited resources to run the necessary algorithms. The system in [9] may be the first multi-UAV cooperative SLAM system. In [10], authors proposed C2TAM system. After merging maps to global map, the server sends the complete global map to agents. However, repeatedly sending the entire map is bound to be a problem in large areas. The solution proposed by Schmuck [7] not only overcomes the disadvantage of heavily relying on the server, but also offloads computationally intensive tasks to the server.

When a drone is used in an unknown environment, tracking loss is common and inevitable. One solution to the tracking loss problem in existing work is that the drone reinitializes the map based on the existing environment [11]. However, such solutions can't effectively use the existing historical information and relocalization in single UAV always fails. As shown in Fig. 2(a), if the tracking is lost, the UAV can't continue to work for a long time (Fig. 2(b)). Another solution [10] is to wait for the server to send the optimized global map to drones. However, global map optimization is a computationally intensive task. When the server sends optimization information to the drone, the UAV has also moved away from the anomaly. To sum up, for the problem of tracking lost, the known work did not provide a solution that can use shared information. So we propose a two-step relocalization utilizing the shared maps, as shown in Fig. 1(a).

The task of map fusion is not only to complete the task of mapping, but more importantly, is to provide drones with more abundant information for localizing. The basic idea of the existing map fusion method, such as [7,12,13], is that each UAV first constructs a local map separately, and the first constructed local map will be used as the initialization of the global map. Once the overlap of the map is detected, the local map and the global map are fused until the map construction task is completed. However, this method has several drawbacks. First of all, in the process of map fusion, as the number of fusion maps increases, the time for global optimization will become longer and longer. UAVs cannot obtain optimization information in time. And it is more likely that the UAV has finished the task, but global optimization has not yet been received. Secondly, as shown in Fig. 2(c), the trajectory of UAV-A may have the overlap with the trajectory of UAV-B at $time_1$, but after this point, two drones move back and their data doesn't need to share. While at $time_2$ right after $time_1$, the trajectories of UAV-A and UAV-C (In Fig. 2(d)) have overlapped part. The overlap of the two drones is increasing over time because the two UAVs move in the same direction. In other words, if UAV-A's map and UAV-C's map are first merged, it is more advantageous for data sharing between two UAVs. Finally, if using the traditional map fusion, there are individual drones that have some part in common but can't be fused into the global map. Aiming at these problems, we present the map fusion method based on hierarchical clustering. As shown in Fig. 2(b), the two UAVs could share information with each other by utilizing the map fusion method based on hierarchical clustering, although their map can't be merged into the other two UAVs' map. Therefore, focusing on the problem of Data Sharing between drones, this paper proposes a centralized cooperative Multi-UAV SLAM (DSM-SLAM) system to enhance the robustness of current SLAM systems. (1) The system off-loads the computational complexity task to the server to support drones in moving autonomously. (2) We propose a two-step relocalization method that supports drones using the shared local map to restart their work in time when tracking is lost. (3) We present a map fusion method based on hierarchical clustering to fuse maps in an order that is more benificial to data sharing.

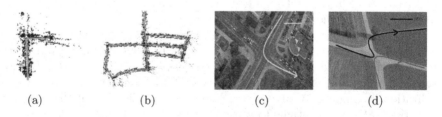

(a)   (b)   (c)   (d)

Fig. 2. Scenes suitable for using the DSM-SLAM

The structure of this paper is as follows. The second part describes the detailed description of the method proposed in this paper. The third part is the experimental result, and the fourth part is the conclusion.

2   Methods

The DSM-SLAM system mainly includes the following two parts, as shown in Fig. 3. The first one is the local map construction and localization mechanism, and the other is the global map construction and sharing mechanism. The local map construction and localization mechanism mainly includes VO, the two-step relocalization and loop detection in the server. VO estimates rough camera motion based on the information between adjacent images and builds a local map synchronously. We only maintain a local map on the UAV to support drones to move autonomously. The high acceleration motion, blurring or lack of visual features in the scene may cause exceptions. The two-step relocalization is designed for constructing local map under abnormal conditions. Local maps generated by the VO are sent to the server at a fixed frequency. The server will process keyframes and map points separately based on different message types. The system runs non-time-critical and computationally expensive loop detection algorithm in the server. At the same time, the server will send the newly received keyframes to the global map construction mechanism. The purpose of the global map construction and sharing mechanism is to fuse the local maps and send the optimization information back to UAVs. The global map construction mechanism mainly includes global map overlap detection, global map fusion, global nonlinear optimization, and global map sharing. If the local map is accepted by the global map overlap detection, the global map fusion module will merge the maps in a certain order. Afterwards, the global optimization module uses the information between the maps to optimize the merged map. Finally, the global optimization information is returned to the drone through the global map sharing. Some parts of the system are based on the ORB-SLAM2 [11] and CORB-SLAM2 [12] projects.

2.1   Local Mapping and Localization Mechanism Based on Two-Step Relocalization

Local map construction mechanism utilizes visual images collected by the camera to construct a local map. The UAV will pass the local map at a fixed frequency to the server. The local map construction mechanism, which provides a good initial value, is the basis of global map construction mechanism. The main task of this part is to estimate the rough camera motion based on the information between adjacent images.

Visual Odometry (VO). VO is mainly composed of tracking module and local mapping module. The tracking module locates the motion of the camera according to the correspondence of the feature points between adjacent images.

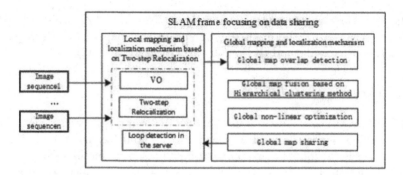

Fig. 3. The overview of the DSM-SLAM system

The local mapping module estimates the spatial position of the feature points in keyframes, thus restores the three-dimensional structure of the environment.

Loop Detection in the Server. In order to conserve bandwidth and improve the real-time performance of the system, we don't deliver images. Instead, we send the keyframes generated by the tracking and local mapping to the server. In addition, the loop detection module is a computational task. The loop detection is used to detect whether the UAV itself has visited this location during the construction of the map. If the UAV have reached this point, the system will use the current and historical information to eliminate accumulated drift errors and improve the accuracy of the local map.

Two-step Relocalization. In the case of abnormal condition such as high acceleration motion or lack of visual features, UAVs always fail to track. We use a two-step relocalization based on local maps in local map construction mechanism to help them continue to work. Two-step Relocalization consists of the first-step relocalization in UAVs and the second-step relocalization in the server. The historical information stored in the UAV is limited and can only help to relocalize near the current location. If the tracking loss is slightly longer, the tracking module will not be able to continue the task. However, relocalization in the server can not only use the historical information of the current UAV, but also use the environmental knowledge collected by other UAVs. In this way, we can effectively improve the accuracy of relocalization. In addition, rich computing resources in the server can improve the efficiency of relocalization. However, there is an unavoidable communication delay in the process of sending information from the server to the UAV. At this time, the UAV may have already left from the position where the frame was just sent, so the relocalization module on the UAV is very necessary. Instead of sending the entire global map, we send an extension information about the local map on the UAV. In this way, the communication and storage requirements are effectively reduced. The DSM-SLAM system can support continuous mapping of large-scale scenes.

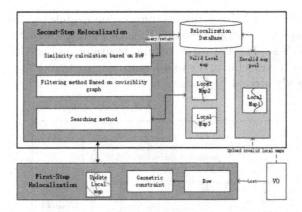

Fig. 4. The two-step relocalization based on sharing local maps

As shown in Fig. 4, when the VO fails to work, it will send the current local map to the invalid map pool in the server. These invalid maps maybe useful for other UAVs. Firstly, the UAV sends the frames acquired after the tracking loss to the server. And then the second-step Relocalization module will query the Relocalization Database to find similar keyframes (KFs) that share a word based on Bag of Words (BoW) method. After that, the second-step relocalization will filter the KFs that share enough words. To find relocalization candidates, the server will calculate similarity score according to the Covisibility Graph [11] and return the candidate keyframes that may be used in the UAV.

For each newly generated or newly received keyframe from the server, the first-step relocalization module queries similar keyframes in the local map and returns some candidate keyframes that share a sufficient number of map points. For these candidate keyframes, the system will measure them according to features or geometric constraints. Finally, the precise relocalization module determine which candidate keyframes match the current frame. Then the UAV will call for more details through searching module in the server according to the matched keyframe. The server will send the associated local maps from other clients to the UAV. In this way, the server can provide more information for relocalization.

2.2   Global Mapping and Sharing Mechanism

This part mainly describes the process of global map construction. It is mainly composed of the following parts: global map overlap detection module, global map fusion module, global non-linear optimization module and global map sharing module. The main storage structure in the server includes some local maps, global map and a place recognition database. The local maps are incrementally built by UAVs and global maps are hierarchically constructed in a certain order. When two maps in the server are detected overlap by global map overlap detection module, they are merged according to the process described at the global map fusion module. Global non-linear optimization module uses Bundle Adjustment (BA) to optimize the global map. The other storage structure is the place recognition database. This database contains keyframes obtained by all UAVs. It mainly provides services for UAVs to query similar keyframes from other clients according to newly arrived keyframes.

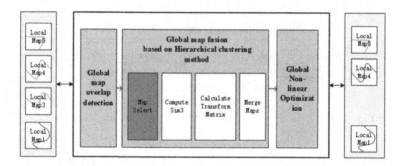

Fig. 5. The global map construction mechanism

Global Map Overlap Detection. It is primarily responsible for detecting the overlap between local maps constructed by UAVs and global maps already stored in the server. The implementation of this module is based on the bag of words theory, the DBoW2 project [14] and the Perspective-n-point (PnP) algorithm [15]. The bag of words model is a technique using visual vocabularies for managing large-scale image collections. First of all, the system deployed on the UAV converts the newly arrived image into sparse digital vectors. Then global map overlap detection will filter out keyframes with good matching according to the BoW vector and PnP algorithm. At the same time, we record the number of all matched keyframes between local maps.

Global Map Fusion Based on Hierarchical Clustering Method. In this paper, we proposes a map fusion algorithm based on hierarchical clustering, as shown in Fig. 5. Map fusion is processed in a bottom-up manner level by level. This method could not only share the keyframes and map points in the overlapped map in time, but also improve the efficiency of constructing maps. In the process of fusion, we take into consideration that the more number of matching keyframes between two local maps, the greater possibility of exchanging information between two UAVs in the future. The process of the algorithm is shown in the Table 1. First of all, the map selector chooses two maps according to the number of matched keyframes. Local maps are built independently, with no global map at the beginning. If a best matching keyframe pair is found in two local maps, the keyframes in a bigger map is used as a reference. We compute the 7 degrees-of-freedom transform matrix between the keyframe pair, which is done in 3D similarity space (Sim3) by the method of Horn [16]. At the same time, all the key frames in another local map are converted using transform matrix, so that the two partial maps have the same scale. Finally, using the matching keyframe as a connection point, two local maps with the same scale are spliced into a global map dynamically. The fused map will replace the original two local maps. And the new fused submap will be the candidate in the next round of iterations. Because of the smaller size of the integrated maps, it is beneficial to provide guidance for the position of UAVs quickly.

Table 1. Global map fusion algorithm based on hierarchical clustering method

Input:	The local maps $M(map_1, map_2, ..., map_n)$
Output:	A global map

	//Initialize the map fuse candidates
Step 1:	Initialize
Loop:	
Step 2:	$Update_{InFixedSequence}$ $(record_{m1}, record_{m2}, ..., record_{mn})$
Step 3:	$(m_i, m_j) = argmax(pair\ (\ record_{mi}, record_{mj}) > \theta\ )$
Step 4:	$(R,t) = argmax\ (MatchedKFs(m_i, m_j))$
Step 5:	$m_{new} \leftarrow merge(m_i, m_j)$
Step 6:	$M \leftarrow delete(m_i, m_j)$
Step 7:	$M \leftarrow add(m_{new})$
Until:	$M = \phi$ or all the local mapping stop

Global Non-linear Optimization. We use the Bundle Adjustment (BA) method to optimize the keyframes contained in the global map. BA takes the spatial position of mappoints as a constraint, and optimizes the map by minimizing the reprojection error of all keyframes contained in the global map. We reduces the errors caused by accumulated drift in this way. We can use the g2o framework [17] to accomplish this optimization.

Global Map Sharing. After the integration of the two submaps, it is necessary to share the information of the fused map to all relevant UAVs. The system sends the updated information of keyframes and map points to UAVs. After receiving the updated map information, the UAV adds the keyframes that are closely related to the current environment from other local UAVs. Thus, UAVs can get more environmental information according to the optimized global map.

3   Experimental Results

The performance of the DSM-SLAM system is analyzed from the three aspects: the loop detection in the server, two-step relocalization, and the process and result of map fusion based on hierarchical clustering method. Experiments are conducted to evaluate the performance of the DSM-SLAM on the KITTI dataset.

The computer that runs experiments is deployed an Ubuntu14.04 64-bit operating system, ROS, an Inter core i5-3470 CPU and 8 GB RAM. Although all clients and the server run on one computer, they can run on an individual UAV.

3.1   Results of the Loop Detection in the Server

The drone built a local map using the pictures collected by the camera and delivered the constructed local map to the server at a fixed frequency. As shown in Fig. 6(a), the left snapshot shows the local map of the drone and the right one is the local map in the server. It can be seen that the map in the server is slightly slower than the map updated on the drone side, but it can basically keep up with the speed of the drone-end mapping. On the right side of Fig. 6(b), the closed loop is detected in the server, but on the left side the local map didn't detect it. As shown in Fig. 6(c), the global optimization in the server had finished, and Fig. 6(d) depicted that the server delivered the optimization information to the UAV. And then UAV updated the result of the pose information of the drone.

3.2   Results of Two-step Relocalization

As shown by Fig. 7(a), VO onboard detected that the tracking of UAV-A was lost and couldn't rely on its own relocalization to continue to work. And the UAV-B had ever gone to the place where UAV-A was lost. When UAV-A was lost, it sent current frame generated from the latest picture to the server and the two-step relocalization module queries the relocalization database to find similarity keyframes and map points from UAV-B. So The server sent the associated keyframes and map points to the UAV-A, as shown by Fig. 7(b). In this way, the UAV-A can begin to work again. Figure 7(c) depict the differnt snapshots when tracking lost and tracking successfully, respectively.

3.3   Results of Global Map Fusion Based on Hierarchical Clustering Method

We selected the map with higher overlap ratio from the sequence 05 of the KITTI dataset, and divided the selected part into three sub-sequences (as shown in Fig. 8(e)). And the map fusion method considers whether the map fusion sequence is more beneficial to the data sharing between two local maps in the next period of time. As shown in Fig. 8(a),(b) and (c), the three UAVs began to construct the local maps at the same time ($t = 0$). And the global map overlap detection recorded accumulated matched keyframes at the fix frequency. The global map overlap detection detected the overlap between the UAV-A's map and the UAV-C's at $t = 10$. But at $t = 12$, the map overlap detection detected more matched keyframes between UAV-A and UAV-B's maps. UAV-A and UAV-B are more likely to exchange data in the next stage. So the global map fusion merged the two maps into a global map, as shown in Fig. 8(d). In this way, the DSM-SLAM has better performance than other systems.

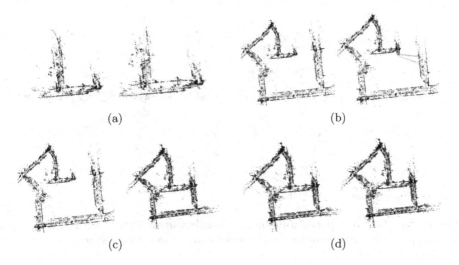

(a)   (b)

(c)   (d)

Fig. 6. The performance of the loop detection in the server. The left snapshots are the local map constructed by UAV, and the right snapshots are the local map in the server.

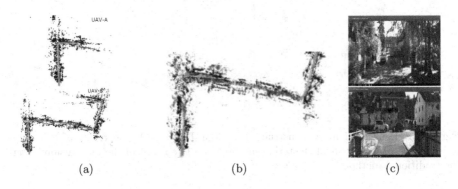

(a)   (b)   (c)

Fig. 7. The performance of the two-step Relocalization in the server. In (a), the green lines mark the poses of keyframes built by 2 different UAVs. In (b), the blue line marks the keyframes that are received from other UAVs to relocalise. (Color figure online)

We selected four sequences of images from the sequence 00 in KITTI dataset. The local maps are constructed by these four sets of image sequences are shown in Fig. 9(a). The global map overlap detection detected overlaps between the local maps of the UAV-A and the UAV-D. And the local maps of UAV-B and UAV-C had overlaps. Therefore, the hierarchical clustering map fusion method can be used to form a global map containing two sub graphs, as shown in Fig. 9(b). If local maps are fused to a global map in a traditional way, two overlapped local maps can not be fused or merged into global map in this scene, as shown in Fig. 9(c). Therefore, map fusion based on hierarchical clustering is more widely applied.

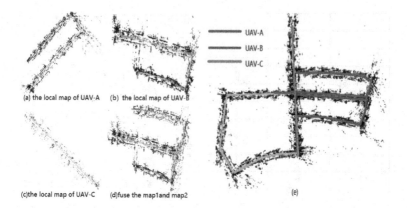

Fig. 8. The result of using map fusion based on hierarchical clustering method. The local maps are fused in an order that is more benificial to the data sharing

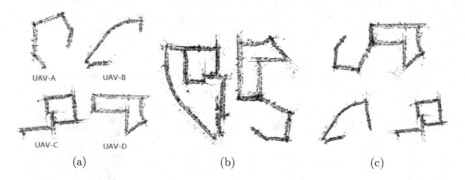

Fig. 9. (a) Four local maps constructed by four UAVs. (b) The result of the map fusion based on hierarchical clustering method. (c) The result of the map fusion using the traditional method.

4   Conclusion

This paper proposes the collaborative visual SLAM system focusing on data sharing for unknown environment explorations. We utilize the Two-step Relocalization based on the shared local map to extend the robustness of the SLAM system. We consider the impact of map fusion order on data sharing among UAVs. Experimental results with public dataset demonstrate the UAVs can work better by using DSM-SLAM. Future research will focus on how to improve the efficiency in optimization step and how to reduce the time that is used to transfer the data between the server and clients.

Acknowledgments. This work was supported by the National Key Research and Development Program of China (2017YFB1001901).

References

1. Davison, A.J., Reid, I.D., Molton, N.D., Stasse, O.: Monoslam: real-time single camera slam. IEEE Trans. Pattern Anal. Mach. Intell. **29**(6), 1052–1067 (2007)
2. Durrantwhyte, H., Bailey, T.: Simultaneous localization and mapping: part i. IEEE Robotic. Amp Autom. Magaz. **13**(2), 99–110 (2017)
3. Cadena, C., et al.: Past, present, and future of simultaneous localization and mapping: toward the robust-perception age. IEEE Trans. Robotic. **32**(6), 1309–1332 (2016)
4. Knuth, J., Barooah, P.: Collaborative localization with heterogeneous inter-robot measurements by riemannian optimization. In: IEEE International Conference on Robotics and Automation, pp. 1534–1539 (2013)
5. Cunningham, A., Paluri, M., Dellaert, F.: DDF-SAM: fully distributed slam using constrained factor graphs. Iros **25**(1), 3025–3030 (2010)
6. Chaimowicz, L.: The next frontier: combining information gain and distance cost for decentralized multi-robot exploration. In: ACM Symposium on Applied Computing, pp. 268–274 (2016)
7. Schmuck, P., Chli, M.: Multi-UAV collaborative monocular slam. In: IEEE International Conference on Robotics and Automation, pp. 3863–3870 (2017)
8. Bai, D., Wang, C., Bo, Z., Xiaodong, Y.I., Yang, X.: CNN feature boosted SEQ SLAM for real-time loop closure detection. Chin. J. Electron. **27**(3), 488–499 (2018)
9. Forster, C., Lynen, S., Kneip, L., Scaramuzza, D.: Collaborative monocular slam with multiple micro aerial vehicles. In: IEEE/RSJ International Conference on Intelligent Robots and Systems, pp. 3962–3970 (2013)
10. Riazuelo, L., Civera, J., Montiel, J.M.M.: C2TAM: a cloud framework for cooperative tracking and mapping. Robot. Auton. Syst. **62**(4), 401–413 (2014)
11. Mur-Artal, R., Tardós, J.D.: ORB-SLAM2: an open-source slam system for monocular, stereo, and RGB-D cameras. IEEE Trans. Robot. **33**(5), 1255–1262 (2017)
12. Li, F., Yang, S., Yi, X., Yang, X.: CORB-SLAM: a collaborative visual slam system for multiple robots. In: EAI International Conference on Collaborative Computing: Networking, Applications and Worksharing (2017)
13. Cui, H., Shen, S., Gao, X., Hu, Z.: CSFM: community-based structure from motion. In: IEEE International Conference on Image Processing, pp. 4517–4521 (2017)
14. Galvez-Lpez, D., Tardos, J.D.: Bags of binary words for fast place recognition in image sequences. IEEE Trans. Robot. **28**(5), 1188–1197 (2012)
15. Moreno-Noguer, F., Lepetit, V., Fua, P.: Accurate non-iterative o(n) solution to the PNP problem. In: IEEE International Conference on Computer Vision, pp. 1–8 (2007)
16. Horn, B.K.P.: Closed-form solution of absolute orientation using unit quaternions. J. Opt. Soc. Am. A **5**(7), 1127–1135 (2016)
17. Kmmerle, R., Grisetti, G., Strasdat, H., Konolige, K., Burgard, W.: G2O: a general framework for graph optimization. In: IEEE International Conference on Robotics and Automation, pp. 3607–3613 (2011)

Comparing Computing Platforms for Deep Learning on a Humanoid Robot

Alexander Biddulph[✉], Trent Houliston, Alexandre Mendes,
and Stephan K. Chalup

School of Electrical Engineering and Computing, The University of Newcastle,
Callaghan, NSW 2308, Australia
Alexander.Biddulph@uon.edu.au

Abstract. The goal of this study is to test two different computing platforms with respect to their suitability for running deep networks as part of a humanoid robot software system. One of the platforms is the CPU-centered Intel® NUC7i7BNH and the other is a NVIDIA® Jetson TX2 system that puts more emphasis on GPU processing. The experiments addressed a number of benchmarking tasks including pedestrian detection using deep neural networks. Some of the results were unexpected but demonstrate that platforms exhibit both advantages and disadvantages when taking computational performance and electrical power requirements of such a system into account.

Keywords: Deep learning · Robot vision · GPU computing
Low powered devices

1 Introduction

Deep learning comes with challenges with respect to computational resources and training data requirements [6,13]. Some of the breakthroughs in deep neural networks (DNNs) only became possible through the availability of massive computing systems or through careful co-design of software and hardware. For example, the AlexNet system presented in [15] was implemented efficiently utilising two NVIDIA® GTX580 GPUs for training.

Machine learning on robots has been a growing area over the past years [4,17,20,21]. It has become increasingly desirable to employ DNNs in low powered devices, among them humanoid robot systems, specifically for complex tasks such as object detection, walk learning, and behaviour learning. Robot software systems that involve DNNs face the challenge of fitting their large computational demands on a suitable computing platform that also complies with the robot's electrical power budget. One way to address these challenges is to develop and

Supported by an Australian Government Research Training Program scholarship and a top-up scholarship through 4Tel Pty.

L. Cheng et al. (Eds.): ICONIP 2018, LNCS 11307, pp. 120–131, 2018.
https://doi.org/10.1007/978-3-030-04239-4_11

use modern software architectures and efficient algorithms for the most resource hungry components, e.g. computer vision [9].

The need for hardware that can run deep convolutional neural networks (DCNNs) on low powered devices can be seen in humanoid robotic systems that only possess CPU resources. Spec et al. [22] developed a DCNN designed to locate soccer balls on a field. The final performance of this system was insufficient for practical use, with an execution time of 0.91s per frame. In a dynamic environment, both the framerate and latency of this detection would reduce its usefulness. Furthermore, the developed architecture consumed too much RAM (> 2 GB) to be usable on the chosen humanoid robot system.

Javadi et al. [14] compared the performance of several well-known DCNN architectures on the task of detecting and classifying other humanoid robots. The performance of these networks was compared both on GPUs and the CPUs of the target robots. Even for the classification task, none of the tested deep network architectures achieved acceptable performance. Only a two-layer network executed in less than 100 ms on their target platform.

While the design of the robot's brain and software system is the most crucial part of a humanoid robot system, every other component of the robot, hardware or software, has to be carefully considered, as well. This paper compares a few currently popular devices that can be used as the main computing device on a low-powered humanoid robot, with the aim to allow usage of deep learning as part of the robot software system.

2   Hardware Platforms

In this section we will describe the robotic platform and two computers that can be used in similar autonomous systems.

The NUgus is a humanoid robotic platform designed to perform human-like activities in real-world environments. It stands 90 cm tall, weighs 7.5 kg, and can serve as an autonomous humanoid robot soccer player. The original robotic platform was developed by the University of Bonn as an open platform in collaboration with the company igus® GmbH in Germany in 2015 [2]. The technical specifications of the NUgus differ only slightly from the original design with the PC being replaced by an Intel® NUC7i7BNH [12], referred to as "NUC" in this paper, and the camera replaced with two FLIR® Flea®3 USB3 cameras fitted with 195° wide angle lenses. The NUC features an integrated GPU, which makes the platform suitable for complex computations, such as deep learning.

The NUgus's 20 servos are the most power hungry component of the robot. Consuming up to 225 W, the power consumption of the computing platform makes up a relatively small portion of this power consumption. Despite this, decreasing power consumption by utilising more efficient algorithms is an effective way of increasing battery life.

The second computer tested in this work is a NVIDIA® Jetson TX2 system, referred to as "Jetson" in this paper, is described in [5]. This system is now becoming a popular alternative for embedded systems that require higher computing capacity, among them autonomous vehicles.

3   The NUgus Robot Software System

The NUgus control software system allows it to play soccer autonomously. It comprises several components; namely vision, behaviour, kinematics and localisation. By far, the most complex component is the vision system, and it is also the most computationally demanding. The four components run in parallel, and information must be shared among them in real time. To achieve this goal the NUClear software framework is used [8]. When running DNNs in the robot vision system, it is important that it does not over-utilise the available computing resources, or the robot will not be able to perform the other activities necessary for playing soccer, e.g. walking while keeping itself balanced, approaching the ball from the right direction and kicking the ball towards the goal.

4   Deep Learning for Robot Vision

The main aim of this work is to trial a DNN in the vision pipeline of a humanoid robotic system to determine the feasibility of using DNNs in battery powered environments. To this end, SSD MobileNet [10,19] pretrained on MS COCO [18], was acquired from the Tensorflow Model Zoo [11]. This network is reported to have a mAP of 21. Testing the network required its integration into the NUgus's vision system and the implementation of an interface to allow communication through the NUbots software architecture.

5   Experiments

In general, while the performance gains of GPU computing over CPU computing are undeniable the performance gain is not always so significant, with GPU computing sometimes providing as little as a 2.5X increase over CPU performance [16]. Gregg et al. [7] highlight the importance of considering data transfer times to and from the GPU, as these transfer times can often dwarf the time spent performing the actual calculations making GPU usage ineffective [1].

Lee et al. [16] and Vanhoucke et al. [23] emphasise that careful implementation and tuning using SIMD intrinsics improves the performance of CPU-based algorithms, especially those using matrix multiplications.

This section details the experiments that were undertaken during the course of this work and is divided into two parts. Section 5.1 focuses on four CPUs and GPUs. Two of the CPUs and GPUs come as part of the NUC and Jetson platforms, while the other two CPUs and GPUs are from a desktop and a laptop. In this section, the task was based on a 2D matrix rotation. Finally, in Sect. 5.2, we present the results for the more complex task of pedestrian detection. This task requires two image pre-processing steps and then the use of a DNN for the detection as a third step. The pedestrian detection benchmark was carried out on the NUC and Jetson platforms only.

It should be noted that the tests presented in this section are not designed to provide a definitive performance indication, but should be seen as a point of reference.

5.1   CPU and GPU Benchmarking

In order to determine a basis for comparing the NUC and the Jetson a simple benchmarking experiment was performed on the CPU and GPU components of both devices. This experiment was also repeated on two consumer grade GPUs – a low-end Intel® HD Graphics 630 and a high-end NVIDIA® GTX1080Ti – to provide an idea of how the devices compare to end-user GPUs, and to ensure differences between OpenCL and CUDA implementations are evaluated on a single device. Two consumer grade CPUs were also used – an Intel ® Core i7-7800X and i7-7920HQ – to provide a similar comparison. Table 1 lists the GPU devices and Table 2 lists the CPU devices that were used in this experiment.

The benchmarking experiment follows the experiment laid out in [3]. The experiment is designed to test the computational power, in terms of floating point operations per second (FLOPS), of a GPU without main memory access.

The FLOPS measure is calculated using a 2D matrix multiplication operation. First, a 2D rotation matrix is created, and a 2D vector is rotated multiple times. Equation (1) shows the iterative formulation used for benchmarking and was implemented in both OpenCL and CUDA for execution on the NUC and Jetson GPUs, respectively.

$$R = \begin{bmatrix} \cos(2) & -\sin(2) \\ \sin(2) & \cos(2) \end{bmatrix} \qquad \begin{aligned} x_0 &= \begin{bmatrix} 1 & 0 \end{bmatrix} \\ x_1^T &= Rx_0^T \end{aligned} \qquad \begin{aligned} x_n^T &= Rx_{n-1}^T \\ out_n &= \langle x_{n,0}, x_{n,1} \rangle \end{aligned} \tag{1}$$

The kernel used in this experiment has allowances for operating with varying dimensionality; mainly 1, 2, and 4-dimensional data. If 1-dimensional data is rotating a single 2D point, then 2-dimensional data would amount to rotating two 2D points, and 4-dimensional data is rotating four 2D points. Only x_0 in Eq. (1) needs to be modified to account for changing dimensionality. The kernel also allows changing the number of iterations that can be performed – in our tests we set the number of iterations to 40,000. The kernels were run with an increasing load on the GPU by changing the number of workers that are performing these calculations. The number of workers ranged from 256 to 1,024,000 in steps of 256.

Analysing Eq. (1) in more depth, each iteration contains 6 floating point operations – 4 multiplications and 2 additions. Factoring in the dimensionality of the data (D), the number of iterations (N), the number of workers (W), and the time taken to complete all iterations (t), we can derive a formula for the number of floating point operations that the GPU can perform per second $FLOPS = \frac{6DNW}{t}$.

To ensure that the compiler does not try to optimise out any of the calculations, the final equation in Eq. (1) is added at the end of the kernel. Strictly speaking, this modifies the formula for $FLOPS$ to $\frac{6DNW}{t} + \frac{(2D-1)W}{t}$. However, this extra term is insignificant and we only use the first term in this work.

For the CPU benchmarking Eq. (1) was implemented using SIMD intrinsics. A varying number of tasks, from 1 up $4,096$ in steps of 32, were created using OpenMP. All other parameters remained the same as in the GPU benchmarking.

5.2 Image Reprojection, Demosaicing, and Pedestrian Detection

This experiment integrates a DNN into the vision system of a humanoid robotic platform. Figure 1 shows the vision pipeline that was used in this experiment.

A modular, multi-language software framework, named NUClear [8], is used as the backbone of this system. Each module in Fig. 1 lists the programming language(s) that were used to implement them.

The camera is fitted with a 195° equidistant field of view lens and streams $1,280 \times 1,024$ images in a Bayer format at up to 60 fps.

The demosaicing and reprojection modules are implemented in OpenCL on the NUC and CUDA on the Jetson and are implemented such that they are performed concurrently. The output of these modules is a $1,280 \times 1,024$ RGB image with a 150° field of view. This means that the reprojection module must also interpolate pixel colour values.

SSD MobileNet [10, 19], pre-trained on MS COCO [18], from the Tensorflow Model Zoo [11] is used in this experiment.

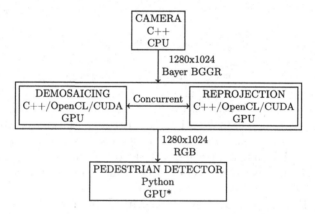

Fig. 1. The software pipeline of the pedestrian detection system. Images from the camera are streamed to the image demosaicing and reprojection module. After demosaicing and reprojection it is passed to the pedestrian detector module. *Unable to run on the GPU of the NUC due to a limitation in Tensorflow*

Due to the use of an equidistant camera lens, the image from the camera has to be projected on to a rectilinear plane before it can be used as an input to the DNN. However, since the camera lens has a field of view greater than π rad, the field of view needs to be restricted, otherwise the resulting rectilinear image would be infinite in size ($\tan\left(\frac{FOV}{2}\right) = \tan\left(\frac{\pi}{2}\right) = \infty$, see Eq. (2)).

Reprojecting an equidistant image to a rectilinear image can be performed by first casting a ray from the center of the camera to a pixel in the rectilinear image and then finding the point that this ray intersects the equidistant image. In this way, the processing time is determined by the resolution and field of view of the rectilinear image. Equation (2) outlines the calculations that are performed for each pixel in the rectilinear image.

$$
f = \frac{\|\boldsymbol{I}_{wh} - 1\|}{2\tan\left(\frac{FOV}{2}\right)} \qquad \theta = \arccos\left(\boldsymbol{v}_x\right) \qquad \boldsymbol{v}_{out} = \begin{cases} \langle 0,0 \rangle & \text{if } \theta = 0, \\ \alpha \boldsymbol{v}_{yz} & \text{otherwise} \end{cases} \tag{2}
$$
$$
\boldsymbol{v} = \|\langle f, \boldsymbol{s}_x, \boldsymbol{s}_y \rangle\| \qquad \alpha = \frac{\theta}{r_{pp}\sin\left(\theta\right)}
$$

Where $\boldsymbol{I}_{wh}$ is the width and height of the rectilinear image, FOV is the field of view of the rectilinear image, f is the focal length of the camera in pixels, $\boldsymbol{s}$ is the screen-centered coordinates of a pixel in the rectilinear image, r_{pp} is a measure of the number of radians spanned by a single pixel in the equidistant image, and $\boldsymbol{v}_{out}$ is the screen-centered coordinates of a pixel in the equidistant image. The screen-centered coordinate system places $(0,0)$ in the middle of the image, with $+x$ to the right and $+y$ up.

Since Tensorflow does not currently support OpenCL based devices, this experiment is performed with the DNN running on the CPU of the NUC and the GPU of the Jetson. Figure 1 details which computing device each module runs on. Moreover, in order to determine the power consumed by the devices, the current drawn from the power supply was recorded during this experiment.

6   Results and Observations

This section details the results of the experiments performed in Sect. 5. Section 6.1 details the results of the CPU and GPU device benchmarking. Section 6.2 details the results of the pedestrian detection tasks.

6.1   CPU and GPU Benchmarking

The results of the GPU benchmarking experiment detailed in Sect. 5.1 provided some surprising results. Figure 2a shows the results of implementing and running the benchmarking equations shown in Sect. 5.1 using a vector of 4 single precision floating point values as the main data type. In terms of performance, we see that the NVIDIA® GTX1080Ti performs best, followed by the NUC GPU, the Intel® HD Graphics 630, and finally the Jetson GPU.

The interesting part of these results is highlighted in Table 1, where we see that the calculated $FLOPS$ value is roughly on par with other reported values for the NVIDIA® GTX1080Ti. However, both of the Intel devices have calculated $FLOPS$ values almost 6 times higher than other reported values. On the other hand, the Jetson is calculated to have a $FLOPS$ value roughly 1.5 times lower than reported.

Unfortunately, it is not currently understood why these values are so wildly out of proportion. It is believed that the Intel OpenCL drivers have found a way to heavily optimise the kernel, while the reported *FLOPS* value for the Jetson is thought to be a theoretical maximum value. This reasoning is backed up by the similar performance of the NVIDIA® GTX1080Ti when running both the OpenCL and CUDA kernels.

Figure 2b shows a comparison between the NUC GPU and the Jetson GPU, using vectors with dimensions of 1, 2 and 4. These results indicate that the Jetson stops providing any performance benefits for vectors with more than 2 dimensions. In comparison, the NUC GPU continues to provide performance benefits for vectors with up to 4 dimensions. It is also interesting that the 1 dimension vector performance of the NUC GPU outperforms the 4 dimension vector performance of the Jetson.

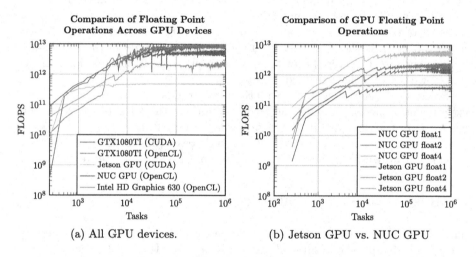

(a) All GPU devices.   (b) Jetson GPU vs. NUC GPU

Fig. 2. Comparison of FLOPS count for the tested GPU devices.

The results of the CPU benchmarking experiment performed in Sect. 5.1 also provided some unexpected results. Figure 3a shows the performance results of the 4 CPU devices and Table 2 lists the calculated FLOPS. Referring to Fig. 3a, the NUC CPU shows the best performance up until 60 tasks are scheduled to run. Above this point, the Intel® Core™ i7-7800X starts outperforming all other devices. The Intel® Core™ i7-7920HQ shows similar performance to the NUC, and the Jetson is approximately 3 times slower than the NUC CPU. The erratic performance of the Intel® Core™ i7-7800X and 7920HQ are due to these desktop/laptop devices running other tasks during the run time of the test.

6.2   Image Reprojection, Demosaicing, and Pedestrian Detection

The results of the experiment run in Sect. 5.2 show that for the purposes of reprojecting and demosaicing a high-resolution image, the GPU in the NUC and

Table 1. GPU devices benchmarked in Sect. 5.1 and their peak TFLOPS.

GPU device	OpenCL	CUDA	Reported[abcd]
NVIDIA® GTX1080Ti	9.14	9.34	10.61
NUC GPU	5.25	Unsupported	0.88
Intel® HD Graphics 630	2.40	Unsupported	0.44
Jetson GPU	Unsupported	0.49	0.75

[a]Wikipedia: GeForce 10 series https://en.wikipedia.org/wiki/GeForcesps-10spsseries.
[b]WikiChip: Intel Iris Plus Graphics 650 https://en.wikichip.org/wiki/intel/irisspsplusspsgraphicssps650.
[c]WikiChip: Intel HD Graphics 630 https://en.wikichip.org/wiki/intel/hdspsgraphicssps630.
[d]Wikipedia: Tegra https://en.wikipedia.org/wiki/Tegra.

Table 2. CPU devices benchmarked in Sect. 5.1 and their peak FLOPS.

CPU device	CPU source	Intrinsics	GFLOPS
Intel® Core™ i7-7800X	Dell Alienware Area-51	AVX/SSE	49.68
Intel® Core™ i7-7920HQ	Apple MacBook Pro 14,3	AVX/SSE	29.33
NUC CPU	Intel® NUC7i7BNH	AVX/SSE	15.54
Jetson CPU	NVIDIA® Jetson TX2	NEON	4.69

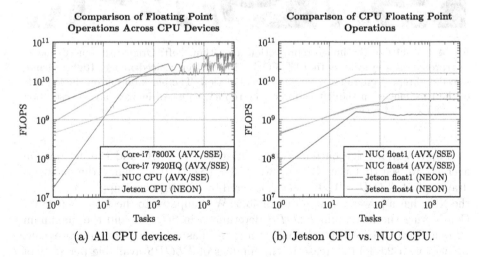

(a) All CPU devices.   (b) Jetson CPU vs. NUC CPU.

Fig. 3. Comparison of the FLOPS count for the tested CPU devices.

the Jetson are closely matched, with the NUC completing the preprocessing tasks in 6.21 ms ($SD = 0.99$) compared to the Jetson's time of 6.48 ms ($SD = 2.43$). However, with respect to running a deep neural network for pedestrian detection on high resolution images, the CPU in the NUC is significantly faster than the GPU of the Jetson (by a factor of 3.5), with the NUC completing the task in 0.17 s ($SD = 0.19$) and the Jetson completing the task in 0.57 s ($SD = 1.10$).

It should be noted that the performance times for the Jetson include the time needed for transferring the image to the GPU and then transferring the results back from the GPU. Such transfer times are not part of the NUC performance times as the pedestrian detection was carried out on the NUC's CPU. Based on the performance times for both the NUC and Jetson on the image demosaicing and reprojection tasks, it would not be unreasonable to conclude that the data transfer times involved in these tasks are not significant enough to drastically alter the results that are seen here.

Figure 4 shows the results of demosaicing and reprojecting an image from the camera. The pinched black shape on the left side of the first two images in Fig. 4 are due to the robot's nose obstructing the field of view. These two images have a 195° equidistant field of view, while the last image has a 150° rectilinear field of view. All images have a resolution of $1,280 \times 1,024$ pixels (not shown to scale).

Fig. 4. Results of the image reprojection algorithm. Left: Original mosaiced image, interpreted as greyscale, with 195° FOV. Middle: Demosaiced image. Right: Demosaiced and reprojected image with 150° field of view. The black border on the left side of the two images on the left is due to the robot's nose. All images have a resolution of $1,280 \times 1,024$ pixels.

Figure 5 shows the current consumption on the NUC and the Jetson. On average, we see that the NUC draws 2.53 A ($SD = 0.35$), while the Jetson is drawing 0.59 A ($SD = 0.09$). When powered by a 16 V power supply we see that the NUC has a power consumption of 40.52 W, compared to the Jetson's 9.48 W. Considering the maximum $FLOPS$ determined in Sect. 6.1 and the maximum current draw, we see that although the NUC has a larger power requirement, we achieve a 2.5-fold increase in the number of $FLOPS$ available per watt of power consumed, with the NUC providing 104.64 GFLOPS/W compared to the Jetson's 41.26 GFLOPS/W.

If we were to power these devices from a 14.8 V 4-cell LiPo battery with a 3850 mA h^{-1} capacity we could expect a battery life of 84.36 min on the NUC and 360.48 min on the Jetson. These figures are calculated based on the average current draw for the two devices, while also assuming that as system voltage decreases current draw will increase in order to keep system power consumption constant.

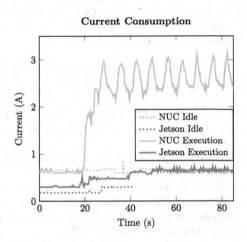

Fig. 5. Comparison of current consumption for the Jetson and NUC.

7   Discussion and Conclusion

This work presented a series of performance tests of computer vision tasks on two computer platforms – the NUC and the Jetson. Despite having a larger power requirement, the NUC proves to be a powerful computing device for an embedded platform. It should be noted that, if Tensorflow was compiled from source for the NUC, further performance improvements are expected as the NUC supports SIMD intrinsics which the pre-built Tensorflow is not compiled to use. Similar performance improvements would not be expected on the Jetson, as Tensorflow is using the GPU device on this platform with very little work being done on the CPU. More significant improvements would be expected on the NUC if Tensorflow was compiled to use OpenCL as a backend, as opposed to CUDA. This would allow Tensorflow to fully utilise the GPU on the NUC. If the intended system has a severely constrained power budget (< 10 W), then the Jetson becomes a very strong candidate for embedded GPU computing tasks.

We conclude that for utilising deep networks on a humanoid robot, like the NUgus, a careful selection of the computing platform is essential. The initial assumption that the GPU focused Jetson would be the best platform for our robot was not supported by this comparative evaluation. It will be interesting to see what opportunities for progress in machine learning on robots the next generation of computing devices may bring.

References

1. Abouelfarag, A.A., Nouh, N.M., ElShenawy, M.: Improving performance of dense linear algebra with multi-core architecture. In: 2017 International Conference on High Performance Computing Simulation (HPCS), pp. 870–874, July 2017. https://doi.org/10.1109/HPCS.2017.132
2. Allgeuer, P., Farazi, H., Schreiber, M., Behnke, S.: Child-sized 3D printed IGUS humanoid open platform. In: IEEE-RAS 15th International Conference on Humanoid Robots (Humanoids) (2015)
3. Bainville, E.: GPU benchmarks (2009). http://www.bealto.com/gpu-benchmarks_flops.html
4. Chalup, S.K., Murch, C.L., Quinlan, M.J.: Machine learning with aibo robots in the four legged league of robocup. IEEE Trans. Syst. Man Cybern. Part C **37**(3), 297–310 (2007)
5. Franklin, D.: NVIDIA Jetson TX2 Delivers Twice the Intelligence to the Edge (2017). https://devblogs.nvidia.com/jetson-tx2-delivers-twice-intelligence-edge/. Accessed 11 Apr 2018
6. Goodfellow, I., Bengio, Y., Courville, A.: Deep Learning. MIT Press, Cambridge (2016)
7. Gregg, C., Hazelwood, K.: Where is the data? why you cannot debate CPU vs. GPU performance without the answer. In: (IEEE ISPASS) IEEE International Symposium on Performance Analysis of Systems and Software, pp. 134–144, April 2011. https://doi.org/10.1109/ISPASS.2011.5762730
8. Houliston, T., et al.: Nuclear: a loosely coupled software architecture for humanoid robot systems. Front. Robot. AI **3**(20) (2016). https://doi.org/10.3389/frobt.2016.00020
9. Houliston, T., Metcalfe, M., Chalup, S.K.: A fast method for adapting lookup tables applied to changes in lighting colour. In: Almeida, L., Ji, J., Steinbauer, G., Luke, S. (eds.) RoboCup 2015. LNCS (LNAI), vol. 9513, pp. 190–201. Springer, Cham (2015). https://doi.org/10.1007/978-3-319-29339-4_16
10. Howard, A.G., et al.: MobileNets: Efficient Convolutional Neural Networks for Mobile Vision Applications. CoRR abs/1704.04861 (2017). http://arxiv.org/abs/1704.04861
11. Huang, J., et al.: Speed/Accuracy Trade-Offs for Modern Convolutional Object Detectors. CoRR abs/1611.10012 (2016). http://arxiv.org/abs/1611.10012
12. Intel Corporation: Product Brief: Intel®NUC Kit NUC7i7BNH (2017). https://www.intel.com.au/content/www/au/en/nuc/nuc-kit-nuc7i7bnh-brief.html. Accessed 11 Apr 2018
13. Jabbar, A., Farrawell, L., Fountain, J., Chalup, S.K.: Training deep neural networks for detecting drinking glasses using synthetic images. In: Proceedings of the The 24th International Conference On Neural Information Processing, ICONIP 2017, Guangzhou, China, 14–18 November 2017
14. Javadi, M., Azar, S.M., Azami, S., Shiry, S., Ghidary, S.S., Baltes, J.: Humanoid robot detection using deep learning: a speed-accuracy tradeoff. In: 21st Annual RoboCup International Symposium (2017, in press)
15. Krizhevsky, A., Sutskever, I., Hinton, G.E.: Imagenet classification with deep convolutional neural networks. In: Pereira, F., Burges, C.J.C., Bottou, L., Weinberger, K.Q. (eds.) Advances in Neural Information Processing Systems (NIPS 2012), vol. 25, pp. 1097–1105. Curran Associates Inc. (2012)

16. Lee, V.W., et al.: Debunking the 100X GPU vs. CPU myth: an evaluation of throughput computing on CPU and GPU. ACM SIGARCH Comput. Archit. News **38**(3), 451–460 (2010)
17. Levine, S., Pastor, P., Krizhevsky, A., Ibarz, J., Quillen, D.: Learning hand-eye coordination for robotic grasping with deep learning and large-scale data collection. Int. J. Robot. Res. **37**(4), 421–436(2017). https://doi.org/10.1177/0278364917710318
18. Lin, T., et al.: Microsoft COCO: Common Objects in Context. CoRR abs/1405.0312 (2014). http://arxiv.org/abs/1405.0312
19. Liu, W., et al.: SSD: Single Shot MultiBox Detector. CoRR abs/1512.02325 (2015). http://arxiv.org/abs/1512.02325
20. Metcalfe, M., Annable, B., Olejniczak, M., Chalup, S.K.: A study on detecting three-dimensional balls using boosted classifiers. In: Interactive Entertainment 2016, at the Australasian Computer Science Week (ACSW 2016), Canberra, Australia. ACM Digital Library, 2–5 February 2016. https://doi.org/10.1145/2843043.2843473
21. Pierson, H.A., Gashler, M.S.: Deep learning in robotics: a review of recent research. Adv. Robot. **31**(16), 821–835 (2017). https://doi.org/10.1080/01691864.2017.1365009
22. Speck, D., Barros, P., Weber, C., Wermter, S.: Ball localization for robocup soccer using convolutional neural networks. In: Behnke, S., Sheh, R., Sarıel, S., Lee, D.D. (eds.) RoboCup 2016. LNCS (LNAI), vol. 9776, pp. 19–30. Springer, Cham (2017). https://doi.org/10.1007/978-3-319-68792-6_2
23. Vanhoucke, V., Senior, A., Mao, M.Z.: Improving the speed of neural networks on CPUs. In: Deep Learning and Unsupervised Feature Learning Workshop, NIPS 2011 (2011)

Min-Max Consensus Algorithm for Multi-agent Systems Subject to Privacy-Preserving Problem

Aijuan Wang, Nankun Mu$^{(\boxtimes)}$, and Xiaofeng Liao

College of Electronic and Information Engineering, Southwest University,
Chongqing 400715, China
nankun.mu@qq.com

Abstract. This paper proposes a privacy-preserving min-max consensus algorithm for discrete-time multi-agent systems, where all agents not only can reach a common state asymptotically, but also can preserve the privacy of their states at each iteration. Based on the proposed algorithm, the detailed consensus analysis is developed, including the impossibility of finite time convergence and the sufficient condition of consensus. Moreover, the privacy-preserving analysis is provided to guarantee the reliability of our privacy-preserving scheme. Finally, a numerical simulation is performed to demonstrate the correctness of our results.

Keywords: Min-max consensus algorithm · Multi-agent systems
Privacy-preserving

1 Introduction

Recently, the consensus of multi-agent systems has attracted considerable interests [1–3] due to their extensively real applications, such as dynamic load balancing [4,5], coordination of groups of mobile autonomous agents [6,7] and cooperative control of vehicle formations [8,9]. One of the fundamental consensus problems is how to design a consensus algorithm that can enable the agents to reach a common state [10,11].

The two common consensus algorithms are average consensus [12–15] and maximum (or minimum) consensus [16,17]. A proportional and derivative-like average consensus algorithm is developed for multi-agent systems with Lipschitz nonlinear dynamics [12]. In [13], the authors propose a distributed finite-time average consensus algorithm for distributed systems. In [14], the authors introduce the event-triggered distributed average consensus algorithm for the digital networks with limited communication bandwidth. In [15], a weighted average consensus algorithm based on unscented kalman filtering (UKF) is developed for the purpose of estimating the true state of interest. On the other hand, the researchers propose a min consensus algorithm with finite-time convergence [16]. In [17], the authors deal with the analysis of the convergence properties of the

© Springer Nature Switzerland AG 2018
L. Cheng et al. (Eds.): ICONIP 2018, LNCS 11307, pp. 132–142, 2018.
https://doi.org/10.1007/978-3-030-04239-4_12

max-consensus protocol in presence of asynchronous updates and time delays. In these consensus algorithms, each agent has to know the others' update laws, then it can infer the others' states by some observability conditions. Especially the state information of each agent is directly available to its neighbors. Therefore, the exact state value of each agent may be computable by the other agents or can be directly obtained by its neighbors, which result in information leakage. From the privacy-preserving perspective, the agents may not want to expose their states values in the process of consensus. An classic example is the multi-agent rendezvous problem [18,19], where the agents are committed to eventually rendezvousing at a certain location and each agent wants to keep its own location secret to the others. This motivates us to propose a privacy-preserving scheme and integrate it into consensus algorithm.

It is worth mentioning that some literatures have successfully applied it into dynamical systems. For example, differentially private filters for dynamical system are designed by adding white Gaussian perturbations to the system [20]. In the context of consensus problems, the differentially private consensus algorithm is firstly proposed in [21]. Then, in order to reach the exact average consensus, the authors [22] propose a privacy preserving average consensus algorithm, where a privacy-preserving scheme is introduced by adding the noises and subtracting all noises added. But they just discuss the case whether the initial state of one agent can be perfectly inferred by the other agents, and they do not provide a quantitative condition that the initial state can be estimated perfectly. Afterwards, in [23], the authors propose a privacy preserving average consensus algorithm to solve this problem, where the covariance matrix of the maximum likelihood estimate on the initial state is derived. In [24], the authors propose a privacy-preserving scheme and apply it to the maximum consensus algorithm.

Motivated by the above discussion, in this paper, we are of interest in developing a privacy-preserving consensus algorithm which can protect the privacy of the states of all agents at each iteration in the whole process of consensus. The major contributions of this paper are summarized as: (1) A privacy-preserving min-max consensus algorithm is developed. (2) The theoretical consensus analysis on our algorithm is developed. (3) We give a detailed analysis on the privacy-preserving degree for each agent.

2   Graph Theory and Min-Max Consensus Algorithm

Consider a connected network of N agents, described by an undirected communication graph $g = (V, \ \varepsilon, \ A)$. The graph is connected if and only if for any two distinct agents $i, j \in V$, there is a path between j and i. The neighbor set of agent i is defined as $N_i = \{j \in V \mid (i, j) \in \varepsilon\}$. The degree d_i of agent i is the cardinality of N_i, namely, $d_i = |N_i|$.

Agent number N of network is defined as the network size. The diameter of graph g is denoted by D. $l_{i,j}$ represents the distance between j and i. Let e_i denote the eccentricity of agent i, i.e., the maximum distance between i and other agents of graph g. $S_{i,j}$ represents the agent set consisting of the agents whose shortest paths to j all include agent i.

In this paper, each agent updates its state to a weighted average of own state together with the minimum and maximum state of whole network as follows:

$$x_i(k+1) = \alpha x_i(k) + \beta \min_{j \in V} x_j(k) + (1 - \alpha - \beta) \max_{j \in V} x_j(k), \tag{1}$$

where $0 < \alpha, \beta < 1$ and $0 < \alpha + \beta < 1$, $\min_{j \in V} x_j(k)$ and $\max_{j \in V} x_j(k)$ denotes the minimum state of whole network at iteration k, respectively.

3   Main Results

3.1   Privacy-Preserving Scheme

Here, we propose a privacy-preserving scheme used in the (1) to protect the state of each agent from disclosure to its neighbors. In the following, the privacy-preserving scheme is established. The steps are listed as follows:

step 1   Initialize $k = 0$.

step 2   For every iteration k, each agent generates two sets of random values. Each set of random has m_i random values for agent i, $(i = 1, 2, ..., N)$. The first group of random numbers d_i^t, $(t = 0, 1, ..., m_i - 1)$ is smaller than $x_i(k)$ for agent i, and the second group of random numbers r_i^t, $(t = 0, 1, ..., m_i - 1)$ is larger than $x_i(k)$.

step 3   Agent i sorts the first group of random numbers in the descending order as is $d_i^0, d_i^1, d_i^2, ..., d_i^{m_i}$ $(d_i^{m_i} = x_i(k))$, and sorts the second of random numbers in the ascending order as is $r_i^0, r_i^1, r_i^2, ..., r_i^{m_i}$ $(r_i^{m_i} = x_i(k))$. Initialize $p = 1, s = 1$.

step 4   Agent obtains the smallest state and the largest state of the whole network agents by executing the following law:

$$y_i(p) = \begin{cases} \min\left\{d_i^p, \min_{j \in N_i} y_j(p-1)\right\}, p \leq m_i \\ \min_{j \in N_i} y_j(p-1), p > m_i \end{cases}, (A)$$

$$z_i(s) = \begin{cases} \max\left\{r_i^s, \max_{j \in N_i} z_j(s-1)\right\}, s \leq m_i \\ \max_{j \in N_i} z_j(s-1), s > m_i \end{cases}, (B)$$

where $z_i(0) = r_i^0$ and $y_i(0) = d_i^0$.

step 5   Let $p = p + 1$ and $s = s + 1$. If $y_i(p) \neq y_i(p-1)$, $z_i(s) \neq z_i(s-1)$, repeat Step 4. Otherwise, proceed to the next step.

step 6   Update state according to the following laws:

$$x_i(k+1) = \alpha x_i(k) + \beta y_i(p) + (1 - \alpha - \beta) z_i(s).$$

step 7   Let $k = k + 1$. If $k \leq 1000$ or $x_i(k+1) \neq x_i(k)$, repeat Step 2-6. Otherwise, end the update process.

In the privacy-preserving scheme, we assume that for $\forall k \in [N], \forall i = 1, 2, ...,$ $2 \leq m_i \leq M$.

Remark 1. In the above privacy-preserving scheme, the random numbers just are used to mix in the states of all agents, preventing the states from being exposed. The random numbers can not alter the update law of each agent. But it can impact the convergence rate of Algorithm (1) that decreases with the increment of random number m_i.

3.2   Finite-Time Convergence Impossibilities

In this subsection, we prove that there is not possible to achieve the finite time consensus under the Algorithm (1).

Theorem 1. *Under the min-max consensus Algorithm (1), finite-time consensus fails for all initial values except for* $\min_{j \in V} x_j(0) \neq \max_{j \in V} x_j(0)$.

Proof. We Define $h(k) = \min_{i \in V} x_i(k), H(k) = \max_{i \in V} x_i(k)$.

Introduce $\varphi(k) = H(k) - h(k)$, then, the asymptotic consensus can be achieved if and only if $\lim_{k \to \infty} \varphi(k) = 0$.

Assume that there exists two agents i, m satisfying that $x_i(t) = h(k), x_m(t) = H(k)$. According to the min-max consensus Algorithm (1), we have

$$x_i(k+1) = (\alpha + \beta) h(k) + (1 - \alpha - \beta) H(k) \tag{2}$$

and

$$x_m(k+1) = \beta h(k) + (1 - \beta) H(k). \tag{3}$$

With (2) and (3), we have

$$\varphi(k+1) = \max_{i \in V} x_i(k+1) - \min_{i \in V} x_i(k+1) \geq x_m(k+1) - x_i(k+1) \\ = \alpha \varphi(k) \tag{4}$$

Therefore, we obtain that

$$\varphi(k) \geq \alpha^k \varphi(0) > 0. \tag{5}$$

Here, $\varphi(0) \neq 0$ is assumed, which is given in the graph theory section. So, $\forall K < \infty, \varphi(K) \geq \alpha^K \varphi(0) > 0$, which means that finite-time consensus is impossible for all initial states $\varphi(0) \neq 0$.

3.3   Asymptotic Consensus Analysis

In this subsection, the consensus results are given. We define $\lambda(k) = \beta \min\limits_{j \in V} x_j(k) + (1 - \alpha - \beta) \max\limits_{j \in V} x_j(k)$.

Lemma 1. *Under the Algorithm* (1), *then, the $\lambda(k)$ is a fixed constant for all $k = 0, 1, 2, \ldots$ and given α, β. The value of $\lambda(k)$ is the following for $\forall k = 0, 1, 2, \ldots$*

$$\lambda(k) = \beta \min_{j \in V} x_j(0) + (1 - \alpha - \beta) \max_{j \in V} x_j(0). \tag{6}$$

Proof. In the following, we prove that for $\forall k = 0, 1, 2, \ldots$

$$\lambda(k) = \beta \min_{j \in V} x_j(0) + (1 - \alpha - \beta) \max_{j \in V} x_j(0)$$

by mathematical induction.

Let us start with $k = 0$. We can clearly see that $\lambda(0) = \beta \min\limits_{j \in V} x_j(0) + (1 - \alpha - \beta) \max\limits_{j \in V} x_j(0)$, which implies that $\lambda(k) = \beta \min\limits_{j \in V} x_j(0) + (1 - \alpha - \beta) \max\limits_{j \in V} x_j(0)$ holds for $k = 0$.

Now, we assume that for k

$$
\begin{aligned}
\lambda(k) &= \beta \min_{j \in V} x_j(k) + (1 - \alpha - \beta) \max_{j \in V} x_j(k) \\
&= \beta \min_{j \in V} x_j(0) + (1 - \alpha - \beta) \max_{j \in V} x_j(0)
\end{aligned} \tag{7}
$$

holds.

Then, under the Algorithm (1), we have for $k + 1$,

$$\lambda(k+1) = \beta \min_{i \in V} \left(\alpha x_i(k) + \beta \min_{j \in V} x_j(k) + (1 - \alpha - \beta) \max_{j \in V} x_j(k) \right)$$

$$+ (1 - \alpha - \beta) \max_{j \in V} \left(\alpha x_i(k) + \beta \min_{j \in V} x_j(k) + (1 - \alpha - \beta) \max_{j \in V} x_j(k) \right). \tag{8}$$

Since $\lambda(k) = \beta \min\limits_{j \in V} x_j(0) + (1 - \alpha - \beta) \max\limits_{j \in V} x_j(0)$ for k holds, the (8) is written

$$\lambda(k+1) = \beta \min_{j \in V} x_j(0) + (1 - \alpha - \beta) \max_{j \in V} x_j(0). \tag{9}$$

The Eq. (9) implies that for $k + 1$, it holds that

$$\lambda(k+1) = \beta \min_{j \in V} x_j(0) + (1 - \alpha - \beta) \max_{j \in V} x_j(0).$$

Hence, from (7)–(9), for $\forall k = 0, 1, 2, \ldots$, we have (6) holds.

Theorem 2. *Suppose that the network topology considered in this paper is connected. Under the min-max consensus Algorithm (1), if $0 < \alpha, \beta < 1$ and $0 < \alpha + \beta < 1$ hold, the states of all agents asymptotic converge to the constant χ, namely, $\lim_{k \to \infty} x_i(k) = \chi$ with*

$$\chi = \frac{\left[\beta \min_{j \in V} x_j(0) + (1 - \alpha - \beta) \max_{j \in V} x_j(0)\right]}{1 - \alpha}. \tag{10}$$

Proof. It follows from Algorithm (1) that

$$x_i(k+1) - \chi = \alpha x_i(k) - \chi + \beta \min_{j \in V} x_j(k) + (1 - \alpha - \beta) \max_{j \in V} x_j(k). \tag{11}$$

Note that

$$\beta \min_{j \in V} x_j(k) + (1 - \alpha - \beta) \max_{j \in V} x_j(k) = (1 - \alpha) \chi.$$

So, the (11) can be written as

$$x_i(k+1) - \chi = \alpha x_i(k) + (1 - \alpha) \chi - \chi = \alpha (x_i(k) - \lambda). \tag{12}$$

By the iteration , we have

$$x_i(k+1) - \chi = \alpha^{k+1} (x_i(0) - \lambda),$$

which means that $\lim_{k \to \infty} x_i(k) = \chi$ due to $0 < \alpha < 1$, .

Remark 2. Under the Algorithm (1), the minimum state owner $h(k)$ is a monotonically non-decreasing sequence, i.e., $h(k) \leq h(k+1)$, and $h(k)$ will always be the minimum state at each iteration $\forall k = 0, 1, 2, \ldots$. Similarly, the maximum state owner $H(k)$ is a monotonically non-increasing sequence, i.e., $H(k) \geq H(k+1)$, and $H(k)$ will always be the maximum state in all states $x_i(k)$ at each iteration $\forall k = 0, 1, 2, \ldots$.

3.4 Privacy Analysis

In this subsection, the privacy of the state for each agent is analyzed with the proposed privacy-preserving scheme. Especially, the disclosure probabilities of the minimum and maximum states are established.

Theorem 3. *With the proposed privacy-preserving scheme, for each pair of neighboring agents i, j. The agent i cannot infer the state of the neighboring agent j, unless the agent j is the maximum state value or the minimum state value.*

Proof. We omit the proof of Theorem 3, since, it is similar with that of [24].

Definition 1. *The maximum expected probability of agent $v_{\max}$, which is recognized by one of its neighbors as the maximum state owner is considered as an index to judge the degree of the privacy leakage of agent $v_{\max}$'s state. And, we have the similar definition for the minimum state owner $v_{\min}$.*

Here, we just discuss the probability calculation of agent $v_{\max}$ recognized by its neighbors.

We give some notations in the following. $J_{i,j}$ denotes the event that agent j is sure that agent i is the maximum state owner. Then, agent j can obtain the probability that agent i is the maximum state when agent j obtains the maximum initial state from agent i. Let R_i denote the R_i-th period at which agent i obtains the maximum state from its some neighbor for the first time. After R_i period, agent j will receive multiple copies of maximum states from its neighbors at the same time. In this case, the probability that agent j is convinced that agent i is the maximum state owner is zero, that is, $P_r(J_{i,j}|R_j + nT) = 0, j \in N_i, n = 1, 2, \dots$. Hence, in the following, we just consider the case where agent j gets the maximum state only from its neighbor i for the first time.

Next, two cases are included as follows:

With Whole Network Topology. In this subsection, we suppose that each agent knows the whole network topology. Namely, the $S_{i,j}$ is known to each agent. In this case, there are several dubious agents in the set $S_{v_{\max},i}, \forall i \in N_{v_{\max}}$ in terms of different q. We let $C_{v_{\max},i}(q)$ denote the sceptical agents set, and it includes the agents in the $S_{v_{\max},i}$ whose distance to i is less than or equal to $q - 1$. Obviously, we have that $C_{v_{\max},i}(q - 1) \subseteq C_{v_{\max},i}(q)$.

Theorem 4. *Suppose that agent i is a neighbor of v_{max}, and it obtains the maximum state in the q-th period for the first time. Then, agent i confirms v_{max} as the agent with maximum state for each iteration $t = k, k = 0, 1, 2, \dots$ with the probability $P_r(J_{v_{\max},i}|R_i = q)$ which is as follows:*

$$P_r(J_{v_{\max},i}|R_i = q) = \frac{P_r(m_{v_{\max}} = q - 1)}{\sum_{s \in C_{v_{\max},i}(q)} P_r(m_s = q - 1 - l_{s,i})}. \tag{13}$$

Proof. $P_r(J_{v_{\max},i}|R_i = q)$ is a conditional probability, which satisfies

$$P_r(J_{v_{\max},i}|R_i = q) = \frac{P_r(R_i = q|J_{v_{\max},i}) P_r(J_{v_{\max},i})}{P_r(R_i = q)}. \tag{14}$$

$P_r(R_i = q|J_{v_{\max},i})$ denotes v_{max} the probability that $q - 1$ generates random numbers, i.e., $P_r(R_i = q|J_{v_{\max},i}) = P_r(m_{v_{\max},i} = q - 1)$.

According to total probability formula, the denominator in (14) can be expressed:

$$P_r(R_i = q) = \sum_{s \in C_{v_{\max},i}(q)} P_r(R_i = q|J_{s,i}) P_r(J_{s,i}). \tag{15}$$

It follows from (14)–(15) that

$$P_r \left(J_{v_{\max},i} | R_i = q \right) = \frac{P_r \left(R_i = q | J_{v_{\max},i} \right) P_r \left(J_{v_{\max},i} \right)}{\sum\limits_{s \in C_{v_{\max},i}(q)} P_r \left(R_i = t | J_{s,i} \right) P_r \left(J_{s,i} \right)}$$

$$= \frac{P_r \left(m_{v_{\max}} = q - 1 \right) P_r \left(J_{v_{\max},i} \right)}{\sum_{s \in C_{v_{\max},i}(q)} P_r \left(m_s = q - l_{s,i} \right) P_r \left(J_{s,i} \right)}. \tag{16}$$

Besides, since we assume that each agent has the same possibility to own the maximum value, we have $P_r \left(J_{v_{\max},i} \right) = P_r \left(J_{s,i} \right) = \frac{1}{|S_{v_{\max},i}|}$.

So, the (16) can be rewritten

$$P_r \left(J_{v_{\max},i} | R_i = q \right) = \frac{P_r \left(m_{v_{\max}} = q - 1 \right)}{\sum_{s \in C_{v_{\max},i}(q)} P_r \left(m_s = q - l_{s,i} \right)}. \tag{17}$$

Corollary 1. *The maximum expected probability $P_r \left(J_{v_{\max},*} \right)$ of agent v_{max} recognized by one of its neighbors as the agent with maximum state in the whole process of consensus iteration is as follows:*

$$P_r \left(J_{v_{\max},*} \right) = \max_{i \in N_{v_{\max}}} \lim_{T \to \infty} \frac{\sum_{t=1}^{T} E \left[P_r \left(J_{v_{\max},i} | R_i \right) \right]}{T}, \tag{18}$$

where

$$E \left[P_r \left(J_{v_{\max},i} | R_i \right) \right] = \sum_{q=2}^{M+1} P_r \left(J_{v_{\max},i} | R_i = q \right) P_r \left(m_{v_{\max},i} = q - 1 \right).$$

With Local or Limited Network Topology. In this subsection, we consider the case where $S_{i,j}$ is unknown to agent i, agent i has to contain as many agents in the set $S_{i,j}$ as possible, that is, $N - d_i$ in all. So, we can take the maximum of all possible cases about the denominator in (13), only obtaining the lower bound of probability.

Corollary 2. *Suppose that each agent just knows the network size N and network diameter D. Then, the lower bound of probability $P_r \left(J_{v_{\max},i} | R_i = q \right)$ that agent i confirms v_{max} as the maximum state owner for any iteration $\forall k, k = 0, 1, 2, \ldots$ is that*

$$\inf \left(P_r \left(J_{v_{\max},i} | R_i = q \right) \right)$$

$$= \frac{P_r \left(m_{v_{\max}} = q - 1 \right)}{P_r \left(m_{v_{\max}} = q - 1 \right) + g \left(q, P_r \left(m \right), d_i, N, D \right)}, \tag{19}$$

where

$$g \left(q, P_r \left(m \right), d_i, N, D \right) = \max_{v=2,\ldots,\min\{q-1,D\}} \left\{ \sum_{s=1}^{v-2} P_r \left(m = q - s - 1 \right) \right.$$

$$\left. + \left(N - d_i - (v - 2) \right) P_r \left(m = q - v \right) \right\}.$$

Theorem 5. *Without the local or limited network topology, the maximum expectation of the lower bound of probability $P_r(J_{v_{\max},*})$ of agent v_{max} referred by one of its neighbors as the agent with maximum state in the whole process of consensus iteration is the following*

$$P_r(J_{v_{\max},*}) = \max_{i \in N_{v_{\max}}} \lim_{T \to \infty} \frac{\sum_{t=1}^{T} E\left[\inf\left(P_r\left(J_{v_{\max},i}|R_i\right)\right)\right]}{T},$$

where

$$E\left[\inf\left(P_r\left(J_{v_{\max},i}|R_i\right)\right)\right] = \left(\sum_{q=2}^{M+1} \inf\left(P_r\left(J_{v_{\max},i}|R_i = q\right)P_r\left(m_{v_{\max},i} = q - 1\right)\right)\right).$$

Remark 3. From (13) and (19), it should be noted that the probability of states' information leakage decreases with the increment of random numbers m_i.

4   Simulation Example

In this section, we verify the consensus results of the proposed Algorithm (1). The network topology graph is shown in Fig. 1, which is connected. And, the weighted matrix is $a_{ij} = 1, i, j \in \varepsilon$, otherwise, $a_{ij} = 0$. We take the initial state as $x_i(0) = \begin{bmatrix} 10.2138 \ 0.4924 \ 1.7239 \ 5.7743 \ -2.7562 \ 8.3814 \end{bmatrix}$ for agent i. Take $\alpha = 0.87$ and $\beta = 0.07$, from the Eq. (10), the consensus value is calculated as $\chi = 3.22995$. As is shown in Fig. 2, state trajectory $x_i(t)$ of each agent converges to value $\chi = 3.22995$, which verifies our results.

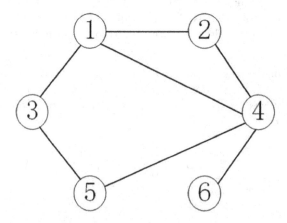

Fig. 1. The undirected and connected network communication topology.

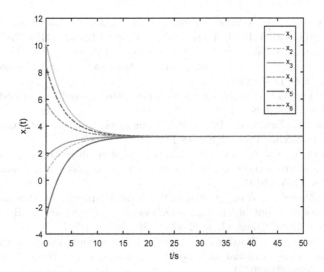

Fig. 2. State trajectory $x_i(t)$ of each agent in networks g under the Algorithm (1)

5   Conclusion

In this paper, we have investigated the privacy-reserving min-max consensus problem. We have proposed a min-max consensus algorithm which not only guarantees that all agents eventually reach a common value but also preserves the privacy of the state of each agent at each iteration. Besides, we have done the privacy-preserving analysis, which exhibits that privacy of the states for all agents (except for the minimum and maximum states) are guaranteed to preserve, and the probability that the minimum and maximum of states are recognized by their neighbors are explicitly given, respectively.

References

1. Zhang, H., Yue, D., Yin, X., Hu, S., xia, D.C.: Finite-time distributed eventtriggered consensus control for multi-agent systems. Inf. Sci. **339**, 132–142 (2016)
2. Wang, A., Dong, T., Liao, X.: On the general consensus protocol in multiagent networks with second-order dynamics and sampled data. Asian J. Control **18**(5), 1914–1922 (2016)
3. Meng, X., Chen, T.: Optimal sampling and performance comparison of periodic and event based impulse control. IEEE Trans. Autom. Control **57**(12), 3252–3259 (2012)
4. Cybenko, G.: Dynamic load balancing for distributed memory multiprocessors. J. Parallel Distrib. Comput. **7**(2), 279–301 (1989)
5. Sun, J., Yang, Q., Liu, X., Chen, J.: Event-triggered consensus for linear continuous-time multi-agent systems based on a predictor. Inf. Sci
6. Jadbabaie, A., Lin, J., Morse, A.S.: Coordination of groups of mobile autonomous agents using nearest neighbor rules. IEEE Trans. Autom. Control **48**(6), 988–1001 (2003)

7. Liu, Q., Wang, J.: A second-order multi-agent network for bound-constrained distributed optimization. IEEE Trans. Autom. Control **60**(12), 3310–3315 (2015)
8. Ren, W., Beard, R.W., Atkins, E.M.: A survey of consensus problems in multi-agent coordination. In: Proceedings of the American Control Conference 2005. IEEE, pp. 1859–1864 (2005)
9. Lu, A.Y., Yang, G.H.: Distributed consensus control for multi-agent systems under denial-of-service. Inf. Sci. **439**, 95–107 (2018)
10. Olfati-Saber, R., Fax, J.A., Murray, R.M.: Consensus and cooperation in networked multi-agent systems. Proc. IEEE **95**(1), 215–233 (2007)
11. Liu, K., Yao, Y., Balakrishnan, V., Ma, H.: Adaptive h8 control of piecewiselinear systems with parametric uncertainties and external disturbances. Asian J. Control **15**(4), 1238–1245 (2013)
12. Wang, D., Zhang, N., Wang, J., Wang, W.: A pd-like protocol with a time delay to average consensus control for multi-agent systems under an arbitrarily fast switching topology. IEEE Trans. Cybern. **47**(4), 898–907 (2017)
13. Oliva, G., Setola, R., Hadjicostis, C.N.: Distributed finite-time averageconsensus with limited computational and storage capability. IEEE Trans. Control Netw. Syst. **4**(2), 380–391 (2017)
14. Li, H., Chen, G., Huang, T., Dong, Z., Zhu, W., Gao, L.: Event-triggered distributed average consensus over directed digital networks with limited communication bandwidth. IEEE Trans. Cybern. **46**(12), 3098–3110 (2016)
15. Li, W., Wei, G., Han, F., Liu, Y.: Weighted average consensus-based unscented kalman filtering. IEEE Trans. Cybern. **46**(2), 558–567 (2016)
16. Zhang, Y., Li, S.: Perturbing consensus for complexity: a finite-time discrete biased min-consensus under time-delay and asynchronism. Automatica **85**, 441–447 (2017)
17. Giannini, S., Petitti, A., Paola, D.D., Rizzo, A.: Asynchronous max-consensus protocol with time delays: convergence results and applications. IEEE Trans. Circuits. Syst. I: Regul. Pap. **63**(2), 256–264 (2016)
18. DeGroot, M.H.: Reaching a consensus. J. Am. Stat. Assoc. **69**(345), 118–121 (1974)
19. Lin, J., Morse, A.S., Anderson, B.D.: The multi-agent rendezvous problem. part 2: the asynchronous case. SIAM J. Control Optim. **46**(6), 2120–2147 (2007)
20. Ny, J.L., Pappas, G.J.: Differentially private filtering. IEEE Trans. Autom. Control **59**(2), 341–354 (2014)
21. Huang, Z., Mitra, S., Dullerud, G.: Differentially private iterative synchronous consensus. In: Proceedings of the 2012 ACM Workshop on Privacy in the Electronic Society. ACM, pp. 81–90 (2012)
22. Manitara, N.E., Hadjicostis, C.N.: Privacy-preserving asymptotic average consensus. In: 2013 European Control Conference (ECC), pp. 760–765. IEEE (2013)
23. Mo, Y., Murray, R.M.: Privacy preserving average consensus. IEEE Trans. Autom. Control **62**(2), 753–765 (2017)
24. Duan, X., He, J., Cheng, P., Mo, Y., Chen, J.: Privacy preserving maximum consensus. In: 2015 IEEE 54th Annual Conference on Decision and Control (CDC), pp. 4517–4522. IEEE (2015)

Robot Navigation on Slow Feature Gradients

Muhammad Haris[2(✉)], Mathias Franzius[2], and Ute Bauer-Wersing[1]

[1] Faculty of Computer Science and Engineering, Frankfurt University of Applied Sciences, 60318 Frankfurt, Germany
[2] Honda Research Institute Europe GmbH, 63073 Offenbach, Germany
muhammad.haris@gmx.de

Abstract. Unsupervised learning with Slow Feature Analysis (SFA) enables an agent to learn spatial representations of its environment from images captured during an exploration phase. In a subsequent application phase, slow features encode the robot's position. The representation is spatially smooth and implicitly encodes the average travel time during exploration. Following the SFA gradient allows the robot to navigate even around obstacles without any planning. Earlier work showed this basic principle in noise-free simulation, using two virtual cameras on a robot. We extend the approach to be more robust and more computationally efficient. We test it on a lawn mower robot with a single camera for navigation in free space and avoiding obstacles.

1 Introduction

One of the most important abilities for autonomous robots is navigation in space. For nontrivial navigation, a mobile agent requires representations of the environment and its location to plan a viable path to a target. Visual simultaneous localization and mapping (vSLAM) is the state of the art approach for generating a map of the environment and simultaneous position estimation based on camera data. Maps generated by SLAM can take various forms, e.g. topological or as occupancy grids, which form the basis for a variety of navigation algorithms [1]. Such navigation approaches fall into a range of computational complexity from simple reactive motion approaches to trajectory planning in metrical maps [2].

Navigation is still a challenging task for mobile agents, whereas rodents, for example, are excellent at navigation. The resulting paths may not be optimal, but very flexibly and quickly generated, leading to an adaptive and robust navigation behavior. One such model is RatSLAM [3]. The agent's position and orientation are encoded as an activity packet in a continuous attractor network. Odometry and image matching introduce energy, shifting the activity peak. A later version of the model [4] allows it to keep spatial representations consistent for longer time periods in a graph-like experience map. Another class of models is based on slowness learning [5,6] and also related to localization as a model of the rodent hippocampus. In slowness learning [7], the resulting representation can

© Springer Nature Switzerland AG 2018
L. Cheng et al. (Eds.): ICONIP 2018, LNCS 11307, pp. 143–154, 2018.
https://doi.org/10.1007/978-3-030-04239-4_13

achieve significant invariances, e.g., to head orientation. Using an uncalibrated catadioptric imaging system, an earlier model successfully allowed a mobile robot to localize in an outdoor environment [8]. The learned representations of such models can be projected into metric space for navigation. While such a step makes quantitative evaluation easier and allows, for example, the integration of Kalman Filtering, it requires additional computational effort and reduces the richness of the representation.

Navigation in topological maps is an established method and can be efficiently done with the graph search algorithm A* [9]. While this algorithm provably finds an optimal path in relevant settings, it can scale unfavorably in large environments with many obstacles. A different navigation approach for metric spaces is based on force fields generated by an attractor at the target and repulsive forces from obstacles [10,11]. While this potential field approach is elegant and simple, it is strongly limited by local minima, e.g. for navigation around non-convex obstacles [12]. This restriction does not apply to the proposed method, as we will show.

Navigation by reinforcement learning based on slow features from visual data has been shown before [13,14]. However, these approaches require extensive additional learning for new target positions.

We present results for robot navigation on the slow feature gradients. The features are orientation-invariant and generated with the model presented in [8,15]. After the learning phase, navigation is performed by following the feature gradients approximated from cost measurements at the target and the current position. The previous work [15] was limited to simulation and required multiple cameras for estimating the slow feature gradients. Here, we extend the approach to become practical in a number of ways. First, we improve robustness, computation speed, and orientation invariance by replacing a part of the Slow Feature hierarchy with a Fourier feature extraction as preprocessing. Second, we present an approach to estimate the navigation gradient with a single camera. Finally, we present quantitative results for navigation with a lawn mower robot in free space and around obstacles.

2   Learning Spatial Representations with SFA

2.1   Slow Feature Analysis

Slow Feature Analysis transforms a set of time varying input signals $x(t)$ into a set of slowly varying output signals $y(t)$. The optimization objective is to find functions $g_j(x)$ such that the output signals [7]

$$y_j(t) := g_j(x(t))$$

minimize

$$\Delta(y_j) := \langle \dot{y}_j^2 \rangle_t$$

under the constraints

$$\langle y_j \rangle_t = 0 \text{ (zero mean)},$$
$$\langle y_j^2 \rangle_t = 1 \text{ (unit variance)},$$
$$\forall i < j : \langle y_i y_j \rangle_t = 0 \text{ (decorrelation and order)}$$

The $\langle \cdot \rangle_t$ and $\dot{y}$ represents time averaging and derivative of y, respectively. The Δ-value defines the temporal variation of the output signal $y_j(t)$ and its minimization is the objective function. A lower value indicates less variation of the signal over time thus means slowly varying signals. The first two constraints normalize all output signals to make their time derivative comparable and avoid trivial solution, $y_j(t) = constant$. The last constraint enforces that different output signals code for distinct features of the input. We employ the MDP implementation of the SFA algorithm [16].

2.2 Learning Spatial Representations

The navigation task requires spatial representations that should code for the position of the robot in space and are independent w.r.t. its orientation. While the objective of SFA enforces temporal slowness on the training trajectory, a suitable intermeshed training trajectory leads to representations that also change smoothly in space [6]. Furthermore, if the environment is suitably rich (i.e., without an identical appearance at different locations) and the function space for training is sufficiently large, each position in space is uniquely encoded by its Slow Feature representation[1]. Together, these properties allow a robot to navigate by following the gradient of the difference between the spatial representations of the current position and a target position. The encoding of position or orientation in the learned slow feature representations depends on the movement statistics of a robot during the training run. If the robot's orientation changes sufficiently faster than its position, the slowest features become orientation invariant [6]. In earlier work, we simulated robot rotation artificially by rotating the images from a 360-degree camera on the robot. [8,15]. Learning orientation invariance, however, generates computational load during training and the results are not perfect for limited training data. Noise on Slow Feature representations deteriorates the navigation gradient and is thus more problematic for gradient based navigation than for localization. As a consequence, we designed the current system with a preprocessing step that explicitly removes orientation dependency from the input representations for SFA. For this purpose, the omnidirectional images captured during the training phase are projected to panoramic images. Each panoramic image consists of 30 rows, which are expanded into Fourier series and only the magnitude part corresponding to the lowest 15 frequency components are stored. This provides a way to achieve rotation invariance as magnitude coefficients of Fourier components are not dependent on robot direction [17,18].

---

[1] Please note that in contrast to PCA, SFA extracts spatial representations cf. [7].

Fourier components obtained for each image by the preprocessing step are used to learn SFA representations. The learning phase has two steps, the first step reduces the dimensionality using a linear SFA while the second step extracts non-linear slow features using a quadratic SFA. The input and output dimensionality for the first step are 450 and 20, respectively. The input and output dimensionality for the second step are 20 and 8, respectively.

We can visualize the SFA representation with so-called spatial firing maps. These maps represent the color-coded function values for each position in a training trajectory. Theory has shown that for localization in open space with a relatively high rotational speed the first two SFA-outputs are orthogonal and monotonic [6], so in this case, there is one global minimum for navigation. The features depicted in Fig. 1 clearly encodes spatial information, with a gradient along the coordinate axes of the two slowest features. In contrast to the optimal solutions in open space as in [6], the representations are distorted around the obstacle. Due to the slowness objective of SFA, views of temporally nearby data during training are encoded with similar slow feature values, while temporally distant data is encoded with distinct values due to the unit variance constraint of SFA. This implicitly encodes the information about obstacles present in a training area as the slow feature values on either side of an obstacle will be highly different. Figure 1 is based on simulation data. Unfortunately, we can not provide such plots for the actual robot scenario, because ground truth data are hard to obtain in our setting and the metric positions we recorded during the experiments are of insufficient quality. Please note, that our approach presented here is unsupervised, hence ground truth data mostly serves the purpose of metric visualization and metric evaluation.

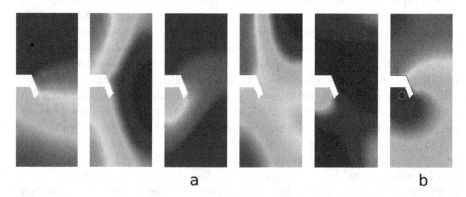

a   b

Fig. 1. Simulated spatial firing maps: (a) Spatial firing maps of the first five SFA units $y_{1..5}$. The maps of the first two SFA units $y_{1,2}$ together uniquely and smoothly encode the robot's position. The maps of functions $y_{3..5}$ show a mixture of the first two units and/or higher modes. (b) The plot shows an example cost surface in a slow feature space when the target position is in the convex region of the obstacle as indicated by the circle. Performing gradient descent on this surface allows a robot to circumnavigate the obstacle.

3   Navigation Method

We use slow features gradients to navigate between arbitrary points in a two-dimensional space. We assume that the target point's slow feature representation is known (e.g. stored during training). We define a cost function C as the Euclidean distance between slow feature representations of points in a 2D space. The approximate gradient of the cost surface determines the navigation direction, thus it is possible to implement navigation by performing gradient descent on the cost surface C. For an n-dimensional slow feature space, the mapping function $f : \mathbb{R}^2 \mapsto \mathbb{R}^n$ maps a position to the slow feature space by processing the associated image. We use Euclidean distance as a cost function $C : \mathbb{R}^n \mapsto \mathbb{R}$, which calculates the cost from current position $p := (x_p, y_p)$ to target position $t := (x_t, y_t)$ with input $f(p)$ [15]:

$$C(f(p)) = \sqrt{\sum_{i=1}^{n}(f(p)_i - f(t)_i)^2}$$

However, we compute a local linear approximation of the gradient as the analytical gradient $\frac{\partial C(f(p))}{\partial p}$ is infeasible to obtain. Here, we use the robotic odometry for estimating the metric position of previous observations. Note that odometry for the recent past is very precise but deteriorates quickly over long distances.

3.1   Implementation

The estimation of the gradient direction at any given position requires the cost measurement at a current and at least two nearby points. These points should not be collinear to fit a plane to a cost surface. In contrast to previous work [15], where two measurements were available at each distinct position, in our current setting, only a single omnidirectional camera is mounted on a mobile robot. Therefore, it is not possible to obtain a gradient estimation directly from the starting point of navigation. For this reason, the robot traverses a fixed V-shaped initial trajectory and captures all the images during traversal. The next step is to compute the cost at each position by transforming all the captured images in the slow feature space. Using the robot's odometry information, an estimation matrix is then created that contains the coordinates and the associated cost value (x, y, C) for each visited position. This is followed by the application of Singular Value Decomposition (SVD) on the estimation matrix to compute the gradient direction. Please note that although three non-collinear points are sufficient to estimate a gradient at any given position p, we use all the intermediate points to increase the robustness of gradient estimation. The constructed estimation matrix for each segment on a navigation path is also retained as a history of past values to use some of them for estimating future gradients depending on the size of the window. The decision of choosing the window size depends on the experiment's scenario (i.e., navigation in an open field or around an obstacle). We parametrize the window size by keeping the full history for free area navigation and a subset of it for obstacle circumnavigation. Using a smaller window size for the latter case allows the robot to take sharp turns–otherwise, it would result in averaging out the gradients.

The odometry quality degrades over time, however, it is still good for local position estimation which the robot uses for motor control. After gradient estimation, the robot moves along the estimated gradient and also captures all the intermediate images. A cost value threshold in slow feature space and a maximum number of iterations serve as stopping criteria. The process is repeated until one of the conditions is met. Each estimated gradient is normalized and multiplied with a constant scaling factor η. Moreover, we used a momentum term γ, which includes information from previous gradients to improve convergence [15]. The navigation algorithm given below uses a previously learned slow feature representation f and a known target representation. (See Algorithm 1).

---

1 Navigation algorithm

1: $\gamma = 0.3$ ▷ momentum term
2: $\eta = 0.5$ ▷ gradient scaling factor
3: previous_gradient = None
4: Traverse a fixed V-shaped trajectory of length 1.96 m and capture images I
5: **do**
6:     **for** each image i in I **do**
7:         k = project omnidirectional image i to a panoramic image
8:         p = row-wise Fourier expansion of k
9:         c = compute cost $C(f(p))$
10:         x,y = retrieve camera position from odometry information
11:         append (x,y,c) to estimation matrix E
12:     **end for**
13:     g = apply SVD on E to compute direction of steepest descent
14:     normalized_gradient = normalize g
15:     **if** previous_gradient is not None **then**
16:         gradient = γ * previous_gradient - η * normalized_gradient
17:     **else**
18:         gradient = η * normalized_gradient
19:     **end if**
20:     previous_gradient = gradient
21:     Move along the gradient direction and capture images I
22:     c = compute cost $C(f(p))$ at the updated position
23: **while** c < 0.3 or number_of_iterations == 200

---

4   Experiments

Experiments were performed in an indoor environment having a training area of size 4×10 m. A V-shaped obstacle was placed inside the training area to later test the navigation performance around obstacles. We used a modified autonomous lawn mower equipped with a single omnidirectional camera for the experiments. The omnidirectional camera recorded images with a resolution of 2880×2880 pixels, which are projected to panoramic images with a resolution of 300×30 pixels. Figure 2 shows the robot, an omnidirectional image, and the associated panoramic image. For the training phase, we put the lawn mower in a free mowing mode and let it run for approximately 50 minutes and captured $15,000$ images to assure that it covers the environment uniformly. In this mode, it drives straight segments until it detects the border, where it turns into a random direction. An example training trajectory is shown in Fig. 3. After the completion of the training phase, slow features were computed directly from Fourier components of a single image. We only used the first two slow feature unit values $s_{1,2}$ for all our experiments as in [15]. The gradient scaling factor and the momentum are set to $\eta = 0.5$ and $\gamma = 0.3$, respectively for all experiments. The maximum number of iterations and cost value threshold in slow feature space are set to 200 and 0.3 as stopping criteria, respectively.

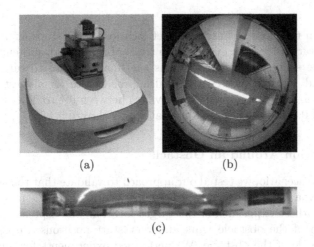

(a)   (b)

(c)

Fig. 2. Robot and Real-world environment. (a) A modified lawn mower robot with an omnidirectional camera is used for experiments. (b) An omnidirectional image. (c) The corresponding panoramic image.

4.1 Navigation in a Free Space Scenario

We start with free area navigation. For this case, we chose start and end positions such that there is no obstacle on the shortest path between them. We tested our approach for five different start and end positions with five trials each.

Results. Figure 4 shows the results of four different start and end positions. We evaluated the goal approach rate as a performance measure. The robot always

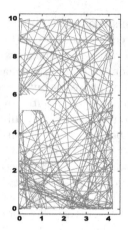

Fig. 3. An example training trajectory, which has straight line segments with random orientation. A V-shaped obstacle is present in the field.

reached within the close vicinity of target location but it stops at different distances w.r.t. target location even for the trials of same start and end positions. We varied the success criterion which is the final distance to target location between 0 and 1 m and plotted the goal approach rate for all test runs. It is shown in Fig. 5. The goal approach rate varies between 0% to 96%. For a target to distance of 0.8 m, for example, the goal approach rate is 85%.

4.2   Navigation Around an Obstacle

For the second scenario, we tested our approach to validate that the slow feature gradients allow a robot to navigate around obstacles without the need of explicit planning. For these experiments, the target location was kept fixed inside the convex region of the obstacle while different start positions are chosen from the opposite side of the obstacle. We performed experiments for four different starting positions with five trials each.

Results. The resulting 20 trajectories are shown in Fig. 6. Here, we repeated the same analysis as for the first scenario. The goal approach rate varies between 0% to 75%. For a success criterion of 0.8 m, the goal approach rate is 70%. The plot for goal approach rate is shown in Fig. 5. The performance decrease comes from the cases of the robot navigating very closely and colliding with the obstacle. In these cases, the robot hit the obstacle and never navigated to the target location. We encountered four such cases out of a total of twenty test runs. For our current analysis, we treated those cases as a failure. Another reason for failed navigation is that the robot got stuck in local minima due to flat gradients as the most variance is concentrated in the regions near to obstacle (see Fig. 1(b)).

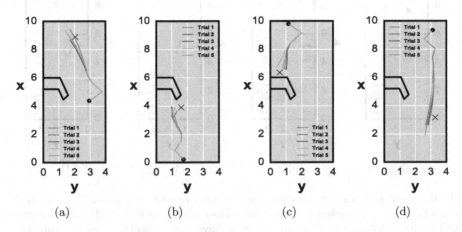

Fig. 4. Navigation in free space. The start and target positions are marked by a dot and cross, respectively. (a)–(d) Results of four different start and end positions using the first two slow feature units. A successful trial is the one that has a final distance of less than 0.8 m from a target location.

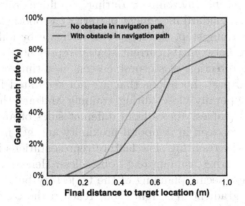

Fig. 5. Goal approach rate: The red line shows the performance tradeoff for free space navigation while the green line shows the performance tradeoff for obstacle circumnavigation. For the criterion of 0.8 m as the final distance to the target location, the robot reached the target location 85% of the trials for free space and 70% of the trials for circumnavigation.

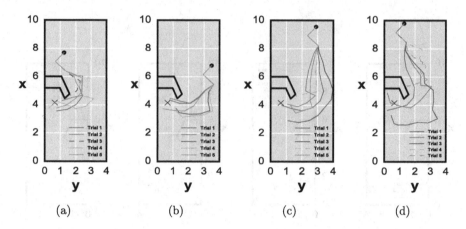

Fig. 6. Resulting trajectories for navigation around obstacles. The start and target positions are marked by a dot and a cross, respectively. (a)–(d) Four different start positions with five trials each. The robot navigated around the obstacle using the first two slow feature unit values. The trajectories for the cases where the robot hit the obstacle are indicated by dashed lines. A successful trial is the one that has a final distance of less than 0.8 m from a target location.

5   Conclusion

We presented a robot navigation system that works in slow feature space by applying gradient descent to a cost surface. The system learned spatial representations for the specific environment during an offline training phase where the robot follows the standard movement pattern of a robotic lawnmower. After the unsupervised learning step, the robot can navigate by following the gradient of the difference between its current spatial representation and the target representation. The direction of steepest descent is estimated from images in the recently traveled past. Locations that are not separated by obstacles were typically visited temporally close during training and are thus encoded with similar Slow Feature values. Views from different sides of an obstacle, on the other hand, were never seen in temporal proximity and get more different representations. Thus, the resulting slow feature representations implicitly encode average travel time during training (e.g., obstacles). Hence, navigating around obstacles requires no specific planning of the trajectory but is achieved by following the steepest gradient. The real robot reached the target in 85% of the trials in an open field scenario and in 70% of the trials when the target was behind the obstacle. Hence, we were able to reproduce the navigation results of extensive statistical experiments from noise-free simulator environment [15]. The variability in each set of trajectories is probably due to small variances in start pose and environmental changes. In some failure cases the robot got stuck in regions with flat gradients. Due to the presence of the obstacle in the training area, the resulting representations distort around it, which leads to relatively flat gradients for most regions of the training area as also depicted in the simulated experiments [15]. An advanced gradient descent algorithm may help with such issues. Further, an extension to the SFA-algorithm made in [19] presents a feasible solution to overcome this issue. Other failures occurred when the robot hit the obstacle. In the future, the robot may fall back to standard behavior in such a case, drive a short distance into a random direction and then resume navigation. Alternatively, an obstacle avoidance sensor on the robot may be used to keep a minimum distance to obstacles. Environmental changes may directly influence the spatial representations, however, we can improve them by using the strategies presented in [20,21].

The approach is not limited to obstacle avoidance by replicating the movement statistics of a training phase. If, for example, the training data is collected from human remote operation, autonomous navigation would replicate the operator's preferences for path planning.

The computational load for our proposed system is orders of magnitudes lower than of standard SLAM or Deep Learning systems, as well as those based on Slow Feature Hierarchies. Learning and application phases are suitable for small and cheap systems based on digital signal processors (DSPs). Our model is especially suited for an application where the robot has a default operation mode with random navigation during which the SFA learning can occur. Over time, the robot's efficiency and capabilities can then improve as described.

References

1. Fuentes-Pacheco, J., Ruiz-Ascencio, J., Rendón-Mancha, J.M.: Visual simultaneous localization and mapping: a survey. Artif. Intell. Rev. **43**(1), 55–81 (2015)
2. Meyer, J., Filliat, D.: Map-based navigation in mobile robots. II. A review of map-learning and path-planning strategies. Cogn. Syst. Res. 4(4), 283–317 (2003)
3. Milford, M., Wyeth, G., Prasser, D.: RatSLAM: a hippocampal model for simultaneous localization and mapping. In: Proceedings of the 2004 IEEE International Conference on Robotics and Automation, ICRA 2004, April 26 - May 1, 2004, New Orleans, LA, USA. pp. 403–408 (2004)
4. Milford, M., Wyeth, G.: Persistent navigation and mapping using a biologically inspired SLAM system. Int. J. Robot. **29**(9), 1131–1153 (2010)
5. Wyss, R., König, P., Verschure, P.F.M.J.: A model of the ventral visual system based on temporal stability and local memory. PLoS Biol **4**, e120 (2006)
6. Franzius, M., Sprekeler, H., Wiskott, L.: Slowness and sparseness lead to place, head-direction, and spatial-view cells. PLoS Comput. Biol. **3**(8), 1–18 (2007)
7. Wiskott, L., Sejnowski, T.: Slow feature analysis: unsupervised learning of invariances. Neural Comput. **14**(4), 715–770 (2002)
8. Metka, B., Franzius, M., Bauer-Wersing, U.: Outdoor self-localization of a mobile robot using slow feature analysis. In: Lee, M., Hirose, A., Hou, Z.-G., Kil, R.M. (eds.) ICONIP 2013. LNCS, vol. 8226, pp. 249–256. Springer, Heidelberg (2013). https://doi.org/10.1007/978-3-642-42054-2_32
9. Hart, P.E., Nilsson, N.J., Raphael, B.: A formal basis for the heuristic determination of minimum cost paths. IEEE Trans. Syst. Sci. Cybern. **SSC-4**(2), 100–107 (1968)

10. Khatib, O.: Real-time obstacle avoidance for manipulators and mobile robots. In: Cox, I.J., Wilfong, G.T. (eds.) Proceedings of the 1985 IEEE International Conference on Robotics and Automation, vol. 2, pp. 500–505. Springer, New York (1985). https://doi.org/10.1007/978-1-4613-8997-2_29

11. Barraquand, J., Latombe, J.: Robot motion planning: a distributed representation approach. Int. J. Robot. Res. **10**(6), 628–649 (1991)

12. Tilove, R.B.: Local obstacle avoidance for mobile robots based on the method of artificial potentials. In: Proceedings of the IEEE International Conference on Robotics and Automation, pp. 566–571, vol. 1, May 1990

13. Legenstein, R., Wilbert, N., Wiskott, L.: Reinforcement learning on slow features of high-dimensional input streams. PLoS Comput. Biol. **6**(8), 1–13 (2010)

14. Böhmer, W., Grünewälder, S., Shen, Y., Musial, M., Obermayer, K.: Construction of approximation spaces for reinforcement learning. J. Mach. Learn. Res. **14**(1), 2067–2118 (2013)

15. Metka, B., Franzius, M., Bauer-Wersing, U.: Efficient navigation using slow feature gradients. In: 2017 IEEE/RSJ International Conference on Intelligent Robots and Systems, IROS 2017, Vancouver, BC, Canada, 24–28 September 2017, pp. 1311–1316 (2017)

16. Zito, T., Wilbert, N., Wiskott, L., Berkes, P.: Modular toolkit for data processing (MDP): a python data processing framework. Front. Neuroinformatics **2**, 8 (2009)

17. Fox, D., Burgard, W., Dellaert, F., Thrun, S.: Monte Carlo localization: efficient position estimation for mobile robots. In: Proceedings of the Sixteenth National Conference on Artificial Intelligence and the Eleventh Innovative Applications of Artificial Intelligence Conference Innovative Applications of Artificial Intelligence, pp. 343–349 (1999)

18. Ishiguro, H., Tsuji, S.: Image-based memory of environment. In: Proceedings of IEEE/RSJ International Conference on Intelligent Robots and Systems, IROS 1996, 4–8 November 1996, Osaka, Japan, pp. 634–639 (1996)

19. Richthofer, S., Wiskott, L.: Global navigation using predictable and slow feature analysis in multiroom environments, path planning and other control tasks. CoRR (2018)

20. Metka, B., Franzius, M., Bauer-Wersing, U.: Improving robustness of slow feature analysis based localization using loop closure events. In: Villa, A.E.P., Masulli, P., Pons Rivero, A.J. (eds.) ICANN 2016. LNCS, vol. 9887, pp. 489–496. Springer, Cham (2016). https://doi.org/10.1007/978-3-319-44781-0_58

21. Haris, M., Metka, B., Franzius, M., Bauer-Wersing, U.: Condition invariant visual localization using slow feature analysis. In: New Challenges in Neural Computation (NC2). Machine Learning Reports, Frank-Michael Schleif, September 2017

Neurodynamics-Based Distributed Receding Horizon Trajectory Generation for Autonomous Surface Vehicles

Jiasen Wang[1,2(✉)] and Jun Wang[1,2]

[1] Department of Computer Science, City University of Hong Kong,
Kowloon, Hong Kong
jiasewang2-c@my.cityu.edu.hk,jwang.cs@cityu.edu.hk
[2] Shenzhen Research Institute, City University of Hong Kong, Shenzhen, China

Abstract. This paper presents a neurodynamics-based distributed algorithm for trajectory generation for a group of autonomous surface vehicles (ASVs). By means of convexification, the trajectory generation problem is formulated as a distributed optimization problem with affine constraints and quadratic objectives. Neurodynamic approach and receding horizon mechanism are used for solving the distributed optimization problem. Simulation results on generating trajectories for four fully-actuated and under-actuated ASVs are reported to substantiate the efficacy of the algorithm.

Keywords: Autonomous surface vehicles
Recurrent neural networks · Receding horizon planning

1 Introduction

Trajectory generation belongs to the guidance system. It is used in many applications such as autopilot (see [5]), search and rescue (see [26]), environmental monitoring (see [9]), etc. Trajectory generation is time dependent and different from path planning, which is usually time independent.

Several methods are available for trajectory generation. For example, a real-time method for the trajectory generation of constrained mechanical systems is presented in [14]. For transforming a constrained trajectory optimization problem into unconstrained one, a barrier function method is proposed in [6]. A nonlinear model predictive control (MPC) framework is presented in [17] for real-time and unconstrained trajectory optimization and tracking control. An MPC-based approach for trajectory generation of autonomous vehicles is presented in [18].

This work was supported in part by the Research Grants Council of the Hong Kong Special Administrative Region of China, under Grants 14207614 and 11208517, and in part by the National Natural Science Foundation of China under grant 61673330.

© Springer Nature Switzerland AG 2018
L. Cheng et al. (Eds.): ICONIP 2018, LNCS 11307, pp. 155–167, 2018.
https://doi.org/10.1007/978-3-030-04239-4_14

Several results are available for trajectory generation of ASVs. For example, a method based on nonlinear sliding mode control for combined trajectory planning and tracking is presented in [21] for ASVs. A trajectory planning method for ASVs in congested traffic area is presented in [19]. A trajectory generation method for communication constrained surface vehicles is presented in [9].

Collision avoidance is a critical issue for trajectory generation for ASVs. An MPC-based collision avoidance system is presented in [10]. The system can avoid multiple dynamic obstacles with sensor or prediction uncertainties. A mid-level collision avoidance method for ASVs based on MPC is presented in [4] for avoiding fixed or moving obstacles.

A distributed system for ASVs is usually more robust than centralized one. Well-designed distributed systems can tolerate part of failures of ASVs. Currently, more attention are paid on distributed control systems of ASVs. There are few works on distributed trajectory generation of ASVs. A cooperative trajectory generation method for ASVs is presented in [8] with emphasis on the ability to handle various constraints and objectives of the algorithm. However, some results are available for general systems and unmanned aerial vehicles (UAVs). For example, for linear systems constrained by interacting and local constraints, a distributed command governor method is proposed in [22]. For distributed trajectory generation for UAVs, an MPC-based method is presented in [16].

In this paper, a neurodynamics-based distributed receding horizon algorithm is presented for the trajectory generation for ASVs. The receding-horizon mechanism is used to reduce the optimization problem size. The distribution of the algorithm allows ASVs to optimize cooperatively and reduces risk of part of failures of ASVs. As pointed in [7], path planning or trajectory generation algorithms for ASVs need to be "lightweight in terms of running time" and capable of handling various constraints. Neurodynamic approaches are equipped with these two properties. For example, neurodynamic approaches are used in [25] for tracking of ASVs. Thus neurodynamic approaches are promising for solving the trajectory generation problems in real-time.

2   Problem Formulation

The problem is to generate collision-free trajectories for N ASVs in a distributed setting. The problem is formulated as decentralized sub-problems for each ASV.

2.1   ASV Dynamics

Let $\eta = [x, y, \psi]^T \in \mathbb{R}^3$ and $\nu = [u, v, r]^T \in \mathbb{R}^3$ denote kinematics vector and kinetics vector, respectively. x and y are coordinates in the earth-fixed reference frame; ψ is the angle of body-fixed reference frame rotated from earth-fixed reference frame in the anticlockwise orientation; u, v, and r denote surge, sway, and yaw velocities in the body-fixed reference frame, respectively. From [5], the kinematics and kinetics equations for an ASV model are

$$\dot{\eta} = R(\psi)\nu,$$
$$M\dot{\nu} = \tau - C(\nu)\nu - D(\nu)\nu, \tag{1}$$

where

$$R(\psi) = \begin{bmatrix} \cos(\psi) & -\sin(\psi) & 0 \\ \sin(\psi) & \cos(\psi) & 0 \\ 0 & 0 & 1 \end{bmatrix},$$

$M^T = M \in \mathbb{R}^{3\times3}$ is an inertial matrix; $C(\nu) = -C(\nu)^T \in \mathbb{R}^{3\times3}$ denotes a Coriolis and centripetal matrix; $D(\nu) \in \mathbb{R}^{3\times3}$ is a nonlinear damping matrix; $\tau = [\tau_u, \tau_v, \tau_r]^T \in \mathbb{R}^3$ is a vector of control input. If the ASV model has an under-actuated configuration, then $\tau_v \equiv 0$; i.e., it is not permitted to directly control the sway velocity. In the sequel, τ is always used as a three dimension vector for convenience. For an under-actuated ASV, this is achieved by setting $\tau = [\tau_u, 0, \tau_r]^T$ manually.

2.2   Nonconvex Constraints

Before going to formulate the optimization problems, the nonconvex constraints are handled in this subsection. In this problem, the nonconvex constraints are: (1) System dynamic equations, and (2) Collision avoidance constraints.

System Dynamic Equations: Suppose that there are N ASVs. Rewrite dynamic equation (1) for the ith ASV as

$$\dot{\eta}_i = R_i(\psi)\nu_i,$$
$$\dot{\nu}_i = M_i^{-1}[-C_i(\nu_i)\nu_i - D_i(\nu_i)\nu_i] + M_i^{-1}\tau_i.$$

Let $\mathrm{x}_i = \mathrm{col}(\eta_i, \nu_i) \in \mathbb{R}^6$ (col is a column concatenation operator), then the above equation becomes

$$\dot{\mathrm{x}}_i = f_i(\mathrm{x}_i)\mathrm{x}_i + g_i(\mathrm{x}_i)\tau_i, \tag{2}$$

where

$$f_i = \begin{bmatrix} 0_{3\times3} & R_i(\psi_i) \\ 0_{3\times3} & -M_i^{-1}(C_i + D_i) \end{bmatrix}, \tag{3}$$

$g_i(\mathrm{x}_i) = [0_{n\times n}, I_n]^T M_i^{-1}$, $0_{n\times n}$ is a zero matrix with dimension $n \times n$, I_n is a identity matrix of dimension n. A discrete time approximation of (2) is

$$\mathrm{x}_i(t+1) = (I_6 + T_s f_i(\mathrm{x}_i(t)))\mathrm{x}_i(t) + T_s g_i(\mathrm{x}_i(t))\tau_i + T_s \alpha_i(\mathrm{x}_i(t)), \tag{4}$$

where T_s is a sampling period, α_i is a term used to rectify the error caused by discretization. If α_i is set to zero vector, (4) reduces to the Euler discretization equation. A quasi-linearization (see [2]) of (4) is

$$x_i(t+1) = A_i(t)x_i(t) + B_i(t)\tau_i(t) + \varepsilon_i(t), \tag{5}$$

where $A_i(t) = I_6 + T_s f_i(\hat{x}_i)$, $B_i(t) = T_s g_i(\hat{x}_i)$, $\varepsilon_i(t) = T_s \alpha_i(\hat{x}_i)$ are coefficients w.r.t a nominal trajectory $\hat{x}_i$. Let

$$\bar{\tau}_i = \mathrm{col}(\tau_i(0), ..., \tau_i(T-1)), \quad \bar{x}_i = \mathrm{col}(x_i(1), ..., x_i(T)),$$

where T (> 1) is an integer time horizon. Note that when used in receding horizon optimization, $\tau_i(0)$ should be understood as the control of current time instant k. It is obtained from (5) that the relation between $\bar{x}_i$ and $\bar{\tau}_i$ is

$$\bar{x}_i = \bar{A}_i \bar{\tau}_i + \bar{b}_i + \bar{\varepsilon}_i, \tag{6}$$

where

$$\bar{A}_i = \begin{bmatrix} B_i(0) & 0 & \cdots & 0 \\ A_i(1)B_i(0) & B_i(1) & \cdots & 0 \\ \vdots & \vdots & \ddots & \vdots \\ \prod_{j=T-1}^{1} A_i(j)B_i(0) & \prod_{j=T-1}^{2} A_i(j)B_i(1) & \cdots & B_i(T-1) \end{bmatrix},$$

$$\bar{b}_i = \begin{bmatrix} A_i(0)x_i(0) \\ A_i(1)A_i(0)x_i(0) \\ \vdots \\ \prod_{j=T-1}^{0} A_i(j)x_i(0) \end{bmatrix}, \quad \bar{\varepsilon}_i = \begin{bmatrix} \varepsilon_i(0) \\ A_i(1)\varepsilon_i(0) + \varepsilon_i(1) \\ \vdots \\ \prod_{j=T-1}^{1} A_i(j)\varepsilon_i(0) + ... + \varepsilon_i(T-1) \end{bmatrix}.$$

Thus the ASV dynamic equation is formulated as an affine constraint (6).

Collision Avoidance Constraints: There are various collision avoidance constraints. Unfortunately, it seems that all of them are nonconvex. In this paper, the collision avoidance constraint (see [1,15]) specified for the ith ASV is as follows:

$$\|G(x_i(t) - x_j(t))\|_2 \geq R_c, \quad j \in \mathcal{P}_i \cap \mathcal{N}_i, \tag{7}$$

where R_c is a pre-set safe distance to avoid collisions, G is a weight matrix, $\mathcal{P}_i$ is a pre-selected set containing the ASVs with higher priority than the ith ASV, $\mathcal{N}_i$ is a set containing those ASVs within the communication scope of the ith ASV. Note that (7) also applies to fixed or moving obstacles by seen $x_j(t)$ as their states. As shown in [15], (7) is ensured if the following affine constraint is satisfied:

$$(\hat{x}_i(t) - \hat{x}_j(t))^T G^T G(x_i(t) - \hat{x}_j(t)) \geq R_c \|G(\hat{x}_i(t) - \hat{x}_j(t))\|_2,$$

where $\hat{x}_i$, $\hat{x}_j$ are some given nominal trajectories. With variable substitutions, above equation is expressed as

$$\hat{a}_{ij}^T(t)G^T G \hat{x}_j(t) + R_c \|G\hat{a}_{ij}(t)\|_2 \leq \hat{a}_{ij}^T(t)G^T G x_i(t), \tag{8}$$

where $\hat{a}_{ij}(t) = \hat{\mathbf{x}}_i(t) - \hat{\mathbf{x}}_j(t)$. Let $\hat{G}_{ij} = \mathrm{diag}(\hat{a}_{ij}^T(1)G^T G, ..., \hat{a}_{ij}^T(T)G^T G)$, $r_{ij}(t) = \hat{a}_{ij}^T(t)G^T G\hat{\mathbf{x}}_j(t) + R_c\|G\hat{a}_{ij}(t)\|_2$, $\hat{r}_{ij} = \mathrm{col}(r_{ij}(1), ..., r_{ij}(T))$, Eq. (8) becomes

$$\hat{r}_{ij} - \hat{G}_{ij}(\bar{b}_i + \bar{\varepsilon}_i) \le \hat{G}_{ij}\bar{A}_i\bar{\tau}_i. \tag{9}$$

An augmented collision avoidance constraint based on (9) is obtained as follows:

$$\hat{r}_i - \hat{G}_i(\bar{b}_i + \bar{\varepsilon}_i) \le \hat{G}_i\bar{A}_i\bar{\tau}_i, \tag{10}$$

where $\hat{r}_i = \mathrm{col}(\hat{r}_{i1}, ..., \hat{r}_{iS})$, $\hat{G}_i = \mathrm{col}(\hat{G}_{i1}, ..., \hat{G}_{iS})$, S denotes the number of elements in $\mathcal{P}_i \cap \mathcal{N}_i$. Thus the collision avoidance constraint (7) is formulated as constraint (10).

2.3 Optimization Problems

Objectives: The objective for N ASVs is as follows:

$$\min J(k) = \sum_{i=1}^{N} J_i(k), \tag{11}$$

where $J_i(k)$ is the objective function of the ith ASV. A choice of $J_i(k)$ (see [15]) is

$$J_i(k) = \sum_{t=0}^{T-1} \|\tau_i(t)\|_R^2, \tag{12}$$

where $\|a\|_R = \sqrt{a^T Ra}$, R is a symmetric and positive definite matrix. This objective is a measure of control effort within time horizon T. Similar to [25], another choice of $J_i(k)$ is

$$J_i(k) = \sum_{t=0}^{T-1} \|\tau_i(t)\|_R^2 + \sum_{t=1}^{T} \|\mathbf{x}_i(t) - \mathbf{x}_{if}\|_W^2, \tag{13}$$

where $\mathbf{x}_{if}$ is a terminal state, W is a symmetric and positive definite weight matrix for terminal cost. This objective is a combination of control effort and reaching terminal state within time horizon T.

Constraints: Apart from ASV dynamic equation and collision avoidance constraint, some other constraints are also needed.

The initial condition is

$$\mathbf{x}_i(0) = \mathbf{x}_{ik}, \tag{14}$$

where $\mathbf{x}_{ik}$ is the system state at time instant k.

The terminal condition is

$$\mathbf{x}_i(T) = \mathbf{x}_{if}. \tag{15}$$

A commonly used velocity constraint is

$$\nu_{\min} \le \nu_i(t) \le \nu_{\max}, \tag{16}$$

where ν_{min} and ν_{max} are the lower and upper bounds of velocities.

The constraints of the control input and its increment are

$$\tau_{min} \leq \tau_i(t) \leq \tau_{max}, \tag{17}$$

$$\varDelta\tau_{min} \leq \varDelta\tau_i(t) \leq \varDelta\tau_{max}, \tag{18}$$

where τ_{min} and τ_{max} are the lower and upper bounds of the control input; $\varDelta\tau_i(t) = \tau_i(t+1) - \tau_i(t)$, $\varDelta\tau_{min}$ and $\varDelta\tau_{max}$ are the lower and upper bounds for the increment of the control input.

Augmented Constraints: Constraints in augmented vector forms are needed for further usage.

Using (6), the terminal equality constraint (15) becomes

$$\bar{A}_{il}\bar{\tau}_i + \bar{b}_{il} + \bar{\varepsilon}_{il} = x_{if}, \tag{19}$$

where $\bar{A}_{il}$, $\bar{b}_{il}$, $\bar{\varepsilon}_{il}$ denote the last six rows of $\bar{A}_i$, $\bar{b}_i$, and $\bar{\varepsilon}_i$, respectively.

The velocity vector ν_i can be written as $\nu_i(t) = Ux_i$, $U = [0_{3\times3}, I_3]$. Let $\bar{U} = \text{diag}(U, ..., U)$, $\bar{\nu}_{min} = \text{col}(\nu_{min}, ..., \nu_{min}) \in \mathbb{R}^{3T}$, $\bar{\nu}_{max} = \text{col}(\nu_{max}, ..., \nu_{max}) \in \mathbb{R}^{3T}$, the constraint (16) becomes

$$\bar{\nu}_{min} - \bar{U}(\bar{b}_i + \bar{\varepsilon}_i) \leq \bar{U}\bar{A}_i\bar{\tau}_i \leq \bar{\nu}_{max} - \bar{U}(\bar{b}_i + \bar{\varepsilon}_i), \tag{20}$$

The input constraint (17) in augmented vector form is

$$\bar{\tau}_{min} \leq \bar{\tau}_i \leq \bar{\tau}_{max}, \tag{21}$$

where $\bar{\tau}_{min} = \text{col}(\tau_{min}, ..., \tau_{min}) \in \mathbb{R}^{3T}$, $\bar{\tau}_{max}$ is defined similarly.

The input increment constraint (18) in augmented form is

$$\varDelta\bar{\tau}_{min} \leq \bar{P}\bar{\tau}_i \leq \varDelta\bar{\tau}_{max}, \tag{22}$$

where $\varDelta\bar{\tau}_{min} = \text{col}(\varDelta\tau_{min}, ..., \varDelta\tau_{min}) \in \mathbb{R}^{3T}$, $\varDelta\bar{\tau}_{max}$ is defined similarly,

$$\bar{P} = \begin{bmatrix} I_3 & 0 & 0 & \cdots \\ -I_3 & I_3 & 0 & \cdots \\ 0 & -I_3 & I_3 & \cdots \\ \vdots & \cdots & \ddots & \cdots \\ 0 & \cdots & -I_3 & I_3 \end{bmatrix} \in R^{3(T-1)\times3T}. \tag{23}$$

Optimizations Problems: An optimization problem with objective function (12) for the ith ASV is

$$\min J_i(k) = \sum_{t=0}^{T-1} \|\tau_i(t)\|_R^2 = \bar{\tau}_i^T \bar{R}\bar{\tau}_i, \text{ s.t. (14), (19), (10), (20), (21),} \tag{24}$$

where $\bar{R} = \text{diag}(R, ..., R) \in \mathbb{R}^{3T \times 3T}$. Similar to [16], a terminal equality constraint (19) is used in (24). It enforces ASVs to terminate at their targets. However, it leads to that the optimization horizon should be chosen sufficiently long to ensure feasibility of the optimization problem. Otherwise the optimization problem may become infeasible. For an under-actuated ASV, infeasibility of the optimization may lead to the ASV to whirl on the surface (see [3]). To solve this problem, another optimization problem with objective function (13) is formulated as follows:

$$
\begin{aligned}
\min\ J_i(k) &= \sum_{t=0}^{T-1} \|\tau_i(t)\|_R^2 + \sum_{t=1}^{T} \|\mathbf{x}_i(t) - \mathbf{x}_{if}\|_W^2 \\
&= \bar{\tau}_i^T (\bar{A}_i^T \bar{W} \bar{A}_i + \bar{R})\bar{\tau}_i + 2(\bar{W}\bar{A}_i)^T(\bar{b}_i + \bar{\varepsilon}_i - \bar{x}_{if})\bar{\tau}_i,
\end{aligned}
$$
$$\text{s.t. (14), (10), (20), (21),} \tag{25}$$

where $\bar{x}_{if} = \text{col}(x_{if}, ..., x_{if}) \in \mathbb{R}^{6T}$, $\bar{W} = \text{diag}(W, ..., W) \in \mathbb{R}^{6T \times 6T}$. Unlike problem (24), there is no terminal equality constraint in problem (25). Instead, an objective function with a terminal cost is used in problem (25) to guide the ASVs to their targets. So the possibility of being feasible of the optimization problem (25) is increased. To take advantages of problems (24) and (25), a switching mechanism can be used as follows: when the ASVs are far away from their targets, problem (25) is adopted to keep feasible of the optimization; when the ASVs are close to their targets, optimization switches to problem (24) to enforce ASVs to terminate at their targets.

Quadratic Optimization Problems: Problems (24) and (25) are rewritten in the following form:

$$\min\ (1/2)\bar{\tau}_i^T Q_i \bar{\tau}_i + c_i^T \bar{\tau}_i, \quad \text{s.t.}\ l_i \le E_i \bar{\tau}_i \le h_i, \tag{26}$$

where for (24), $Q_i = 2\bar{R}$, $c_i = 0_T$, $E_i = \text{col}(\hat{G}_i \bar{A}_i, \bar{U}\bar{A}_i, I_T, \bar{A}_{il}, \bar{P})$, $l_i = \text{col}(\hat{r}_i - \hat{G}_i(\bar{b}_i + \bar{\varepsilon}_i), \bar{\nu}_{\min,i} - \bar{U}(\bar{b}_i + \bar{\varepsilon}_i), \bar{\tau}_{\min}, \mathbf{x}_{if} - \bar{b}_{il} - \bar{\varepsilon}_{il}, \Delta\bar{\tau}_{\min})$, $h_i = \text{col}(\infty, \bar{\nu}_{\max,i} - \bar{U}(\bar{b}_i + \bar{\varepsilon}_i), \bar{\tau}_{\max}, \mathbf{x}_{if} - \bar{b}_{il} - \bar{\varepsilon}_{il}, \Delta\bar{\tau}_{\max})$; for (25), $Q_i = 2(\bar{A}_i^T \bar{W} \bar{A}_i + \bar{R})$, $c_i = 2(\bar{W}\bar{A}_i)^T(\bar{b}_i + \bar{\varepsilon}_i - \bar{x}_{if})$, $E_i = \text{col}(\hat{G}_i \bar{A}_i, \bar{U}\bar{A}_i, I_T, \bar{P})$, $l_i = \text{col}(\hat{r}_i - \hat{G}_i(\bar{b}_i + \bar{\varepsilon}_i), \bar{\nu}_{\min,i} - \bar{U}(\bar{b}_i + \bar{\varepsilon}_i), \bar{\tau}_{\min}, \Delta\bar{\tau}_{\min})$, $h_i = \text{col}(\infty, \bar{\nu}_{\max,i} - \bar{U}(\bar{b}_i + \bar{\varepsilon}_i), \bar{\tau}_{\max}, \Delta\bar{\tau}_{\max})$.

3 Trajectory Generation

3.1 Neurodynamic Optimization

Several neurodynamic approaches are available for solving (26); see e.g., [11–13]. In particular, a recurrent neural network (RNN) model is proposed in [24]. For the ith ASV, its dynamic and output equations are respectively

$$
\begin{aligned}
\epsilon\frac{dz_i}{dt} &= -W_i z_i + P_{\mathcal{Z}_i}(W_i z_i - z_i + q_i) - q_i, \\
\bar{\tau}_i &= Q_i^{-1}(E_i^T z_i - c_i),
\end{aligned}
\tag{27}
$$

Algorithm 1. Neurodynamics-based receding horizon trajectory generation algorithm for the ith ASV.

Input: K, K', T, $T' \leq T$, x_{i0}, x_{if}, $\mathcal{Q}_i$, $\mathcal{P}_i$;
Output: $x_i(t)$, $t = 0, ..., K$;
1: SwitchingFlag$_i \leftarrow 0$; // store switching information;
2: $k \leftarrow 0$; // k is the current time instant;
3: $x_i(t) \leftarrow x_{i0}$, $t = 0, ..., K$; // initialize memory;
4: **while** $k \leq K$ **do** //this "while" is the receding horizon procedure;
5:     $x_i^*(t) \leftarrow x_i(t)$, $t = k, ..., k + T - T'$;
6:     $x_i^*(t) \leftarrow x_i(k + T - T')$, $t = k + T - T' + 1, ..., k + T$;
7:     Detect neighboring ASVs and save them to $\mathcal{N}_i$;
8:     **for** LoopID $= 1$ **to** K' **do** // Run this "for loop" in parallel;
9:         Send $x_i^*(t)$ $(t = k, ..., k + T)$ to ASVs in $\mathcal{N}_i \cap \mathcal{Q}_i$;
10:        Receive $x_j^*(t)$, $t = k, ..., k + T$, $j \in \mathcal{N}_i \cap \mathcal{P}_i$;
11:        // above "Receive" should synchronize LoopID;
12:        $\hat{x}_j(t) \leftarrow x_j^*(t)$, $t = k, ..., k + T$, $j \in \mathcal{N}_i \cap \mathcal{P}_i$;
13:        Compute SwitchingFlag$_i$;
14:        **if** SwitchingFlag$_i = 1$ **then**
15:            Solve (24) via (27) to get $\bar{\tau}_i^*$;
16:        **else**
17:            Solve (25) via (27) to get $\bar{\tau}_i^*$;
18:        **end if**
19:        Reset $x_i^*(t)$, $t = k, ..., k + T$ using $\bar{\tau}_i^*$ and (6);
20:    **end for**
21:    $x_i(t) \leftarrow x_i^*(t)$, $t = k, ..., k + T$;
22:    $k \leftarrow k + T'$; // move T' time steps forward
23: **end while**

---

where ϵ is a time constant, $W_i = E_i Q_i^{-1} E_i^T$, $q_i = -E_i Q_i^{-1} c_i$, $\mathcal{Z}_i = \{z \in \mathbb{R}^n | l_i \leq z \leq h_i\}$, n is the dimension of l_i, the activation function $P_{\mathcal{Z}_i}(z_i) = [P_{\mathcal{Z}_i}(z_{i0}), ..., P_{\mathcal{Z}_i}(z_{in})]^T$, z_{ij} is the jth element of z_i,

$$P_{\mathcal{Z}_i}(z_{ij}) = \begin{cases} h_{ij} & z_{ij} > h_{ij}, \\ z_{ij} & l_{ij} \leq z_{ij} \leq h_{ij}, \\ l_{ij} & z_{ij} < l_{ij}, \end{cases}$$

l_{ij} and h_{ij} are the jth elements of l_i and h_i, respectively. As analyzed in [24], the output of the RNN is globally convergent to the optimal solution of (26) provided that Q_i is symmetric and positive definite.

3.2   A Distributed Algorithm

Based on the works in [16, 22], the pseudocode of a neurodynamics-based receding horizon algorithm deployed in the ith ASV is proposed in Alg. 1. In Alg. 1, $\mathcal{Q}_i$ is a set containing all the ASVs having lower priority than the ith ASV. In line 11, "synchronize LoopID" means that current ASV only receive information of those ASVs whose "LoopIDs" have the same value as its. This requires that in communication, "LoopID" should be sent also. In line 13, "Compute SwitchingFlag$_i$" could be done by checking whether the current state is close enough to the target state: if the distance is smaller than some preset bound, then "SwitchingFlag$_i$ =1", otherwise, "SwitchingFlag$_i$ =0". Note that in Alg. 1, no communication is needed in the RNN computation procedure. This renders the RNNs possibility to compute rapidly without being influenced by wireless communication bottleneck.

4   Simulation Results

Consider a problem of generating trajectories for four ASVs. As in [20], the parameters used for ASV model are

$$
M = \begin{bmatrix} 25.8 & 0 & 0 \\ 0 & 33.8 & 1.0948 \\ 0 & 1.0948 & 2.76 \end{bmatrix}, \quad C = \begin{bmatrix} 0 & 0 & c_{13} \\ 0 & 0 & 25.8u \\ -c_{13} & -25.8u & 0 \end{bmatrix}, \quad D = \begin{bmatrix} d_{11} & 0 & 0 \\ 0 & d_{22} & d_{23} \\ 0 & d_{32} & d_{33} \end{bmatrix},
$$

where $c_{13} = -33.8v - 1.0948r$, $d_{11} = 0.72 + 1.33|u| + 5.87u^2$, $d_{22} = 0.8896 + 36.5|v| + 0.805|r|$, $d_{23} = 7.25 + 0.845|v| + 3.45|r|$, $d_{32} = 0.0313 + 3.96|v| + 0.13|r|$, $d_{33} = 1.90 - 0.08|v| + 0.75|r|$. Let $x_{10} = [1, 1, 0.78, 0, 0, 0]^T$, $x_{20} = [2.5, 1, 2.36, 0, 0, 0]^T$, $x_{30} = [5, 1, 2.36, 0, 0, 0]^T$, $x_{40} = [4, 0, 0.78, 0, 0, 0]^T$ be the initial states for the four ASVs. Let $x_{f1} = [5, 5, 0.78, 0, 0, 0]^T$, $x_{f2} = [-2, 5.5, 2.36, 0, 0, 0]^T$, $x_{f3} = [1, 5, 2.36, 0, 0, 0]^T$, $x_{f4} = [8.2, 4.7, 0.78, 0, 0, 0]^T$ be four target states. Suppose that after using neurodynamics-based method in [23], the assigned target states for four ASVs are x_{f1}, x_{f2}, x_{f3}, and x_{f4}, respectively. Suppose that there are two fixed location obstacles in the X-Y plane with coordinates $(2, 3)$ and $(4.5, 2.5)$, respectively. Other parameters in the simulations are $R = \text{diag}(1, 1, 1)$, $W = \text{diag}(10, 10, 800, 1, 1, 1)$, $R_c = 1$, $\nu_{max} = -\nu_{min} = \text{col}(1, 1, 1)$, $\tau_{max} = -\tau_{min} = \text{col}(10, 10, 10)$, $T_s = 0.1$, $T' = 1$, $T = 50$, $K' = 2$, $K = 200$ ($K = 250$) for fully-actuated (under-actuated) ASVs, $\epsilon = 10^{-6}$. The effective communication distance for two ASVs is set to five

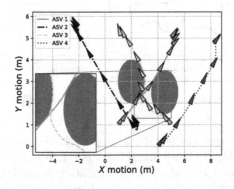

Fig. 1. Generated trajectories of four fully-actuated ASVs.

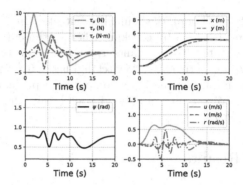

Fig. 2. Control and controlled variables for the first fully-actuated ASV.

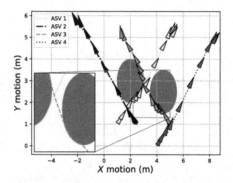

Fig. 3. Generated trajectories of four under-actuated ASVs.

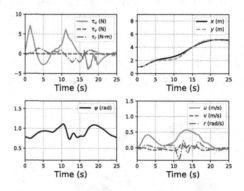

Fig. 4. Control and controlled variables for the first under-actuated ASV.

meters. The priority relation is $1 > 2 > 3 > 4$; i.e., the first ASV has the highest priority. In Alg. 1, "SwitchingFlag$_i$" is set to one if the Euclidean distance between current location and target location is smaller than 1.1 meters; otherwise, "SwitchingFlag$_i$" is set to zero. Figures 1 and 2 depict the generated trajectories for fully-actuated ASVs. Figures 3 and 4 depict the generated

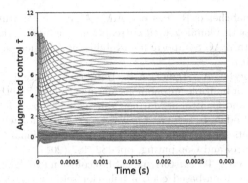

Fig. 5. A snapshot of the output convergence of the RNN deployed in the first under-actuated ASV.

trajectories for under-actuated ASVs. From Figs. 1 and 3, it is noted that there is no collision between ASVs and obstacles. From Figs. 2 and 4, it is observed that the under-actuated ASV needs more time than the fully-actuated ASV to reach its target. Figure 5 depicts the output convergent behaviors of the RNN in a snapshot.

5   Conclusion

In this paper, a distributed algorithm is presented for generating collision-free trajectories for a group of ASVs based on neurodynamic approach and receding-horizon mechanism. The receding-horizon mechanism can reduce the problem size, whereas neurodynamic optimization is suitable for real-time optimization. Simulation results for four ASVs with under-actuated and fully-actuated configurations are discussed to substantiate the efficacy of the algorithm.

References

1. Beard, R.W., Lawton, J., Hadaegh, F.Y.: A feedback architecture for formation control. In: Proceedings of American Control Conference, vol. 6, pp. 4087–4091 (2000)
2. Boyd, S.: Sequential convex programming (2018). http://stanford.edu/class/ee364b/
3. Do, K.D., Jiang, Z.P., Pan, J.: Underactuated ship global tracking under relaxed conditions. IEEE Trans. Autom. Control **47**(9), 1529–1536 (2002)
4. Eriksen, B.H., Breivik, M.: MPC-based mid-level collision avoidance for ASVs using nonlinear programming. In: IEEE Conference on Control Technology and Applications, pp. 766–772 (2017)
5. Fossen, T.I.: Handbook of Marine Craft Hydrodynamics and Motion Control. Wiley, New York (2011)
6. Hauser, J., Saccon, A.: A barrier function method for the optimization of trajectory functionals with constraints. In: Proceeding of 45th IEEE Conference on Decision and Control, pp. 864–869 (2006)

7. Häusler, A.J., Ghabcheloo, R., Pascoal, A.M., Aguiar, A.P.: Multiple marine vehicle deconflicted path planning with currents and communication constraints. In: Proceedings of 7th IFAC Symposium on Intelligent Autonomous Vehicles, vol. 43, pp. 491–496 (2010)
8. Häusler, A.J., Saccon, A., Aguiar, A.P., Hauser, J., Pascoal, A.M.: Cooperative motion planning for multiple autonomous marine vehicles. In: 9th IFAC Conference on Manoeuvring and Control of Marine Craft, vol. 45, pp. 244–249 (2012)
9. Hervagault, Y., Prodan, I., Lefevre, L.: Trajectory generation with communication-induced constraints for surface vehicles. In: 21st International Conference on System Theory, Control and Computing, pp. 482–487 (2017)
10. Johansen, T.A., Perez, T., Cristofaro, A.: Ship collision avoidance and colregs compliance using simulation-based control behavior selection with predictive hazard assessment. IEEE Trans. Intell. Transp. Syst. **17**(12), 3407–3422 (2016)
11. Li, G., Yan, Z., Wang, J.: A one-layer recurrent neural network for constrained nonconvex optimization. Neural Netw. **61**, 10–21 (2015)
12. Liu, Q., Wang, J.: A one-layer recurrent neural network with a discontinuous hard-limiting activation function for quadratic programming. IEEE Trans. Neural Netw. **19**(4), 558–570 (2008)
13. Liu, Q., Wang, J.: A one-layer recurrent neural network for constrained nonsmooth optimization. IEEE Trans. Syst. Man Cyber. Part B (Cybern.) **41**(5), 1323–1333 (2011)
14. Milam, M.B., Mushambi, K., Murray, R.M.: A new computational approach to real-time trajectory generation for constrained mechanical systems. In: Proceeding of 39th IEEE Conference on Decision and Control, vol. 1, pp. 845–851. IEEE (2000)
15. Morgan, D., Chung, S.J., Hadaegh, F.Y.: Model predictive control of swarms of spacecraft using sequential convex programming. J. Guid. Control. Dyn. **37**(6), 1725–1740 (2014)
16. Morgan, D., Subramanian, G.P., Chung, S.J., Hadaegh, F.Y.: Swarm assignment and trajectory optimization using variable-swarm, distributed auction assignment and sequential convex programming. Int. J. Robot. Res. **35**(10), 1261–1285 (2016)
17. Neunert, M., et al.: Fast nonlinear model predictive control for unified trajectory optimization and tracking. In: IEEE International Conference on Robotics and Automation, pp. 1398–1404 (2016)
18. Nolte, M., Rose, M., Stolte, T., Maurer, M.: Model predictive control based trajectory generation for autonomous vehicles-an architectural approach. In: IEEE Intelligent Vehicles Symposium, pp. 798–805 (2017)
19. Shah, B.C., et al.: Trajectory planning with adaptive control primitives for autonomous surface vehicles operating in congested civilian traffic. In: IEEE/RSJ International Conference on Intelligent Robots and Systems, pp. 2312–2318 (2014)
20. Skjetne, R., Fossen, T.I., Kokotović, P.V.: Adaptive maneuvering, with experiments, for a model ship in a marine control laboratory. Automatica **41**(2), 289–298 (2005)
21. Soltan, R.A., Ashrafiuon, H., Muske, K.R.: State-dependent trajectory planning and tracking control of unmanned surface vessels. In: American Control Conference, pp. 3597–3602 (2009)
22. Tedesco, F., Raimondo, D.M., Casavola, A., Lygeros, J.: Distributed collision avoidance for interacting vehicles: a command governor approach. In: 2nd IFAC Workshop on Distributed Estimation and Control in Networked Systems, vol. 43, pp. 293–298 (2010)

23. Wang, J., Wang, J.: Task assignment based on a dual neural network. In: Huang, T., Lv, J., Sun, C., Tuzikov, A.V. (eds.) ISNN 2018. LNCS, vol. 10878, pp. 686–694. Springer, Cham (2018). https://doi.org/10.1007/978-3-319-92537-0_78
24. Xia, Y., Feng, G., Wang, J.: A recurrent neural network with exponential convergence for solving convex quadratic program and related linear piecewise equations. Neural Netw. **17**(7), 1003–1015 (2004)
25. Yan, Z., Wang, J.: Model predictive control for tracking of underactuated vessels based on recurrent neural networks. IEEE J. Ocean. Eng. **37**(4), 717–726 (2012)
26. Zhao, W., Meng, Q., Chung, P.W.: A heuristic distributed task allocation method for multivehicle multitask problems and its application to search and rescue scenario. IEEE Trans. Cybern. **46**(4), 902–915 (2016)

Adaptive Finite-Time Synchronization of Inertial Neural Networks with Time-Varying Delays via Intermittent Control

Lin Cheng, Yongqing Yang$^{(\boxtimes)}$, Xianyun Xu, and Xin Sui

School of Science, Wuxi Engineering Research Center for Biocomputing,
Jiangnan University, Wuxi 214122, People's Republic of China
yongqingyang@163.com

Abstract. In this paper, the adaptive finite-time synchronization is investigated for inertial neural networks with time-varying delays. The second-order inertial systems can be transformed into two first-order differential systems by selecting the appropriate variable substitution. Using the adaptive periodically intermittent controllers, the slave system can realize synchronization with the master system in finite time. By the several differential inequalities and finite-time stability theory, some simple finite-time synchronization criteria for an array of inertial neural networks are derived. A numerical example is finally provided to illustrate the effectiveness of the obtained theoretical results.

Keywords: Finite-time synchronization · Inertial neural networks
Time-varying delays · Adaptive intermittent control

1 Introduction

Over the last few decades, periodic oscillation, chaotic behaviors, and stability analysis for neural neural network has aroused the discussion and research of many scholars. Meanwhile, neural networks play a significant role in different areas, since neural networks can be applied to image processing, combinatorial optimization, secure communication, and pattern recognition [1–6]. Synchronization means agreement or correlation of different processes in time. Among many dynamical behaviors of neural networks, synchronization is one of the most significance ones that has aroused widespread attentions of many researchers. Research on synchronization phenomena has been an active subject, such as the analysis of synchronization of chaotic system. So far, there are many different types of synchronization, for example, projective synchronization [7], lag synchronization [8], cluster synchronization [9], complete synchronization [10], phase synchronization [11] etc.

Y. Yang—This work was supported by the Natural Science Foundation of Jiangsu Province of China under Grant No. BK20170171, BK20161126.

© Springer Nature Switzerland AG 2018
L. Cheng et al. (Eds.): ICONIP 2018, LNCS 11307, pp. 168–179, 2018.
https://doi.org/10.1007/978-3-030-04239-4_15

Most previous literature has mainly been devoted to the stability analysis and periodic oscillations of different kinds of neural networks. The problem about delayed Hopfield neural networks with global exponential stability has been discussed in [12]. The authors in [13] investigated exponential stability for stochastic BAM networks with discrete and distributed delays. Exponential stability of complex-valued memristor-based neural networks with time-varying delays have been studied in [14]. Note that many of the studies focused on neural networks, and only the first derivative of states is important for introducing inertial terms into neural networks. The inertial terms are considered as key tools for generating complex bifurcations and chaos. Up to now, the inertial neural networks have attracted the attention of many researchers. The authors in [15] discussed the robust stability of inertial BAM neural networks with time delays and uncertainties via impulsive effect. In [16], the inertial Cohen-Grossberg-type neural networks with time delays was proposed and its stability analysis were discussed.

Lately, a volume of the existing research on inertial neural networks were mainly focused on exponential synchronization or asymptotical synchronization of networks [17,18]. That it to say, as the time goes to infinity, the dynamical systems only can achieve stability. However, in many actual situations, the dynamical system might be hoped to be stabilised as speedy as possible in a finite time. Since then, problems related to finite time synchronization for networks becomes a hot topic [19,20]. The work in [21] only investigated finite-time stability for inertial neural networks. As it is well known, delays are ubiquitous in the real world, and the introduction of delays may make neural networks their dynamical behaviors much more complicated, even in causing instability [22,23]. However, the authors have ignored the time-varying delays in [21]. The problem on finite-time and fixed-time synchronization analysis for inertial memristive neural networks via state feedback control has been investigated in [24]. The advantages of discontinuous control with different continuous control strategies are non control sections. As far as we know, there are few results on finite-time synchronization of inertial neural networks with time-varying delays.

Motivated by the aforementioned discussion, this paper addresses the problem of adaptive finite-time synchronization of inertial neural networks with time-varying delay via periodically intermittent control. Rather, by the finite-time stability analysis techniques and the linear matrix inequalities, some effective criteria are derived, which can guarantee the master system synchronizes to the slave system in finite time. Meanwhile, the general continuous feedback control is discussed with inertial neural networks. In the end, an example is given to demonstrate the effectiveness of the proposed synchronization criteria.

The rest of this paper is organized as follows. In Sect. 2, the model description and some preliminaries are proposed. In Sect. 3, the main results and remark for finite time synchronization of inertial neural networks with time-varying delay. Moreover, in Sect. 4, an example is given to show the effectiveness of our results. Finally, in Sect. 5, conclusions are given.

2   Model Description and Preliminaries

Considering the inertial neural network with time-varying delays. The model is described by the following equations:

$$\frac{d^2 x_i(t)}{dt^2} = -c_i \frac{dx_i(t)}{dt} - d_i x_i(t) + \sum_{j=1}^{n} a_{ij} f_j(x_i(t)) + \sum_{j=1}^{n} b_{ij} f_j(x_i(t-\tau(t))) + I_i, \quad (1)$$

where $x_i(t)(i = 1, 2, ..., n)$ is the state vector of the ith neuron at time t, the second derivative of $x_i(t)$ is called an inertial neural term of system (1), c_i and d_i are positive constants. The nonlinear function f_j denotes activation function of the jth neuron at time t, a_{ij} and b_{ij} are constants and denotes the connection strengths, I_i is an external inputs for the ith neuron, $\tau(t)$ is the time-varying delay of inertial neural network that satisfies $0 \leq \tau(t) \leq \tau_1, \dot{\tau}(t) \leq \mu_m < 1$, τ_1 and μ_m are constants.

Remark 1. The chaotic neural network is a highly nonlinear dynamic system. The research on chaotic neural network mainly lies in recognizing the chaotic characteristics of individual neurons and the behavior analysis of simple chaotic neural networks. The second-order inertial neural networks (1) has chaos, which is different from the first order neural network chaos, such as chen's system and chua's system.

Next, let the following variable transformation be: $y_i(t) = \frac{dx_i(t)}{dt} + \theta_i x_i(t)$, $i = 1, 2, ..., n$. Denote $x(t) = (x_1(t), x_2(t), ..., x_n(t))^T$, $y(t) = (y_1(t), y_2(t), ..., y_n(t))^T$, then inertial neural network (1) can be written as:

$$\begin{cases} \frac{dx(t)}{dt} = -\Theta x(t) + y(t), \\ \frac{dy(t)}{dt} = -Cy(t) - Dx(t) + Af(x(t)) + Bf(x(t - \tau(t))) + I, \end{cases} \quad (2)$$

where $\bar{c}_i = c_i - \theta_i$, $\bar{d}_i = \theta_i(\theta_i - c_i) + d_i$, $C = diag(\bar{c}_1, \bar{c}_2, ..., \bar{c}_n)$, $D = diag(\bar{d}_1, \bar{d}_2, ..., \bar{d}_n)$, $A = (a_{ij})_{n \times n}$, $B = (b_{ij})_{n \times n}$, $\Theta = diag(\theta_1, \theta_1, ..., \theta_n)$, $f(x(t)) = (f_1(x_1(t)), f_2(x_2(t))...f_n(x_n(t)))^T$, $f(x(t - \tau(t))) = (f_1(x_1(t - \tau(t))), f_2(x_2(t - \tau(t))), ..., f_n(x_n(t - \tau(t))))^T$, $I = (I_1, I_2, ..., I_n)^T$. For simplicity, we choose (3) as the master system, the corresponding slave system is formulated as follows:

$$\begin{cases} \frac{dv(t)}{dt} = -\Theta v(t) + w(t) + u_1(t), \\ \frac{dw(t)}{dt} = -Cw(t) - Dv(t) + Af(v(t)) + Bf(v(t - \tau(t))) + I + u_2(t), \end{cases} \quad (3)$$

where $v(t) = (v_1(t), v_2(t), ..., v_n(t))^T$, $w(t) = (w_1(t), w_2(t), ..., w_n(t))^T$, are the state variables of the slave system, $u_1(t), u_2(t)$ are the appropriate control inputs to be designed later.

Denote the synchronization error $e(t) = v(t) - x(t)$, $\bar{e}(t) = w(t) - y(t)$, we can get the following error system

$$\begin{cases} \frac{de(t)}{dt} = -\Theta e(t) + \bar{e}(t) + u_1(t), \\ \frac{d\bar{e}(t)}{dt} = -C\bar{e}(t) - De(t) + A(f(v(t)) - f(x(t))) + B(f(v(t - \tau(t))) \\ \qquad - f(x(t - \tau(t)))) + u_2(t). \end{cases} \quad (4)$$

In order to realize adaptive finite-time synchronization of inertial neural networks between the master system (3) and slave system (4), the intermittent control $u_i(t)$ is defined by:

$$
\begin{cases}
u_1(t) = k \odot e(t) - \lambda sign(e(t)) \\
\quad - \lambda (\int_{t-\tau(t)}^{t} e^T(s)e(s)ds)^{\frac{1}{2}} \frac{e(t)}{\|e(t)\|^2}, & lT \leq t \leq lT + \delta \\
u_2(t) = \varepsilon \odot \overline{e}(t) - \lambda sign(\overline{e}(t)), & lT \leq t \leq lT + \delta, \\
u_1(t) = u_2(t) = 0, & lT + \delta \leq (l+1)T,
\end{cases}
\tag{5}
$$

where $k = (k_1, k_2, ..., k_n)^T$, $\varepsilon = (\varepsilon_1, \varepsilon_2, ..., \varepsilon_n)^T$ are the adaptive laws, and the $\odot$ is defined as $k \odot e(t) = [k_1 \cdot e_1(t), k_2 \cdot e_2(t), ..., k_n \cdot e_n(t)]^T$, $\lambda > 0$ is real constant. $\mathcal{L} = \{1, 2, ..., l\}$ is a finite natural number set. $T > 0$ is called the control period. $\theta = \frac{\delta}{T}$ denote the control rate.

At the same time, the adaptive rule defined as follows:

$$
\dot{k}_i =
\begin{cases}
-\alpha_i \left(e^T(t)e(t) + \frac{\lambda}{\sqrt{\alpha_i}} sign(k_i) + \frac{\eta_i e(t)^T e(t)}{k_i} \right), & lT \leq t \leq lT + \delta, \\
0, & lT + \delta \leq (l+1)T,
\end{cases}
\tag{6}
$$

$$
\dot{\varepsilon}_i =
\begin{cases}
-\alpha_i \left(\overline{e}^T(t)\overline{e}(t) + \frac{\lambda}{\sqrt{\alpha_i}} sign(\varepsilon_i) + \frac{\epsilon_i \overline{e}(t)^T \overline{e}(t)}{\varepsilon_i} \right), & lT \leq t \leq lT + \delta, \\
0, & lT + \delta \leq (l+1)T,
\end{cases}
\tag{7}
$$

where $\alpha_i > 0$ is a positive constant. $\eta_i > 0, \epsilon_i > 0$ are nonnegative constants denotes the control gain.

Assumption 1. For all x, $y \in \mathbb{R}^n$, suppose that the activation function $f(\cdot)$ satisfies the following condition,

$$
\| f(x) - f(y) \| \leq \| J(x - y) \| .
$$

where $J \in \mathbb{R}^{n \times n}$ is a known constant matrix.

Definition 1 ([25]). *The slave system (4) is said to reach finite-time synchronization with the master system (3), if there exists a constant $t_1 \geq 0$ such that*

$$
\lim_{t \to t_1} \| e(t) \| = 0,
$$

and $\| e(t) \| = 0$ for $t \geq t_1$, where t_1 denotes the settling time.

Lemma 1 ([26]). *If $b_1, b_2, ..., b_n \geq 0$, $0 < k \leq 1$, after that*

$$
(\sum_{i=1}^{n} b_i)^k \leq \sum_{i=1}^{n} b_i^k .
$$

Lemma 2 ([27]). *If X, Y, and Q are real matrices with appropriate dimensions, there exists a constant $\sigma > 0$ and $Q = Q^T > 0$ such that*

$$
2X^T Y \leq \sigma X^T Q X + \sigma^{-1} Y^T Q^{-1} Y.
$$

Lemma 3 ([28]). *If there exist a continuous, positive definite $V(t)$ satisfies the following inequality:*

$$\dot{V}(t) \leq -\alpha V^\eta(t) + hV(t), \forall t \geq t_0, V^{1-\eta}(t_0) \leq \frac{\alpha}{h},$$

where $\alpha > 0, 0 < \eta < 1, h > 0$ are three constants, then the settling time t_1 is given by

$$t_1 \leq \frac{\ln(1 - \frac{h}{\alpha}V^{1-\eta}(0))}{h(\eta - 1)}.$$

3   Main Results

Now, we are in a position to present our results. We will introduce the synchronization criteria between the master system and the salve system in finite time with time-varying delay via adaptive intermittent controllers.

Theorem 1. *Suppose that Assumption 1 hold. For given positive constants ρ, σ, δ, if there exist two diagonal positive definite matrices $\Xi = diag(\eta_1, \eta_2, ..., \eta_n)$ and $\Lambda = diag(\epsilon_1, \epsilon_2, ..., \epsilon_n)$ such that the following conditions hold.*

$$\Phi = \begin{pmatrix} \Phi_{11} & \frac{1}{2}(I_n - D) & 0 \\ * & \Phi_{22} & 0 \\ * & * & \rho J^T J - \frac{1}{2}(1 - \mu_m)I_n \end{pmatrix} < 0, \tag{8}$$

$$(\frac{1}{2} + \delta - \beta)I_n + \sigma J^T J - \Theta \leq 0, \tag{9}$$

$$\sigma^{-1}AA^T + \rho^{-1}BB^T + \delta^{-1}(I_n - D)(I_n - D)^T - \beta I_n - C - \Lambda \leq 0, \tag{10}$$

$$\rho J^T J - \frac{1}{2}(1 - \mu_m)I_n \leq 0. \tag{11}$$

Then the master system (3) and slave system (4) can be finite-time synchronized under the adaptive periodically intermittent control:

$$t_1 = \frac{V^{1-\eta}(0)e^{(1-\eta)(1-\theta)\gamma t}}{\alpha\theta(1 - \eta)}, \tag{12}$$

where $\Phi_{11} = \sigma J^T J + \frac{1}{2}I_n - \Theta - \Xi$, $\Phi_{22} = \sigma^{-1}AA^T + \rho^{-1}BB^T - C - \Lambda$, let $e(t) = (e_1^T(t), e_2^T(t), ..., e_n^T(t))^T \in \mathbb{R}^n$.

Proof. Constructing the following Lyapunov-Krasovskii function: $V(t) = V_1(t) + V_2(t) + V_3(t)$,

$V_1(t) = \frac{1}{2}e^T(t)e(t) + \frac{1}{2}\bar{e}^T(t)\bar{e}(t), V_2(t) = \frac{1}{2}\int_{t-\tau(t)}^t e^T(s)e(s)ds,$
$V_3(t) = \frac{1}{2}\sum_{i=1}^n \frac{1}{\alpha_i}k_i^2 + \frac{1}{2}\sum_{i=1}^n \frac{1}{\alpha_i}\varepsilon_i^2.$

When $t \in [lT, lT + \theta T]$, the time derivative of $V(t)$ along the trajectories of the error system (5) and using Assumption 1, one have

$$\dot{V}_1(t) = e^T(t)\Big(-\Theta e(t) + \bar{e}(t) + k \odot e(t) - \lambda sign(e(t))$$

$$-\lambda\Big(\int_{t-\tau(t)}^{t} e^T(s)e(s)ds\Big)^{\frac{1}{2}}\frac{e(t)}{\|e(t)\|^2}\Big) + \bar{e}^T(t)\Big(-C\bar{e}(t) - De(t)$$

$$+\overline{A}(f(v(t)) - f(x(t))) + B(f(v(t-\tau(t))) - f(x(t-\tau(t))))$$

$$+\varepsilon \odot \bar{e}(t) - \lambda sign(\bar{e}(t))\Big)$$

$$\leq -e^T(t)\Theta e(t) + e^T(t)\bar{e}(t) + e^T(t)k \odot e(t) - \lambda \mid e^T(t) \mid$$

$$-\lambda\Big(\int_{t-\tau(t)}^{t} e^T(s)e(s)ds\Big)^{\frac{1}{2}} - \bar{e}^T(t)C\bar{e}(t) - \bar{e}^T(t)De(t)$$

$$+\sigma^{-1}\bar{e}^T(t)AA^T\bar{e}(t) + \sigma e^T(t)J^T Je(t) + \rho^{-1}\bar{e}^T(t)BB^T\bar{e}(t)$$

$$+\rho e^T(t-\tau(t))J^T Je(t-\tau(t)) + \bar{e}^T(t)\varepsilon \odot \bar{e}(t) - \lambda \mid \bar{e}^T(t) \mid .$$

$$\dot{V}_2(t) + \dot{V}_3(t) = \frac{1}{2}\Big(e^T(t)e(t) - (1 - \dot{\tau}(t))e^T(t-\tau(t))e(t-\tau(t))\Big)$$

$$+\sum_{i=1}^{n}\frac{1}{\alpha_i}k_i\Big[-\alpha_i\Big(e^T(t)e(t) + \frac{\lambda}{\sqrt{\alpha_i}}sign(k_i) + \frac{\eta_i e(t)^T e(t)}{k_i}\Big)\Big]$$

$$+\sum_{i=1}^{n}\frac{1}{\alpha_i}\varepsilon_i\Big[-\alpha_i\Big(\bar{e}^T(t)\bar{e}(t) + \frac{\lambda}{\sqrt{\alpha_i}}sign(\varepsilon_i) + \frac{\epsilon_i \bar{e}(t)^T \bar{e}(t)}{\varepsilon_i}\Big)\Big]$$

$$\leq \frac{1}{2}\Big(e^T(t)e(t) - (1 - \mu_m)e^T(t-\tau(t))e(t-\tau(t))\Big)$$

$$-\sum_{i=1}^{n}k_i e^T(t)e(t) - \sum_{i=1}^{n}\frac{\lambda}{\sqrt{\alpha_i}} \mid k_i \mid - \sum_{i=1}^{n}\eta_i e^T(t)e(t)$$

$$-\sum_{i=1}^{n}\varepsilon_i \bar{e}^T(t)\bar{e}(t) - \sum_{i=1}^{n}\frac{\lambda}{\sqrt{\alpha_i}} \mid \varepsilon_i \mid - \sum_{i=1}^{n}\epsilon_i \bar{e}^T(t)\bar{e}(t).$$

$$(13)$$

By lemma 1 and combine (10) and (11), we get

$$
\begin{aligned}
\dot{V}(t) \leq\ & e^T(t)\Big(\sigma J^T J + \frac{1}{2}I_n - \Theta - \Xi\Big)e(t) \\
& + \bar{e}^T(t)\Big(\sigma^{-1}AA^T + \rho^{-1}BB^T - C - \Lambda\Big)\bar{e}(t) \\
& + e^T(t-\tau(t))\Big(\rho J^T J - \frac{1}{2}(1-\mu_m)I_n\Big)e(t-\tau(t)) \\
& + e^T(t)(I_n - D)\bar{e}(t) - \lambda \mid e^T(t) \mid -\lambda \mid \bar{e}^T(t) \mid \\
& - \lambda\Big(\int_{t-\tau(t)}^{t} e^T(s)e(s)ds\Big)^{\frac{1}{2}} - \sum_{i=1}^{n}\frac{\lambda}{\sqrt{\alpha_i}}\mid k_i \mid - \sum_{i=1}^{n}\frac{\lambda}{\sqrt{\alpha_i}}\mid \varepsilon_i \mid \\
\leq\ & \xi^T(t)\Phi\xi(t) - \sqrt{2}\lambda\Big(\frac{1}{2}e^T(t)e(t) + \frac{1}{2}\bar{e}^T(t)\bar{e}(t) \\
& + \frac{1}{2}\int_{t-\tau(t)}^{t} e^T(s)e(s)ds + \frac{1}{2}\sum_{i=1}^{n}\frac{1}{\alpha_i}k_i^2 + \frac{1}{2}\sum_{i=1}^{n}\frac{1}{\alpha_i}\varepsilon_i^2\Big)^{\frac{1}{2}}
\end{aligned}
$$

$$(14)$$

where $\xi(t) = (e^T(t), \bar{e}^T(t), e^T(t-\tau(t)))^T$, based on the condition (9) that $\Phi \leq 0$, we can obtain as $V(t) \leq -\sqrt{2}\lambda V^{\frac{1}{2}}(t)$.

When $lT + \theta T \leq (l+1)T$, based on the conditions (10)–(12), then the time derivative of V for $t > 0$ is given by

$$
\begin{aligned}
\dot{V}(t) =\ & e^T(t)\Big(-\Theta e(t) + \bar{e}(t)\Big) + \bar{e}^T(t)\Big(-C\bar{e}(t) - De(t) + A(f(v(t)) - f(x(t))) \\
& + B[f(v(t-\tau(t))) - f(x(t-\tau(t)))]\Big) + \frac{1}{2}e^T(t)e(t) \\
& - \frac{1}{2}(1-\dot{\tau}(t))e^T(t-\tau(t))e(t-\tau(t)) \\
\leq\ & -e^T(t)\Theta e(t) + e^T(t)\bar{e}(t) - \bar{e}^T(t)C\bar{e}(t) - \bar{e}^T(t)De(t) \\
& + \sigma^{-1}\bar{e}(t)AA^T\bar{e}(t) \\
& + \sigma e^T(t)J^T Je(t) + \rho^{-1}\bar{e}^T(t)BB^T\bar{e}(t) + \rho e^T(t-\tau(t))J^T Je(t-\tau(t)) \\
& + \frac{1}{2}e^T(t)e(t) - \frac{1}{2}(1-\mu_m)e^T(t-\tau(t))e(t-\tau(t)) \\
\leq\ & e^T(t)\Big((\frac{1}{2}+\delta-\beta)I_n + \sigma J^T J - \Theta\Big)e(t) \\
& + e^T(t-\tau(t))\Big(\rho J^T J - \frac{1}{2}(1-\mu_m)I_n\Big)e(t-\tau(t)) \\
& + \bar{e}^T(t)\Big(\sigma^{-1}AA^T + \rho^{-1}BB^T + \delta^{-1}(I_n - D)(I_n - D)^T - \beta I_n\Big)\bar{e}(t) \\
& + \beta(e^T(t)e(t) + \bar{e}^T(t)\bar{e}(t)) \leq \beta V_1(t).
\end{aligned}
$$

$$(15)$$

Then $\dot{V}(t) \leq \beta V(t)$, by Theorem 1, it follows from (18) and (20), let $\alpha = \sqrt{2}\lambda, \eta = \frac{1}{2}$, we get

$$\begin{cases} \dot{V}(t) \leq -\sqrt{2}\lambda V^{\frac{1}{2}}(t), & lT \leq t \leq lT + \delta T. \\ \dot{V}(t) \leq \beta V(t), & lT + \delta T \leq (l+1)T. \end{cases} \tag{16}$$

By Lemma 3, we obtain $t \leq \dfrac{V^{\frac{1}{2}}(0)e^{\frac{1}{2}(1-\theta)\beta t}}{\sqrt{2}\lambda\theta} = t_1$. therefore, the synchronization of inertial neural networks for the master system (3) and the slave system (4) is achieved in a finite time. So the Theorem 1 is proved.

The adaptive periodically intermittent controller is degenerated to continuous feedback control strategy when $\theta = 1$. A new controller (6)–(8) are proposed as follows:

$$\begin{cases} u_1(t) = k \odot e(t) - \lambda sign(e(t)) - \lambda(\int_{t-\tau(t)}^{t} e^T(s)e(s)ds)^{\frac{1}{2}} \frac{e(t)}{\|e(t)\|^2}. \\ u_2(t) = \varepsilon \odot \overline{e}(t) - \lambda sign(\overline{e}(t)). \end{cases} \tag{17}$$

$$\begin{cases} \dot{k}_i = -\alpha_i\left(e^T(t)e(t) + \frac{\lambda}{\sqrt{\alpha_i}}sign(k_i) + \frac{\eta_i e(t)^T e(t)}{k_i}\right). \\ \dot{\varepsilon}_i = -\alpha_i\left(\overline{e}^T(t)\overline{e}(t) + \frac{\lambda}{\sqrt{\alpha_i}}sign(\varepsilon_i) + \frac{\epsilon_i \overline{e}(t)^T \overline{e}(t)}{\varepsilon_i}\right), \end{cases} \tag{18}$$

then the following corollary can be obtained.

Corollary 1. *Under Assumptions 1, for given positive constants ρ, σ, δ, if there exist two diagonal positive definite matrices $\Xi = diag(\varepsilon_1, \varepsilon_2, ..., \varepsilon_n)$ and $\Lambda = diag(\epsilon_1, \epsilon_2, ..., \epsilon_n)$ such that the following conditions hold:*

$$\Psi = \begin{pmatrix} \Psi_{11} & 0 & 0 \\ * & \Psi_{22} & 0 \\ * & * & \rho J^T J - \frac{1}{2}(1 - \mu_m)I_n \end{pmatrix} < 0, \tag{19}$$

afterwards the master system (3) and the slave system (4) can achieve finite time synchronization with the continuous controller in the setting time: $t_2 = \dfrac{2V^{\frac{1}{2}}(0)}{\sqrt{2}\lambda}$. where $\Psi_{11} = \sigma J^T J + (\frac{1}{2} + \delta)I_n - \Theta - \Xi$, $\Psi_{22} = \sigma^{-1}AA^T + \rho^{-1}BB^T + \delta^{-1}(I_n - D)(I_n - D)^T - C - \Lambda$.

4   Numerical Example

In this section, an example is given to verify the effectiveness of the synchronization for inertial neural networks scheme obtained in the previous section. Considering the following inertial neural networks:

$$\frac{d^2 x_i(t)}{dt^2} = -c_i \frac{dx_i(t)}{dt} - d_i x_i(t) + \sum_{j=1}^{2} a_{ij} f_j(x_i(t)) + \sum_{j=1}^{2} b_{ij} f_j(x_i(t - \tau(t))) + I_i, \quad i = 1, 2. \tag{20}$$

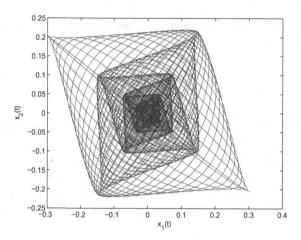

Fig. 1. Phase portrait of the system (20).

then, the corresponding slave system is formulated as follows:

$$\begin{cases} \frac{dv(t)}{dt} = -\Theta v(t) + w(t) + u_1(t), \\ \frac{dw(t)}{dt} = -Cw(t) - Dv(t) + Af(v(t)) + Bf(v(t - \tau(t))) + I + u_2(t), \end{cases} \quad (21)$$

we can get the corresponding matrix

$$C = \begin{pmatrix} 0.1 & 0 \\ 0.1 & 0 \end{pmatrix}, D = \begin{pmatrix} 0.11 & 0 \\ 0 & 0.11 \end{pmatrix}, A = \begin{pmatrix} -0.95 & 0.01 \\ 0.01 & -1 \end{pmatrix}, B = \begin{pmatrix} 0.6 & -0.5 \\ 1.8 & 0.5 \end{pmatrix},$$

where $\Theta = diag(0.5, 0.5)$, $f(x) = tanh(x)$, let $\tau(t) = 0.5\sin(t) + 0.3$, Obviously, $\dot{\tau}(t) \leq \mu_m = 0.5$. Moreover, the initial values are given as: $x(0) =$

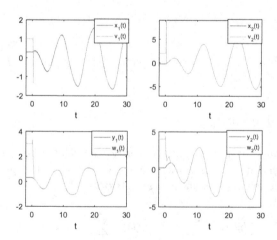

Fig. 2. Trajectory of the synchronization $x(t), v(t), y(t), w(t)$ with the intermittent controller.

$(0.3, -0.2)^T$, $y(0) = (0.3, 0.2)^T$, $v(0) = (1, 2)^T$, $w(0) = (3, 4)^T$. Given $\theta = 0.6$, $T = 3$, $\lambda = 3$, $k = 2.1$. By the Matlab LMI Control Toolbox to solve the LMI in the Theorem 1. We have a set of feasible solutions: $\Xi = diag(13.3641, 13.3641)$, $\Lambda = diag(24.7574, 24.7574)$. Then, the Fig. 1 show the phase portrait of the system (33). Then, trajectory of the synchronization $x(t), v(t)$ and $y(t), w(t)$ with the adaptive periodically intermittent strategy in Fig. 2. Finally, the synchronization errors are shown for the systems by using the adaptive intermittent controller in Fig. 3.

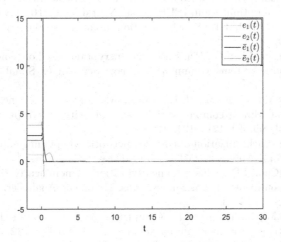

Fig. 3. Trajectory of the synchronization errors $e(t), \overline{e}(t)$ with the intermittent controller.

5   Conclusion

In this paper, the finite-time synchronization for a class of inertial neural networks with time-varying delay was studied. By selecting suitable variable substitution, the original system can changed to two first-order differential equations. The discontinuous intermittent controller was proposed to adjust the system to realize synchronization with finite time. By the some adequate conditions and finite-time stability theory, we have proposed the finite-time synchronization of master-slave systems. In the end, the numerical simulation given to demonstrate the effectiveness the proposed method. In the future, a mixed intermittent controller with different control rates may be researched. Hence, it is worth learning the finite-time synchronization for coupled inertial neural networks under aperiodically intermittent control, impulsive control, sampled-data control and so on.

References

1. Wang, D.L.: Pattern recognition: neural networks in perspective. IEEE Expert. **8**, 52–60 (1993)
2. Cochocki, A., Unbehauen, R.: Neural Networks for Optimization and Signal Processing. Wiley, Chichester (1993)
3. Kwok, T., Smith, K.A.: Experimental analysis of chaotic neural network models for combinatorial optimization under a unifying framework. Neural Netw. **13**, 731–744 (2000)
4. Kwok, T., Smith, K.A.: A unified framework for chaotic neural-network approaches to combinatorial optimization. IEEE Trans. Neural Netw. **10**, 978–981 (1999)
5. Pecora, L.M., Carroll, T.L.: Synchronization in chaotic systems. Phys. Rev. Lett. **64**, 821 (1990)
6. Xu, Y., Wang, H., Li, Y.G., Pei, B.: Image encryption based on synchronization of fractional chaotic systems. Commun. Nonlinear Sci. Numer. Simul. **19**, 3735–3744 (2014)
7. Mahmoud, G.M., Mahmoud, E.E., Arafa, A.A.: Projective synchronization for coupled partially linear complex-variable systems with known parameters. Math. Methods Appl. Sci. **40**, 1214–1222 (2017)
8. Guo, W.: Lag synchronization of complex networks via pinning control. Nonlinear Anal. Real World Appl. **12**, 2579–2585 (2011)
9. Wang, Y.L., Cao, J.D.: Cluster synchronization in nonlinearly coupled delayed networks of non-identical dynamic systems. Nonlinear Anal. Ser. B Real World Appl. **14**, 842–851 (2013)
10. Mahmoud, G.M., Mahmoud, E.E.: Complete synchronization of chaotic complex nonlinear systems with uncertain parameters. Nonlinear Dyn. **62**, 875–882 (2010)
11. Mahmoud, G.M., Mahmoud, E.E.: Phase and antiphase synchronization of two identical hyperchaotic complex nonlinear systems. Nonlinear Dyn. **61**, 141–152 (2010)
12. Chen, T.P.: Global exponential stability of delayed Hopfield neural networks. Neural Netw. **14**, 977–980 (2001)
13. Bao, H.B., Cao, J.D.: Exponential stability for stochastic BAM networks with discrete and distributed delays. Appl. Math. Comput. **218**, 6188–6199 (2012)
14. Shi, Y.C., Cao, J.D., Chen, G.R.: Exponential stability of complex-valued memristor-based neural networks with time-varying delays. Appl. Math. Comput. **313**, 222–234 (2017)
15. Zhang, W., Huang, T.W. Li, C.D., Yang, J.: Robust stability of inertial BAM neural networks with time delays and uncertainties via impulsive effect. Neural Process. Lett. **48**, 1–12 (2017)
16. Ke, Y.Q., Miao, C.F.: Stability analysis of inertial Cohen-Grossberg-type neural networks with time delays. Neurocomputing **117**, 196–205 (2013)
17. Zhang, W., Li, C.D., Huang, T.W., Tan, J.: Exponential stability of inertial BAM neural networks with time-varying delay via periodically intermittent control. Neural Comput. Appl. **26**, 1781–1787 (2015)
18. Cao, J.D., Wan, Y.: Matrix measure strategies for stability and synchronization of inertial BAM neural network with time delays. Neural Netw. **53**, 165–172 (2014)
19. Sowmiya, C., Raja, R., Cao, J.D., Rajchakit, G.: Enhanced robust finite-time passivity for Markovian jumping discrete-time BAM neural networks with leakage delay. Adv. Differ. Equ. **2017** Article no. 318 (2017)

20. Bao, H.B., Cao, J.D.: Finite-time generalized synchronization of nonidentical delayed chaotic systems. Nonlinear Anal. Model. Control **21**, 306–324 (2016)
21. Cui, N., Jiang, H.J., Hu, C., Abdujelil, A.: Finite-time synchronization of inertial neural networks. J. Assoc. Arab. Univ. Basic Appl. Sci. **24**, 300–309 (2017)
22. Li, Y.N., Sun, Y.G., Meng, F.W.: New criteria for exponential stability of switched time varying systems with delays and nonlinear disturbances. Nonlinear Anal. Hybrid Syst. **26**, 284–291 (2017)
23. Guo, Y.X.: Globally robust stability analysis for stochastic Cohen-Grossberg neural networks with impulse control and time-varying delays. Ukr. Math. J. **69**, 1220C–1233 (2017)
24. Wei, R.Y., Cao, J.D., Ahmed, A.: Finite-time and fixed-time synchronization analysis of inertial memristive neural networks with time-varying delays. Cogn. Neurodyn. **12**, 1–14 (2017)
25. Shen, J., Cao, J.D.: Finite-time synchronization of coupled neural networks via discontinuous controllers. Cogn. Neurodyn. **5**, 373–385 (2011)
26. Wang, L., Xiao, F.: Finite-time consensus problems for networks of dynamic agents. IEEE Trans. Autom. Control **55**, 950–955 (2010)
27. Yang, Z.C., Xu, D.Y.: Stability analysis and design of impulsive control systems with time delay. IEEE Trans. Autom. Control **52**, 1448–1454 (2007)
28. Shen, Y.J., Xia, X.H.: Semi-global finite-time observers for nonlinear systems. Automatica **44**, 3152–3156 (2008)

Adaptive Critic Designs of Optimal Control for Ice Storage Air Conditioning Systems

Zehua Liao and Qinglai Wei[✉]

The State Key Laboratory of Management and Control for Complex Systems,
Institute of Automation, Chinese Academy of Sciences, Beijing 100190, China
qinglai.wei@ia.ac.cn

Abstract. In this paper, the optimal control scheme for ice storage air conditioning system is solved via an adaptive critic design method. Adaptive critic design is also called adaptive dynamic programming (ADP). First, the operation of the air conditioning system is analyzed. Next, adaptive critic method is designed to realize the optimal control for the air conditioning system. Numerical results show that using the data-based ADP optimal control method can reduce the operation costs.

Keywords: Ice storage air conditioning · Adaptive critic design
Adaptive dynamic programming · Neural network · Optimal control

1 Introduction

Recently, ice storage air conditioning has been widely used in the world due to its outstanding characteristics [1–3]. It has become an important issue on how to realize the optimal control for the air conditioning system and give full play to its advantages of shifting peak and filling valley, so as to help the users achieve the greatest benefit in the economy.

Adaptive critic design (ACD) is one of the effective methods to solve the problems of nonlinear systems optimal control, and it can also efficiently conquer the "curse of dimensionality" in general dynamic programming. Adaptive dynamic programming (ADP) is another name of adaptive critic design. ADP was first proposed by Werbos to solve the forward-in-time problems of optimal control [4]. The main principle of ADP is to approximate the control law and the performance index function in dynamic programming equation by using function approximation structure, and obtain the system's optimal performance [5]. Werbos used two neural networks in function approximation structure, to implement the ADP method. So ADP is often called neuro-dynamic programming [6]. Now ADP has been introduced in the field of energy management [7–11].

Developing a new self-learning scheme of optimal control for ice storage air conditioning via a data-based ADP method is the main focus of this paper. A self-learning optimal control method is designed to manage ice storage of

© Springer Nature Switzerland AG 2018
L. Cheng et al. (Eds.): ICONIP 2018, LNCS 11307, pp. 180–188, 2018.
https://doi.org/10.1007/978-3-030-04239-4_16

the air conditioning system, in order to save money and meet the cooling load demand, simultaneously. Compared with the current control strategies [12,13], it is emphasized that the air conditioning system can realize self-learning by the load demand and real-time electricity rate, without requiring a mathematical model of the system. Numerical results by the method will show the effectiveness.

2 Self-learning Optimal Control Scheme for Ice Storage Air Conditioning System via Data-Based ADP

2.1 Ice Storage Air Conditioning System

The system is made up of air conditioning refrigerator, cooling load demand, cold storage equipment system (including cooling converter) and cooling management system. The management system is connected to the cold storage equipment through the converter. In this ice storage air conditioning system, the cold storage equipment adopts different control strategies to meet the cooling load demand. For the air conditioning system, three operational modes are considered.

(1) Store mode: when the electricity rate is high and the cooling load demand is low, the air conditioning refrigeration system will meet the load demand directly and store cooling into the cold storage equipment.
(2) Idle mode: the air conditioning refrigeration system will meet the load demand directly at certain hours, while the quantity of cold storage keeps constant.
(3) Release mode: considering load demand and electricity rate, when the cost rate is high, the cold storage equipment releases cooling to supply the load at hours.

2.2 Air Conditioning Dynamics

The cooling capacity stored in the cold storage equipment can be expressed as

$$P_I(t+1) = P_I(t) - p_I(t)\eta(p_I(t)), \; p_I(t) < 0. \tag{1}$$

Let t denote the time index. Let $P_I(t)$ (kWh) denote the residual cooling capacity in the cold storage equipment. Let $p_I(t)$ (kW) denote the cooling capacity output of the cold storage equipment, and let $\eta(p_I(t))$ denote the conversion efficiency. The cooling capacity released from the cold storage equipment can be expressed as

$$P_I(t+1) = P_I(t) - p_I(t)\eta(p_I(t)), \; p_I(t) > 0. \tag{2}$$

The load demand is shared between the air conditioning refrigerator and the cold storage equipment, which is expressed as

$$p_L(t) = p_I(t) + p_C(t). \tag{3}$$

Let $p_L(t)$ denote the cooling load demand, and let $p_C(t)$ denote the cooling capacity output of the refrigerator. The optimization problem can be described as minimizing the performance index function, shown as

$$\begin{cases} \min J(t) = \alpha \sum_{t=1}^{\infty} C(t) \times p_C(t) \\ \text{s.t. Physical Constrains} \end{cases} \tag{4}$$

where α is the coefficient of power consumption. Let $J(t)$ denote the performance index function, and let $C(t)$ denote the electricity rate.

2.3 Data-Based Adaptive Dynamic Programming

The delays in $p_L(t)$ and $p_C(t)$ are introduced for convenience of analysis, so the load balance as $p_L(t-1) = p_I(t-1) + p_C(t)$ can be defined. Then, we let $x_1(t) = p_C(t)$ and $x_2(t) = P_I(t)$. Let $x(t) = [x_1(t), x_2(t)]^T$ and $u(t) = p_I(t)$. According to the air conditioning model, the discrete nonlinear system is given by

$$x(t+1) = F[x(t), u(t), t] = \begin{pmatrix} p_L(t) - u(t) \\ x_2(t) - u(t)\eta(u(t)) \end{pmatrix}. \tag{5}$$

We let

$$J[x(t), t] = \sum_{k=t}^{\infty} \gamma^{k-t} U[x(k), u(k), k] \tag{6}$$

denote the performance index function, where the utility function is defined as $U[x(k), u(k), k] = \alpha \times C(t) \times x_1(t)$, and γ is the discount factor between pre and post stages of the system with $0 < \gamma \leq 1$. Finding the sequence of control actions $u(k)$ is the research object of dynamic programming. And it helps to minimize $J[x(t), t]$ in (6). The optimal performance index function according to Bellmans̀ principle is equals to

$$J^*[x(t), t] = \min_{u(t)} \left(\gamma J^*[x(t+1), t+1] + U[x(t), u(t), t] \right). \tag{7}$$

We let $u^*(t)$ denote the optimal control actions. It can be expressed as

$$u^*(t) = \arg\min_{u(t)} \left(\gamma J^*[x(t+1), t+1] + U[x(t), u(t), t] \right). \tag{8}$$

A data-based ADP is employed to obtain the optimal control without constructing the dynamic of the cold storage system in this work. This kind of ADP is called action-dependent heuristic dynamic programming (ADHDP). Figure 1 shows the scheme of ADHDP, which minimizes the following error to train the critic network.

$$\|E_p\| = \sum_t E_p(t) = \sum_t \frac{1}{2} [Q(t-1) - U(t) - \gamma Q(t)]^2. \tag{9}$$

Let $Q(t)$ denote the critic network output.

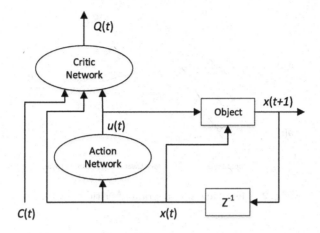

Fig. 1. The ADHDP scheme

If for all time t, there is $E_p(t) = 0$, then it implies from (9) that

$$Q(t-1) = \gamma Q(t) + U(t)$$
$$= \gamma[\gamma Q(t+1) + U(t+1)] + U(t)$$
$$= \cdots \tag{10}$$
$$= \sum_{k=t}^{\infty} \gamma^{k-t} U[k].$$

2.4   Self-Learning Scheme for Air Conditioning System

The optimal control scheme for air conditioning system is based on ADHDP. It includes action module and critic module. In the critic module, the performance index function is approximated by a BP neural network. The BP neural network has 3 layers with 4 nodes of input layer, 1 node of output layer and 9 nodes of hidden layer. As Fig. 2 shows, $x(t)$, $C(t)$ and $u(t)$ are the inputs of the network, while $x(t)$ denotes the system state that includes the residual cooling capacity $P_I(t)$ and the cooling capacity output of refrigerator $p_C(t)$, $u(t)$ denotes the control action and $C(t)$ denotes the electricity rate. Besides, $Q(t)$ is the critic network output, W_{c1} and W_{c2} are the corresponding weight matrices. The object of the critic module is to minimize its error function in (9).

In the action module, three actions are defined as follow: release with $u(t) = -1$, idle with $u(t) = 0$, or store with $u(t) = 1$. The value of $p_I(t)$ is related to $u(t)$ and the parameters of the cold storage equipment, which needs to be discussed. Figure 3 shows the self-learning process of the optimal control scheme. When a load demand is received, it will find which action can minimize the critic network output, and then select this action as the current control action. The above process is based on the successful training of the critic network.

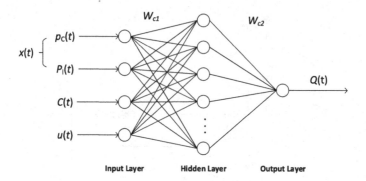

Fig. 2. The schematic diagram of critic network

Based on the above preparation, *Algorithm 1* describes the optimal control scheme for the ice storage air conditioning system.

Algorithm 1. ADP Implementation

① Data collecting. In this stage, when a cooling load demand happens, the random action $u(t)$ with same probability will be adopted. At the same time, corresponding to each action, the electricity rate $C(t)$ and the state $x(t)$ should be collected, and the utility function will be calculated. In this work, the normalized utility function can be given by

$$U(t) = \frac{the\ cost\ of\ power\ consumption}{the\ possible\ maximum\ cost}. \tag{11}$$

② Critic network training. In this stage, the data collected in the last stage is used to train the critic network.

③ Applying the critic network that is trained successfully into the process as in Fig. 3. And the actions which minimizes the output will be selected in the system. The progress will continue until the number of total iterations is reached.

3   Numerical Experiment

3.1   Experiment Preparation

The target of experiment is to minimize the cost of the air conditioning system by using the self-learning optimal control scheme, meeting the load demand and system constraints. Before the implementation, some settings are required.

(1) The air conditioning system should meet the demand of cooling load at any time.
(2) The capacity of the cold storage equipment used in the simulation is 50000 kWh and a minimum of 20% of the storage is required to maintain cooling load. The maximum rate of store/release is 8000 kW. The initial cold storage capacity is 20000 kWh.

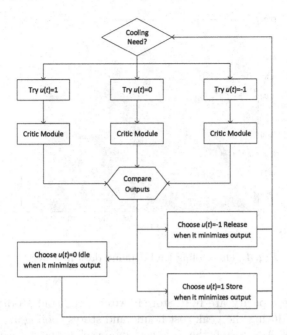

Fig. 3. Block diagram of the self-learning scheme

(3) It is assumed that the cold storage equipment and the air conditioning refrigerator will not supply cooling at the same time. The demand of cooling load is supplied by either cold storage equipment or refrigerator at any time.

(4) Using the load prediction model to predict the hourly cooling load in a certain day.

The training set for the cooling load demand data is chosen in [14, 15], which are the load data of Yonyou Software Park No. 2 R & D Center in 2004. The cooling load demand data is given in Fig. 4. The daily real-time electricity rate is chosen by Beijing commercial electricity [16], which is shown in Fig. 5.

3.2   Results and Analysis

Based on the daily real-time electricity rate and the cooling load demand, the optimal control scheme can be implemented by Algorithm 1, where the structure of the critic network is set as 4–9–1. The optimal cooling storage/release control law for the air conditioning system in a week is shown in Fig. 6.

From the daily real-time price, we can see that, in the period of 11:00–15:00 and 19:00–21:00, there are two price peaks occurring. According to the cooling load curve, it can be found that the peak of the load occurs in the afternoon while the real-time price is high. So it is obvious that the peak of the cost happens when both peaks of load demand and real-time price occur. As can be seen form Fig. 6, when the real-time price is cheap and the load demand is

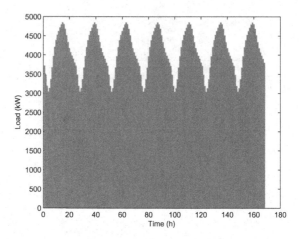

Fig. 4. The cooling load demand data in a week

low, the cooling capacity are fully stored. After that, cold storage equipment releases cooling during the peak cost hours, and stores cooling again during the valley cost hours. As a result, the total cost in a week without optimal control is 187881.04$. Implementing the optimal control via data-based ADP method, the corresponding total cost reduces to 116487.71$. So the optimal control scheme help the user to save 71393.33$ in a week, and the saving rate is 38.0%.

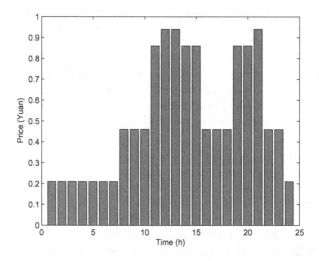

Fig. 5. The daily real-time electricity rate

According to the numerical results, the saving rate has reached a very high standard. In a real-time application, it is not easy to implement the optimal control scheme because it has heavy neural network computations, which require fast microcontrollers.

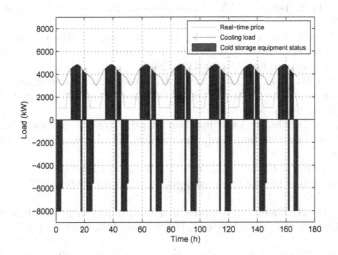

Fig. 6. Optimal scheduling of cold storage equipment in one week

4   Conclusion

In this paper, a self-learning optimal control scheme is developed based on ADP for ice storage air conditioning system. The main contents of this work include data-based ADP design and numerical experiment. The numerical results indicate that the developed ADP method is effective in minimizing the system cost. Compared with the current control strategies, it is of higher economy efficiency and does not need a mathematical model of the system.

References

1. B. H, Y. Tang, X. Liu, and W. Cheng, Ice storage air-conditioning system design and evaluation for Taizhou electric power control center, in Proceedings of the International Conference on Electric Technology and Civil Engineering, Lushan, China, pp. 2296–2299 (2011)
2. Hu, W., Chai, C.C., Yang, W., Yu, R.: Comprehensive modeling and joint optimization of ice thermal and battery energy storage with provision of grid services. In: Proceedings of TENCON 2017: IEEE Region 10 Conference, Penang, Malaysia, pp. 528–533 (2017)
3. Rahnama, S., Bendtsen, J.D., Stoustrup, J., Rasmussen, H.: Robust aggregator design for industrial thermal energy storages in smart grid. IEEE Trans. Smart Grid **8**(2), 902–916 (2017)

4. Werbos, P.: Advanced forecasting methods for global crisis warning and models of intelligence. General Syst. Yearbook, **22**, 25–38 (1977)
5. Xu, X., Yang, H., Lian, C., Liu, J.: Self-learning control using dual heuristic programming with global Laplacian eigenmaps. IEEE Trans. Ind. Electron. **64**(12), 9517–9526 (2017)
6. Bertsekas, D.P., Tsitsiklis, J.N.: Neuro-Dynamic Programming. Athena Scientfic, MA (1996)
7. Wei, Q., Liu, D., Lewis, F.L., Liu, Y.: Mixed iterative adaptive dynamic programming for optimal battery energy control in smart residential microgrids. IEEE Trans. Ind. Electron. **64**(5), 4110–4120 (2017)
8. Wei, Q., Shi, G., Song, R., Liu, Y.: Adaptive dynamic programming-based optimal control scheme for energy storage systems with solar renewable energy. IEEE Trans. Ind. Electron. **64**(7), 5468–5478 (2017)
9. Wei, Q., Liu, D., Lin, H.: Value iteration adaptive dynamic programming for optimal control of discrete-time nonlinear systems. IEEE Trans. Cybern. **46**(3), 840–853 (2016)
10. Wei, Q., Lewis, F.L., Sun, Q., Yan, P., Song, R.: Discrete-time deterministic Q-learning: a novel convergence analysis. IEEE Trans. Cybern. **47**(5), 1224–1237 (2017)
11. Yang, X., He, H., Zhong, X.: Adaptive dynamic programming for robust regulation and its application to power systems. IEEE Trans. Ind. Electron. **65**(7), 5722–5732 (2018)
12. Zhang, M., Gu, Y.: Optimization of ice-storage air conditioning system with ASAGA. In: Proceedings of IEEE Workshop on Advanced Research and Technology in Industry Applications, Ottawa, Canada, pp. 1042–1046, September 2014
13. Huang, J.A., Ha, T.T., Zhang, Y.J.: An optimal control method of ice-storage air conditioning based on reducing direct cooling cost sequentially. In: IEEE International Conference on Power and Renewable Energy, Shanghai, China, pp. 264–268 (2016)
14. Zhang, W.: The load prediction and the rooms decoupling control for ice storage air conditioning system, Ph.D. thesis, Lanzhou University of Technology, Lanzhou, China (2006)
15. National Meteorological Information Center of China: Special Meteorological Data Set for Analysis of Thermal Environment in China, Beijing. China Architecture & Building Press, China (2005)
16. Beijing commercial electricity. https://wenku.baidu.com/view/f3b1624676eeaea-ad1f330e4.html

Impulsive Constraint Control of Coupled Neural Network Model with Actual Saturation

Deqiang Ouyang[1(✉)], Tingwen Huang[2], Chuandong Li[3], Caiping Chen[3], and Hongfei Li[3]

[1] Center for Future Media, School of Computer Science and Engineering, University of Electronic Science and Technology of China, Chengdu 611731, People's Republic of China
ouyangdeqiang@std.uestc.edu.cn
[2] Department of Mathematics, Texas A and M University at Qatar, Doha, Qatar
tingwen.huang@qatar.tamu.edu
[3] National and Local Joint Engineering Laboratory of Intelligent Transmission and Control Technology (Chongqing), College of Electronic and Information Engineering, Southwest University, Chongqing, People's Republic of China
licd@cqu.edu.cn, c_p_chen@163.com, hongfli@126.com

Abstract. In this paper, the exponential synchronization of a class of coupled neural network model under impulsive constraint control is presented. Under impulsive constraint control, several useful linear matrix inequalities (LMIs) are derived by applying Lyapunov function and generalized sector condition. Moreover, under impulsive partial constraint control, a novel sufficient condition guaranteeing exponential synchronization of coupled neural network model is presented. Finally, a numerical simulation is presented to verify the validity of the theoretical analysis results.

Keywords: Coupled neural network model
Impulsive constraint control · Exponential synchronization
Actual saturation

1 Introduction

The dynamical analysis of a class of coupled neural network model have been attracted many scholars to study. The dynamical behaviors of coupled neural network model have been considered, such as periodicity [9], dissipativity [12], stability [5,6,8,10], synchronization [2,7,18] and so on. Moreover, a lot of results about coupled neural network model have been considered [2,18], for example, in [2], under impulsive pinning control, pinning synchronization of coupled neural network model with distributed-delay was obtained. In [18], by applying nonsmooth analysis theorem, several discontinuous or continuous controllers were established to ensure finite-time synchronization of coupled neural network model.

© Springer Nature Switzerland AG 2018
L. Cheng et al. (Eds.): ICONIP 2018, LNCS 11307, pp. 189–199, 2018.
https://doi.org/10.1007/978-3-030-04239-4_17

Moreover, impulsive control is an interesting topic, and has been considered for stability of chaos systems [11, 13–15]. Impulsive control not only has a simple structure but also has great control effect. For example, in [11], under impulsive control, the fixed-time stability of Cohen-Grossberg BAM neural network was presented. In [13–15], based on polytopic representation method, stability of nonlinear dynamical systems under impulsive control with input saturation was obtained. At the same time, actual saturation is one of the most common limiting device, and the capability of hardware devices may be limited to a specific area. At present, there have two approaches to deal with input saturation items [1, 4, 16, 19–21], i.e., sector nonlinearity model approach and polytopic representation approach. Sector nonlinearity model approach is the most common method. For example, in [19], based on sector nonlinearity model approach and under sampled-data control, exponential synchronization of chaotic neural networks was established. In [17], stability of model with partial state saturation nonlinearities was investigated. It is very meaningful to study impulsive constraint control of coupled neural network model, and there are not many related results. Therefore, it is necessary to study that the dynamical behavior of coupled neural network model under impulsive constraint control.

In the paper, exponential synchronization of coupled neural network model with impulsive constraint control is obtained. Moreover, based on sector nonlinearity model approach, it is easy to deal with actual saturation items. Furthermore, under impulsive partial constraint control and same assumption, several LMIs conditions to guarantee exponential synchronization of model are investigated. Finally, under unconstraint impulsive control, sufficient conditions are established to ensure exponential synchronization of coupled neural network model.

The paper is organized as follows. In Sect. 2, a class of coupled neural network model and some useful lemmas are given. In Sect. 3, two different impulsive constraint controls are presented, and some theorems and corollary are established. In Sect. 4, an example to show the validity of the theoretical analysis results is given. In Sect. 5, the conclusion and future prospects are presented.

Notations: Let $\mathbf{M} = \{1, 2, \ldots, n\}$, $\mathbf{N} = \{1, 2, \ldots, N\}$, $\mathbf{R}$, $\mathbf{R}_+$ and $\mathbf{N}_+$ denote the sets of real numbers, nonnegative real numbers and positive integer respectively. Moreover, $\mathbf{R}^m$ denotes the m-dimensional real column vector space endowed with the norm $\|\cdot\|$. For matrix $\mathcal{D} = (d_{ij})_{m \times m} \in \mathbf{R}^{m \times m}$, $\mathcal{D}^{\mathbf{T}}$ and $\mathcal{D}^{-1}$ represent the transpose of and the inverse of matrix $\mathcal{D}$, and $\mathcal{D} > 0$ means positive definite matrix.

2   Problem Formulation and Preliminaries

Consider the following a class of coupled neural network model

$$\frac{\mathbf{d}(x_i(t))}{\mathbf{d}t} = -\mathbf{C}x_i(t) + \mathbf{A}f(x_i(t)) + \mathbf{I} + h\sum_{j=1}^{N} d_{ij}x_j(t) + u_i(t), \qquad (1)$$

where $x_i(t) = (x_{i1}(t), x_{i2}(t), \ldots, x_{in}(t))^{\mathbf{T}} \in \mathbf{R}^N$ is the state variable, $\mathbf{A} = (a_{ij})_{n \times n}$ is the connection weight matrix, $\mathbf{C} = \mathrm{diag}(c_1, c_2, \ldots, c_n)$ represents a diagonal matrix, $c_i > 0$ represents the neuron self-inhibitions, $h > 0$ is a coupling strength, $\mathbf{D} = (d_{ij})_{N \times N}$ is the configuration matrix, in which $d_{ij} = d_{ji} \neq 0$, otherwise, $d_{ij} = d_{ji} = 0$ ($i \neq j$), and the diagonal elements are defined by $d_{ii} = -\sum_{j=1, j \neq i}^{N} d_{ij}$, $u_i(t) \in \mathbf{R}^n$ is control input, and $f(x_i(t)) = (f_1(x_{i1}(t)), f_2(x_{i2}(t)), \ldots, f_n(x_{in}(t)))^{\mathbf{T}}$ is continuous function, and satisfies there is a constant l_i such that $0 \leq \frac{f_i(x_{ir}(t)) - f_i(z_{ir}(t))}{x_{ir}(t) - z_{ir}(t)} \leq l_i$ holds, for all $x_{ir}(t) \neq z_{ir}(t) \in \mathbf{R}$, where $\mathbf{L} = \mathrm{diag}(l_1, l_2, \ldots, l_n)$.

The isolate node in the network (1) is established as follows

$$\frac{\mathbf{d}(y(t))}{\mathrm{d}t} = -\mathbf{C}y(t) + \mathbf{A}f(y(t)) + \mathbf{I}, \tag{2}$$

where $y(t) = (y_1(t), y_2(t), \ldots, y_n(t))^{\mathbf{T}} \in \mathbf{R}^n$ is the state vector of the isolate neuron.

In order to synchronize model (1), the impulsive control with actual saturation is proposed as follows:

$$\mathbf{R}_i(t) = \sum_{\kappa=1}^{\infty} \mathrm{sat}(\chi_i(t))\delta(t - t_\kappa) \quad t \in [t_\kappa, t_{\kappa+1}), \tag{3}$$

where the saturation function $\mathrm{sat}(\chi_i(t)) = (\mathrm{sat}(\chi_{i1}(t)), \mathrm{sat}(\chi_{i2}(t)), \ldots, \mathrm{sat}(\chi_{in}(t)))^{\mathbf{T}} \in \mathbf{R}^n$ with $\mathrm{sat}(\chi_{is}(t)) = \mathrm{sign}(\chi_{is}(t))\min\{\chi_{ois}, |\chi_{is}(t)|\}$, where χ_{ois} is the known saturation level. $\chi_i(t) = \mathbf{K}e_i(t)$, where the error signal $e_i(t) = x_i(t) - y(t)$, $\mathbf{K} \in \mathbf{R}^{n \times n}$ is the control gain, and $\delta(\cdot)$ is the Dirac delta function with sequence $\{t_\kappa\}_{\kappa \in N_+}$ satisfy $0 = t_0 < t_1 < \ldots < t_\kappa < \ldots$, $\lim_{n \to \infty} t_\kappa = \infty$. Suppose that there is a constant $\upsilon > 0$ such that $0 < t_\kappa - t_{\kappa+1} < \upsilon$.

Consider impulsive constraint control (3), subtracting (2) from (1), then the error models are obtained as follows

$$\begin{cases} \dfrac{\mathbf{d}(e_i(t))}{\mathrm{d}t} = -\mathbf{C}e_i(t) + \mathbf{A}g(e_i(t)) + h\sum_{j=1}^{N} d_{ij}e_j(t), \\ \Delta e_i(t_\kappa) = \mathrm{sat}(\mathbf{K}e_i(t_\kappa^-)), \quad \kappa \in N_+, \\ e_i(0) = e_{i0} \in \mathbf{R}^n, \end{cases} \tag{4}$$

where $\Delta e_i(t_\kappa) = e_i(t_\kappa) - e_i(t_\kappa^-)$, $e_i(t_\kappa) = e_i(t_\kappa^+)$ and $e_i(t_\kappa^-) = \lim_{t \to t_\kappa^-} e_i(t)$, and $g(e_i(t)) = f(y(t) + e_i(t)) - f(y(t))$.

Let us define the dead-zone nonlinearity $\Omega(\mathbf{K}e_i(t))$ by

$$\Omega(\mathbf{K}e_i(t)) = \mathrm{sat}(\mathbf{K}e_i(t)) - \mathbf{K}e_i(t), \tag{5}$$

then, model (4) is rewritten as

$$
\begin{cases}
\dfrac{d(e_i(t))}{dt} = -\mathbf{C}e_i(t) + \mathbf{A}g(e_i(t)) + h\sum_{j=1}^{N} d_{ij}e_j(t), \\
e_i(t_\kappa) = (I_n + \mathbf{K})e_i(t_\kappa^-) + \varOmega(\mathbf{K}e_i(t_\kappa^-)), \quad \kappa \in N_+, \\
e_i(0) = e_{i0} \in \mathbf{R}^n,
\end{cases}
\tag{6}
$$

where $I_n \in \mathbf{R}^{n \times n}$ is an identity matrix.

Lemma 1. [3] Let $A_i = (\alpha_{i1}, \alpha_{i2}, \ldots, \alpha_{in})^{\mathbf{T}} \in \mathbf{R}^n$ and $B_i = (\beta_{i1}, \beta_{i2}, \ldots, \beta_{in})^{\mathbf{T}} \in \mathbf{R}^n$. Suppose $-\chi_{oi} \le A_i - B_i \le \chi_{oi}$ where $\chi_{oi} \in \mathbf{R}_+^n$, there is a diagonal matrix $\mathbf{F} > 0 \in \mathbf{R}^{n \times n}$ such that the nonlinearity $\varOmega(\cdot)$ satisfies the following inequality:

$$
\varOmega(A_i)^{\mathbf{T}}\mathbf{F}(\varOmega(A_i) + B_i) \le 0.
$$

By applying Lemma 1 to model (3), for any matrix $\mathbf{G} \in \mathbf{R}^{n \times n}$, if $-\chi_{ois} \le (\mathbf{K}_{(s)} - \mathbf{G}_{(s)})e_i(t_\kappa^-) \le \chi_{ois}$, $e_i(t_\kappa^-) \in \mathbf{R}^n$, where $\mathbf{K}_{(s)} - \mathbf{G}_{(s)} \in \mathbf{R}^{1 \times n}$, $\chi_{ois} \in \mathbf{R}_+$, $i \in N$, then

$$
\varOmega(\mathbf{K}e_i(t_\kappa^-))^{\mathbf{T}}\mathbf{F}[\varOmega(\mathbf{K}e_i(t_\kappa^-)) + \mathbf{G}e_i(t_\kappa^-)] \le 0,
\tag{7}
$$

where $\mathbf{K}_{(s)}$ and $\mathbf{G}_{(s)}$ represent the sth row of matrices $\mathbf{K}$ and $\mathbf{G}$, respectively.

Moreover, suppose that $\forall e_i(t_\kappa^-) \in \varUpsilon(|\mathbf{K} - \mathbf{G}|, \chi_{oi}) = \{e_i(t_\kappa^-) \in \mathbf{R}^n; -\chi_{oi} \le (\mathbf{K} - \mathbf{G})e_i(t_\kappa^-) \le \chi_{oi}\}$, where $\chi_{oi} \in \mathbf{R}_+^n$, $\mathbf{K} - \mathbf{G} \in \mathbf{R}^{n \times n}$, $i \in N$, then

$$
\varOmega(\mathbf{K}e_i(t_\kappa^-))^{\mathbf{T}}\mathbf{F}[\varOmega(\mathbf{K}e_i(t_\kappa^-)) + \mathbf{G}e_i(t_\kappa^-)] \le 0.
\tag{8}
$$

3 Main Results

In this section, several LMIs to ensure exponential synchronization of coupled neural network model via impulsive constraint control are obtained as follows.

Theorem 1. *Let the given three scalars $\sigma > 0$, $\upsilon > 0$ and $0 < \gamma < 1$ such that $\sigma + \frac{\ln \gamma}{\upsilon} < 0$. Suppose that there are a positive definite matrix $\mathbf{H} \in \mathbf{R}^{n \times n} > 0$ and two diagonal matrices $\mathbf{F} \in \mathbf{R}^{n \times n} > 0$, $\varGamma \in \mathbf{R}^{n \times n} > 0$ such that the following LMIs hold:*

$$
\begin{bmatrix} \mathbf{H} & (\mathbf{K}_{(s)} - \mathbf{G}_{(s)})^{\mathbf{T}} \\ \star & \chi_{ois}^2 \end{bmatrix} \ge 0, \quad s \in \mathbf{M}, \quad i \in \mathbf{N},
\tag{9}
$$

$$
\begin{aligned}
I_N \otimes [-(2\mathbf{C} + \sigma)\mathbf{H} + \mathbf{H}\mathbf{A}\varGamma^{-1}\mathbf{A}^{\mathbf{T}}\mathbf{H} \\
+ \mathbf{L}^{\mathbf{T}}\varGamma\mathbf{L}] + 2h\mathbf{D} \otimes \mathbf{H} \le 0,
\end{aligned}
\tag{10}
$$

$$
\varXi = \begin{bmatrix} \varTheta & (I_n + \mathbf{K})^{\mathbf{T}}\mathbf{H} - \mathbf{G}^{\mathbf{T}}\mathbf{F} \\ \star & \mathbf{H} - 2\mathbf{F} \end{bmatrix} < 0,
\tag{11}
$$

where $\varTheta = -\gamma\mathbf{H} + (I_n + \mathbf{K})^{\mathbf{T}}\mathbf{H}(I_n + \mathbf{K})$, then model (1) with the impulsive constraint control (2) is exponentially synchronization.

Proof. Consider the following Lyapunov function:

$$\mathbf{V}(t) = \sum_{i=1}^{N} e_i^{\mathbf{T}}(t)\mathbf{H}e_i(t). \tag{12}$$

Then the derivative of $\mathbf{V}(t)$ along the trajectories of system (4) can be obtained as follows,

$$\mathbf{D}^+\mathbf{V}(t) = 2\sum_{i=1}^{N} e_i^{\mathbf{T}}(t)\mathbf{H}[-\mathbf{C}e_i(t) + \mathbf{A}g(e_i(t)) + h\sum_{j=1}^{N} d_{ij}e_j(t)]. \tag{13}$$

Note that for a positive definite diagonal matric Γ, one gets

$$2e_i^{\mathbf{T}}(t)\mathbf{H}\mathbf{A}g(e_i(t)) \leq e_i^{\mathbf{T}}(t)\mathbf{H}\mathbf{A}\Gamma^{-1}\mathbf{A}^{\mathbf{T}}\mathbf{H}e_i(t) + g^{\mathbf{T}}(e_i(t))\Gamma g(e_i(t))$$
$$\leq e_i^{\mathbf{T}}(t)[\mathbf{H}\mathbf{A}\Gamma^{-1}\mathbf{A}^{\mathbf{T}}\mathbf{H} + \mathbf{L}^{\mathbf{T}}\Gamma\mathbf{L}]e_i(t). \tag{14}$$

Let $e(t) = [e_1^{\mathbf{T}}(t), e_2^{\mathbf{T}}(t), \ldots, e_N^{\mathbf{T}}(t)]^{\mathbf{T}} \in \mathbf{R}^{N \times n}$, we can get the following coupled term

$$2he_i^{\mathbf{T}}(t)\mathbf{H}\sum_{j=1}^{N} d_{ij}e_j(t) \leq 2he^{\mathbf{T}}(t)(\mathbf{D} \otimes \mathbf{H})e(t). \tag{15}$$

From (14)–(15), then (13) becomes the following inequality

$$\mathbf{D}^+\mathbf{V}(t) \leq e^{\mathbf{T}}(t)\Big[I_N \otimes \Big(-(2\mathbf{C} + \sigma)\mathbf{H} + \mathbf{H}\mathbf{A}\Gamma^{-1}\mathbf{A}^{\mathbf{T}}\mathbf{H}$$
$$+\mathbf{L}^{\mathbf{T}}\Gamma\mathbf{L}\Big) + 2h\mathbf{D} \otimes \mathbf{H}\Big]e(t) + \sigma e^{\mathbf{T}}(t)\mathbf{H}e(t)$$
$$= \sigma\mathbf{V}(t),$$

which implies

$$\mathbf{V}(t) \leq e^{\sigma(t-t_\kappa)}\mathbf{V}(t_\kappa), \quad t \in [t_\kappa, t_{\kappa+1}).$$

When $t = t_\kappa$, from (11), one gets

$$\mathbf{V}(t_\kappa) = \sum_{i=1}^{N} e_i^{\mathbf{T}}(t_\kappa)\mathbf{H}e_i(t_\kappa)$$

$$\leq \sum_{i=1}^{N} \Big[(I_n + \mathbf{K})e_i(t_\kappa^-) + \Omega(\mathbf{K}e_i(t_\kappa^-))\Big]^{\mathbf{T}}$$

$$\times \mathbf{H}\Big[(I_n + \mathbf{K})e_i(t_\kappa^-) + \Omega(\mathbf{K}e_i(t_\kappa^-))\Big]$$

$$-2\sum_{i=1}^{N} \Omega(\mathbf{K}e_i(t_\kappa^-))^{\mathbf{T}}\mathbf{F}[\Omega(\mathbf{K}e_i(t_\kappa^-)) + \mathbf{G}e_i(t_\kappa^-)]$$

$$\leq \sum_{i=1}^{N} \Phi^{\mathbf{T}}(t_\kappa^-)\Xi\Phi(t_\kappa^-) + \gamma\sum_{i=1}^{N} e_i^{\mathbf{T}}(t_\kappa^-)\mathbf{H}e_i(t_\kappa^-))$$

$$\leq \gamma\mathbf{V}(t_\kappa^-),$$

where $\Phi(t) = \begin{bmatrix} e_i(t) & \Omega(\mathbf{K}e_i(t)) \end{bmatrix}^{\mathbf{T}}$. Therefore,

$$\mathbf{V}(t) \leq \gamma^\kappa e^{\sigma t}\mathbf{V}(0), \quad \text{for} \quad t \in [t_\kappa, t_{\kappa+1}). \tag{16}$$

Based on $0 < \gamma < 1$, when $t \in [t_\kappa, t_{\kappa+1})$, one gets

$$t \leq t_{\kappa+1} = t_{\kappa+1} - t_\kappa + \ldots + t_1 - t_0 \leq (\kappa + 1)\upsilon, \tag{17}$$

which implies

$$\kappa \geq \frac{t}{\upsilon} - 1. \tag{18}$$

From (16) and (18), one gets

$$\mathbf{V}(t) \leq \gamma^\kappa e^{\sigma t}\mathbf{V}(0) \leq \gamma^{\frac{t}{\upsilon}-1}e^{\sigma t}\mathbf{V}(0)$$
$$= \gamma^{-1}e^{(\sigma + \frac{\ln \gamma}{\upsilon})t}\mathbf{V}(0). \tag{19}$$

The proof is completed.

In the follows, the impulsive control with partial input saturation is established as follows

$$u_i(t) = \begin{cases} \displaystyle\sum_{\kappa=1}^{\infty} \mathbf{sat}(\chi_i(t))\delta(t - t_\kappa), & i \in \{1, 2, \ldots, w\}, \\ \displaystyle\sum_{\kappa=1}^{\infty} \chi_i(t)\delta(t - t_\kappa), & i \in \{w + 1, w + 2, \ldots, N\}. \end{cases} \tag{20}$$

Consider the impulsive control with partial input saturation (20), model (1) is rewritten as

$$\begin{cases} \dfrac{d(e_i(t))}{dt} = -\mathbf{C}e_i(t) + \mathbf{A}g(e_i(t)) + h\displaystyle\sum_{j=1}^{N} d_{ij}e_j(t), \\ \Delta e_i(t_\kappa) = \begin{cases} \mathbf{sat}(\mathbf{K}e_i(t_\kappa^-)), & i \in \{1, 2, \ldots, w\}, \\ \mathbf{K}e_i(t_\kappa^-) & i \in \{w + 1, w + 2, \ldots, N\}, \end{cases} \\ e_i(0) = e_{i0} \in \mathbf{R}^n. \end{cases} \tag{21}$$

Based on (5), model (21) is converted into

$$\begin{cases} \dfrac{d(e_i(t))}{dt} = -\mathbf{C}e_i(t) + \mathbf{A}g(e_i(t)) + h\displaystyle\sum_{j=1}^{N} d_{ij}e_j(t), \\ e_i(t_\kappa) = \begin{cases} (I_n + \mathbf{K})e_i(t_\kappa^-) + \Omega(\mathbf{K}e_i(t_\kappa^-)), i = \{1, 2, \ldots, w\}, \\ (I_n + \mathbf{K})e_i(t_\kappa^-), i = \{w + 1, w + 2, \ldots, N\}, \end{cases} \\ e_i(0) = e_{i0} \in \mathbf{R}^n. \end{cases} \tag{22}$$

Under impulsive partial constraint control (20), several LMIs to guarantee exponential synchronization of model (1) are presented as follows.

Theorem 2. *Let the given three scalars* $\sigma > 0$, $\upsilon > 0$ *and* $0 < \gamma < 1$ *such that* $\sigma + \frac{\ln \gamma}{\upsilon} < 0$. *Suppose that there are a positive definite matrix* $\mathbf{H} \in \mathbf{R}^{n \times n} > 0$ *and two diagonal matrices* $\mathbf{F} \in \mathbf{R}^{n \times n} > 0$, $\Gamma \in \mathbf{R}^{n \times n} > 0$ *such that the LMIs* (9)–(11) *hold and the following LMI holds:*

$$\begin{bmatrix} \gamma \mathbf{H} & (I_n + \mathbf{K})^{\mathbf{T}} \\ \star & \mathbf{H}^{-1} \end{bmatrix} < 0, \tag{23}$$

then model (1) *under the impulsive partial constraint control* (20) *is exponentially synchronization.*

Proof. Due to $\mathbf{H}^{-1} > 0$, by applying Schur complements, from (23) one gets

$$- \gamma \mathbf{H} + (I_n + \mathbf{K})^{\mathbf{T}} \mathbf{H}(I_n + \mathbf{K}) \leq 0. \tag{24}$$

When $t = t_\kappa$, from (11) and (23) that

$$\begin{aligned}
\mathbf{V}(t_\kappa) &= \sum_{i=1}^{w} e_i^{\mathbf{T}}(t_\kappa)\mathbf{H}e_i(t_\kappa) + \sum_{i=w+1}^{N} e_i^{\mathbf{T}}(t_\kappa)\mathbf{H}e_i(t_\kappa) \\
&\leq \sum_{i=1}^{w} \left[(I_n + \mathbf{K})e_i(t_\kappa^-) + \Omega(\mathbf{K}e_i(t_\kappa^-)) \right]^{\mathbf{T}} \\
&\quad \times \mathbf{H}\left[(I_n + \mathbf{K})e_i(t_\kappa^-) + \Omega(\mathbf{K}e_i(t_\kappa^-)) \right] \\
&\quad -2\sum_{i=1}^{w} \Omega(\mathbf{K}e_i(t_\kappa^-))^{\mathbf{T}}\mathbf{F}[\Omega(\mathbf{K}e_i(t_\kappa^-)) + \mathbf{G}e_i(t_\kappa^-)] \\
&\quad + \sum_{i=w+1}^{N} e_i^{\mathbf{T}}(t_\kappa^-)(I_n + \mathbf{K})^{\mathbf{T}}\mathbf{H}(I_n + \mathbf{K})e_i(t_\kappa^-) \\
&\leq \sum_{i=1}^{w} \Phi^{\mathbf{T}}(t_\kappa^-)\Xi\Phi(t_\kappa^-) + \gamma \sum_{i=1}^{N} e_i^{\mathbf{T}}(t_\kappa^-)\mathbf{H}e_i(t_\kappa^-)) \\
&\leq \gamma \mathbf{V}(t_\kappa^-).
\end{aligned}$$

The proof is completed.

If all impulsive controls are not subjected to the input saturation i.e., $w = 0$, the corollary can be obtain as follows.

Corollary 1. *Let the given three scalars $\sigma > 0$, $\upsilon > 0$ and $0 < \gamma < 1$ such that $\sigma + \frac{\ln \gamma}{\upsilon} < 0$. Suppose that there are a positive definite matrix $\mathbf{H} \in \mathbf{R}^{n \times n} > 0$ and two $m \times m$ diagonal matrices $\mathbf{F} > 0$ and $\Gamma > 0$ such that the LMIs (9)–(10), and (23) hold, then model (1) with the impulsive unconstrain control (20) (w = 0) is exponentially synchronization.*

4   Numerical Example

In this section, an example is given to illustrate the effectiveness of proposed design schemes.

Example 1. Consider the following coupled neural network model with three nodes:

$$\frac{d(x_i(t))}{dt} = -\mathbf{C}x_i(t) + \mathbf{A}f(x_i(t)) + h \sum_{j=1}^{N} d_{ij}x_j(t), \tag{25}$$

where

$$\mathbf{A} = \begin{pmatrix} 1.25 & -3.2 & -3.2 \\ -3.2 & 1.1 & -4.4 \\ -3.2 & 4.4 & 1.0 \end{pmatrix}, \mathbf{D} = \begin{pmatrix} -2 & 1 & 1 \\ 1 & -1 & 0 \\ 1 & 0 & -1 \end{pmatrix},$$

$\mathbf{C} = 2I_3$, $h = 0.5$, $x_i(t) = (x_{i1}(t), x_{i2}(t), x_{i3}(t))^{\mathbf{T}} \in \mathbf{R}^3$, $f(x_i(t)) = (f_1(x_{i1}(t)), f_2(x_{i2}(t)), f_3(x_{i3}(t)))^{\mathbf{T}}$, and $f_r(\xi_{ir}(t)) = \frac{1}{2}(|x_{ir}(t) + 1| - |x_{ir}(t) - 1|)$ $(r = 1, 2, 3)$. According to the condition of activation function, it is easy to see that $\mathbf{L} = \text{diag}(1, 1, 1)$. Let $\sigma = 11$, $\gamma = 0.3$, $\upsilon = 0.1$ and $\chi_{oi1} = \chi_{oi2} = \chi_{oi3} = 1$, it is easy to check that $\sigma + \frac{\ln \gamma}{\upsilon} \approx -1.0397 < 0$.

Note that, the dynamical behavior of the isolated node can be described by

$$\frac{d(y(t))}{dt} = -\mathbf{C}y(t) + \mathbf{A}f(y(t)),$$

which has a chaotic attractor shown in [22].

The impulsive constraint control (2) is

$$\mathbf{K} = \begin{pmatrix} -1 & 0 & 0 \\ 0 & -1.1 & 0 \\ 0 & 0 & -1.03 \end{pmatrix}.$$

From Theorem 1, by solving the LMIs (9)–(11), we can get $\mathbf{H} = \text{diag}(1, 1, 1)$, $\mathbf{F} = \text{diag}(3, 3, 3)$, $\Gamma = \text{diag}(10, 10, 10)$. Let $E(t) = \sum_{i=1}^{3} e_i^{\mathbf{T}}(t)e_i(t)$, from Fig. 1, we can see that model (25) can not be synchronized without impulsive full constraint control (2) or (20). From Fig. 2, we can see that model (25) can be synchronized with impulsive full constraint control (2).

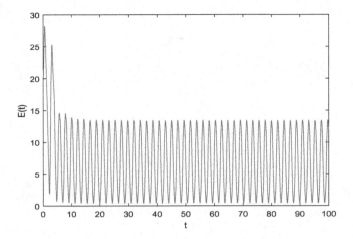

Fig. 1. The synchronization error of model (25) without impulsive full constraint control (2) or (20).

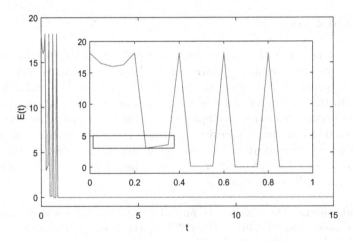

Fig. 2. The synchronization error of model (25) with impulsive full constraint control (2).

From Theorem 2, by solving the LMIs (9)–(11) and (23), we can get $\mathbf{H} =$ diag$(3, 3, 3)$, $\mathbf{F} =$ diag$(8, 8, 8)$, $\Gamma =$ diag$(11, 11, 11)$. From Fig. 3, we can see that model (25) can be synchronized with impulsive partial constraint control (20) (*i.e.*, $w = 1$). From Figs. 2 and 3, it is easy to see that the state trajectories under impulsive full constraint controller (2) is faster than the one under the impulsive partial constraint controller (20).

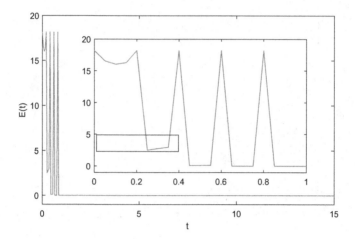

Fig. 3. The synchronization error of model (25) with impulsive partial constraint control (20).

5   Conclusions

Exponential synchronization of a class of coupled neural network model under impulsive control with actual saturation is obtained. Based on generalized sector condition, it is easy to see that saturation item is easy to replace and more easier to handle. In the future, we will study exponential synchronization of discrete-time neural network model with impulsive constraint control, moreover, taking actual saturation term into impulsive constraint controller is more conducive to stability of model. Furthermore, the study of stabilization of discrete-time neural network model based on TS-fuzzy model under impulsive constraint control= is still a big very challenging problem.

Acknowledgments. This study was funded by National Natural Science Foundation of China (Nos. 61633011, 61374078), Qatar National Research Fund (No. NPRP 8-274-2-107), Graduate Student Research Innovation Project of Chongqing (No. CYB17076), Chongqing Research Program of Basic Research and Frontier Technology (No. cstc2015jcyjBX0052).

References

1. Ding, L., Zheng, W.X., Guo, G.: Network-based practical set consensus of multi-agent systems subject to input saturation. Automatica **89**, 316–324 (2018)
2. He, W., Qian, F., Cao, J.: Pinning-controlled synchronization of delayed neural networks with distributed-delay coupling via impulsive control. Neural Networks **85**, 1–9 (2017)
3. Hu, T., Lin, Z.: Control Systems with Actuator Saturation: Analysis and Design. Springer Science & Business Media, New York (2001)

4. Huang, H., Li, D., Lin, Z., Xi, Y.: An improved robust model predictive control design in the presence of actuator saturation. Automatica **47**(4), 861–864 (2011)
5. Huang, T., Chan, A., Huang, Y., Cao, J.: Stability of cohen-grossberg neural networks with time-varying delays. Neural Networks **20**(8), 868–873 (2007)
6. Huang, T., Li, C., Duan, S., Starzyk, J.A.: Robust exponential stability of uncertain delayed neural networks with stochastic perturbation and impulse effects. IEEE Trans. Neural Netw. Learning Syst. **23**(6), 866–875 (2012)
7. Huang, T., Li, C., Yu, W., Chen, G.: Synchronization of delayed chaotic systems with parameter mismatches by using intermittent linear state feedback. Nonlinearity **22**(3), 569 (2009)
8. Li, C., Feng, G., Huang, T.: On hybrid impulsive and switching neural networks. IEEE Trans. Syst. Man Cybern. Part B **38**(6), 1549–1560 (2008)
9. Li, H., Li, C., Huang, T.: Periodicity and stability for variable-time impulsive neural networks. Neural Networks **94**, 24–33 (2017)
10. Li, H., Li, C., Huang, T., Ouyang, D.: Fixed-time stability and stabilization of impulsive dynamical systems. J. Frankl. Inst. **354**(18), 8626–8644 (2017)
11. Li, H., Li, C., Huang, T., Zhang, W.: Fixed-time stabilization of impulsive cohen-grossberg BAM neural networks. Neural Networks **98**, 203–211 (2018)
12. Li, H., Li, C., Zhang, W., Xu, J.: Global dissipativity of inertial neural networks with proportional delay via new generalized halanay inequalities. Neural Process. Lett., 1–19 (2018)
13. Li, L., Li, C., Li, H.: An analysis and design for time-varying structures dynamical networks via state constraint impulsive control. Int. J. Control, 1–9 (2018)
14. Li, L., Li, C., Li, H.: Fully state constraint impulsive control for non-autonomous delayed nonlinear dynamic systems. Nonlinear Anal. Hybrid Syst. **29**, 383–394 (2018)
15. Li, Z., Fang, J., Huang, T., Miao, Q., Wang, H.: Impulsive synchronization of discrete-time networked oscillators with partial input saturation. Inf. Sci. **422**, 531–541 (2018)
16. Lin, X., Li, X., Zou, Y., Li, S.: Finite-time stabilization of switched linear systems with nonlinear saturating actuators. J. Frankl. Inst. **351**(3), 1464–1482 (2014)
17. Liu, D., Michel, A.N.: Stability analysis of systems with partial state saturation nonlinearities. IEEE Trans. Circuits Syst. I Fundam. Theory Appl. **43**(3), 230–232 (1996)
18. Liu, X., Cao, J., Yu, W., Song, Q.: Nonsmooth finite-time synchronization of switched coupled neural networks. IEEE Trans. Cybern. **46**(10), 2360–2371 (2016)
19. Rakkiyappan, R., Latha, V.P., Zhu, Q., Yao, Z.: Exponential synchronization of markovian jumping chaotic neural networks with sampled-data and saturating actuators. Nonlinear Anal. Hybrid Syst. **24**, 28–44 (2017)
20. Seuret, A., da Silva Jr., J.M.G.: Taking into account period variations and actuator saturation in sampled-data systems. Syst. Control Lett. **61**(12), 1286–1293 (2012)
21. Zhou, B., Gao, H., Lin, Z., Duan, G.: Stabilization of linear systems with distributed input delay and input saturation. Automatica **48**(5), 712–724 (2012)
22. Zou, F., Nossek, J.A.: Bifurcation and chaos in cellular neural networks. IEEE Trans. Circuits Syst. I Fundam. Theory Appl. **40**(3), 166–173 (1993)

Value Iteration Algorithm for Optimal Consensus Control of Multi-agent Systems

Qichao Zhang[1,2]([✉]) and Dongbin Zhao[1,2]

[1] The state Key Laboratory of Management and Control for Complex Systems, Institute of Automation, Chinese Academy of Sciences, Beijing 100190, China
{zhangqichao2014,dongbin.zhao}@ia.ac.cn
[2] University of Chinese Academy of Sciences, Beijing 100049, China

Abstract. In this paper, we investigate the optimal consensus control problem for the multi-agent systems by utilizing the Heuristic Dynamic Programming (HDP) algorithm under the centralized learning and decentralized execution framework, which is a kind of value iteration algorithms in reinforcement learning. Different from independent learning framework, a centralized value function which is shared for all the agents is defined. To approach the Nash equilibrium, we prove the equivalence relationship between the Bellman optimality equation and the discrete-time Hamilton-Jacobi-Bellman (DTHJB) equation. For the implementation purpose, the actor-critic structure with NN approximators is proposed to approach the solution of DTHJB equation, where the critic network for all the agents is centralized using the global information, and each actor network for the corresponding agent is decentralized using the local information. Finally, the simulation results are provided, which demonstrates the effectiveness of the proposed HDP algorithm under the centralized learning and decentralized execution framework.

Keywords: Reinforcement learning · Multi-agent systems
Value iteration

1 Introduction

With the rapid development of artificial intelligence, reinforcement learning (RL) [1] and adaptive dynamic programming (ADP) [2,3] techniques have been extended from the single agent environment to multi-agent systems (MAS) field in the last two decades. Compared with the RL in single agent environment, there are two main difficulties for the multi-agent reinforcement learning. First, the external environment is partial observability for each agent [4], which means

This work is supported by the Beijing Science and Technology Plan under Grant Z181100008818075, National Natural Science Foundation of China (NSFC) under Grants No. 61803371, No. 61573353, and No. 61873268.

© Springer Nature Switzerland AG 2018
L. Cheng et al. (Eds.): ICONIP 2018, LNCS 11307, pp. 200–208, 2018.
https://doi.org/10.1007/978-3-030-04239-4_18

that agents lack full information about the environment and other agents. They have to learn the individual optimal policies based on observed local information including its own information and the neighbors' information. Second, the external environment for each agent is nonstationary [5]. As the behaviors of other agents are constantly changing and the agent does not achieve the policies of all the other agents during the learning process, the environment for each agent is no longer stationary. Considering the two characteristics of MAS, three learning schemes are proposed for multi-agent RL. The first one is called as centralized learning [6], where a centralized policy is learned to map the global observation to a joint action. The second one is independent learning [7], where each agent learns its own independent policy based on its local observation. The third one is centralized learning and decentralized execution [5], which is proposed and widely used recently.

It should be mentioned that consensus problem, which is a hot topic in MAS, allows each agent to track the leader's state or reach a common state by designing the control protocols [8]. Recently, RL has been extensively used to solve the optimal consensus control problem based on the independent learning scheme. However, the system stability is hard to be guaranteed under this framework. Here, we focus on the optimal consensus for MAS based on the centralized learning and decentralized execution framework. During the learning process, the observation and policy of each agent are fed into the centralized critic network. During the execution process, the control policy of each agent is obtained using the trained actor networks based on individual local observation.

2   Problem Statement

2.1   Basic Graph Theory

Define a digraph with information interactions between the agent as $Gr = (V, E)$. The non-empty set of nodes denoting N agents is expressed by $V = (1, 2, ..., N)$, and $E \subseteq v \times v$ is the set of edges. A weighted matrix $L = [l_{ij}] \in \mathbf{R}^{N \times N}(i \neq j)$ denotes the graph's topology. A directed edge of Gr is represented by (v_j, v_i), which means that the node i receives the transmitted information from node j. Thus, node i is considered as a neighbor of node j. If $(v_j, v_i) \in E$, $l_{ij} > 0$, otherwise $l_{ij} = 0$. The neighbors of node i are represented by the set $N_i = \{j \mid (v_j, v_i) \in E\}$. The in-degree of node i is denoted by $d_i = \sum_{j=1}^{N} l_{ij}$. The Laplacian matrix of graph Gr is denoted by $L = D - E$. Note that the diagonal matrix $D = \text{diag}(d_1, d_2, ..., d_N)$. A digraph is strongly connected, if there is a accessible path from an arbitrary node i to node j. A spanning tree is a digraph, as there is a root which has a directed edge from the root to the other nodes.

2.2   Optimal Consensus Control of Multi-agent Systems

Consider the following discrete-time MAS

$$x_i(k + 1) = Ax_i(k) + B_i u_i(k), \quad i = 1, ..., N \tag{1}$$

where $x_i \in \mathbf{R}^n$ denotes the state vector, $u_i \in \mathbf{R}^{m_i}$ denotes the control input. For all the agents, we consider the matrix A is the same and unknown.

The dynamics of leader agent is described by

$$x_0(k+1) = Ax_0(k) \qquad (2)$$

where $x_0 \in \mathbf{R}^n$ is the measurable state of the leader.

Denote the local neighborhood error vector for the graph Gr as

$$\delta_i(k) = \sum_{j \in N_i} l_{ij}(x_i(k) - x_j(k)) + g_i(x_i(k) - x_0(k)), i = 1, ..., N \qquad (3)$$

where $\delta_i \in \mathbf{R}^n$, and the pinning gain $g_i \geq 0$. Then, we have the global neighborhood error vector

$$\begin{aligned} \delta(k) &= (L \otimes I_n)x(k) + (G \otimes I_n)(x(k) - x_0(k)) \\ &= ((L+G) \otimes I_n)(x(k) - \underline{x}_0(k)) \end{aligned} \qquad (4)$$

where $\delta(k) = [\delta_1^T(k), \delta_2^T(k), ..., \delta_N^T(k)]^T \in \mathbf{R}^{nN}$ is the global error state, $x(k) = [x_1^T(k), x_2^T(k), ..., x_N^T(k)]^T \in \mathbf{R}^{nN}$, $\underline{x}_0(k) = \underline{I}x_0(k) \in \mathbf{R}^{nN}$ with $\underline{I} = \underline{1} \otimes I_n \in \mathbf{R}^{nN \times n}$, and $\underline{1}$ the vector of ones with the length of N. $L \in \mathbf{R}^{N \times N}$ is the Laplacian matrix for the graph Gr, $G \in \mathbf{R}^{N \times N}$ is a diagonal matrix with the pinning gains g_i as its diagonal entries, $G = \text{diag}(g_1, g_2, ..., g_N)$. $I_n \in \mathbf{R}^n$ is an identity matrix and $\otimes$ denotes the Kronecker product operator. Note that the consensus error vector is defined as $\eta(k) = x(k) - \underline{x}_0(k)$.

Definition 1: The consensus of MAS (1) with the leader (2) can be achieved, if their states reach agreement, i.e., $\lim_{k \to \infty} (x_i(k) - x_0(k)) = 0$, $i = 1, ..., N$

To obtain a stabilizing controller, the following assumption is given.

Assumption 1: a. The pair (A, B_i) is reachable for all $i \in \mathbf{N}$.
b. The graph Gr contains at least a spanning tree and $g_i \neq 0$ for a root node.

Lemma 1: Assume that the graph Gr has a spanning tree and $G \neq 0$ for at least one root node. Then $\|\eta(k)\| \leq \|\delta(k)\|/\underline{\sigma}(L+G)$, where $\underline{\sigma}(L+G)$ is the minimum singular value of matrix $(L+G)$.

It is shown in [9] that $L+G$ is nonsingular under Assumption 1. Then, the consensus error η can approach to zero by making the local neighborhood error δ arbitrarily small, which means the consensus of MAS can be guaranteed. Based on (3), the local neighborhood error dynamics (3) is

$$\begin{aligned} \delta_i(k+1) &= \sum_{j \in N_i} l_{ij}(x_i(k+1) - x_j(k+1)) + g_i(x_i(k+1) - x_0(k+1)) \\ &= A\delta_i(k) + (d_i + g_i)B_iu_i(k) - \sum_{j \in N_i} l_{ij}B_ju_j(k) \end{aligned} \qquad (5)$$

where l_{ij} denotes the element which is the i_{th} row and j_{th} column of the Laplacian matrix L. Note that the agent i can only exchange information with all the neighbors, i.e., agents $\{j|j \in N_i\}$, which is a distributed control problem.

Under the centralized learning framework, we have

$$\delta(k+1) = (A \otimes I_N)\delta(k) + ((L+G) \otimes I_n)\,\bar{B}u(k) \tag{6}$$

where $\bar{B} = \text{diag}\{B_1, ..., B_N\}$, $u(k) = [u_1^T(k), ..., u_N^T(k)]^T$.

Define the centralized value function for the global consensus error corresponding to the state dependent policies as

$$
\begin{aligned}
V(\delta(k)) &= \sum_{t=k}^{\infty} \frac{1}{2}\Big(\delta^T(t)Q\delta(t) + \sum_{i=1}^{N} u_i^T(t)R_i u_i(t)\Big) \\
&= r\big(\delta(k), u_i(k), u_{Gr-i}(k)\big) + V(\delta(k+1))
\end{aligned} \tag{7}
$$

where $Q = \text{diag}(Q_1, ..., Q_N)$, $Q_i > 0 \in \mathbf{R}^{n \times n}$, $R_i > 0 \in \mathbf{R}^{m \times m}$ are symmetric matrices, and u_{Gr-i} denote the control inputs of the other nodes except for node i. For the MAS with centralized learning, it is aimed to learn the optimal policies to minimize the shared centralized value function, such that

$$V^*(\delta(k)) = \min_{u_i, u_{Gr-i}} \big(r(\delta(k), u_i(k), u_{Gr-i}(k)) + V^*(\delta(k+1))\big) \tag{8}$$

Definition 2 [10]: The policy set $\{u_1, ..., u_i, ..., u_N\}$ is admissible if it can stabilize (6) and make sure the centralized value (8) is finite.

Define the centralized Hamiltonian function [2] of MAS as

$$
\begin{aligned}
H(\delta(k), \nabla V(\delta(k+1)), u_i(k), u_{Gr-i}(k)) &= \nabla V^T\big(\delta(k+1)\big)\big((A \otimes I_N)\delta(k) \\
+ ((L+G) \otimes I_n)\,\bar{B}u(k)\big) + \frac{1}{2}\Big(\delta^T(k)Q\delta(k) &+ \sum_{i=1}^{N} u_i^T(k)R_i u_i(k)\Big)
\end{aligned} \tag{9}
$$

where $\nabla V(\delta(k+1)) = \nabla V_{k+1} = [\nabla V_{1,k+1}^T, ..., \nabla V_{N,k+1}^T]^T \in \mathbf{R}^{nN}$ with initial condition $V(0) = 0$, and $\nabla V_{i,k+1} = \nabla V_i\big(\delta(k+1)\big) = \partial V\big(\delta(k+1)\big)/\partial \delta_i(k+1)$.

Lemma 2: Let $0 < V^*(\delta_k) \in C^2$ satisfies the DTHJB equation

$$
\begin{aligned}
H\big(\delta(k), \nabla V^*(\delta(k+1)), u_i^*(k), u_{Gr-i}^*(k)\big) &= \nabla V^{*T}(\delta(k+1)) \\
\times \delta(k+1) + \frac{1}{2}\Big(\delta^T(k)Q\delta(k) + \sum_{i=1}^{N} u_i^{*T}(k)R_i u_i^*(k)\Big) &= 0
\end{aligned} \tag{10}
$$

with initial condition $V^*(0) = 0$, where

$$u_i^*(k) = -(d_i + g_i)R_i^{-1}B_i^T \nabla V_{i,k+1}^* \tag{11}$$

Then, $V^*(\delta(k+1))$ satisfies the following Bellman optimality equation

$$V^*(\delta(k)) = V^*(\delta(k+1)) + \frac{1}{2}\big(\delta^{\mathrm{T}}(k)Q\delta(k)$$
$$+ \sum_{i=1}^{N}(d_i + g_i)^2 \nabla V_{i,k+1}^{*\mathrm{T}} B_i R_i^{-1} B_i^{\mathrm{T}} \nabla V_{i,k+1}^*\big) \tag{12}$$

Lemma 3: Let (A, B_i) satisfies Assumption 1. If $0 < V^*(\delta_k) \in C^2$ satisfies the Bellman optimality Eq. (12), it satisfies the DTHJB Eq. (10).

Note that Lemmas 2 and 3 reveal the relationship between the DTHJB equation and Bellman optimality equation. For the centralized learning and decentralized execution framework, all agents are in best response to their neighbors if and only if they can approach the Nash equilibrium during the learning process, which is defined in the following.

Definition 3: A sequence of control laws $\{u_1^*, ..., u_N^*\}$ is said to constitute a global Nash equilibrium for the multi-agent differential games, if the optimal policy u_i^* of agent i with response to the other agents satisfied

$$V(u_i^*, u_{Gr-i}^*) \leq V(u_i, u_{Gr-i}^*), i \in \mathbf{N} \tag{13}$$

To obtain the global Nash equilibrium solution (11), the optimal centralized value function V^* should be obtained by solving the DTHJB equation firstly.

3   HDP Algorithm

3.1   HDP Algorithm

To approach the solution of the DTHJB Eq. (10) for the fully cooperative multi-agent systems, we propose a novel HDP algorithm under the centralized learning and decentralized execution framework.

HDP for multi-agent systems

1. Find $V^{m+1}(\delta)$ by solving the following equation:

$$V^{m+1}(\delta(k)) = r\big(\delta(k), u_i^m(k), u_{Gr-i}^m(k)\big) + V^m(\delta(k+1)) \tag{14}$$

with the iteration index m.
2. Update the control policies simultaneously using

$$u_i^{m+1}(\delta) = -(d_i + g_i)R_i^{-1} B_i^{\mathrm{T}} \nabla V_i^{m+1}(\delta(k+1)) \tag{15}$$

It should be mentioned that only a centralized Eq. (14) is included for all the agents in the proposed algorithm, which means only one single-critic NN should be constructed to approach the iterative value function. It should be mentioned that the control policy u_i^{m+1} (15) is guided by the centralized value function. During the centralized learning process, the control policy (15) is the target value of the actor NN for agent i. Then, we can design the actor NN for each agent with partially observable local information δ_i to approach the target value, which can guarantee the distributed control during the decentralized execution process.

3.2 Actor-Critic Structure for HDP

In this subsection, neural networks (NNs) are constructed to solve (14) for $V^m(\delta(k))$ for the implementation of the proposed HDP algorithm.

Design of Critic NN: Define the critic NN approximator as

$$\hat{V}(\delta(k)) = \delta^{\mathrm{T}}(k)\hat{w}_c^{\mathrm{T}}\delta(k) \tag{16}$$

where $\delta(k)$ is the global error, and $\hat{w}_c \in \mathbf{R}^{nN \times nN}$ is the unknown coefficient vectors of critic NN. Let $\bar{V}(\delta(k))$ be the target of the critic NN at step m such that

$$\bar{V}(\delta(k)) = r\big(\delta(k), \hat{u}_i^m(k), \hat{u}_{Gr-i}^m(k)\big) + \hat{V}^m(\delta(k+1)) \tag{17}$$

The critic NN approximation error is defined as

$$e_c(k) = \bar{V}(\delta(k)) - \hat{V}^m(\delta(k)) \tag{18}$$

In order to minimize the error performance index

$$E_c(k) = \frac{1}{2}e_c^{\mathrm{T}}(k)e_c(k) = \frac{1}{2}\|\bar{V}(\delta(k)) - \hat{V}^m(\delta(k))\|^2, \tag{19}$$

the gradient descent method is applied to update the weights of critic NN

$$\hat{w}_c(k+1) = \hat{w}_c(k) - \alpha_c\delta(k)\delta(k)^{\mathrm{T}}\big(V(\delta(k)) - \hat{V}^m(\delta(k))\big) \tag{20}$$

where α_c denotes the learning rate of the critic NN.

Design of Actor NNs: We define the actor networks for each agent as

$$\hat{u}_i(\delta_i(k)) = \hat{w}_{ai}^{\mathrm{T}}\delta_i(k) \tag{21}$$

where $\delta_i(k)$ is the local neighborhood error of agent i, and $\hat{w}_{ai} \in \mathbf{R}^{nN_i \times m_i}$ is the unknown coefficient vectors of actor NN for agent i.

Then, we give the updating laws for the actor NNs. For the actor NN of each agent, its target is the iterative control policy by (15), i.e., $\bar{u}_i(k) = (d_i + g_i)R_i^{-1}B_i^{\mathrm{T}}\nabla\hat{V}_i^{m+1}(\delta(k+1))$. Based on (16), we can obtain

$$\bar{u}_i(k) = (d_i + g_i)R_i^{-1}B_i^{\mathrm{T}}O_i\hat{w}_c^{\mathrm{T}}\delta(k) \tag{22}$$

where $O_i = 2 \times [0, \cdots [I]_{ii}, \cdots 0] \in \mathbf{R}^{n \times nN}$. As a result, the actor NN error is

$$e_{ai}(k) = \hat{u}_i(\delta(k)) - \bar{u}_i(k) = \hat{w}_{ai}^{\mathrm{T}}\delta_i(k) - (d_i + g_i)R_i^{-1}B_i^{\mathrm{T}}O_i\hat{w}_c^{\mathrm{T}}\delta(k) \tag{23}$$

Similar with the critic NN, the updating law for actor NNs is

$$\hat{w}_{ai}(k+1) = \hat{w}_{ai}(k) - \beta_i\delta_i k\big(\hat{w}_{ai}^{\mathrm{T}}\delta_i(k) - \bar{u}_i(k)\big) \tag{24}$$

where β_i is the learning rate of actor NN for agent i.

4   Simulation

Consider the four-node communication graph. The leader is pinned to node 4 with $g_4 = 1$ and $g_i = 0 (i \neq 4)$ with the following graph Laplacian matrix L

$$L = \begin{bmatrix} 1.6 & -0.4 & -0.3 & -0.9 \\ -0.8 & 1.4 & -0.6 & 0 \\ -0.75 & -0.6 & 1.35 & 0 \\ 0 & -0.2 & 0 & 0.2 \end{bmatrix}$$

The dynamics of the agents $i(i = 1, 2, 3, 4)$ is described as

$$\begin{bmatrix} \dot{x}_{i1} \\ \dot{x}_{i2} \end{bmatrix} = \begin{bmatrix} 0 & 1 \\ -1 & 0 \end{bmatrix} \begin{bmatrix} x_{i1} \\ x_{i2} \end{bmatrix} + \begin{bmatrix} 1 \\ 0.5 \end{bmatrix} u$$

Define the weight matrices of the shared value function as $Q = I, R_1 = R_2 = R_3 = R_4 = I$. The learning rates are $\alpha_c = 0.2$ and $\beta_i = 0.1$. Let the initial states of each agent in the games be

$$x_1(0) = \begin{bmatrix} 0.3 \\ 0.4 \end{bmatrix}, \quad x_2(0) = \begin{bmatrix} 0.2 \\ 0.6 \end{bmatrix}, x_3(0) = \begin{bmatrix} -0.4 \\ -0.3 \end{bmatrix}, \quad x_4(0) = \begin{bmatrix} -0.2 \\ -0.6 \end{bmatrix}.$$

The initial states of the leader is chosen as $x_0(0) = [0.5 - 0.5]^T$. The initial parameters of critic NN is chosen randomly in $[0, 1]$, and the parameters of actor NNs are both initialized as zeros. The sampling time interval is 0.01s. The trajectories of the consensus errors between the agents $i(i = 1, 2, 3, 4)$ with the leader during the learning process are given in Fig. 1, where the consensus errors are converged to zero. Then, the centralized learning phase is terminated.

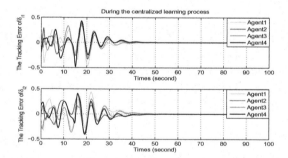

Fig. 1. The consensus errors $\delta_i, i = 1, 2, 3, 4$ during the centralized learning process

Then, we apply the obtained control policies during the decentralized execution process with different initial states.

$$x_1(0) = \begin{bmatrix} 0.2 \\ -0.2 \end{bmatrix}, x_2(0) = \begin{bmatrix} 0.3 \\ -0.1 \end{bmatrix},$$

$$x_3(0) = \begin{bmatrix} 0.1 \\ -0.6 \end{bmatrix}, x_4(0) = \begin{bmatrix} -0.2 \\ -0.6 \end{bmatrix}, x_0(0) = \begin{bmatrix} 0.4 \\ -0.3 \end{bmatrix}$$

Then, we execute the trained actor NN for the MAS only based on the local information, From Fig. 2, we can see that the consensus error between the agents with the leader can be converged to the equilibrium state quickly.

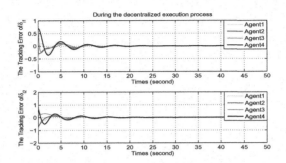

Fig. 2. The consensus error $\delta_i, i = 1, 2, 3, 4$ during the execution process

5 Conclusion

In this paper, we investigate the value iteration scheme for the optimal consensus of MAS under the centralized learning and decentralized execution framework. The HDP algorithm for MAS is proposed with the actor-critic structure. Finally, we give the simulation results to verify the effectiveness of proposed HDP algorithm.

References

1. Sutton, R. S., Barto, A. G.: Reinforcement Learning: An Introduction. MIT Press, Cambridge (1998)
2. Lewis, F. L., Vrabie, D., Syrmos, V. L.: Optimal Control. Wiley, Hoboken (2012)
3. Zhu, Y., Zhao, D., Li, X.: Iterative adaptive dynamic programming for solving unknown nonlinear zero-sum game based on online data. IEEE Trans. Neural Networks Learn. Syst. **28**(3), 714–725 (2017)
4. Zhang, Z., Zhao, D., Gao, J., et al.: FMRQ-A multiagent reinforcement learning algorithm for fully cooperative tasks. IEEE Trans. Cybern. **47**(6), 1367–1379 (2017)
5. Lowe, R., Wu, Y., Tamar, A., et al.: Multi-agent actor-critic for mixed cooperative-competitive environments. In: Advances in Neural Information Processing Systems, pp. 6382–6393 (2017)
6. Tesauro, G.: Extending Q-learning to general adaptive multi-agent systems. In: Advances in Neural Information Processing Systems, pp. 871–878 (2004)
7. Vamvoudakis, K.G., Lewis, F.L., Hudas, G.R.: Multi-agent differential graphical games: Online adaptive learning solution for synchronization with optimality. Automatica **48**(8), 1598–1611 (2012)
8. Zhu, Y., Zhao, D., Zhong, Z.: Adaptive optimal control of heterogeneous CACC system with uncertain dynamics. IEEE Trans. Control Syst. Technol. (2018) https://doi.org/10.1109/TCST.2018.2811376

9. Khoo, S., Xie, L., Man, Z.: Robust finite-time consensus tracking algorithm for multirobot systems. IEEE/ASME Trans. Mechatron. **14**(2), 219–228 (2009)
10. Pedregal, P.: Optimal Control. Introduction to Optimization. TAM, vol. 46, pp. 195–236. Springer, New York (2004). https://doi.org/10.1007/0-387-21680-4_6

Potential and Sampling Based RRT Star for Real-Time Dynamic Motion Planning Accounting for Momentum in Cost Function

Saurabh Agarwal, Ashish Kumar Gaurav, Mehul Kumar Nirala, and Sayan Sinha(✉)

Indian Institute of Technology Kharagpur, Kharagpur, India
{shourabhagarwal,ashishkg0022,mehulkumarnirala,sayan.sinha}@iitkgp.ac.in

Abstract. Path planning is an extremely important step in every robotics related activity today. In this paper, we present an approach to a real-time path planner which makes use of concepts from the random sampling of the Rapidly-exploring random tree and potential fields. It revises the cost function to incorporate the dynamics of the obstacles in the environment. Not only the path generated is significantly different but also it is much more optimal and rigid to breakdowns and features faster replanning. This variant of the Real-Time RRT* incorporates artificial potential field with a revised cost function.

Keywords: Path planning · Robotics · RRT · Potential energy
Wavefront · Doppler effect

1 Introduction

Robotics challenges today involve the rapid motion of robots in a highly dynamic environment where agents try to outperform their opponents with superior planning and strategy. At the lowest level of operation, path planning has a major contribution to the performance of an agent. In this paper, we describe the steps and methods involved in the making of a dynamic path planner suited for such challenging environments. We take into consideration the real-time knowledge of the two-dimensional coordinate representing the position of various robots. We also assume that the data available is correct and do not question its accuracy.

Random sampling path planners have been highly used in various fields of robotics, such as aerial robotics [6] and robosoccer [1]. Many variants of Rapidly-exploring Random Trees (RRTs) have been used in such areas. Along with that, path planning based on potential fields has also been popular in robotics and are pretty commonly used in areas such as self-driving cars [3]. In this paper, we describe the use of a variant of the RRT algorithm and discuss incorporation of the concept of artificial potential field (APF) into it.

S. Agarwal, A.K. Gaurav, M.K. Nirala and S. Sinha—Equal contribution.

© Springer Nature Switzerland AG 2018
L. Cheng et al. (Eds.): ICONIP 2018, LNCS 11307, pp. 209–221, 2018.
https://doi.org/10.1007/978-3-030-04239-4_19

2   Previous Works

Extensive research has been done on the path-planning problem in the past few years. Traditional approaches fail to provide estimates before the motion begins, and hence, it becomes difficult to provide proper control on motion of agent. Moreover, their instantaneous nature is prone to making the velocity graph abstract. This makes the motion of the robot unstable, bringing in more error into the scenario, leading to the violation of kinodynamic constraints. Such planners include MergeSCurve [10] and Dynamic Window [4]. Visibility Graphs [7] is another widely used algorithm for obstacle avoidance. But, it generates paths that are very close to the obstacles, and hence is not suitable for a dynamic environment. Other models such as Graph Plan [2] prefer concentrating on fields of robotics with a large number of degrees of freedom, and is difficult to be developed into generic algorithms for path planning.

Path planning algorithm for the dynamic environment needs to be real-time and robust. The paths need to be regenerated or updated from time to time based on various factors. The RRT too is not suitable for dynamic environments. As soon as a new path needs to be generated, a tree growth is performed from the source to the destination, which can be pretty costly. However, various variants of the RRT have been proposed to make them suitable according to specific needs. The variant of RRT suitable in our case is the RT-RRT* [9]. Similarly, the concept of a potential field for path planning was initially applied for static situations only. Though later on, various research work has been performed to introduce dynamism into Artificial Potential Field (APF) [15] algorithms and create ameliorative APF models [11] making them suitable for a rapidly changing environment. In this section, we first discuss the previously proposed algorithms of the RRT. Several graph and potential based algorithms have been discussed.

2.1   Rapidly-Exploring Random Tree

Rapidly-exploring Random Tree (RRT) [8] finds path between $x_{init} \subset X$ and $x_{goal} \subset X$. It is assumed that an obstacle region $X_{obs} \subset X$ is given and we can check if a point lies in this region or not. If any point or edge lies in this region, that should not be used to extend our tree to avoid collision with any obstacle. It is a simple and fast algorithm for finding a path between two points, but it has several limitations. It does not guarantee convergence to an optimal path. Furthermore, no measure is taken to make this algorithm compatible with the dynamic environment. Later, various extensions have been proposed to address these problems. Below is a brief description of RRT and Algorithm 1 presents its pseudo code. In this algorithm, $SampleFree()$ samples a node $x_{rand} \subset X$ randomly in space. Then $x_{nearest}$ is found using $Nearest(V, x)$ which finds a node nearest to x among a list of nodes V. After this $Steer(x_t, x_o)$ is used to give point on path which originates from x_{rand} and terminates to $x_{nearest}$. If $(x_{nearest}, x_{node})$, a path joining $x_{nearest}$ and x_{node}, belong to X_{free} then x_{node} is inserted into V and similarly $(x_{nearest}, x_{node})$, edge joining $x_{nearest}$ and x_{node}, is inserted in E.

Algorithm 1. RRT algorithm

1: **procedure** RRT(x_{init})   ▷ The source node
2:     $V \leftarrow \{x_{init}\}$
3:     $E \leftarrow \phi$
4:     **for** $i = 1, 2, ..., n$ **do**
5:         $x_{rand} \leftarrow SampleFree(i)$
6:         $x_{nearest} \leftarrow Nearest(V, x_{rand})$
7:         $x_{node} \leftarrow Steer(x_{nearest}, x_{rand})$
8:         **if** ObstacleFree($x_{nearest}, x_{node}$) **then**
9:             $V \leftarrow V \cup \{x_{node}\}$
10:            $E \leftarrow E \cup \{(x_{nearest}, x_{node})\}$
11:    **return** $G = (V, E)$

2.2 RRT*

RRT* [5] algorithm proposes a way which tries to minimise the distance of the root from the nodes at each new iteration. After the random sampling of a new node, this algorithm allows rewiring of the tree to reduce the distance from root to child node. It inspects each of the nodes of the tree which are within a neighbourhood of the newly generated child node. This child node is then reconnected to that node, which traces up to the tree root at the shortest distance.

2.3 A Real-Time Path Planning Algorithm Based on RRT*

RT-RRT* [9] algorithm is similar to RRT*, but the tree growth takes place just once, and the nodes are rewired from time to time based on a variety of factors. Since the tree growth does not take place again and again, RT-RRT* is suitable for a dynamic environment as the overhead costs are low. Moreover, rewiring of nodes makes sure the tree has been modified to suit the changes in the environment.

2.4 Potential Guided Directional-RRT*

Potential guided directional-RRT*(PDG-RRT*) [12] is a modification of RRT* which uses Artificial Potential Field [15] for guiding the random sampling more towards the goal. This enhances the rate of convergence and provides a more optimal solution. At each iteration, a random node x_{rand} is generated. This node x_{rand} is then moved a fixed distance α along the direction of the potential gradient to give a new node z. This node z is then added to the tree by adding an edge from this node to another node in the tree such that the distance from the source to z is minimised, and we define $nodes_{max}$ is the maximum number of nodes.

Algorithm 2. Potential Guided Directional-RRT* algorithm

1: **procedure** POTENTIAL GUIDED DIRECTIONAL-RRT*(x_{init}) ▷ The source node
2:     $V \leftarrow \{x_{init}\}$
3:     $E \leftarrow \phi$
4:     **for** $i = 1, 2, 3 \ldots nodes_{max}$ **do**
5:         $x_{rand} \leftarrow SampleFree(i)$
6:         $x_{prand} \leftarrow GradientDescent(x_{rand})$
7:         $x_{nearest} \leftarrow Nearest(V, x_{prand})$
8:         $x_{node} \leftarrow Steer(x_{nearest}, x_{prand})$
9:         **if** ObstacleFree$(x_{nearest}, x_{node})$ **then**
10:             $V \leftarrow V \cup \{x_{node}\}$
11:             $E \leftarrow E \cup \{(x_{nearest}, x_{new})\}$
12:     **return** $G = (V, E)$

---

3   Proposed Path Planner

The previous works describe various ways in which the RRTs have been modified to improve the optimality, speed and dynamism of path generation. The PGD-RRT* [12] promises more optimal paths, and we attempt to extend it to the dynamic environment. The RT-RRT* is a variant of the RRT* which features dynamism. In this paper, the RT-RRT* has been made more optimal, using concepts from PGD-RRT*. Along with that, some novel concepts have been introduced to bridge the gap, such as a new formula for computing potential and a new cost function.

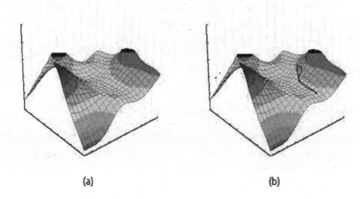

(a)   (b)

Fig. 1. (a) The graph of superimposition of potential and frequency. (b) Straight line path is taken as cost in other planners marked in grey; actual path taken by our planner marked in black. The black patches mark infinity.

3.1   Tree Growth

The RT-RRT* [9] proposes to grow the tree just once during the lifetime of the path planner, and reuses the nodes and the edges to predict the path. We use a similar concept here, with the difference being in the fact that the random sampling is influenced by the potential field created.

Potential Field: The potential field generated can be compared to an electrostatic field. The obstacles are assumed to be positively charged, and the destination to be negatively charged. This places the obstacles at a relatively higher positive potential. The aim is to descend towards lower potential as much as possible. All charges are taken as point charges. At every point x in the space, the potential is calculated for a particular charge as $V_i = \frac{kQ_i}{r_i}$ where Q_i is the magnitude of the ith point charge (obstacle or destination), and r_i is the distance of the point x from it. The net potential at a point is given by $\sum_{i=1}^{n} V_i = k\sum_{i=1}^{n} \frac{Q_i}{r_i}$. Here k is a proportionality constant, and n is the total number of charges. We determine the values of Q and k by validation.

The sampling method used in the RT-RRT* [9] to generate new nodes is applicable here as well. Briefly, a random number between 0 and 1 is generated, and on the basis of that, it is determined where random sampling is to be performed. $LineTo(x_{goal})$ is invoked when the random number generated is greater than $1 - \alpha$, else if the number is less than $\frac{1-\alpha}{\beta}$ $Uniform(X)$ is invoked. Otherwise $Ellipsis(x_0, x_{goal})$ is called. Here β is a real number used for making a distinction between the last two function calls. Next, taking inspiration from PGD-RRT* [12], the randomly generated point is allowed to descend according to the potential gradient for a certain amount of time. The resulting point is treated as the initially generated node for an RRT*. The best parent to this node is determined within the surrounding grids of the node. From that parent, at a particular step length, the new node is created. The entire procedure, in a nutshell, is provided in the Algorithm 3.

3.2   Hexagonal Grid

The entire area is divided into a set of grids, and all the nodes are mapped to a specific grid. In another way, every node is assigned a particular grid ID, which is developed in such a manner that the entire area gets divided into a set of hexagonal grids. Hexagon ensures most efficient packing in 2D space and is quite popular in path planning domain [13]. The potentials developed correspond to those of the points at the centres of the respective grids.

3.3   Edge Rewiring

The RT-RRT* proposes to rewire edges under three categories. We explain the changes we make to each of them:

Algorithm 3. Random sampling

1: **procedure** RANDOM SAMPLING(U, G, α, β, γ, $step\_size$)   ▷ Graph G is the existing graph, Field U is the generated field
2:     $A \leftarrow rand(0,1)$ ▷ Random number generation between 0 and 1.
3:     **if** $A \geq 1 - \alpha$ **then**
4:         $x_{rand} = LineTo(x_{goal})$
5:     **else**
6:         **if** $\frac{1-\alpha}{\beta}$ **then**
7:             $x_{rand} = Uniform(X)$
8:         **else**
9:             $x_{rand} = Ellipsis(x_0, x_{goal})$
10:     $x_{parent} = BestParent(x_{rand})$▷ Best parent is decided as per the recursive cost function
11:     $x_{random} = NodeAtDist(Angle(x_{parent}, x_{rand}), step\_size)$ ▷ Finding the node on the line joining x_{parent} and x_{rand} at a distance $step\_size$ from x_{parent}
12:     $x_{final} = x_{random} - \gamma \frac{\delta(U)}{\delta(x)}|_{x=x_{final}}$
13:     $U = AddToGraph(U, x_{final})$
14:     **if** Graph has been saturated **then**
15:         return
16:     **else**
17:         $RANDOMSAMPLING(U, G, \alpha, \beta, \gamma, \_size)$

Rewire in the Presence of an Obstacle: When a path is being approached by an obstacle, a rewiring is necessary. According to the RT-RRT* algorithm, any path which goes through the obstacles attain infinite cost. Similarly, when obstacles have charge, they provide an infinite potential at their point of existence. Hence, both of them seamlessly integrate with one another. But, here, the infinite potential is assumed for a greater radius, as deemed fit for the situation. Accordingly, rewiring is prioritised in such regions.

Rewire from Root: Rewiring begins at the root and keeps proceeding throughout the tree. Hence, it starts rewiring from the position where the agent lies. Then, it continues to assign nodes which are at greater distance from the agent to their corresponding best-matching parents.

Random Rewiring: This paper proposes to remove random rewiring from RT-RRT* and introduces a way of determining edges which are to be rewired. At every point of the field, these wavefronts would be perceivable. In case the robot is moving, the frequency detected at a point would be different from that of what had been emitted. This is known as the Doppler Effect. The frequency due to Doppler effect is expressed as $f = \left(\frac{v+v_r}{v+v_s}\right) f_0$ where f is the perceived frequency, f_0 is the constant frequency emitted by obstacles, v is the constant velocity of wave in space, v_r is receiver's velocity and v_s is the source's velocity which will be equal to the obstacle's velocity in our case. The frequency at every point in the field is calculated using $\sin(\omega_1 t) + \sin(\omega_2 t) = 2\sin\left(\frac{\omega_1+\omega_2}{2}t\right)\cos\left(\frac{\omega_1-\omega_2}{2}t\right)$ in a binary fashion. This has been illustrated in Fig. 2(a). Thus, we obtain the net frequency of every point in the field. After plotting this across the field, we obtain the local minima. Finding out the maximum among these local minima, we rewire edges randomly in that grids.

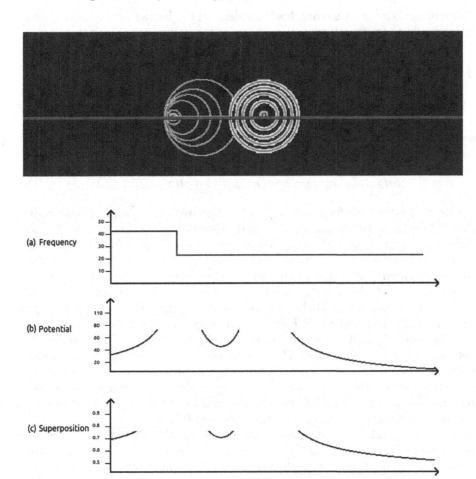

Fig. 2. Wavefronts due to robot while stationary and while moving. This is described mathematically by the Doppler Effect. Along the violet line, the (a) frequency graph, (b) the potential graph and (c) the overall graph have been shown (Color figure online)

3.4   Boundary Conditions

An exponential potential control barrier function is defined in 3D space. Let the boundary of the field/environment be defined by $\mathbf{B}(\mathbf{x})$ and let $\mathbf{x}_b$ be a point on the boundary. The growth of the potential function around boundary is defined by $B(x) = \frac{1}{1-e^{-||x_b-x||}}$. Thus, the potential field around the boundary of the environment proliferates as,

$$\lim_{x \to x_b} \frac{1}{1 - e^{-||x_b-x||}} \to \infty$$

$||x||$ represents $x^T.x$. A moving body incident on the barrier gets reflected, and hence would restrict path generation.

4   Cost Function

The cost function in the RT-RRT* [9] algorithm is used to find the best parent of a node while performing edge-rewiring. Edge rewiring, as described in previous sections, helps in optimising an already generated path, by providing nodes with better parents. The RT-RRT* defines cost recursively as:

$$cost_{distance}(x_{new}) = cost(x_{closest}) + dist(x_{closest}, x_{new}) \tag{1}$$

Where x_{new} is the node for which we wish to compute the cost, $x_{closest}$ is the node closest to x_{new} and $dist(x_{closest}, x_{new})$ is the distance between x_{new} and $x_{closest}$. This paper proposes to make some changes to the existing cost function, making PGD-RRT* suitable for a dynamic environment. The recursive approach to the cost function remains the same. The difference lies in the way the distance (using $dist$) is computed. Hence $x_{closest}$ shall be modified as well. Nevertheless, $x_{closest}$ remains the closest node via the distance metric, but in this paper, we attempt to replace the commonly used Euclidean distance with geodesic distance. The two-dimensional graph of the area initially developed using an artificial potential field is taken into consideration. The straight line path which gives the Euclidean distance between x_{new} and $x_{closest}$ is projected on this graph. The length of the curve thus obtained is taken as the distance. Thus, the formula remains the same, but the approach to computing the distance is a bit different. It is explained in a better fashion through the illustration provided (Fig. 1).

Let the potential surface be represented as S and the position vectors of the points $x_{closest}$ and x_{new} be a and b. The distance is measured along the curve formed by intersection of S with a plane which is parallel to the z axis and passes through the straight line path between a and b. We assign $\hat{n}$ to be the cross product between $a - b$ and a unit vector along the z axis. We obtain a plane P as $(r - a) \cdot \hat{n} = 0$, where r is an arbitrary point on the plane. Thus, the curve formed along the path between two nodes becomes $C: S - P = 0$. $dist(x_{closest}, x_{new})$ is the arclength along C from a to b. If we represent C parametrically as $p(t) = < x(t), y(t), z(t) >$ with t_1 and t_2 being the parametric values for a and b, then the arclength is given by

$$dist(x_{closest}, x_{new}) = \int_{t_1}^{t_2} \sqrt{\left(\frac{dx}{dt}\right)^2 + \left(\frac{dy}{dt}\right)^2 + \left(\frac{dz}{dt}\right)^2} \, dt$$

Doppler Effect: The cost function incorporates the velocity using the Doppler effect. This accounts for the dynamic nature of obstacles. The shift in frequency helps in the prediction of approach/separation of the body concerning obstacles. The frequency at every point in the field is calculated by superimposition of waves from all the obstacles, as illustrated in Fig. 2. This shift in frequency is plotted on a two dimensional axis for all obstacles. The obtained map explains the relative velocity of approach/separation of obstacles with respect to the moving body.

$$\mathbf{F}(\hat{\mathbf{x}}, t) = \sum_{i=0}^{|obstacles|} f_i \tag{2}$$

where, $|obstacles|$ represents the cardinality of the obstacle vector, f_i denotes the observed frequency of i_{th} obstacle and t is time. The surface so obtained corresponds to obstacles' activity. The minima on the surface correspond to low activity regions of obstacles. To extract the local minima the surface is descended iteratively. $\mathbf{x}_{n+1} = \mathbf{x}_n - \gamma_n \nabla F(\mathbf{x}_n)$, $n \geq 0$. where $\mathbf{x}_n$ are points in Cartesian plane with $\mathbf{x_0}$ as initial guess. To ensure convergence step size γ is kept small. However an effective γ via Barzilai-Borwein method [14] yields promising results. The Barzilai-Borwein method is shown below

$$\gamma_n = \frac{(\mathbf{x}_n - \mathbf{x}_{n-1})^T [\nabla F(\mathbf{x}_n) - \nabla F(\mathbf{x}_{n-1})]}{||\nabla F(\mathbf{x}_n) - \nabla F(\mathbf{x}_{n-1})||^2} \tag{3}$$

The cost is defined as

$$cost_{velocity}(x_{new}) = \lambda |\mathbf{F}(\hat{\mathbf{x}}, t)|_{at\ \mathbf{x}=x_{new}} \tag{4}$$

where λ is a constant close to unity. Thus, low activity regions would be associated with low costs. Effective or total cost is computed as a linear combination of given by,

$$cost(x) = c_0 cost_{distance}(\mathbf{x}) + c_1 cost_{velocity}(\mathbf{x}) \tag{5}$$

Where c_0 and c_1 are two constants which depend on the environment and can be tuned to achieve promising results. Incorporating boundary condition would change effective cost to

$$cost(x)_{effective} = cost(x)B(x)$$

5   Experimentation and Analysis

We conducted an experiment based on tree generation time given a set of obstacles. The experimental setup included ROS based communication nodes and a GUI based interactive platform for setting up the environment. This was performed on Linux 4.4.0-127-generic with 8 GiB memory and 2.3 GHz x 8 CPU. Thus, it can be seen that our planner takes nearly the same time to generate the

Algorithm 4. Rewiring random edges

1: **procedure** REWIRE RANDOM(U, G, D, $grid$, ϵ, $root$)   ▷ Graph G is the existing graph, Field U is the generated field
2:    $F = U + \epsilon D$
3:    $grid_{min} = rand(1, length(grid))$
4:    $max\_min = grid_{min}$
5:    **for** each i; i in $length(grid)$ **do**
6:        $flag = False$
7:        **if** $F[i] < F[$each $neighbour$ of $grid[i]]$ **then**
8:            **if** $max\_min < F[i]$ **then** $max\_min < F[i]$
9:    $cost_{org} = cost(Parent(max\_min), max\_min)$
10:    $node_{final} = max\_min$
11:    **for** each node i in current and neighbouring grids **do**
12:        $cost_{here} = cost(i, max\_min) + cost(i, root)$
13:        **if** $cost_{here} < cost_{org}$ **then**
14:            $cost_{org} = cost_{here}$
15:            $node_{final} = i$
16:    $max\_min.setParent = node_{final}$

---

tree given similar set of environmental conditions. Along with that, our planner provides a more optimised path. The average number of iterations required to find the path to each goal was 10.36 for our method and 18.43 for RT-RRT*.

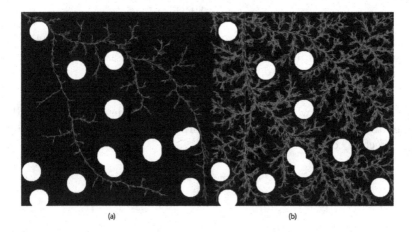

(a)   (b)

Fig. 3. The simulation environment, showing (a) our path planner and (b) RT-RRT*

Replanning: In a dynamic environment, the replanning of path is an important aspect. Moving obstacles when come along the path, a new path has to be found incorporating the new position of the obstacles. For better performance of agents, this replanning of path should take place in less amount of time.

Table 1. Comparison of time (in milli seconds) in generation of path

Obstacles	Average Traversal Time			
	PGD-RRT*	RRT*	RT-RRT*	Our planner
5	4.1996	3.7078	3.7951	2.7958
7	4.2315	3.7912	3.8105	2.8110
9	4.2555	3.9105	3.8492	2.8862
10	4.2658	3.9824	3.8908	2.8873
12	4.3655	4.1473	4.1335	3.1080
15	4.4662	4.2194	4.3264	3.1866
18	4.6905	4.4505	4.5155	3.4242
20	4.9696	4.9191	4.9583	3.9823

Our proposed path planner outperforms other planners in this, hence making it suitable for a real-time dynamic environment. With respect to the PGD-RRT*, our path planner is very quick in replanning as the tree growth does not take place again and again, but gets rewired when required. With respect to the RT-RRT*, our planner provides a more optimised path, which is clear from the illustration provided (Fig. 3) and Table 1. Our planner is based on a dynamic environment only and can be outperformed by other planners in a static environment (Table 2).

Table 2. Comparison of time (in milli seconds) in replanning of path

Obstacles	Average Replanning Time			
	PGD-RRT*	RRT*	RT-RRT*	Our planner
5	8.1996	8.7078	3.7958	3.7951
7	8.2758	8.8213	3.8501	3.8491
10	8.2783	8.8582	3.8614	3.8519
11	8.2811	8.8607	3.8688	3.8564
15	8.2902	8.9078	3.8891	3.8888
18	8.3215	8.9303	3.9011	3.8997
21	8.3414	8.9489	3.9114	3.9114

Our planner takes lesser tree generation time compared to the RT-RRT*, as it is biased in an intelligent way to reach the destination faster, using the concept of potential fields. Moreover, our planner features faster replanning as compared to PGD-RRT* as the tree growth does not take place again and again and is rewired. A similar advantage is seen over RRT* as well. The path generated is more optimised as compared to the RT-RRT* due to intelligent rewiring, which is evident from Table 1.

6   Conclusion

Our work focuses on a path planner with concepts from a variant of rapidly exploring random trees and artificial potential fields. They are merged innovatively, with concepts from various other domains, and a cost function redefined to suit a dynamic environment. This remains efficient in time with respect to tree growth and requires a lesser number of iterations for path generation, compared to that of similar planners.

Future Work

In the future, we propose to research on how the acceleration of a robot can help in computing a more optimal path. Though the mathematics proposed in this paper can easily be extended to take acceleration into consideration, acceleration is not that easy to measure and a lot of noise comes in the way of the location data of the robots being received. Hence, our research aims to find an efficient solution to denoising the data post which we can take into consideration the acceleration, and if possible, other forms of kinetics.

Acknowledgement. We thank Manjunath Bhatt (manjunathbhat9920@iitkgp.ac.in), Rahul Kumar (vernwalrahul@iitkgp.ac.in) and Shubham Maddhashiya (shubhamsipah@iitkgp.ac.in) for assisting us in this project and supporting us as and when required.

References

1. Bhushan, M., Agarwal, S., Gaurav, A.K., Nirala, M.K., Sinha, S., et al.: KgpKubs 2018 team description paper. In: RoboCup 2018 (2018)
2. Blum, A.L., Furst, M.L.: Fast planning through planning graph analysis. Artif. Intell. **90**(1–2), 279–298 (1997)
3. Dolgov, D., Thrun, S., Montemerlo, M., Diebel, J.: Practical search techniques in path planning for autonomous driving. In: Proceedings of the First International Symposium on Search Techniques in Artificial Intelligence and Robotics (STAIR-08) (2008)
4. Fox, D., Burgard, W., Thrun, S.: The dynamic window approach to collision avoidance. IEEE Robot. Autom. Mag. **4**, 23–33 (1997)
5. Karaman, S., Frazzoli, E.: Incremental Sampling-based Algorithms for Optimal Motion Planning. Robotics: Science and Systems. arXiv preprint:1005.0416 (2010)
6. Kim, J., Ostrowski, J.P.: Motion planning of aerial robot using rapidly-exploring random trees with dynamic constraints. In: IEEE International Conference on Robotics and Automation (Cat. No.03CH37422), vol. 2, pp. 2200–2205 (2003)
7. Kunigahalli, R., Russell, J.S.: Visibility graph approach to detailed path planning in CNC concrete placement. In: Proceedings of the 11th ISARC, pp. 141–147 (1994)
8. LaValle, S.M.: Rapidly-exploring random trees: A new tool for path planning. Report No. TR 98–11. Computer Science Department, Iowa State University (1998)
9. Naderi, K., Rajamki, J., Hmlinen, P.: RT-RRT*: a real-time path planning algorithm based on RRT*. In: 8th ACM SIGGRAPH Conference on Motion in Games (MIG 2015), pp. 113–118 (2015)

10. Nguyen, K.D., Ng, T.C., Chen, I.M.: On algorithms for planning S-curve motion profiles. Int. J. Adv. Robot. Syst. **5**(1), 99–106 (2008)
11. Qixin, C., Yanwen, H., Jingliang, Z.: An evolutionary artificial potential field algorithm for dynamic path planning of mobile robot. In: International Conference on Intelligent Robots and Systems, pp. 3331–3336 (2006)
12. Qureshi, A.H., et al.: Potential guided directional-RRT* for accelerated motion planning in cluttered environments. In: IEEE International Conference on Mechatronics and Automation, Takamatsu, pp. 519–524 (2013)
13. Sinha, S., Nirala, M.K., Ghosh, S., Ghosh, S.K.: Hybrid path planner for efficient navigation in urban road networks through analysis of trajectory traces. In: 24th International Conference on Pattern Recognition (2018, in Press)
14. Tan, C., Ma, S., Dai, Y., Qian, Y.: Barzilai-Borwein step size for stochastic gradient descent. arXiv preprint:1605.04131 (2016)
15. Vadakkepat, P., Lee, T.H., Xin, L.: Application of evolutionary artificial potential field in robot soccer system. In: Joint 9th IFSA World Congress and 20th NAFIPS International Conference, vol. 5, pp. 2781–2785 (2008)

Dynamic Control of Storage Bandwidth Using Double Deep Recurrent Q-Network

Kumar Dheenadayalan$^{(\boxtimes)}$, Gopalakrishnan Srinivasaraghavan,
and V. N. Muralidhara

International Institute of Information Technology, Bangalore, India
d.kumar@iiitb.org

Abstract. We propose a novel approach to optimize the performance of
a large scale physical system by mapping the performance optimization
problem into a reinforcement learning framework. A reasonably efficient
manual bandwidth control for large storage servers seems to be a difficult
task for system administrators, but a dynamic bandwidth control can
be effectively learned by a reinforcement learning agent. We adopt a
combination of Double Deep Q-Network and a Recurrent Neural Network
as our function approximator to identify the extent of bandwidth control
(actions) given the state representation of a storage server. Allowing the
agent to control the amount of allowable bandwidth to each logical unit
within a filer has shown to enhance throughput as-well-as reduce the
overload duration of storage servers.

1 Introduction

In an era of big-data analytics, the explosion of data has imposed tremendous
stress on various entities of a large distributed infrastructure. Attempts to predict and forecast the performance of such entities provide limited advantages
in terms of reducing the load on such infrastructure. Beyond understanding the
state or the extent of load, there is lack of literature on enhancing the online
performance of these systems. Self-managing and self-tuning mechanisms are
required to manage such large scale systems. Thus, we present one of such self-
tuning mechanisms for a cluster-mode storage server popularly referred to as
filers. They are responsible for servicing high capacity I/O request from a large
number of clients. A modern filer has PetaBytes (PB) of storage capacities with
thousands of underlying disks. Self-tuning is a challenging research problem due
to the varying effects of different components.

The prediction of anomalies and potential problem has been attempted in
the past, especially for small scale systems (i.e. disk level). In most cases, the
solutions have an implicit assumption of availability of accurate device models
[1] which is impractical given the scale of the filer. The choice of actions to be
taken based on these predictions is another key task and is difficult even for an
experienced system administrator. It is further complicated and constrained by
the presence of numerous Service Level Agreements (SLAs). The automation of

© Springer Nature Switzerland AG 2018
L. Cheng et al. (Eds.): ICONIP 2018, LNCS 11307, pp. 222–234, 2018.
https://doi.org/10.1007/978-3-030-04239-4_20

implementing potential action in case of an anomalous or undesired high load on a filer is an open problem.

Research related to the application of Reinforcement learning (RL) is popular in the fields of gaming, robotics as well as other domains. Reinforcement learning has been applied in diverse domains such as telecommunications and networking [2,12], physical system optimization and maintenance [3,9] and data migration in multi-tier storage systems [14]. Hence, RL-algorithms are an interesting choice for optimizing a large scale system such as a cluster-mode filer to understand its state and choose appropriate actions (if such an action exists) so as to achieve optimal storage performance. If an RL-agent is able to successfully learn to optimize the storage performance, then transferring the RL-agent to similar filers is possible and this can reduce the time taken for convergence to occur.

Environment: The environment for this problem consists of not only the filer but all the entities (i.e. applications and clients) that communicate with the filer for I/O processing. Even if we assume the predictability of application and user behavior, the filer as a single entity is known to have a complex interaction with various components at physical, logical and software levels. Hence, the application of RL is not trivial due to the complexity posed by the large state space. This is especially true for large-scale storage systems where the number of state variables/features is of the order of 10^5. On the other hand, the action space can also be large due to the wide range of components which require different actions (e.g. modifying TCP packet size, throttling logical volumes, online backup rescheduling).

Given the fact that compressed search space also approximately models the physical state of a filer [4,5], it provides a meaningful alternative for incorporating only a selected number of features in the state representation. In addition to this, the action space can also be reduced to focus on a single component that has a high impact on the performance of the filer. Although our primary aim is to maximize the performance of a filer completely driven by an RL-agent, reducing the action space by limiting it to those actions having a direct positive impact on the performance is also important. While the search space can be reduced by an order of 10^3, the problem still looks very challenging and as such, it warrants the use of a Deep Reinforcement Learning algorithm.

In this present work, we use Double Deep Recurrent Q-Network (DRQN) with Dueling architecture. We also as well as reduce the action space to a single logical component within a filer. A logical component is a logical boundary defining the storage space available for a subset of users. A crude analogy of a volume in the filer is the logical partition on a hard disk in the modern computer system. The task of an RL-agent is to apply an action, for example, throttle the number of operations performed on one of the volumes at any particular time. Due to the dynamic nature of the filer environment, it poses various challenges in practically applying reinforcement learning algorithms, much of which will be discussed in Sect. 3. We also argue that this dynamism in the environment makes reinforcement learning a better choice for performance optimization (Fig. 1).

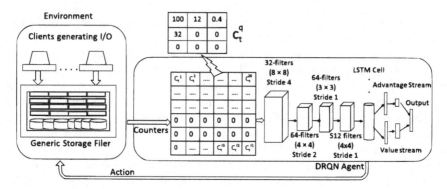

Fig. 1. System architecture for RL controlled filers.

1.1 Performance Counters for a Filer

Definition 1. c_t^q *represent a vector of counters related to component q at time t, then* $\mathcal{C}$ *represents the matrix of n components vectors of possibly varying length collected over a period of time to form the dataset* $\mathcal{D}$ *for performance analysis.*

Let us consider a well-known component like a CPU, which consists of a number of counters. For example: $c_t^q = \{\text{system\_cpu\_busy} = 99\%, \text{raid\_percent} = 15\%, \text{system\_cpu\_nwk\_percent} = 30\%, ...\}$. Each counter is important as it provides different type of information about the CPU. There may be tens to hundreds of such counters for each component and the number of components can be of the same scale. Nevertheless, not all the counters are useful for an RL-agent to make a decision since the number of implementable actions is limited due to fewer configurable components for performance enhancement. At the same time, there are global level components like the CPU whose performance influences the entire storage system. As a result, such counters need to be retained.

2   Related Work

The only known application of RL algorithms for any storage system was presented in [14]. A multi-tier storage system may host files of different priorities at multiple storage levels. If the most frequently accessed files called "hot" files are dynamically placed on the fastest storage tiers, it can minimize the response of the storage system to user requests. Temporal-difference learning was used to achieve the objectives in [14]. Our proposed solution, to the best of our knowledge, is the first attempt to solve the problem of dynamic storage performance optimization using RL algorithm to improve throughput and minimize overload scenarios.

Work related to storage load prediction either at a small scale (disk-level) [15] or large-scale (storage filers or a cluster of filers) [13] have been solved using various machine learning algorithms. But *how do we use these predictions?* Are there any useful means of implementing some action given these predictions is more relevant than knowing the load on the storage system? There has been minimal work on how these predictions can be used in performance improvement. Cluster storage load prediction and fixed time period throttling were recently proposed in [5]. The intrusive retraining mechanism as well as the loss of throughput due to rigid throttling mechanism proposed in [5] limits their applicability. We adopt the idea of using performance counters as a representative for the state of the filer as used in [5] and this forms the state of the environment in our RL problem. Given the state of the filer, the objective of an RL-agent is to learn the appropriate choice of action which results in throughput maximization.

2.1   Reinforcement Learning Background

We briefly introduce the Reinforcement Learning setting that consists of an *agent* interacting with the *environment, E*. The state information of E is available to the *agent* at each time step t based on which the agent chooses an action from the set of valid actions in the action space $\mathcal{A}$. The effect of an action on the environment is measured by the reward r_t received along with the transition in the state from s_t to s_{t+1}. The aim of the agent is to learn a policy π that can maximize the expected cumulative reward which is also called the return G. The policy maps the states to actions that are learned over a period of time through the experience of the agents' interaction with the environment. Let $\gamma \in (0, 1]$ represent the discount factor. The expected return from time step t is $G_t = \sum_{k=0}^{\infty} \gamma^k \ r_{t+k}$. There are many popular choices of RL algorithms available in existing literatures. Q-Learning [17] is one of such popular off-policy algorithms which is also a model-free algorithm. Q-values represent the long-term expected return as a result of executing an action from a given state. At each time step, Q-values are iteratively updated using $Q(s,a) := Q(s,a) + \alpha(r + \gamma \max_{a'} Q(s', a') - Q(s,a))$.

An optimal value function $Q^*(s,a) = \max_\pi Q^\pi(s, a)$ gives the maximum action value achievable by an optimal policy. In practical applications, a parameterized function for approximating the Q-Value estimation is learnt especially where the state space is huge [10]. The Deep Q-Network proposed in [10] uses a separate target Q-network parameterized by θ^- to give consistent targets during temporal difference backups. Nevertheless, this method is known to be overoptimistic in its evaluation of Q-values. However, many variants of DQN have been proposed to reduce the estimation errors and induce stability. One such variation is the Double DQN that decomposes the *max* operation in the target into action selection and action evaluation. $y_i = (r + \gamma Q(s', \text{argmax}_a Q(s', a; \theta); \theta^-))$.

As a result of this change, a significant decrease in the overestimation of state action values were reported in [6]. Another variation to Double DQN that produces much faster convergence was presented in [16], which involves using Duelling architecture that depends on an advantage function along with the traditional value function. The advantage function A^π, is defined as the difference between the action value function and the state value function: $A^\pi(s,a) = Q^\pi(s,a) - V^\pi(s)$. Here $V^\pi(s)$ is the value function that measures the importance of being in a particular state s. The Dueling DQN explicitly splits the representation of state values and advantage values into two streams and later combining them using a special aggregating layer to produce an estimate of the state-action value function Q.

$$Q(s,a;\theta,\alpha,\beta) = V(s;\theta,\beta) + (A(s,a;\theta,\alpha) - \frac{1}{|A|}\sum_{a'} A(s,a';\theta,\alpha)) \qquad (1)$$

where α, β, and θ are individual parameters related to the advantage stream, value stream and network parameters, respectively. Thus, the dueling architecture is known to give better value estimates with significantly faster convergence especially with the increase in the number of actions. Besides, since the number of actions for the current problem is likely to be high, Dueling DQN is more appropriate.

Performance counters measured over a period of time represent a time series with significant temporal properties that need to be accommodated in the decision-making process. The choice of the recurrent neural network which can be combined with any of the DQN architecture was proposed in [7]. The use of similar architecture for physical system performance enhancement through a combination of Duelling architecture with Double Deep Recurrent Q-Network is the first of its kind.

3   Mapping to Deep-RL Problem

3.1   Action Space

The action space to be explored consists of three choices for each logical unit (l), which include: (1) increase the bandwidth by 10% of the average bandwidth observed since the last action (10_l^+), (2) decrease the bandwidth by 10% of the average bandwidth observed since the last action (10_l^-), and (3) no change in the bandwidth (N_l).

$$\mathcal{A} \in (10_1^+, 10_1^-, N_1, \ldots, 10_l^+, 10_l^-, N_l)$$

As these actions are implemented for each logical unit and the action space can change dynamically with the addition of newer logical units, which is another challenge that we address by resetting the rate of annealing ϵ to the value used at the beginning of the experiment. This encourages the RL-agent to explore more and in the process accommodate all new actions included, which can be considered as a means of retraining or updating the policy π. Three actions are considered for each value of l and the number of actions increases by 3 units per l. Since the maximum value of l is restricted by the practical limitation of the filer itself, the number of actions is also bounded [11]. A typical industrial setup will have workloads spread across 10 to 30 different logical units per node. Another important consideration is the customer SLAs which guarantees a minimum bandwidth for each logical layer. Such SLAs restrict the number of valid actions similar to the restriction in actions observed in the boundary cells of a grid world problem.

We consider discrete action space instead of a continuous action space because the rate of change in bandwidth has to be small. *A continuous action space implies uncontrolled variance in the level of throttling which can hamper the user experience.* For example, let's consider a scenario where the agent decides that the optimal bandwidth for a logical unit is 50% of the current mean bandwidth. A sudden drop of 50% can seriously hamper the application performance, which needs to be avoided. Another choice of action would be to control all logical units at each time step, which is an even bigger challenge as the number of actions would be combinatorial in nature.

3.2 Reward Function

The design of a reward function should be in line with the objective of enhancing the storage throughput. The actions explored are based on the principle of enhancing the probability of I/O bandwidth access for the needy without starving any other logical unit. As multiple users are associated with a single logical unit, starvation should always be avoided. Balancing the load through bandwidth management does not only provide better storage performance from an application or a clients' perspective but also prevents overload scenarios that occur when a large number of clients launch multiple I/O intensive jobs. For each action a_t, the reward r_t is evaluated as the sum of throughput observed on all logical units normalized by the maximum achievable throughput (T_{max}). $r_t = \frac{1}{T_{max}} \sum_{l=1}^{l} T_t^l$ The range of reward values output by the environment is $r \in \{0, 1\}$. A reward value of 1 represents the theoretical maximum throughput achievable and a zero reward represents no I/O operation. Throughput maximizing reward function enables appropriate action to be taken only when the filer malfunctions due to its inability to process different workloads and not due to the complexity of the data placements on disks or network issues.

Challenges with the Reward Function: Maximizing the rewards and in turn the throughput depends on balancing the amount of bandwidth among all logical units when the load is high. If the I/O generated by clients is low, then the throughput itself will be low and as a result, the reward will also be low. This can affect the type of actions chosen in different states of the environment. The agent is supposed to learn the fact that actions leading to {low latency, high throughput} as well as {low latency, low throughput} are good actions. The variance in the actions for such states and the convergence of the approximating function will be faster if latency related information is explicitly provided as features.

Definition 2. *Loading effect of a filer refers to the variation in throughput observed due to either slow data generation at the source or various latencies at client, network or disk level. Due to this throughput of the filer may be low but no meaningful actions can be implemented to improve the situation.*

For the problem being discussed, there is no guarantee of rewards showing any increasing or decreasing trend. The improvements achieved through the RL algorithm should only be assessed by comparing the performance of an RL-agent based filer with the default filer behavior. This is purely because of the *loading effect* defined previously.

Performance counters have enough features that accommodate the latencies of all the components within the filer but due to preprocessing of features, some may not be included. As a result, such excluded latencies are explicitly included before feeding the input to the neural network. Preprocessing is necessary because we are dealing with at least 100,000 features, which is expected to increase with the addition of disks and creation of newer logical units. Furthermore, the nature of features (performance counters) can be treated as time series which can reveal the overall I/O pattern. If this time series aspect is adequately captured during modeling, then appropriate throttling levels can be implemented for a fixed time interval.

3.3   Input Space

Since the dimension of input space is very high ($\geq$100,000), which is also expected to increase with the addition of newer logical layers, it is necessary to remove redundant counters at the component level. This is achieved by running an inter-correlation test for all counters belonging to a particular component. An inter-correlation test is performed on randomly queried performance counters belonging to each component q over a period of time. In this process, if a strong correlation exists between two features belonging to the same component, one of the features is eliminated [5]. This enhances the diversity of features as well as reduces their number.

We represent the state information as a combination of multiple blocks of 3×3 cells with each cell representing one counter of a component. A maximum of 9 counters are retained per component and a minimum of 1. If there are less than 9 counters, then zero-fill is applied while the top 9 counters are used otherwise. Each of these 3×3 cell blocks are placed inside a matrix of size 60×60 that represent the state information of the environment and it has the capacity to accommodate 400 different components. In case filers with large number of logical units or additional components are part of the state information, they can be accommodated without changing the shape of the input data. As many as 40 different components are placed inside this matrix in our experiment and the remaining cells are zero-filled to accommodate various additional logical layers that may be present on different filers.

4   Algorithm

The dueling architecture proposed in [16] can be used for Q-value estimation with any of the standard RL algorithms such as DQN and its variations or SARSA. Due to the various advantages of dueling, we use Dueling architecture for Double Deep Recurrent Q-Network where the Q-value function is approximated by using a combination of CNN and LSTM. The ouput layer of CNN are linked to an LSTM cell as in [7] and LSTM cells have the ability to save temporal information that can handle the time series aspect of performance counters. The output of LSTM is split into the V^π and A^π streams. A replay buffer, N_r is used to periodically train the network about temporal dependencies. As we are dealing with Recurrent Neural Networks (LSTM), we feed fixed length consecutive (s, a, r, s') tuples from a random episode to retain the time-series in the state representation. The algorithm used for training the RL-agent is similar to the Double DQN with modifications in dueling architecture. Details pertaining to the algorithm can be found in [16] and the details related to the architecture of CNN and LSTM are presented in the next section.

5   Experiment Setup

From a system perpspective, the basic framework of our experimental setup centers on generating workload similar to those observed in real industrial environment. Whereas, as regard Reinforcement learning, the design of the network and hyperparameters are the important components. In order to generate workload similar to those observed in industry, we used the 'Standard Performance Evaluation Corporations' (SPEC) Solution File Server (SFS-2014) tool, developed by a consortium of storage vendors for standard industrial workload. For the sake of comparison with similar dynamic load control, we generate workload patterns identical to that used in [5]. With this environment setup in place, we deployed an RL-agent on a separate server to observe the state information (performance counters) of the filer and apply the throttling mechanism to control the load so as to maximize throughput.

We also used NetApp FAS series filer with the RL-agent trying to optimize a single node of the cluster mode filer. NetApp provides an administrative interface responsible for fetching all possible c^q that is used to generate the state information of the environment. Throttling was applied through the Quality Of Service (QOS) interface available to the administrators who are responsible for increasing or decreasing the level of I/O available to each logical unit. Due to the high computational cost of a Deep Neural Network, we do not embed the RL-agent within the filer but rather, it is present on a separate server with 8-CPUs and 32 GB of RAM. Although we did not encounter any practical difficulties during the learning process of the RL-agent, learning outside the environment is much more preferable since the computational cost can compound and hamper the behavior of the filer itself.

The Q-Network consists of four convolution layers and LSTM layer. The first convolution layer has 32 8×8 filters with stride 4, the second layer has 64 4×4 filters with stride 2, the third layer comprises 64 3×3 filters with stride 1 and the fourth convolution layer has 512 4×4 filters with stride 1. Recurrence effects were extracted with a fully connected recurrent LSTM layer of the same size as the number of filters in the previous layer. This output is further split into two streams with one computing V^π and the other A^π. The final hidden layers of the value and advantage streams were both fully connected, with the value stream having one output and the advantage stream having as many outputs as the number of valid actions. All filters were initialized with uniformly sampled random values. In addition, rectifier non-linearities [8] were inserted between all adjacent layers. We test this setup on 5 logical layers and hence the number of actions were set to 15.

Analysis of RL-agent from a System Perspective: The *length of the episodes are restricted* to 30 min based on the average completion time of various applications launched by clients. We also *reset the maximum allowable operations* per logical unit to 4,000 operations at the start of each episode, i.e., every 30 min. This value should be high such that it exhibits a behavior similar to unrestricted I/O permitted by the default behavior of a filer. A number of unique start states were automatically generated due to the fixed length of episodes and varying number of applications/processes encountered at the start of each episode. Actions were implemented at a frequency of 1 min and a total of 30 actions were applied in each episode with a view to maximizing the reward.

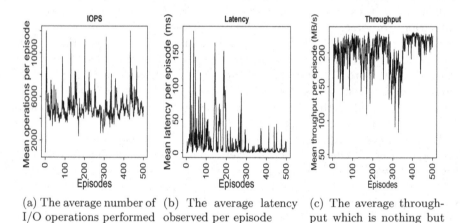

(a) The average number of I/O operations performed on the filer.

(b) The average latency observed per episode

(c) The average throughput which is nothing but the scaled reward

Fig. 2. Performance of filer under the supervision of an RL-agent.

The mean number of operations performed by various clients is shown in Fig. 2a. In order to negate the loading effect and accurately measure the performance improvements, we analyzed the average throughput per episode and the average latency observed per episode in Fig. 2c and b respectively. The average latency was restricted to less than 150 ms throughout the training period, which is significant in terms of providing assured service to clients. The mean throughput increases as the mean latency decreases with increase in the number of episodes. However, it is not clear if the performance achieved by the RL-agent is better than the normal filer behavior. In episodes 250 to 500, which are of interest to us, we observe a general decrease in the latency (Fig. 2b) even though the number of operations tend to gradually increase (Fig. 2a). Similarly, the utilization of the throughput shows an upward trend.

The number of IOPS handled in each episode should be high due to the reasons given below:

- The number of iops, each having a size not more than 64 KB, were high
- The block size of data generated by applications were more than 64 KB.

While the first reason listed above ensures that the number IOPS generated at the source and the number of IOPS handled within the filer are identical, the second reason shows a deviation in the number of IOPS measured in the source and the destination. Filers typically break down large blocks of data into 64 KB size before processing the them and this is true for the NetApp filer that was used in our experiment. Hence, a variation in the throughput can be observed for the same number of IOPS in Fig. 2.

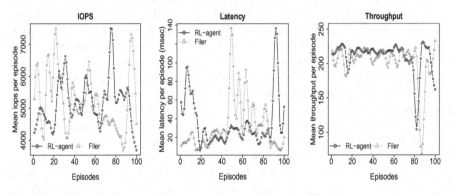

(a) The average number of I/O operations performed on the filer.

(b) The average latency observed per episode

(c) The average throughput observed per episode

Fig. 3. Comparison of performance of the filer under the supervision of an RL-agent post the training duration of 500 episodes.

Comparision of RL-agents' Performance with Default Filer Behavior:
Ideally, limits on logical units that dynamically vary with the requirements of the workload should be implemented by the storage administrators. However, the scale of the problem as well as lack of understanding of the workload patterns restricts the implementation of such dynamic changes. A fixed number of IOPS are enabled whenever a filer is created and the frequency of revisiting these limits is very rare. In our comparison, we applied a limit of 10,000 IOPS for all the logical units during the data collection for a default filer behavior.

After the completion of 500 episodes of learning by the RL-agent, we compared the performance of the default filer behavior with the performance induced by actions chosen by the RL-agent. Applications were initiated on a random number of clients ranging from 5 to 20 clients. Figure 3 shows the performance of the RL-agent aided filer compared with that of the default filer behavior with respect to IOPS, latency and throughput. All 20 clients were used in the load generation during the testing phase and throttling was implemented at various stages across the 100 episodes considered for comparison. Because identical operations may not be performed in each episode, a point-to-point comparison should not be considered. The average gain in throughput over all the 100 episodes is about 2.84% and the mean increase in the number of IOPS performed by the RL-agent aided filer is 1.8%. Previous attempts at throttling I/O to control load on the filer presented in [5] had reported a 1.8% loss of throughput. Hence, the performance achieved here is an improvement with reasonable practical benefits in various industries depending on I/O intensive applications.

Influence of RL-agent: So far, the results presented are the most significant performance improvements achieved during our testing. When the number of clients generating I/O were low (i.e. 5 clients with each running 5 to 20 different processes), the results were identical to the results of a default filer behavior with no change in the average throughput or latency and the most frequent action chosen by the agent N_l indicating no change in the bandwidth.

As the number of clients were gradually increased, the performance fluctuated though significant gains were observed when the number of clients were high. Based on the results from the experiment, an interesting observation was made regarding the amount of load to be allowed at the end of each episode. A high value of 4000 IOPS was used which led to N_l being chosen more frequently. Furthermore, bandwidth changes were implemented only when the number of clients were higher than 10. It must be noted that these results are peculiar to the NetApp filer and the performance trends may be different for other filers.

6   Conclusion

We have been able to introduce a new challenging application of RL for optimizing filer behavior especially in a high load scenario. A combination of two different classes of neural network architecture (i.e. deep neural network and recurrent neural network) was applied to align the approximating function with the temporal patterns present in the state information. The challenges due to the scale of the problem restricts the advantages only to high load scenarios with a little drawback in the performance under normal load behavior. In addition, other aspects of RL such as the ease of transferring the model to multiple similar systems with faster convergence makes the proposed method more attractive than other alternative strategies available in existing literatures.

References

1. Anderson, E., Spence, S., Swaminathan, R., Kallahalla, M., Wang, Q.: Quickly finding near-optimal storage designs. ACM Trans. Comput. Syst. **23**(4), 337–374 (2005)
2. Boyan, J.A., Littman, M.L.: Packet routing in dynamically changing networks: a reinforcement learning approach. In: Proceedings of the 6th International Conference on Neural Information Processing Systems, NIPS 1993, pp. 671–678 (1993)
3. Crites, R., Barto, A.: Improving elevator performance using reinforcement learning. In: Advances in Neural Information Processing Systems, pp. 1017–1023 (1996)
4. Deshpande, S., Dheenadayalan, K., Srinivasaraghavan, G., Muralidhara, V.N.: Filer response time prediction using adaptively-learned forecasting models based on counter time series data. In: 15th IEEE International Conference on Machine Learning and Applications, ICMLA 2016, pp. 13–18 (2016)
5. Dheenadayalan, K., Srinivasaraghavan, G., Muralidhara, V.N.: Self-tuning filers — overload prediction and preventive tuning using pruned random forest. In: Kim, J., Shim, K., Cao, L., Lee, J.-G., Lin, X., Moon, Y.-S. (eds.) PAKDD 2017. LNCS (LNAI), vol. 10235, pp. 495–507. Springer, Cham (2017). https://doi.org/10.1007/978-3-319-57529-2_39
6. van Hasselt, H., Guez, A., Silver, D.: Deep reinforcement learning with double Q-learning. In: Proceedings of the Thirtieth AAAI Conference on Artificial Intelligence, February 2016, pp. 2094–2100 (2016)
7. Hausknecht, M.J., Stone, P.: Deep recurrent Q-learning for partially observable MDPs. CoRR abs/1507.06527 (2015)
8. Maas, A.L., Hannun, A.Y., Ng, A.Y.: Rectifier nonlinearities improve neural network acoustic models. In: Proceedings of ICML, vol. 30 (2013)
9. Mahadevan, S., Marchalleck, N., Das, T.K., Gosavi, A.: Self-improving factory simulation using continuous-time average-reward reinforcement learning. In: 14th International Conference on Machine Learning, pp. 202–210 (1997)
10. Mnih, V., et al.: Human-level control through deep reinforcement learning. Nature **518**(7540), 529–533 (2015)
11. NetApp Inc.: Storage limits. https://library.netapp.com/ecmdocs/ECMP1196906/html/GUID-AA1419CF-50AB-41FF-A73C-C401741C847C.html
12. Singh, S., Bertsekas, D.: Reinforcement learning for dynamic channel allocation in cellular telephone systems. In: Proceedings of the 9th International Conference on Neural Information Processing Systems, NIPS 1996, pp. 974–980 (1996)
13. Tang, H., Gulbeden, A., Zhou, J., Strathearn, W., Yang, T., Chu, L.: A self-organizing storage cluster for parallel data-intensive applications. In: SC 2004: Proceedings of the 2004 ACM/IEEE Conference on Supercomputing, p. 52, November 2004
14. Vengerov, D.: A reinforcement learning framework for online data migration in hierarchical storage systems. J. Supercomput. **43**(1), 1–19 (2008)
15. Wang, M., Au, K., Ailamaki, A., Brockwell, A., Faloutsos, C., Ganger, G.R.: Storage device performance prediction with cart models. In: Proceedings of the Joint International Conference on Measurement and Modeling of Computer Systems, pp. 412–413 (2004)

16. Wang, Z., Schaul, T., Hessel, M., van Hasselt, H., Lanctot, M., de Freitas, N.: Dueling network architectures for deep reinforcement learning. In: Proceedings of the 33rd International Conference on Machine Learning, ICML 2016, pp. 1995–2003 (2016)
17. Watkins, C.J.C.H., Dayan, P.: Technical note: Q-learning. Mach. Learn. **8**(3–4), 279–292 (1992)

Adaptive Modeling and Control of an Upper-Limb Rehabilitation Robot Using RBF Neural Networks

Liang Peng[1], Chen Wang[1,2], Lincong Luo[1,2], Sheng Chen[1,2],
Zeng-Guang Hou[1,2,3(✉)], and Weiqun Wang[1]

[1] State Key Laboratory of Management and Control for Complex Systems,
Institute of Automation, Chinese Academy of Sciences, Beijing 100190, China
{liang.peng,wangchen2016,luolincong2014,chensheng2016,
zengguang.hou,weiqun.wang}@ia.ac.cn
[2] University of Chinese Academy of Sciences, Beijing 100049, China
[3] CAS Center for Excellence in Brain Science and Intelligence Technology,
Beijing 100190, China

Abstract. Robot-assisted rehabilitation following neurological injury is most successful when subject participation is maximized in the training tasks. Developing control strategies that can provide subject-specific assistance is accordingly an active area of research. For robot-assisted rehabilitation training, it is challenging to adapt the robotic assistance to each patient's impairment, and model-based control methods in previous studies are difficult to implement because of the computational complexity of human-robot interaction dynamics and changes of human active efforts during rehabilitation exercises. This study implements adaptive modeling and control for an two-DOF upper-limb rehabilitation robot by combining an RBF-based feedforward controller with a feedback impedance controller. Simulation and experiment results show that, the RBF neural network is able to adaptively establish the human-robot dynamics as well as estimating the human efforts, and the impedance controller guarantees compliant human-robot interaction and regulates the maximal tolerated tracking error. Besides, the proposed controller is defined in the robot workspace, thus is easy to be generalized to be used for multi-DOFs exoskeleton-type rehabilitation robots.

Keywords: Rehabilitation robot · Adaptive control
RBF neural networks · Assist-as-needed

1 Introduction

Stroke has become one of the leading causes of death over the world, and most post-stroke patients have certain kinds of sensory-motor deficits such as

This research is supported by in part by National Natural Science Foundation of China (Grants #61603386, U1613228, 61720106012, 61533016, 61421004) and Beijing Natural Science Foundation (Grant L172050, Z161100001516004).

L. Cheng et al. (Eds.): ICONIP 2018, LNCS 11307, pp. 235–245, 2018.
https://doi.org/10.1007/978-3-030-04239-4_21

hemiparalysis. Long-term rehabilitation of patients with neurological is essential for patients to improve muscle strength and movement coordination. Conventional rehabilitation training is performed by human therapists, which is time consuming and labor intensive, and robot-assisted therapy has drawn more attention over the last decades. Robotic devices are well suited to offer multiple training sessions with consistent delivery of therapy, and holds great promise of increasing training efficiency [1,2]. Furthermore, robot can precisely perform objective and quantitative performance evaluation of subjects during therapy, and the integration of virtual reality games adds more entertainment and promotes more patient motivation than in conventional manual therapy [3].

In addition to rehabilitation robot design, lots of researches focus more on robot control strategies, which is a key factor for training effectiveness [4]. Hogan et al. [5] demonstrated that continuous passive motion-based therapy did not produce significant improvements in poststroke patients. Generally speaking, active training is better than passive training for post-acute patients, as the former motivates more active participation of the patient, which is beneficial for brain plasticity and motor relearning [6]. Besides, safe and compliant human-robot interaction is essential, and a secondary injury is unacceptable.

One of the key principles for active training is "assist-as-needed" [7], which emphasizes that the robot only provides necessary assistance according to the patient's condition and performance, and the patient exploits most of his/her residual efforts to complete the task. Bio-signals like EEG [8] and EMG [9] provide solutions to directly measure human brain and muscle activations, but the most widely used pattern recognition can only separate between several actions, and there still lacks a simple and precise computational model relating EMG to muscle force [10]. To obtain the desired robot assistance for a certain training task, in [11] human active effort was estimated by subtracting the human-robot dynamics from the measured interaction force/torque. However, the model calibration process is time-consuming, and modeling errors may cause over intervention of the robot during training.

With a rehabilitation robot named CASIA-ARM [12] developed by us, an adaptive controller based on RBF neural networks is proposed, where RBFs are located over the workspace and adaptively learn both the human-robot dynamics and human efforts. Furthermore, an impedance controller provides compliant human-robot interaction and regulates the tolerated tracking error.

Figure 1 shows the robotic training scenario, where the robot has a five-bar structure, and only two base joints are driven by two DC motors respectively, thus it has two DOFs in the horizontal plane. The design of parallel structure and cable transmission system make the robot have small reflected inertia and friction at the end-effector, which is beneficial to compliant interaction between the robot and patients. Moreover, virtual reality training games are developed based on OpenGL. The visual/audio and haptic feedback of the robot produce significant improvements in patients' hand-eye coordination abilities [13].

Fig. 1. Training scenario using CASIA-ARM.

Table 1. Technical specifications of CASIA-ARM.

Items	Characteristics
DOF	2
Actuation	2 DC motors
Sensors	2 rotary encoders
Range of joint motion	$80°$–$220°$, $-40°$–$100°$
Workspace	500 mm * 416 mm
Motor torque	$\sim$ 450 mNm
Reduction ratio	20:1
Force capability	>32.8 N

The rest of this paper is organized as follows. Section 2 introduces details of the proposed method. Section 3 verifies the method through simulation. The experiment results using CASIA-ARM for the validation of the developed controller are presented in Sect. 4. Finally, Sect. 5 concludes this paper and present the remarks for future work (Table 1).

2   Methods

Suppose a patient is asked to follow a training trajectory with assistance from the robot, and the tracking error in the two-dimensional workspace is defined as:

$$\tilde{X} = X - X_d \tag{1}$$

where $X = [x, y]^T$ is the Cartesian coordinates of the robot endpoint, and X_d is the desired trajectory.

Define the sliding variable also in the workspace:

$$s = \dot{\tilde{X}} + \Lambda\tilde{X} \tag{2}$$

and the reference trajectory of s:

$$v = \dot{X}_d - \Lambda\tilde{X} \tag{3}$$

where Λ is a diagonal positive definite matrix. Then we have:

$$s = \dot{X} - v \tag{4}$$

The human-robot system dynamics can be expressed in the workspace as:

$$D(X)\ddot{X} + C\left(X, \dot{X}\right)\dot{X} + f\left(X, \dot{X}\right) + G(X) = F_r + F_h \tag{5}$$

where $D(X)$ and $C\left(X, \dot{X}\right)$ are inertial and Centripetal/Coriolis terms, respectively; $f\left(X, \dot{X}\right)$ represents the friction and $G(X)$ is the gravity term, as our robot moves in the horizontal plane, the gravity term remains zero; the terms F_r and F_h represent robot force and human efforts, respectively.

Then the controller in this method is defined as:

$$F_r = Y\left(X, \dot{X}, s, \dot{s}\right)\hat{a} - K_D s \tag{6}$$

where $Y\hat{a}$ is an estimation model of the human-robot dynamics and human force, which can be defined as:

$$Y\hat{a} = \hat{D}\ddot{x} + \hat{C}\dot{x} + \hat{f} - \hat{F}_h \tag{7}$$

where $\hat{D}$, $\hat{C}$, and $\hat{f}$ express the estimated human-robot dynamics.

As a special property, robot dynamics can be expressed in a linearized form in terms of system parameters, where Y contains generalized variables, and a is the parameter vector about both kinematic and dynamic parameters of the robot. On the other hand, Eq. (7) also includes the patient active efforts $\hat{F}_h$, which is also approximated using a linear controller, and the performance is tested in following experiments.

Besides, the term of $K_D s$ in Eq. (6) is an impedance controller. By considering the sliding variable, K_D and $K_D \Lambda$ represent stiffness and damping factors, and $K_D = \begin{bmatrix} k_{dx} & 0 \\ 0 & k_{dy} \end{bmatrix}$ is a symmetric, positive definite gain matrix. This impedance controller guarantees compliant interaction between human and the robot, and the selection of K_D determines the maximal tolerated tracking error.

According to [14], the tracking error tends to be zero by adopting the following adaption law:

$$\dot{\hat{a}} = -\Gamma^{-1}Y^T s \tag{8}$$

where Γ is a symmetric positive definite matrix.

Instead of using the linearly parameterized robot-human dynamics as in [11], in this study, $Y\hat{a}$ is implemented using linear combinations of Gaussian RBFs, which can uniformly approximate any continuous function, and the regressor matrix Y is defined as:

$$Y^{2\times 50} = \begin{bmatrix} g^T & 0 \\ 0 & g^T \end{bmatrix} \tag{9}$$

where the Gaussian RBF vector is defined as 25 spatially distributed radial basis functions:

$$g = [g_1\ g_2\ \dots g_{25}]^T \tag{10}$$

and each Gaussian RBF is defined as:

$$g_n = \exp\left(-|X - \mu_n|^2/2\sigma^2\right) \tag{11}$$

where μ_n is the location of the nth Gaussian RBF, and σ is a scalar smoothing constant. For CASIA-ARM, we have implemented a two-dimensional grid of RBFs, with 5 grid divisions in the X direction, and 5 grid in the Y direction across the workspace, and $\sigma = 6$ cm. The RBFs are evenly spaced at 5 cm apart, and there are enough overlaps of RBFs over the workspace so as to achieve good approximation performance.

3   Simulation

This section verifies the effectiveness of the proposed method by simulation. The dynamics of the human-robot system is introduced, and a spatial circular tracking task is tested by simulation in MATLAB.

3.1   Human-Robot Dynamics

As shown in Fig. 2, the human arm is simplified as a two-DOF serial linkage (l_5 and l_6) in horizontal plane, and the five-bar parallel robot has four articulated links from l_1 to l_4.

Parameter values in simulation are shown in Table 2, including *length* of each link, *mass*, *center of mass* and *moment of inertia*. The length of A_1A_2 (distance between two base joints) is 12 cm. The robot parameters are obtained using CAD software (SolidWorks 2012), and the human parameters are estimated based on the anthropometric data of a person who is 180 cm tall and 70 Kg in weight [15].

As shown in Fig. 2, the human-robot system is a redundantly actuated dynamic system, as there are four actuators (two robot motors and two human joints) but only two DOFs. The dynamics derivation is a difficult problem by directly using Lagrangian method, and in [16] we propose a novel method, then the human-robot dynamics defined in the workspace can be expressed as:

$$D\ddot{X} + C\dot{X} = F_h + F_r \tag{12}$$

where D is the Cartesian inertia term, C is the Cartesian Coriolis/centrifugal term, and please refer to [16] for more details.

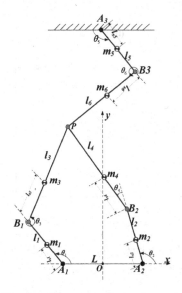

Fig. 2. Diagram of the human-robot system, and definition of the dynamics parameters.

Table 2. Simulation Parameters. Of the parameters in the table, l is the length, COM is the distance between the center of mass and the proximal end, m is the mass, and I is the moment of inertia of the link about its center of mass.

Links	$l(m)$	$COM(m)$	$m(Kg)$	$I(kg.m^2)$
1	0.3	0.15	0.629	0.0047
2	0.3	0.15	0.629	0.0047
3	0.4	0.2	0.815	0.0109
4	0.4	0.2	0.815	0.0109
5	0.35	0.15	1.96	0.025
6	0.37	0.25	1.54	0.046

3.2   Simulation Results

The simulation was implemented in MATLAB using ODE45 function, and the system states (acceleration, velocity, joint angles, endpoint position, RBF weight, etc.) were calculated based on the system dynamics (12) and robot control (6). During the simulation, the matrix K_d was set to be $\begin{bmatrix} 100 & 0 \\ 0 & 100 \end{bmatrix}$, Γ was selected as $\begin{bmatrix} 0.5 & 0 & \cdots & 0 \\ 0 & 0.5 & \cdots & 0 \\ \vdots & \vdots & \ddots & \vdots \\ 0 & 0 & \cdots & 0.5 \end{bmatrix}_{50 \times 50}$. The task was selected as a circular trajectory tracking training, where the center was at $(0, 0.4\,\mathrm{m})$, the period was $6\,\mathrm{s}$ and the radius was $0.15\,\mathrm{m}$.

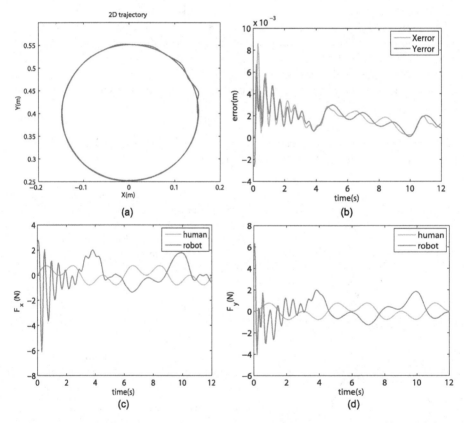

Fig. 3. Simulation results: (a) actual 2D trajectory, (b) tracking errors, (c) and (d) are force outputs of human and robot in X and Y directions.

In this simulation, the human voluntary force was set to be

$$
\begin{cases}
F_{hx} = A\left(t\right) \cdot \cos\left(\frac{2\pi}{T_d}t\right) \\
F_{hy} = A\left(t\right) \cdot \sin\left(\frac{2\pi}{T_d}t\right)
\end{cases}
\tag{13}
$$

therefore we have:

$$
F_h^T \dot{X}_d = 0
$$

which means the human force was perpendicular to the velocity, and the amplitude was selected as:

$$
A\left(t\right) = \sin\left(\frac{4\pi}{T_d}t\right)
\tag{14}
$$

so the human force worked as a varying disturbance (maximum: 5 N, period: $\frac{T_d}{2} = 3\,\text{s}$) perpendicular to the desired motion.

As shown in Fig. 3, there were relatively large errors at the beginning, which produces oscillations of the robot force output. The tracking errors reduced to less than 4 mm after one period, and the robot outputs both in X and Y directions become stable.

4 Experiment

4.1 Experiment Setup

In the experiment, our proposed method was implemented on the CASIA-ARM, and the task was set to be repeated reaching training, which is widely used in rehabilitation training. As shown in Fig. 4, the subject was guided to reach between two points A and B, which took turns to become a large red dot indicating the target, and the trajectory follows the minimum-jerk principle [17].

The coordinates of A and B were $(-0.14, 0.4)$ and $(0.14, 0.4)$ (unit: m), respectively, and the desired trajectory in the workspace was:

$$
\begin{cases}
x_{ref}\left(t\right) = x_i + \left(x_d - x_i\right)\left[10(t/\tau)^3 - 15(t/\tau)^4 + 6(t/\tau)^5\right] \\
y_{ref}\left(t\right) = y_i + \left(y_d - y_i\right)\left[10(t/\tau)^3 - 15(t/\tau)^4 + 6(t/\tau)^5\right]
\end{cases}
\tag{15}
$$

where (x_i, y_i) and (x_d, y_d) were the starting point and target point, respectively. As target A and target B were in the X direction, thus $y_{ref} = 0.4$, the movement time $\tau = 2\,\text{s}$.

As the task was constrained in the X direction, different from the RBF grid setup (5 * 5 over the workspace) for simulation in Sect. 3, there were 20 gaussian basis functions distributed uniformly between A and B, which was beneficial for smoothness of the approximating dynamics. Besides, the robot moved in different directions (A to B, B to A) in the experiment, while only in counter clockwise direction in the simulation.

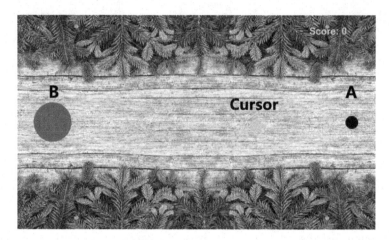

Fig. 4. Visual display of the task, where the subject controlled the cursor using the robot to reach B from A.

Therefore, 20 gaussian basis functions were set in two directions, respectively, which means the regressor Y were implemented via two RBF neural networks i.e. Y_R2L (right->left) and Y_L2R (left->right) according to the motion directions. The weight vector of Y_R2L was defined as $\left(\hat{a}_1 \ \hat{a}_2 \ ... \ \hat{a}_{40} \right)^T$, and the weight vector of Y_R2L was $\left(\hat{a}_{41} \ \hat{a}_{42} \ ... \ \hat{a}_{80} \right)^T$.

The controller was realized by programming in C++, and the control period was 1 ms. During each control period, following operations were performed:

1. Calculate the sliding variable s according to the error between the minimum jerk trajectory and current position.
2. Update the regressor matrix Y_L2R or Y_R2L according to the current position and the direction.
3. Update the weight vector $\hat{a}$ according to Eq. (8).
4. Calculate the desired robot force F_r according to Eq. (6).
5. Transform F_r to joint torques via Jacobian matrix $(\tau = J^T F)$ and send commands to the robot motor drivers.

4.2   Experiment Results

In order to validate the controller's adaptation to human efforts, the participant (healthy, male, 30 years old) imposed a resisting force of about 10 N during the motion, which was measured by a 6-axis force sensor and is shown by F_{sensor} in Fig. 5(b).

In the experiment, the weights of RBFs $\hat{a}$ were zero-initialized, therefore only the impedance controller was working in the beginning. In order to test the adaptation performance of the RBF controller, the impedance parameters in the X direction were set to a relatively small value ($k_{dx} = 100\,\text{N/m}$, $\Lambda =$

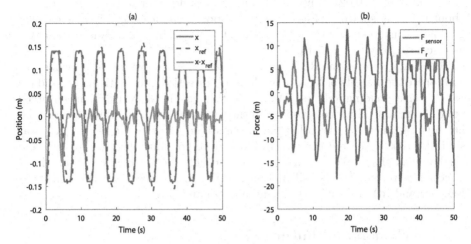

Fig. 5. Experiment results: (a) tracking performance in the X direction, (b) the robot force F_r and the human force F_{sensor} measured by the force sensor.

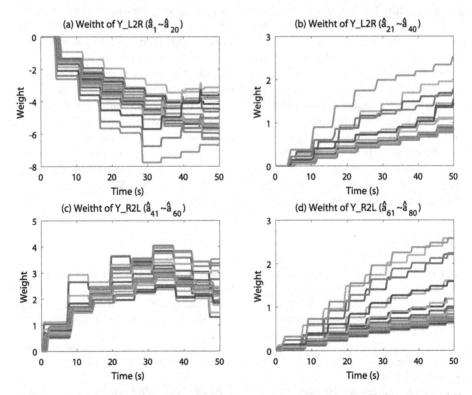

Fig. 6. Adaptation of RBFs: (a) and (b) are weights of Y_{L2R} in the X direction and Y direction, respectively, (c) and (d) are the corresponding results of Y_{R_L}.

$diag(0.02, 0.02, ..., 0.02))$, and the maximal tracking error in the beginning was about 0.1 m as shown in Fig. 5(a). After about 7 periods, the RBF controller gradually converged and the errors reduced to less than 0.03 m, and the changes of F_r can be seen in Fig. 5(b).

The adaptation of the RBF weights $\hat{a}$ over time is shown in Fig. 6, where (a) and (b) are the weights of Y_{L2R} along X and Y directions, respectively, (c) and (d) are the corresponding weights of Y_{R2L}. Most weights in the X direction converged after about 30 s, which contributed to the reduction of the tracking error under human resistance. On the other hand, the stiffness in the Y direction K_{dy} was set to be 1000 N/m, and the human forces were not controlled deliberately and smaller than that in the X direction. As a result, the tracking errors on the Y direction were relatively small (0.01 m). It also can be seen that the weights in Fig. 6(c) and (d) are smaller than those in (a) and (b).

5   Conclusion and Future Work

In this study, an RBF based adaptive controller is proposed to control an upper-limb rehabilitation robot. The experimental and simulation results show that this algorithm has several merits over other methods. Firstly, it doesn't need to establish an accurate dynamic model of the human-robot system, which is generally difficult for a complex robot. Secondly, this controller can learns both the system dynamics and human active efforts adaptively, thus making it unnecessary to estimate the human efforts. Thirdly, it's implemented in the robot workspace thus is easy to be generalized to be used in multi-joint robots.

However, though this controller can adapt to different human conditions, advanced neural-rehabilitation therapy emphasizes more on human active participation and immediate feedbacks, thus more improvements are necessary. Therefore, future modifications based on this study may include robot trajectory planning according to the human intention instead of a predefined trajectory, and haptic rendering to encouraging his/her motivations, etc. Besides, studies of clinical feasibility on stroke patients will be conducted.

References

1. Riener, R., Nef, T., Colombo, G.: Robot-aided neurorehabilitation of the upper extremities. Med. Biol. Eng. Comput. **43**(1), 2–10 (2005)
2. Lo, A.C., Guarino, P.D., Richards, L.G., Haselkorn, J.K.: Robot-assisted therapy for long-term upper-limb impairment after stroke. N. Engl. J. Med. **362**(19), 1772–1783 (2010)
3. Maciejasz, P., Eschweiler, J., Gerlach-Hahn, K., Jansen-Troy, A., Leonhardt, S.: A survey on robotic devices for upper limb rehabilitation. J. Neuroeng. Rehabil. **11**(1), 3 (2014)
4. Marchal-Crespo, L., Reinkensmeyer, D.J.: Review of control strategies for robotic movement training after neurologic injury. J. Neuroeng. Rehabil. **6**(1), 20 (2009)

5. Hogan, N., Krebs, H.I., Rohrer, B., Palazzolo, J.J.: Motions or muscles? Some behavioral factors underlying robotic assistance of motor recovery. J. Rehabil. Res. Dev. **43**(5), 605 (2006)

6. Prange, G.B., Jannink, M.J., Groothuis-Oudshoorn, C.G., Hermens, H.J., IJzerman, M.J.: Systematic review of the effect of robot-aided therapy on recovery of the hemiparetic arm after stroke. J. Rehabil. Res. Dev. **43**(2), 171 (2006)

7. Reinkensmeyer, D.J., Wolbrecht, E., Bobrow, J.: A computational model of human-robot load sharing during robot-assisted arm movement training after stroke. In: Proceedings of Annual International Conference of the IEEE Engineering in Medicine and Biology Society, pp. 4019–4023 (2007)

8. Lotte, F., et al.: A review of classification algorithms for eeg-based brain–computer interfaces: a 10 year update. J. Neural Eng. **15**(3), 031005 (2018)

9. Nazmi, N., Abdul Rahman, M.A., Yamamoto, S.-I., Ahmad, S.A., Zamzuri, H. Mazlan, S.A.: A review of classification techniques of EMG signals during isotonic and isometric contractions. Sensors 6(8), 1304 (2016)

10. Sartori, M., Reggiani, M., Farina, D., Lloyd, D.G.: EMG-driven forward-dynamic estimation of muscle force and joint moment about multiple degrees of freedom in the human lower extremity. PloS one **7**(12), e52618 (2012)

11. Wang, W., et al.: Toward patients motion intention recognition: dynamics modeling and identification of ilegan LLRR under motion constraints. IEEE Trans. Syst. Man Cybern. Syst. **46**(7), 980–992 (2016)

12. Peng, L., Hou, Z.G., Peng, L., Wang, W.: Design of CASIA-ARM: a novel rehabilitation robot for upper limbs. In: Proceedings of IEEE/RSJ International Conference on Intelligent Robots and Systems (IROS), pp. 5611–5616 (2015)

13. Peng, L., Hou, Z.G., Peng, L., Luo, L., Wang, W.: Robot assisted rehabilitation of the arm after stroke: prototype design and clinical evaluation. Sci. China Inf. Sci. **60**(7), 073201 (2017)

14. Slotine, J.J.E., Li, W.: Applied Nonlinear Control. Prentice hall, Englewood Cliffs (1991)

15. Winter, D.A.: Biomechanics and Motor Control of Human Movement. Wiley, New York (2009)

16. Peng, L., Hou, Z.G., Wang, W.: Dynamic modeling and control of a parallel upper-limb rehabilitation robot. In: Proceedings of IEEE International Conference on Rehabilitation Robotics (ICORR), pp. 532–537 (2015)

17. Hogan, N.: An organizing principle for a class of voluntary movements. J. Neurosci. **4**(11), 2745–2754 (1984)

Modelling Predictive Information
of Stochastic Dynamics in the Retina

Min Yan[1(✉)], Yiko Chen[2], C. K. Chan[2], and K. Y. Michael Wong[1]

[1] Department of Physics, Hong Kong University of Science and Technology,
Kowloon, Hong Kong SAR, China
{myanaa,phkywong}@ust.hk
[2] Institute of Physics, Academia Sinica, Taipei, Taiwan
ykchen12@gmail.com, ckchan@gate.sinica.edu.tw

Abstract. Many experiments showed that the retina processes information before transmitting them to the visual cortex. We propose a model to elucidate the predictive effect of the amacrine cells and ganglion cells in the retina. We generate the input signals with OU (Ornstein-Uhlenbeck) and HMM (Hidden Markov model) process, and compare the mutual information calculated from the simulations with those of the moving bar experiments on bullfrog retina mounted on a multi-electrode array, illustrating that the model agrees with the experiments.

Keywords: Retina · Predictive information · Direction selectivity
Amacrine cells · Ganglion cells

1 Introduction

The brain receives huge quantity of information everyday from the outside world. Among all senses, vision occupies a decisive position, which is crucial for animals to hunt and survive. The retina is the starting point to receive visual information from the surrounding environment. Many experiments showed that the beginning step of processing information is in the retina rather than the visual cortex [1,2]. In addition, the retina also makes predictions about the state of the stimulus in the future [3–5], usually 50 to 100 ms in advance, according to experiments on bullfrog retina and larval tiger salamander retina [3,6,7]. This period of prediction is significant for animals to make responses to avoid potential dangers or to launch attacks [8,9].

The neuron populations responsible for the prediction effect are mainly amacrine cells and ganglion cells in the retina. They also have the ability to recognize moving directions [10–15]. As verified in many experiments, it is the inhibitory couplings from the amacrine cells to the ganglion cell that give rise to the direction selectivity and prediction effect [16]. Amacrine cells release inhibitory neurotransmitters (γ-Aminobutyric acid (GABA)) to the corresponding receptors located at the ganglion cells. Each amacrine cell has different preferred moving directions in each sector. Each sector connects to the ganglion cells asymmetrically, which result in the direction selectivity [17,18].

© Springer Nature Switzerland AG 2018
L. Cheng et al. (Eds.): ICONIP 2018, LNCS 11307, pp. 246–257, 2018.
https://doi.org/10.1007/978-3-030-04239-4_22

Modelling the prediction effect with respect to the amacrine cells and ganglion cells has been explored previously. Berry et al. introduced the concept of contrast-gain control, according to which a moving high-contrast stimulus somehow desensitizes the response of ganglion cells after a short time delay [2]. By introducing the spatial and temporal filters that desensitize short-term neural response, they were able to generate retinal neural response that anticipates external stimuli moving with constant speed. Similar to this desensitization phenomenon, the encoded information from the retina can adapt to changes in environment, resulting in the so-called dynamic predictive coding [19]. The model of Berry et al. further predicted that the anticipation is weaker when the stimulus contrast is low, or when the stimulus speed is too high. These predictions agreed with experimental results. In this approach, the direction selectivity of the amacrine cells does not play a role, and the issue of neuronal mechanisms remained open, as long as they produce the required temporal filtering function.

On the other hand, Nijhawan et al. modelled the prediction effect by the direction selectivity of the amacrine cells [5]. Since a stimulus moving at a constant speed is considered, the direction selectivity is well defined. The amacrine cells exert an asymmetric inhibition on the ganglion cells, resulting in the anticipation of the future position of the moving stimulus. However, it is not clear whether direction selectivity plays a role in the prediction of stochastic movements of stimuli.

Experimentally, Palmer et al. measured the prediction effect of ganglion cells in the presence of a stimulus consisting of a bar with stochastic movement [3]. By calculating the mutual information between visual stimuli and responses of salamander ganglion cells at different time shifts, they found that the ganglion cells carry the strongest information of the stimulus at around 80 ms in the past, although they also carry considerable information in the future sector. On the other hand, Chen et al. performed experiments by probing a bullfrog retina with spatially uniform light pulses of correlated stochastic intervals, and found that the peak position of the mutual information is located in the future time when the correlation times of the stochastic inputs are sufficiently long, meaning that ganglion cells can have a predominantly predictive behavior [6]. However, since light was shining on an extensive area of the retina, no spatial prediction was required. Only the temporal prediction was involved, and the direction selectivity of the amacrine cells was not relevant. Furthermore, they found that the predictive behaviour depends on the process generating the stochastic movement. Stochastic movements generated by the Hidden Markov Model (HMM) are predictable, whereas those generated by the Ornstein-Uhlenbeck (OU) process are not.

In this paper, we consider a model in which the inhibitory interaction of the amacrine cells contributes to the predictive response of the retina subject to an external stimulus of a stochastically moving bar. In contrast to the previous experiment in [6] with stimuli that are spatially uniform and temporally stochastic, the moving bar experiment is stochastic in both space and time, and the local responses of the amacrine cells have to be taken into account. In both our simulation and experiment, we found that the mutual information can be predominantly predictive. Our results show that the predictive response of the retina arise from the inertial dynamics due to the local inhibition of the amacrine cells. The issue of whether direction selectivity of these cells is necessary for prediction will also be discussed.

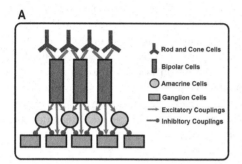 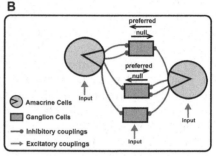

Fig. 1. The schematic diagram of the information pathway in the retina and the corresponding model structure. (A) The hierarchical structure of the information flow pathway in the retina. (B) The model of amacrine and ganglion cells.

2   The Model

Figure 1(A) schematically shows the hierarchical structure of the information flow pathway in the retina [14]. The photoreceptor layer (rod and cone cells) receives information, then transmits them to the bipolar cells. Then the bipolar cells convey the signals to both the amacrine cells and ganglion cells simultaneously. Afterwards the amacrine cells generate asymmetric inhibitory inputs to the ganglion cells, which also give rise to prediction effect and direction selectivity.

Our model involves both populations of amacrine and ganglion cells. There are different sectors with different preferred directions distributed on single amacrine cells, which link to different populations of ganglion cells [10,11,15]. The inhibitory couplings have an asymmetric distribution from the amacrine cells to the ganglion cells [5,20–23]. For ganglion cells, which have preferred directions and null directions (opposite to the preferred directions), the quantity of couplings from amacrine cells in the preferred directions are less than those in their null directions. The asymmetric couplings from amacrine to ganglion cells objectively achieve the direction selectivity, as shown in Fig. 1(B). Due to the inhibitory inputs (GABA), one side of the responses of ganglion cells is inhibited, and at the same time the opposite side of responses are pushed forward, resulting in the prediction effect.

Let $U(x,t)$ be the synaptic inputs to the ganglion cells at time t and position x. The external inputs delivered by the bipolar cells, denoted as $E(x,t)$, are transmitted to amacrine and ganglion cells simultaneously, and the centers of inputs are represented by $z(t)$. In the moving bar experiment [3,6], $z(t)$ denotes the bar positions at each instant. In the ganglion cells, the signals are coupled with the spatio-temporal filter, $K_E(x-x', t-t')$, then multiplied by an inhibitory gain function $g(x,t)$, given by amacrine cells. Therefore the dynamics of the ganglion cells can be written as:

$$U(x,t) = g(x,t) \int_{-\infty}^{\infty} dx' \int_{-\infty}^{t} dt' K_E(x - x', t - t') E(x', t'), \qquad (1)$$

where

$$E(x,t) = E_0 \exp\left[-\frac{(x - z(t))^2}{2a^2}\right], \qquad (2)$$

$$K_E(x - x', t - t') = J_g(x - x') \cdot (t - t') e^{-\alpha(t-t')}. \qquad (3)$$

Here the $J_g(x - x')$ denotes the connections from bipolar cells to ganglion cells, taking the form of a Gaussian function. ω_g is the coupling strength and σ_g is the coupling width,

$$J_g(x - x') = \omega_g \exp\left[-\frac{(x - x')^2}{2\sigma_g^2}\right]. \qquad (4)$$

The firing rate $r(x,t)$ of ganglion cells is determined by the ReLu function of $U(x,t)$, (which reduces to a linear function in our model),

$$r(x,t) = \text{ReLu}[U(x,t)]. \qquad (5)$$

The gain function $g(x,t)$ denotes outputs of amacrine cells. Similar to ganglion cells, the inputs to the amacrine cells are coupled with the spatio-temporal filter, $K_I(x - x', t - t')$:

$$g(x,t) = \exp\left[-\mathscr{A} \int_{-\infty}^{\infty} dx' \int_{-\infty}^{t} dt' K_I(x - x', t - t') I(x', t')\right]. \qquad (6)$$

Here the inputs to the amacrine cells are denoted as $I(x,t)$ instead of $E(x,t)$. Actually, the $I(x,t)$ consists of two components, one is the inputs from the bipolar cells, and the other one is the feedback from the ganglion cells. Since we are studying the behaviors of the whole system, we approximate $I(x,t) \approx E(x,t)$. This amounts to approximating that the feedback from the ganglion cells is also proportional to the strength of the stimulus. Further refinement of the approximation will be left for future work.

For the spatio-temporal filter $K_I(x - x', t - t')$, we take two different situations into account, moving in positive and negative directions. The neural mechanism of detecting velocity direction locally will not be discussed here, but it is sufficient to note that it can be deduced from a comparison of the spatial and temporal gradients of the profile of $U(x,t)$. When the stimulus moves in the positive direction, the amacrine cells release inhibitory neurotransmitters (GABA) behind the leading edge of the stimulus. According to the spatio-temporal features, we propose:

$$K_I(x - x', t - t') = \Theta\left[v(x', t')(x' - x)\right] \exp\left[-\frac{|x - x'|}{\lambda} - \frac{p(t - t')}{\lambda}\right], \quad (7)$$

where Θ represents the step function.

Figure 2(A) shows the response of the ganglion cells in the presence of a moving stimulus. When there are no inputs from amacrine cells, the peak response is just located at the same position as the peak of the excitatory input. However, when the ganglion cells simultaneously receive inhibitory inputs from amacrine cells, the responses are altered. Since the GABA is released by amacrine cells after the stimulus, the tail of the responses of ganglion cells is inhibited strongly. Meanwhile, the leading edge of the responses are not inhibited. Therefore 'inhibiting the tail' gives rise to the direction selectivity. Due to the inhibitory inputs from amacrine cells, the bump of responses of ganglion cells is protruded forward, and the peak of the responses is also ahead of the stimulus, resulting in the prediction effect. Hence, facilitated by the amacrine cells, no matter the stimulus moves towards positive or negative direction, the ganglion cells are always able to recognize the moving direction and make predictions.

In order to study the influences of the parameters on the prediction effect, we tested different sets of parameters shown in Fig. 2(B)–(D).

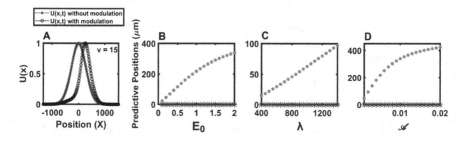

Fig. 2. (A) The profile of the synaptic input $U(x, t)$ of the ganglion cells. Blue-star lines: Response of ganglion cells without inputs from amacrine cells. Black-circle lines: Response of ganglion cells with inputs from amacrine cells. (B)–(D) Effects of different parameters on the predictive positions. Red points are the predictive positions, and blue stars denote the positions without the prediction effect. (B) is for varying intensity E_0, (C) for varying the transmission range λ, (D) for varying the magnitude of the gain coupling $\mathscr{A}$. (Color figure online)

It can be seen from Fig. 2(B) that the prediction effect is enhanced as the intensity E_0 increases. However, the decay rate of neurotransmitters GABA, p/λ, is inversely proportional to the prediction effect, as shown in Fig. 2(C). In Fig. 2(D), the prediction effect will also get more pronounced when the magnitude of the gain coupling $\mathscr{A}$ from amacrine cells to ganglion cells increases.

3   Simulations of the Network Model

After exploring the properties and dynamics of the network model, we fit the experiment to the model.

3.1   Inputs to the Network Model

In previous sections we have introduced predictive effects reported by other groups under constant moving velocities [2–6,8,9]. Now we consider the prediction effect with respect to stochastically moving stimulus. There has been verified that the signals generated from OU process do not contain sufficient information for predictions [6], but the signals generated from HMM carry information that can be utilized to make predictions [3,6]. In this paper, we take the moving-bar experiment as an example, and according to experiments [3,6], for HMM, we consider a damped harmonic oscillator driven by noise. The position and moving velocity of the moving bar at each time step are iterated by the following equations [3,6]

$$x_{t+\Delta\tau} = x_t + v_t\Delta\tau, \tag{8}$$

$$v_{t+\Delta\tau} = [1 - \Gamma\Delta\tau]v_t - \omega^2 x_t\Delta\tau + \xi_t\sqrt{D\Delta\tau}, \tag{9}$$

where the natural frequency $\omega = 2\pi f$ rad/s, and $\Gamma/2\omega = 1.06$ to keep the dynamics always slightly overdamped. The time step $\Delta\tau = 1/60s$, and $D = 2.7 \times 10^6$ pixel$^2/s^3$. ξ_t is a Gaussian random variable with zero mean and unit variance, independently of each time step. For the OU Process, there is only one equation denoting the moving bar positions at each time step:

$$x_{t+\Delta\tau} = (1 - \theta\Delta\tau)x_t + \xi_t\sqrt{D\Delta\tau}. \tag{10}$$

The positions iterated from Eqs. (8) and (10) are inserted as input positions in Eq. (2) ($z(t)$) to the network model, which correspond to the positions of moving bar at each instant in the experiment. Eventually, by calculating the mutual information between peak positions of the stimulus and the retina response, the behaviors and competence of the prediction effect of the network model can be measured.

3.2   Mutual Information

To evaluate the prediction effect, we calculate the mutual information between the input positions and the output positions, defined by

$$I_m(S, R, \delta t) = \sum_i p(s_i, r_{i-k}) \log_2 \frac{p(s_i, r_{i-k})}{p(s_i)p(r_{i-k})}, \tag{11}$$

in which the S denotes the moving stimulus states, given by $S = \{s_1, s_2, \ldots\}$ and R denotes the responses of the ganglion cells, denoted as $R = \{r_1, r_2, \ldots\}$. δt is the time shift between the signals and responses, and $k = \delta t / \Delta t$ is the difference in time indices between s and r. $p(s_i)$ is the probability of having a state s_i, and $p(s_i, r_{i-k})$ is the joint probability of the state (s_i, r_{i-k}).

3.3   Comparisons of Mutual Information

We compare the mutual information between the experiments and simulations in HMM and OU Process respectively. For experimental details, please refer to reference [7]. We find that the predictive behavior depends on the correlation time τ of the input time series.

HMM. In HMM, the theoretical value of the correlation time is $\tau = (\Gamma^2 - \omega^2)/(\omega^2 \Gamma)$. In Fig. 3, we present the mutual information from HMM calculated respectively from simulations and experiments [7]. In experiment it has been found that the signals generated from HMM have predictive behaviors owing to the hidden variables, as shown in Fig. 3(B). In the simulation results shown in Fig. 3(B), it can be seen that the network model works very well with respect to the prediction effect. As the correlation time increases, both of the experiments and simulations show improvement of the predictions, and the peak of the mutual information gradually shift to the positive, and the bump of the mutual information becomes wider. When the correlation time is too small, e.g., $\tau = 0.16$ s, both of the experiments and simulations do not show prediction effect. When the correlation time reaches a high value, e.g., $\tau = 1.28$ s, the mutual information in both simulations and experiments become more predictive, namely including more available information for making prediction.

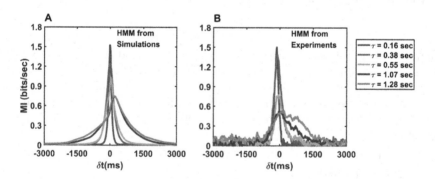

Fig. 3. The mutual information curves for different correlation times τ of HMM. (A) The mutual information calculated from simulations. (B) The mutual information measured from experiment.

In order to illustrate the predictive effect of the network model, we plot the peak positions and the center of mass of the mutual information under different correlation times in both experiments and simulations, as shown in Fig. 4.

For the peak positions of the mutual information, we can see the simulations and experiments exhibit similar trends. When the correlation time is small, the available information in the past is too poor for the ganglion cells to make reliable predictions, therefore the peak positions are around 0 in the simulations, and even some lag behind in the experiments, as shown in Fig. 4(B). When the correlation time is sufficiently large, the network can make use of the information in the past to make reliable predictions, hence the peak positions become more positive. The center of mass of the mutual information is shown in Figs. 4(C) and (D), which show a general picture of the prediction effect. In the network model, when the correlation time is small, the network model is not able to make predictions, hence the center of mass mainly stays around 0. In the experiments, the center of mass is even located at negative time values, which verifies that the ganglion cells are not capable of making predictions on small correlation times. However, when the correlation time is large, e.g., $\tau = 1.07\,\mathrm{s}$ and $\tau = 1.28\,\mathrm{s}$, the center of mass of mutual information increases prominently, which can be seen in both simulations and experiments.

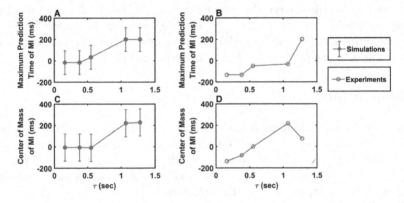

Fig. 4. The peak positions and center of mass of the mutual information curves of HMM. (A) The peak positions for various correlation times calculated from simulations. (B) The peak positions for various correlation times measured from experiments. (C) The center of mass for distinct correlation times calculated from simulations. (D) The center of mass for distinct correlation times measured from experiments. Solid circles with error bars denote simulation results, and hollow circles indicate the measurements from experiments.

OU Process. The theoretical value of the correlation time in the OU Process is $\tau = 1/\theta$. Similar to the HMM, we also explored the prediction effect under the moving inputs generated by OU Process.

In the Fig. 5, we show the mutual information calculated from experiments and simulations under different correlation times, respectively. Contrary to the

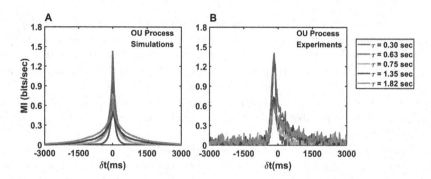

Fig. 5. The mutual information curves for different correlation times τ of OU Process. (A) The mutual information calculated from simulations. (B) The mutual information measured from experiment.

HMM, OU Process as input signals have no predictive behaviors, which are verified both by experiments and simulations. In the simulations shown in Fig. 5(A), no matter how small or large the correlation time is, the peak positions of the mutual information are always lied on 0, meaning no prediction effect found. As for the experiments shown in Fig. 5(B), the peak positions of mutual information not only fail to make predictions, but also lag behind the 0 point.

4 Conclusion

We compare the mutual information calculated from simulations with those measured in experiments of HMM and OU process plotted in Figs. 3, 4 and 5. The results show that our model agrees with experiments very well, especially as the correlation time τ increases, the prediction effect also becomes more and more prominent in both experiments and simulations.

To explain the experimental results, our model involves amacrine cells and ganglion cells, in which the amacrine cells provide inhibitory couplings to the ganglion cells. As the stimulus moves, the amacrine cells transmit GABA neurotransmitters after the stimulus passes by. This inhibition destabilizes the firing region of the population of ganglion cells, increasing its tendency to shift the firing profile to the neighborhood. Hence when the population profile moves in a certain direction, its tendency to move forward is stronger than backward, since the concentration of GABA neurotransmitters is higher in the tail region.

It is interesting to note that while both HMM and OU processes are stochastic in nature, only HMM contains predictive information detectable by neural systems, as confirmed in both simulations and experiments. The difference between HMM and OU is that the dynamics of the former is inertial, or in other words, the motion carries momentum. For the neural system to be predictive, its dynamics must also be inertial, and this is achieved in the retina by the local inhibition of the amacrine cells, in the sense that the response to a continuously moving stimulus will be stronger in the forward direction.

The above dynamical description is operationally equivalent to the spatial and temporal filters proposed by Berry et al. [2]. The spatial filter is a center-surround one, which serves to stabilize the activity profile of the ganglion population as a bump shape, and the temporal profile is biphasic with short-term inhibition relaxing to zero after a period of time. As shown in Fig. 2(b), the anticipation in our model is weaker when the intensity is low, consistent with the weakening effect of low contrast in [2]. In our model, the anticipation is also weaker when the stimulus speed is too high, consistent with [2]. (This result is not shown here, but similar speed-dependent effects can be found in the cases of other local inhibitory mechanisms to be discussed in the second last paragraph.)

However, the local inhibitory mechanism described above does not preclude the contribution of the direction selectivity of the amacrine cells to anticipatory tracking. These cells asymmetrically release inhibitory neurotransmitters to the ganglion cells [5, 14]. This endows the ganglion cells the ability to recognize the moving directions of the stimuli. When these inhibitory inputs enter ganglion cells, they inhibit the tail of the responses of the ganglion cells, which leads to the forward bias of the responses, giving rise to the prediction effect. This mechanism is effective in anticipating stimuli moving in a constant direction, since in this case the head and tail regions are well defined. It is expected that the mechanism is not effective in predicting stimuli with stochastic motion, since the head and tail regions are stochastically changing. Hence, we expect that the anticipatory behaviors for constant and stochastic stimulus directions may have qualitative difference. Comparison of these two behaviors is an interesting issue to be explored.

It is worthwhile to note that there are other neural mechanisms, such as short-term synaptic depression, spike frequency adaptation and inhibitory feedback from upper layers, that will exhibit the same universal anticipative effect [24]. Short-term synaptic depression is the reduction of synaptic efficacy due to the overconsumption of neurotransmitters after prolonged firing activities of the neurons. Spike frequency adaptation is the adaptation of the firing threshold of neurons due to factors such as the accumulation of calcium ions after prolonged firing activities [4]. Both have the same effect of destabilizing the activity profile of the ganglion population and contributing to the buildup of inertia. It would be interesting to combine these mechanisms with the action of the amacrine cells to give a fuller picture of the population dynamics.

In the present model, we have not considered the feedback couplings from ganglion cells to amacrine cells. This feedback circuitry is found in the form of gap junctions for OFF ganglion cells that exhibit predictive behavior, but not in ON ganglion cells that do not exhibit such behavior [25]. As a part of connections between the amacrine cells and ganglion cells, adding the feedback couplings from ganglion cells will be one direction for further study. Already without the feedback couplings, we have sufficient drives to yield the predictive behavior. We anticipate that the inclusion of feedback couplings will further enhance the effects.

Acknowledgements. This work is supported by the Research Grants Council of Hong Kong (grant numbers 16322616 and 16306817).

References

1. Jessell, T., Siegelbaum, S., Hudspeth, A.J.: Learning and memory. In: Kandel, E.R., Schwartz, J.H., Jessell, T.M. (eds.) Principles of Neural Science, vol. 4, pp. 1227–1246. McGraw-hill, New York (2000)
2. Berry II, M.J., et al.: Anticipation of moving stimuli by the retina. Nature **398**(6725), 334 (1999)
3. Palmer, S.E., Marre, O., Berry, M.J., Bialek, W.: Predictive information in a sensory population. Proc. Natl. Acad. Sci. U.S.A. **112**(22), 6908–6913 (2015)
4. Mi, Y., Fung, C.C.A., Wong, K.Y.M., Wu, S.: Spike Frequency Adaptation Implements Anticipative Tracking in Continuous Attractor Neural Networks. In: NIPS, pp. 505–513 (2014)
5. Nijhawan, R., Wu, S.: Compensating time delays with neural predictions: are predictions sensory or motor? Philos. Trans. Royal Soc. A **367**(1891), 1063–1078 (2009)
6. Chen, K.S., Chen, C.C., Chan, C.K.: Characterization of predictive behavior of a retina by mutual information. Front. Comput. Neurosci. **11**, 66 (2017)
7. Chen, Y.: Anticipative Responses of a Retina to a Stochastically Moving Bar. Master thesis, National Tsing Hua University (2018)
8. Fung, C.C.A., Wong, K.Y.M., Wang, H., Wu, S.: Dynamical synapses enhance neural information processing: gracefulness, accuracy, and mobility. Neural Comput. **24**(5), 1147–1185 (2012)
9. Marre, O., Botella-Soler, V., Simmons, K.D., Mora, T., Tkačik, G., Berry II, M.J.: High accuracy decoding of dynamical motion from a large retinal population. PLOS Comput. Biol. **11**(7), e1004304 (2015)
10. Briggman, K.L., Helmstaedter, M., Denk, W.: Wiring specificity in the direction-selectivity circuit of the retina. Nature **471**(7337), 183 (2011)
11. Euler, T., Detwiler, P.B., Denk, W.: Directionally selective calcium signals in dendrites of starburst amacrine cells. Nature **418**(6900), 845 (2002)
12. Wei, W., Hamby, A.M., Zhou, K., Feller, M.B.: Development of asymmetric inhibition underlying direction selectivity in the retina. Nature **469**(7330), 402 (2011)
13. Fried, S.I., Münch, T.A., Werblin, F.S.: Mechanisms and circuitry underlying directional selectivity in the retina. Nature **420**(6914), 411 (2002)
14. Masland, R.H.: The neuronal organization of the retina. Neuron **76**(2), 266–280 (2012)
15. Venkataramani, S., Taylor, W.R.: Orientation selectivity in rabbit retinal ganglion cells is mediated by presynaptic inhibition. J. Neurosci. **30**(46), 15664–15676 (2010)
16. Manu, M., Baccus, S.A.: Disinhibitory gating of retinal output by transmission from an amacrine cell. Proc. Natl. Acad. Sci. U.S.A. **108**(45), 18447–18452 (2011)
17. Baden, T., Berens, P., Franke, K., Rosón, M.R., Bethge, M., Euler, T.: The functional diversity of retinal ganglion cells in the mouse. Nature **529**(7586), 345 (2016)
18. Zhang, A.J., Wu, S.M.: Responses and receptive fields of amacrine cells and ganglion cells in the salamander retina. Vision Res. **50**(6), 614–622 (2010)
19. Hosoya, T., Baccus, S.A., Meister, M.: Dynamic predictive coding by the retina. Nature **436**, 71–77 (2005)
20. de Vries, S.E., Baccus, S.A., Meister, M.: The projective field of a retinal amacrine cell. J. Neurosci. **31**(23), 8595–8604 (2011)

21. Lin, B., Wang, S.W., Masland, R.H.: Retinal ganglion cell type, size, and spacing can be specified independent of homotypic dendritic contacts. Neuron **43**(4), 475–485 (2004)
22. Vaney, D.I., Sivyer, B., Taylor, W.R.: Direction selectivity in the retina: symmetry and asymmetry in structure and function. Nat. Rev. Neurosci. **13**(3), 194 (2012)
23. Dong, W., Sun, W., Zhang, Y., Chen, X., He, S.: Dendritic relationship between starburst amacrine cells and direction-selective ganglion cells in the rabbit retina. J. Physiol. **556**(1), 11–17 (2004)
24. Fung, C.C.A., Wong, K.Y.M., Mao, H.Z., Wu, S.: Fluctuation-response relation unifies dynamical behaviors in neural fields. Phys. Rev. E **92**(2), 022801 (2015)
25. Pang, J.J., Gao, F., Wu, S.M.: Light-evoked excitatory and inhibitory synaptic inputs to ON and OFF α ganglion cells in the mouse retina. J. Neurosci. **23**(14), 6063–6073 (2003)

Local Tracking Control for Unknown Interconnected Systems via Neuro-Dynamic Programming

Bo Zhao[1]([✉]), Derong Liu[2], Mingming Ha[3], Ding Wang[1], Yancai Xu[1], and Qinglai Wei[1]

[1] The State Key Laboratory of Management and Control for Complex Systems, Institute of Automation, Chinese Academy of Sciences, Beijing 100190, China
{zhaobo,ding.wang,yancai.xu,qinglai.wei}@ia.ac.cn
[2] School of Automation, Guangdong University of Technology, Guangzhou 510006, China
derongliu@foxmail.com
[3] School of Automation and Electrical Engineering, University of Science and Technology Beijing, Beijing 100083, China
hamingming_0705@foxmail.com

Abstract. This paper develops a neuro-dynamic programming based local tracking control (LTC) scheme for unknown interconnected systems. By using the local input-output data and the desired states of coupling subsystems, a local neural network (NN) identifier is established to obtain the local input gain matrix online. By introducing a modified local cost function, the Hamilton-Jacobi-Bellman equation is solved by a local critic NN with asymptotically convergent weight vector, which is obtained by nested update law, and the LTC can be derived with the desired state aided augmented subsystem. The stability of the closed-loop system is shown by Lyapunov's direct method. The simulation on the parallel inverted pendulum system illustrates that the developed LTC scheme is effective.

Keywords: Adaptive dynamic programming
Neuro-dynamic programming · Local tracking control
Optimal control · Unknown interconnected systems

1 Introduction

Modern systems requires more advanced control performance since they are becoming increasingly large-scale, nonlinear and complex, such as multi-agent systems [1], reconfigurable modular robotic systems [2]. Decentralized control or local control which uses only local subsystem information is a reasonable approach to solve the corresponding control problems. It is worth pointing out that the optimal control is also desired besides the excellent control quality.

© Springer Nature Switzerland AG 2018
L. Cheng et al. (Eds.): ICONIP 2018, LNCS 11307, pp. 258–268, 2018.
https://doi.org/10.1007/978-3-030-04239-4_23

To address the optimal control problems for aforementioned systems, the adaptive dynamic programming (ADP) [3–5] is an effective way with the help of neural networks (NNs) [6]. Throughout the related literature, considerable efforts have been made on ADP-based approaches for nonlinear systems [7–12]. For trajectory tracking control of interconnected systems, Mehraeen *et al.* [13] assumed that the input gain matrix was available and the unknown interconnection was weak, and proposed a near optimal decentralized control method for nonlinear systems with available dynamics by using online updated action NN and critic NN. By using an online updated single critic NN, Qu *et al.* [14] proposed a decentralized adaptive tracking control strategy which consisted of a steady-state controller and a modified optimal feedback controller for uncertain systems. For unknown nonlinear systems, Zhao *et al.* [15] presented an observer-critic structure based decentralized tracking control scheme, which was composed of local desired control, local tracking error control and a compensator.

We can observe that from the aforementioned literature that the critic NNs were guaranteed to be ultimately uniformly bounded (UUB). Motivated from it, this paper develops a local tracking control (LTC) scheme for unknown interconnected systems by using the local critic NN with nested update laws, which ensures the weight vector of local critic NN to asymptotic convergence.

2   Problem Statement

Considering unknown interconnected systems, whose ith $(i = 1, 2, \ldots, N)$ subsystem dynamics can be described by

$$\dot{x}_i(t) = f_i(x_i(t)) + g_i(x_i(t))u_i(x_i(t)) + h_i(x(t)), \tag{1}$$

where $x_i(t) = [x_{i1}(t), x_{i2}(t), \ldots, x_{i(n_i)}(t)]^{\mathsf{T}} \in \mathbb{R}^{n_i}, i = 1, \ldots, N$ and $u_i(x_i(t)) \in \mathbb{R}^{m_i}$ are the state vector and local control input of the ith subsystem, respectively; $x(t) = [x_1^{\mathsf{T}}(t), \ldots, x_N^{\mathsf{T}}(t)]^{\mathsf{T}} \in \mathbb{R}^n$ is the entire system state vector with $n = \sum_{i=1}^{N} n_i$; $f_i(x_i(t)), g_i(x_i(t))$ and $h_i(x(t))$ are unknown nonlinear internal dynamics, input gain matrix and interconnection term, respectively.

In order to improve the modeling accuracy, the desired states of the coupling subsystems are substituted into the interconnection term, we thus rewrite the interconnection term as

$$h_i(x) = h_i(x_i, x_{kd}) + \Delta h_i(x, x_{kd}), \tag{2}$$

where x_{kd} denotes the desired states of the kth coupled subsystems, $\Delta h_i(x, x_{kd}) = h_i(x) - h_i(x_i, x_{kd})$ denotes the substitution error. Thus, (1) can be rewritten as

$$\dot{x}_i = f_i(x_i, x_{kd}) + g_i(x_i)u_i(t) + \Delta h_i(x, x_{kd}), \tag{3}$$

where $f_i(x_i, x_{kd}) = f_i(x_i(t)) + h_i(x_i, x_{kd})$.

Assumption 1. *The nonlinear functions $f_i(x_i, x_{kd})$, $g_i(x_i)$ and $\Delta h_i(x, x_{kd})$ are Lipschitz and continuous in their arguments with $f_i(0) = 0$, and the subsystem (3) is controllable.*

Assumption 2. *The desired trajectory x_{id} is twice differentiable and the vector $X_{id} = [x_{id}, \dot{x}_{id}, \ddot{x}_{id}]^\mathsf{T}$ is norm-bounded as $\|X_{id}\| \leq q_{iM}$, where q_{iM} is an available positive constant.*

Since the interconnection term $\Delta h_i(x, x_{kd})$ satisfies the Lipschitz condition, which thus implies $\|\Delta h_i(x, x_{kd})\| \leq \sum_{j=1}^{n} d_{ik} E_k$, where $E_k = \|x_k - x_{kd}\|$, and $d_{ik} \geq 0$ is an unknown Lipschitz constant.

To achieve the desired output trajectory, we use the exosystem $\dot{x}_{id} = f_{id}(x_{id})$ for the ith subsystem, where $f_{id}(x_{id})$ is a Lipschitz continuous function.

For the ith subsystem, define the local tracking error as $e_i = x_i - x_{id}$. Thus, the tracking error dynamics can be expressed as $\dot{e}_i = f_i(x_i, x_{kd}) + g_i(x_i)u_i(t) + \Delta h_i(x, x_{kd}) - f_{id}(x_{id})$.

Defining the augmented subsystem states as $x_{ia} = [e_i, x_{id}]^\mathsf{T} \in \mathbb{R}^{2n}$, and the augmented subsystem can be expressed as

$$\dot{x}_{ia} = f_{ia}(x_{ia}) + g_{ia}(x_i)u_i(t) + \Delta h_{ia}(x_a), \tag{4}$$

where $f_{ia}(x_{ia}) = \begin{bmatrix} f_i(e_i + x_{id}) - f_{id}(x_{id}) \\ f_{id}(x_{id}) \end{bmatrix}$, $g_{ia}(x_{ia}) = \begin{bmatrix} g_i(e_i + x_{id}) \\ 0 \end{bmatrix}$,

$\Delta h_{ia}(x_a) = \begin{bmatrix} \Delta h_i(x, x_{kd}) \\ 0 \end{bmatrix}$, and $x_a = \begin{bmatrix} x \\ x_{kd} \end{bmatrix}$.

Similar to [8], the nominal augmented subsystem (4) is described as

$$\dot{x}_{ia} = f_{ia}(x_{ia}) + g_{ia}(x_i)u_i(t). \tag{5}$$

The local optimal cost function is defined as

$$V_i^*(x_{ia}(t), u_i^*) = \int_t^\infty \left(\hat{\eta}_i \|\nabla V_i^*\| E_i + x_{ia}^\mathsf{T} Q_{ia} x_{ia} + u_i^{*\mathsf{T}}(\tau) R_i u_i^*(\tau) \right) d\tau, \tag{6}$$

where $\hat{\eta}_i$ is the estimation of later defined positive constant η_i, $\nabla V_i^*(x_{ia}) = \partial V_i^*(x_{ia})/\partial x_{ia}$ is the partial derivative of $V_i^*(x_{ia})$ with respect to the subsystem state x_{ia}, $Q_{ia} = \begin{bmatrix} Q_i & 0_{n \times n} \\ 0_{n \times n} & 0_{n \times n} \end{bmatrix} \in \mathbb{R}^{2n}$, $R_i \in \mathbb{R}^n$ and $Q_i \in \mathbb{R}^n$ are positive definite matrices, and $0_{n \times n}$ is a square matrix of zeros.

If (6) is continuously differentiable, then the infinitesimal version of it is the so-called local Lyapunov equation

$$\nabla V_i^*(x_{ia})\dot{x}_{ia} + \hat{\eta}_i \|\nabla V_i^*(x_{ia})\| E_i + x_{ia}^\mathsf{T} Q_{ia} x_{ia} + u_i^{*\mathsf{T}} R_i u_i^* = 0. \tag{7}$$

Thus, the Hamiltonian of the optimal problem is defined as

$$H_i(x_{ia}, \nabla V_i^*, u_i^*) = \nabla V_i^* \dot{x}_{ia} + x_{ia}^\mathsf{T} Q_{ia} x_{ia} + \hat{\eta}_i \|\nabla V_i^*(x_{ia})\| E_i + u_i^{*\mathsf{T}} R_i u_i^*. \tag{8}$$

And the local optimal cost function $V_i^*(x_{ia})$ is derived by solving the local Hamilton-Jacobi-Bellman (HJB) equation

$$\min_{u_i^*} H_i(x_{ia}, \nabla V_i^*, u_i^*) = 0 \tag{9}$$

with $V_i^*(0) = 0$. Therefore, the local optimal tracking control is expressed as

$$u_i^* = -\frac{1}{2} R_i^{-1} g_{ia}^\mathsf{T}(x_{ia}) \nabla V_i^*(x_{ia}). \tag{10}$$

Thus, we have

$$\nabla V_i^{*\mathsf{T}} g_{ia}(x_{ia}) = -2u_i^{*\mathsf{T}} R_i. \tag{11}$$

The objective of this paper is to find a neuro-dynamic programming (NDP) based LTC policy $u_i(x_{ia})$ to minimize the local cost function such that the actual states of the ith subsystem follow its desired trajectories.

3   Neuro-Dynamic Programming-Based Local Tracking Control

With the modern systems become complex and large-scale, it is difficult or even unable to design the controller in the centralized architecture, since the complexity of controller design procedure and computational burden become heavier. Thus, LTC which uses only local subsystem information is reasonable to be implemented in such systems. In this section, the detailed LTC design procedure is provided based on NDP for unknown interconnected systems.

3.1   Local NN Identifier

For the ith subsystem (2), let $F_i(x_i, x_{kd}, u_i) = f_i(x_i, x_{kd}) + g_i(x_i)u_i(x_i)$, which is identified by the radial basis function (RBF) NN as

$$F_i(x_i, x_{kd}, u_i) = W_{iF}^\mathsf{T} \sigma_{iF}(x_i, x_{kd}, u_i) + \varepsilon_{iF}, \tag{12}$$

where $W_{iF} \in \mathbb{R}^{l_{iF}}$, $\sigma_{iF}(x_i, x_{kd}, u_i) \in \mathbb{R}^{l_{iF}}$ is the RBF, l_{iF} is the number of neurons in the hidden layer, ε_{iF} is the approximation error, which is assumed to be norm-bounded as $\|\varepsilon_{iF}\| \leq \varepsilon_{iFM}$, where ε_{iFM} is an unknown positive constant.

Define the identification error as

$$e_{io} = x_i - \hat{x}_i. \tag{13}$$

From (2), the local NN identifier can be established as

$$\dot{\hat{x}}_i = \hat{F}_i(\hat{x}_i, x_{kd}, u_i) + l_{io}e_{io}, \tag{14}$$

where $\hat{x}_i$ is the identification of subsystem state x_i, l_{io} is the identification gain matrix with all positive elements, $\hat{F}_i(\hat{x}_i, x_{kd}, u_i) = \hat{W}_{iF}^\mathsf{T} \hat{\sigma}_{iF}(\hat{x}_i, x_{kd}, u_i)$ is the identification of $F_i(\hat{x}_i, x_{kd}, u_i)$, where $\hat{W}_{iF}$ and $\hat{\sigma}_{iF}(\hat{x}_i, x_{kd}, u_i)$ are the estimations of W_{iF} and $\sigma_{iF}(x_i, x_{kd}, u_i)$, respectively.

Therefore, the local input gain matrix $\hat{g}_{ia}(x_{ia})$ can be derived by

$$\hat{g}_{ia}(x_{ia}) = \frac{\partial \hat{F}_i(\hat{x}_i, x_{kd}, u_i)}{\partial u_i}. \tag{15}$$

Theorem 1. *Considering the ith subsystem dynamics (2), the local input-output data and desired states of coupling subsystems based local NN identifier (14) guarantees the local identification error e_{io} to be ultimately uniformly bounded (UUB) with the updated law*

$$\dot{\hat{W}}_{iF} = \Gamma_{iF} e_{io} \hat{\sigma}_{iF}(\hat{x}_i, x_{kd}, u_i), \tag{16}$$

where Γ_{iF} is a positive constant.

Proof: Define the weight estimation error as $\tilde{W}_{iF} = W_{iF} - \hat{W}_{iF}$. Select a Lyapunov function candidate as

$$L_{i1} = \frac{1}{2} e_{io}^{\mathsf{T}} e_{io} + \frac{1}{2} \tilde{W}_{iF}^{\mathsf{T}} \Gamma_{iF}^{-1} \tilde{W}_{iF}. \tag{17}$$

Its time derivative is

$$\begin{aligned}
\dot{L}_{i1} &= e_{io} \dot{e}_{io} - \tilde{W}_{iF}^{\mathsf{T}} \Gamma_{iF}^{-1} \dot{\hat{W}}_{iF} \\
&= -\lambda_{\min}(l_{io}) e_{io}^2 + e_{io} \tilde{W}_{iF}^{\mathsf{T}} \hat{\sigma}_{iF} + e_{io} \delta_{iF} - \tilde{W}_{iF}^{\mathsf{T}} \Gamma_{iF}^{-1} \dot{\hat{W}}_{iF}, \tag{18}
\end{aligned}$$

where $\lambda_{\min}(\cdot)$ presents the minimum eigenvalue of the matrix, $\delta_{iF} = W_{iF}^{\mathsf{T}} \tilde{\sigma}_{iF} + \varepsilon_{iF}$. By assuming that $\|\delta_{iF}\| \leq \delta_{iFM}$ with δ_{iFM} a positive constant. Substituting (16) into (18), (18) becomes

$$\begin{aligned}
\dot{L}_{i1} &\leq -\lambda_{\min}(l_{io}) e_{io}^2 + \|e_{io}\| \delta_{iFM} \\
&\leq -(\lambda_{\min}(l_{io}) \|e_{io}\| - \delta_{iFM}) \|e_{io}\|. \tag{19}
\end{aligned}$$

Hence, $\dot{L}_{i1} \leq 0$ when e_{io} lies outside the compact set $\Omega_{e_{io}} = \left\{ e_{io} : \|e_{io}\| \leq \frac{\delta_{iFM}}{\lambda_{\min}(l_{io})} \right\}$. Therefore, according to Lyapunov's direct method, the identification error e_{io} can be guaranteed to be UUB. This completes the proof.

3.2 Local Critic Neural Network with Asymptotically Convergent Weight Vector

Since the universal approximation property, the local optimal cost function (6) is constructed by local critic NN on a compact set Ω_i as

$$V_i^*(x_{ia}) = W_{ic}^{*\mathsf{T}} \sigma_{ic}(x_{ia}) + \varepsilon_{ic}(x_{ia}), \tag{20}$$

where $W_{ic}^* \in \mathbb{R}^{l_{ic}}$ is ideal weight vector, $\sigma_{ic}(x_{ia}) \in \mathbb{R}^{l_{ic}}$ is the activation function, l_{ic} is the number of neurons in the hidden layer, and $\varepsilon_{ic}(x_{ia}) \in \mathbb{R}$ is the reconstruction error.

The partial gradient of (20) with respect to the corresponding state x_{ia} is

$$\nabla V_i^*(x_{ia}) = \nabla \sigma_{ic}^{\mathsf{T}}(x_{ia})W_{ic}^* + \nabla \varepsilon_{ic}(x_{ia}). \tag{21}$$

In order to approximate the unknown ideal weight vector W_{ic}^*, the local critic NN can be approximated as

$$\hat{V}_i(x_{ia}) = \hat{W}_{ic}^{\mathsf{T}}\sigma_{ic}(x_{ia}), \tag{22}$$

where $\hat{W}_{ic} \in \mathbb{R}^{l_{ic}}$ is the approximate weight vector, and its partial gradient is

$$\nabla \hat{V}_i(x_{ia}) = \nabla \sigma_{ic}^{\mathsf{T}}(x_{ia})\hat{W}_{ic}. \tag{23}$$

Thus, the LTC policy (10) is approximated as

$$\hat{u}_i(x_{ia}) = -\frac{1}{2}R_i^{-1}\hat{g}_{ia}^{\mathsf{T}}(x_{ia})\nabla \hat{V}_i(x_{ia}). \tag{24}$$

The Hamiltonian is approximated as

$$\hat{H}_i(x_{ia}, \nabla \hat{V}_i(x_{ia}), \hat{u}_i(x_{ia})) = \nabla \hat{V}_i(x_{ia})\dot{x}_{ia}(t) + \hat{\eta}_i \left\|\nabla \hat{V}_i(x_{ia})\right\| E_i$$
$$+ x_{ia}^{\mathsf{T}}Q_{ia}x_{ia} + \hat{u}_i^{\mathsf{T}}(x_{ia})R_i\hat{u}_i(x_{ia}) = e_{ic}, \tag{25}$$

where $\hat{\eta}_i$ is updated as $\dot{\hat{\eta}}_i = \Gamma_{i\eta} \left\|\nabla \hat{V}_i(x_{ia})\right\| E_i$, where $\Gamma_{i\eta}$ is a positive constant.

Comparing (25) with the optimal local Hamiltonian (8), and defining the weight error vector as $\tilde{W}_{ic} = W_{ic}^* - \hat{W}_{ic}$, we have

$$H_i(x_{ia}, \nabla V_i^*(x_{ia}), u_i^*) - \hat{H}_i\left(x_{ia}, \nabla \hat{V}_i(x_{ia}), \hat{u}_i\right) = \tilde{W}_{ic}^{\mathsf{T}}\nabla \sigma_{ic}(x_{ia})\dot{x}_{ia} + \delta_i, \tag{26}$$

where $\delta_i = \hat{\eta}_i \left(\|\nabla V_i^*(x_{ia})\| - \left\|\nabla \hat{V}_i(x_{ia})\right\|\right) E_i + \varepsilon_{ic}\dot{x}_{ia}(t) + (u_i^*(x_{ia}) + \hat{u}_i(x_{ia}))^{\mathsf{T}}$
$R_i(u_i^*(x_{ia}) - \hat{u}_i(x_{ia}))$.

Assumption 3. δ_i is norm-bounded, i.e., $\|\delta_i\| \leq \delta_{icM}$, where δ_{icM} is a positive constant.

Noticing the fact that $H_i(x_{ia}, \nabla V_i^*(x_{ia}), u_i^*(x_{ia})) = 0$, from (25), we have $\frac{\partial e_{ic}}{\partial \hat{W}_{ic}} = \nabla \sigma_i(x_{ia})\dot{x}_{ia}(t) = \theta_i$, where $\theta_i \in \mathbb{R}^{l_{ic}}$. Thus, from (26), we have $e_{ic} = -\delta_i - \tilde{W}_{ic}^{\mathsf{T}}\theta_i$. In order to tune the weight vector $\hat{W}_{ic}$ of the local critic NN, the objective function $E_{ic} = \frac{1}{2}e_{ic}^2$ should be minimized by the steepest descent algorithm as

$$\dot{\hat{W}}_{ic} = -\Gamma_{ic1}\frac{\theta_i}{(1 + \theta_i^{\mathsf{T}}\theta_i)^2}\left(e_{ic} - \hat{\delta}_{icM} - \eta_{ic}\text{sgn}(\tilde{W}_{ic}^{\mathsf{T}}\theta_i)\right), \tag{27}$$

where Γ_{ic1} and η_{ic} are the designed positive constants, the robust term $\hat{\delta}_{icM}$, which is the estimation of the upper-bound of δ_{icM}, is updated by

$$\dot{\hat{\delta}}_{icM} = \Gamma_{ic2}\frac{\tilde{W}_{ic}^{\mathsf{T}}\theta_i}{(1 + \theta_i^{\mathsf{T}}\theta_i)^2}, \tag{28}$$

where Γ_{ic2} is a positive constant. We can observe that (27) is derived by updating δ_{icM} as (28). Thus, (27) and (28) are the so-called nested update laws.

Then, the weight error dynamics of the local critic NN can be expressed as

$$\dot{\tilde{W}}_{ic} = -\dot{\hat{W}}_{ic}$$
$$= -\Gamma_{ic1}\frac{\theta_i}{(1+\theta_i^{\mathsf{T}}\theta_i)^2}\left(\delta_i + \delta_{icM} - \tilde{\delta}_{icM} + \tilde{W}_{ic}^{\mathsf{T}}\theta_i + \eta_{ic}\mathrm{sgn}(\tilde{W}_{ic}^{\mathsf{T}}\theta_i)\right), \quad (29)$$

where $\tilde{\delta}_{icM} = \delta_{icM} - \hat{\delta}_{icM}$, which is updated by

$$\dot{\tilde{\delta}}_{icM} = -\Gamma_{ic2}\frac{\tilde{W}_{ic}^{\mathsf{T}}\theta_i}{(1+\theta_i^{\mathsf{T}}\theta_i)^2}. \quad (30)$$

Remark 1. *It is worth mentioning that $\dot{\tilde{W}}_{ic}$ is used in (27) and (28). Since W_{ic} is a constant, its time derivative is zero. Thus, with the help of (29), (27) and (28) can be obtained.*

Theorem 2. *Considering the dynamics of the subsystem (1), the local weight error vector dynamics of the local critic NN $\tilde{W}_{ic}$ can be guaranteed to convergent asymptotically with the nested update laws (27) and (28).*

Proof: Choose the Lyapunov function candidate as

$$L_{i2} = \frac{1}{2}\tilde{W}_{ic}^{\mathsf{T}}\Gamma_{ic1}^{-1}\tilde{W}_{ic} + \frac{1}{2}\Gamma_{ic2}^{-1}\tilde{\delta}_{icM}^2. \quad (31)$$

Substituting (29) and (30) into the time derivative of (31), we have

$$\dot{L}_{i2} = \tilde{W}_{ic}^{\mathsf{T}}\Gamma_{ic1}^{-1}\dot{\tilde{W}}_{ic} + \Gamma_{ic2}^{-1}\tilde{\delta}_{icM}\dot{\tilde{\delta}}_{icM}$$
$$= -\tilde{W}_{ic}^{\mathsf{T}}\frac{\theta_i}{(1+\theta_i^{\mathsf{T}}\theta_i)^2}\left(\delta_i + \delta_{icM} + \tilde{W}_{ic}^{\mathsf{T}}\theta_i + \eta_{ic}\mathrm{sgn}(\tilde{W}_{ic}^{\mathsf{T}}\theta_i)\right)$$
$$\quad - \frac{\tilde{W}_{ic}^{\mathsf{T}}\theta_i}{(1+\theta_i^{\mathsf{T}}\theta_i)^2}\tilde{\delta}_{icM} + \Gamma_{ic2}^{-1}\tilde{\delta}_{icM}\dot{\tilde{\delta}}_{icM}$$
$$= \frac{1}{(1+\theta_i^{\mathsf{T}}\theta_i)^2}\left(-\tilde{W}_{ic}^{\mathsf{T}}\theta_i(\delta_i + \delta_{icM}) - \left\|\tilde{W}_{ic}^{\mathsf{T}}\theta_i\right\|^2 - \eta_{ic}\left\|\tilde{W}_{ic}^{\mathsf{T}}\theta_i\right\|\right)$$
$$\leq -(\eta_{ic} - 2\delta_{icM})\frac{\left\|\tilde{W}_{ic}^{\mathsf{T}}\theta_i\right\|}{(1+\theta_i^{\mathsf{T}}\theta_i)^2} - \frac{\left\|\tilde{W}_{ic}^{\mathsf{T}}\theta_i\right\|^2}{(1+\theta_i^{\mathsf{T}}\theta_i)^2}. \quad (32)$$

Therefore, $\dot{L}_{i2} < 0$ as long as the designed parameter $\eta_{ic} > 2\delta_{icM}$. It indicates the weight error vector $\tilde{W}_{ic}$ of the local critic NN is ensured to asymptotic convergence. This completes the proof.

Theorem 3. *Considering the subsystem dynamics of unknown interconnected systems (4), the local critic NN whose weight vector is tuned by the nested update laws (27) and (28), and the Assumptions 1–3, the closed-loop system of unknown interconnected systems can be guaranteed to asymptotic convergence with the developed LTC.*

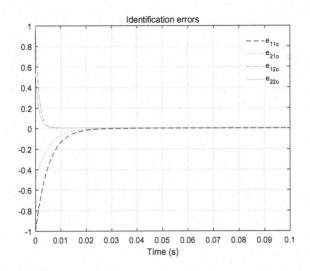

Fig. 1. Identification errors of each subsystem

Proof: The procedure of prove is similar to that in [15], it is omitted for the reason of length limitation.

4 Simulation Study

A hard spring connected parallel inverted pendulum system [7] is employed in our simulation.

The initial values of the parallel inverted pendulum and local NN identifiers are $[1; 1; 0; 0]$ and $[2; 1.5; -1; -0.5]$, respectively. The desired trajectories are given as $x_{11d} = 0.5\cos(0.5t)$ and $x_{21d} = 0.8\sin(0.3t + \pi/6)$. The local optimal cost function (6) for each subsystem is approximated by establishing the local critic NN with 4–10–1 structure, which has 4 input neurons, 10 hidden neurons and 1 output neuron, the weight vector is chosen as $\hat{W}_{ic} = [\hat{W}_{ic1}, \hat{W}_{ic2}, \dots, \hat{W}_{ic10}]^{\mathsf{T}}$, the activation function is chosen as $\sigma_{ic}(e_i, x_i) = [e_{i1}^2, e_{i1}e_{i2}, e_{i1}x_{i1}, e_{i1}x_{i2}, e_{i2}^2, e_{i2}x_{i1}, e_{i2}x_{i2}, x_{i1}^2, x_{i1}x_{i2}, x_{i2}^2]$. The related control parameters are set as $l_{io} = diag[200; 800]$, $\Gamma_{iF} = 1$, $\Gamma_{i\eta} = 0.1$, $\eta_{ic} = 0.1$, $R_i = 0.001I$, $Q_i = 10I_2$, $\Gamma_{ic1} = 0.01$, $\Gamma_{ic2} = 0.01$, where I_n denotes the identity matrix with n dimensions.

The simulation results are depicted in Figs. 1, 2, 3 and 4. Figure 1 illustrates the identification errors converge to a small region of equilibrium point. The derived local input gain matrix is used to develop the LTC for each subsystem. Figure 2 illustrates the trajectory tracking performance by using the developed LTC policy (24). The trajectory tracking error curves are displayed in Fig. 3. We can see from these two figures that the actual trajectories can follow their desired ones within $4s$. Figure 4 shows the LTC input curves.

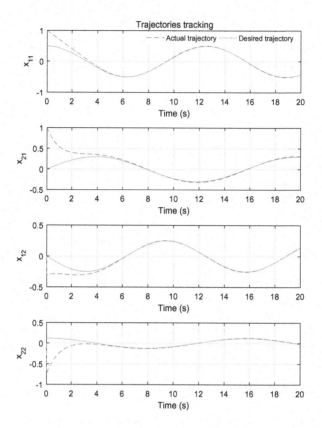

Fig. 2. Subsystem trajectory tracking performance

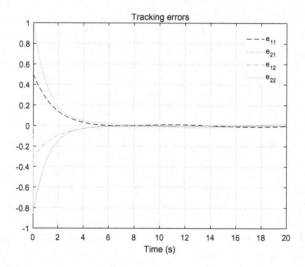

Fig. 3. Subsystem trajectory tracking errors

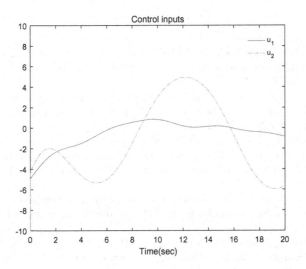

Fig. 4. LTC inputs

Remark 2. *In the simulation, the reasonable control parameters are selected by "trial and error" cooperatively.*

5   Conclusion

In this paper, the NDP based LTC scheme is proposed for unknown interconnected systems. In order to identify the local input gain matrix, which is used to derive the LTC, the local NN observer is constructed based on local input-output data and desired information of the interconnected subsystems. The interconnection is inserted to establish the modified local cost function, and the local HJB equation is solved by using the local critic NN with asymptotically convergent weight vector. The simulation results show that the proposed LTC scheme ensures the actual trajectories of each subsystem follow their desired ones successfully. In the future work, the event-triggered control strategy is combined to solve the optimal control problems for nonlinear systems, since it can reduce the energy cost and computational burden further such that the developed scheme is more feasible to be implemented in practice.

Acknowledgments. This work was supported in part by the National Natural Science Foundation of China under Grants 61603387, 61773075, 61533017 and 61773373, and in part by the Early Career Development Award of SKLMCCS under Grant 20180201.

References

1. Wen, G., Chen, C.L.P., Liu, Y.J.: Formation control with obstacle avoidance for a class of stochastic multiagent systems. IEEE Trans. Industr. Electron. **65**(7), 5847–5855 (2018)
2. Zhao, B., Li, Y.: Model-free adaptive dynamic programming based near-optimal decentralized tracking control of reconfigurable manipulators. Int. J. Control Autom. Syst. **16**(2), 478–490 (2018)
3. Werbos, P.J.: Approximate dynamic programming for real time control and neural modeling. In: White, D.A., Sofge, D.A. (eds.) Handbook of Intelligent Control: Neural, Fuzzy, and Adaptive Approaches. Van Nostrand Reinhold, New York (1992)
4. Abu-Khalaf, M., Lewis, F.L., Huang, J.: Neurodynamic programming and zero-sum games for constrained control systems. IEEE Trans. Neural Netw. **19**(7), 1243–1252 (2008)
5. Liu, D., Wei, Q., Wang, D., Yang, X., Li, H.: Adaptive Dynamic Programming with Applications in Optimal Control. Springer, Cham (2017). https://doi.org/10.1007/978-3-319-50815-3
6. Wei, Y., Park, J.H., Karimi, H.R., Tian, Y.C.: Improved stability and stabilization results for stochastic synchronization of continuous-time semi-Markovian jump neural networks with time-varying delay. IEEE Trans. Neural Netw. Learn. Syst. **29**(6), 2488–2501 (2018)
7. Zhao, B., Wang, D., Shi, G., Liu, D., Li, Y.: Decentralized control for large-scale nonlinear systems with unknown mismatched interconnections via policy iteration. IEEE Trans. Syst. Man. Cyber. Syst. **48**(10), 1725–1735 (2018)
8. Gao, W., Jiang, Y., Jiang, Z.P., Chai, T.: Output-feedback adaptive optimal control of interconnected systems based on robust adaptive dynamic programming. Automatica **72**, 37–45 (2016)
9. Wu, H., Li, M., Guo, L.: Finite-horizon approximate optimal guaranteed cost control of uncertain nonlinear systems with application to mars entry guidance. IEEE Trans. Neural Netw. Learn. Syst. **26**, 1456–1467 (2015)
10. Zhao, B., Liu, D., Li, Y.: Online fault compensation control based on policy iteration algorithm for a class of affine non-linear systems with actuator failures. IET Control Theory Appl. **10**(15), 1816–1823 (2016)
11. Wen, G., Ge, S.S., Tu, F.: Optimized backstepping for tracking control of strict-feedback systems. IEEE Trans. Neural Netw. Learn. Syst. **29**(8), 3850–3862 (2018)
12. Fu, Y., Fu, J., Chai, T.: Robust adaptive dynamic programming of two-player zero-sum games for continuous-time linear systems. IEEE Trans. Neural Netw. Learn. Syst. **26**(12), 3314–3319 (2015)
13. Mehraeen, S., Jagannathan, S.: Decentralized optimal control of a class of interconnected nonlinear discrete-time systems by using online Hamilton-Jacobi-Bellman formulation. IEEE Trans. Neural Netw. **22**(11), 1757–1769 (2011)
14. Qu, Q., Zhang, H., Feng, T., Jiang, H.: Decentralized adaptive tracking control scheme for nonlinear large-scale interconnected systems via adaptive dynamic programming. Neurocomputing **225**, 1–10 (2017)
15. Zhao, B., Liu, D., Yang, X., Li, Y.: Observer-critic structure-based adaptive dynamic programming for decentralised tracking control of unknown large-scale nonlinear systems. Int. J. Syst. Sci. **48**(9), 1978–1989 (2017)

Optimal Control for Dynamic Positioning Vessel Based on an Approximation Method

Xiaoyang Gao, Tieshan Li[(⊠)], and Qihe Shan

Navigation College, Dalian Maritime University, Dalian, China
gaoxiaoyang@dlmu.edu.cn, tieshanli@126.com,
shanqihe@163.com

Abstract. The paper investigates an approximation method of dynamic positioning (DP) vessel optimal control problem. The approximation method is used for sequential improvement of the control law which converges to the optimal by designing a recursive algorithm. It is proved that the designed control law can maintain vessel's position and heading at desired values, while guaranteeing the asymptotical stability in the control system. The optimal control problem of DP vessel can be solved by this method. Finally, simulation results involving a supply vessel demonstrate the validity of the proposed control law.

Keywords: An approximation method · Optimal control
Dynamic positioning vessel

1 Introduction

According to the International Maritime Organization (IMO) and the certifying class societies (DNV, ABS, etc.), the definition of vessel dynamic positioning (DP) is a vessel that can maintain its position and heading state (fixed location or pre-determined track) exclusively by means of active thrusters [1]. Dynamic positioning is easy to operate to avoid damage of the seabed, the cost will not increase with depth's increasing, which has been widely used in drilling platform, laying ships, supply ships, etc. [2] There exists a practical significance to the exploitation and utilization of marine resources by studying the dynamic positioning technology deeply.

Researchers did a large amount of in-depth research in DP vessel motion control problem where many methods of control theory were used in. In the early 1960s, DP control adopted PID control method [3], which controlled the single degree of freedom (DOF) after decoupled the model. Then, optimal control theory was formed and developed in the 1950s under the promotion of space technology. The pioneering work of optimal control mainly depended on dynamic programming and maximal principle by R. E. Bellman and Pontryagin, etc. In the late 1970s, Kalman filtering methods and Linear Quadratic Gaussian (LQG) started to be used in the control design of DP vessel [4–6]. Then, Fu et al. presented a control method based on the optimal control theory and a vector guidance strategy [7]. Veremey, E. et al. proposed a special unified multipurpose control law structure constructed on the basis of nonlinear asymptotic observers that allows the decoupling of a synthesis into simpler particular optimization problems [8]. However, the traditional optimal control method is difficult to be applied

© Springer Nature Switzerland AG 2018
L. Cheng et al. (Eds.): ICONIP 2018, LNCS 11307, pp. 269–278, 2018.
https://doi.org/10.1007/978-3-030-04239-4_24

in some nonlinear systems or the performance index is not in a quadratic form. George N. Saridis et al. proposed an optimal control approximation method could be used in a class of nonlinear system and linear system which was applied to control the trainable manipulators [9]. This paper uses this approximation method to solve the linear optimal control problem of DP vessel through designing a recursive algorithm with an initial admissible control law to approximate the optimal control law and performance index instead of solving Hamilton-Jacobi-Bellman (HJB) equation. The optimal control problem of DP vessel can be solved by this approximation method.

The rest of the paper is organized as follows. Section 2 gives the preliminary knowledge about the approximation method of optimal control, as well as the linear mathematical model of the DP vessel. In the Sect. 3, the approximation method is employed for the model of DP vessel. Section 4 is the simulation to prove the feasibility of the control law. Conclusion is given in Sect. 5.

2 Preliminaries

2.1 The Optimal Control Problem and the Approximation Method

Consider a class of continuous system as follows:

$$\dot{x} = f(x) + Bu \tag{1}$$

$x(t_0) = x_0$, $u \in \Omega_u \subset R^m$, where $x(t) \in R^n$ is the state vector, Ω_u is a compact set of admissible controls, u is control input and B is an $n \times m$ matrix. The system can be controlled by a continuous control on Ω_u.

Define a performance index

$$V(x_0) = \int_{t_0}^{T} (Q(x) + R(u))dt \tag{2}$$

where $Q(x)$ and $R(u)$ could be designed as $Q(x) = x^T Q x$, $R(u) = u^T R u$, $Q \in R^{n \times n}$ and $R \in R^{m \times m}$ are positive definite matrices. Calculate derivative of Eq. (2) with respect to t, we obtain:

$$-\frac{\partial V}{\partial t} = Q(x) + R(u) + \frac{\partial V^T}{\partial x} f(x) + \frac{\partial V^T}{\partial x} Bu \tag{3}$$

Define the Hamiltonian as:

$$H(x, u, \frac{\partial V}{\partial x}) = Q(x) + R(u) + \frac{\partial V^T}{\partial x} f(x) + \frac{\partial V^T}{\partial x} Bu \tag{4}$$

Then,

$$\frac{\partial V}{\partial t} + H(x, u, \frac{\partial V}{\partial x}) = 0 \tag{5}$$

It can be found that:

$$\frac{dV}{dt} = -Q(x) - R(u) \tag{6}$$

It is well known that $Q(x)$ and $R(u)$ are all positive definite functions. Thus, $\frac{dV}{dt} \leq 0$ can be a Lyapunov function and if $Q(x) + R(u) \neq 0$ on a segment of a trajectory, the system is asymptotically stable.

The optimal u^* can be found as follows:

$$u^* = -\frac{1}{2} R^{-1} B^T \frac{\partial V^*}{\partial x} \tag{7}$$

Substituting Eq. (7) into Eq. (5), the HJB equation is obtained as follows:

$$\frac{\partial V^*}{\partial t} + Q(x) + \frac{\partial V^*}{\partial x} f(x) - \frac{1}{4} \frac{\partial V^{*T}}{\partial x} BR^{-1} B^T \frac{\partial V^*}{\partial x} = 0 \tag{8}$$

If the performance index in a quadratic form, the optimal control law is a form of state feedback which can be obtained by solving Riccati equations. If the performance index is not a quadratic form with state and control input, then the optimal control law can be found by solving HJB equation. However, it is hard to solve this kind of partial differential equation. Then, an approximation method is applied which may serve to obtain controllers will make the system stable and approximate the optimal solution. The design procedure is given as follows:

Step 1: Select a feedback admissible control law u_i for system Eq. (1). Then, we can obtain a V_i by satisfying Eq. (5).

Step 2: Design the update control law

$$u_{i+1} = -\frac{1}{2} R^{-1} B^T \frac{\partial V_i}{\partial x} \tag{9}$$

Through Eq. (9), V_{i+1} can be obtained by satisfying Eq. (5).

Step 3: If $\|V_i - V_{i+1}\| < \delta$, δ is an extremely small positive constant, stop updating and the u_{i+1} is an approximation to the optimal control u^*. If $\|V_i - V_{i+1}\| \geq \delta$, repeat step 2 by increasing update time by one and go on.

The following lemma can be found in [9].

Lemma 1: If a sequence of pairs $\{u_{i+1}, V_{i+1}\}$ satisfying Eq. (5) is generated by selecting the control u_{i+1} to minimize Eq. (4) associated with the previous performance index V_i as Eq. (9), then the corresponding performance index satisfies the inequality

$$V_i \geq V_{i+1}$$

Thus by selecting the pairs $\{u_{i+1}, V_{i+1}\}$ in the above manner sequentially, the resulting sequence V_{i+1} converges monotonically to the optimal V^* associated with the optimal control u^*.

2.2 The Linear Model of DP Vessel

As shown in [10], the 3 DOF linear model of low velocity DP vessel without external disturbance is expressed as:

$$
\begin{aligned}
\dot{\eta}_p &= v \\
\dot{v} &= -M^{-1}Dv + M^{-1}\tau
\end{aligned}
\tag{10}
$$

$\eta_p = [x, y, \psi]^T$ are vectors obtained by decomposing the position of North-east-down coordinate into body-fixed coordinate which indicate the vessel's position and heading angle. $v = [u, v, r]^T$ represented surge, sway and yaw velocity of the ship. While M was initial matrix include hydrodynamic added mass, D was damping coefficient matrix. The vessel-fixed propulsion forces and moments were represented by $\tau = [\tau_1, \tau_2, \tau_3]^T$. The linear model can also be written as follows:

$$
\begin{aligned}
\dot{x} &= Ax + B\tau \\
y &= Cx
\end{aligned}
\tag{11}
$$

$x = [\eta_p \quad v]^T$, $A = \begin{bmatrix} 0 & I_{3\times3} \\ 0 & -M^{-1}D \end{bmatrix}$, $B = \begin{bmatrix} 0 \\ M^{-1} \end{bmatrix}$, $C = [I_{3\times3} \quad 0]$, where $I_{3\times3}$ a 3×3 identity matrix, y is output of system.

Assume that the reference state is $z_d = [\eta_d, v_d]^T$. Sometimes z_d is considered as a constant vector in dynamic positioning let $X = x - z_d$ and the model which contains reference state can be written as:

$$\dot{X} = AX + B\tau \tag{12}$$

Define a performance index of DP vessel tracking system as:

$$V(X) = \int_{t_0}^{\infty} (X^T Q X + \tau^T R \tau) dt \tag{13}$$

For the linear model, the performance index can be written as:

$$V(X) = \frac{1}{2} X^T P X \tag{14}$$

P is a positive definite symmetric matrix.

3   DP Vessel Optimal Control with the Approximation Method

In this section, based on the content above, the approximation method is applied to solve the optimal tracking control problem of DP vessel.

Substituting Eq. (12) into Eq. (5), Eq. (15) is obtained as follows:

$$\frac{1}{2}[(AX + B\tau)^T PX + X^T P(AX + B\tau)] + X^T QX + \tau^T R\tau = 0 \qquad (15)$$

Then the approximation method is used to obtain the optimal control law and optimal performance index

Step 1: Select an admissible feedback control

$$\tau_i = K_i X \qquad (16)$$

Substituting Eq. (16) into Eq. (15) and Eq. (17) is obtained as follows:

$$\frac{1}{2}[(AX + BK_iX)^T P_iX + X^T P_i(AX + BK_iX)] + X^T QX + (K_iX)^T R(K_iX) = 0 \qquad (17)$$

which is the same as

$$\frac{1}{2}[(A^T + K_i^T B^T)P_i + P_i(A + BK_i)] + Q + K_i^T RK_i = 0 \qquad (18)$$

The initial P_i can be obtained by solving the Eq. (18).

Step 2: Design the update control law as follows:

$$\tau_{i+1} = -R^{-1}B^T P_iX \qquad (19)$$

By satisfying Eq. (15), then P_{i+1} can be obtained with τ_{i+1}.

Step 3: Comparing P_i and P_{i+1}. If $\|P_i - P_{i+1}\| < \delta$, it can be considered that obtain the optimal control law. If not, repeat step 2 by increasing update time by one and continue.

According to Lemma 1, it can be proved that the final control law converges to the optimal supposing that the initial u_i is chosen to be admissible and the following obtained control laws are admissible as well at every step.

A piece of pseudo-code which contains the steps above to make the procedure clearly is established as follows:

Algorithm 1:	Calculate optimal τ^*
Input:	τ_i : Initial admissible control law
Output:	τ^*, V^* : Optimal control and performance index
1:	$\tau_i \leftarrow K_i X$, $i = 1$, $\delta = 10^{-5}$
2:	**while** $e > \delta$ **do**
3:	$P_i \leftarrow$ solve Eq.(5) with known τ_i
4:	$\tau_{i+1} \leftarrow -R^{-1} B^T P_i X$
5:	$P_{i+1} \leftarrow$ solve Eq.(5) with known τ_{i+1}
6:	$e \leftarrow \left\| P_{i+1} - P_i \right\|$
7:	$i = i + 1$
7:	**end while**
8:	**return** $P^* \leftarrow P_{i+1}$, $u^* \leftarrow -R^{-1} B^T P^* X$

4 Simulation and Analysis

In this section, the feasibility of the proposed controller was verified in a simulation environment in MATLAB using the surface ship model of supply vessel [11]. The mass of vessel was 6.4×106 kg, length was 76.2 m, and inertia matrix and hydrodynamic damping matrix, respectively was

$$M = \begin{bmatrix} 1.1274 & 0 & 0 \\ 0 & 1.8902 & -0.0744 \\ 0 & -0.0744 & 0.1278 \end{bmatrix}, \quad D = \begin{bmatrix} 0.0414 & 0 & 0 \\ 0 & 0.1775 & -0.0141 \\ 0 & -0.1073 & 0.0568 \end{bmatrix}$$

In the simulation, the initial state $x_0 = [\,30\,\text{m}\ 35\,\text{m}\ (20°/180°) \times \pi\,\text{rad}\ 0\,\text{m/s}$ $0\,\text{m/s}\ 0\,\text{rad/s}]^T$, the reference state $z_d = [\,5m\ 7m\ (10°/180°) \times \pi\,\text{rad}\ 0m/s\ 0m/s$ $0rad/s]^T$, and the running time $T = 50s$. Algorithm is simulated via MATLB. The optimal control law calculated by this approximation method is shown as:

$$\tau^* = \begin{bmatrix} -14.1421 & 0 & 0 & -15.1865 & 0 & 0 \\ 0 & -14.1421 & 0.0016 & 0 & -15.7441 & 0.0576 \\ 0 & -0.0016 & -14.1421 & 0 & -0.0390 & -14.2125 \end{bmatrix} X$$

The simulation results are shown in Figs. 1, 2, 3, 4 and 5. Figure 1 shows that the approximate algorithm have iterated for 10 times and obtained 9 groups of performance index through which it can be found that the value of performance index decreasing by times of approximation and the last approximation can be seemed as the optimal performance index. The position and heading angle transformation of vessel are shown in Figs. 2 and 3 which indicated that vessel reach the reference state by designed control law. Figures 4 and 5 show the control forces and control moment of vessel when performance index becomes optimal. Therefore, from the simulation results, it is seen that the proposed optimal control law can achieve the control target quickly and effectively.

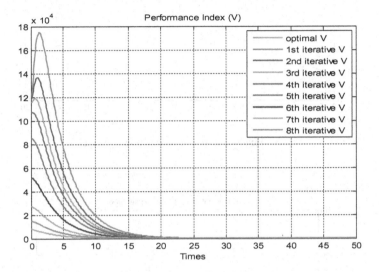

Fig. 1. The changes of performance index in approximation process

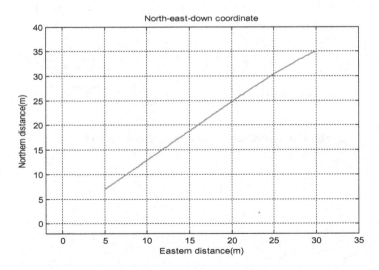

Fig. 2. Position transformation of vessel

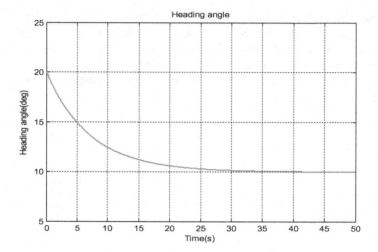

Fig. 3. Heading angle transformation of vessel

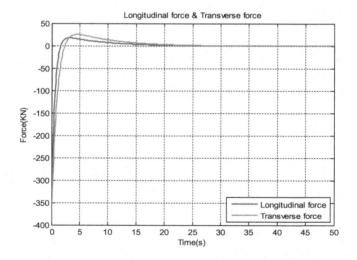

Fig. 4. Longitudinal and transverse control force of vessel

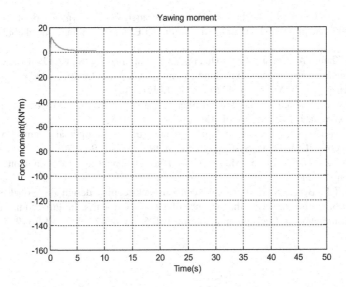

Fig. 5. Control force moment of vessel

5 Conclusion

In this paper, a control law combining an approximation method of optimal problem has been proposed for the DP vessel. The approximation method can make the control law and performance index converge and overcome the optimal control problem instead of solving HJB equation. It has been proved that the proposed optimal control law can maintain vessel's position and heading at desired values. Simulation results demonstrate the effectiveness of the control law.

Acknowledgments. This work was supported in part by the National Natural Science Foundation of China (61751202).

References

1. Sørensen, A.J.: A survey of dynamic positioning control systems. Annu. Rev. Control **35**(1), 123–136 (2011)
2. El-Hawary, F. (ed.): The ocean Engineering Handbook. CRC Press, Boca Raton (2000)
3. Morgan, M.J.: Dynamic Positioning of Offshore Vessels. Petroleum Pub. Co., Tulsa (1978)
4. Balchen, J.G., Jenssen, N.A., Sælid, S.: Dynamic positioning using Kalman filtering and optimal control theory. In: IFAC/IFIP Symposium on Automation in Offshore Oil Field Operation, vol. 183, p. 186 (1976)
5. Grimble, M., Patton, R., Wise, D.: The design of dynamic ship positioning control systems using extended Kalman filtering techniques. In: OCEANS 1979, San Diego, CA, USA, pp. 488–497. IEEE (1979)

6. Saelid, S., Jenssen, N., Balchen, J.: Design and analysis of a dynamic positioning system based on Kalman filtering and optimal control. IEEE Trans. Autom. Control **28**(3), 331–339 (1983)
7. Fu, M., Zhang, A., Liu, J., Xu, J., Huang, Y.: The optimal control based on vector guidance for dynamic positioning vessel rotating around a fixed point. In: 2013 32nd Chinese Control Conference (CCC), Xi'an, pp. 4370–4375. IEEE (2013)
8. Veremey, E., Sotnikova, M.: Optimal filtering correction for marine dynamical positioning control system. J. Mar. Sci. Appl. **15**(4), 452–462 (2016)
9. Saridis, G.N., Lee, C.-S.G.: An approximation theory of optimal control for trainable manipulators. IEEE Trans. Syst. Man Cybern. **9**(3), 152–159 (1979)
10. Fossen, T.I.: Handbook of Marine Craft Hydrodynamics and Motion Control. Wiley, Hoboken (2011)
11. Fossen, T.I., Berge, S.P.: Nonlinear vectorial backstepping design for global exponential tracking of marine vessels in the presence of actuator dynamics. In: Proceedings of the 36th IEEE Conference on Decision and Control, vol. 5, pp. 4237–4242. IEEE (1997)

Interactive Incremental Online Learning of Objects Onboard of a Cooperative Autonomous Mobile Robot

Stephan Hasler[1]([✉]), Jennifer Kreger[2], and Ute Bauer-Wersing[2]

[1] Honda Research Institute Europe GmbH,
Carl-Legien-Str. 30, 63073 Offenbach am Main, Germany
`stephan.hasler@honda-ri.de`
[2] Frankfurt University of Applied Sciences,
Nibelungenplatz 1, 60318 Frankfurt am Main, Germany

Abstract. Detecting objects and referring to them in a dialog is a crucial requirement for robotic systems that cooperate with humans. For this, in an unrestricted natural environment the innate concepts of the robot must be extended and adapted over time. In this paper we describe an autonomous mobile robot system that performs online interactive incremental learning of objects. We argue that this combination strongly contributes to the variation of appearance, context, and labels under which visual concepts are encountered and thus overcomes limitations of existing databases and robotic systems where one or more of these aspects are missing. In the current prototype version, objects are shown to the robot in hand and are learned by a standard classifier on top of pre-trained CNN features. We evaluate the basic feasibility of the current approach on an existing database of hand-held objects, show how it performs online on the robot, and discuss extensions of the system towards life-long learning and data acquisition.

Keywords: Interactive online learning · Autonomous robot

1 Introduction

To cooperate with humans in a natural way, assistant robots must be able to detect and discriminate the task-relevant objects and uniquely refer to them in changing contexts. In recent years deep learning led to a strong gain in performance for various object recognition tasks and shows remarkable performance on various vision benchmarks. However, the applicability of models pre-trained on such benchmarks for cooperative robots in unrestricted real-world settings is limited.

First of all, even a rich offline database can never cover all the necessary categories that might be relevant for all possible tasks and environments an agent might be confronted with. Second of all, also the known categories might

© Springer Nature Switzerland AG 2018
L. Cheng et al. (Eds.): ICONIP 2018, LNCS 11307, pp. 279–290, 2018.
https://doi.org/10.1007/978-3-030-04239-4_25

not generalize well to the diverse newly encountered instances and viewing conditions. Both aspects can partially be addressed with domain adaptation and transfer strategies. But in general there will always exist cases when the system must learn a new category or adapt to new instances of an existing one, which reflects the general motivation for incremental learning.

A third point to make is, that in current popular internet image collections like ImageNet [5] and MS COCO [22] the variations in viewpoint, lighting, and object instance of a category are distributed randomly over the large set of samples. Hence there is no physical object concept where temporal and spatial stability over different views can be exploited by a robot during training and recognition. Therefore many benchmarks focus on single view recognition. This might not be required for a robotic system that can actively integrate information from several view points. Moreover recognition can also be very difficult if not impossible for certain views, thus confusing the classifier or wasting its resources.

There are databases with natural or rendered images where categories are represented in a more structured way, each containing several object instances with several views. The views can originate from different sessions that reflect changing environmental conditions or object transformations. This does better reflect how robots encounter objects in the real world. However, these existing datasets with natural images are obtained using a quasi static robot in dedicated acquisition sessions like in [30,31]. As a result, the viewing conditions are discrete and controlled by the human and thus prone to lack sufficient context variation. For the rendered datasets a wide variety of viewing conditions can be generated artificially. However, the set of object instances is also limited and the robust transfer of the models trained on such data to natural images is still an open research topic [32].

We argue that a very natural and unbiased way to train/test a recognition system and acquire data is to use a cooperative autonomous mobile robot with onboard interactive incremental online learning.

If a robot is mobile it can see the same object from different poses. But only if it has a certain degree of long-term autonomy it can explore its environment by its own, decide on which place to be at which time, and which object to focus on and with which person to interact. This drastically increases the appearance variations under which objects and the categories they belong to are encountered. Optimally this autonomy would be driven by the robot's will to cooperate with humans and to learn how to do so for various tasks and contexts.

There are systems that autonomously navigate indoors and predict the presence of new objects by comparing 3D measurements with stored reference frames [6]. Driving around the object several views can be collected and merged to a 3D or appearance based model, which eventually is presented to a human for getting label information. This post-labeling is a valid approach but getting labels in direct situated interaction with the human close by gives potentially richer information, as similar objects are described differently by different persons, in different contexts and tasks.

Only in certain situations it is necessary that a robot must immediately learn and apply a new visual concept after it has been presented to it by a teacher. However, the more learning a robot can perform online, the more actively and economically it can exploit the information in the current situation including the teacher and the more intelligent it will appear. So the system could potentially request/acquire further samples for a concept and ask for alternative labels or relations to other concepts and objects. In general, the availability of strong onboard resources increases the online learning capabilities of the robot and also its autonomy.

We have implemented a first reference system that fulfills these requirements with a focus on smooth interactive incremental online learning of objects in unrestricted settings. We have started using a standard incremental classifier alternative to demonstrate the benefits of our approach over systems in the literature that miss some of the said issues.

2   Related Work

Already before the great attention on deep learning there were robust online systems for discrimination of objects [12] and detection on a mobile robot [1]. These were based on hierarchical architectures where lower layers used hand-crafted features and only the kernels of an intermediate layer and the parameters of the final classifier on top were optimized by learning. Nowadays with deep learning all layers of the representation can be optimized, yielding highly robust recognition for large-scale benchmarks [4,20,33]. With the use of modern GPUs several of these models can be used online.

Also there were early systems for incremental online learning. So first in [19] hand-held natural objects were learned in front of a black background and associated with a pre-defined text label. This was later extended in [18] towards multi-class object categorization in cluttered background using a heuristic strategy for prototype and feature selection. In [8] models for spatial concepts and speech labels were learned simultaneously and associated to each other. By replacing feature extraction by a pre-trained CNN, these approaches gained in performance. This was e.g. evaluated in [7] for the system of [18]. However, naive long-term incremental training of a standard CNN is prone to gradually forgetting concepts that are not presented regularly.

For several classifiers there exist stable incremental versions. A comprehensive evaluation of some standard approaches on some standard benchmarks was published in [24]. In general, the optimal classifier choice strongly depends on the order in which data of different classes is encountered and also which changes a class undergoes over time. A prototype-based combination of short-term and long-term memory was presented in [23] to deal with heterogeneous concept drift of known classes. In [3] an incremental Regularized Least Squares Classifier was proposed, where new classes could be added efficiently. A general Nearest Neighbor based memory module is proposed in [16]. This can be used to increase stability during incremental training of deep architectures by remembering rare events. In [15] continuous fine-tuning of a CNN on new classes was shown to be practical. But the experiments focused on very short-term effects only. More long-term stability is obtained for CNN approaches that explicitly distribute resources for remembering old concepts and acquiring new ones as e.g. in [21]. This emphasizes that incremental learning is an open research field and there is not yet any gold standard. For building our reference system the choice of the optimal classifier is not the initial focus and will be addressed in the next project stage.

As motivated in the introduction, the popular large internet image collection category benchmarks like MS COCO [5] and ImageNet [22] have strong limitations for the robotic context. There are more dedicated robotic databases containing several views per object. The iLab-20M [2] is a large collection of toy vehicles with varying background recorded on a turn-table. In [30,31] the iCub robot is used to record hand-held objects, and in [29] 300 Baxter robots were used to automatically scan and get physical experience with millions of objects. However, in non of these approaches the robot could autonomously explore its environment. Thus the variation of objects and contexts is limited. With our moving autonomous robot system we establish the basis for collecting a much wider range of data.

3   System

In this section we will shortly describe the basic setup and capabilities of our robot and later in more detail how the behavior for interactive incremental online learning was implemented and integrated.

3.1   Robot Hardware

The used platform is the wheel-based MetraLabs SCITOS G5 robot [25] as shown in Fig. 1. The base system is equipped with a front and a back Laser scanner that can be used to detect obstacles, and localize and navigate in recorded maps. Our robot additionally has a Kinect 2 camera and a microphone array mounted on a pan-tilt unit. A small touch screen is attached to the chest of the robot. Furthermore a KINOVA JACO arm [17] is located at the right side of the robot. This arm can be used for pointing gestures and grasping of objects and was also tested to have enough force to operate door handles. There is a base computer onboard for processing the Laser data and doing navigation. Additionally we integrated a computer with a GTX1070 GPU that is mainly used for image acquisition and processing. The base system is safety certified. It can drive up to 1.4 m/s. Loading of the onboard battery is done by driving onto a charging station and a full load takes 6 h. The base system can move and navigate for up to 18 h. Using additional equipment, especially the GPU, the run time is considerably reduced to several hours. In our laboratory setting we have three identical robots that are located in separated office areas. In general each robot acts independently but they can exchange information over a central server/WiFi.

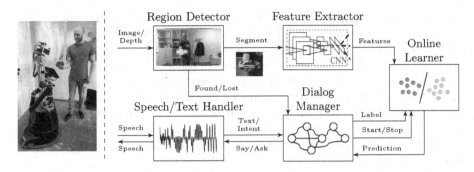

Fig. 1. Left: User showing an object to the robot. Right: Online learning system. Based on proximity to the camera a segment is cropped from an image. For this segment features are computed using a CNN, and classified and collected by the online learner. The dialog manager regularly communicates the predicted class to the user. The user can confirm or correct the prediction. After presentation of the object the dialog manager sends the label to the online learner to trigger an update of the representation.

3.2 Basic Robot Software System

The middle-ware that we use on the robot is ROS [34] Indigo. Only the Laser processing and basic navigation uses the MIRA middle-ware of MetraLabs, which offers a ROS interface.

We took over the basic behavior system used in the STRANDS project [26], which mainly is a task scheduler/executor. The basic task for the robot is to reach a certain location in its known map. On the way the robot is avoiding obstacles and has to pass doors. If a location is not reachable the robot cancels the current goal. This basic task is what the robot is doing all the time currently, each time choosing a new random target location from a predefined set distributed over several office rooms and floors. In the background the robot observes its battery state and on demand can add a high-priority task to go to the charging station to the scheduler. Taken this together, the robot has an autonomous behavior, although driving around randomly is not very purposeful at the moment.

On the way to or at a target location a user can call the robot. The intent of the user utterance is classified and triggers a certain action. Such an action can be the response to a question ("How are you?") or the start of an incremental object learning session ("I want to show you an object."). The object learning can also be started by showing an object in a certain distance to the robot for one second.

3.3 Online Learning Action

The components that are relevant for the online learning action are shown in Fig. 1 and described in the following.

Region Detector. The region detector determines the closest depth blob to the camera. If this blob fulfills constraints on minimal size and distance range, it is cropped from the HD image (1920 × 1080) of the Kinect 2 and normalized to a given segment size, here 227 × 227 corresponding to the input size of the CNN feature detector. This strategy follows the concept of the peri-personal space proposed in [9] which was already used in other online interaction systems [12,18,31]. However, we neglect the figure-ground segregation step of [12,18] as the study in [7] showed that this might be less important in combination with deep features. If a region is found in five consecutive frames an *object-found* signal is send to the dialog manager which signals the online-learner to start collecting feature vectors. For five consecutive frames without a region an *object-lost* signal is send respectively.

Feature Extractor. As features we use the 1000-dimensional activation of the 'fc8' layer of the BVLC Reference CaffeNet [14] trained on ImageNet, which is a version of the architecture proposed in [20]. We did not do any adaptation to our camera but used the image mean subtraction corresponding to ImageNet which is provided with the CaffeNet.

Online Learner. When the online learner receives the *start* signal from the dialog manager, it begins collecting feature vectors and predicts a class label for each using its internal classifier. When the object is lost the dialog manager closes the feature gate. After object presentation, the label that was given or confirmed by the user is send to the online learner, which it uses together with the collected data to update the internal classifier. As our reference system should enable research on incremental learning the choice or implementation of the ultimate incremental classifier was not in focus at this point in time. For simplicity we first use a Nearest Neighbor Classifier that stores all feature vectors as prototypes. Class predictions are integrated over all views of a presented object by counting how often each class was the winner. The contribution of this is evaluated in Sect. 4.

Speech/Text Handler. In the outward direction this component synthesizes speech from text that it receives from the dialog manager. For this the Nuance Vocalizer Expressive TTS [28] is used. In the inward direction the speech signal is acquired using the microphone-array and enhanced using HARK sound source localization [27]. After this the signal is translated to text using Google's Cloud Speech-To-Text streaming API [10]. This is followed by intent recognition that is performed using the IBM Watson Assistant service [13]. The intents used for the online learning are 'accept', 'decline', 'goodbye', 'label', 'forgetit', and 'train'. For the 'label' intent it is necessary to extract the words that make up the label from a sentence. Currently we solve this by only allowing the phrase "It is a ...". With the help of a natural language understanding service it will be possible to extract the label also from less restricted sentences. We also plan to show the understood labels on the touch screen. In case of misunderstanding or ambiguities the user would then have a chance to correct or select an alternative. Speech-to-text and intent recognition are the only parts of the system that are not done onboard but are currently realized using cloud services.

Dialog Manager. The dialog manager controls the whole dialog using a state machine that reacts to the signals received from the other components. When the region detector finds or loses the object the manager tells the online learner to start or stop collecting data. During presence of the object it asks the online learner regularly for its accumulated class prediction, reports this to the user via speech, and waits for him to confirm or correct the prediction. If the true label was not given during presentation of the object the manager will asks the user afterwards. Finally, the label is send to the online learner where it triggers an update of the classifier. After this the system is waiting for a new object. The state machine is implemented using the SMACH package that comes with ROS. SMACH also provides a helpful tool to visualize state machines online.

The raw data for each presented object is stored in a separate directory. This currently comprises the RGB and depth segments, the computed features and the given label. This can potentially be extended with various information like the name of the user, the room, the robot's position/orientation, and the time of day. During interaction the user can trigger the 'forgetit' intent, in which case all data of current object will be deleted and no training will be performed. The collected raw data can later be used for large-scale offline evaluation/benchmarks and training. A framework to manage this ever growing data has to be developed.

Currently the system runs at 6 Hz. The limiting factor is not the feature extraction but the use of the HD Kinect images that have to be sent to several processes.

4   Results

When looking at the activation of the 'fc8' layer of CaffeNet, we observed that there was a strong variation in the set of highly activated neurons and for several of them the relation to the presented object was unclear. A reason for this might be that the context under which objects are shown in hand in our office is quite different to and less natural than the context in ImageNet, e.g. comparing real cars and toy cars. This might 'confuse' the CNN and increase the influence of background variation. To get a first impression of how well standard classifiers can cope with this effect, we tested several of them on the iCubWorld28 database [30], which has similar characteristics as our online system.

The iCubWorld28 database contains 28 objects. For each object on each of four days there is a separate train and test session with 200 segments. The objects are rotated in hand in front of an office background. Each segment has a size of 128×128. We simply train on all training views of the objects and test on all test views. Before feature extraction each view is up-sampled to the CaffeNet input dimension of 227×227.

For comparison we tested a Nearest Neighbor Classifier (NNC), a Regularized Least Squares Classifier (RLSC) as used in [30], a Support Vector Machine with RBF kernel (SVM), a Linear SVM (LSVM), Generalized Learning Vector Quantization (GLVQ) [35], and Generalized Relevance Learning Vector Quantization (GRLVQ) [11]. The classifiers are listed again in the top part of Fig. 2

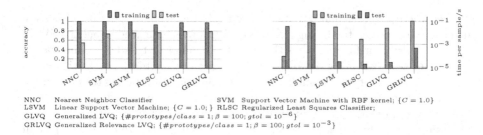

Fig. 2. Left: Accuracy of different classifiers on the iCubWorld28 database. Right: Average training and test time per sample.

together with their parameter settings. For the two LVQ approaches the optimal parameters were determined with a grid search in parameter space, keeping part of the training data as validation set. The classification accuracies are given in the lower left-hand part of Fig. 2. The NNC yields a test accuracy of only 54.4% by simply storing all training views as prototypes. The NNC mostly confuses similar objects inside the same category. So there are e.g. several cups, sprayers, and sponges whose difference in appearance seems not to be prominent enough compared to the effect of unseen background and pose variation. All other tested classifiers have means to weight object information over background noise. This is not so obvious but also true for the GLVQ, that can assign the same value to a dimension inside prototypes of different classes, thus making the dimension irrelevant during nearest neighbor search. These approaches yield a substantially higher accuracy of 73.5% to 78.7%, which basically reproduces the 80.8% reported in [30] where a RLSC was used on CaffeNet features. The small deviation might be caused by a different pre-processing of images and the use of a different layer of CaffeNet. Both is not described in [30]. Surprisingly, even the very low-parametric classifiers like LSVM, RLSC, GLVQ, and GRLVQ show a strong performance compared to the kernel SVM, which suggests that the objects cluster nicely in a weighted feature space. Still there is a residual error of about 20%, caused by very similar appearance of objects in several views. However, in [30] they showed that this error can strongly be reduced by temporal integration, which is natural to do in a robotic setting. We confirmed this finding for our own dataset.

In Fig. 2 also the training and test time per sample is shown. For our offline classifiers we see two extreme cases. The NNC can be trained very quickly but is slow during test, while the other approaches are quick in predicting a class label, but much slower to train. RLSC and GLVQ are among the most accurate and allow rapid classification, but a full re-training is too slow for smooth online interaction. A next step for us is to investigate in how far the existing incremental versions of these approaches trade online training speed against test performance. For now we decided to use the NNC on the robot, because it shows a very smooth training and test behavior for up to 20000 training views. This corresponds to one hour non-stop interaction given our current frame rate of 6 Hz.

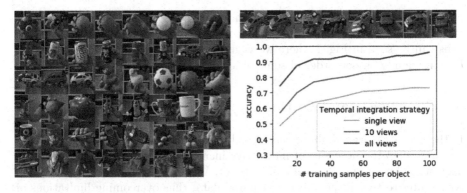

Fig. 3. Left: The 47 objects of the Johnny47 database. Each object has a training session with 100 views and a test session with 100 views. Right top: Views of one object. Right bottom: Accuracy during incremental learning for temporal integration strategies.

To convey the current impression of the incremental online system, we acquired a first small database which we call the Johnny47 database. This database contains 47 objects with one training (100 segments) and one test session (100 segments). The segments have a size of 227×227 pixels. The objects of this database and the variation of views for one object are shown in Fig. 3. To mimic an incremental training we split the training session into several steps. In each step ten consecutive views of each object are added as prototypes to the NNC and the performance is evaluated on the whole test session. As outlined in the introduction, a robot can usually integrate evidence from several viewpoints. We evaluated the effect of this by accumulating class votes alternatively over a single, ten, and all 100 consecutive views of an object. A similar voting was also done in [12,18] and evaluated in [30]. The results in Fig. 3 show that the performance seems to converge from 60 training views per object onwards. The simple temporal integration leads to a very strong gain in performance and confirms the good impression that people have when testing/teaching the robot online, where predictions are also integrated over all seen views of the presented

Fig. 4. Interaction example with robots predictions and user confirmation or correction. Often unknown objects are confused with known objects of the same basic category, like balls, cars, mugs, and animals.

object. Furthermore, it is interesting and motivating for the user that usually the predictions for unknown objects are somehow explainable, even for the NNC. This can be seen in Fig. 4 where we visualize user input and robot predictions for a randomly chosen order of some objects.

5   Conclusion and Outlook

In this paper we argued about the special nature of object recognition on robots and described the necessity of interactive incremental object learning and data acquisition on an autonomous mobile robot. We outlined how this combination contributes to the richness of the acquired data, thus overcoming limitations of existing databases.

To our knowledge we have the first running system that fulfills all these properties and is ready to operate 24-7. In particular we have a smoothly running dialog for interactive incremental object learning, currently using a baseline classifier on top of pre-trained CNN features.

We evaluated the general applicability of CNN features pre-trained on ImageNet in combination with standard classifiers in a robotic setting on the iCubWorld28 database. The results show that classifiers that can weight feature dimensions are able to realize an adaptation to the robotic target domain. Furthermore we confirmed the expectation that temporal integration strongly improves recognition performance in a robotic context on our own database Johnny47.

Next we will start a first autonomous long-run of the robot. Observing how people interact with the system will help us to make the dialog more intuitive. We will also get an impression about the variation of objects, contexts, and labels under long-term conditions. Based on this we plan to evaluate the suitability of incremental learning approaches, especially towards dealing with label noise in multi-class object recognition. We will also have to develop general means for consistent data management for the given acquisition strategy.

Currently, the objects can only be shown by hand and are selected by depth criteria. A natural extension is to add other attention mechanisms, e.g. a CNN region detector. With this the robot can collect different object views by itself e.g. by driving around a table. This can also be used to remember locations of objects in the office and gives us the opportunity to realize more meaningful robot services on top of this.

Acknowledgments. We thank MetraLabs for the setup and support of the robots. We got a quick start with our robots by being able to use the software developed in the STRANDS project. For this we thank the whole project team, especially Lenka Mudrová and Nick Hawes. We also thank Manuel Mühlig for establishing and maintaining the basic robot software system at our institute.

References

1. Andreopoulos, A., Hasler, S., Wersing, H., Janssen, H., Tsotsos, J.K., Körner, E.: Active 3D object localization using a humanoid robot. IEEE Trans. Robot. **27**(1), 47–64 (2011)
2. Borji, A., Izadi, S., Itti, L.: iLab-20M: a large-scale controlled object dataset to investigate deep learning. In: IEEE Conference on Computer Vision and Pattern Recognition, pp. 2221–2230. IEEE Computer Society (2016)
3. Camoriano, R., Pasquale, G., Ciliberto, C., Natale, L., Rosasco, L., Metta, G.: Incremental robot learning of new objects with fixed update time. In: IEEE International Conference on Robotics and Automation, pp. 3207–3214. IEEE (2017)
4. Chatfield, K., Simonyan, K., Vedaldi, A., Zisserman, A.: Return of the devil in the details: delving deep into convolutional nets. In: Valstar, M.F., French, A.P., Pridmore, T.P. (eds.) British Machine Vision Conference. BMVA Press (2014)
5. Deng, J., Dong, W., Socher, R., Li, L., Li, K., Li, F.: ImageNet: a large-scale hierarchical image database. In: IEEE Conference on Computer Vision and Pattern Recognition, pp. 248–255. IEEE (2009)
6. Fäulhammer, T., et al.: Autonomous learning of object models on a mobile robot. IEEE Robot. Autom. Lett. **2**(1), 26–33 (2017)
7. Fischer, L., Hasler, S., Schrom, S., Wersing, H.: Improving online learning of visual categories by deep features. In: Future of Interactive Learning Machines workshop at the Conference on Neural Information Processing Systems (2016)
8. Goerick, C., et al.: Interactive online multimodal association for internal concept building in humanoids. In: 9th IEEE-RAS International Conference on Humanoid Robots, pp. 411–418. IEEE (2009)
9. Goerick, C., Wersing, H., Mikhailova, I., Dunn, M.: Peripersonal space and object recognition for humanoids. In: 5th IEEE-RAS International Conference on Humanoid Robots, pp. 387–392. IEEE (2005)
10. Google: Cloud Speech-To-Text. https://cloud.google.com/speech-to-text/
11. Hammer, B., Villmann, T.: Generalized relevance learning vector quantization. Neural Netw. **15**(8–9), 1059–1068 (2002)
12. Hasler, S., Wersing, H., Kirstein, S., Körner, E.: Large-scale real-time object identification based on analytic features. In: Alippi, C., Polycarpou, M., Panayiotou, C., Ellinas, G. (eds.) ICANN 2009. LNCS, vol. 5769, pp. 663–672. Springer, Heidelberg (2009). https://doi.org/10.1007/978-3-642-04277-5_67
13. IBM: Watson Assistant. www.ibm.com/watson/services/conversation/
14. Jia, Y., et al.: Caffe: convolutional Architecture for Fast Feature Embedding. In: Hua, K.A., Rui, Y., Steinmetz, R., Hanjalic, A., Natsev, A., Zhu, W. (eds.) Proceedings of the ACM International Conference on Multimedia, pp. 675–678. ACM (2014)
15. Käding, C., Rodner, E., Freytag, A., Denzler, J.: Fine-tuning deep neural networks in continuous learning scenarios. In: Chen, C.-S., Lu, J., Ma, K.-K. (eds.) ACCV 2016. LNCS, vol. 10118, pp. 588–605. Springer, Cham (2017). https://doi.org/10.1007/978-3-319-54526-4_43
16. Kaiser, L., Nachum, O., Roy, A., Bengio, S.: Learning to remember rare events. CoRR abs/1703.03129 (2017)
17. Kinovarobotics: Kinova. www.meetjaco.com/about/
18. Kirstein, S., Denecke, A., Hasler, S., Wersing, H., Gross, H., Körner, E.: A vision architecture for unconstrained and incremental learning of multiple categories. Memetic Comput. **1**(4), 291–304 (2009)

19. Kirstein, S., Wersing, H., Körner, E.: Rapid online learning of objects in a biologically motivated recognition architecture. In: Kropatsch, W.G., Sablatnig, R., Hanbury, A. (eds.) DAGM 2005. LNCS, vol. 3663, pp. 301–308. Springer, Heidelberg (2005). https://doi.org/10.1007/11550518_38
20. Krizhevsky, A., Sutskever, I., Hinton, G.E.: ImageNet classification with deep convolutional neural networks. In: Bartlett, P.L., Pereira, F.C.N., Burges, C.J.C., Bottou, L., Weinberger, K.Q. (eds.) Advances in Neural Information Processing Systems 25: Proceedings of the 26th Annual Conference on Neural Information Processing Systems, pp. 1106–1114 (2012)
21. Lee, J., Yoon, J., Yang, E., Hwang, S.J.: Lifelong learning with dynamically expandable networks. CoRR abs/1708.01547 (2017)
22. Lin, T.-Y., et al.: Microsoft COCO: common objects in context. In: Fleet, D., Pajdla, T., Schiele, B., Tuytelaars, T. (eds.) ECCV 2014. LNCS, vol. 8693, pp. 740–755. Springer, Cham (2014). https://doi.org/10.1007/978-3-319-10602-1_48
23. Losing, V., Hammer, B., Wersing, H.: KNN classifier with self adjusting memory for heterogeneous concept drift. In: Bonchi, F., Domingo-Ferrer, J., Baeza-Yates, R.A., Zhou, Z., Wu, X. (eds.) IEEE 16th International Conference on Data Mining, pp. 291–300. IEEE (2016)
24. Losing, V., Hammer, B., Wersing, H.: Incremental on-line learning: a review and comparison of state of the art algorithms. Neurocomputing 275, 1261–1274 (2018)
25. MetraLabs: Scitos G5. www.metralabs.com/en/mobile-robot-scitos-g5/
26. Mudrová, L., Lacerda, B., Hawes, N.: An integrated control framework for long-term autonomy in mobile service robots. In: 2015 European Conference on Mobile Robots, pp. 1–6. IEEE (2015)
27. Nakadai, K., Takahashi, T., Okuno, H.G., Nakajima, H., Hasegawa, Y., Tsujino, H.: Design and implementation of robot audition system 'Hark' - open source software for listening to three simultaneous speakers. Adv. Robot. 24(5–6), 739–761 (2010)
28. Nuance: Vocalizer Text-To-Speech. www.nuance.com/mobile/mobile-solutions/vocalizer-expressive.html
29. Oberlin, J., Meier, M., Kraska, T., Tellex, S.: Acquiring object experiences at scale. In: AAAI-RSS Special Workshop on the 50th Anniversary of Shakey: The Role of AI to Harmonize Robots and Humans (2015)
30. Pasquale, G., Ciliberto, C., Odone, F., Rosasco, L., Natale, L.: Real-world object recognition with off-the-shelf deep Conv Nets: how many objects can iCub learn? CoRR abs/1504.03154 (2015)
31. Pasquale, G., Ciliberto, C., Rosasco, L., Natale, L.: Object identification from few examples by improving the invariance of a Deep Convolutional Neural Network. In: IEEE/RSJ International Conference on Intelligent Robots and Systems, pp. 4904–4911. IEEE (2016)
32. Peng, X., Usman, B., Kaushik, N., Hoffman, J., Wang, D., Saenko, K.: VisDA: the visual domain adaptation challenge. CoRR abs/1710.06924 (2017)
33. Redmon, J., Farhadi, A.: YOLO9000: better, faster, stronger. In: IEEE Conference on Computer Vision and Pattern Recognition, pp. 6517–6525. IEEE Computer Society (2017)
34. ROS: Robot Operating System. www.ros.org
35. Sato, A., Yamada, K.: Generalized learning vector quantization. In: Touretzky, D.S., Mozer, M., Hasselmo, M.E. (eds.) Advances in Neural Information Processing Systems 8, pp. 423–429. MIT Press (1995)

Resilient Consensus for Multi-agent Networks with Mobile Detectors

Haofeng Yan[1], Yiming Wu[2], Ming Xu[2(✉)], Ting Wu[2], Jian Xu[1],
and Tong Qiao[2]

[1] School of Computer Science and Technology,
Hangzhou Dianzi University, Hangzhou 310018, China
[2] School of Cyberspace, Hangzhou Dianzi University, Hangzhou 310018, China
mxu@hdu.edu.cn

Abstract. This paper investigates the problem of resilient consensus for multi-agent systems under malicious attacks. Compared with most of existing works, a more flexible network topology scheme is considered, where a kind of specific agents as the mobile detectors and builders of network robustness are adopted. Specifically, the mobile agents can perceive the message of their nearby agents in the dynamic network, and acquire both in-degree and state information of each node as characteristics to judge the network state as well as communication links between nodes. It is shown that even in poor network robustness, the non-faulty agents can still achieve a consensus in finite time with the help of mobile agents. Finally, the simulation results show the effectiveness of the proposed method.

Keywords: Resilient consensus · Network security · Mobile detector

1 Introduction

With high robustness and strong flexibility, distributed computation plays a key role in multi-agent systems [1–4]. As one of the most effective methods for distributed computation, consensus means that nodes in the network achieve an agreement on a certain state variable by using local information. Most of existing works assume that all agents perform the algorithm faithfully with the prescribed update rules. However, the multi-agent systems are usually deployed in a real-world environment, nodes may update with outliers due to failures or cyber attacks, thus these existing consensus algorithms could become vulnerable or even invalid.

Recently, a family of consensus algorithms named Mean Subsequence Reduced (MSR) algorithms is proposed in [5–7]. In the MSR algorithms, each node disregards the smallest and largest F values collected from its neighbors and then updates its own state to be an average of the remaining values. However, MSR algorithms need to be run on a system satisfying a particular network

© Springer Nature Switzerland AG 2018
L. Cheng et al. (Eds.): ICONIP 2018, LNCS 11307, pp. 291–302, 2018.
https://doi.org/10.1007/978-3-030-04239-4_26

topology property called network robustness [16]. Some works analyze and develop this topology property for multi-agent systems in the presence of misbehaving agents [8,17].

In this paper, we attempt to implement MSR algorithm in more general topologies. Specifically, we design a novel method to reduce the dependence of complex graph topology by using clustering method. We first analyze the characteristics of running MSR algorithm in general networks. According to these characteristics, networks are decomposed into some subunits, and then, we adopt a mobile node to identify different subunits and act as links between them. Finally, non-faulty nodes receive the information from the mobile nodes and add it to their own update data set on the basis of the weight.

The rest of paper is organized as follows. Section 2 describes related works. In Sect. 3, we give the problem formulation and the background related to this paper. And in Sect. 4, the details of our method are given. Section 5 shows some simulations. Finally, we conclude this paper with a short review in Sect. 6.

2   Related Work

Consensus control in multi-agent systems with malicious agents has attracted increasingly research interests [10,11]. There are two types of methods to solve this problem: one is fault detection and isolation, for example, Zhao et al. [12] and Mi et al. [13] respectively adopted different detection mechanisms to identify abnormal nodes as a security solution of distributed consistency. The other is fault (or attack) tolerance algorithms, such as the MSR-type algorithms. These algorithms are able to mitigate the effects of malicious agents without the need for non-faulty nodes to explicitly identify the sources of attacker [6,8–10].

2.1   Analyze MSR-Type Algorithms in General Topologies

In [14], the authors consider a large and sparsely connected network, and give some expressions of local convergence in local networks under two distinct fault models. In addition, the authors give a methodology for analyzing global network convergence properties. In [15], the authors give the necessary and sufficient conditions for the MSR-type algorithm to achieve resilient Byzantine consensus in arbitrary directed graphs. And in [16], authors propose a concept of r-robust graph and show that this concept provides the condition for achieving distributed resilient consensus goals. Authors in [8] summarize the work of the predecessors and exploit a novel graph-theoretic property, named *network robustness*. They indicates that traditional properties such as connectivity are not sufficient to support the operation of MSR algorithms. Moreover, in [17], the authors prove that determining the robustness of the given network is NP-complete.

2.2 Community Detection in Networks

Community detection can help us to discover the topics of information networks or cyber-communities of social networks. In [18], Radicchi et al. introduce a divisive algorithm that detect inter-community links and then remove these links from the graph. In [19], authors propose a fast hierarchical agglomeration algorithm for optimizing the modularity of networks. Another agglomerative, algorithm which merges similar nodes recursively is proposed by Pons et al. in [20]. In addition, Vincent et al. [21] propose a method based on modularity optimization to extract the community structure of large networks.

3 Preliminary and Problem Statement

In this section, we introduce some fundamental matters related to graph theory, the scope of threats and the concepts of resilient consensus.

Notations. A directed graph is given by $\mathcal{G} = (\mathcal{V}, \mathcal{E})$, where $\mathcal{V} = 1, ..., n$ is the node set, and $\mathcal{E} \subseteq \mathcal{V} \times \mathcal{V}$ is the set of edges. The edge $(j, i) \in \mathcal{E}$ indicates that information flows from node j to node i, which is called an incoming edge of node i. The node with the edge pointing to node i is referred to as a neighbor of node i, and the set of the entire neighbors of the node i is denoted by $\mathcal{J}_i = \{j : (j, i) \in \mathcal{E}\}$. The number of neighbors that node i has is called in-degree, which is denoted as $d_i = |\mathcal{J}_i|$.

As mentioned in the Introduction, some works focus on a graph property known as r-robust, which is given by the following definitions from [8] and [16] for analysis of resilient consensus of multi-agent systems.

Definition 3.1: For digraph $\mathcal{G} = (\mathcal{V}, \mathcal{E})$ is (r, s)-robust $(r, s < n)$ if for every pair of nonempty disjoint subsets $S_1, S_1 \subset \mathcal{V}$, at least one of the following conditions holds:

(1) $\mathcal{X}_{S_1}^r = S_1$,
(2) $\mathcal{X}_{S_2}^r = S_2$,
(3) $|\mathcal{X}_{S_1}^r| + |\mathcal{X}_{S_2}^r| \geq s$,

where $\mathcal{X}_{S_l}^r$ is the entire set of nodes in S_l which have at least r incoming edges from outside S_l. In particular, graphs which are $(r, 1)$-robust are called r-robust.

The following lemma shows the basic properties of the robust graphs [8].

Lemma 1: For an (r, s)-robust graph $\mathcal{G}$, the following holds:

(i) $\mathcal{G}$ is (r', s')-robust, where $0 \leq r' \leq r$ and $1 \leq s' \leq s$, and in particular, it is r-robust.
(ii) $\mathcal{G}$ has a directed spanning tree.
(iii) $r \leq \lceil n/2 \rceil$, where $\lceil \cdot \rceil$ is the ceiling function. Also, if $\mathcal{G}$ is a complete graph, then it is (r', s)-robust for all $0 < r' \leq \lceil n/2 \rceil$ and $1 \leq s \leq n$.

Moreover, a graph $\mathcal{G}$ is (r, s)-robust if it is $(r + s - 1)$-robust.

It is clear that (r, s)-robustness is more restrictive than r-robustness. Consider a network with five agents as shown in Fig. 1, which satisfies a $(2, 1)$-robust graph. We can also name it a 2-robust graph. And taking a closer look at this graph, for any pairs of disjoint, nonempty subsets of nodes in the graph, we can see that at least one node in the subset would be sufficiently influenced by two nodes outside its set (thus we could only remove one node which value is abnormal compared with its own). This would drive it away from the values of its subset, and thereby allow it to lead its subset to the values of the other set. Moreover, the consensus will fail if more than one node is abandoned. This causes no node has enough neighbors in the outside set, every node throws away all information from outside of its set.

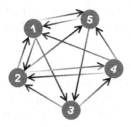

Fig. 1. A (2,1)-robust graph with five nodes.

3.1   Threat Model

In this paper, we consider there should be an upper bound on the number of malicious nodes either in each nodes' neighborhood (f-local) or in whole network (f-total), the definitions are as follows:

Definition 3.2 (f-local model): A node $i \in V$ is called f-local if the number of malicious nodes in the neighborhood $\mathcal{J}_i$ of each node i is no greater than f, $\forall f \in \mathbb{Z}_{\geq 0}$.

Definition 3.3 (f-total model): A node $i \in V$ is called f-total if the number of malicious nodes in the network is no greater than f, $\forall f \in \mathbb{Z}_{\geq 0}$.

3.2   Problem Formulation

Now we define the concept of resilient consensus as follows:

Definition 3.4: It is called reaching a resilient consensus if all normal nodes satisfy the following two conditions, for any initial values and malicious inputs:

(1) Safety condition: There exists a bounded interval S defined by the initial α of the normal nodes, and $\alpha_i[t] \in S$, $\forall i \in V \setminus \mathcal{M}, t \in \mathbb{Z}_{\geq 0}$;

(2) Consensus condition: The state values of all normal nodes agree on a constant c which satisfies $\lim\limits_{t\to\infty} \alpha_i[t] = c$, $\forall i \in \mathcal{V} \setminus \mathcal{M}, t \in \mathbb{Z}_{\geq 0}$.

When applying update MSR-type algorithms to node in a network which satisfies r-connected but not r-robust, we found that all normal nodes will unable to reach consensus under the f-local threat model $(2f + 1 \leq r)$. Thus, our goal is to design a method to ensure that nodes can reach consensus when the network only satisfies r-connected.

4 Algorithm Description

We now introduce our method to achieve resilient consensus for multi-agent systems in adversarial environment. The algorithm can be able to apply to a more general network topology situation, which is composed of Average Iteration Algorithm, MSR algorithms for normal nodes and Mobility Detection Algorithms for the mobile nodes. Specifically, the framework of our method is shown in Fig. 2, First of all, after receiving the message from its neighbors, each normal node first uses MSR algorithm to eliminate outliers. Then, the remaining message is utilized for the update according to the iteration rule. In addition, mobile nodes move around randomly according to the algorithm, which collect information from node in network and finds high modularity partitions of large networks. Finally, the mobile nodes act as link between subunits. The detailed information is shown in Algorithm 1.

The lower half of the algorithm is allocated to each node to run separately, and the detailed process about this part will be shown in Sect. 4.1. The other part of the algorithm is run by mobile nodes, and the detailed process well be shown in Sect. 4.2, which space and time complexities are both $O(e)$.

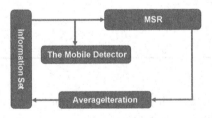

Fig. 2. The framework of the proposed method.

4.1 Average Iteration Algorithm and MSR Algorithm

Each normal node receives the values of neighbors at every time-step t, and according to our attack model, there are at most f of node's neighbors may be malicious nodes. In traditional MSR, each node is unaware of which neighbors may be attackers. Therefore, node simply removes the extreme values with respect to its own value when updates its value. The details are as follows:

Algorithm 1. Network topology compensation algorithm

Input: T, $\mathcal{G} = \{\mathcal{V}, \mathcal{E}, \mathcal{L}\}$, F
Output: a stable set of values $\mathcal{L}$

1: $\mathcal{L}_{error} \leftarrow \{\}$;
2: **while** *The values in the L set are unlikeness* **do**
3:      **if** *time.equals(T)* **then**
4:          community detection;
5:          **foreach** *community* **do**
6:              **if** *community.size* $< F$ **then**
7:                  $\mathcal{L}_{error}\{\} \leftarrow$ for each i in this community$\{\}$;
8:              **else**
9:                  continue;
10:              **end if**
11:          **end foreach**
12:          $\mathcal{G}.\mathcal{E} \leftarrow$ new $\mathcal{G}.\mathcal{E}$
13:      **end if**
14:      **foreach** $i \in \mathcal{V}$ **do**
15:          $i_{value} \leftarrow \mathcal{L}[i]$;
16:          $\mathcal{J}_i \leftarrow \mathcal{G}.\mathcal{E}$;
17:          **foreach** j *in* $\mathcal{J}_i$ **do**
18:              **if** j_{value} *in* $\mathcal{L}_{error}$ **then**
19:                  Delete $\mathcal{G}.\mathcal{L}[j]$;
20:                  $d_i \leftarrow d_i - 1$;
21:              **else**
22:                  continue;
23:              **end if**
24:          **end foreach**
25:          $\mathcal{J}_i.value \leftarrow$ remove the outliers that differ greatly from i;
26:          Average Iteration;
27:          $\mathcal{L}[i] \leftarrow$ new $\mathcal{L}[i]$;
28:      **end foreach**
29: **end while**
30: **return** $\mathcal{L}$

---

1 At each update time t, each normal node i obtains and sorts $x_j[t]$ $(j \in \mathcal{J}_i)$, which is received from its neighbors.
2 If there are less than f neighbors' value larger than its own value, $x_i[t]$, then normal node i removes all values which are larger than its own. Otherwise, the largest f values are removed. Similarly, if there are less than f neighbors' value less than its own value, node i removes all values that are less than its own. Otherwise, the least f values are removed.
3 Let $R_i[t]$ denote the set of nodes who were removed by i in step 2. Each value of node i at this time-step is updated as:

$$x_i[t+1] = \sum_{j \in \mathcal{J}_i[t] \setminus \mathcal{R}_i[t]} \{w_{ij}x_j[t]\} \tag{1}$$

where w_{ij} is the weight of edge from j to i.

As we can see that MSR does not require any node to have knowledge of the identities of non-neighbor nodes. However, it turns out that it needs a specific structure characterized by graph robustness rather than simply possess enough neighbors. Consider this problem, a question we need to answer next is: how to extend this situation to a series of more general networks.

4.2   The MDA Mechanism

The main objective of MDA mechanism is to use community detection method to identify subunits in a multi-agent system and establish links between them.

Critical Phenomena of Networks: We first try to run MSR algorithm on a low-robust network (which set f that $r < 2f + 1$) and observe the changes of all normal nodes' values. From Fig. 3, we can see that as the number of iterations increases, normal nodes in the network form different subunits based on their value and the edge. Nodes in the same subunit have the same value that node 1 and node 2 have the same value 5 and other nodes' value is approximately 7.5. Intuitively, we can find that when each node runs the MSR algorithm, due to the network robustness is insufficient, there will be nodes in one region that cannot communicate with other area. Thus we can acquire a new network topology after running MSR algorithm based on these features and values of the nodes. This network can be divided into areas of densely connected nodes, with the nodes belonging to different areas only sparsely connected.

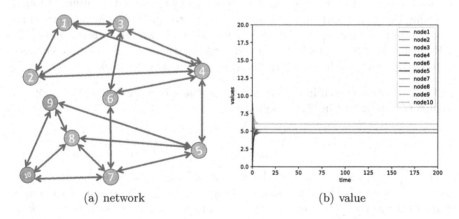

(a) network   (b) value

Fig. 3. Each node's value trajectory in the network with MSR algorithm.

Clustering: In response to the need outlined above, we need a way to find reasonably good partitions in a fast way and establish contact for these partitions. Each agent in a multi-agent system can be regarded as an individual of a social

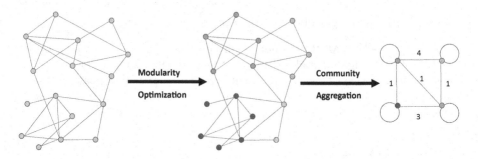

Fig. 4. Visualization of the steps of clustering.

network. The links between the agents can be regarded as communication in the social network. Therefore, we use the community detection method [19] to find a reasonable network partition after using the MSR algorithm. This method is divided in two phases that are repeated iteratively. This method is described as follows:

1 Each node in the network is assigned a different subunit.
2 For each node i consider its neighbors j, try removing i from its subunit and placing it in the subunit of j, evaluate their gain of modularity, and the node i finally placed in the subunit for which this gain is maximum (this gain is positive). If the gain for all neighbors j is negative, i stays in its original subunit.
3 Repeat step 2 until the subunit to which all nodes belong does not change.
4 Build a new network whose nodes whose nodes are now the subunits found during the step 2 and step 3. The weight of the edges between the new nodes is the sum of the weights of the edges between the two previous subunits [22].
5 Repeat the above steps in the new network until there are no more changes and attain the maximum of modularity of the entire network.

The process is shown in Fig. 4, and the formula of gain that a node move in subunit C is provided in [21] as follows:

$$\Delta Q = [\frac{\sum_{in} + k_{i,in}}{2m} - (\frac{\sum_{tot} + k_i}{2m})^2] - [\frac{\sum_{in}}{2m} - (\frac{\sum_{tot}}{2m})^2 - (\frac{k_i}{2m})^2] \qquad (2)$$

where $\sum_{in}$ denotes the sum of the weights of the edges whose starting and ending points are both in the same subunit C, $\sum_{tot}$ is the sum of the edges which incident to C, k_i denotes the sum of the in-degree with node i, $k_{i,in}$ denotes the sum of the weights of the edges from i to nodes in C and m denotes the sum of the weights of all edges in the network.

In this paper, we do not really need to find the exact value of modularity and just need to know how much the current operating module grows relative to other operations (operations mean move node i into a subunit). So we use the formula to determine relative gain. This formula can reduce the time complexity of the algorithm greatly.

$$\Delta Q = k_{i,in} - \frac{\sum_{tot} \times k_i}{m} \tag{3}$$

And we set the weight of the edge to the reciprocal of the absolute values of difference between two node values.

Mobility Model: We divide the network into multiple regions, in each region, there are mobile nodes that move around randomly according to the protocol in [23]. The mobile node has a large powerful receiver that can receive information broadcast by nodes in the vicinity. In addition, there are mobile nodes in different areas to contact each other. Mobile nodes run the community detection algorithm at a certain period of time based on the information it collects. When it found different subunits, it will help them deliver message. Then we can assume that each mobile node has sufficient mobility so that the nodes in multi-agent network can be contacted with a positive probability in each time-step, which is used in [24]. The contact probability relates to the number of mobile nodes and the frequency of movement. We use the following assumption to simplify the statement.

Assumption 4.1: The probability that each node in the network is in contact with the mobile node is the same and equals $\rho, 0 < \rho < 1$, at each time-step.

The network model with mobile nodes is widely used in many fields. For example, mobile nodes in wireless sensor networks can be considered as a secondary node to improve network performance, in addition to wireless charging, sensor coverage, data collection, etc.

5   Simulation

In this section, we design simulations to illustrate the convergence and effectiveness of our method in a more general network.

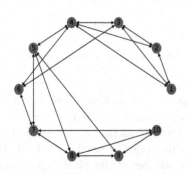

Fig. 5. Network topology.

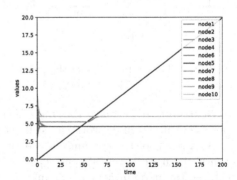

Fig. 6. MSR under 2-local threat model.

We built a $(1,4)$-robust graph as show in Fig. 5, in which the node set is $\mathcal{V} = \{1, 2, ..., 10\}$ and node $i \in \mathcal{V}$ has initial value $x_i(0)$, and $x(0) = [3, 5, 9, 7, 0, 0, 7, 4, 6, 8]^T$. According to [8], the node in this network can run MSR with f-local ($f <= 1$) threat model and reach resilient consensus, if $f > 1$, this network may break down into different subunits and not reach consensus. Now let's set 5 and 6 nodes to be malicious nodes that can inject any false data into its neighborhood at each time-step to break consistency. And next we tried several different forms of data injection to effectiveness. In addition, according to our setup, we can get that h (the number of malicious nodes which neighbor and collude with each other) is 2. The mobile node is set to randomly appear around the nodes in the network with a probability of 0.1 and collects the messages to implement community detection. And when the time period $t = 100$, the mobile node acts as a link between subunits based on the results of the community detection.

As shown in Figs. 6 and 7, the value injected by malicious nodes increases over time, and both of them are not affected by malicious nodes. However, in Fig. 6, we can see that the value of each normal node under the traditional MSR method will be affected by the network robustness. Intuitively, values of normal node tend to be two different values, so that the network can not reach resilient consensus. But average consensus can still be achieved under our method as shown in Fig. 7(a).

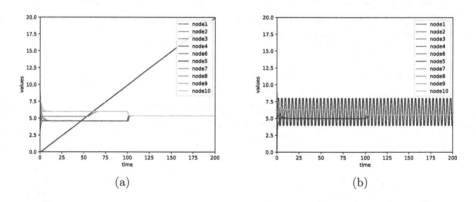

Fig. 7. The performance of the proposed method under 2-local threat model.

Next we tried to make the injected data floating up and down. Figure 7(b) shows that nodes in network can still achieve consensus with our method.

As a result of these examples, we observe that our method can be effective even if the node have different types of malicious nodes as neighbors. By this method, the node in network maintains the advantages of the MSR algorithm, that the node does not detect the received information from neighbors, just needs to eliminate the largest outliers. Moreover, it has lower requirements on the network topology to achieve resilient consensus.

6   Conclusions

In this paper, we proposed a novel MSR-based algorithm to reach resilient consensus for multi-agent systems under attacks. Specifically, we first observed and analyzed the phenomenon of traditional MSR algorithms running on a poor robust network, then we effectively utilized mobile agents to improve the MSR algorithm, and made the algorithm can apply to a more general network. We designed an effective mechanism to protect the information in each agent' broadcast data, and then we analyzed the convergence of the proposed control law under f-local attack model. Simulation results showed the effectiveness of our method.

Acknowledgment. This work is supported by the cyberspace security Major Program in National Key Research and Development Plan of China under grant 2016YFB0800201, Natural Science Foundation of China under grants 61572165, 61702150 and 61803135, State Key Program of Zhejiang Province Natural Science Foundation of China under grant LZ15F020003, Key Research and Development Plan Project of Zhejiang Province under grants 2017C01062 and 2017C01065, and Zhejiang Provincial Basic Public Welfare Research Project under grant LGG18F020015.

References

1. Cheng, L., Wang, Y., Ren, W., Hou, Z.G., Tan, M.: On convergence rate of leader-following consensus of linear multi-agent systems with communication noises. IEEE Trans. Autom. Control. **61**(11), 3586–3592 (2016)
2. Cheng, L., Wang, Y., Ren, W., Hou, Z.G., Tan, M.: Containment control of multi-agent systems with dynamic leaders based on a PI^n-type approach. IEEE Trans. Cybern. **46**(12), 3004–3017 (2016)
3. Zheng, Y., Ma, J., Wang, L.: Consensus of hybrid multi-agent systems. IEEE Trans. Neural Netw. Learn. Syst. **29**(4), 1359–1365 (2018)
4. Zhu, Y., Li, S., Ma, J., Zheng, Y.: Bipartite consensus in networks of agents with antagonistic interactions and quantization. IEEE Trans. Circ. Syst. II Express Briefs (2018). https://doi.org/10.1109/TCSII.2018.2811803
5. Dolev, D., Lynch, N.A., Pinter, S.S., Stark, E.W., Weihl, W.E.: Reaching approximate agreement in the presence of faults. J. ACM (JACM) **33**(3), 499–516 (1986)
6. LeBlanc, H.J., Koutsoukos, X.D.: Consensus in networked multi-agent systems with adversaries. In: 14th International Conference on Hybrid Systems: Computation and Control, pp. 281–290. ACM (2011)
7. Kieckhafer, R.M., Azadmanesh, M.H.: Reaching approximate agreement with mixed-mode faults. IEEE Trans. Parallel Distrib. Syst. **5**(1), 53–63 (1994)
8. LeBlanc, H.J., Zhang, H., Koutsoukos, X., Sundaram, S.: Resilient asymptotic consensus in robust networks. IEEE J. Sel. Areas Commun. **31**(4), 766–781 (2013)
9. Wu, Y., He, X., Liu, S., Xie, L.: Consensus of discrete-time multi-agent systems with adversaries and time delays. Int. J. Gen. Syst. **43**(3–4), 402–411 (2014)
10. Dibaji, S.M., Ishii, H.: Resilient multi-agent consensus with asynchrony and delayed information. IFAC-Pap. OnLine **48**(22), 28–33 (2015)
11. Wu, Y., He, X.: Secure consensus control for multi-agent systems with attacks and communication delays. IEEE/CAA J. Autom. Sin. **4**(1), 136–142 (2017)

12. Zhao, C., He, J., Chen, J.: Resilient consensus with mobile detectors against malicious attacks. IEEE Trans. Signal Inf. Process. Netw. **4**(1), 60–69 (2018)
13. Mi, S., Han, H., Chen, C., Yan, J., Guan, X.: A secure scheme for distributed consensus estimation against data falsification in heterogeneous wireless sensor networks. Sensors **16**(2), 252 (2016)
14. Kieckhafer, R., Azadmanesh, M.: Low cost approximate agreement in partially connected networks. J. Comput. Inf. **3**(1), 53–85 (1993)
15. Vaidya, N.H., Tseng, L., Liang, G.: Iterative approximate byzantine consensus in arbitrary directed graphs. In: 2012 ACM Symposium on Principles of Distributed Computing, pp. 365–374. ACM (2012)
16. Zhang, H., Sundaram, S.: Robustness of information diffusion algorithms to locally bounded adversaries. In: 2012 American Control Conference (ACC 2012), pp. 5855–5861. IEEE (2012)
17. Zhang, H., Fata, E., Sundaram, S.: A notion of robustness in complex networks. IEEE Trans. Control. Netw. Syst. **2**(3), 310–320 (2015)
18. Radicchi, F., Castellano, C., Cecconi, F., Loreto, V., Parisi, D.: Defining and identifying communities in networks. Proc. Natl. Acad. Sci. U. S. A. **101**(9), 2658–2663 (2004)
19. Clauset, A., Newman, M.E., Moore, C.: Finding community structure in very large networks. Phys. Rev. E **70**(6), 066111 (2004)
20. Pons, P., Latapy, M.: Computing communities in large networks using random walks. J. Graph Algorithms Appl. **10**(2), 191–218 (2006)
21. Blondel, V.D., Guillaume, J.L., Lambiotte, R., Lefebvre, E.: Fast unfolding of communities in large networks. J. Stat. Mech. Theory Exp. **2008**(10), P10008 (2008)
22. Arenas, A., Duch, J., Fernández, A., Gómez, S.: Size reduction of complex networks preserving modularity. New J. Phys. **9**(6), 176 (2007)
23. Ma, C.Y., Yau, D.K., Chin, J.c., Rao, N.S., Shankar, M.: Matching and fairness in threat-based mobile sensor coverage. IEEE Trans. Mob. Comput. **8**(12), 1649–1662 (2009)
24. Duan, X., He, J., Cheng, P., Chen, J.: Exploiting a mobile node for fast discrete time average consensus. IEEE Trans. Control. Syst. Technol. **24**(6), 1993–2001 (2016)

Multi-feature Fusion for Deep Reinforcement Learning: Sequential Control of Mobile Robots

Haotian Wang, Wenjing Yang$^{(\boxtimes)}$, Wanrong Huang, Zhipeng Lin,
and Yuhua Tang

National University of Defense Technology, Changsha, China
{wanghaotian13,wenjing.yang,linzhipeng13,yhtang}@nudt.edu.cn,
huangwr1990@163.com

Abstract. Compared with traditional motion planners, deep reinforcement learning has been applied more and more widely to achieving sequential behaviours control of mobile robots in indoor environment. However, the state of robot in deep reinforcement learning is commonly obtained through single sensor, which lacks accuracy and stability. In this paper, we propose a novel approach called multi-feature fusion framework. The multi-feature fusion framework utilizes multiple sensors to gather different scene images around the robot. Once environment information is gathered, a well-trained autoencoder achieves the fusion and extraction of multiple visual features. With more accurate and stable states extracted from the autoencoder, we train the mobile robot to patrol and navigate in 3D simulation environment with an asynchronous deep reinforcement learning algorithm. Extensive simulation experiments demonstrate that the proposed multi-feature fusion framework improves not only the convergence rate of training phase but also the testing performance of the mobile robot.

Keywords: Mobile robot · Deep reinforcement learning
Feature fusion

1 Introduction

Sequential behaviours control of mobile robots has received extensive attention in the last two decades [11]. In this paper, we only consider the high-level sequential behaviours of mobile robots, which include patrol and navigation in an indoor environment with some obstacles. Traditional motion planners for mobile robots often rely on highly precise sensors together with hand-crafted observations of

---

H. Wang—This work was funded through National Science Foundation of China (No. 91648204).

© Springer Nature Switzerland AG 2018
L. Cheng et al. (Eds.): ICONIP 2018, LNCS 11307, pp. 303–315, 2018.
https://doi.org/10.1007/978-3-030-04239-4_27

environment. However, it is suggested that traditional motion planners need hand-crafted guidance and thus are lack of generalization to new scenarios.

Recently, owing to the development of learning-based methods, Deep Reinforcement Learning (deep-RL) has achieved success over many aspects [9]. Moreover, successful robot patrol and navigation requires learning the relationships between actions and environment [17] and that is exactly what deep reinforcement learning concentrates on. Thus we take deep reinforcement learning as an unsupervised learning strategy to control continuous and sequential behaviours of mobile robots in indoor environment.

For deep reinforcement learning, state is the function expression of the history and it is commonly observed from for the environment. In this paper, state represents the location of the robot in the world. However, traditional approaches often utilize single sensor such as one lidar or one camera to observe the environment, causing the state trends to be isolated and one-sided. Thus it is difficult for the inaccurate and unstable state observed through single sensor to sustain the learning process in a comparatively complex environment. In order to improve the accuracy and stability of state, we propose the multi-feature fusion framework. This framework consists of four components: observation, learning process, simulation platform and reward function. We take Deep Deterministic Policy Gradient (DDPG) as the learning strategy and gym-gazebo[1] as the simulation platform. Additionally, we design two different reward functions for patrol and navigation respectively. For observation, we design a multi-sensor system with multiple camera sensors from different angles to capture the scene images around the mobile robot. Once scene images are captured, the framework will perform feature fusion and extraction through an autoencoder. The extracted state can help the mobile robot learn from environment more effectively and accurately.

To verify the proposed approach, we build the testing scenes in gym-gazebo and make extensive comparison experiments to fully demonstrate the improvement of our multi-feature fusion framework. The results of experiments including patrol and navigation show that the deep reinforcement learning with our multi-feature fusion framework achieves more accurate and stable sequential behaviours control of mobile robots. Moreover, reward curves indicate that the multi-feature fusion framework not only improves the convergence rate of learning process in the training phase but also facilitate the performance of the mobile robot in the testing phase.

In brief, the contributions of our research can be summarized as follows:

- We propose the multi-feature fusion framework and facilitate the accuracy and stability of sequential behaviours control of the mobile robot.
- We achieve robot patrol and navigation in 3D simulation environment, with the help of the deep reinforcement learning and our multi-feature fusion framework.
- The experimental results demonstrate that our multi-feature fusion framework improves both convergence rate and testing performance.

---

[1] https://github.com/erlerobot/gym-gazebo/.

2 Related Work

Sequential behaviours control of mobile robot in indoor environment concentrates on map-building-based approach and learning-based approach in general [3]. Traditional map-building-based approach of mobile robots control has strict requirement for devices and human guidance. Although it has been used widely, map-building-based approach is still lack of generalization to new scenarios [10, 14].

Learning-based approach endows the robot the ability to learning from the interaction with environment without human guidance. Deep reinforcement learning is the most popular and suitable method to achieve this goal [7]. Tai *et al.* proposed a navigation planner of mobile robots on the basis of Asynchronous Deep Reinforcement Learning (ADDPG) [13]. To enable a mobile robot to patrol and navigate across similar scenes, Zhang *et al.* proposed a successor feature-based deep reinforcement learning algorithm [16]. In 2017, Zhu *et al.* built a target-driven navigation model, which makes a combination of ResNet [6] and Siamese Network [2] in order to further enhance the scene generalization ability of robots [17]. Recently, Yang *et al.* proposed a deep reinforcement learning algorithm to learn multiple tasks concurrently [15].

Additionally, traditional feature fusion techniques such as Kalman Filtering and variants have already been widely applied to robot control, such as object localization [8, 12] and robot navigation [4]. Recently, deep networks have been focused on achieving feature fusion. Eitel *et al.* [5] proposed a Convolutional Neural Network (CNN) for object recognition with RGB-D inputs. Bohez *et al.* built a multi-layer fusion CNN with multiple lidar sensors to achieve robot navigation based on Deep Q-Networks (DQNs) [1]. The study in this study has fundamental difference with their research. First, instead of lidars, we achieve feature fusion through multiple cameras. Second, we utilize the auto-encoder to achieve the multi-feature fusion.

3 Multi-feature Fusion Framework

In this section, we give detailed elaboration of our multi-feature fusion framework. The framework is divided into four components: observation, learning strategy, experiment platform and reward function. Among these components, we concentrate on presenting the observation part as the main innovation point in this paper.

3.1 Observation: Multi-feature Fusion and Extraction

Deep reinforcement learning concentrates on endowing the robot the ability to learn from the interaction with the environment. Thus the state observed from environment is critical for the robot to identify its current position in learning sequence, which represents the location of robot in the world in this paper. Different from utilizing single camera or lidar sensor in traditional approach, we propose a novel approach called multi-feature fusion and extraction to help the mobile robot observe and extract the environment information.

Autoencoder is widely applied to extract the stable and representative features of images. Thus we build an autoencoder to achieve feature fusion and extraction. The autoencoder consists of two parts: an encoder and a decoder. The encoder part is a Convolutional Neural Network (CNN) with two convolution layers, two pooling layers and two dense layers. The structure of decoder part is similar to the encoder, where its deconvolutional and upsample layers are applied to recover the feature map to the raw feature input. More detailed parameters of autoencoder are elaborated in Fig. 1. The autoencoder is already trained before the learning behaviours start. To pre-train the autoencoder, we gather 10000 scene images in the gazebo environment as the training dataset. When finishing training for 20000 epoches, we save the model of the autoencoder.

After pre-training of the Autoencoder, we take 5 synchronized monocular cameras as the raw input from the environment as shown in Fig. 1. All the monocular cameras are fixed at the same position but with different orientations, where the angular distance between two adjacent cameras is 15°. Then the 5 scene images captured from the cameras will be concatenated on the axis of channel. The 128-dimensional output of the encoder will be taken as the state for DDPG. For the Autoencoder, the loss function is defined as the square distance between the input of encoder and the output of the decoder. In fact, here we only train the Autoencoder to achieve extraction and dimensionality reduction of the raw inputs.

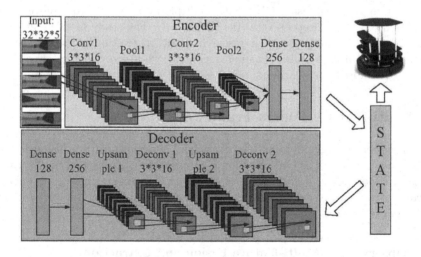

Fig. 1. Multi-feature fusion and extraction

3.2   Learning Strategy

Deep Deterministic Policy Gradient (DDPG) is proposed based on a combination of Actor-Critic and Policy-Gradient [7]. Different from traditional Q-learning and Deep Q Network which choose discretized actions based on Q values, DDPG generates continuous and deterministic actions. As our mobile robot requires continuous actions and DDPG has better performance on convergence speed, we choose DDPG as the learning strategy for the mobile robot.

As shown in Fig. 2, we build the deep networks for actor and critic separately. For actor, we use four dense layers to generate the continuous actions with the input as the processed state. Considering the requirement for the format of the robot's speed in gazebo simulator, we take the output of actor network as the mergence of linear.x and angular.z of geometry message[2]. Moreover, we constrain the linear speed in $[0.0, 1.0]$ through taking *sigmoid* as activation function. Similarly, an *tanh* activation function is applied to keep the angular speed in $[-1.0, 1.0]$. For critic, we use a dense layer with 512 nodes to extract the state and merge the processed state with the actions in the second layer as a common criterion. Then three dense layers with each of 512 nodes are used to generate the Q value to instruct the actor.

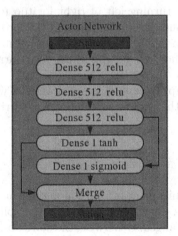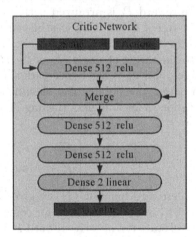

Fig. 2. Network structure of DDPG

3.3   Experiment Platform: Gym-gazebo

In order to simulate the mobile robot in a realistic and well-designed 3D environment, we choose gym-gazebo as our experiment platform. Gym-gazebo is a

---

[2] http://wiki.ros.org/geometry_msgs.

combination of gazebo[3] and gym[4], which can utilize various deep reinforcement learning algorithms to drive the mobile robot in a highly realistic simulation environment. In gym-gazebo, the simulation environment is built in gazebo and the interaction between the robot and the environment is achieved in gym. As a core, the control of robot in environment is completed by Robot Operating System (ROS)[5], through the communication among ROS nodes.

3.4 Reward Function

We design two different reward functions for the navigation and patrol respectively. For navigation, the reward function can be given as follow:

$$
Reward_n = \begin{cases} P_a & if\ D_t < \Phi_d \\ P_c & if\ collision \\ P_t & if\ turning \\ P_n * (D_t - D_{t-1}) & else \end{cases} \tag{1}
$$

In the reward function (1), a positive reward P_a is arranged if the mobile robot arrives to the target, but if the robot collides with the obstacle, a negative reward P_c is arranged. Moreover, to prevent the robot to fall into the loop, a negative reward P_t is arranged if the robot turning left of right for too many times in recent steps. This is achieved through keeping a queue to record the recent directions of the robot. Additionally, D_t is the real distance from the robot to the target of step t, Φ_d is the threshold for judging if the robot arrives the target, and P_n is the hyper-parameter.

For patrol, the mobile robot should attempt to explore the environment without collision or meaningless turning. Thus the reward function can be written as follow:

$$
Reward_p = \begin{cases} P_c & if\ collision \\ P_r * (MaxAng - |V_{Ang}| + P_b) & else \end{cases} \tag{2}
$$

In the reward function (2), we give a punishment of collision and encourage the robot to go straight except for some necessary turning. If the robot collides with the obstacles, a negative reward P_c is arranged. Otherwise, the more positive reward is arranged if the robot drives more straightly. Here P_r and P_b are the hyper-parameters, where $MaxAng$ and V_{Ang} are the maximum angular speed and the real angular speed of mobile robot respectively.

4   Experiment

In this section we will give a detailed elaboration of experiments. Our experiments make comparison between DDPG with multi-feature fusion and DDPG

---

[3] http://www.gazebosim.org/.
[4] http://gym.openai.com/.
[5] http://www.ros.org/.

with single camera sensor, according to the robot's performance of patrol and navigation respectively. We build different simulated environments for patrol and navigation, as shown in Figs. 3 and 4. The mobile robot is the turtlebot in ROS, which is same for patrol and navigation.

Fig. 3. Patrol   **Fig. 4.** Navigation

In the experiments, the configuration of the Autoencoder keeps the same for patrol and navigation. The initialization of weight follows normal distribution with mean 0 and standard deviation 0.01. The learning rate of the Autoencoder is fixed to 0.001 and the batch-size is 32 for training. Additionally, an Adam-optimizer is utilized as the learning algorithm.

4.1 Patrol Configuration and Training Performance

The patrol task is comparatively easier to achieve, thus we firstly endow the mobile robot the ability to patrol in the environment in Fig. 3. We firstly train the turtlebot utilizing DDPG strategy with single monocular camera as visual input and record the reward curve in the training phase. Then we install 5 monocular cameras with different orientations on the turtlebot and use our multi-feature fusion framework with DDPG strategy to train the robot. Similarly, the reward curve of the learning process with multi-feature fusion framework is also recorded.

In patrol, we use a discount rate $\gamma = 0.995$ to reduce the rate of exploration in ϵ-greedy strategy. The RMSProp learning algorithm is run for 1000 episodes in DDPG and each episode has a maximum length of 550, which requires the robot to patrol in the maze for nearly two loops stably. The learning rate is fixed at 0.0001 for actor and 0.001 for critic, where the TAU for updating target networks is fixed at 0.001. During training phase, the batch-size is kept at 32 for all the time. Additionally, the steps of exploration before learning is 2000 steps.

The training reward curves of single camera and multi-feature fusion are reported as follow:

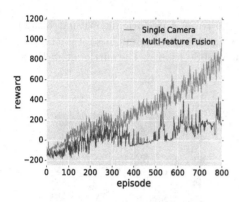 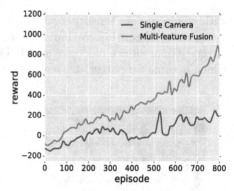

Fig. 5. Raw reward curve   **Fig. 6.** Smooth reward curve

As shown in Figs. 5 and 6, we present the reward curves, which contain the changing trend of episode reward over the number of training episodes, with the same random seed. The two reward curves in the Fig. 5 indicates the raw reward, where the another two in the Fig. 6 are smoothed by the Gaussian Kernel in order to make convenience for visualization.

The reward curves above indicates the improvement of training using multi-feature fusion compared with using traditional single camera sensor in the patrol process. Firstly, our multi-feature fusion achieves faster convergence rate than single camera sensor, which is obvious from 300 to 650 episodes in the Fig. 6. Secondly, the multi-feature fusion is more stable than single camera, with less fluctuation during training phase. We suggest that the reasons for the improvement of performance of multi-feature fusion method can be analyzed from two aspects: more accurate states for location and more stable features for training. The fusion of multi-feature from multiple camera sensors improves the accuracy of state, which represents the location of the robot in the environment. The autoencoder achieves the feature extraction and provides stable states for the deep network of critic and actor in DDPG to train the robot.

4.2   Navigation Configuration and Training Performance

In the navigation scene, we set three targets for the turtlebot in the start point, as shown in Fig. 7. We record the rewards of each episode during the training phase for navigation processes of three targets, utilizing multi-feature fusion and single camera respectively. The installation of 5 monocular cameras on the turtlebot is the same as the approaches in patrol.

In navigation, the discount rate is set as $\gamma = 0.9995$ to reduce the rate of exploration in ϵ-greedy strategy. The RMSProp learning algorithm is utilized in DDPG strategy and each episode has a maximum length of 300. The learning rate is fixed at 0.0002 for actor and 0.0001 for critic, where the TAU for updating target networks is fixed at 0.001. During training phase, the batch-size is kept at 64 for all the time. Additionally, the steps of exploration before learning is

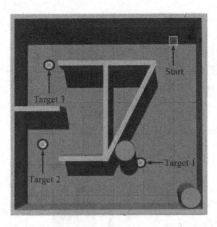

Fig. 7. Targets for navigation

3000 steps. For three targets, we respectively report the raw curves of training reward of the navigation process utilizing single camera and multi-feature fusion, together with their smooth curves to make comparisons. For target 1, we train the robot for 4000 episodes. Considering the training curves almost remain stable in the later stages, we only display the first 3000 episodes for each approach and 2500 episodes for comparison as follow:

For target 2, we train the turtlebot for 2000 episodes and only display the first 1500 episodes of the training rewards as follow:

For target 3, for the reason that the target is close to the wall in the environment and the location is more challenging to the robot, we train the turtlebot for 8000 episodes and display only the first 6000 episodes:

In the three figures above, the reward curves increase dramatically in some specific episodes when the mobile robot starts to navigate to the target stably, for example, the 1500-th episode in (a) and the 600-th episode in (b) of Fig. 8. It is caused by the positive reward P_a in the reward function (1), which is comparatively higher to the other positive reward. This positive reward is set a high value in order to promote the convergence of training process. Moreover, the

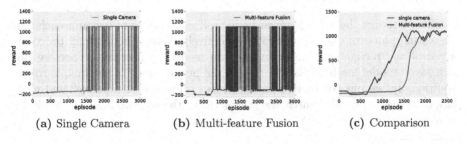

(a) Single Camera   (b) Multi-feature Fusion   (c) Comparison

Fig. 8. Target 1

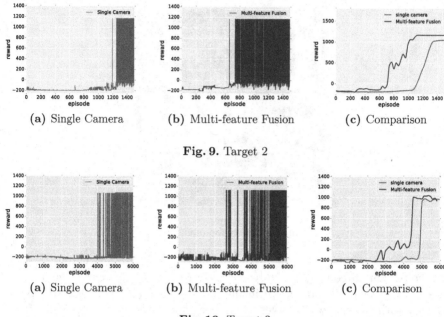

(a) Single Camera　　　　(b) Multi-feature Fusion　　　　(c) Comparison

Fig. 9. Target 2

(a) Single Camera　　　　(b) Multi-feature Fusion　　　　(c) Comparison

Fig. 10. Target 3

negative reward P_c which is set to punish the collision also leads to a comparative low episode reward before the robot arrives to the target.

As shown in the Figs. 8, 9 and 10, we can find that the navigation with multi-feature fusion has improvement on the convergence rate of the training process, compared with the single camera. Specifically, in (a) and (b) of Fig. 8, the training process with multi-feature fusion begins to converge in the 600-th episode, while the training process with single camera starts converging until 1500-th episode. The disparity of two approaches is more obvious in the smooth curves displayed in (c) of Fig. 8. Additionally, as shown in Figs. 9 and 10, the episode reward of multi-feature fusion is higher than single camera when the robot can navigate to the target stably. This indicates the better testing performance of multi-feature fusion and will be analyzed later.

4.3　Testing Performance

Finally, we report the average reward and task success rate of two approaches in the testing phase of our experiments respectively. When the training processes of patrol and navigation finish, we save the models respectively. Then we run each saved model for 100 episodes to test the performance of the mobile robot. Additionally, we put the related testing videos on website for patrol[6] and navigation[7] respectively. (To watch the patrol video, the website address should be input manually.)

---

[6] https://youtu.be/9V2_QTS_aJo.
[7] https://youtu.be/EdqG-W-V3pE, https://youtu.be/5LcTV67ZmfE.

Firstly, the average reward in 100 episodes of multi-feature fusion and single camera are reported as follow:

Table 1. Testing reward

Approaches	Multi-feature fusion	Single camera
Patrol	373.61	315.52
Navigation: Target1	1108.94	1031.76
Navigation: Target2	1092.30	1025.8
Navigation: Target3	1168.15	1124.28

Secondly, we report the task success rate in 100 episodes of multi-feature fusion and single camera as follow. Here we note that the success of task in each episode refers that the mobile robot can accomplish the patrol or navigation task in the episode without collisions.

Table 2. Task success rate

Approaches	Multi-feature fusion	Single camera
Patrol	88%	84%
Navigation: Target1	86%	75%
Navigation: Target2	92%	87%
Navigation: Target3	89%	81%

From the Table 1, we find that multi-feature fusion achieves an average testing reward of 373.61 in patrol, while the single camera sensor only achieves 315.52. Similarly, for three targets, the average testing reward of multi-feature fusion is higher than single camera. From the Table 2, we find that the multi-feature achieves higher success rate than single camera. Additionally, some failure tasks in the test can be considered that the 3D environment in gazebo is noisy and not stable enough.

We suggest that the significant difference in the average rewards is due to the stability and accuracy of state are enhanced effectively through multi-feature fusion. The multi-feature fusion framework gathers more abundant scene images and performs stable and accurate feature fusion and extraction. The task success rate in Table 2 also reflects the improvement on testing performance through multi-feature fusion framework.

5   Conclusion and Future Work

In this paper, in order to improve the sequential behaviours control of mobile robot in indoor environment, we propose the multi-feature fusion framework with gathering more abundant observation and obtain more accurate and stable state. Based on multiple visual sensors and an autoencoder, we achieve mutli-feature fusion and extraction. With the mutli-feature fusion framework, we perform the patrol and navigation of mobile robot through deep reinforcement learning. Finally, experiments in 3D simulation environment demonstrate that our mutli-feature fusion framework improves both on the convergence rate and the testing performance. Moreover, our future work plans to develop our methods on the hardware experiments with a more complex real-world environment.

References

1. Bohez, S., Verbelen, T., De Coninck, E., Vankeirsbilck, B., Simoens, P., Dhoedt, B.: Sensor fusion for robot control through deep reinforcement learning. arXiv Preprint arXiv:1703.04550 (2017)
2. Chopra, S., Hadsell, R., LeCun, Y.: Learning a similarity metric discriminatively, with application to face verification. In: IEEE Computer Society Conference on Computer Vision and Pattern Recognition, CVPR 2005, vol. 1, pp. 539–546. IEEE (2005)
3. DeSouza, G.N., Kak, A.C.: Vision for mobile robot navigation: a survey. IEEE Trans. Pattern Anal. Mach. Intell. **24**(2), 237–267 (2002)
4. Dobrev, Y., Flores, S., Vossiek, M.: Multi-modal sensor fusion for indoor mobile robot pose estimation. In: 2016 IEEE/ION Position, Location and Navigation Symposium (PLANS), pp. 553–556. IEEE (2016)
5. Eitel, A., Springenberg, J.T., Spinello, L., Riedmiller, M., Burgard, W.: Multimodal deep learning for robust RGB-D object recognition. In: 2015 IEEE/RSJ International Conference on Intelligent Robots and Systems (IROS), pp. 681–687. IEEE (2015)
6. He, K., Zhang, X., Ren, S., Sun, J.: Deep residual learning for image recognition. In: Proceedings of the IEEE Conference on Computer Vision and Pattern Recognition, pp. 770–778 (2016)
7. Lillicrap, T.P., et al.: Continuous control with deep reinforcement learning. arXiv Preprint arXiv:1509.02971 (2015)
8. Malyavej, V., Kumkeaw, W., Aorpimai, M.: Indoor robot localization by RSSI/IMU sensor fusion. In: 2013 10th International Conference on Electrical Engineering/Electronics, Computer, Telecommunications and Information Technology (ECTI-CON), pp. 1–6. IEEE (2013)
9. Mnih, V., et al.: Human-level control through deep reinforcement learning. Nature **518**(7540), 529 (2015)
10. Richter, M., Sandamirskaya, Y., Schöner, G.: A robotic architecture for action selection and behavioral organization inspired by human cognition. In: 2012 IEEE/RSJ International Conference on Intelligent Robots and Systems (IROS), pp. 2457–2464. IEEE (2012)
11. Rimon, E., Koditschek, D.E.: Exact robot navigation using artificial potential functions. IEEE Trans. Robot. Autom. **8**(5), 501–518 (1992)
12. Stroupe, A.W., Martin, M.C., Balch, T.: Distributed sensor fusion for object position estimation by multi-robot systems. In: Proceedings of 2001 IEEE International Conference on Robotics and Automation, ICRA 2001, vol. 2, pp. 1092–1098. IEEE (2001)

13. Tai, L., Paolo, G., Liu, M.: Virtual-to-real deep reinforcement learning: continuous control of mobile robots for mapless navigation. In: 2017 IEEE/RSJ International Conference on Intelligent Robots and Systems (IROS), pp. 31–36. IEEE (2017)
14. Wu, Y., Zhang, W., Song, K.: Master-Slave Curriculum Design for Reinforcement Learning. In: IJCAI, pp. 1523–1529 (2018)
15. Yang, Z., Merrick, K.E., Abbass, H.A., Jin, L.: Multi-task deep reinforcement learning for continuous action control. In: IJCAI, pp. 3301–3307 (2017)
16. Zhang, J., Springenberg, J.T., Boedecker, J., Burgard, W.: Deep reinforcement learning with successor features for navigation across similar environments. arXiv Preprint arXiv:1612.05533 (2016)
17. Zhu, Y., et al.: Target-driven visual navigation in indoor scenes using deep reinforcement learning. In: 2017 IEEE International Conference on Robotics and Automation (ICRA), pp. 3357–3364. IEEE (2017)

Dynamics Based Fuzzy Adaptive Impedance Control for Lower Limb Rehabilitation Robot

Xu Liang[1,2], Weiqun Wang[1], Zengguang Hou[1,2(✉)], Zihao Xu[3], Shixin Ren[1,2], Jiaxing Wang[1,2], and Liang Peng[1]

[1] Institute of Automation, Chinese Academy of Sciences, Beijing 100190, China
liangxucsu@163.com, {weiqun.wang,zengguang.hou}@ia.ac.cn,
1027872138@qq.com, renshixincn@gmail.com, 2837939728@qq.com
[2] University of Chinese Academy of Sciences, Beijing 100149, China
[3] Lanzhou Jiaotong University, Lanzhou 730070, China
1325195510@qq.com

Abstract. Human-robot interaction control plays a significant role in the research and clinical application of rehabilitation robots. A fuzzy adaptive variable impedance control strategy is proposed in this paper. Firstly, a dynamic model is established by using the Lagrangian method and the traditional friction model, which can be used to predict human-robot interaction forces. Then, a fuzzy adaptive variable impedance control strategy based on the human-robot system dynamic model is designed. In the designed control strategy, the interaction forces, position and velocity errors are taken as the system inputs, and a fuzzy adaptive law is used to adjust the damping and stiffness coefficients. Finally, the dynamics identification experiments and simulation of the fuzzy adaptive variable impedance control strategy are carried out, by which performance of the proposed method is validated.

Keywords: Fuzzy adaptive impedance · Dynamics
Parameter identification · Assist-as-needed · Rehabilitation robot

1 Introduction

Rehabilitation treatment of patients suffering from stroke can be improved with the aid of rehabilitation robots significantly [1–5]. It is well known from the neuron plasticity theory that, recovery of patients' motor function and compensation and reorganization of the nervous system can be promoted through exercise training [6]. The active training based on patients' motion intention is

This research is supported in part by the National Natural Science Foundation of China (Grants 61720106012, 91648208), and the Beijing Municipal Natural Science Foundation (Grant L172050, 3171001).

an effective training strategy in the rehabilitation therapy, which can be used to raise patients' enthusiasm for participation in training, and improve the rehabilitation effect.

The key premise of active rehabilitation training is to accurately recognize the human movement intention. The system dynamics and the measurement from force sensors can be used to calculate the human-robot interaction forces, which are an intuitive embodiment of human movement intention. The system dynamic model can be obtained by accurate modeling and identification methods.

In human-robot interaction systems, impedance control strategies are widely used. Mechanical impedance, including stiffness, damping and inertia, reflects the relationship between motion and force. Human joint impedance can be continuously modulated to accommodate environment variation and maintain stable interaction either voluntarily or reflexively during movement. Lee and Hogan characterized human ankle mechanical impedance in the frontal and sagittal planes simultaneously [7]. Mendoza et al. proposed a wave-based bilateral teleoperation scheme for robot assisted rehabilitation therapies [8]. A pair of motion-based impedance controllers are integrated into the scheme to enhance human-robot interaction.

It is time consuming for human operators to precisely adjust the impedance control parameters. Automatic processing adjustment program will make rehabilitation robots more economic and accessible for clinic application. Cartesian impedance control is employed by Ficuciello to realize the robot's end executor's supple behavior, in response to force imposed by human operators. In addition, different variable impedance control strategies have been assessed, in which the impedance parameters are revised according to human movement intention. The associated experimental results have shown that the variable impedance control with an appropriate regulation strategy for parameters' adjustment is better than constant impedance, because it improves the patients perceived comfort during human instruction [9]. Finite state impedance control and fuzzy logic inference are the widely used approaches for tuning impedance parameters. Liu developed a set of dempster-shafer theory based transition rules to resolve the uncertainty-caused challenge [10]. Huang designed a network expert system which used fuzzy logic inference to code a specialist's experience and technology into program to effectively adjust the impedance coefficients of a powered knee prosthesis during level-ground walking [11].

In this paper, we concentrate on modeling dynamics and design of fuzzy adaptive impedance control strategies for patients' active exercises on the lower limb rehabilitation robot. A fuzzy adaptive impedance control strategy is proposed to adjust the impedance parameters based on the human-robot interaction terms (forces, positions and speeds). Based on the system dynamic model, the human-robot interaction forces can be estimated. Then, the obtained interaction forces and the position and speed tracking errors are used as inputs of the fuzzy system. Finally, a fuzzy adaptive law is designed to adjust the impedance parameters to create a compliant interface.

The rest of this paper is organized as follows. Section 2 gives the method that mainly includes establishment of dynamics of the lower limb rehabilitation robot, the parameter identification method and the fuzzy adaptive impedance control strategy. Dynamics identification and validation experiments for the fuzzy adaptive impedance control are performed in Sect. 3. In Sect. 4, the methods and experiments are summarized, and the future work is briefly introduced.

2 Method

In this study, we proposed nonlinear dynamics based fuzzy adaptive impedance control strategy for the lower limb rehabilitation robot. The experiment equipment is given in Fig. 1.

Fig. 1. Experiment platform for identification and validation of dynamics

2.1 Dynamic Model

The lower limb rehabilitation robot can be simplified as a two degrees of freedom linkage as shown in Fig. 2. The dynamic model can be obtained by using the following equations.

$$\frac{d}{dt}\left(\frac{\partial L}{\partial \dot{\theta}}\right) - \frac{\partial L}{\partial \theta} = \tau \tag{1}$$

where L represents the difference between the kinetic and potential energy of the system.

$$L = K - P \tag{2}$$

where K and P represents the kinetic and potential energy respectively.

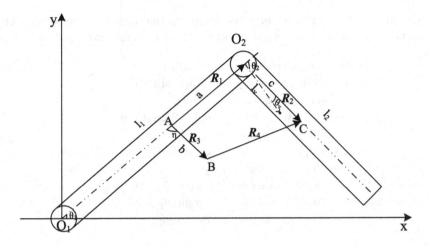

Fig. 2. Simplification of the lower limb rehabilitation robot

A closed loop vector relationship shown in Fig. 2 can be expressed as follows:

$$\mathbf{R_1} + \mathbf{R_2} = \mathbf{R_3} + \mathbf{R_4} \tag{3}$$

According to the virtual work principle, the force f_2, along the direction of R_{25}, can be calculated by the equation (4) as:

$$f_2 = k\tau_2 \tag{4}$$

$$k = \frac{\sqrt{a^2 + b^2 + c^2 + 2ab\cos(\eta) + 2ac\cos(\theta_2) + 2bc\cos(\eta - \theta_2)}}{bc\sin(\eta - \theta_2) - ac\sin(\theta_2)} \tag{5}$$

Joint friction is related to the mechanical mechanism of transmission system, lubrication, inertia of each component, etc. It often exhibits a high degree of nonlinearity, which poses a great challenge to the modeling of joint friction. In this paper, the classic joint friction model is adopted, the form of which is given by [12].

$$\tau_{i,f} = c_{i,1}\dot{\theta}_i + c_{i,2}\text{sign}(\dot{\theta}_i), i = 1, 2 \tag{6}$$

where $c_{i,1}$ and $c_{i,2}$ represent the viscosity and coulomb friction coefficient respectively. This friction model could reflect the commutation characteristics of coulomb friction when joint steering is switched. After getting rid of the similar terms and recombining the parameters, the dynamics of the lower limb rehabilitation robot is gained as follows:

$$\Phi_r P_r = \tau_r \tag{7}$$

where Φ_r is a 2*10 regressor matrix, of which the items are composed of the joint angular and its accelerations and velocities of lower limb rehabilitation robot.

P_r is a 10*1 vector which denotes the unknown parameters to be solved. The elements of Φ_r and P_r are defined respectively by the following equation.

$$
\begin{aligned}
&\phi_{11}= 2\ddot{\theta}_1, \quad \phi_{12}=\cos{(\theta_1)}, \quad \phi_{13}= 2(\ddot{\theta}_1+\ddot{\theta}_2),\\
&\phi_{14}= 2\ddot{\theta}_1\cos{(\theta_2)}+\cos{(\theta_2)}\ddot{\theta}_2-2\dot{\theta}_1\dot{\theta}_2\sin{(\theta_2)}\\
&\qquad -\dot{\theta}_1^2\sin{(\theta_2)}+g\cos{(\theta_1+\theta_2)}/l_2,\\
&\phi_{15}= -\cos{(\theta_2)}\ddot{\theta}_2^2-2\dot{\theta}_1\dot{\theta}_2\cos{(\theta_2)}-2\sin{(\theta_2)}\ddot{\theta}_1\\
&\qquad -\sin{(\theta_2)}\ddot{\theta}_2^2-g\sin{(\theta_1+\theta_2)}/l_2,\\
&\phi_{16}= -\sin{(\theta_1)}, \quad \phi_{17}=\dot{\theta}_1, \quad \phi_{18}=\text{sign}(\dot{\theta}_1), \quad \phi_{19}= \phi_{10}= 0,\\
&\phi_{21}= \phi_{22}= 0, \quad \phi_{23}= 2k(\ddot{\theta}_1+\ddot{\theta}_2),\\
&\phi_{24}=k[\cos{(\theta_2)}\ddot{\theta}_1+\sin{(\theta_2)}\dot{\theta}_1^2+g\cos{(\theta_1+\theta_2)}/l_2],\\
&\phi_{25}=k[\cos{(\theta_2)}\ddot{\theta}_1^2-\sin{(\theta_2)}\ddot{\theta}_1^2-g\sin{(\theta_1+\theta_2)}/l_2],\\
&\phi_{26}=\phi_{27}=\phi_{28}= 0, \quad \phi_{29}=k\ddot{\theta}_2, \quad \phi_{20}=k\text{sign}(\dot{\theta}_2),
\end{aligned}
\tag{8}
$$

and

$$
\begin{aligned}
&p_1= \int_{\nu_1} \tfrac{1}{2}\rho_\nu l_\nu^2 d\nu + \int_{\nu_2} \tfrac{1}{2}\rho_\nu l_2^2 d\nu,\\
&p_2= \int_{\nu_1} \rho_\nu g l_\nu \cos{(\theta_\nu)}\, d\nu + \int_{\nu_2} \rho_\nu g l_2 d\nu,\\
&p_3= \int_{\nu_2} \tfrac{1}{2}\rho_\nu l_\nu^2 d\nu, \quad p_4= \int_{\nu_2} \rho_\nu l_\nu \cos{(\theta_\nu)}\, d\nu,\\
&p_5= \int_{\nu_2} \rho_\nu l_\nu \sin{(\theta_\nu)}\, d\nu, \quad p_6= \int_{\nu_1} \rho_\nu g l_\nu \sin{(\theta_\nu)}\, d\nu,\\
&p_7=c_{1,1}, \quad p_8=c_{1,2}, \quad p_9=c_{2,1}, \quad p_{10}=c_{2,2},
\end{aligned}
\tag{9}
$$

where ν_1 and ν_2 are the lower limb rehabilitation robot thigh and calf's volume respectively; ρ_ν, l_ν and θ_ν represent material density, the distance between the related joint axis and $d\nu$, and the angle offset of $d\nu$ from the central line respectively, and they are connected with the position; $d\nu$ is a differential term; $p_1 - p_{10}$ are the uncertain coefficients to be recognized.

2.2   Fuzzy Adaptive Impedance Control

Impedance control is intended to regulate the dynamic relationship between the patient's active applied force and the robot's motion trajectory deviation. The basic idea is to permit the experimental subject to deviate from the predetermined reference path instead of forcing the subject to move on a fixed path. The degree of deviation depends on the magnitude of the moment applied by the patient and its behavioral pattern so as to create a comfortable and natural compliant tactile interface for the patient.

The impedance control method can be represented by the spring-damping-inertia model:

$$
F_h = M\ddot{X}_e + B\dot{X}_e + KX_e
\tag{10}
$$

where K, B and M denote the rigidity, damping and inertial coefficient matrices respectively. M reflects the smoothness of the system response, B reflects the energy consumption of the system, and K reflects the system's rigidity. X_e, $\dot{X}_e$ and $\ddot{X}_e$ represent the position, velocity and acceleration deviation between the robot's actual and reference trajectory respectively. F_h represents the mutual haptic force imposed on robot by patients.

In the course of rehabilitation treatment, the robot system usually needs to adjust the impedance parameters in real time to provide appropriate assistance or resistance force according to the training form and the patient's condition. The mechanical and human body impedance are constantly changing, in order to obtain a better rehabilitation effect, dynamics based fuzzy adaptive impedance control strategy is proposed to adjust stiffness and damping coefficients on line to approximate the ideal impedance parameters. Figure 3 gives the diagram of controlling system, where X, X_d are the robot's actual and desired location separately. As seen in Fig. 3, the position errors and human-robot interaction forces are taken as inputs of stiffness fuzzy adaptive regulators to adjust $K_d(x)$, and the human-robot interaction forces and velocity errors are taken as inputs of damping fuzzy adaptive regulators to adjust $B_d(x)$.

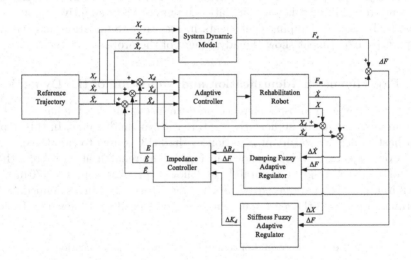

Fig. 3. The diagram of fuzzy adaptive impedance control system

To represent the time-varying uncertainties of variables $B_d(x)$ and $K_d(x)$, fuzzy variables $\tilde{B}_d(x)$ and $\tilde{K}_d(x)$ that reflect the state of the human-robot system are used instead. $\tilde{B}_d(x)$ is constructed as follows:

(1) Define l_i fuzzy sets $H_i^{k_i}(k_i = 1, 2, ...l_i)$ for variable $x_i(i = 1, 2, ...n)$;

(2) $\prod\limits_{i=1}^{n} l_i$ fuzzy rules is adopted to construct fuzzy system $\tilde{B}_d(x)$: if x_1 is $H_1^{k_1}$ and ... and x_n is $H_n^{k_n}$, then $\tilde{B}_d$ is $C^{k_1 \cdots k_n}$.

Where fuzzy sets $H_i^{k_i}$ is defined as {NB, NS,O,PS,PB}, namely $l_i = 5$. The fuzzy variables $\tilde{B}_d(x)$ can be obtained by using the plane inference machine, the single-value fuzzy and the center average deblurring.

$$\tilde{B}_d(x) = \frac{\sum\limits_{k_1=1}^{l_1} \cdots \sum\limits_{k_n=1}^{l_n} \bar{w}_f^{k_1\cdots k_n} (\sum\limits_{i=1}^{n} \mu H_i^{k_i}(x_i))}{\sum\limits_{k_1=1}^{l_1} \cdots \sum\limits_{k_n=1}^{l_n} (\sum\limits_{i=1}^{n} \mu H_i^{k_i}(x_i))} \tag{11}$$

where $\mu H_i^{k_i}(x_i)$ is membership function, $\bar{w}_f^{k_1\cdots k_n}$ is free parameters.

3   Experiments

In this section, we designed several experiments to verify the validity of the methodology mentioned above. The human-robot system dynamics identification experiments are carried out on the robot designed by our laboratory, which can supply various active exercise pattern, so that the forces imposed on the robot by the subject can be obtained by calculation. Then the force and velocity and position errors will be taken as simulation inputs of fuzzy adaptive regulator to adjust stiffness and damping coefficients in real time. Simulation analysis and comparative experiment show the advantages of the proposed method.

3.1   Experiment for Identification and Validation of the Dynamics

The convergence and anti-interference performance of the least squares estimation can be improved by sufficiently exciting the system dynamic. In this paper, a stochastic particle swarm optimization method [13] is used to gain the optimal excitation trajectory, which is shown in Fig. 4. The rehabilitation robot's thigh and calf's length are given separately as follows: $l_1 = 420$ mm, $l_2 = 470$ mm, as shown in Fig. 2. The parameters of human-robot system dynamics defined by (8) is estimated by using the least square method. The results are shown in Table 1.

Table 1. The results of dynamics parameters identification

Parameters	Unit	Value	Parameters	Unit	Value
P_1	$Kg*m^2$	828.4950	P_6	Nm	9380.8
P_2	Nm	−2984.2	P_7	Nm	−2084.3
P_3	$Kg*m^2$	1644.1	P_8	Nm	−102.8052
P_4	$Kg*m^2$	486.5616	P_9	Nm	−12537
P_5	$Kg*m^2$	−81.2372	P_{10}	Nm	−392.5691

In the verification experiment, a trajectory different from the optimal excitation trajectory was designed and carried out on the lower limb rehabilitation robot under the same condition as the parameter identification experiment. The estimated and measured torques are given in Fig. 5. From Fig. 5, we can see that the estimation errors of knee and hip joints are small , which indicates that the

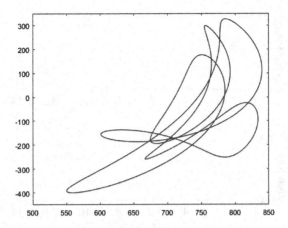

Fig. 4. The OET for identification of dynamics

dynamics of human-robot system is appropriate for torque estimation. Therefore, the patient's active torque imposed on robot can be calculated as follows:

$$\tau_h = \tau_{ms} - \tau_{er} \tag{12}$$

where τ_h represents the active torque applied by human body, which is used as inputs of fuzzy adaptive regulators. τ_{ms} represents the measured torque from force transducers installed on the knee and hip joint, τ_{er} represents the estimated torque which is calculated from the human-robot system's dynamic model.

3.2 Experiment for Validation of the Fuzzy Adaptive Impedance Control

The proposed fuzzy adaptive impedance control strategy was test with MATLAB to perform the numerical simulations. τ_h, position and velocity errors were taken as inputs of fuzzy adaptive regulators to adjust the stiffness and damping coefficients in real time. Sine wave was used as reference trajectory as treadmill and gait are similar to sine waves. The proposed method is compared with traditional impedance control of fixed parameters in the simulation, and the results are given in Figs. 6 and 7.

From Figs. 6 and 7, we can see that the fuzzy adaptive impedance control method has better compliance and flexibility. When the system is subjected to the same external force interference, the fuzzy adaptive impedance control method can track the trajectory well without a sudden increase in velocity. However, the position tracking of the comparison method fluctuates greater, and the velocity also changes obviously, as shown in Fig. 6. The experiment results indicate that the proposed method has a good control effect and enhances the compliance of human-robot interaction system.

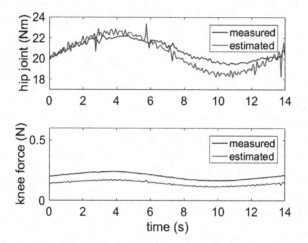

Fig. 5. Results of identification experiments of the dynamics

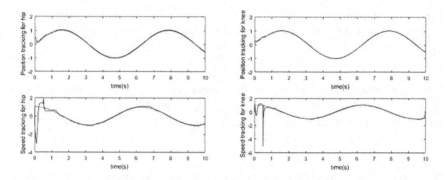

Fig. 6. Results of impedance control with fixed parameters

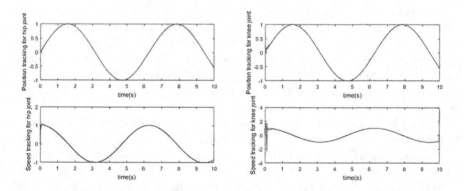

Fig. 7. Results of fuzzy adaptive impedance control method

4 Conclusion

Active rehabilitation training based on the patients' ability is an effective way to raise their enthusiasm for participating in exercise and improve the therapeutic effect. As a result, the impedance parameters should be tuned online based on the human-robot interactive forces and the state of motion. A fuzzy adaptive impedance control method is proposed in this paper. The human-robot interaction forces can be obtained through the method of human-robot system dynamic model identification. Then, the interaction force, position and velocity errors are taken as inputs of fuzzy adaptive regulators to adjust the stiffness and damping coefficients online.

The identification and validation experimental results show that the established system dynamic model is suitable for estimation of human-robot interaction torque. The simulation results show that the fuzzy adaptive impedance control method can increase the robustness and stability of the human-robot system, and the system's compliance is obviously better than traditional control method. The further investigation will be carried out to reduce the estimation error for hip joint. Meanwhile, the proposed method will be performed on lower limb rehabilitation robot and then applied to patient experiments to further validate its effectiveness in promoting recovery of patient's motor function.

References

1. Von Schroeder, H.P., Coutts, R.D., Lyden, P.D., Billings Jr., E., Nickel, V.L.: Gait parameters following stroke: a practical assessment. J. Rehabil. Res. Develop. **32**, 25–31 (1995)
2. Olney, S.J., Griffin, M.P., Monga, T.N., McBride, I.D.: Work and power in gait of stroke patients. Arch. Phys. Med. Rehabil. **72**, 309–314 (1991)
3. Banala, S.K., Kim, S.H., Agrawal, S.K., Scholz, J.P.: Robot assisted gait training with active leg exoskeleton (ALEX). IEEE Trans. Neural Syst. Rehabil. Eng. **17**, 2–8 (2009)
4. Cai, L.L., et al.: Implications of assist-as-needed robotic step training after a complete spinal cord injury on intrinsic strategies of motor learning. J. NeuroSci. **26**, 10564–10568 (2006)
5. Riener, R., Lunenburger, L., Jezernik, S., Anderschitz, M., Colombo, G., Dietz, V.: Patient-cooperative strategies for robot-aided treadmill training: First experimental results. IEEE Trans. Neural Syst. Rehabil. Eng. **13**, 380–394 (2005)
6. Pons, T.P., Garraghty, P.E., Ommaya, A.K., Kaas, J.H., Taub, E., Mishkin, M.: Massive cortical reorganization after sensory deafferentation in adult macaques. Science **252**, 1857–1860 (1991)
7. Lee, H., Hogan, N.: Time-varying ankle mechanical impedance during human locomotion. IEEE Trans. Neural Syst. Rehabil. Eng. **23**, 755–764 (2015)
8. Mendoza, M., Bonilla, I., González-Galván, E., Reyes, F.: Impedance control in a wave-based teleoperator for rehabilitation motor therapies assisted by robots. Comput. Meth. Prog. Bio. **123**, 54–67 (2016)
9. Ficuciello, F., Villani, L., Siciliano, B.: Variable impedance control of redundant manipulators for intuitive human-robot physical interaction. IEEE Trans. Robot. **31**, 850–863 (2015)

10. Liu, M., Zhang, F., Datseris, P., Huang, H.: Improving finite state impedance control of active-transfemoral prosthesis using dempster-shafer based state transition rules. J. Intell. Robot. Syst. **76**, 461–474 (2014)
11. Huang, H., Crouch, D.L., Liu, M., Sawicki, G.S., Wang, D.: A cyber expert system for auto-tuning powered prosthesis impedance control parameters. Ann. Biomed. Eng. **44**, 1613–1624 (2016)
12. Wit, C.C.D., Noel, P., Aubin, A., Brogliato, B., Drevet, P.: Adaptive friction compensation in robot manipulators: low velocities. Int. J. Robot. Res. **10**, 1352–1357 (1991)
13. Wei, L.Y., Qi, H., Ren, Y.T., Sun, J.P., Wen, S., Ruan, L.M.: Application of hybrid SPSO-SQP algorithm for simultaneous estimation of space-dependent absorption coefficient and scattering coefficient fields in participating media. Int. J. Therm. Sci. **124**, 424–432 (2018)

A New Overvoltage Control Method Based on Active and Reactive Power Coupling

Guangbin Li, Yanhong Luo[✉], and Dongsheng Yang

Northeastern University, Shenyang 110819, China
luoyanhong@ise.neu.edu.cn

Abstract. With the development of photovoltaic policy, a large number of PVs will be connected to rural networks, which cause the higher photovoltaic permeability in rural networks in the future. However, the PV of high-permeability will cause very serious reverse power flow in the rural networks, which will lead to the problems of overvoltage and the increasement of network loss, and which also seriously affect the safety of users. The relationship between the active power and reactive power of the inverter is analyzed. The quantization's relationship between the voltage and the active power or the reactive power in the rural networks is analyzed according to the capacity characteristics of the inverter. In order to minimize the reduction of photovoltaic active power, an active and reactive power coupling control strategy is proposed. A multi-objective particle swarm optimization algorithm is adopted considering the characteristics of economy and the topology of rural networks. The results of simulation show that the proposed control strategy and the optimal control algorithm can not only guarantee the efficient usage of inverter's active and reactive power, but also realize the optimization of network loss.

Keywords: Rural power grid · Active and reactive power coupling
Overvoltage · Multi-objective particle swarm optimization algorithm

1 Introduction

With the depletion of energy sources and the increasing destruction of the environment, the development and use of clean energy has become the mainstream trend of the society, and PV is a typical representative of clean energy [1]. With the country's strong support for the use of PV, rural PV will become the mainstream trend in the future. However, the increasing of PV will cause a serious impact for the stable operation of rural networks [2].

The rural networks have the characteristics of long power supply lines, wide distribution areas, small and dispersed loads, strong seasonal power consumption, low utilization rate of equipment and weak grid structures [3]. Therefore, with the connection of a large number of PV, it will lead to more electrical energy into the rural networks. And it will cause a phenomenon that the power was reversed to the network and create the risk of overvoltage, which will bring a great risk to the safe and stable operation of the rural networks [4, 5]

© Springer Nature Switzerland AG 2018
L. Cheng et al. (Eds.): ICONIP 2018, LNCS 11307, pp. 327–338, 2018.
https://doi.org/10.1007/978-3-030-04239-4_29

In order to solve the problem of overvoltage caused by PVs, some solutions have been proposed to solve the overvoltage. The reference [6, 7] proposed that used the reactor to compensate the node voltage, but this solution had the problems of large-transient conflict, slow response and possible system resonance. In reference [8, 9], it was proposed to adjust the voltage by changing the tap of the transformer, but the adjustment method was too simple and the flexibility was not enough, in addition the frequent adjustment of the transformer might shorten the service life of the transformer. In reference [10], the energy storage system was proposed to store the excess energy of PV, but the cost of energy storage equipment was too high, which was not economical for users.

In recent years, some scholars have made some breakthroughs in the control of photovoltaic inverter. In reference [11–13], an overvoltage solution of active power reduction in PV high-permeability and low-voltage feeders was proposed. However, by reducing the output of PV to achieve voltage control, serious abandonment will occur, which results in some unnecessary economic losses. With the improvement of inverter technology, more and more photovoltaic inverters have the ability of reactive power regulation. Compared with the regulation methods of active power reduction, the reactive power regulation of the inverter will have a greater economic advantage. The reference [14] uses the BP network in artificial neural networks to optimize the reactive voltage of increasingly complex electric power systems. And it can avoid the problem of solving multivariable nonlinear mixed constraint equations. In reference [15], a scheme of reactive power control based on droop control was proposed, in which segmented the droop curve and reduced the flow of reactive power in the circuit and helped to reduce the network loss. The reference [16] proposed a new adaptive neural network DC/AC converter-based control. Aiming at voltage control and reactive power compensation, explore the effect of reliability and power quality, and propose coherent active and reactive power distribution in order to alleviate this problem.

However, the above control schemes are based on the sufficient reactive power capacity of the inverter, and don't consider the situation when the reactive power capacity is insufficient. Because of the large impedance ratio of rural distribution lines, reactive power has a weak effect on voltage. In order to solve these problems, the mechanism of active power and reactive power on voltage in rural network is studied at first. On this basis, a voltage control strategy based on coupling of active and reactive power is proposed. And combined with the control strategy proposed in this paper, a multi-objective particle swarm optimization algorithm is further adopted to the control strategy.

2   Analysis of the Mechanism of Voltage Action in Rural Power Network

2.1   Analysis of Reactive Power Capacity of PV

PV system has the advantages of no environmental and noise pollution, unlimited use of geographical environment, long service life, easy maintenance and construction. The grid-connected photovoltaic system consists of photovoltaic panels, a DC/DC converter and a DC/AC inverter. Figure 1 is a typical grid-connected photovoltaic system structure. The active power of PV can be adjusted by adjusting the DC bus voltage at the outlet of the DC/DC converter.

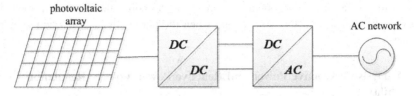

Fig. 1. Topological structure of grid-connected PV system

The relationship between adjustable reactive power capacity and inverter capacity is as follows [17]:

$$Q_{PVOUT} = \sqrt{S^2 - P_{PVOUT}^2},\qquad(1)$$

In the formula (1), Q_{PVOUT} is reactive power capacity of the inverter, P_{PVOUT} is the actual output power of photovoltaic, and S is the capacity of the inverter, which is generally taken as 110% of rated power [18]. According to the maximum inverter capacity of 110%, the inverter's controllable reactive power capacity is 46% of the rated power. It can be seen that the inverter can provide reactive power to adjust the voltage. However, due to the large impedance ratio of rural distribution lines, the effect of reactive power on voltage is weak. So the reactive power capacity is not enough to adjust overvoltage to a reasonable range, so it is necessary to use some active power to prevent overvoltage.

According to formula (1), it is known that the relationship between active power and reactive power of inverter can be expressed by Fig. 2.

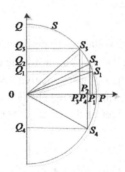

Fig. 2. Diagram of relationship between active and reactive power of inverter

It can be seen from Fig. 2 that the point on the semicircle is the capacity of the inverter. When the point is on S_1 position, the actual output is active power P_1 of the inverter, and Q_1 is the available reactive power capacity of the inverter. When the reactive power capacity is insufficient, the inverter reduces the active power from P_1 to P_2, and increases its reactive power capacity $Q_2 - Q_1$, and so on until the voltage is adjusted to a safe and stable operating range. On S_4 point, the inverter absorbs the reactive power to prevent the low voltage when the photovoltaic generation does not produce power.

2.2 Analysis of Reactive Power and Reactive Power Voltage Regulation Ability

From power flow calculation of the power system, the commonly used Newton-Raphson power flow calculation in the distribution network satisfies the following equation:

$$\begin{bmatrix} \Delta P \\ \Delta Q \end{bmatrix} = \begin{bmatrix} H & N \\ M & L \end{bmatrix} \begin{bmatrix} \Delta\theta \\ \Delta V/V \end{bmatrix}, \tag{2}$$

It is possible to change the matrix above by:

$$\begin{bmatrix} \Delta\theta \\ \Delta V \end{bmatrix} = \begin{bmatrix} S^{\theta-P} & S^{\theta-Q} \\ S^{V-P} & S^{V-Q} \end{bmatrix} \begin{bmatrix} \Delta P \\ \Delta Q \end{bmatrix}, \tag{3}$$

Thus, for a system with N PQ nodes, the voltage variation ΔV is related to the active power variation ΔP and the reactive power variation ΔQ:

$$\Delta V = S^{V-P}\Delta P + S^{V-Q}\Delta Q, \tag{4}$$

The above formula can also be expressed as:

$$\Delta V_i = \sum_{j=1}^{n} S_{ij}^{V-P}\Delta P_j + \sum_{j=1}^{n} S_{ij}^{V-Q}\Delta Q_j, \tag{5}$$

In the expression, S_{ij}^{V-P} is the voltage-active power sensitivity of node i to node j, that is, the change in voltage of node i is caused by the change of unit active power of node j. S_{ij}^{V-Q} is the voltage-reactive power sensitivity of node i to node j, that is, the change in the voltage of node i is caused by the change of unit reactive power of node j. ΔP_j is the variation of active power at the node j. ΔQ_j is the change of the reactive power at the node j.

3 Photovoltaic Inverter Active and Reactive Coupling Control Strategy

By formula (5), it can be seen that the change in the active power and reactive power of each node on the line will cause the voltage variation of the node itself and other nodes. Therefore, the voltage variation of the node can be controlled by controlling the active power and reactive power of the inverter.

In rural networks, the line exhibits a high-impedance line, that is, the unit resistance of the line is much larger than the unit reactance. Comparing the reduction of the PV's active power and the adjustment of the PV's reactive power, the reduction of the PV's active power has a greater advantage (in terms of regulating voltage). However, adjusting the inverter's reactive power has greater advantages (in terms of economy). Photovoltaics are dispersedly connected to each user in rural power grids. When the PV is insufficient, the power flow's direction on the line is from the head end to the tail end. When the PV is sufficient and surplus, the power on the line flows from the user to the first node, and the line voltage increases gradually from the first node. It can be seen that the phenomenon of overvoltage is most likely to occur in the end nodes of rural power network. Therefore, controlling the terminal node's voltage on the line can ensure that the voltage of each node on the line work within the normal range [19].

Therefore, the control idea of this paper is to regulate reactive power and active power to prevent overvoltage. The adjustment process is shown in Fig. 3:

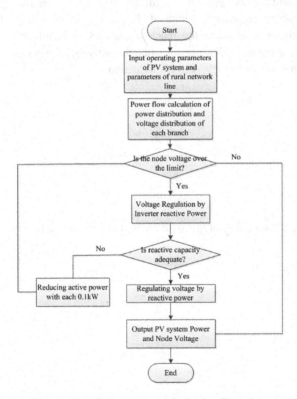

Fig. 3. Sensitivity power optimization flow chart

3.1 Control Strategy in the Case of Sufficient Reactive Power Capacity

Figure 4(a) shows the node voltage control curve was adopted by the rural networks when the reactive power capacity is sufficient. In the figure, Q is the reactive output capacity of photovoltaic. V^1_{cri} is the threshold of reactive initial control when the voltage starts to rise. When the node voltage is greater than V^1_{cri}, PVs will absorb reactive power from the network to prevent overvoltage, and when the node voltage is less than V^1_{cri}, the reactive power output of the inverter is zero. V^2_{cri} is the voltage's boundary threshold corresponding to the maximum reactive output of the inverter, and $-Q_{max}$ is the maximum reactive power when the PVs are operated at rated power.

3.2   Control Strategy in the Case of Insufficient Reactive Power Capacity

When the reactive power capacity of the inverter reaches the maximum reactive power, the node voltage is still rising. In this case, the voltage cannot be controlled by adjusting the reactive power of the photovoltaic generation, and the overvoltage can only be prevented by reducing the active power of the photovoltaic generation. the reduction of PV can further release the inverter's reactive power control capability.

Figure 4(b) shows the active and reactive coupling control curve under the condition of insufficient reactive capacity. In the figure, P_{MPPT} is the maximum output of PV when the reactive capacity is sufficient. P_{cut} is the actual output of PV after the reduction of active power when the reactive capacity is insufficient; $-Q^{add}_{max}$ is the increase of reactive power of the inverter when the active power is reduced to P_{cut}. V^3_{cri} is the value corresponding to PV active's reduction.

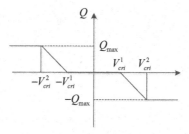

(a) Control strategy with sufficient reactive capacity

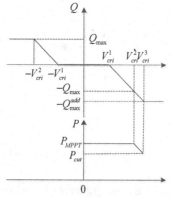

(b) Control strategy with sufficient reactive capacity

Fig. 4. Active/reactive power control strategy of inverter

3.3   Multi-objective Optimization Algorithm

Based on the above analysis, this paper optimizes the proposed control strategy based on economics.

Objective function:

$$\min P_{loss} = \Delta P_{loss,1} + \Delta P_{loss,2} + \cdots + \Delta P_{loss,n} = \sum_{i=1}^{n} R_i \frac{P_i^2 + Q_i^2}{V_i^2}, \tag{6}$$

$$\min \Delta V = \Delta V_1 + \Delta V_2 + \cdots + \Delta V_n = \sum_{i=1}^{n} \left(\sum_{j=1}^{n} S_{ij}^{V-P} \Delta P_j + \sum_{j=1}^{n} S_{ij}^{V-Q} \Delta Q_j \right), \tag{7}$$

Where, n is the total number of rural network's branches. R_i is the resistance of the ith branch, Ω. P_i is the active power which flows at the end of the ith branch, kW. Q_i is the reactive power which flows at the end of the ith branch, kVAr. V_i is the voltage at the end of the ith branch.

Restrictions:

$$\begin{cases} P_{Gi} + P_{DGi} - P_{Di} = U_i \sum_{j=1}^{n} U_j (G_{ij} \cos \theta_{ij} + B_{ij} \sin \theta_{ij}) \\ Q_{Gi} + Q_{DGi} - Q_{Di} = U_i \sum_{j=1}^{n} U_j (G_{ij} \sin \theta_{ij} - B_{ij} \cos \theta_{ij}) \end{cases}, \tag{8}$$

Formula (8) is the active and reactive power flow equation, θ_{ij} is the phase difference between node i and node j. P_{Gi}, P_{DGi} and P_{Di} are the active power output of the generator, the active output of PVs, and the active load at node i. Q_{Gi}, Q_{DGi} and Q_{Di} are the reactive power output of the generator, the reactive power generated by PVs, and the reactive power load at node i.

$$0 \leq P_{DGi} \leq P_{DGmax}, \tag{9}$$

Equation (9) represents the constraint of the active output P_{DGi} of the ith PV. Among them, P_{DGmax} is the maximum active output of PVs.

4   Simulation Study

By MATLAB, this paper uses a 220 low-voltage single-phase distribution feeder for simulation. As shown in Fig. 5, the system includes a distribution transformer and a low-voltage feeder. There are 8 nodes on the feeder, that is, 8 users installed with PV. The basic parameters of the line are shown in Table 1.

Table 1. Line basic parameter settings.

Line length (m)	40
Line type	LGJ-25/4
Headend voltage (V)	220

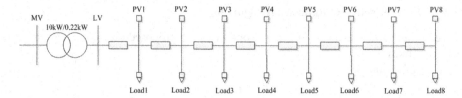

Fig. 5. Feeder diagram of low voltage distribution in rural power network

In the rural network, the geographical distance between the user and the user is relatively close, just like the assumed distance of Fig. 5, so its external environment is close. Therefore, PV production among users is basically the same, and assumed that each PV production's Rated power is 5 kW.

When the PV production is zero and the PV is operating at the rated power, the PV curve is shown in Fig. 6. From the figure, it can be seen that the voltage of each user on the line gradually decreases from the head to the end when the PV production is zero. when the PV operates at the rated power, PV's curve shows an increasing trend from point to point, and the node voltage is the worst case at the end. According to the notice issued by the State Grid Corporation of China about the issue of distributed power grid interconnection opinions and specifications, the upper and lower limits of the rural network's voltage are +5% and –10%, respectively, i.e. the upper limit is 231 V. As can be seen from Fig. 6, the voltage has started to appear overvoltage phenomenon since the second user. The voltage at the end of the line can reach 248 V, which seriously exceed the voltage's stable operation range specified by the state.

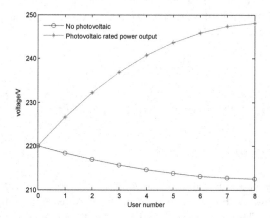

Fig. 6. Voltage at each point of the line without photovoltaic and full photovoltaic transmission

Using the above active and reactive power coupling control strategy, the over-voltage's nodes in Fig. 6 can be controlled. The control chart is shown in Fig. 7.

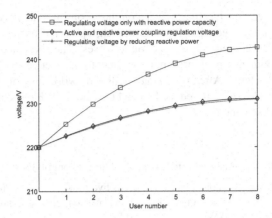

Fig. 7. Voltage at different points of the Line under different control schemes

As shown in Fig. 7, when only using the reactive capacity of the inverter to adjust voltage, it can be seen that although the voltage at each point of the line is reduced, there is still an overvoltage phenomenon on the line, which starts from the third user, the maximum voltage is 242.7 V. In view of the shortage of reactive power capacity of the inverter, the active and reactive power coupling control strategy is proposed in this paper. As can be seen from Fig. 7, compared with the method of controlling reactive power of inverter, the method proposed in this paper can solve the problem of overvoltage very well. Although the line voltage is still on the increasing trend, the maximum voltage on the line is 231 V, which does not reach the upper limit of voltage stable operation. Compared with the method of only reducing the active power of the inverter, the method proposed in this paper can reduce the power of all the inverters by the same value and ensure the economy at the same time, which can ensure that the income of each user is basically the same.

Table 2 lists the network loss and PV's active power reduction and reactive power compensation under different control schemes. From the table, it can be seen that its network loss increases with the access of PVs, which is due to the reverse power flow caused by a large amount of PVs. So it causes the increase of network loss, and The more PV power is connected, the larger the network loss will be. When PV is not connected (scheme 1), the network loss is 0.191 kW. when PV is connected without any control (scheme 2), PVs are outputting by MPPT, and the network loss is 3.021 kW, although the network loss of scheme 2 is smaller than scheme 1, it doesn't solve the overvoltage problem. In the comparison of scheme 3, scheme 4, scheme 5, although the network loss of scheme 4 (only reducing active power) is relatively small, which is only 0.429 kW, but the amount of active power reduction is relatively large, reaching 20 kW, and the power factor of PV is relatively small, it is only 0.454. Compared the method of only controlling reactive power (scheme 3) with the active and reactive power coupling control strategy (scheme 5) proposed in this paper, the network loss of scheme 3 is smaller, but the reactive power capacity is insufficient, so the expected control purpose cannot be achieved. Compared with scheme 4 and scheme 5, it can be seen that the network loss caused by only reducing active power is lower than the active and reactive power coupling control strategy proposed in this paper, but the amount of reducing PV is too much, up to half of the maximum active power of PV, which caused serious economic losses. The presented multi-objective particle swarm optimization algorithm is applied to the active and reactive power coupling control strategy. After optimization, the network loss of the scheme 6 is obviously suppressed, which is reduced by 1.11% and 2.49% respectively compared with scheme 3 and scheme 5.

Table 2. Analysis of results under different control Scenario.

Control scheme	Network loss (kW)	Photovoltaic reduction (kW)	Reactive power compensation (kW)	Photovoltaic power factor
Scheme 1	0.191	0	0	\
Scheme 2	3.021	0	0	0.91
Scheme 3	4.06	0	18.4	0.91
Scheme 4	0.429	20	0	0.454
Scheme 5	4.55	32	31.02	0.727
Scheme 6	4.015	32	31.02	0.727

From the above control results, it can be seen that the proposed active and reactive power coupling control strategy can effectively control the node voltage within the safe and stable operation range. By multi-objective particle swarm optimization algorithm, the proposed active and reactive power coupling control strategy can be optimized to avoid unnecessary reactive power flow in the network, thereby reducing the network loss and ensuring the economic operation of the network.

5   Conclusion

The work of this paper mainly includes the following aspects:

(1) Combining the line parameters and topology of the rural network, the relationship between active power or reactive power and voltage in the rural network is studied, and the expression between voltage-active and voltage-reactive is given.
(2) Aiming at the previous methods of only controlling the reactive power of the inverter and only reducing the active power, a control strategy based on the coupling of active and reactive power is proposed in this paper.
(3) Based on the above research, a multi-objective particle swarm optimization algorithm is adopted to ensure the economic operation of the network on the basis of the effectiveness of the voltage control in the agricultural network.

In future work, it will be interesting to study the effect of photovoltaic power forecasting for reducing active power. Because in this paper, the proposed control strategy doesn't consider the effect of frequent switch between active and reactive power on the inverter. So forecasted by the active power to be reduced, it can reduce the amount of switch and reduce the impact on the inverter. It can accurately reduce the active power and utilize the reactive power.

References

1. Worthmann, K., Braun, P., et al.: Distributed and decentralized control of residential energy systems incorporating battery storage. IEEE Trans. Smart Grid 6(4), 1914–1923 (2015)
2. Fan, Y., Zhao, B., Jiang, Q., Cao, Y.: Peak capacity calculation of distributed photovoltaic source with constraint of over-voltage. Autom. Electr. Power Syst. 36(17), 40–44 (2012)
3. Sun, Y., Zhang, L., Liu, D.: Research on the consumption capability of distributed photovoltaic access in rural areas. Zhejiang Electr. Power 36(11), 45–50 (2017)
4. Wang, Y.: Influence of grid-connected photovoltaic generation system on feeder voltage of rural power grid and its protection. Technol. Wind (24), 169 (2017)
5. Yang, C.: Study on influence of grid-connected photovoltaic distributed generation system on rural network feeder voltage and self-protection. Sci. Technol. Inf. 12(10), 102 (2014)
6. Xu, X., Huang, Y., Liu, C., Wang, W.: Influence of distributed photovoltaic generation on voltage in distribution network and solution of voltage beyond limits. Power Syst. Technol. 34(10), 140–146 (2010)
7. Chen, X., Zhang, Y., Huang, X.: Review of reactive power and voltage control method in the background of active distribution network. Autom. Electr. Power Syst. 40(1), 143–151 (2016)
8. Long, C., Procopiou, A.T., et al.: Performance of OLTC-based control strategies for LV networks with photovoltaics. In: 2015 IEEE Power & Energy Society General Meeting, pp. 1–5. IEEE Press, Denver (2015)
9. Choi, J.H., Kim, J.C.: Advanced voltage regulation method of power distribution systems interconnected with dispersed storage and generation systems. IEEE Trans. Power Delivery 16(2), 329–334 (2001)
10. Zhao, B., Wei, L., Xu, Z., et al.: Photovoltaic accommodation capacity determination of actual feeder based on stochastic scenarios analysis with storage system considered. Autom. Electr. Power Syst. 39(9), 34–40 (2015)
11. Reinaldo, T., Luiz, A., Lopes, C., Tarek, H.M.: Coordinated active power curtailment of grid connected PV inverters for overvoltage prevention. IEEE Trans. Sustain. Energ. 2(2), 139–147 (2011)
12. Tonkoski, R., Lopes, L.A.C.: Impact of active power curtailment on overvoltage prevention and energy production of PV inverters connected to low voltage residential feeders. Renewable Energy 36(12), 3566–3574 (2011)
13. Wai, K.Y., Havas, L., Overend, E., et al.: Neural network-based active power curtailment for overvoltage prevention in low voltage feeders. Expert Syst. Appl. 41(4), 1063–1070 (2014)
14. Xiu, W., Xia, L., Hai, L.: Application of artificial neural network in reactive voltage optimization of power system. J. Shenyang Agric. Univ. 39(06), 713–717 (2008)
15. Demirok, E., Gonzalez, P.C., Frederiksen, K.H.B., et al.: Local reactive power control methods for overvoltage prevention of distributed solar inverters in low-voltage grids. IEEE J. Photovoltaics 1(2), 174–182 (2011)
16. Xiong, C., Li, C., Yang, L., et al.: Study of reactive power compensation based on neural network. In: Proceedings of the 35th Chinese Control Conference, pp. 323–326. IEEE Press, Chengdu (2016)

17. Kabiri, R., Holmes, D.G., McGrath, B.P., et al.: LV grid voltage regulation using transformer electronic tap changing with PV inverter reactive power injection. IEEE J. Emerg. Sel. Top. Power Electron. **3**(4), 1182–1192 (2015)

18. Alam, M.J.E., Muttaqi, K.M., Sutanto, D.: A multi-mode control strategy for VAr support by solar PV inverters in distribution networks. IEEE Trans. Power Syst. **30**(3), 1316–1326 (2015)

19. Weckx, S., Driesen, J.: Optimal local reactive power control by PV inverters. IEEE Trans. Sustain. Energy **7**(4), 1624–1633 (2016)

A Neural Network Compensation Technique for an Inertia Estimation Error of a Time-Delayed Controller for a Robot Manipulator

Seul Jung[✉]

Intelligent Systems and Emotional Engineering (ISEE) Laboratory,
Department of Mechatronics Engineering, Chungnam National University,
99 Daehak-ro, Yuseong-gu, Daejeon 34134, Korea
jungs@cnu.ac.kr
http://isee.cnu.ac.kr

Abstract. In this paper, a neural network is added to compensate for the deviation error of an estimated inertia matrix of the time-delayed controller for a robot manipulator. The time-delayed control (TDC) method is known as a simple and practical control method for controlling robot manipulators. The previously sampled information is used to cancel uncertainties for the current control using a time- delay. One of the problems of TDC is the constant inertia selected for simplicity and how to deal with the error of the inertia model estimation. In this paper, a neural network is used to compensate for the deviated inertia error. Simulation studies of position tracking control performances of a three link rotary robot are presented.

Keywords: Robot manipulator · Neural network · Time-delayed control

1 Introduction

Recently, research and development of small and compact manipulators are demanded for conducting the cooperation tasks in the smart manufacturing cell in industries, which is a core of Industry 4.0. The size of robot arms is minimized and the safety protection mechanisms are added for the cooperating task in the small workspace. A pair of robot arms is required to perform sophisticated tasks such as part assembly, peg in hole, and part dismantling.

Commercial robot arms such as Baxter [1], Ridgeback [2], RB-1 [3], Fetch [4], TIAGo [5], and AUBO [6] are available in the market and designed as a size of a human arm for the task although the cost is still high. Some of them are equipped with compliant joints for the safety as well as rigid bodies for the accurate position control task.

Many industrial robot manipulators still use a simple PID (proportional–integral-derivative) control with the gravity compensation for the joint control algorithm. Since PID control methods are simple and practical, the methods are still used dominantly in the industries. However, compact and compliant robot arms have to deal with many sophisticated tasks such as contact force control in the constrained space.

© Springer Nature Switzerland AG 2018
L. Cheng et al. (Eds.): ICONIP 2018, LNCS 11307, pp. 339–346, 2018.
https://doi.org/10.1007/978-3-030-04239-4_30

Control algorithms for those cooperative robots should deal with disturbances to perform accurate position tracking control. Many advanced control algorithms for robot manipulators have been developed in the literature.

The time-delayed control (TDC) method is one of the robust control algorithms dealing with disturbances from either internal and external input. The concept of TDC is quite simple that it can be easily realized in practice [7]. The advanced TDC methods with the sliding mode control method have been presented to improve the performance [8, 9]. TDC has been used as a robust controller for force control application of a robot manipulator as well [10].

However, there are three problems to overcome for implementing the TDC scheme. Firstly, the fast sampling time is required since it uses the previous information. The sampling time should be fast enough to make the control action be continuous. Secondly, an accurate estimation of acceleration signals is required. Lastly, the constant values of an inertia matrix should be selected with care to satisfy both the performance and the stability [7]. The sampling issue can be solved with the state-of-the-art hardware technology. The problem of the acceleration estimation can be solved by the direct measurement of signals using sensors such as accelerometers or a rate gyro directly. The last problem of selecting the inertia matrix to satisfy the stability still remains.

Therefore, in this paper, a neural network technique is applied to compensate for the deviated error between the real and the estimated inertia matrices in the TDC framework. This scheme is called a neural network TDC scheme (NN-TDC). The purpose of using a neural network is to generate the compensating signals for the inertia matrix in association with TDC to improve the tracking control performance. Train and control of the neural network compensator are done in on-line fashion. Learning algorithm of neural network is derived. Simulation studies of controlling a three link rotary robot manipulator are presented to confirm the performance of the proposed control method.

2 Robot Manipulator Dynamics for TDC

The robot dynamic equation is defined as

$$D\ddot{q} + C + G + \tau_u = \tau, \tag{1}$$

where q is the n × 1 joint angle vector, D is the n × n inertia matrix, C is the n × 1 Coriolis and centrifugal force, G is the n × 1 gravity force, τ is the n × 1 input torque, and τ_u is the n × 1 other unmodeled torque.

Equation (1) can be modified as

$$\bar{D}\ddot{q} + D\ddot{q} - \bar{D}\ddot{q} + C + G + \tau_u = \tau, \tag{2}$$

where the constant inertia matrix $\bar{D} = \alpha I$, α is a constant. Then Eq. (2) can be denoted as a new dynamic equation of a robot manipulator. The dynamic equation forms with a constant inertia matrix and all other dynamic terms as

$$\bar{D}\ddot{q} + h = \tau, \tag{3}$$

where h includes all the terms except an inertial force such as $h = D\ddot{q} - \bar{D}\ddot{q} + C + G + \tau_u$.

From (3), the model h can be represented as

$$h(t) = \tau(t) - \bar{D}\ddot{q}(t). \tag{4}$$

Since $h(t)$ is not available at the same sampling time, it can be obtained by using the delayed information as $\bar{h}(t)$

$$\bar{h}(t) = h(t - \lambda) = \tau(t - \lambda) - \bar{D}\ddot{q}(t - \lambda), \tag{5}$$

where λ is the time delay.

Then a new TDC law becomes

$$\tau = \bar{D}u + \bar{h}. \tag{6}$$

The control input u is represented as PD type control.

$$u = \ddot{q}_d + K_D\dot{e} + K_P e, \tag{7}$$

where K_D, K_P are control gains, q_d is the n × 1 desired joint angle vector, and the tracking error becomes $e = q_d - q$.

The combined control law from (5), (6), and (7) becomes

$$\tau(t) = \bar{D}(\ddot{q}_d + K_D\dot{e} + K_P e) + \tau(t - \lambda) - \bar{D}\ddot{q}(t - \lambda). \tag{8}$$

To implement TDC, we need $\bar{D}$, $\ddot{q}(t - \lambda)$ and the sampling time λ in (8). The sampling time and the acceleration estimation can be done by the selection of suitable hardware, but the inertia problem still remains on the control designer. Therefore, in this paper, our goal is to remedy the inertia problem in association with $\bar{D}$.

Figure 1 shows the TDC block diagram to implement (8).

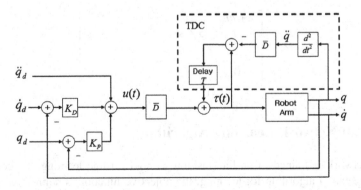

Fig. 1. TDC block diagram

3   Neural Network Compensator

In the framework of TDC, a neural network is added as an auxiliary controller to compensate for the deviated error in the inertia matrix to improve the control performance. Figure 2 shows the NN-TDC block diagram. Neural network outputs are added to the diagonal elements of the inertia matrix such as

$$d_{ii} = \alpha + dn_i \tag{9}$$

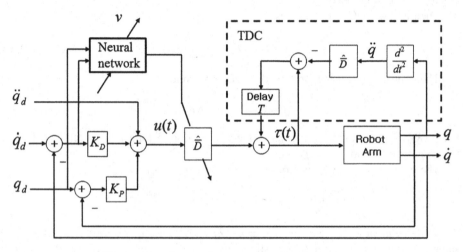

Fig. 2. Neural network compensator for TDC scheme

where d_{ii} is the diagonal element of $\bar{D}$, α is a constant, and dn_i is the *ith* output of neural network. The updated inertia matrix becomes

$$\hat{\bar{D}} = \begin{bmatrix} \alpha + dn_1 & 0 & \cdots & 0 \\ 0 & \alpha + dn_2 & \cdots & 0 \\ \vdots & & \ddots & \vdots \\ 0 & 0 & \cdots & \alpha + dn_n \end{bmatrix} \tag{10}$$

4   Neural Network Learning Algorithm

Neural network is trained in on-line fashion without off-line learning. To derive the back-propagation algorithm for learning, the objective function is defined as

$$E = \frac{1}{2} v^T v, \tag{11}$$

where $v \in R^{n \times 1}$ is considered as the training signal.

The gradient can be obtained as

$$\Delta w = -\eta \frac{\partial E}{\partial w} = -\eta \frac{\partial E}{\partial v} \frac{\partial v}{\partial w}, \tag{12}$$

where η is the learning rate and w is the weight vector.

Note that we have the closed loop error equation from (3) and (6) as

$$\bar{D}\ddot{q} + h = \hat{\bar{D}}u + \overline{h}. \tag{13}$$

Arranging (13) yields

$$\bar{D}v = D_n u + \Delta h. \tag{14}$$

where D_n is the neural network compensation signal, $v = \ddot{e} + K_D \dot{e} + K_P e$ and $\Delta h = h - \bar{h}$.

The gradient to complete (12) can be obtained from (14) as

$$\Delta w = \eta \bar{D}^{-1} v^T \frac{\partial D_n}{\partial w} u \tag{15}$$

Weights are updated at every sampling time.

$$w(t+1) = w(t) + \Delta w(t) + \beta \Delta w(t-1) \tag{16}$$

where β is the momentum constant.

Figure 3 shows the neural network structure. It has one input- one hidden layer, and one output layer. The output layer is linear while the hidden layer is nonlinear.

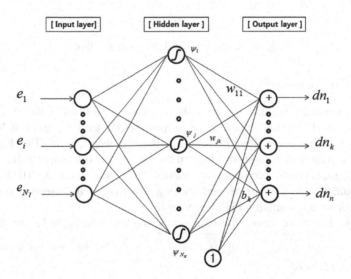

Fig. 3. Neural network structure

5   Simulation Studies

In this section, we verify the proposed control method by simulation studies. The circular trajectory tracking tasks of a 3 link robot manipulator are conducted and the performances are compared. The circular trajectory is slanted so that all directional component can be tested. Robot parameters for the simulation are copied from the first three joints of the PUMA robot.

Two control schemes are tested: one is TDC and another is NN-TDC. PD controller gains are set to $K_D = 20, K_P = 100$ for both schemes. The sampling time is set to 0.001 s.

5.1   TDC Scheme

Figure 4 shows the tracking result of the circular trajectory by TDC. We select the constant inertial value as $\alpha = 0.02$.

We see from Fig. 4 that the tracking error is notable. This error is caused from the ill estimation of a contant inertia value.

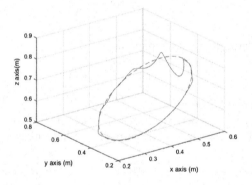

Fig. 4.   Performance of TDC when $\alpha = 0.02$

5.2   NN-TDC Scheme

In order to minimize the tracking error shown in Fig. 4, a neural network is used to compensate for the inertial values. Figure 5(a) shows the tracking result of NN-TDC. We clearly see that the tracking performance is better than that of TDC. Figure 5(b) shows the corresponding joint tracking errors. Neural network output values are also plotted in Fig. 5(c) and the compensating values at the convergence are 0.0740, 0.1809, 0.0991. Note that the patterns of joint tracking errors and neural network outputs are quite similar when comparing Fig. 5(b) with (c).

For the clear comparison, the total errors are listed in Table 1. The errors are calculated as $Es = \sqrt{(x_d - x)^2 + (y_d - y)^2 + (z_d - z)^2}$. We see that the error of NN-TDC is much smaller.

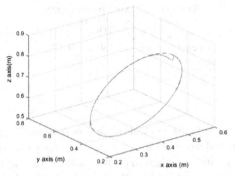

(a) Circular tracking performace

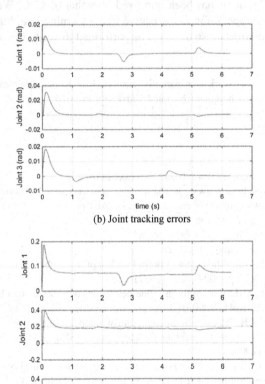

(b) Joint tracking errors

(c) Neural network outputs

Fig. 5. Performance of NN-TDC

Table 1. Tracking performances

	TDC	NN-TDC
Tracking errors, Es	1.4717	0.4196

6 Conclusions

In this paper, the neural network control scheme is presented to improve the control performance of the time-delayed control scheme for robot manipulators. Neural network adapts the constant inertia values in on-line fashion. Learning algorithm has been derived for on-line learning and control. We have confirmed that the tracking performance by NN-TDC scheme has been improved over that of TDC. We also found that the neural network compensator was sensitive to the learning rate. Stability proof will be analyzed and experimental studies will be performed to verify the proposal on a real robot manipulator.

Acknowledgements. This work was partially supported by Korea Environmental Industry & Technology Institute and Chungnam National University in 2018.

References

1. Rethink Robotics. https://www.rethinkrobotics.com/baxter/
2. Clearpath Robotics. https://www.clearpathrobotics.com/ridgeback-indoorrobot-platform/
3. Robotnik. https://www.robotnik.eu/manipulators-2/rb-one/
4. Fetch Robotics. https://fetchrobotics.com/
5. PAL Robotics. http://tiago.pal-robotics.com/
6. AUBO Rotics. https://aubo-robotics.com/
7. Hsia, T.C.: On a simplified joint controller design for robot manipulators. In: IEEE Conference on CDC, pp. 1024–1025 (1987)
8. Chang, P.H., Park, S.H.: On improving time-delay control under certain hard nonlinearities. Mechatronics **13**, 393–412 (2003)
9. Lee, S.U., Chang, P.H.: The development of anti-windup scheme for time delay control with switching action using integral sliding surface. Trans. ASME J. Dyn. Syst. Meas. Control **125**(4), 630–638 (2003)
10. Jung, S., Hsia, T.C., Bonitz, R.G.: Force tracking impedance control of robot manipulators under unknown environment. IEEE Trans. Control. Syst. Technol. **12**(3), 474–483 (2004)

Biomedical Applications

Estimating Criticality of Resting-State Phase Synchronization Network Based on EEG Source Signals

Li Zhang[1,2], Bo Shi[1], Mingna Cao[1], Sai Zhang[1], Yiming Dai[1], and Yanmei Zhu[2(✉)]

[1] School of Medical Imaging, Bengbu Medical College,
Bengbu 233030, Anhui, China
{shibo,mingna_mit,xmxu,daiyiming}@bbmc.edu.cn
[2] Research Center for Learning Science, Southeast University,
Nanjing 210096, Jiangsu, China
{li_zhang,zhuyanmei}@seu.edu.cn

Abstract. EEG phase synchrony is an important signature in estimating functional connectivity of brain network, in which criticality of phase-locking state has been viewed as the key factor in facilitating dynamic reorganization of functional network. Based on source trace of resting-state EEG signals recorded from 24 subjects, this study extracted phase-locking intervals (PLIs) between pairwise source signals and constructed PLI sets with size higher than 10^5 for each subject, from frontal-parietal, frontal-temporal, and temporal-parietal cortical areas, respectively. Through further data fitting in power-law model, this study finds that θ- (4–8 Hz) and α-band (8–13 Hz) activities have longer phase-locking duration in a broader power-law distribution interval, compared to those in high frequency bands, indicating higher temporal stability of functional coupling between brain areas. In contrast, the probability density of PLIs oscillating in β (13–30 Hz) and γ (30–60 Hz) bands has less data fitting errors and bigger power-law exponent, suggesting higher criticality and flexibility of reorganization of phase synchronization networks. The findings are expected to provide effective neural signatures for comparison and recognition of neural correlations of cognition, emotion, disease etc. in the future.

Keywords: EEG resting-state source networks · Phase-locking interval
Power-law model · Criticality

1 Introduction

Electroencephalography (EEG) based phase synchronization is an important analytical approach to dynamic functional connectivity of brain network. Phase-locking activity between neuroelectric signals is particularly viewed as a representative measurement to obtain a statistical quantification of frequency-specific synchronization strength [1]. Previous study has suggested that the communication windows for input and output between phase-locked neuronal groups are open at the same time, which supports effective information communication [2]. Though further decomposing phase signals of

© Springer Nature Switzerland AG 2018
L. Cheng et al. (Eds.): ICONIP 2018, LNCS 11307, pp. 349–357, 2018.
https://doi.org/10.1007/978-3-030-04239-4_31

low-frequency EEG, Thatcher et al. have pointed out that spontaneous EEG phase synchronization is actually a mixture of alternative episodes of phase-locking and phase-shift duration over time course. While continuous phase-locks mean the emergence of functional connection between brain areas, phase-shifts mark the "disconnection" state and the beginning of a different set of connections in neuronal network, i.e., the occurrence of functional network reconfiguration [3].

By extracting phase-locking intervals (PLIs) from computer simulation data and fMRI signals respectively, Kitzbichler et al. have found that these episodic PLIs have non-uniform time length, with probability density conforming to power-law rule, which has been widely accepted as a typical empirical signature of non-equilibrium systems in self-organized critical state [4]. Particularly, Bassett et al. have found that small-world network of the brain is actually operated in a "metastable" connecting state, i.e., a highly critical edge of synchronizing activity in transit to desynchronization through "neuronal avalanche" [5, 6]. The criticality of synchronization has been considered to account for rapid network reorganization in response to endogenous perturbation and external event [4, 5]. However, previous fMRI study lacked temporal accuracy in assessing alternative change of transient phase-locking and phase-shift activities, and criticality estimation was restricted in low-frequency (0.05–0.45 Hz) oscillations of brain activity [4]. Therefore, EEG is a necessary approach to criticality assessment on multiband neuronal activities. In this study, we extracted PLIs from resting-state source EEG in different frequency bands and then fitted them to power-law model, to estimate and discuss the criticality of phase synchronizations between different cortical areas.

2 Materials and Methods

2.1 Participants

EEG data were recorded from 24 normal participants composed of 10 males and 14 females aged 22.4 ± 2.3 (mean $\pm$ SD), during eyes-open resting state with the time length of about 3–5 min. The study was approved by the Academic Committee of the Research Center for Learning Science, Southeast University, China. All the subjects have signed the informed consent. EEG data were recorded by 60-channel Neuroscan international 10–20 system, with a sampling rate of 500 Hz.

2.2 EEG Preprocessing and Cortical Source Estimation

For the raw EEG signals, the preprocessing procedure of the Scan 4.3 software included band-pass filtering, baseline correction, artifact rejection etc. After that, the EEG signals were transformed into source currents at the cortical surface with equal time length, through a cortical source estimation procedure in Brainstorm toolbox [7]. Through downsampling the original sources, 248 current dipoles covering frontal, parietal, temporal and occipital regions were generated (Fig. 1).

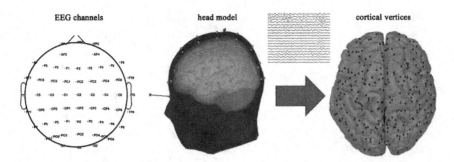

Fig. 1. EEG electrode placement of the Neuroscan international 10–20 system, head model and cortical vertices transformed through a source estimation procedure.

2.3 Extracting PLIs from Multi-frequency Cortical Source Signals

For each subject, the resting-state time series were then filtered into θ (4–8 Hz), α (8–12 Hz), β (12–30 Hz) and γ (30–60 Hz) frequency bands, respectively.

The analytical signal of a time series $x(t)$ is defined as

$$H(t) = x(t) + i\tilde{x}(t) \tag{1}$$

Here the Hilbert transform (HT) is used to obtain the phase of a signal, and $\tilde{x}(t)$ represents the HT of $x(t)$, which is obtained by

$$\tilde{x}(t) = \frac{1}{\pi} PV \int_{-\infty}^{\infty} \frac{x(t')}{t - t'} dt' \tag{2}$$

PV is the Cauchy principal value. The instantaneous phase of the signals $x(t)$ is

$$\varnothing_x(t) = arctan\frac{\tilde{x}(t)}{x(t)} \tag{3}$$

The phase $\varnothing_y(t)$ of a time series $y(t)$ with the same time length can be acquired as well

According to [4], PLI is defined as the time interval containing continuous phase-synchronized activity within a limited range, i.e., the phase difference $\Delta\varnothing_{xy}(t)$ between $x(t)$ and $y(t)$ satisfies the following condition at each time point:

$$\Delta\varnothing_{xy}(t) = |\varnothing_x(t) - \varnothing_y(t)| < const \tag{4}$$

where *const* represents a threshold between phase lock and phase shift. While this condition does not hold true, a phase-locking process is interrupted and replaced by phase shift.

Figure 2 exemplifies the alternative process of phase locks and phase shifts of a pair of source signals from frontal-parietal cortical vertices. By setting the threshold of phase difference as $\pi/4$ and counting the continuous time points of phase locks within the threshold, we can get multiple episodes of PLIs.

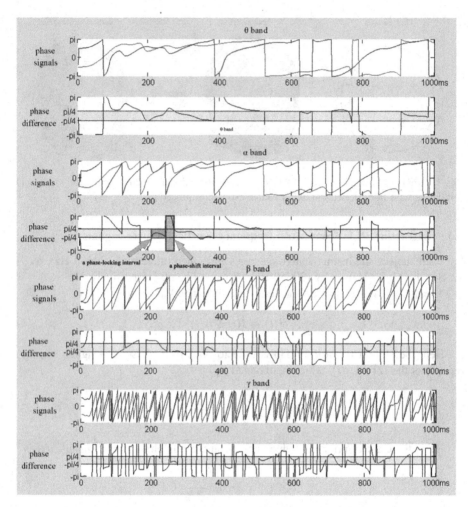

Fig. 2. An illustration of PLIs between a pair of frontal-parietal signals. The horizontal axis is the time course, and the vertical axis represents the phase signals of the two cortical signals and their phase difference respectively. The yellow boxes refer to the phase differences within threshold $\pi/4$. The red and blue boxes are examples of a PLI and a phase shift interval. (Color figure online)

2.4   Criticality Estimation of Resting-State Phase-Locking Duration

In each frequency band, we fitted the PLIs extracted between different brain areas to power-law model, which is an acknowledged indicator of self-organization criticality of dynamical system that has a critical point as an attractor [8].

It should be noted that the changes in inter-connections among frontal, parietal and temporal areas are especially crucial for self-reorganization of functional networks. Therefore, we selected 30 cortical vertices uniformly distributed in bilateral frontal area, 30 vertices in parietal area and 10 vertices at temporal lobes, to construct inter-node PLI sample sets by concatenating all the episodes of PLIs.

In this study, the parameter fitting method proposed by [9] was applied to each PLI dataset for quantifying its power-law distribution characteristic. Here x represents a set composed of discrete PLIs, with its probability density defined as

$$p(x) = P_r(X = x) = Cx^{-\alpha} \tag{5}$$

where X is the observed PLI value, C indicates a normalization constant, and α represents the power-law exponent. However, in natural phenomenon with power-law characteristic, not all values abide by the rule and only the values greater than a minimum value x_{min} conform to power-law distribution, so the first step is to estimate x_{min} for each data set. If the data in the region $x \geq x_{min}$ can accurately obey power-law rule, the scaling parameter α can be estimated accordingly. Particularly, we used the maximum likelihood estimator to approximately estimate α for the special case of $x_{min} = 1$, through the solution to the transcendental equation $\frac{\zeta'(\hat{\alpha})}{\zeta(\hat{\alpha})} = -\frac{1}{n}\sum_{i=1}^{n} \ln x_i$ (ζ is the Riemann zeta function). While $x_{min} > 1$, the generalized zeta $\frac{\zeta'(\hat{\alpha}, x_{min})}{\zeta(\hat{\alpha}, x_{min})} = -\frac{1}{n}\sum_{i=1}^{n} \ln x_i$ was used to replace the zeta function. Then α could be estimated by the maximum likelihood estimator for each x_{min}.

The Kolmogorov-Smirnov goodness-of-fit statistic is given by

$$D = max_{x \geq x_{min}} |S(x) - P(x)| \tag{6}$$

where $S(x)$ and $P(x)$ represent the cumulative distribution function of the observations and that of the data best fitting to power-law model in $x \geq x_{min}$, respectively. What gives the minimum value of D is the optimal estimation of x_{min}. To assess the goodness-of-fit of the power-law scaling, root-mean-square error (RMSE) indicated by $R_e = \sqrt{[\sum d_i^2 / n]}$ was computed, where d_i refers to the deviation between the observed and estimated values.

3 Results and Discussions

For each subject, relevant scaling parameters used for criticality estimation were obtained. Figure 3 illustrates the probability density distributions of 12 sample sets of PLIs from frontal-parietal, frontal-temporal and temporal-parietal areas in θ, α, β and γ frequency bands, from a randomly selected individual subject.

It can be seen that, after determining the minimum value x_{min} of PLIs, the cumulative probability density $P(x)$ in the region $x \geq x_{min}$ is very close to the distribution of $P(x) = Cx^{-\alpha}$, showing a dramatic fall-off in exponential form with the increase of PLI, especially in β and γ frequency bands.

The statistic bars in Fig. 4 give data concerning the basic scaling parameters of PLIs in power-law model. From the resting-state EEG source signals with the duration of 3–5 min, we extracted episodic PLIs more than 10^5 for each subject. Specifically, β- and γ-band PLIs show smaller estimated minimum $\hat{x}_{min}$ than those in θ- and α-band, while low-frequency oscillations show bigger mean PLIs. Besides, compared to θ- and α-band activities, β- and γ-band PLIs have less x_{max} and shorter interval in which data distribution conforms to power-law rule. Correspondingly, the cumulative probability density of high-frequency PLIs presents larger power-law exponent $\hat{\alpha}$, i.e., more significant decline of $P(x)$, which indicates lower probability of long phase-locking duration.

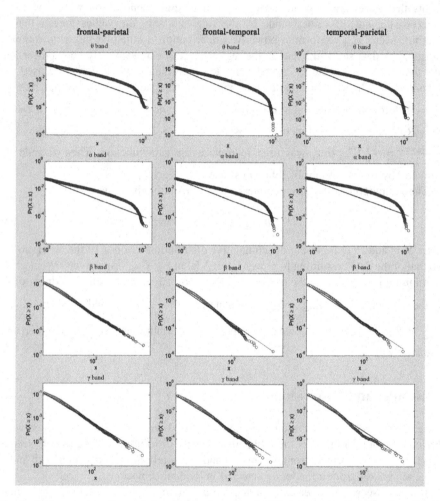

Fig. 3. Fitting curves of cumulative distribution of real PLIs ($x \geq x_{min}$), plotted on logarithmic axes. The horizontal axis is the time length of PLI and the vertical axis refers to its cumulative probability density. The black dotted line represents the ideal power-law distribution with the same estimated exponent $\hat{\alpha}$.

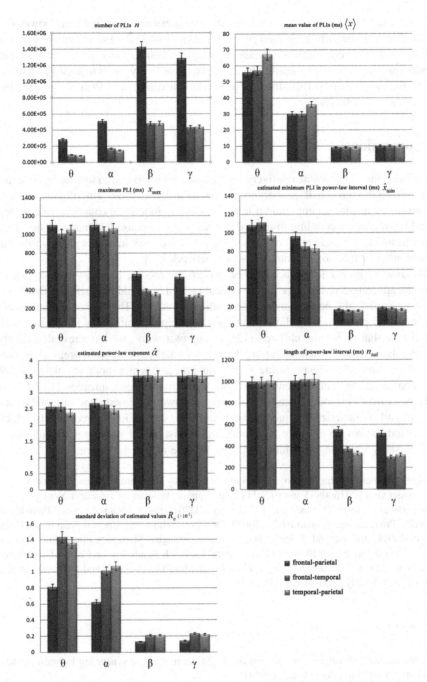

Fig. 4. Statistical bars of scaling parameters of PLIs in power-law model for 24 subjects. n-sample size; x-mean value of PLIs; x_{max}-maximum of PLI; $\hat{x}_{min}$-estimated minimum of PLIs in power-law distribution interval; $\hat{\alpha}$-estimated power-law exponent; $n_{tail} - [x_{min}, x_{max}]$; R_e-standard deviation of estimated values ($\times 10^{-2}$).

Additionally, there are smaller fitting errors between real data and ideal power-law distribution with the same estimated power-law exponent $\hat{\alpha}$ in β and γ frequency bands, indicating higher criticality of phase synchronization toward desynchronization. Besides, in goodness-of-fit statistic, the fitting errors are very small in each PLI sample set (<2%), suggesting the robustness of power-law distribution of PLIs across different brain areas and subjects.

4   Conclusions

The critical interval contains relatively long phase-locking duration among brain areas. Although taking up small proportion in the total PLIs, these long-duration phase synchronizations play an important role in maintaining functional coupling state among distant brain areas. According to critical dynamics, these phase-locking activities have been adjusted to a critical point of state transition, which would access phase shifts and redistribution of functional connections in a network [10].

Because of higher fitting accuracy in power-law model, β- and γ-band synchronizations are considered to have higher criticality, which would intensify the flexibility and adaptiveness of functional network reconfiguration [11]. Thatcher et al. have found longer α-band phase-locking duration at parietal-occipital brain areas in children with autism than that of normal children [12]. In our past work, we have revealed highly critical γ-band phase synchronizations between frontal-parietal brain areas in mathematically gifted adolescents during reasoning process, which both maintains stable functional coupling state and enhances lability of network reorganization [13]. In this study, we further extended parameter scaling of power-law mode and criticality estimation to all characterized frequency bands in resting-state EEG source signals, which is expected to provide effective references and robust neural signatures for functional network study involving cognition, emotion, disease etc. in the future.

Acknowledgements. This work was supported in part by the Natural Science Foundation of China under Grants 31600862 and 61501115, the Support Program of Excellent Young Talents in Universities of Anhui Province under Grant gxyqZD2017064, the Natural Science Foundation of Jiangsu Province under Grant BK20150633, the China Scholarship Council Fund under Grant 201808340011, the Natural Science Foundation of Bengbu Medical College under Grant BYKY1604ZD and BYKY1638, the Fundamental Research Funds for the Central Universities under Grant CDLS-2018-04, and Key Laboratory of Child Development and Learning Science (Southeast University), Ministry of Education.

References

1. Lachaux, J.P., Rodriguez, E., Martinerie, J.: Measuring phase synchrony in brain signals. Hum. Brain Mapp. **8**, 194–208 (1999)
2. Fries, P.: A mechanism for cognitive dynamics: neuronal communication through neuronal coherence. Trends Cogn. Sci. **9**, 474–480 (2005)
3. Thatcher, R.W.: Coherence, phase differences, phase shift, and phase lock in EEG/ERP analyses. Dev. Neuropsychol. **37**, 476–496 (2012)

4. Kitzbichler, M.G., Smith, M.L., Christensen, S.R.: Broadband criticality of human brain network synchronization. PLoS Comput. Biol. **5**, e1000314 (2009)
5. Werner, G.: Metastability, criticality and phase transitions in brain and its models. Biosyst. **90**, 496–508 (2007)
6. Bassett, D.S., Meyer-Lindenberg, A., Achard, S.: Adaptive reconfiguration of fractal small-world human brain functional networks. Proc. Natl. Acad. Sci. **103**, 19518–19523 (2006)
7. Tadel, F., Baillet, S., Mosher, J.C.: Brainstorm: a user-friendly application for MEG/EEG analysis. Comput. Intell. Neurosci. **2011**, 8 (2011)
8. Thatcher, R.W., North, D.M., Biver, C.J.: Self-organized criticality and the development of EEG phase reset. Hum. Brain Mapp. **30**, 553–574 (2009)
9. Clauset, A., Shalizi, C.R., Newman, M.E.J.: Power-law distributions in empirical data. SIAM Rev. **51**, 661–703 (2009)
10. Levina, A., Herrmann, J.M., Geisel, T.: Phase transitions towards criticality in a neural system with adaptive interactions. Phys. Rev. Lett. **102**, 118110 (2009)
11. Levina, A., Herrmann, J.M., Geisel, T.: Dynamical synapses causing self-organized criticality in neural networks. Nat. Phys. **3**, 857–860 (2007)
12. Thatcher, R.W., North, D.M., Neubrander, J.: Autism and EEG phase reset: deficient GABA mediated inhibition in thalamo-cortical circuits. Dev. Neuropsychol. **34**, 780–800 (2009)
13. Zhang, L., Gan, J.Q., Wang, H.: Optimized gamma synchronization enhances functional binding of fronto-parietal cortices in mathematically gifted adolescents during deductive reasoning. Front. Hum. Neurosci. **8**, 430 (2014)

A Spatio-Temporal Fully Convolutional Network for Breast Lesion Segmentation in DCE-MRI

Mingjian Chen[1], Hao Zheng[1], Changsheng Lu[2], Enmei Tu[1],
Jie Yang[1(✉)], and Nikola Kasabov[3]

[1] Institute of Image Processing and Pattern Recognition,
Shanghai Jiao Tong University, Shanghai, China
mjchensjtu@gmail.com, {zhenghaobs,jieyang}@sjtu.edu.cn,
hellotem@hotmail.com
[2] Key Laboratory of System Control and Information Processing,
Shanghai Jiao Tong University, Shanghai, China
ChangshengLuu@gmail.com
[3] Knowledge Engineering and Discovery Research Institute,
Auckland University of Technology, Auckland, New Zealand
nkasabov@aut.ac.nz

Abstract. Breast lesion segmentation result has a huge impact on the subsequent clinical analysis, and therefore it is of great importance for image-based diagnosis. In this paper, we propose a novel end-to-end network utilizing both spatial and temporal features for fully automated breast lesion segmentation from dynamic contrast-enhanced magnetic resonance imaging (DCE-MRI). Our network is based on a modified convolutional neural network and a recurrent neural network, and it is capable of unearthing rich spatio-temporal features. In our network, a multi-pathway structure and a fusion operator are introduced to acquire 3D information of different tissues, which is helpful for reducing false positive segmentation while boosting accuracy. Experimental results demonstrate that the proposed network produces a significantly more accurate result for lesion segmentation on our evaluation dataset, achieving 0.7588 dice coefficient and 0.7390 positive predictive value.

Keywords: Breast DCE-MRI · Lesion segmentation
Recurrent neural network · Convolutional neural network
Spatio-temporal features

1 Introduction

Breast cancer is one of the most common malignancies in the world [23], and early detection can significantly reduce mortality. Dynamic contrast-enhanced magnetic resonance imaging (DCE-MRI), which possesses more sensitivity than mammography for the early lesion detection in dense breasts [24], is an important

© Springer Nature Switzerland AG 2018
L. Cheng et al. (Eds.): ICONIP 2018, LNCS 11307, pp. 358–368, 2018.
https://doi.org/10.1007/978-3-030-04239-4_32

complementary imaging modality in the clinical examination of breast lesions [13]. A typical breast DCE-MRI study involves several series of images, containing dynamic contrast-enhanced series acquired before and after the injection of a contrast agent at different time points. In breast DCE-MRI, lesions and healthy tissues own different temporal representations, which are very useful in clinical analysis. However, it is a complex and time-consuming task for manual diagnosis due to the high dimensionality (3D over time) of DCE-MRI study [17]. In order to improve the efficiency of diagnosis, breast MRI computer-aided diagnosis (CAD) is developed, where the main tasks can be divided into detection, segmentation and classification of lesions. In this paper, we focus on breast lesion segmentation because of its direct influences on the classification results of suspicious lesions.

It still remains a challenging task for lesion segmentation due to the fact that breast lesions vary in size, texture and shape [27]. Traditional methods mainly depend on hand-crafted features [8,10] and classification algorithms. [18] extracts pixel-wise dynamic features for distinguishing suspicious pixels and true lesion pixels, yet neglects spatial correlations of breast lesions. [9] models dynamics of each voxel in the DCE-MRI by time-series analysis that heavily relies on DCE-MRI intensity and ignores texture and shape features of breast lesions. Recently, deep learning methods, such as convolutional neural networks (CNN), have achieved great successes in image classification [7,12], detection [21] and segmentation [15] without manually designed features. Several deep learning methods have been proposed for breast lesion detection and segmentation. [2] uses saliency analysis and CNN-based classification to reject false positive detections. [16] utilizes a globally optimal inference and multi-scale CNN semantic segmentation as shape prior to improve the accuracy of lesion segmentation. [25] uses a trained deep artificial neural network to acquire lesion segmentation by classifying overlapping tiles in breast DCE-MRI.

Although the CNN-based methods have achieved a considerable success in breast lesion segmentation, they do not make full use of the temporal information of breast lesion, which is widely used by radiologists and clinicians. Inspired by the fact that recurrent neural network (RNN) [6] specializes in processing temporal sequence data, we propose a novel end-to-end network for automatic breast lesion segmentation by extracting both spatial and temporal features from DCE-MRI sequences. Our network incorporates both CNN and RNN architecture. The RNN extracts temporal features from DCE-MRI sequences, and meanwhile 3D information is acquired through a multi-pathway structure and a fusion operator. Then a fully convolutional network utilizes context information of the extracted features to inference the probability map of lesions (Fig. 2). Our main contributions are summarized as follows: (1) We propose an effective lesion segmentation network taking temporal information into consideration that has not been well exploited by other CNN-based methods; (2) A multi-pathway structure allows the network to capture 3D information of different tissues, which is important to reduce false positive segmentation while boost accuracy; (3) Our novel architecture provides promising results on the evaluation dataset that achieves 0.7588 dice coefficient and 0.7390 positive predictive value.

The rest of this paper is organized as follows. Section 2 gives the details of our automatic segmentation architecture. Section 3 introduces the evaluation dataset and presents some experiments to validate the effectiveness of the proposed network. In Sect. 4 we conclude this paper.

2   Method

Our method can be summarized by a series of processes on DCE-MRI shown in Fig. 1. The subtraction sequences are acquired by subtracting pre-contrast from post-contrast sequences to eliminate the influences of the tissues without enhancement after the injection of a contrast agent. Then, breast region extraction and lesion segmentation network are applied to obtain lesion segmentation results.

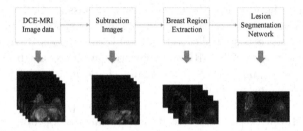

Fig. 1. Pipeline of the proposed method

2.1   Breast Region Extraction

In this stage, breast region will be extracted in each slice, and then normalization will be applied since our dataset acquires from different devices and protocols. The typical MRI image includes air background, breast, heart and other organs. Since the main focus is breast region, breast region extraction is indispensable for reducing the computation complexity of lesion segmentation network and eliminating noises in breast DCE-MRI. Firstly, otsu thresholding [20] is applied to get breast region boundary closed to the muscle from subtraction sequences. The breast-air boundary can be also obtained by otsu thresholding in pre-contrast sequences. Secondly, morphological refinements are applied to smoothen the boundary and to fill internal holes. For each patient's breast region, the images are transformed to the same dynamic range by contrast stretching. Finally, the breast region is resized to 256×144 to unify various sizes of breast regions.

2.2   Lesion Segmentation Network

In this subsection, the proposed network architecture will be introduced completely. Our network is a hybrid of a CNN and a RNN. The RNN has been proved to be a powerful tool for generalizing temporal correlations. However, it is not suitable to process 2D images due to lack of ability to encode spatial relationship. On the contrary, CNN excels at processing two dimension images due to its convolutional structure.

U-net [22] is a variant of CNN, which demonstrates promising performance in bio-medical image segmentation tasks. However, there are two main drawbacks in U-net for DCE-MRI breast lesion segmentation. Firstly, the standard U-net architecture includes four downsampling operations for encoding the input images to high-level representations. Large downsampling factor can be beneficial to classification task, but the accurate location information will gradually decrease while downsampling factor increases, which causes a dilemma situation [14]. On one hand, the information of small lesions will be weakened even removed if downsampling factor is too large. On the other hand, small downsampling factor means small receptive field and less context information in feature maps, which prejudices the classification accuracy. Secondly, U-net fails to capture temporal information which is crucial in breast DCE-MRI analysis. To overcome these shortcomings, we propose a spatio-temporal network by modifying the standard U-net and adding a novel ConvLSTM [26] structure, which extracts deep spatio-temporal information and preserves high spatial resolution simultaneously. The specific architecture of the proposed network is shown in Fig. 2.

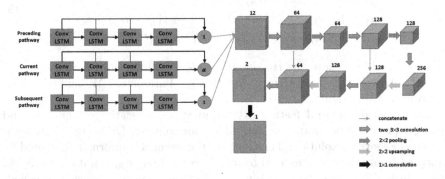

Fig. 2. The specific architecture of the proposed network. Each blue rectangle and blue cube refers to a single channel feature map and a multi-channel feature map respectively, whereas the blue circle represents concatenation and element-wise production. (Color figure online)

The ConvLSTM is proposed for precipitation nowcasting that takes 3D tensor (2D radar map with time dimension) as input and returns future radar maps, and shows great performance in processing spatio-temporal data. We adapt the ConvLSTM architecture to fit segmentation task in DCE-MRI sequences by adding three parallel ConvLSTM pathways that offer multi-view features of the lesion, including 3D information and temporal features. Given one slice of a certain position, the preceding, current and subsequent slices are fed into the three

ConvLSTM pathways respectively. In this way, our network can capture 3D information which can facilitate distinguishing the lesion from mammary ducts and mammary glands, those two are the major obstacles to obtain accurate lesion segmentation results since they present similar temporal features to the lesions. A 3×3 convolution with padding for the same size is set for state-to-state and input-to-state transition in all ConvLSTM cells, in which the number of hidden states is set to one. The outputs of all ConvLSTM cells are concatenated to fuse more temporal information and to take full advantage of different sequences. A hyper parameter α greater than one is introduced in the fusion operation to highlight current slice features since the predicted segmentation result corresponds to the current slice. The fusion operation can be formulated as:

$$h^{(i)} = f^{(i)}_{\text{convLSTM}}(x_1^{(i)}, x_2^{(i)}, \ldots, x_N^{(i)}), i = 1, 2, 3 \qquad (1)$$

$$h_{\text{fusion}} = [h^{(1)}, \alpha h^{(2)}, h^{(3)}] \qquad (2)$$

where $[h^{(1)}, \alpha h^{(2)}, h^{(3)}]$ refers to the concatenation of the feature maps. i denotes three different positions of slices and N represents the temporal number of subtraction sequences. $f_{\text{convLSTM}}(\cdot)$ represents the ConvLSTM function. The main equations in the ConvLSTM function are shown below:

$$
\begin{aligned}
i_t &= \sigma(W_{xi} * x_t + W_{hi} * h_{t-1} + b_i) \\
f_t &= \sigma(W_{xf} * x_t + W_{hf} * h_{t-1} + b_f) \\
c_t &= f_t \circ c_{t-1} + i_t \circ tanh(W_{xc} * x_t + W_{hc} * h_{t-1} + b_c) \\
h_t &= \sigma(W_{xo} * x_t + W_{ho} * h_t - 1) \circ tanh(C_t)
\end{aligned}
\qquad (3)
$$

where 'o' and '*' denote the Hadamard product and the convolution operator respectively. 'σ' is the hard sigmoid function [5]. Parameters 'W' and 'b' will be determined after training.

After spatio-temporal feature extraction stage, a relatively high-level and strong representation would be obtained from complex DCE-MRI data without loss of spatial resolution. Then, the context semantic information should be captured from the fusion features to inference the final segmentation result. As shown in Fig. 2, the network for semantic segmentation part consists of encoding path and decoding path. In our network, two downsampling operators are adequate to capture rich context information for lesion segmentation. Therefore our network is lightweight and has less tendency to overfitting compared with the original U-net. In encoding path, a 3×3 convolutional layer followed by a rectified linear unit repeats twice, then a 2×2 max pooling operation with stride 2 is applied for downsampling. After two max pooling operations, the size of the feature map is reduced to a quarter of the input image (e.g. 64×36). In decoding part, a 2×2 upsampling operation is used to enlarge feature maps twice for concatenating the corresponding feature maps from skip connection. The skip connection assembles relatively low-level features and high-level features, yielding more precise localization. After two upsampling operations, a 1×1 convolutional layer followed by a sigmoid activation function is applied to generate a probability map, the size of which is the same as that of the input image (e.g. 256×144). The value of each pixel in the probability map is between 0 and 1.

In order to effectively train the spatio-temporal network, we choose dice coefficient loss as the objective function. Dice coefficient loss [19] is defined as:

$$L(\bar{b}, b) = -\frac{\sum (2 \times \bar{b} \circ b)}{\sum \bar{b} + \sum b} \tag{4}$$

where the summations run over every pixel in the binary ground truth $\bar{b}$, the probability map b in the denominator and the Hadamard product of them in the numerator.

3   Experiments

3.1   Experimental Dataset

In our evaluation dataset, a total of 73 DCE-MRI studies are acquired from two different devices and protocols, the GE (18 studies) and Philips (55 studies). Each DCE-MRI study includes one pre-contrast sequences and at least four post-contrast sequences. Two experienced breast radiologists from Shanghai Renji Hospital manually annotate pixel-wise labels on 73 lesions (benign = 45 and malignant = 28) as the ground truth in this paper, with total 679 annotated slices. The entire dataset is randomly patient-wise split, where the training set consists of 48 MRI studies and the testing set consists of 25 MRI studies.

3.2   Experimental Setup

Since the number of slices containing breast lesions is much smaller than that without breast lesions, we select the slices with at least one lesion for training. With this training strategy, our proposed network can focus more on the essential features of breast lesions. The first three and the last subtraction sequences are chosen to train and test due to the temporal number of subtraction sequences varying with acquisition devices. Finally, the entire evaluation dataset consists of 468 training slices and 211 testing slices. We set $\alpha = 2.5$ in Eq. 2 for the following experiments.

Our network is trained by Adam stochastic gradient descent method [11] with a learning rate 5×10^{-5}. The network is trained 100 epochs with a batch size of 16. All experiments in this paper are implemented in keras framework [4] with Tensorflow [1] backend, and run on a NVIDIA TitanX GPU with 12GB memory.

3.3   Experiments of Spatio-Temporal Feature Extraction

To validate that our ConvLSTM structure can extract more abundant spatio-temporal features, three contrast networks are designed as follows: (1) The four-channel images consisted of the first three and the last subtraction sequences are fed into semantic part of our network without three convLSTM pathways, denoted as Semantic net; (2) The same subtraction sequences are used as sequential input for the network only consisting of current convLSTM pathway followed with semantic part, denoted as ST2D-net; (3) The proposed network shown in Fig. 2 is denoted as ST3D-net.

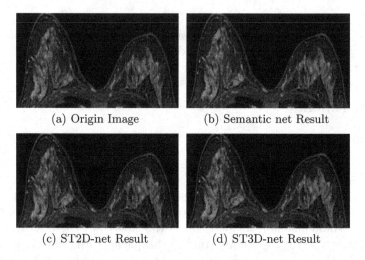

(a) Origin Image   (b) Semantic net Result

(c) ST2D-net Result   (d) ST3D-net Result

Fig. 3. Segmentation results of the lesion with marked background parenchymal enhancement that is defined as the enhancement of normal mammary glands [3]. Red line refers to the ground truth segmentation boundary, while blue line refers to the segmentation boundary predicted by the corresponding networks. The image shown in (a) is from the last subtraction sequences. (Color figure online)

Figure 3 illustrates a representative example of segmentation results predicted by three contrast networks on the testing set. Semantic net only segments part of the breast lesion (Fig. 3(b)). Semantic net including only two pooling layers lacks context information which is vital for complete segmentation. The improvement from Semantic-net to ST2D-net demonstrates that it is reasonable to treat the DCE-MRI data as sequential data and our designed network structure combining the advantage of RNN and CNN has a strong representation ability of spatio-temporal data. Through observing the segmentation results of the three contrast networks, we can see that our ConvLSTM structure can not only extracts the spatio-temporal features but also enhances the network's ability of combining context information. In addition, a satisfactory lesion segmentation result is obtained by ST3D-net (Fig. 3(d)), which therefore demonstrates that our ST3D-net is capable of segmenting the breast lesion accurately from the patient with marked background parenchymal enhancement.

To obtain quantitative comparison, we calculate the pixel-wise dice coefficient and positive predictive value on the testing set as a metric of segmentation

Table 1. Quantitative results of three contrast networks on the testing set

Model	Dice coefficient	Positive predictive value
Semantic net	0.6553	0.5945
ST2D-net	0.6963	0.6655
ST3D-net	0.7588	0.7390

accuracy and false positive rate. A higher dice coefficient means higher segmentation accuracy, whereas a higher positive predictive value indicates less false positives. As Table 1 shows, ST3D-net has a higher positive predictive value and dice coefficient compared with ST2D-net, proving that 3D features can help to reduce false positives and produce more accurate results.

3.4 Comparisons with Other Approaches

In order to compare our proposed method with other baseline segmentation approaches including Pixel-based method [18], ANN [25], U-net [22] and PSPnet [28], we train all networks on the training set and compare the segmentation performance on the testing set. The PSPnet is a novel deep convolutional network architecture for semantic segmentation, which performs very well in many scenarios. In detail, U-net and PSPnet adopt three-channel images as input, therefore three-channel images comprising the first two and the last subtraction sequences are used for training these networks. We also compare the number of parameters and test time of different methods as well as dice coefficient and positive predictive value.

Table 2. Comparisons with other approaches

Model	Dice coefficient	Positive predictive value	#Parameter (Million)	Test time (/slice)
Pixel-based [18]	0.5213	0.5022	-	98.35 s
ANN [25]	0.5343	0.4203	**0.002**	1.82 s
U-net [22]	0.6715	0.5419	31.0	**21 ms**
PSPnet [28]	0.6952	0.6259	46.8	107 ms
ST3D-net(ours)	**0.7588**	**0.7390**	1.9	22 ms

As Table 2 shows, our method achieves the best segmentation accuracy compared to other baseline methods. In addition, the number of parameters of our network is much less than that of U-net and PSPnet. Large improvement in segmentation accuracy with relatively less parameters indicates that our network struct is more suitable for breast segmentation task in DCE-MRI. Our network unearths more essential temporal features from DCE-MRI data, while other networks do not make full use of them. The test time in the last column in Table 2 is calculated by testing the same number of breast slices. Our network and U-net cost almost the same amount of time for inferring the segmentation result per slice, while our network yields much better dice coefficient and positive predictive value. In general, our method is not only lightweight but also effective for handling breast segmentation tasks.

Figure 4(a)–(g) shows some segmentation results obtained by the comparative methods. Other methods segment false positive regions (e.g. mammary ducts) which are rejected by our proposed method. Note that we do not apply any post-processing to the segmentation results predicted by our network. By observing all segmentation results obtained by our method on the testing set, we discover that some slices located in the edge of lesions (Fig. 4(h)) will be missed by our method as well as other methods. The possible reason is that it is hard to learn useful features for the lesion that has such low intensity and small size.

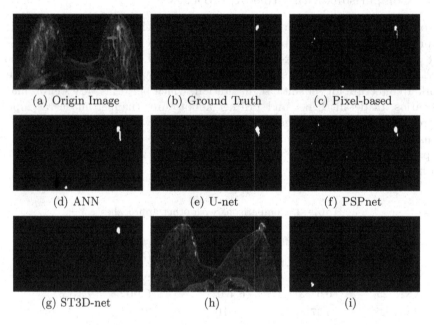

Fig. 4. (a)–(g): origin image with the red arrow pointing to the mammary duct, its ground truth and segmentation results predicted by the methods shown in Table 2 respectively. (h)–(i): a slice located in the edge of a lesion and its ground truth. (Color figure online)

4   Conclusion

In this paper, we propose a spatio-temporal convolutional network for fully automatic breast lesion segmentation. With the advantage of three convLSTM pathways and the fusion operator, our network not only extracts the spatio-temporal features but also possesses the ability to acquire more context information. Experimental results demonstrate that our method apparently improves the segmentation accuracy with relatively less parameters compared with other baseline methods, suggesting its potential of clinical applications. In future work, we would like to explore how to combine our network with 3D CNN to process all DCE-MRI sequences in single inference to further improve segmentation performance.

Acknowledgments. This research is partly supported by NSFC, China (No: 61572315, 6151101179) and 973 Plan, China (No. 2015CB856004).

References

1. Abadi, M., et al.: Tensorflow: a system for large-scale machine learning. In: Proceedings of the 12th USENIX Symposium on Operating Systems Design and Implementation, vol. 16, pp. 265–283 (2016)
2. Amit, G., et al.: Hybrid mass detection in breast MRI combining unsupervised saliency analysis and deep learning. In: Descoteaux, M., Maier-Hein, L., Franz, A., Jannin, P., Collins, D.L., Duchesne, S. (eds.) MICCAI 2017. LNCS, vol. 10435, pp. 594–602. Springer, Cham (2017). https://doi.org/10.1007/978-3-319-66179-7_68
3. Arslan, G., Çelik, L., Çubuk, R., Çelik, L., Atasoy, M.M.: Background parenchymal enhancement: is it just an innocent effect of estrogen on the breast? Diagn. Interv. Radiol. **23**(6), 414 (2017)
4. Chollet, F., et al.: Keras (2015). https://github.com/fchollet/keras
5. Courbariaux, M., Hubara, I., Soudry, D., El-Yaniv, R., Bengio, Y.: Binarized neural networks: Training deep neural networks with weights and activations constrained to +1 or −1. arXiv preprint arXiv:1602.02830 (2016)
6. Graves, A.: Supervised Sequence Labelling With Recurrent Neural Networks. Studies in Computational Intelligence, vol. 385. Springer, Heidelberg (2008). https://doi.org/10.1007/978-3-642-24797-2
7. Huang, G., Liu, Z., Der Maaten, L.V., Weinberger, K.Q.: Densely connected convolutional networks. In: Computer Vision and Pattern Recognition, pp. 2261–2269 (2017)
8. Huang, Q., Yang, J.: A multistage target tracker in IR image sequences. Infrared Phys. Technol. **65**(7), 122–128 (2014)
9. Jayender, J., Chikarmane, S., Jolesz, F.A., Gombos, E.: Automatic segmentation of invasive breast carcinomas from DCE-MRI using time series analysis. J. Magn. Reson. Imaging JMRI **40**(2), 467–75 (2013)
10. Jin, Q., Grama, I., Kervrann, C., Liu, Q.: Nonlocal means and optimal weights for noise removal. SIAM J. Imaging Sci. **10**(4), 1878–1920 (2017)
11. Kingma, D.P., Ba, J.: Adam: A method for stochastic optimization. arXiv preprint arXiv:1412.6980 (2014)
12. Krizhevsky, A., Sutskever, I., Hinton, G.E.: Imagenet classification with deep convolutional neural networks. In: Advances in Neural Information Processing Systems, pp. 1097–1105 (2012)
13. Lehman, C.D., et al.: MRI evaluation of the contralateral breast in women with recently diagnosed breast cancer. N. Engl. J. Med. **356**(13), 1295–1303 (2007)

14. Li, Z., Peng, C., Yu, G., Zhang, X., Deng, Y., Sun, J.: Detnet: A backbone network for object detection. arXiv preprint arXiv:1804.06215 (2018)
15. Long, J., Shelhamer, E., Darrell, T.: Fully convolutional networks for semantic segmentation. In: Computer Vision and Pattern Recognition, pp. 3431–3440 (2015)
16. Maicas, G., Carneiro, G., Bradley, A.P.: Globally optimal breast mass segmentation from DCE-MRI using deep semantic segmentation as shape prior. In: IEEE International Symposium on Biomedical Imaging, pp. 305–309 (2017)
17. Maicas, G., Carneiro, G., Bradley, A.P., Nascimento, J.C., Reid, I.: Deep reinforcement learning for active breast lesion detection from DCE-MRI. In: Descoteaux, M., Maier-Hein, L., Franz, A., Jannin, P., Collins, D.L., Duchesne, S. (eds.) MICCAI 2017. LNCS, vol. 10435, pp. 665–673. Springer, Cham (2017). https://doi.org/10.1007/978-3-319-66179-7_76
18. Marrone, S., Piantadosi, G., Fusco, R., Petrillo, A., Sansone, M., Sansone, C.: Automatic lesion detection in breast DCE-MRI. In: Petrosino, A. (ed.) ICIAP 2013. LNCS, vol. 8157, pp. 359–368. Springer, Heidelberg (2013). https://doi.org/10.1007/978-3-642-41184-7_37
19. Milletari, F., Navab, N., Ahmadi, S.A.: V-net: fully convolutional neural networks for volumetric medical image segmentation. In: Fourth International Conference on 3D Vision (3DV) 2016, pp. 565–571. IEEE (2016)
20. Otsu, N.: A threshold selection method from gray-level histograms. IEEE Trans. Syst. Man Cybern. $9(1)$, 62–66 (1979)
21. Ren, S., He, K., Girshick, R., Sun, J.: Faster R-CNN: towards real-time object detection with region proposal networks. In: Advances in Neural Information Processing Systems, pp. 91–99 (2015)
22. Ronneberger, O., Fischer, P., Brox, T.: U-Net: convolutional networks for biomedical image segmentation. In: Navab, N., Hornegger, J., Wells, W.M., Frangi, A.F. (eds.) MICCAI 2015. LNCS, vol. 9351, pp. 234–241. Springer, Cham (2015). https://doi.org/10.1007/978-3-319-24574-4_28
23. Siegel, R.L., Miller, K.D., Jemal, A.: Cancer statistics, 2016. CA Cancer J. Clin. $54(1)$, 8–29 (2015)
24. Siu, A.L.: Screening for breast cancer: us preventive services task force recommendation statement. Ann. Intern. Med. $164(4)$, 279–296 (2016)
25. Wu, H., Gallego-Ortiz, C., Martel, A.: Deep artificial neural network approach to automated lesion segmentation in breast. In: Proceedings of the 3rd MICCAI Workshop on Breast Image Analysis, pp. 73–80 (2015)
26. Xingjian, S., Chen, Z., Wang, H., Yeung, D.Y., Wong, W.K., Woo, W.c.: Convolutional LSTM network: a machine learning approach for precipitation nowcasting. In: Advances in Neural Information Processing Systems, pp. 802–810 (2015)
27. Yuan, Y., Giger, M.L., Hui, L., Suzuki, K., Sennett, C.: A dual-stage method for lesion segmentation on digital mammograms. Med. Phys. $34(11)$, 4180–4193 (2007)
28. Zhao, H., Shi, J., Qi, X., Wang, X., Jia, J.: Pyramid scene parsing network. In: Computer Vision and Pattern Recognition, pp. 6230–6239 (2017)

Glomerulus Detection on Light Microscopic Images of Renal Pathology with the Faster R-CNN

Ying-Chih Lo[1], Chia-Feng Juang[2(✉)], I-Fang Chung[3],
Shin-Ning Guo[2], Man-Ling Huang[3], Mei-Chin Wen[4],
Cheng-Jian Lin[5], and Hsueh-Yi Lin[5]

[1] Section of Nephrology, Department of Medicine,
Taichung Veterans General Hospital, Taichung, Taiwan
[2] Department of Electrical Engineering,
National Chung-Hsing University, Taichung, Taiwan
cfjuang@dragon.nchu.edu.tw
[3] Institute of Biomedical Informatics,
National Yang-Ming University, Taipei, Taiwan
[4] Department of Pathology, Taichung Veterans General Hospital,
Taichung, Taiwan
[5] Department of Computer Science and Information Engineering,
National Chin-Yi University of Technology, Taichung, Taiwan

Abstract. Glomerulus is an important component in human kidney. The appearance of the glomeruli on light microscopic image can provide abundant information for disease diagnosis. Due to the importance of glomeruli on accurate renal disease diagnosis, this paper proposes an automatic method to detect glomeruli in light microscopy images with Periodic Acid Schiff (PAS) or hematoxylin and eosin (H&E) stains at 100x, 200x, or 400x optical magnification. The faster region-based convolutional neural network (R-CNN) is applied to the detection task. The proposed detection approach performs an end-to-end glomerulus detection without any *a priori* information of the stains and magnifications of the images. The training dataset contains 2,511 images with 3,956 glomeruli. The test dataset contains 482 images with 563 glomeruli. The recall and precision of the test result are 91.54% and 86.50%, respectively, which shows the effectiveness of the proposed detection method.

Keywords: Glomerulus detection · Renal pathology · Deep learning
Convolutional neural networks · Region-based convolutional neural networks

1 Introduction

As the advancement of the healthcare and increased longevity, many countries had gradually become the aging societies. This results in an increase of patients with chronic diseases, such as diabetes, hypertension, and hyperlipidemia. In addition to these well-known diseases, chronic kidney disease (CKD) is also a prevalent chronic disease nowadays. According to a previous study [1], there were about 12% of the

© Springer Nature Switzerland AG 2018
L. Cheng et al. (Eds.): ICONIP 2018, LNCS 11307, pp. 369–377, 2018.
https://doi.org/10.1007/978-3-030-04239-4_33

citizens in Taiwan had CKD and there were only 3.5% of them aware of their diseases. In order to improve the care for CKD patients in Taiwan and decrease the medical expense, the National Health Insurance Administration launched a comprehensive care program for early CKD patients since 2011 to facilitate the early detection and treatment for CKD. Even though, there is still some CKD patients having poor disease control due to that the etiology of the underlying kidney disease is unknown after routine laboratory survey. For these patients, the clinician may suggest renal biopsy for further evaluation.

Renal biopsy is a valuable and irreplaceable diagnostic tool for renal diseases. For patients with unknown cause CKD, heavy proteinuria and unexplainable rapid renal function deterioration, the clinicians may suggest renal biopsy, an invasive examination for further investigation. After local anesthesia, the clinician will obtain a small proportion of kidney tissue with needle biopsy gun under real-time sonography guide. The renal specimen will receive a series of preprocessing and staining techniques before it can be read by pathologist with microscopy. In general, there are three different kinds of image studies, including light microscopy, immunofluorescene, and electron microscopy, need to be thoroughly evaluated before the final diagnosis is made. Among these studies, light microscopy is the most fundamental one and can reveal much important information such as the basic microstructure of the kidney.

In general, the human kidney is composed of four basic microstructures, including glomerulus, renal tubule, interstitial tissue and vessel. Each component plays a different role in the control of renal regulatory mechanism. Glomerulus is the basic functional unit for waste and excess water removal. The appearance of the glomeruli on light microscopic image can provide abundant information for disease diagnosis. Besides, the number of the glomeruli in a renal specimen is also an indicator of the sampling quality. For a sample with only limited glomeruli, the value for further interpretation is minimal. Besides, the appearance of glomeruli alone provides a clue for the underlying renal disease and long-term outcome. During the process of renal pathological image interpretation, the glomeruli definitely play a crucial role.

Due to the importance of glomeruli on accurate renal disease diagnosis, many scholars had devoted themselves to the study of glomerular classification and detection [2–6]. The early approach in this topic is the combination of handcrafted features with a classifier. Yoshihiro *et al.* proposed the combination of Rectangular Histogram of Oriented Gradient (R-HOG) features with linear support vector machine (SVM) classifiers to detect glomeruli [2]. The dataset used were whole slide renal pathological images of Spontaneously Diabetic Torii (SDT) rats [7]. A new descriptor called Segmental HOG (S-HOG) which can be adaptively fitted to input renal images and perform a comprehensive detection of glomeruli in kidney sections images was proposed in [3]. The result showed that the S-HOG method achieved the higher F-measure of 0.866 in comparison with 0.838 of the R-HOG-based method in the rat dataset. A novel system was proposed to detect the renal corpuscle objects and measure the glomerulus diameter and Bowman's space width of renal albino rats in [4]. The system used the particles analyzer technique to extract three descriptors (aspect ratio, circularity, and solidity) to detect glomerulus on green channel images.

Due to the rise of big data, deep learning models have been widely used in medical image diagnosis in recent years. Different from hand-crafted features, deep learning techniques enable machines to learn the best features by themselves to classify or detect objects [8, 9]. Gallego *et al.* applied Convolutional Neural Networks (CNNs) to classify human glomerulus and non-glomerulus patches manually cropped from whole slide images (WSI) [5]. The CNN-based classification approach was further applied to detect glomeruli in WSI by using a sliding window with overlapping [6]. Detection was performed on WSI stained with Periodic Acid Schiff (PAS) at 20x magnification. A total of 98 WSI (with around 4000 glomeruli) and 10 WSI (with 275 glomeruli) were used for training and test, respectively.

This paper considers the detection of human glomeruli in whole light microscopy images. The images were stained with PAS or hematoxylin and eosin (H&E) and taken at 100x, 200x, or 400x optical magnification. The variations in glomeruli colors due to different stains and sizes due to different magnifications making the detection task much more challenging than that studied in [6]. The faster region-based convolutional neural network (R-CNN) is a popular method for end-to-end objection detection with different sizes and shapes [10, 11]. This paper applies the faster R-CNN to this detection task. Experimental results show that an F-measure of 0.889 is achieved for the detection of 563 glomeruli in 482 test images with two stains and three magnifications using a single faster R-CNN.

This paper is organized as follows. Section 2 introduces the collection of training and test data. Section 3 introduces glomerulus detection using the faster R-CNN. Section 4 presents experimental results. Finally, Sect. 4 presents the conclusions.

2 Materials

This study was approved by the Institutional Review Board of Taichung Veterans General Hospital, Taiwan (TCVGH-1073605D). In this study, we use renal biopsy pathological images collected for the purpose of routine case discussion during 2011 to 2017 in Taichung Veterans General Hospital, Taiwan. The images were taken with Olympus Digital Camera System and the image resolution was 4080×3072 pixels. We choose light microscopy images with H&E and PAS stains as the material in this study. The images were taken with optical magnifications of 100x, 200x, or 400x on light microscopy. Figure 1 shows the appearances of renal pathology images with different stains and magnifications. In general, the PAS stain can highlight the basement membrane of glomeruli and renal tubules. Images with different stain methods and different magnifications can provide different information to the pathologist. Table 1 summarizes the numbers of images collected for training, validation, and test. The total numbers of glomeruli in the training and test images are 3956 and 563, respectively.

Figure 2 shows different patches in the images. A normal glomerulus in the light microscopy images is roughly circular on pathological images, as shown in Fig. 2(a). It is observed that the glomerulus is surrounded by Bowman's capsule, and this unique feature makes glomerulus isolated from other structures. In addition to glomeruli, there are other circular structures in the biopsy images, including renal tubules and blood vessels, as shown in Fig. 2(b), which may cause the model to misinterpret as false positives. If glomeruli have abnormalities, such as global/segmental glomerulosclerosis and crescentic glomerulonephritis, as shown in Fig. 2(c), the shape of glomerular changes, which may result in difficulty for the model to identify it as the glomerulus (false negative). Overall, the variations in glomerulus structures, stains, and sizes in the images make the automatic detection of glomeruli a challenging task.

Table 1. The distribution and numbers of renal biopsy pathological images collected for training, validation, and test datasets.

Dataset	Magnifications	100x	200x	400x	Total
Training	H&E	377	302	753	1,432
	PAS	170	204	705	1,079
	Subtotal	547	506	1,458	**2,511**
Validation	H&E	63	59	176	**298**
	PAS	22	38	122	**182**
	Subtotal	85	97	298	**480**
Test	H&E	64	60	176	300
	PAS	22	38	122	182
	Subtotal	86	98	298	**482**

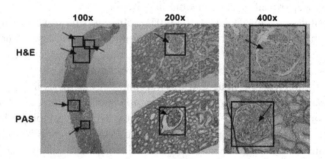

Fig. 1. Appearances of whole renal pathology images of different stains and resolutions, where the arrows indicate the locations of glomeruli and the black bounding boxes are manually annotated glomeruli.

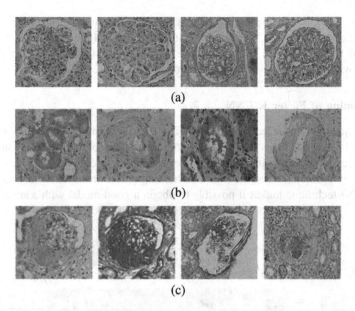

Fig. 2. Patches of (a) normal glomerulus, (b) other structures that mimic glomerulus, and (c) abnormal glomerulus in the dataset.

3 Glomerulus Detection Using Faster R-CNN

3.1 Structure of Faster R-CNN

This section introduces the use of a single faster R-CNN to automate the detection of glomeruli in the whole images with H&E or PAS stain at different magnifications. Figure 3 shows the structure of the faster R-CNN used in the glomerulus detection task. The VGG-16 model is utilized in the faster R-CNN. The feature extraction part in the VGG-16 consists of 13 convolution layers and 4 pooling layers. The classification part of the VGG-16 consists of three fully connected layers. The region proposal network (RPN) is performed on each 3x3 spatial sliding window over the last convolution layer (feature maps) of the VGG-16. The RPN provides the classification scores and region coordinates of proposed objects in the feature maps. The region of interest (ROI) pooling layer adjusts the dimension of the feature maps to ensure a constant dimension size to the fully connected layer. In the end, the score and position adjustments are generated by the fully connected layer.

In the collection of the training dataset of the faster R-CNN, only the positive rectangular patches (glomeruli) need to be annotated in the images. Figure 1 shows some of the annotated glomeruli in the whole images with different stains and magnifications. The faster R-CNN automatically and randomly samples negative patches (non-glomerulus) from the whole image, which greatly reduces the effort on data collection. Both the PAS and H&E stained images are collected in the training dataset so that the single faster R-CNN can detect glomeruli in the images with different stains. The sizes of the glomeruli in the images taken with the same optical magnification are similar but have a large variation in the images with different magnifications, as shown in Fig. 1. For this problem, the training dataset also mixes the images at the magnifications of 100x, 200x, and 400x so that the single faster R-CNN is robust in detecting glomeruli with different sizes from different magnifications.

3.2   Training of Faster R-CNN

The technique of transfer learning is used. The transfer learning technique uses a pre-trained model and fine tunes it based on the training dataset of a new classification task. This technique has shown its merit in model training, especially for the classification and detection problems when the available training data is insufficient such as medical images. This technique makes it possible to obtain a good model with a small number

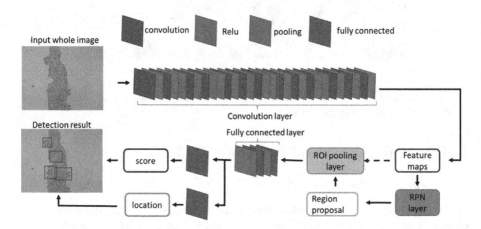

Fig. 3. The architecture of the faster R-CNN, where the VGG16 model is used.

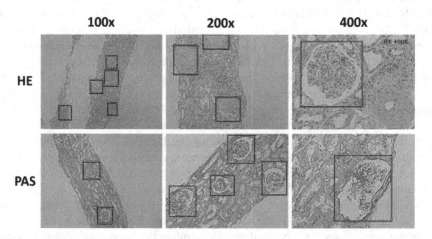

Fig. 4. The correction detections using the faster R-CNN, where the yellow boxes indicate the true positives. (Color figure online)

of data and is, therefore, used in the glomerulus detection problem. In this paper, the VGG-16 pre-trained using the 2014 ImageNet Large Scale Visual Recognition Challenge (ILSVRC) dataset is utilized. Due to the use of transfer learning, the weights in the convolution and fully connected layers are fine tuned. On this side, the stochastic gradient descent is used. The learning rate and the momentum coefficient are set to 0.001 and 0.9, respectively. The total iteration number is set to 30000. These coefficients are selected according to the performance of the validation dataset.

4   Experimental Result and Discussions

This section presents the experimental result using the faster R-CNN. To evaluate the performance of the proposed detection method, the recall (R) and the precision (P) were used. The R measure is defined as the ratio between the number of glomeruli correctly detected (true positives) and the actual number of glomeruli. The P measure is defined as the ratio between the detected glomeruli (true positives) and the total number of detections (true positives + false positives). In addition, the following F-measure (F_1) [6] was also used to measure the detection performance

$$F_1 = 2\frac{R \cdot P}{R + P} \tag{1}$$

The R, P, and F_1 were 91.54%, 86.50%, and 0.889, respectively. Figure 4 shows some of the correct detections from the whole images with H&E or PAS stains at the three magnifications. The single faster R-CNN succeeded in detecting most glomeruli in the images without *a priori* information of the stains and magnifications. It should be emphasized that the training and test datasets were different from those in [6] and contained higher variations in the glomerulus colors (two stains) and sizes (three magnifications). The window-based scanning method in [6] cannot be directly applied to detect glomeruli with such variations in sizes.

Table 2. The test result (#false positives/#false negatives) distributed over the images with two stains and three magnifications by using the faster R-CNN.

Dataset	100x	200x	400x	Total
H&E	26/18	15/9	10/3	51/30
PAS	10/7	9/2	6/6	25/15
Total	36/25	24/11	16/9	**76/45**

To analyze the false detections, Table 2 shows the numbers of false positives and false negatives distributed over the images with different stains and magnifications. The result showed that the false positives were nearly evenly distributed over different stains and magnifications with no bias to a particular type of image. Most of the false negatives occurred at the images with 100x and 200x magnifications with no bias to the stains. The images with 400x magnification showed a smaller number of false negatives mainly because of the large glomerulus size in the image. Figure 5(a) shows examples of the false negatives, where it was observed that some of them were at the image boundaries and some were abnormal in structure. Figure 5(b) shows examples of the false positives, where some of them were glomerulus-like in structure and some of them were caused by improper detection sizes.

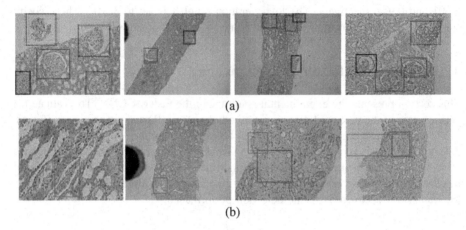

(a)

(b)

Fig. 5. (a) False negatives (red boxes) and (b) false positives (blue boxes) in the images, where the yellow boxes indicate the true positives. (Color figure online)

5   Conclusion

In this study, we have demonstrated the feasibility of automatic glomerulus detection on renal pathology images with different stains and magnifications using deep learning. The use of a single faster R-CNN succeeds in detecting most glomeruli in the images with H&E or PAS stains at 100x, 200x, or 400 magnifications. In the future, the collection of more abnormal glomeruli in the training dataset to improve the detection rate will be studied. To automate the whole medical diagnosis process, deep-learning based automatic disease diagnosis following the detection of the glomeruli from the renal pathology images will be studied as well. The proposed approach may improve the efficiency of the pathological diagnosis process if widely used in the clinical settings in the future.

References

1. Wen, C.P., et al.: All-cause mortality attributable to chronic kidney disease: a prospective cohort study based on 462293 adults in Taiwan. Lancet **371**(9631), 2173–2182 (2018)
2. Hirohashi, Y., et al.: Automated quantitative image analysis of glomerular desmin immunostaining as a sensitive injury marker in Spontaneously Diabetic Torii rats. J. Biomed. Image Process. **1**(1), 20–28 (2014)
3. Kato, T., et al.: Segmental HOG: new descriptor for glomerulus detection in kidney microscopy image. BMC Bioinform. **16**(316), 1–16 (2015)

4. Kotyk, T., et al.: Measurement of glomerulus diameter and Bowman's space width of renal albino rats. Comput. Methods Programs Biomed. **126**, 143–153 (2016)
5. Pedraza, A., Gallego, J., Lopez, S., Gonzalez, L., Laurinavicius, A., Bueno, G.: Glomerulus classification with convolutional neural networks. In: Valdés Hernández, M., González-Castro, V. (eds.) MIUA 2017. CCIS, vol. 723, pp. 839–849. Springer, Cham (2017). https://doi.org/10.1007/978-3-319-60964-5_73
6. Gallego, J., et al.: Glomerulus classification and detection based on convolutional neural networks. J. Imaging **4**(20), 1–19 (2018)
7. Sasase, T., Ohta, T., Masuyama, T., Yokoi, N., Kakehashi, A., Shinohara, M.: The spontaneously diabetic torii rat: an animal model of nonobese type 2 diabetes with severe diabetic complications. J. Diabetes Res. **2013**, 1–12 (2013)
8. Krizhevsky, A., Sutskever, I., Hinton, G.: Imagenet classification with deep convolutional neural networks. In: Proceedings of the Advances in Neural Information Processing Systems, pp. 1097–1105 (2012)
9. Szegedy, C., et al.: Going deeper with convolutions. In: Proceedings of the IEEE Conference on Computer Vision and Pattern Recognition, Boston, pp. 1–9 (2015)
10. Ren, S., He, K., Girshick, R., Sun, J.: Faster R-CNN: towards real-time object detection with region proposal networks. IEEE Trans. Pattern Anal. Mach. Intell. **39**(6), 1137–1149 (2017)
11. Zhao, X., Li, W., Zhang, Y., Gulliver, T.A., Chang, S., Feng, Z.: A faster RCNN-based pedestrian detection system. In: Proceedings of the IEEE 84th Vehicular Technology Conference (VTC-Fall), pp. 1–5 (2016)

One-Bit DNA Compression Algorithm

Deloula Mansouri and Xiaohui Yuan[(✉)]

School of Computer Science and Technology,
Wuhan University of Technology, Wuhan, China
{deloula,yuanxiaohui}@whut.edu.cn

Abstract. Recently, the ever-increasing growth of genomic sequences DNA or RNA stored in databases poses a serious challenge to the storage, process and transmission of these data. Hence effective management of genetic data is very necessary which makes data compression unavoidable. The current standard compression tools are insufficient for DNA sequences compression. In this paper we proposed an efficient lossless DNA compression algorithm based One-Bit Compression method (OBComp) that will compress both repeated and non-repeated sequences. Unlike direct coding technique where two bits are assigned to each nucleotide resulting compression ratio of 2 bits per byte (bpb), OBComp used just a single bit 0 or 1 to code the two highest occurrence nucleotides. The positions of the two others are saved. To further enhance the compression, modified version of Run Length Encoding technique and Huffman coding algorithm are then applied respectively. The proposed algorithm has efficiently reduced the original size of DNA sequences. The easy way to implement our algorithm and the remarkable compression ratio makes its use interesting.

Keywords: DNA sequences · Redundancies · Lossless compression
One-Bit algorithm

1 Introduction

Bioinformatics is an interdisciplinary field that combines computer science and biology to research, develop, and apply computational tools and approaches to solve the problems that biologists face in the agricultural, medical science and study of the living world [1]. The most important element in bioinformatics is the study of biomolecules present in DNA. Deoxyribonucleic Acid (DNA) is a molecule that encodes the genetic information, it helps a lot of medical treatments, identify individuals, detect the risk for certain diseases and diagnose genetic disorders [2]. DNA sequences are the combinations of only four bases adenine, cytosine, guanine, and thymine (A, C, G, T) [1]. Around 3 billion characters and over 23 pair of chromosome are there in human genome and can even reach more than 100 billion nucleotides for certain amphibian species [3]. One million Terabytes (TB) is required to store one million genomes, equivalent to 1000 Petabyte (PB). All of this amount of data is stored in special databases also known as databanks which has been developed by the scientific community where biologists can also analyze their data [4, 5].

From 1982 to the present, the number of bases generate in databases has doubled approximately every 18 months. Approximately 231 842 951 552 bases are stored in

© Springer Nature Switzerland AG 2018
L. Cheng et al. (Eds.): ICONIP 2018, LNCS 11307, pp. 378–386, 2018.
https://doi.org/10.1007/978-3-030-04239-4_34

GenBank in April 2017 [6]. When database becomes full, adding new data to the databases will be impossible, subsequently this problem can lead to faults or inability to work with the current web application. Storing, sharing and performing operations on this huge amount of genomic data will require enormous space and very longtime [1–7]. As well as the high level of similarity within DNA sequences and specially between individuals of same species is the peculiarity of genomic information and can be used for efficient compression. To solve the previous problems in an efficient way, compression becomes very necessary [7, 8].

Data compression is the process of encoding the information in a reduced form using fewer bits than the original representation by eliminating the redundancies which can not only save the storage capacity but also helpful in fast searching, retrieval, analysis and transmitting data over the internet with limited bandwidth [9, 10]. Two main categories for compression of data based on the requirements of reconstruction, lossless and lossy compression [11]. In lossless compression techniques, the original data can be exactly produced from the compressed data. Which is means that there is no loss of information. It is used for applications that cannot accept any difference between the original and recovered data [12]. Lossy compression means that there is some loss of information and the original data generally cannot be exactly recovered from the compressed one. In many situations such as images and sound, the production of no identical copy of the original one after decompression is not a necessary [13]. It's known that the DNA datasets cannot afford to lose any part of their data, then lossless compression techniques are more suitable for their compression [2].

In this article, the proposed research is a lossless DNA compression algorithm based on new compression method OBComp. It had mainly three phases, pre-processing phase, coding phase and post-processing phase. OBComp is used like pre-processing step when Modified Run Length Encoding algorithm (Modified RLE) proposed in [14] and Huffman coding are used in coding step. However, not all the generated files after using Modified RLE are compressed. We implement the same idea of Modified RLE, but with the compression of all generated outputs files. One of these files is compressed by using Huffman coding algorithm. In post-processing step, the binary files generated in the previous step are divided into blocs of 8 bits then each bloc is converted on its decimal value. To enhance more the compression, all the final output files are compressed by using one of existing compression tool.

This paper is organized in 5 sections, we will start by presenting the background and the previous work related to the research (Sect. 2). In Sect. 3, we will present the proposed methodology and working principle. We will illustrate, discuss and analyzes the results in Sect. 4. Finally, Sect. 5 includes conclusion and some future works.

2   Related Works

Gzip or bzip2 are the firstest simple compression tools, they failed to compress the biological sequences well [15]. Because these algorithms are designed mainly for English text compression and doesn't use the special structures of biological sequences [16]. Gzip tool [15] cannot compress DNA sequences but only expand them in size, it has a compression ratio of 2.217 bit per base (bpb). A compression algorithm provides significant results only if the compression ratio is lower than two bpb because the DNA sequences only consist of four nucleotide bases {A, C, G, T}, thus 2 bits are enough to store each base. Huffman's code [17] also fails to compress the DNA sequences because the frequencies of {A, C, G, T} are not very different [17, 18]. Ultimately, tailored DNA compression methods are needed and several methods have been proposed.

The first DNA compression algorithm found in the literature is BioCompress developed in 1993 [19]. It is one of the algorithms based on finding repetition of DNA nucleotides, such as repeats, palindromes, complement repeats etc. BioCompress [19] and BioCompress-2 [20] two lossless compression algorithms based on Lempel-Ziv compression method. First, it detects exact and reverse complement repeats in the DNA and encoded them by the repeat length and the position of a previous repeat occurrence. For non-repeat regions, it is encoded by 2 bpb. BioCompress-2 is the extended version of BioCompress, the difference between them is in the encoding of non-repeat regions. In BioCompress 2, if no significant repetition is found the arithmetic coding of order 2 is applied. GenBit compress tool divides the input DNA sequence into blocs of four characters (8 bits). A specific bit "1" is introduced as a 9th bit if the consecutive blocs are same. Else, a specific bit "0" is introduced as a 9th bit. It achieves the best compression ratios for Entire Genome (DNA sequences) when the repetitive blocs are maximum. otherwise, When the repetitive fragments are less, or nil, the compression ratio is higher than two [21].

DNABIT Compress Algorithm [22] divides the sequence into small blocs and compresses them taking into consideration if they existed before or not using binary bits in bit-pre-processing stage to make the difference between exact and reverse repeats fragments of DNA sequences. It significantly improves the running time and achieves better compression for larger genomes. SBVRLDNAComp [23] is sequential compression algorithm designed only for the compression of DNA and RNA sequences. This algorithm compressed any sequence firstly by searching the repeats in four different ways using four different methods. After getting the bits sequence form these methods, it compares the bit length and chooses the optimal one dynamically before going to final compression using LZ77.

Recently, Combination between some compression methods was proposed [24]. Firstly, Burrows-Wheeler Transform (BWT) takes a block of data and rearranges it lexicographically. Move-To-Front (MTF) and Run Length Encoder (RLE) are then respectively applied. In coding step, Arithmetic Coding is used. Modified HuffBit algorithm [25] is the extended version of HuffBit algorithm, the use of R language in DNA compression and adding the six bits at the end of file (to know the code of each base) are the main potentiality of this research work. Testing the real biological sequences using these two algorithms could not be performed because the frequencies of A, C, G and T in the real DNA sequences are almost the same which represent here the worst case.

3   Proposed Methodology

DNA sequence is a combination of four nucleotides bases {A, C, G and T} which are both repetitive and non-repetitive in nature. The proposed method is a combination between three compression algorithms. We propose a new pre-processing method based on one-bit representation of nucleotides named OBComp. It compresses DNA sequences by replacing just the two highest occurrence nucleotides by single bit 0 or 1 and the positions of the two others nucleotides is saved. Thus, three output files are generated in this step, binary file which represent the input file of next step and position files.

In coding step, we take the modified version of Run Length Encoding algorithm (Modified RLE) used in [14] to compress the binary file. Two files will be the output of this step, the numeric file is compressed by using Huffman coding algorithm [17]. Every 8 bits in the binary files is converted into its decimal value. To enhance the compression furthermore, we are using one of existing LZ-family compressor to compress position files generated in pre-processing step. OBComp and Modified RLE methods are more detailed in this section.

3.1   One-Bit Compression Method

The following algorithm explain OBComp.

   Begin

- Calculate the frequency Fr of each nucleotide in DNA input file.
- Put the frequency in descending order, then we have:

$$Fr(X) > Fr(Y) > Fr(Z) > Fr(Q)$$

When each of: X, Y, Z and Q represent a specific DNA base.
- For each occurrence of Z in DNA input file, save its position in PosZ file.
- For each occurrence of Q in DNA input file, save its position in PosQ file.
- Remove each occurrence of Z and Q in the original file.
- Replace each X by 0 and each Y by 1 in DNA input file.

   End.

3.2   Modified RLE

Specially applied to compress binary sequences in which runs of data (that is, replacing any sequence (of one's ending with zero and any sequence of zero's ending with one) with the repeat length) are stored in Run_File. And the data value stored in another file Order_File. Table 1 shows an example of Modified RLE.

3.3   Algorithm Description

In this part, we explain encoding and decoding procedures.

Encoding Procedure. Simplified approach is the art of the proposed compression method. Three phases are presented, OBComp explaining before is applied in the first

Table 1. Example of the modified RLE algorithm.

Binary sequence	11110	01	0001	000001	01	01	0000001	11110	1110	110
Run_File	4	1	3	5	1	1	6	4	3	2
Order_File	1	0	0	0	0	0	0	1	1	1

step. In the second step we use Modified RLE then Huffman algorithm to reduce the size of the binary output of the first step. Third step represent post-processing step, decimal value for every 8 bits in binary files is calculated. All other output files of the first step are then more compressed by using one of existing compressor. The flow of proposed model for encoding procedure is given in the figure Fig. 1.

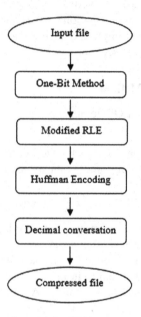

Fig. 1. Proposed model for DNA sequences compaction.

Decoding Procedure. Decoding algorithm is important to get the original file. The proposed algorithm is a lossless compression algorithm. So, the decoding method is just the reversal of encoding procedure. The compressed file will be the input file of decompression procedure, then post processing step of encoding procedure will be pre-processing step and pre-processing method will be applied in post processing step in decoding procedure.

3.4   Example and Calculation

In order to illustrate our algorithm's approach, throughout this section, we will use the following sequence (Đ) as an example:

ATGTACGTGATCGCTAAGAATCCGTGCGAAAACTAGGCGAGCTA
GACCTTGTCGATCATTAGATCG

Our proposed algorithm is applied to the DNA sequence (Đ) above.

Pre-processing Steps. Using OBComp method.
Step 1: Sequence length = 70.
Frequency of A Fr (A)= 20, Frequency of T Fr (T)= 16,
Frequency of G Fr (G)= 19, Frequency of C Fr (C)=15.
Such that, $Đ = \Sigma$ (Fr (A) + Fr (T) + Fr (G) + Fr (C)) = 70.
Since: Fr (A) > Fr (G) > Fr (T) > Fr (C).
Step 2: Save the position of the bases C and T.
PosZ = {5,11,13,21,22 ...}, PosQ = {1,3,7,10,14 ...}.
Step 3: Delete the occurrence of C and T from Đ then assigning A=0 and G=1, a partially compressed sequence is obtained.
01011010010011100000 ... (1)

Coding Steps. Double compression is used.
Step 1: Modified RLE algorithm is applied in (1):
111112242111321 ... (2)
001100101111100
Step 2: Huffman coding algorithm is using to code (2), the result is:
111110101001011111000011
The final output of this step:
11111010 10010111 10000110 01100101 11110000.

Post-processing Steps Decimal conversation is used, then we have:
250 151 134 101 240

Remark: Position files generated in pre-processing step are compressed by using one of based LZ-family compressor.

4   Results and Discussions

4.1   Performance Evaluation

For practical purpose, we tested our algorithm on standard benchmark data used in [2]. DNA input data for experimentations is taken from GenBank. GenBank is a popular public genetic sequence database. The performance of proposed technique was verified on various real DNA sequences. We used five human genes, HUMDYSTROP (human dystrophin gene), HUMHDABCD (human sequence of contig), HUMHBB (human globin gene) whose size is about 73000 nucleotides. We also used Mitochondria genome (MPOMTCG) and the complete genome of Vaccinia Virus (VACCG) whose size is about 191000 nucleotides. The above input data has been taken from the website of NCBI [26].

To quantify the efficiency of a given compression run, several measures can be applied such as compressed size and compression ratio. Table 2 shows the sequence size before compression, sequence size after using the proposed compression algorithm and the percentage of reduced size for the above stated DNA sequences.

To test the performance of our algorithm, we compared our approach to existing

Table 2. Performance of our proposed algorithm in different DNA sequences and genomes.

Sequence name	Number of nucleotides	Sequence size before compression (byte)	Size after compression (byte)	The percentage of reduced size (%)
Humdystrop	38770	39876	9538.375	76.08
Humhprtb	56737	58060	13901.125	76.06
Humghcsa	66495	68096	13212.5	80.60
Humhbb	73308	74357	17804,375	76.06
Mpomtcg	186609	189274	46706.25	75.33
Vaccg	191737	194476	45023	76,85

DNA compression algorithms in terms of binary representation rate per nucleotide (bpb) using the real biological sequences [2]. When the compression ratio is defined as follow:

Compression ratio = (Size after compression/Size before compression) *8

The Table 3 shows the comparison among the various existed DNA compression algorithms like FS4CP [27], BWT+MTF+ARI [24], Modified HuffBit [25], Genbit [21] and also compare the proposed algorithm with standard compression tool Gzip [15]. To test the performance of Modified HuffBit algorithm, the author in [25] didn't use the real data benchmarks. In this case we test it using the real DNA datasets.

Table 3. Compression with other existing methods (Bits/Base).

Sequence name	Gzip	Genbit	FS4CP	Modified HuffBit	BWT+MTF +ARI	Proposed method
Humdystr-op	2.24	2.23	2.06	2.04	–	**1.91**
Humhprtb	2.27	2.24	1.98	2.23	2.01	**1.91**
Humghcsa	–	–	2.04	1.68	1.60	**1.55**
Humhbb	2.21	2.23	2.07	2.16	2.02	**1.91**
Mpomtcg	2.21	2.24	1.98	2.13	–	**1.97**
Vaccg	2.17	2.24	1.97	2.00	2.07	**1.85**

The Table 2 indicate that the proposed technique gives better result than existed compression algorithms with average compression ratio equal to 1.83 bpb.

5 Conclusion

In this paper, new pre-processing compression method is proposed. One-Bit technique is used for compressing repetitive and non-repetitive DNA sequences without using complex dynamic programming. The main idea is replacing the two highest occurrence nucleotides by single unlike a 2-bit based compression technique where two unique bits are assigned to each nucleotide. Then using combination between two compression algorithms like Modified RLE and Huffman helps to reduce the size of input file to less than the quarter. Not only the average compression ratio of the proposed method is better than the existing compression algorithms but also it is simple, fast, flexible and very easy to implement make it more interesting.

In the future, we will try to compress position files using an efficient compression method specially for integer number to get better compression, we will try also to associate our new method to other vertical compression algorithms based on statistical approaches to achieve better result.

References

1. Saada, B., Zhang, J.: Vertical DNA sequences compression algorithm based on hexadecimal representation. In: Proceedings of the World Congress on Engineering and Computer Science, pp. 21–25. WCECS, San Francisco (2015)
2. Jahaan, A., Ravi, T., Arokiaraj, S.: A comparative study and survey on existing DNA compression techniques. Int. J. Adv. Res. Comput. Sci. **8**, 732–735 (2017)
3. Majumder, A.B., Gupta, S.: CBSTD: a cloud based symbol table driven DNA compression algorithm. In: Bhattacharyya, S., Sen, S., Dutta, M., Biswas, P., Chattopadhyay, H. (eds.) Industry Interactive Innovations in Science, Engineering and Technology. LNNS, vol. 11, pp. 467–476. Springer, Singapore (2018). https://doi.org/10.1007/978-981-10-3953-9_45
4. Aly, W., Yousuf, B., Zohdy, B.: A Deoxyribonucleic acid compression algorithm using auto-regression and swarm intelligence. J. Comput. Sci. **9**, 690–698 (2013)
5. Kuruppu, S., Puglisi, S.J., Zobel, J.: Reference sequence construction for relative compression of genomes. In: Grossi, R., Sebastiani, F., Silvestri, F. (eds.) SPIRE 2011. LNCS, vol. 7024, pp. 420–425. Springer, Heidelberg (2011). https://doi.org/10.1007/978-3-642-24583-1_41
6. GenBank. https://www.ncbi.nlm.nih.gov/genbank/
7. Behzadi, B., Le Fessant, F.: DNA compression challenge revisited: a dynamic programming approach. In: Apostolico, A., Crochemore, M., Park, K. (eds.) CPM 2005. LNCS, vol. 3537, pp. 190–200. Springer, Heidelberg (2005). https://doi.org/10.1007/11496656_17
8. Keerthy, A.S., Appadurai, A.: An empirical study of DNA compression using dictionary methods and pattern matching in compressed sequences. Int. J. Appl. Eng. Res. **10**, 35064–35067 (2015)
9. Al-Okaily, A., Almarri, B., Al Yami, S., Huang, C.H.: Toward a better compression for DNA sequences using huffman encoding. J. Comput. Biol. **24**, 280–288 (2017)

10. Arya, G.P., Bharti, R.K., Prasad, D., Rana, S.S.: An Improvement over direct coding technique to compress repeated & non-repeated nucleotide data. In: 2016 International Conference on Computing, Communication and Automation, pp. 193–196. IEEE Press, Noida (2016)

11. Rastogi, K., Segar, K.: Analysis and performance comparison of lossless compression techniques for text data. Int. J. Eng. Comput. Res. **3**, 123–127 (2014)

12. Singh, A.V., Singh, G.: A survey on different text data compression techniques. Int. J. Sci. Res. **3**, 1999–2002 (2014)

13. Brar, R., Singh, B.: A survey on different compression techniques and bit reduction algorithm for compression of text/lossless data. Int. J. Adv. Res. Comput. Sci. Softw. Eng. **3**, 579–582 (2013)

14. Priyanka, M., Goel, S.: A compression algorithm for DNA that uses ASCII values. In: 2014 IEEE International Advance Computing Conference, pp. 739–743. IEEE Press, Gurgaon (2014)

15. Gzip. http://www.gzip.org/

16. Nour, S.B., Amr, A.S.: DNA lossless compression algorithms: review. Am. J. Bioinform. Res. **3**, 72–81 (2013)

17. Khalid, S.: Introduction to Data Compression. Morgan Kaufmann, San Francisco (2006)

18. Mark, N., Jean-Loup, G.: The Data Compression Book. Morgan Kaufmann, New York (2012)

19. Grumbach, S., Tahi, F.: Compression of DNA Sequences. In: Proceedings of the Data Compression Conference, DCC 1993, pp. 340–350. IEEE Press, Snowbird (1993)

20. Korodi, S., Tabus, I., Rissanen, J., Astola, J.: DNA sequence compression based on the normalized maximum likelihood model. IEEE Signal Process. Mag. **24**, 47–53 (2007)

21. Rajeswari, P.R., Apparao, A., Kumar, V.K.: Genbit compress tool (GBC): a Java based tool to compress DNA sequences and compute compression ratio (BITS/BASE) of genomes. Int. J. Comput. Sci. Inf. Technol. **2**, 181–191 (2010)

22. Rajeswari, P.R., Apparao, A.: DNABIT compress - genome compression algorithm. Bioinformation **5**, 350–360 (2011)

23. Roy, S., Bhagot, A., Sharma, K., Khatua, S.: BVRLDNAComp: an effective DNA sequence compression algorithm. Int. J. Comput. Sci. Appl. **5**, 73–85 (2015)

24. Rexline, S.J., Aju, R.G., Trujilla, L.F.: Higher compression from burrows-wheeler transform for DNA sequence. Int. J. Comput. Appl. **173**, 11–15 (2017)

25. Habib, N., Ahmed, K., Jabin, I., Rahman, M.M.: Modified HuffBit compress algorithm – an application of R. J. Integr. Bioinform. **15**, 1–13 (2018)

26. National Center for Biotechnology Information. https://www.ncbi.nlm.nih.gov/

27. Roy, S., Khatua, S.: DNA data compression algorithms based on redundancy. Int. J. Found. Comput. Sci. Technol. **4**, 49–58 (2014)

Robust Segmentation of Overlapping Cells in Cervical Cytology Using Light Convolution Neural Network

Shusong Xu[1], Chen Sang[1], Yulan Jin[2], and Tao Wan[1(✉)]

[1] School of Biomedical Science and Medical Engineering, Beijing Advanced Innovation Centre for Biomedical Engineering, Beihang University, Beijing, China
taowan@buaa.edu.cn
[2] Department of Pathology, Beijing Obstetrics and Gynecology Hospital, Capital Medical University, Beijing, China

Abstract. Automated segmentation of cells in cervical cytology images poses a great challenge due to the presence of fuzzy and overlapping cells, noisy background, and poor cytoplasmic contrast. We present an improved method for segmenting nuclei and cytoplasm from a cluster of cervical cells using convolutional network and fast multi-cell labeling. A light convolutional neural network (CNN) model is employed to generate nuclei candidates, which can serve as accurate initializations for the subsequent level set segmentation and provide a priori knowledge for the cytoplasm segmentation. A fast multi-cell labeling method based on the superpixel map is devised to roughly segment clumped and inhomogeneous cytoplasm before applying a cell boundary refinement approach. A shape constraint in conjunction with boundary and region information drive a level set formulation to perform a robust cell segmentation. The qualitative and quantitative evaluations demonstrated that the presented cellular segmentation method is effective and efficient.

Keywords: Nuclei detection · Cytoplasm segmentation
Cervical cytology · Convolutional Neural Network · Multi-cell labeling

1 Introduction

According to the World Health Organization, cervical cancer is the fourth most common cause of cancer death in women worldwide [9]. Currently, pap smear test or Thinprep cytologic test is an important routine screening in the early detection of cervical cancer [12]. During the test, a sample of cells from the cervix is smeared, or spread, onto a glass slide, and examined under a microscope by a cytologist or pathologist for nuclear and cytoplasmic atypia to detect precancerous abnormalities. However, this manual process is time consuming, prone

T. Wan—This work was partially supported by the National Natural Science Foundation of China under award No. 61876197.

to errors, intra- and inter-observer variability, and leads to a large variation in false negative results [7]. The issues involved in the manual analysis have motivated the development of automated systems for the analysis of cervical cytology images. Among them, the automated segmentation of overlapping cells remains one of the most critical challenges.

Recently, many approaches for cervical cell segmentation have been proposed in the literature [10]. Most of the current automated methods mainly aim to segment overlapping nuclei and cellular masses [8]. For example, Tareef *et al.* [14] developed a nuclei and cytoplasm segmentation in a cluster of cervical cells based on distinctive local features and guided sparse shape deformation. Lu *et al.* [7] devised a two-stage method using a joint optimization of multiple level set functions for the segmentation of cytoplasm and nuclei. Learning based methods have become popular in tackling the cell segmentation on cervical cytology images. Song *et al.* [13] combined multiscale convolutional neural network (CNN) and graph partitioning to segment cervical cytoplasm and nuclei of overlapping cells. In their latest work [12], multiscale deep CNN was adopted to learn the cell appearance features and shape information was considered to guide cell segmentation. Although learning process usually requires a longer computational time, these methods have demonstrated superior performance in segmenting overlapping cells. This motivates our work on building an accurate and efficient learning-based segmentation method for partitioning overlapping cervical cells.

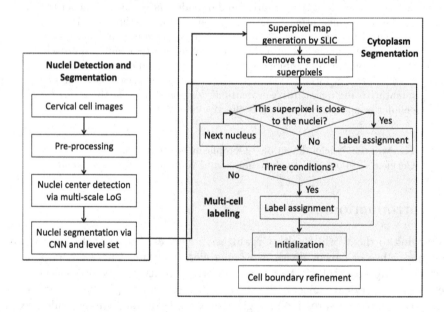

Fig. 1. Workflow of the presented cell segmentation method

In this paper, we present an automated segmentation method for accurately splitting the touching or overlapping cells in digitalized cervical cytology images.

The framework consists of two main modules shown in Fig. 1. In the nuclei detection process, a light CNN model is employed to distinguish nuclei and non-nuclei patches with reduced computational burden. The nuclei candidates can serve as initializations for the subsequent level set segmentation and provide a priori knowledge for the cytoplasm segmentation. We generate a superpixel map via a simple linear iterative clustering (SLIC) method [1]. A derived fast multi-cell labeling method is applied to accurately segment clumped cytoplasm before a cell boundary refinement process. Further, we incorporate a shape constraint in addition to the region and boundary information in a level set formulation to provide a successful solution to tackle the difficulties of segmenting clustered cells with large shape variations and their high degree of overlapping.

This paper is organized as follows. The presented methods are described in Sect. 2. Experimental results and conclusions are demonstrated in Sect. 3 and Sect. 4, respectively.

2 Methods

2.1 Nuclei Detection and Segmentation

Pre-processing. To suppress image artifacts due to the variations in sample preparation, staining process, and scanning condition, we applied an anisotropic filtering to the cervical cell image, which was able to smooth the image meanwhile preserving important edge information.

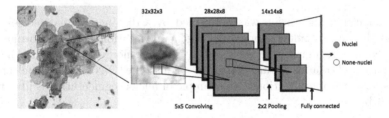

Fig. 2. Illustration of the CNN architecture

Nuclei Center Detection via Multi-scale LoG. The Laplacian of Gaussian (LoG) filter is often used for detecting elliptical blob structures appearing in the images, which can be defined as [4]: $\nabla^2 G = -\frac{1}{2\pi\sigma^4}[1 - \frac{x^2+y^2}{\sigma^2}] \cdot \exp(-\frac{x^2+y^2}{2\sigma^2})$, where G is a Gaussian filter with standard derivation of σ, and (x, y) are image pixel coordinates. A 2-D LoG filter is a circularly symmetric function. Therefore, after convolving with a digitalized cervical cell image, the response of the filter would be maximum at a scale that approximately matched the actual shape of nucleus. The matched nucleus radius (r) and the scale (σ) meet the condition of $\sigma = (r - 1)/3$. In order to identify all the different sizes of nuclei, a multi-scale

LoG filter was employed to obtain nuclei center detection. The grayscale values of center pixels were used by the LoG operator. The image patches with size of 32×32 pixels centered with the detected nuclear pixels were extracted and would serve as potential candidates containing real nuclei.

Nuclei Segmentation via CNN and Level Set. A CNN classifier was trained to distinguish nuclei and non-nuclei within the nuclei candidates identified using the multi-scale LoG filtering. Since CNN approach is a supervised classification method, we trained the CNN model with manually labeled nuclei patches. A light CNN architecture was designed with a convolutional layer, a pooling layer, and a fully-connected layer (shown in Fig. 2) [15]. The classification is as follows:

Input: The nuclei patches with RGB color channels were sized into 32×32 pixels.

Convolutional layer: A 2-D convolution of the input feature maps with a 5×5 convolution kernel. The output of this layer was eight 2-D 28×28 feature maps. The activation function of rectified linear unit (ReLU) was used [2].

Pooling layer: We applied a subsampling pooling operation over a 2×2 non-overlapping window on each output feature map, allowing to learn the invariant features.

Fully connected layer: Each neuron in this layer would be connected to the output feature maps of pooling layer. The total number of neurons is $14 \times 14 \times 8$.

Output: Two neurons (nuclei and non-nuclei) in this layer are activated by a logistic regression model.

The nuclei segmentation was performed by a distance regularized level set method [6]. In the segmentation model, a general variational level set formulation with a distance regularization term and an external energy term drive the motion of the zero level contour toward desired locations. The CNN detected nuclei patch centers served as initializations. An example of nuclei segmentation is illustrated in Fig. 3(b).

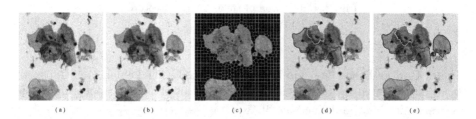

(a)   (b)   (c)   (d)   (e)

Fig. 3. Illustration of cervical cell segmentation. (a) Original image; (b) Nuclei detection; (c) Superpixel map; (d) Cell segmentation after multi-cell labeling on the overlapping cell mass; (e) Final segmentation after cell boundary refinement.

2.2 Cytoplasm Segmentation

We classified cell regions into single cells and overlapping or cell masses by measuring the region attributes, i.e., ratio of region area and average area across all the segmented regions, eccentricity, and ratio of region diameter and perimeter [3]. For isolated cells, the segmentation can be obtained by directly performing the level set method. For a cell mass containing touching or overlapping cells, we devised a computerized method involving the superpixel partitioning, multi-cell labeling, and cell boundary refinement.

Superpixel Map Generation. We first over-segmented the cell image using the simple linear iterative clustering method [1]. To speed up the process, the background was assigned to 0. We noted that some nuclei would also be partitioned into several superpixels. Since these nuclei superpixels would affect the subsequent multi-cell labeling, we discarded these superpixels. The grayscale value of a superpixel was computed by averaging the grayscale values of all the containing pixels. An example of a superpixel map is shown Fig. 3(c).

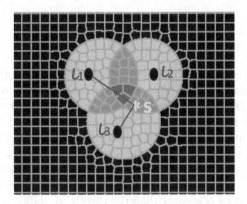

Fig. 4. Illustration of multi-cell labeling. The circle represents a cell, the dark ellipse is a nucleus, and the star is a superpixel. l_1, l_2, and l_3 are the cell labels.

Multi-cell Labeling Based on Superpixel. Assume a cell mass $C = \{c_1, c_2, ..., c_N\}$ consisting of a number of N cells (i.e., containing N nuclei). Each cell has a label $l(c)$. Let a superpixel set denote as $S = \{s_1, s_2, ..., s_M\}$, and M is the total number of superpixels in C. Each superpixel can be assigned with one or more labels belonging to the multiple cells. Clearly, a superpixel belongs to a cell if it is close to the cell's nucleus. It is more complicated if there are many superpixels around the cell. The superpixel can be assigned a label by meeting the following three conditions: (i) The distance between the superpixel's center and the nucleus center is within a specific range, which is related to the radius

---

Algorithm 1. The multi-cell labeling method

   Input: Superpixel s, thresholdings $\{T_1, T_2, T_3\}$
   Output: Superpixel's label $l(s)$
   for $i = 1 : M$ **do**
     for $j = 1 : N$ **do**
       if s_i is close to the nucleus l_j **then**
         Assigning l_j to s_i
       else
         Computing $Dist(s_i, l_j)$
         if $(Dist(s_i, l_j) < T_1)$ && $(grayscale(s_i) < T_2)$ **then**
           Finding the superpixels between s_i and l_j;
           Computing the minimum of these superpixels' mean grayscale value (λ);
           if $grayscale(s_i) - \lambda < T_3$ **then**
             Assigning l_j to s_i
           end if
         end if
       end if
     end for
     if s_i has no label **then**
       Assigning the closest nuclear label to s_i
     end if
   end for

---

of the cell; (ii) The superpixel's mean grayscale value is relatively low because the overlapping region between cells is darker than the non-overlapping region; (iii) The change of grayscale values from dark to bright is less than a pre-defined threshold, ensuring the superpixel almost located within the cell or the overlapping region. Figure 4 gives a better interpretation on the three conditions of multi-cell labeling process. In the figure, a cell is represented as a circle, and the dark ellipse is nucleus. The star s is a superpixel. The superpixel s can be easily assigned to label l_1 if using only conditions (i) and (ii). However, the grayscale change between l_1 and s is greater than the threshold, indicating that the superpixel s is out of the overlapping region of l_1. Therefore, it is useful in considering the constraint of the grayscale value change between superpixel and nucleus. The detailed implementation is shown in Algorithm 1.

Cell Boundary Refinement. We connected the superpixels with the same label to generate a cell segmentation in the cell mass. Since the superpixel is a small region, the boundary of cell is curvy and imprecise. For each cell, we ran an improved distance regularized level set method to refine the cell contour. Since the cells are roughly elliptical in shape, we introduced a prior shape constraint in the energy function, which can be expressed as: $\mathcal{E}(\phi) = \omega_1 \mathcal{R}(\phi) + \omega_2 \mathcal{L}(\phi) + \omega_3 \mathcal{A}(\phi) + \omega_4 \mathcal{S}(\phi)$, where ϕ denotes level set function in the image domain, $(\omega_1, \omega_2, \omega_3, \omega_4)$ are the weights to balance the four terms. $\mathcal{R}$ is the regularization term, $\mathcal{L}$ is the distance term, $\mathcal{A}$ is the region term, and $\mathcal{S}$ is the elliptical shape prior. $\mathcal{R}$, $\mathcal{L}$, $\mathcal{A}$, and $\mathcal{S}$ are defined as [11]: $\mathcal{R}(\phi) \triangleq$

$$\int_\phi R(\nabla\phi)\,dx,\ R(\phi) = \begin{cases} \frac{1}{(2\pi)^2}(1-\cos(2\pi\phi)) & \text{if } \phi < 1 \\ \frac{1}{2}(\phi-1)^2 & \text{if } \phi \geq 1 \end{cases},\ \mathcal{L}(\phi), \triangleq \int_\Omega g\delta(\phi)|\nabla\phi|\,dx,$$

$\mathcal{A}(\phi) \triangleq \int_\Omega gH(\phi)\,dx$, $\mathcal{S}(\phi) \triangleq \int_\Omega g(x)H(l(\phi)\,dx$, where H is the Heaviside function, δ is the derivative of H. $g = 1/(1+|1+\nabla G_\sigma * I|^2)$, where G_σ is the Gaussian kernel with standard deviation of σ, encourages the evolving contours to align with edges. $l(\phi)$ returns the signed distance map of the elliptical shape approximation of ϕ. The rough cell contour from the multi-cell labeling was used as an initialization for the level set segmentation method. We combined region, boundary, and local shape information to drive the level set evolution for a better segmentation of overlapping cervical cells in the microscopic images.

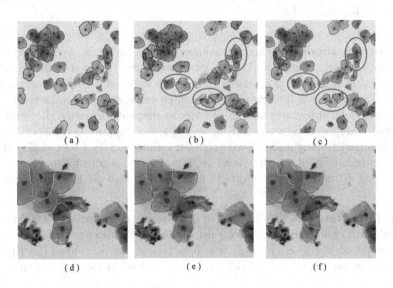

Fig. 5. Two examples of segmentation results. (a)(d) Original images; (b)(e) The presented cell segmentation method; (c)(f) Lu's method [7]. The first row is from ISBI, and the second row is from the in-house dataset. The thick red circles in the first row indicate the comparison differences. (Color figure online)

3   Experimental Results

3.1   Experimental Design

Data Description. We collected three independent datasets, including two public datasets provided in the first and second overlapping cervical cell segmentation challenges (ISBI2014 and ISBI2015) [7], and a in-house data cohort. ISBI2014 dataset contained 45 synthetic cervical cytology images in the training set and 90 synthetic images in the test set. ISBI2015 had 4 images along with their volume images in the training set and 5 images with their volume images in

the test set. Our dataset contained 14 scanned cervical cytology images at 20×
magnification. We used 4 images for training and 10 for testing. All the images
were accompanied by nucleus and cytoplasm annotations except for the ISBI2015
test dataset, therefore we did not include these images in the evaluation.

Parameter Settings. We used a learning rate of 0.01 and epoch of 30 for training the CNN model. A stochastic gradient descent was used to optimize the entire
training process. For both ISIB2014 and ISBI2015 datasets, three thresholding
values $\{T_1, T_2, T_3\} = \{60, 40, 5\}$ in the multi-cell labeling, and weights in the
improved level set method were assigned as $\{\omega_1, \omega_2, \omega_3, \omega_4\} = \{0.2, 0.5, 0.2, 0.1\}$.
For the in-house dataset, we set $\{T_1, T_2, T_3\} = \{60, 50, 15\}$ and $\{\omega_1, \omega_2, \omega_3,$
$\omega_4\} = \{0.2, 0.5, 0.1, 0.1\}$.

Reference Methods. We compared our segmentation method with three alternative approaches. Song *et al.* [12] developed a framework based on deep learning
technique and a deformation model to segment cervical cells from overlapping
clumps. Tareef *et al.* [14] devised an automated cell segmentation method based
on distinctive local features and guided sparse shape deformation. Lee and Kim
presented a segmentation method for multiple overlapping cervical cells using
superpixel partitioning and cell-wise contour refinement [5]. Lu *et al.* [7] utilized
a joint optimization of multiple level set functions to segment clump cells. The
original quantitative results reported in the papers were used for comparison.

Evaluation Metrics. The original evaluation metrics used in both ISBI2014
and ISBI2015 challenges, including dice similarity coefficient (DSC), false negative rate at object level (FNRo), true positive rate at pixel level (TPRp) and
false positive rate at pixel level (FPRp), were utilized in the experiments.

3.2 Qualitative Results

For the visual evaluation, we demonstrated two examples shown in Fig. 5, in
which two images were selected from the ISBI challenges and our in-house
dataset, respectively. We compared the method with the manual annotation and
Lu *et al.*'s method using the authors' original implementation [7]. By observing
the figures, we noted that our segmentation provided more consistent results with
the manual delineations compared with Lu's method. In addition, our method
was robust and effective in segmenting the clumped cells with inhomogeneous
cytoplasm due to the uneven straining.

3.3 Quantitative Results

The segmentation performance was quantitatively evaluated using four measures, including DSC, FNRo, TPRp, and FPRp. A segmented cell in the ground
truth was considered to be missed if there is no region in the segmentation result

Table 1. Performance comparison of cervical cell segmentation using DSC, FNRo, TPRp, and FPRp.

Dataset	Method	DSC	FNRo	TPRp	FPRp
ISBI2014	Tareef et al. [14]	0.89 ± 0.07	0.27 ± 0.28	0.91 ± 0.09	0.004 ± 0.005
	Lu's method [7]	$0.88 \pm N/A$	$0.21 \pm N/A$	$0.92 \pm N/A$	$0.002 \pm N/A$
	Lee and Kim [5]	0.90 ± 0.08	0.14 ± 0.19	0.88 ± 0.10	0.002 ± 0.002
	The presented method	$\mathbf{0.91 \pm 0.09}$	$\mathbf{0.13 \pm 0.14}$	$\mathbf{0.96 \pm 0.04}$	$\mathbf{0.002 \pm 0.003}$
ISBI2015	Song et al. [12]	$0.89 \pm N/A$	$0.26 \pm N/A$	$\mathbf{0.92 \pm N/A}$	$0.002 \pm N/A$
	Lee and Kim [5]	0.88 ± 0.09	0.43 ± 0.17	0.88 ± 0.12	0.001 ± 0.001
	The presented method	$\mathbf{0.90 \pm 0.03}$	$\mathbf{0.21 \pm 0.07}$	0.89 ± 0.02	$\mathbf{0.001 \pm 0.002}$
In-house dataset	Lu's method [7]	0.87 ± 0.05	0.32 ± 0.14	0.88 ± 0.05	0.004 ± 0.004
	The presented method	$\mathbf{0.90 \pm 0.03}$	$\mathbf{0.24 \pm 0.05}$	$\mathbf{0.90 \pm 0.04}$	0.004 ± 0.003

that has a DSC greater than 0.7 with it. FNRo is the rate of cells missed in the ground truth and TPRp and FPRp are the average of true positive and false positive rates at the pixel level of those regions that are not missed. The higher values of DSC and TPRp and the lower values of FNRo and FPRp indicate better segmentation performance. The experimental results are listed in Table 1. Tareef et al. and Lu et al. used the ISBI2014 dataset [7,14]. Song's method [12] was assessed using the ISBI2015 dataset. Lee and Kim utilized both ISBI2014 and ISBI2015 datasets [5].

By examining the results shown in Table 1, our method achieved the best segmentation performance in terms of DSC, FNRo, and FPRp for both ISBI2014 and ISBI2015 datasets, and highest TPRp for ISBI2014 and second best TPRp for ISBI2015. In Lee and Kim's method, they also adopted superpixel partitioning and cell contour refinement technique for the clump cell segmentation. Our method outperformed Lee and Kim's method in all the metrics. This is mainly because that we excluded the nuclei superpixels in the initialization and set a constraint of grayscale change between the nucleus and superpixel during the multi-cell labeling, which could improve overlapping cytoplasm segmentation. The presented method yielded the DSC value of 0.90 for the in-house data cohort, which was comparable with the other two datasets. Our method outperformed Lu's method [7] in both ISBI2014 and in-house dataset, suggesting that the integration of CNN model and multi-cell labeling technique could provide good segmentation for largely clustered and highly overlapping cells. Moreover, an evaluation on three different datasets demonstrated the superiority of the presented algorithm over the other state-of-the-art methods.

3.4 Computational Time

Our method was performed under MATLAB R2017a platform on a machine with a 3.40 GHz Intel Core i5-7500 CPU and 8 GB RAM. The running time is approximately 12 s for segmenting a image patch with size of 512×512 pixels. This indicates that our method is at least 10 times faster than the fastest method reported in [7].

4   Conclusion

We presented an automated computerized method to segment the overlapping cells in cervical cytology images. A combination of the CNN technique and multi-cell labeling allows to provide superior performance in segmenting inhomogeneous cell cytoplasm. Moreover, we incorporated a shape prior in addition to the region and boundary information in the cell boundary refinement process to provide a successful solution to tackle the difficulties of segmenting clustered cells due to the large variation in the cell shapes and their high degree of overlapping. Our approach offered improvements over existing techniques in accuracy and speed of computation.

References

1. Achanta, R., Shaji, A., Smith, K., Lucchi, A., Fua, P., Susstrunk, S.: SLIC superpixels compared to state-of-the-art superpixel methods. IEEE Trans. Pattern Anal. Mach. Intell. **34**, 2274–2282 (2012)
2. Hinton, G.E.: A practical guide to training restricted Boltzmann machines. In: Montavon, G., Orr, G.B., Müller, K.-R. (eds.) Neural Networks: Tricks of the Trade. LNCS, vol. 7700, pp. 599–619. Springer, Heidelberg (2012). https://doi.org/10.1007/978-3-642-35289-8_32
3. Jing, J., Wan, T., Cao, J., Qin, Z.: An improved hybrid active contour model for nuclear segmentation on breast cancer histopathology. In: IEEE International Symposium on Biomedical Imaging, pp. 1155–1158 (2016)
4. Kong, H., Akakin, H., Sarma, S.: A generalized Laplacian of Gaussian filter for blob detection and its applications. IEEE Trans. Cybern. **43**(6), 1719–1733 (2013)
5. Lee, H., Kim, J.: Segmentation of overlapping cervical cells in microscopic images with superpixel partitioning and cell-wise contour refinement. In: the IEEE Conference on Computer Vision and Pattern Recognition Workshops, pp. 1367–1373 (2016)
6. Li, C., Xu, C., Gui, C., Fox, M.: Distance regularized level set evolution and its application to image segmentation. IEEE Trans. Image Process. **19**(12), 3243–3254 (2010)
7. Lu, Z., Carneiro, G., Bradley, A.: An improved joint optimization of multiple level set functions for the segmentation of overlapping cervical cells. IEEE Trans. Image Process. **24**(4), 1261–1272 (2015)
8. Lu, Z., et al.: Evaluation of three algorithms for the segmentation of overlapping cervical cells. IEEE J. Biomed. Health Inform. **21**, 441–450 (2017)
9. McGuire, S.: World cancer report 2014 Geneva, Switzerland: world health organizaiton, international agency for research on cancer. Adv. Nutr. **7**(2), 418–419 (2016)
10. Phoulady, H., Goldgof, D., Hall, L., Mouton, P.: A framework for nucleus and overlapping cytoplasm segmentation in cervical cytology extended depth of field and volume images. Comput. Med. Imaging Graph. **59**, 38–49 (2017)
11. Rousson, M., Paragios, N.: Shape priors for level set representations. In: Heyden, A., Sparr, G., Nielsen, M., Johansen, P. (eds.) ECCV 2002. LNCS, vol. 2351, pp. 78–92. Springer, Heidelberg (2002). https://doi.org/10.1007/3-540-47967-8_6
12. Song, Y., et al.: Accurate cervical cell segmentation from overlapping clumps in pap smear images. IEEE Trans. Med. Imaging **36**(1), 288–300 (2017)

13. Song, Y., Zhang, L., Chen, S., Ni, D., Lei, B., Wang, T.: Accurate segmentation of cervical cytoplasm and nuclei based on multiscale convolutional network and graph partitioning. IEEE Trans. Biomed. Eng. **62**(10), 2421–2433 (2015)
14. Tareef, A., et al.: Automatic segmentation of overlapping cervical smear cells based on local distinctive features and guided shape deformation. Neurocomputing **221**, 94–107 (2017)
15. Wan, T., Cao, J., Chen, J., Qin, Z.: Automated grading of breast cancer histopathology using cascaded ensemble with combination of multi-level image features. Neurocomputing **229**, 34–44 (2017)

Semantic Similarity Measures to Disambiguate Terms in Medical Text

Kai Lei, Jiyue Huang, Shangchun Si, and Ying Shen[✉]

Shenzhen Key Lab for Information Centric Networking and Blockchain Technology
(ICNLAB), School of Electronics and Computer Engineering,
Peking University, Shenzhen, China
{leik,shenying}@pkusz.edu.cn,
{huangjiyue,shangchunsi}@sz.pku.edu.cn

Abstract. Computing the semantic similarity accurately between words is an important but challenging task in the semantic web field. However, the semantic similarity measures involve the comprehensiveness of knowledge learning and the sufficient training of words of both high and low frequency. In this study, an approach MedSim is presented for semantic similarity measures to identify synonym terms in medical text with effectiveness and accuracy well-balanced. Experimental results on Chinese medical text demonstrate that our proposed method has robust superiority over competitors for synonym identification.

Keywords: Semantic similarity · Word embedding
Feature selection · Chinese medical text

1 Introduction

Semantic similarity measures between terms or concepts, also known as synonym identification, becoming intensively used for most applications in natural language processing and information retrieval, such as sense disambiguation [11] and document classification [17]. Given a knowledge sources, semantic similarity measures compute the similarity between words in order to perform estimations.

Semantic similarity measures are usually performed using some kind of metrics. Some use statistical models [16] or lexical pattern, and other represent contexts using the vector space model [3], or make use of conceptual hierarchies [2]. Several similarity measures have been developed, being given the existence of a structured knowledge representation offered by ontologies and corpus which enable semantic interpretation of terms [6].

Despite effectiveness of previous studies, the study of semantic similarity measures in medical field in Chinese language is still a relatively new territory. There are several challenges: (1) The global context information and Chinese linguistic information, which play crucial roles in semantic comprehension, are underexplored in recent work for synonym identification. (2) Limited performance for learning the low frequency words in a large-scale corpus, compared with the high frequency words, resulting in the inaccurate entity knowledge presentation.

© Springer Nature Switzerland AG 2018
L. Cheng et al. (Eds.): ICONIP 2018, LNCS 11307, pp. 398–409, 2018.
https://doi.org/10.1007/978-3-030-04239-4_36

To alleviate these limitations, we propose a learning method MedSim that determines the intended concept associated with an ambiguous word using semantic similarity measures. In specific, we first explore the linguistic and contextual semantic features, so as to augment the amount of information of the low frequency words and relieve the demand of a corpus with complete knowledge. Then, we alleviate the limitations of plain word information by involving global context extracted from search engine. Finally, we concatenate all the information that has well-balanced the effectiveness and accuracy and compare its performance with the state-of-art approaches.

The contribution of this papers can be summarized as follows: (1) We learn the linguistic and contextual semantic features, which improve the knowledge representation of low frequency words; (2) We exploit the global context information to capture useful topical information of pairwise word, which improve the limitations of plain word information learning; (3) The experimental results show that Medsim consistently outperforms the state-of-the-art methods.

2   Related Works

Existing methods that have been proposed to automatically identify synonyms in text can be classified into three groups: supervised, unsupervised and knowledge-based methods. Supervised methods [15] use machine learning algorithms to assign concepts/terms to instances containing the ambiguous word. To employ these methods, it is inevitably to create training data for each target word to be disambiguated, which is usually labor-intensive and time-consuming. Knowledge-based methods mainly use information from an external knowledge source or a corpus of text [8]. The accuracy of semantic similarity measures heavily relies on the degree of knowledge completeness and structure sparseness of the adopted knowledge base.

In this study, we apply unsupervised methods which adopt the distributional characteristics of a corpus to compute the semantic similarity. Several studies have attempted to explore the text corpus information with unsupervised methods. A text corpus typically contains two types of context information: global context and local context. Global context carries topical information which can be utilized by topic models, e.g. Latent Dirichlet Allocation (LDA) and GloVe [13], to discover topic structures from the text corpus. Local context can train word embeddings such as NPLM [1] and Word2Vec [10] to capture semantic regularities reflected in the text corpus [8]. As a distributed representation of words, the basic idea of word embedding is to convert a word into a vector then project into a low-dimensional vector space, enabling the similar vectors in the same space share higher relevance.

3   Methodology

In this section, we propose an extensible word embedding model MedSim to conduct synonym identification of Chinese medical characters in detail, as Fig. 1 has briefly illustrated the general architecture. MedSim firstly transforms the input Chinese pairwise words into vectors using Word2vec. Then, features are extracted and mapped into low dimension vectors by feature-adjusted mapping function. Finally, word embeddings and feature embeddings are concatenated to identify synonyms. Components will be presented in detail in this section.

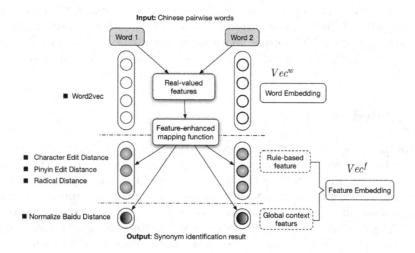

Fig. 1. MedSim architecture.

3.1   Feature Selection

Following our previous work [7], 2^{13} experiments with or without one or more specific features were performed to observe the contribution of each feature and to identify which combinations of features are more effective. We employ the SVM classifier to classify several pairs of words based on the selected features to implement synonym identification, and unified measurements are precision, recall, and $F1$ values. A total of 1,526 pairs of synonyms are included as positive samples. Negative pairs are constructed from optional words in the modern Chinese dictionary. In the 2^{13} combinations, we select the top 10% (sorted by $F1$ score) to observe the frequency of each feature (see Fig. 2). Features with higher frequency means they perform better in the combinations. Thus, we notice that the most useful features for our task include the Chinese cosine similarity, Radical, Normalized Baidu distance, Chinese edit distance, and the Pinyin Edit Distance.

3.2 Feature Embedding

Rule-Based Features. Rule-based features such as radicals and pronunciation can effectively reflect the similarities between Chinese characters and pronunciation. The application of Rule-based features which related to the linguistic information can reduce the requirement of the knowledge completeness of the adopted knowledge base, thus improving the synonym identification of low frequency words.

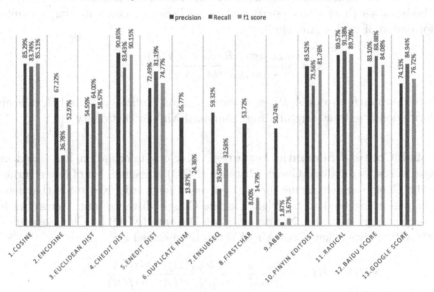

Fig. 2. Precision, recall and F1 of single feature in combinations.

Chinese Character Edit Distance. Shorter editing distance tend to imply synonym such as "维生素B2 缺乏病" and "维生素B2 缺乏症" (they share an editing distance of 1). In this paper, we define relative Chinese Character Edit Distance as:

$$EditDist(A, B) = \frac{editDistance(A, B)}{maxLength(A, B)} \tag{1}$$

where $editDistance(A, B)$ is the minimum number of edit operations from one string to another, and $maxLength(A, B)$ is the max length of A, B.

Pinyin Edit Distance. Pinyin refers to pronunciation of Chinese characters, in the medical domain, Pinyin can eliminate the difference brought by transliteration. Such as "埃博拉病毒" and "埃播拉病毒" . This paper extracts pinyin from standard Xinhua Dictionary, we define relatively Pinyin Edit Distance based on Edit Distance:

$$pinyinEditDist = \frac{pinyinEditDistance(A, B)}{maxLength(A, B)} \qquad (2)$$

an example for $pinyinEditDistance(A, B)$ is pinyin for both "埃博拉病毒" and "埃播拉病毒" is "ai'bo'la'bing'du", thus, making $pinyinEditDist = 0$.

Radical Distance. As a pictographic language, radicals of the Chinese characters has its their semantic meaning. Specifically, radicals which appear more frequently in the medical domain distinct from other domains such as " "月" (part of body), "艹" (bacterial related) and "疒" (diseases), compensating semantic information of words. And similarly:

$$CR = \frac{commonRadicals(A, B)}{maxLength(A, B)} \qquad (3)$$

where $commonRadicals(A, B)$ represents common Radicals of A and B, radicals information is extract carefully from standard Xinhua Dictionary.

Global Context Semantic Features. Previous information has based on single words, Normalized Google Distance (NGD) is the number of hits returned by the Google search engine for a given set of keywords. Keywords that have similar meaning tend to be "tight" in the Google distance unit. According to pre-experiment, we utilize Normalize Baidu Distance (NBD) for Chinese medical text, NBD between keywords A and B can be specified as:

$$NBD(A, B) = \frac{max(logf(A), logf(B) - logf(A, B))}{logM - min(logf(A), logf(B))} \qquad (4)$$

where M is the total numbers of web pages by searching, $f(A)$ and $f(B)$ is the hit times of A and B respectively, while $f(A, B)$ is time of co-occurrence.

In each combination of corpus and frequency extractors, the relative frequency of words or phrases is defined. The results returned by the search engine can be approximated as the relative usage frequency in practice. Thus, NBD approximately capture the semantic relevance of the two lexical items, and all web page retrieved by Baidu can be utilized by NBD.

3.3 Feature-Adjusted Mapping Function

Feature-adjusted mapping function is designed to adjust the impact of different features. By projecting the Rule-based and Global context semantic features into real-valued vectors, the mapping function $y = f(x)$ is given by:

$$y_i = \begin{cases} \frac{x_i}{thr} & \text{if } x_i < \text{thr} \\ \frac{1 - 2thr + x_i}{2(1 - thr)} & \text{if } x_i \geq \text{thr} \end{cases} \qquad (5)$$

where x_i is a real number feature with a range of $[0, 1]$, y is the mapping output with a range of $[-1, 1]$. The threshold thr is set to 0.3.

In this piecewise function, when x is greater than thr, its corresponding y value tends to 1, indicating that the two words are very similar; when x is smaller than thr, y value tends to -1, which indicates that the two words are with lower similarity. A fuzzy similarity of 0.5 is set when x equals to thr.

For a pairwise word $(\mathbf{w1}, \mathbf{w2})$, given a feature x as a real number:

$$Vec_{w1} = [f(x), f(x), ..., f(x)] \tag{6}$$

$$Vec_{w2} = [1, 1, ..., 1] \tag{7}$$

where Vec_{w1} represents vector concatenated by transforming y from real-valued feature x following (5), and Vec_{w2} is an all 1 vector in respected to Vec_{w1}. Length of Vec_{w1} and Vec_{w2} can be determined manually.

3.4 Extensible Embedding Model

Different features are transformed to vectors and then extended to existing word embedding vectors. The specific approach is to convert the selected features into vectors based on feature embedding representations by expanding in the vector dimension. For each feature extracted above, our model translates it into a real-valued vector, which is concatenated to the existing word vector representation by the following format:

$$word = [vec^w, vec^{f1}, vec^{f2}, ..., vec^{fm}] \tag{8}$$

where vec^w is word vectors by the existing Word2vec model, and vec^{fi} is the ith feature representation as illustrated before. Hence, MedSim can capture linguistic information and global context information based on the existing word vector which focus on local context information. Our model can better handle the problems of contextual solidification in texts in the medical domain as well as the low accuracy for words appears less frequently.

4   Experiment

4.1   Dataset and Experimental Setups

Datasets. The proposed approach is evaluated on synonym datasets extracted from ve authoritative Chinese medical websites and knowledge bases: A+ Medical Encyclopedia (A+ Medical)[1], Baidu Encyclopedia (Baidu)[2], Hudong Encyclopedia (Hudong)[3], Xywy-A Doctor-Patient interaction website[4] and China Disease Knowledge Total Database (CDD)[5]. The five data sources can be described in detail in Table 1:

---

[1] http://cht.a-hospital.com/.
[2] https://baike.baidu.com/.
[3] http://www.baike.com/.
[4] http://www.xywy.com/.
[5] http://www.lib.whu.edu.cn/dc/viewdc.asp?id=536.

Table 1. Description of dataset sources in detail.

Source	A+ Medical	CDD	Baidu	Hudong	Xywy
Quantity	8493	17517	5238	3095	8126

Implementation Details. We use a subset of our dataset mentioned before, which includes 42469 pairs of annotated examples of synonym. We adopt 7666 pairs out of them for the task of synonym identification, with unfamiliar entities removed. Trained Word2vec embeddings of 100 dimensions are adopted as word embeddings. When features are involved as combination information, each feature corresponds to a 10-dimensional vector. Thr of feature-enhanced mapping function was set as 0.4. In our experiment, the ratio of positive and negative samples is 1:1, and all negative samples are randomly selected from our dataset.

Evaluation. In this paper, the evaluation metrics are described as follows:

Correlation Coefficient. Spearman's Rank Correlation Coefficient (ρ) and Pearson Correlation Coefficient (r) were used to evaluate the improvement effect of the model. These evaluation methods are widely used to evaluate the consistency between the results of automatic prediction and the manually labeled standard results.

$$\rho = 1 - \frac{6 \sum_{i=1}^{n} (R_{X_i} - R_{Y_i})^2}{n(n^2 - 1)} \tag{9}$$

The Spearman correlation coefficient shows the relative directions of X (independent variables) and Y (dependent variables). If Y tends to increase when X increases, the Spearman correlation coefficient is positive. if on the contrary, the Spearman correlation coefficient is negative. A Spearman correlation coefficient of zero indicates that Y does not have any tendency when X increases.

$$r = \frac{\sum_{i=1}^{n} (X_i - \bar{X})(Y_i - \bar{Y})}{\sqrt{\sum_{i=1}^{n} (X_i - \bar{X})} \sqrt{\sum_{i=1}^{n} (Y_i - \bar{Y})}} \tag{10}$$

The Pearson correlation coefficient ranges from -1 to 1. A value of 1 means that X and Y are well described by the equation of a line, all data points follow a straight line, and Y increases as X increases; the value of -1 means that Y decreases as X increases; a value of 0 means no linear relationship between the two variables.

Precision, Recall and F1 Score. The Confusion Matrix is a widely used in the classification problem. According to the combination of the real category and the predicted category of the model, the sample is divided into four cases: True Positive, False Positive, True Negative, and False Negative. Precision (P) indicates whether predicted positive samples are true, While Recall (R) indicates how many positive examples are correctly predicted.

$$P = \frac{TP}{TP + FP} \tag{11}$$

$$R = \frac{TP}{TP + FN} \tag{12}$$

since they are two contradictory measures, we introduce $F1\ Score$ as:

$$F1 = \frac{2 * P * R}{P + R} \tag{13}$$

ROC Curve. Points on the Receiver Operating Characteristic Curve reflects the susceptibility to the same input. According to the results predicted by the model, samples were ordered as positive examples. The False Positive rate (FP) and True Positive rate (TP) were obtained, and they were plotted as the horizontal and the vertical axis. Defined as:

$$FP = \frac{FP}{TN + FP} \tag{14}$$

$$TP = \frac{TP}{TP + FN} \tag{15}$$

Accuracy. The Accuracy can reflect the ability to determine True Positive and True Negative. And can be calculated by the following formula:

$$Accuracy = \frac{TP + TN}{TP + FN + FP + TN} \tag{16}$$

4.2 Experimental Results

For all implemented method, we apply the same parameter settings as comparison, GloVe, Cilin+W2V [12], SVM-uni [9], BiNB [14], STS [4] and SOC-PMI [5] are selected and Table 2 compares MedSim proposed in this paper with other existing methods of synonym identification using our dataset. We observe that MedSim achieves the best results in both correlation coefficient and accuracy. Additionally, ablation test is reported to analysis the effectiveness of different features of MedSim in term of discarding features (w/o rule/global). Generally, both types of features contribute.

From the results we can see that, MedSim has achieved a highest accuracy compared with state-of-art method. Multiple features added in this model is interpretable.

As shown in the result for correlation: (1) The PMI and PPMI model show poor correlation; (2) A comparison of the results of the two widely used word vector models shows that the results obtained by Word2vec are better correlated than GloVe; (3) Our base models, i.e., adding linguistic information and global context information separately, outperforms the baseline Word2vec, indicating that our method can aid in the performance of low-frequent words; (4) The utilization of search engines improves the performance of our model. This is within our expectation since search engines can provide background information and hidden relations beyond the context, which play crucial roles in human text comprehension. Therefore, our model has robust superiority over competitors (Table 3).

Table 2. Result of synonym identification (using accuracy)

Model	Accuracy
PMI	0.756
STS (2008)	0.726
SOC-PMI (2006)	0.763
SVM-uni (2011)	0.794
GloVe (2014)	0.637
W2V (2013)	0.807
BiNB (2013)	0.831
MedSim	**0.897**
w/o global	0.818
w/o rule	0.849

Table 3. Result of synonym identification (using correlation coefficient)

Model	Spearman	Pearson
PMI	0.41482	0.42197
PPMI	0.49546	0.28410
Cilin+W2V (2016)	0.45700	0.45500
GloVe (2014)	0.54086	0.52131
W2V (2013)	0.54086	0.60303
MedSim	**0.76860**	**0.74286**
w/o global	0.74362	0.65688
w/o rule	0.64461	0.64106

4.3 Analysis

Effect of Adding Rule-Based Features. In this section, we verified the distribution of similarity for both positive and negative samples in the model after combining the linguistic information. As in Fig. 3, the left panel shows the experimental result of the positive samples, and the right panel for negative samples. Comparing with the red curve, adding linguistic information can lead to a noticeable shift of the similarity distribution of the blue curve, indicating the improvement of the disambiguation identification. However, the homophonic word with different meanings cannot be well distinguished by the application of local rule-based feature (i.e. pinyin), thus the model (Word2vec+linguistic) does not operate well in certain situation (e.g. in x axis less than 0.0).

Effective of Adding Global Context Features. Based on the previous rule-based feature embeddings, global feature of Normalized Baidu Distance is added to verify the effect of adding global context semantic features.

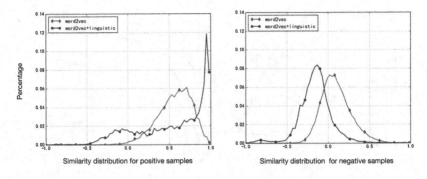

Fig. 3. Effect of adding rule-based features

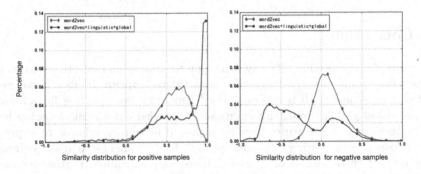

Fig. 4. Effect of adding global context semantic features

Similarly, we can see from Fig. 4 that: (1) For the blue curve, its peak shifts to the right and exceeds the given threshold obviously in the positive samples, while its peak shifts to the left for negative ones, indicating that words can be better distinguished; (2) Specifically, the left panel shows that adding global context semantic information can solve the problem caused by linguistic information alone presented in Fig. 3.

Model presented in this paper is labeled as MedSim (MedSimc for model based on CBOW and MedSims for Skip-gram). Experimental results above prove that adding global context information based on linguistic information can improve performance of our synonym identification model. *Precision* for adding rule-based features is 0.79246, and 0.77602 for Recall, which indicates that the global context information can improve Recall without reducing *Precision*. Global context information ensures the model to make better use of contextual semantic information, better solving the problem of different morphologies or pronunciations with same semantics. It can improve Precision and complement linguistic information (Fig. 5).

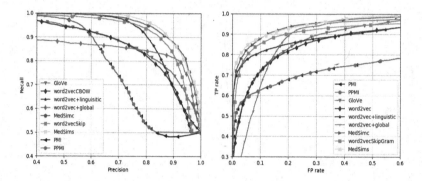

Fig. 5. Comparison for *PR* and *ROC Curve* with ablation

5   Conclusion

Motivated by the need to identify synonym accurately and effectively so as to reduce the requirement of large-scale knowledge-complete databases, we presented a novel approach to measure word similarity by capturing fine-grained linguistic and global context information. We carried out our model using feature embedding, such features have been proved by our experiment to improve the performance of existing methods especially for handling low frequency words, showing that these sources of information are complementary. Further research will include entity linking based on identified synonyms, contributing to the ontology completion.

Acknowledgements. This work has been financially supported by Guangdong Prenational project (Grant No.: 2014GKXM054), the National Natural Science Foundation of China (No. 61602013), and the Shenzhen Key Fundamental Research Projects (Grant No.: JCYJ20170818091546869).

References

1. Bengio, Y., Schwenk, H., Senécal, J.S., Morin, F., Gauvain, J.L.: Neural probabilistic language models. In: Holmes, D.E., Jain, L.C. (eds.) Innovations in Machine Learning. Studies in Fuzziness and Soft Computing, vol. 194, pp. 137–186. Springer, Heidelberg (2006). https://doi.org/10.1007/3-540-33486-6_6
2. Christensen, J., Pasca, M.: Instance-driven attachment of semantic annotations over conceptual hierarchies. In: Proceedings of the 13th Conference of the European Chapter of the Association for Computational Linguistics, pp. 503–513. Association for Computational Linguistics (2012)
3. Gábor, K., Zargayouna, H., Tellier, I., Buscaldi, D., Charnois, T.: Exploring vector spaces for semantic relations. In: Proceedings of the 2017 Conference on Empirical Methods in Natural Language Processing, pp. 1814–1823. Association for Computational Linguistics (2017)
4. Islam, A., Inkpen, D.: Semantic text similarity using corpus-based word similarity and string similarity. ACM Trans. Knowl. Discov. Data **2**(2), 1–25 (2008)

5. Islam, M.A., Inkpen, D.: Second order co-occurrence PMI for determining the semantic similarity of words. In: Proceedings of the International Conference on Language Resources and Evaluation, pp. 1033–1038 (2006)
6. Krishnan, R., Hussain, A., Sherimon, P.C.: Retrieval of semantic concepts based on analysis of texts for automatic construction of ontology. In: Huang, T., Zeng, Z., Li, C., Leung, C.S. (eds.) ICONIP 2012. LNCS, vol. 7663, pp. 524–532. Springer, Heidelberg (2012). https://doi.org/10.1007/978-3-642-34475-6_63
7. Lei, K., Si, S., Wen, D., Shen, Y.: An enhanced computational feature selection method for medical synonym identification via bilingualism and multi-corpus training. In: IEEE International Conference on Big Data Analysis, pp. 909–914 (2017)
8. Medrouk, L., Pappa, A.: Deep learning model for sentiment analysis in multilingual corpus. In: International Conference on Neural Information Processing, pp. 205–212 (2017)
9. Mendes, A.C., Wichert, A.: From symbolic to sub-symbolic information in question classification. Artif. Intell. Rev. **35**(2), 137–154 (2011)
10. Mikolov, T., Yih, W.T., Zweig, G.: Linguistic regularities in continuous space word representations. In: HLT-NAACL (2013)
11. Pasini, T., Navigli, R.: Train-o-matic: large-scale supervised word sense disambiguation in multiple languages without manual training data. In: Proceedings of the 2017 Conference on Empirical Methods in Natural Language Processing, pp. 78–88. Association for Computational Linguistics (2017)
12. Pei, J., Zhang, C., Huang, D., Ma, J.: Combining word embedding and semantic lexicon for chinese word similarity computation. In: Lin, C.-Y., Xue, N., Zhao, D., Huang, X., Feng, Y. (eds.) ICCPOL/NLPCC -2016. LNCS (LNAI), vol. 10102, pp. 766–777. Springer, Cham (2016). https://doi.org/10.1007/978-3-319-50496-4_69
13. Pennington, J., Socher, R., Manning, C.: Glove: global vectors for word representation. In: Conference on Empirical Methods in Natural Language Processing, pp. 1532–1543 (2014)
14. Socher, R., et al.: Recursive deep models for semantic compositionality over a sentiment treebank (2013)
15. Virk, S.M., Muller, P., Conrath, J.: A supervised approach for enriching the relational structure of frame semantics in framenet. In: Proceedings of COLING 2016, the 26th International Conference on Computational Linguistics: Technical Papers, pp. 3542–3552. The COLING 2016 Organizing Committee (2016)
16. Wang, W., Alkhouli, T., Zhu, D., Ney, H.: Hybrid neural network alignment and lexicon model in direct HMM for statistical machine translation. In: Proceedings of the 55th Annual Meeting of the Association for Computational Linguistics, vol. 2: Short Papers), pp. 125–131. Association for Computational Linguistics (2017)
17. Yin, Y., Song, Y., Zhang, M.: Document-level multi-aspect sentiment classification as machine comprehension. In: Proceedings of the 2017 Conference on Empirical Methods in Natural Language Processing, pp. 2044–2054. Association for Computational Linguistics (2017)

Age Estimation from MR Images via 3D Convolutional Neural Network and Densely Connect

Qi Qi[1,2](✉) ⓘ, Baolin Du[1,2](✉) ⓘ, Mingyong Zhuang[1,2](✉) ⓘ,
Yue Huang[1,2](✉) ⓘ, and Xinghao Ding[1,2](✉) ⓘ

[1] Fujian Key Laboratory of Sensing and Computing for Smart City,
Xiamen University, Xiamen 361005, Fujian, China
18150076754@163.com, 747127841@qq.com, tozhuang2012@qq.com
[2] School of Information Science and Engineering, Xiamen University,
Xiamen 361005, Fujian, China
{yhuang2010,dxh}@xmu.edu.cn

Abstract. The estimation of brain age from magnetic resonance (MR) images is useful for computer-aided diagnosis (CAD) in neurodegenerative diseases. Some deep learning methods has been proposed for age estimation from MR images recently. These methods release the burden of pre-processing dramatically, and they outperform the methods with hand-crafted features as well. However, the existing models of brain age estimation simply stack several convolution layers together, whose fitting ability is still limited. In this paper, we propose a deep learning framework based on 3D convolution neural network and dense connections to predict brain ages from MR images. The densely connect block in the proposed framework has a stronger fitting ability. Besides, combined with the domain knowledge of brain age estimation, the high-frequency structures of brain MR images are extracted and then are sent into the deep network. The proposed method is evaluated on a public brain MRI dataset. With the comparisons with existing methods, the experimental results demonstrated that our method achieved the state-of-the-art performances with the accuracy of 4.28 years on mean absolute error (MAE).

Keywords: MR images · Age estimation · CNN · 3D convolution
Densely connect

The work is supported in part by National Natural Science Foundation of China under grants of 81671766, 61571382, 61571005, 61172179 and 61103121, in part by CCF-Tencent Open Fund, in part by Natural Science Foundation of Guangdong Province under grant 2015A030313007, in part by the Fundamental Research Funds for the Central Universities under Grants 20720180059, 20720160075, in part of the Natural Science Foundation of Fujian Province of China (No. 2017J01126).

© Springer Nature Switzerland AG 2018
L. Cheng et al. (Eds.): ICONIP 2018, LNCS 11307, pp. 410–419, 2018.
https://doi.org/10.1007/978-3-030-04239-4_37

1   Introduction

In the process of human aging, the changing of brain structures, is recommended as an indicator of individual aging and some neurodegenerative diseases [1,2]. Senior people are under a higher risk of neurodegenerative diseases, such as Alzheimer's Disease (AD) and Parkinson's Disease (PD) [3]. The early aging of brains due to these diseases may cause the age of the brain older than the chronological age. The differences between the brain age and actual age of a person could be helpful to detect some brain disease at the early stage [4]. Therefore, a precise age predicting method from brain MR images is prerequisite for diagnosing brain-related diseases as early as possible.

In the recent years, some works had been reported on age estimation from MR images. Franke et al. [5] took the gray matter MR images as input, and then used the relevance vector regression (RVR) to estimate the brain age. Wang et al. [6] used the Hidden Markov Models (HMMs) to estimate the brain age. Kondo et al. [7] used the brain local features of T1-weighted MR images to estimate the brain age. Wang et al. [8] combined the cortical thickness with curvatures to design a cortical surface pattern (CSP), and then estimated the brain age with RVR. The age estimation methods based on traditional machine learning require complex and time-consuming data pre-processing, and strongly rely on hand crafted features, which limits its generalization and robustness.

Deep learning methods have demonstrated good performances on many medical image processing task such as tumor segmentation and disease classification [9]. Recently, some age estimation methods based on deep learning have been proposed. Huang et al. [10] used the stacked convolutional neural networks (CNNs) to predict the brain age. Cole et al. [11] put simple pre-processed MR data into a 3D CNNs to predict the brain age from MR images. Though the deep learning methods have demonstrated a better performances than traditional methods with hand-crafted features, there are still some limitations. That is, these methods simply stack the nonlinear layers together without considering any domain knowledge. Besides, they has insufficient capacity for feature extraction with a small training dataset.

Therefore, in the proposed work, we develop a 3D CNNs framework with densely connect to estimate brain ages from MR images. Firstly, based on the domain knowledge that the brain aging usually cause morphological changes, we extra the high-frequency structures of brain as the input to enforce the morphological changes in the brain. Secondly, to fully utilize the space information between slices, the three-dimensional convolution is used to extract the three-dimensional features of the brain MRI image in order to encode the rich spatial information in the stereo data and improve the accuracy of age estimation. Finally, we use the dense connect method to reduces the gradient vanishing and improves fitting ability of the network. The parameters are reduced, which makes up for the shortcoming of the large computational resources required for three-dimensional convolutions.

The proposed framework has been demonstrated in a public dataset called IXI. The experimental results demonstrated that with the proposed network, the mean absolute error (MAE) in age estimation decreases to 4.28 years, which is a better performances than that of most state-of-the-art methods.

In this article, the main contributions are summarized as: (1) For the domain knowledge in the age estimation task, the edge information of the brain MR data is extracted, and the three-dimensional convolution method is used to better utilize the information between the slices of the MR image by sharing the weight parameters in the third dimension of the data, fully tapping the features of the data, thus the age estimation result can be improved. (2) Due to limited training samples in medical image analysis applications, densely connect is used as the basic module of the network, which can well reuse the output features of different convolutional layers, improve the fitting ability of the network and prevent overfitting to some extent.

2   Proposed Method

The proposed network architecture is shown in Fig. 1. We will give a detail description of the framework as follows.

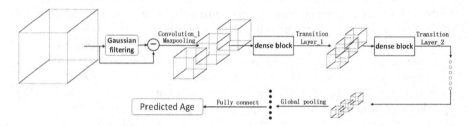

Fig. 1. The proposed age estimation framework. The details of dense block (red) will be shown in Fig. 2 (Color figure online)

The structures of human brain could be segmented into white matter (WM), gray matter (GM) and cerebrospinal fluid (CSF) region according to different voxel value and shape on images. In some related research [1,12], brain aging causes morphological changes of brain structure, which results in the changing of the shape and bulk of WM, GM and CSF. That is, the changing of shape and volume of the WM, GW and CSF is of great significance to the age estimation task. To better making use of morphological information of brain, we extra the high-frequency of the brain data, and then extract the features to make the prediction, instead of directly extract the features from the raw data. MR images are divided into slices along the Z axis. We get the high-pass part of MR slices by subtract the low-pass part, which is gained by Gaussian Filtering.

The brain MR images are three-dimensional data. Learning feature representations from all the three dimensions is very important for biometric estimation tasks from 3D medical data [13]. In some previous works, 2D convolution is used for extracting features of brain MR image [6] without considering convolutional kernels sharing on the third dimension, thus the space coding information of the third dimensional is inevitably lost. So we use 3D convolution to learn the feature representation in all three dimensions. At this point, the feature is a three-dimensional data block instead of two-dimensional plane data map, which is called a feature block. The network employs the entire stereo data [14], as the input, and then three-dimensional convolution kernels slide through the entire three-dimensional topology, in order to encode the rich spatial information in the stereo data and improve the accuracy of age estimation. By utilizing the sharing of the convolution kernel in all three dimensions, the network can make full use of spatial inter-pixel information.

Compared to the 2D convolution, the computing resources required for three-dimensional convolution increase exponentially. So a network structure with fewer parameters and high-efficiency fitting ability is urgently required. Moreover, due to the privacy and difficulty of collecting MR data, the training set is of small sample size, thus it is prone to suffer overfitting during training. To solve these problems, we introduce the densely connect [15]. The detail architecture of dense connect is shown in Fig. 2. In each densely connect block, the output of the l-th layer can be represented as:

$$x_l = H_l(x_0, x_1, \ldots, x_{l-1}) \tag{1}$$

$x_0, x_1, \ldots, x_{l-1}$ denotes the output feature maps of the layer 0 (input) to layer l-1, H_l denotes the nonlinear mapping of l-th layer.

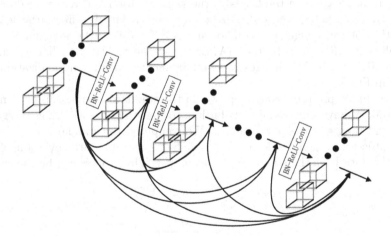

Fig. 2. The details of dense block in Fig. 1.

In the dense module, the input of each layer makes up of the outputs of all the previous layers. The densely connect method makes each layer directly connected to the input of the dense block and the back-propagated gradient of the loss function, which greatly relieve the gradient vanishing. Thus, the network can be stacked deeper to improve the fitting ability. The number of output feature maps at each layer is indicated as growth rate. In order to avoid the overfitting, the growth rate is always set to be a small value [15]. Therefore, dense connection is suitable for applications with the limited size of training samples, such as medical image processing. The transition layer between each dense block consists of a convolution layer with kernel size of 1×1 and a pooling layer, which can reduce the data dimension and combine the feature information.

The loss function can be denoted as:

$$loss = \frac{1}{N} \sum_{i-1}^{N} (\tilde{y}_i - y_i)^2 \tag{2}$$

N denotes the number of samples, $\tilde{y}_i, y_i$ denotes the output of the age estimation result and the age label of sample i.

3   Experiments

3.1   Dataset and Pre-processing

We train and test our network on brain MR images from the publicly accessible IXI database (http://brain-development.org/ixi-dataset/). There are 581 T1-images collected on three different scanners of different normal people aged 19–86 years in the IXI database. The data of each subject is composed of $256 \times 256 \times 150$ voxels. In order to align the heads to the same position and remove redundant voxels to save the computational resources, which largely affects the feature extraction of the image in the following process, we do the realignment and normalize of MR images by the MATLAB package SPM12 [16]. The examples of raw MR images, MR images after realignment and after normalization are shown in Fig. 3.

As a simple pre-processing step, we do the standard normalization to make every subject have the similar distribution, so that the network can convergence faster and then get a better accuracy. We computed the standard deviation of every subject and then divided each sample by its own standard deviation, which made the data have the standard deviation of 1. The process can be shown as:

$$\hat{M}^s = M^s - mean(M^s) \tag{3}$$

$$\tilde{M}^s = \frac{\hat{M}^s}{std(\hat{M}^s)} \tag{4}$$

where M_s denotes the s-th subject of the training data, and $std(\cdot)$ denotes the standard deviation.

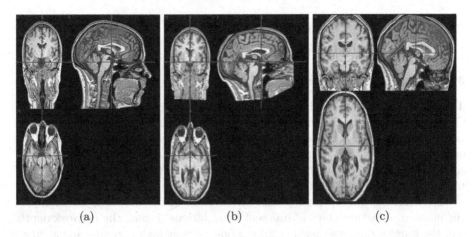

Fig. 3. (a) The raw MR image. (b) MR image after realignment. (c) MR image after normalize.

3.2 Data Augmentation

After the processing, the subjects are of the same pose and position. The size of every subject becomes $189 \times 157 \times 130$ voxels after the data preprocessing. We randomly choose 90 subjects for testing, and the others for training. Since training samples are less than five hundred, in order to reduce the overfitting in the experiments, we apply the training data augmentation procedure as follows: (1) shift the image in the same slice horizontally or vertically for pixels randomly sampled from $[-15, 15]$, and then pad the images to the original size with zeros; (2) flip the images horizontally or vertically; (3) rotate the images for 180 degrees and (4) combine any two types of above procedures together. In order to keep the entirety morphology of each subject, the slices of the same brain share the same process parameter in one process. The training dataset is expanded to more the 6000 samples after the augmentation.

3.3 Assessment Criteria

The MAE (mean absolute error) are used to measure the accuracy of age estimation results [7,8,11].

$$MAE = \frac{1}{n} \sum_{i=1}^{n} |f_i - y_i| \tag{5}$$

f_i, y_i denotes the predicted age and age label of sample i.

In addition, we use the ME (mean error) and RMSE (root mean square error) to describe the distribution of the age estimation error.

$$ME = \frac{1}{n} \sum_{i=1}^{n} (f_i - y_i) \tag{6}$$

$$RMSE = \frac{1}{n} \sum_{i=1}^{n} (f_i - y_i)^2 \tag{7}$$

3.4 Experimental Results

The experiments were implemented in a server with NIVDIA GeForce 1080Ti GPU, 64G RAM and 56 Intel(R)Xeon(R) CPU E5-2683 V3 @ 2.00 GHz. The network weights were trained by minimizing the mean square loss using an Adam optimization algorithm [17]. The batch size was set to 15. The learning rate was set to 0.0001, and then was reduced to 0.00001 after 10000 times training. The weight decay was set to 0.0001.

To determine the optimal parameter settings of the network, for example, the number of feature maps of different convolutional layers, the network depth and the growth rate, we do some validations without high-pass filtering at first. The network parameters and the corresponding results on the test set are shown in Tables 1 and 2, where Conv_1 denotes the number of output feature maps of layer convolution_1, nb_layers_1, nb_layers_2, nb_layers_3, and nb_layers_4 denote the depth of the every dense block respectively.

Table 1. The parameters of the network (no high-pass filtering).

Model index	Conv_1	nb_layers_1	nb_layers_2	nb_layers_3	nb_layers_4	Growth rate
1	30	6	6	6	6	12
2	30	6	12	12	12	12
3	30	6	12	24	16	12
4	30	6	12	32	12	12
5	20	6	12	12	12	12
6	40	6	12	12	12	12
7	30	6	12	12	12	8
8	30	6	12	12	12	16

Table 2. The corresponding result of models in Table 1

Model index	1	2	3	4	5	6	7	8
MAE	5.22	**4.63**	4.98	5.00	5.08	4.69	4.87	4.79
ME	−0.07	**−0.04**	0.03	−0.01	0.35	−0.47	−0.16	−0.75
RMSE	6.38	**5.39**	5.68	5.94	6.30	5.54	6.33	5.68

As the result shows, model 2 has the lowest MAE of 4.63 years. That is, when the layer convolution_1 generated 30 feature maps, the layers in every dense block was set to 6, 12, 12 and 12 respectively, and the growth rate was set to 12, the model got the best result.

Based on the result of Table 2, to find out the best Gaussian filter kernel size of the proposed framework, we set the different Gaussian kernel sizes and the result of age estimation on the test set are shown in Table 3.

Table 3. The parameter and the result of the proposed framework

Kernel sizes of Gaussian filter	3×3	5×5	7×7	9×9
MAE	4.64	4.30	**4.28**	4.42
ME	0.21	0.11	**0.16**	−0.18
RMSE	5.55	4.98	**4.98**	5.17

As we can see in the Table 3, when the kernel size of the Gaussian filter was set to 7×7, the proposed age estimation framework received the best result of 4.28 years on MAE.

The visualised result of the proposed framework is show in Fig. 4. In the figure, the estimation result of each subject is represented as a point. The green line shows the ideal result where predicted age equals real age. All points are very close to the green line, which demonstrates the strong correlation between the predicted age and the real age ($r = 0.94$), thus illustrates that the proposed method can model the brain aging process and predict the age.

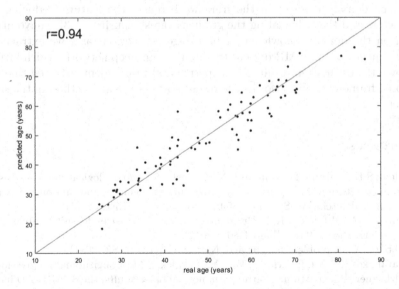

Fig. 4. The visualised result of the proposed method. Every point in the figure represents a subject. The green line shows the ideal case that the predicted age is equal to the real age. The predicted age and the real age is strongly related ($r = 0.94$) (Color figure online)

For a comparison with other methods, the results of different methods are shown in Table 4. The result of Franke et al. [5] method and Wang et al. [8] method were also train and tested on IXI dataset with a very similar validation setting. For a fair comparison, we reproduced the state-of-the-art method of Cole et al. [11] by ourselves on IXI with the same validation setting as our experiment. As we can see in the Table 4, our method has the smallest MAE compared to the previous methods.

Table 4. The result of different age estimation methods

Method	Samples	Training samples	Test samples	MAE
Franke et al. [5]	547	410	137	4.61
Wang et al. [8]	360	324	36	4.57
Cole et al. [11]	581	491	90	5.07
Proposed	581	491	90	4.28

4   Conclusion

In this paper, to estimate age from MR T1-images, we proposed a deep network architecture based on 3D convolution and dense connect. The 3D convolution operations learn the feature representation in all dimensions of brain MR images. And then densely connect in the framework reuses the feature, reducing the parameters and then alleviating the gradient vanish causing by 3D convolution. Based on the domain knowledge of brain age estimation task, we extract the high-frequency of brain MR data as the input of our deep network, which further improves the estimation result. The experimental results demonstrated that the proposed framework outperforms the recent state-of-the-art method in the same dataset.

References

1. Alam, S.B., Nakano, R., Kamiura, N., Kobashi, S.: Morphological changes of aging brain structure in MRI analysis. In: 15th International Symposium on Soft Computing and Intelligent Systems (SCIS), pp. 683–687. IEEE (2014)
2. Bunge, S.A., Whitaker, K.J.: Brain imaging: your brain scan doesn't lie about your age. Curr. Biol. **22**(18), R800–R801 (2012)
3. Abbott, A.: A problem for our age. Nature **475**(7355), S2 (2011)
4. Davatzikos, C., Xu, F., An, Y., Fan, Y., Resnick, S.M.: Longitudinal progression of Alzheimer's-like patterns of atrophy in normal older adults: the SPARE-AD index. Brain **132**(8), 2026–2035 (2009)
5. Franke, K., Ziegler, G., Klöppel, S., Gaser, C., Alzheimer's Disease Neuroimaging Initiative: Estimating the age of healthy subjects from T1-weighted MRI scans using kernel methods: exploring the influence of various parameters. Neuroimage **50**(3), 883–892 (2010)

6. Wang, B., Pham, T.D.: MRI-based age prediction using hidden Markov models. J. Neurosci. Methods **199**(1), 140–145 (2011)
7. Kondo, C., et al.: An age estimation method using brain local features for T1-weighted images. In: 37th Annual International Conference of the IEEE Engineering in Medicine and Biology Society, pp. 666–669. IEEE (2015)
8. Wang, J., Li, W., Miao, W., Dai, D., Hua, J., He, H.: Age estimation using cortical surface pattern combining thickness with curvatures. Med. Biol. Eng. Comput. **52**(4), 331–341 (2014)
9. Litjens, G., et al.: A survey on deep learning in medical image analysis. Med. Image Anal. **42**, 60–88 (2017)
10. Huang, T.W., et al.: Age estimation from brain MRI images using deep learning. In: IEEE 14th International Symposium on Biomedical Imaging, pp. 849–852. IEEE (2017)
11. Cole, J.H., et al.: Predicting brain age with deep learning from raw imaging data results in a reliable and heritable biomarker. NeuroImage **163**, 115–124 (2017)
12. Good, C.D., Johnsrude, I.S., Ashburner, J., Henson, R.N., Friston, K.J., Frackowiak, R.S.: A voxel-based morphometric study of ageing in 465 normal adult human brains. Neuroimage **14**(1), 21–36 (2001)
13. Dou, Q., et al.: Automatic detection of cerebral microbleeds from MR images via 3D convolutional neural networks. IEEE Trans. Med. Imaging **35**(5), 1182–1195 (2016)
14. Ji, S., Xu, W., Yang, M., Yu, K.: 3D convolutional neural networks for human action recognition. IEEE Trans. Pattern Anal. Mach. Intell. **35**(1), 221–231 (2013)
15. Huang, G., Liu, Z., Weinberger, K.Q., van der Maaten, L.: Densely connected convolutional networks. In: Proceedings of the IEEE Conference on Computer Vision and Pattern Recognition, vol. 1, p. 3 (2017)
16. Friston, K.J., Holmes, A.P., Worsley, K.J., Poline, J.P., Frith, C.D., Frackowiak, R.S.: Statistical parametric maps in functional imaging: a general linear approach. Hum. Brain Mapp. **2**(4), 189–210 (1994)
17. Kingma, D.P., Ba, J.: Adam: a method for stochastic optimization. arXiv preprint arXiv:1412.6980 (2014)

Low-Shot Multi-label Incremental Learning for Thoracic Diseases Diagnosis

Qingfeng Wang[1,2], Jie-Zhi Cheng[3(✉)], Ying Zhou[4], Hang Zhuang[1],
Changlong Li[1], Bo Chen[2], Zhiqin Liu[2], Jun Huang[2], Chao Wang[1],
and Xuehai Zhou[1(✉)]

[1] School of Software Engineering,
University of Science and Technology of China, Hefei, China
qfwangyy@mail.ustc.edu.cn, xhzhou@ustc.edu.cn
[2] School of Computer Science and Technology,
Southwest University of Science and Technology, Mianyang, China
[3] Shanghai United Imaging Intelligence, Shanghai, China
jzcheng@ntu.edu.tw
[4] Mianyang Central Hospital, Mianyang, China

Abstract. Despite promising results of 14 types of diseases continuously reported on the large-scale NIH dataset, the applicability on real clinical practice with the deep learning based CADx for chest X-ray may still be quite elusive. It is because tens of diseases can be found in the chest X-ray and require to keep on learning and diagnosis. In this paper, we propose a low-shot multi-label incremental learning framework involving three phases, i.e., representation learning, low-shot learning and all-label fine-tuning phase, to demonstrate the feasibility and practicality of thoracic disease abnormalities of CADx in clinic. To facilitate the incremental learning in new small dataset situation, we also formulate a feature regularization prior, say multi-label squared gradient magnitude (MLSGM) to ensure the generalization capability of the deep learning model. The proposed approach has been evaluated on the public ChestX-ray14 dataset covering 14 types of basic abnormalities and a new small dataset MyX-ray including 6 types of novel abnormalities collected from Mianyang Central Hospital. The experimental result shows MLSGM method improves the average Area-Under-Curve (AUC) score on 6 types of novel abnormalities up to 7.6 points above the baseline when shot number is only 10. With the low-shot multi-label incremental learning framework, the AI application for the reading and diagnosis of chest X-ray over-all diseases and abnormalities can be possibly realized in clinic practice.

Keywords: Chest X-ray · Thoracic diseases diagnosis
Low-shot learning · Multi-label learning · Incremental learning

© Springer Nature Switzerland AG 2018
L. Cheng et al. (Eds.): ICONIP 2018, LNCS 11307, pp. 420–432, 2018.
https://doi.org/10.1007/978-3-030-04239-4_38

1 Introduction

Recent progress of deep learning techniques has revolutionized the computer-aided diagnosis (CADx) on the clinical reading of chest X-rays, since the release of public chest X-ray dataset by National Institutes of Health (NIH) in 2017 [1,2]. The NIH dataset contains more than 110,000 front-view chest X-ray images and covers 14 types of abnormalities. In particular, the CheXNet developed by Rajpurkar et al. [3] was shown to even outperform the radiologists' performance on the diagnosis of pneumonia. Although promising results have been continuously reported on the NIH dataset, the applicability on real clinical practice with the deep learning based CADx for chest X-ray may still be quite elusive.

It is because tens of diseases can be found with the chest X-ray, and therefore training the neural network with limited 14 abnormalities can be far from enough to reach promising performance for the AI-based chest X-ray diagnosis. Due to the difficulty of accessing medical annotation resource and regional dependence of disease popularization, the complete annotation for chest X-ray images with all possible diseases and abnormalities can be barely available. Accordingly, the training of a useful neural networks may be more reasonably realized in an incremental way. Specifically, the number of classes/labels can be increased from time to time, and hence the neural network shall be retrained correspondently for the recognition of newly added abnormality labels. To minimize the effort of re-training neural network with the availability of new types with small samples, the low-shot learning scheme seems to be a suitable learning framework [4].

In this study, we exploit a low-shot learning scheme for a deep learning model to incrementally learn new abnormalities in chest X-ray images. Specifically, we train a deep learning model on the NIH dataset to recognize 14 abnormalities optimizing with multi-label squared gradient magnitude (MLSGM) regularization. Afterward, a new small dataset where all chest X-ray images were annotated with unseen abnormalities of pulmonary tuberculosis (PTB), senile cardiopulmonary change (SCC), degenerative thoracic spine (DTS), pulmonary infection (PI), interstitial lung disease (ILD), calcification (CAL), etc. is introduced. The network equipped with knowledge of 14 abnormalities from NIH dataset is then further trained with the new small dataset in the low-shot learning framework. It will be shown that with the proposed low-shot learning framework, the incrementally trained network can effectively learn knowledge from the newly added small dataset. Therefore, it may shed a light for the possibility of incremental learning for the artificial intelligent medical applications.

The contributions of this paper can be three-fold. First, we demonstrate the feasibility of low-shot learning for the learning of multiple thoracic diseases depicted in chest X-ray images by minimizing the effort of re-training deep learning model. This may potentially address the current reality of medical AI bottleneck that the availability of medical annotations for the chest X-ray images is always limited. With the low-shot learning framework, the AI application for the reading and diagnosis of chest X-ray over-all diseases and abnormalities can be possibly realized. To our best knowledge, the low-shot learning has been less explored for the medical image analysis. Second, we formulate a feature regularization prior, say multi-label squared gradient magnitude (MLSGM), which can be carried out in low-shot multi-label fashion to facilitate the incremental learning in new small dataset scenario. The efficacy of the MLSGM is corroborated with our experimental results. Third, the experiments are conducted on the 14 types of abnormalities on the NIH dataset as well as the newly small dataset collected from Mianyang Central Hospital. It will be suggested that the significant improvement for the new abnormalities in the small dataset will not compromise the performance of basic 14 abnormalities.

2   Related Work

Recently, deep learning techniques have been applied on various CADx applications with promising performance, such as the differential diagnosis of benign and malignant pulmonary nodules in computed tomography (CT) scans [5–7], computerized prognosis for Alzheimer's disease and mild cognitive impairment in magnetic resonance imaging (MRI) [8], breast anatomy segmentation in whole breast ultrasound [9], fetal standard plane retrieval [10] and quality assessment in ultrasound (US) images [11], etc. However, these applications typically address one abnormality or lesion. CADx for multiple abnormalities has been less explored. A more practical AI application for medical images shall be capable of identifying multiple abnormalities and diseases, but is quite challenging to develop as medical image data with annotations of more than 1 abnormality are rarely available.

In these two years, several studies have been conducted for the computer-aided diagnosis of 14 abnormalities in chest X-ray images since the introduction of the chest X-ray dataset by NIH. Specifically, Rajpurkar et al. developed a CheXNet to detect the 14 types of abnormalities and achieved state-of-the-art performance [3]. Yao et al. exploited the dependencies among labels to predict pathologic patterns by combing the technique of densenet and recurrent neural network [12]. Li et al. attempt to address the problems of thoracic disease identification and localization with the multiple instance learning framework [13]. Although promising performances have been reported in these studies, the extension of network model trained with 14 abnormalities on other abnormalities is unknown. Accordingly, this paper explores the low-shot learning framework for the extension possibility without large re-training effort.

Low-shot learning for computer vision was introduced by Hariharan et al. to address multi-class natural images classification with limit novel examples [4]. Dave et al. further proposed a novel progressive multi-label classifier for class-incremental data, whereas it needed to retrain neural network when novel labels were arrived [14]. However, there is less work that exploits the low-shot learning for the medical image analysis. Accordingly, we introduce the representation regularization method into the multi-label scenario to attain the low-shot learning on the diagnosis of thoracic abnormalities in chest X-ray images. Since the annotations on the chest X-ray images of the NIH dataset were simply on image level. The problem is a weakly-supervised learning problem and therefore, the difficulty can be further exacerbated. It will be shown that although the aforementioned difficulties are given, our low-shot learning can still achieve satisfactory performance for the learning of new types of abnormalities.

3 Methods

3.1 Framework Overview

Our goal is to build a low-shot multi-label learning framework to facilitate the incremental learning of a deep learning model with new small data, i.e., MyX-ray introduced in Sect. 4.1 that includes unseen abnormalities in chest X-ray images. In this section, the proposed low-shot learning method is elaborated. In this study, the involved basic and new groups of abnormalities are used to evaluate the effectiveness of the low-shot learning of deep learning model.

The low-shot multi-label incremental learning framework can be carried out in three phases, i.e., representation learning phase, low-shot learning phase and all-label fine-tuning phase. The flowchart of proposed low-shot learning is summarized in Fig. 1. We first train a deep learning model for multi-label learning of 14 abnormalities from the large NIH dataset. The 14 abnormalities constitutes the basic group of labels in this study, and is used for the phase of representation learning. The representation learning aims to obtain features with higher generalization capability, and is optimized with the multi-label squared gradient magnitude (MLSGM) regularization. The MLSGM feature regularization term, which will be introduced in Sect. 3.3, aims to ensure the generalization capability of the deep learning model. After the representation learning on the basic group of abnormalities, the network structure before the final fully connected layer is treated as the feature extractor. The feature extractor network is then further connected with a fully connected layer with 6 output neurons for the low-shot learning from the new added MyX-ray dataset. Afterward, the sub-network without the fully connected layer with the low-shot learning is then connected with a new fully connected layer with 20 output neurons, which include 14 abnormality labels and 6 new labels from the MyX-ray dataset for the all-label fine-tuning phase.

3.2 Multi-label Classification Loss

Our multi-label learning network is constructed with a ConvNet as feature extractor ϕ and a multi-label classifier W to recognize C types of abnormalities. Multiple abnormalities can possibly co-exist in one chest X-ray image x and are not mutually exclusive [13]. Accordingly, W is implemented by a fully connected layer with C independent output neurons. The loss function of the multi-label learning network is defined as the sum of binary cross entropy loss of each independent output

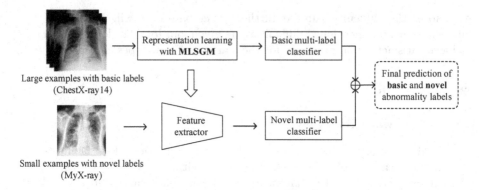

Fig. 1. Overview of low-shot multi-label incremental learning framework.

neuron that corresponds to each abnormality with the ground truth label vector $y = [y_1, y_2, \ldots, y_c, \ldots, y_C], y_c \in \{0, 1\}$ as shown in Eq. (1).

$$F_{loss}(W, \phi(x), y) = -\sum_{c=1}^{C} y_c \log P_c(W, \phi(x)) + (1 - y_c) \log(1 - P_c(W, \phi(x))) \quad (1)$$

where y_c indicates the presence of abnormality c in x. The predicted probability of abnormality c is applied by sigmoid nonlinearity as in Eq. (2).

$$P_c(W, \phi(x)) = \frac{1}{1 + \exp(-w_c^T \phi(x))} \quad (2)$$

Let D represent the whole set of samples in ChestX-ray14 dataset. Therefore, the joint training the feature extractor ϕ and the multi-label classifier W on D can be realized with the minimization of the objective function:

$$\min_{W, \phi} L_D(\phi, W) = \min_{W, \phi} \frac{1}{|D|} \sum_{(x,y) \in D} F_{loss}(W, \phi(x), y) \quad (3)$$

3.3   MLSGM Feature Regularization

Since novel labels are agnostic in learning process, one more generalized feature representation training is a good solution for the facilitation of incremental learning of new abnormalities in chest X-ray images. In order to enable better learning for small samples with novel multi-label abnormalities, we reformulate the work of Hariharan et al. w.r.t squared gradient magnitude (SGM) loss [4] to multi-label SGM (MLSGM) for medical image chest X-ray analysis scenario. The motivation of MLSGM is to reduce the difference between the multi-label classifier W trained on large dataset D and the multi-label classifiers V trained on small datasets S ($S \subset D$) so that those trained on new small datasets can generalize better. **We simulate low-shot multi-label learning experiments on the** basic labels by considering several tiny training sets $S \subset D, |S| \ll |D|$ as in [4]. The objective of V trained on S can be depicted as in Eqs. (4–6).

$$F_{loss}(V, \phi(x), y) = -\sum_{c=1}^{C} y_c \log P_c(V, \phi(x)) + (1 - y_c) \log(1 - P_c(V, \phi(x))) \quad (4)$$

$$P_c(V, \phi(x)) = \frac{1}{1 + \exp(-v_c^T \phi(x))} \quad (5)$$

$$\min_{V,\phi} L_S(\phi, V) = \min_{V,\phi} \frac{1}{|S|} \sum_{(x,y) \in S} F_{loss}(V, \phi(x), y) \quad (6)$$

In order to reduce the difference between the multi-label classifier W trained on the large dataset D and the multi-label classifiers V trained on these small datasets S, W is used to minimize $L_S(\phi, V)$. Therefore, the gradient of $L_S(\phi, V)$ at $V = W$, denoted as $\nabla_V L_S(\phi, V)|_{V=W}$, shall be closed to zero. That is to say, the closer W is to the global minimum of $L_S(\phi, V)$, the lower the magnitude of this gradient. Thus, the objective function can be represented as the minimization of Eq. (7):

$$\tilde{L}_S(\phi, W) = \| \nabla_V L_S(\phi, V)|_{V=W} \|^2 \quad (7)$$

where the gradient vector $\nabla_V L_S(\phi, V)$ of C types of abnormalities can be represented as:

$$\nabla_V L_S(\phi, V) = [\nabla_{v_1} L_S(\phi, V), \ldots, \nabla_{v_c} L_S(\phi, V), \ldots, \nabla_{v_C} L_S(\phi, V)] \quad (8)$$

According to Eqs. (4–6), $\nabla_{v_c} L_S(\phi, V) = \frac{1}{|S|} \sum_{(x,y) \in S} (P_c(V, \phi(x)) - y_c) \cdot \phi(x)$. Thus, Eq. (7) can be rewritten as:

$$\tilde{L}_S(\phi, W) = \frac{1}{|S|^2} \sum_{c=1}^{C} \| \sum_{(x,y) \in S} (P_c(W, \phi(x)) - y_c) \cdot \phi(x) \|^2 \quad (9)$$

Considering an extreme version of Eq. (9) where S is a single example (x, y):

$$\tilde{L}_S(\phi, W) = \sum_{c=1}^{C} (P_c(W, \phi(x)) - y_c)^2 \| \phi(x) \|^2 \quad (10)$$

where ϕ, W, P_c and y_c are the same as defined in Eqs. (1–2). Note that y_c indicates the presence of abnormality c in one chest X-ray. Thus, We have deduced a feature regularization term as in Eq. (10), denoted as multi-label SGM (MLSGM). Let $\alpha(W, \phi(x), y) = \sum_{c=1}^{C} (P_c(W, \phi(x)) - y_c)^2$, where $\alpha(W, \phi(x), y) \in [0, 2C]$. Intuitively, $\alpha(W, \phi(x), y)$ is a weighted term of per example's convolutional features, which will be high when abnormality labels are erroneously identified. Therefore, our goal is to optimize the feature representation of basic abnormalities by minimizing the incorporation of MLSGM loss in standard multi-label classification loss formulation:

$$\min_{W,\phi} L_D(\phi, W) + \lambda \frac{1}{|D|} \sum_{(x,y) \in D} \alpha(W, \phi(x), y) \| \phi(x) \|^2 \quad (11)$$

Algorithm 1. Low-Show Multi-label Incremental Learning

Input: X-ray images x in ChestX-ray14; X-ray images x' in MyX-ray(dataset M);
   basic multi-label vector $y = [y_1, \ldots, y_c, \ldots, y_C], y_c \in \{0, 1\}$;
   novel multi-label vector $y' = [y'_1, \ldots, y'_n, \ldots, y'_N], y'_n \in \{0, 1\}$.
Output: Probability scores of $C + N$ types of abnormalities.

1: Jointly training feature extractor ϕ and multi-label classifier W, and optimizing representation learning of basic types of abnormalities with MLSGM by Eq. (11);
2: Using the trained ϕ to extract convolutional features $\phi(x')$ of novel abnormalities, training novel multi-label classifier W', and optimizing low-shot learning by $\min_{W'} \frac{1}{|M|} \sum_{(x',y') \in M} F_{loss}(W', \phi(x'), y')$; (Note that M is the MyX-ray dataset)
3: Merging W and W' into W'', and generating an incremental model by connecting ϕ and W''.

This equation can be viewed as the aggregation of the standard multi-label classification loss and the weighted L_2 regularization on the feature activations of multiple labels, where λ is a weighting factor which is empirically set as $1e-5$. Minimizing the norm of the feature activations can constrain the feature encoding in the network and drive to find useful features [4]. The incorporation of MLSGM in the training process can endow the network with more generalization capability. Note that the notation V in this section only exists in the formula derivation of MLSGM, whereas W' introduced in Sect. 3.4 is different from V.

3.4 Training Scheme

The low-shot learning framework is composed of three learning phases, i.e., representation learning phase, low-shot learning phase and all-label fine-tuning phase. The three learning phases are summarized in **Algorithm 1**. In this study, the ResNet-50 [15] architecture pre-trained from the ImageNet [16,17] without the final full-connected layer for the 1000 output neurons is used as an initialized feature extractor ϕ in the representation learning phase. Then, we construct a sub-network of a fully connected layer with C output neurons, which are independently operated with element-wise sigmoid nonlinearity, as a multi-label classifier W, where C is the number of the basic types of abnormalities (here $C = 14$). Afterwards, ϕ and W are jointly trained with ChestX-ray14 dataset optimizing with MLSGM by Eq. (11) to predict the basic 14 types of abnormalities. In low-shot learning phase, we use the ϕ trained in representation learning phase to extract convolutional features $\phi(x')$ of X-ray images x' in MyX-ray. A new multi-label classifier W' is constructed with a fully connected layer and N output neurons, where N is the number of novel types of abnormalities in MyX-ray (here $N = 6$). Then we concatenate $\phi(x')$ and W' to train W' optimized by $\min_{W'} \frac{1}{|M|} \sum_{(x',y') \in M} F_{loss}(W', \phi(x'), y')$ with MyX-ray dataset M, where y' is the N-dimensional multi-label vector of novel abnormalities. We have obtained a multi-label classifier W' that is able to identify N types of novel abnormalities. Finally, we fine-tune all basic and novel labels by fusing W and W' to concatenate their fully connected layers and merging their weights into a new multi-classifier W''. Thus, an incremental model is generated through connecting ϕ and W'' in total equiped with the ability to recognize $N + C$ types of abnormalities.

4 Experiments

4.1 Datasets

ChestX-ray14 [1] is a recently released large-scale public dataset of thoracic abnormalities, which contains 112,120 frontal-view X-ray images. The ChestX-ray was released by NIH. This dataset is annotated with 14 abnormalities found in chest X-ray images. The 14 abnormalities are Atelectasis, Cardiomegaly, Effusion, Infiltration, Mass, Nodule, Pneumonia, Pneumothorax, Consolidation, Edema, Emphysema, Fibrosis, Pleural Thickening (PT), and Hernia.

MyX-ray. In order to equip the deep learning model with the ability of identifying more abnormalities in chest X-ray images, we collected a new small dataset of 1657 front-view chest X-rays from Mianyang Central Hospital. We called the new dataset as MyX-ray. The MyX-ray dataset includes 6 types of novel abnormalities, i.e., pulmonary tuberculosis (PTB), senile cardiopulmonary change (SCC), degenerative thoracic spine (DTS), pulmonary infection (PI), interstitial lung disease (ILD) and calcification (CAL). The MyX-ray dataset also includes some abnormalities defined in the NIH dataset. The quantitative statistics of abnormalities in MyX-ray dataset are shown in Table 1.

Table 1. The statistical illustration of abnormality labels in MyX-ray.

Novel label	#	Basic label	#	Basic label	#
PTB	802	Atelectasis	79	Pneumothorax	53
SCC	311	Cardiomegaly	197	Consolidation	92
DTS	271	Effusion	167	Edema	14
PI	271	Infiltration	419	Emphysema	74
ILD	224	Mass	30	Fibrosis	283
CAL	165	Nodule	530	PT	260
		Pneumonia	40	Hernia	10

4.2 Experimental Settings

We employ the model ResNet-50, which was pre-trained from ImageNet dataset, as an initialized learning framework. All experiments in this study are conducted on a linux server with 4 NVIDIA Titan X GPUs. The Area-Under-Curve (AUC) score of each abnormality are used as evaluation metrics [1].

Representation Learning Phase Setup. We adopt the data split[1] published by Yao et al. for this study. The 70% and 10% of data are treated as training and validation samples respectively and the remaining 20% data are used for testing, All X-ray images are resized to 224×224 and normalized. We also perform data augmentation with randomly resized cropping and randomly horizontal flipping for training set. We optimize the model using stochastic gradient descent (SGD) with a mini-batch size of 32. We train the model for 90 epochs, while the learning rate starts at 0.01 and is decreased by a factor of 10 at every 30 epochs. The momentum parameter is set as 0.9 and weight decay is fixed at 1e−4. The weighted factor λ of MLSGM term is set as 1e−5.

Low-Shot Learning Phase Setup. In order to illustrate the efficiency of MLSGM for different low-shot number settings, we randomly select n positive examples of each novel label from MyX-ray dataset as training set every time, where $n \in \{10, 20, 50, 100\}$, and then the remainder examples are used for testing. That is we will randomly split MyX-ray into training and testing set for four times, and each time the corresponding total numbers of training examples for 6 types of novel abnormalities are 60, 120, 300, 600 respectively. We train the multi-label classifier sub-network using SGD for 30 epochs, while the learning rate starts at 0.1 and is decreased by a factor of 10 at every 10 epochs. The momentum is set as 0.9 and weight decay is fixed at 0.001.

4.3 Results and Analysis

In this study, we perform multi-label classification for 14 types of basic and 6 types of novel thoracic disease abnormalities in chest X-ray images. To illustrate the efficiency of MLSGM feature regularization for low-shot multi-label incremental learning, we use a scheme without optimizing by MLSGM, denoted as **baseline** method. Table 2 shows the comparison of AUC scores of 6 types of novel labels between baseline and MLSGM approaches. As can be observed, MLSGM performs much better than baseline for each novel abnormality with different shot number settings. It suggests that the incorporation of MLSGM could improve the novel labels learning with small samples.

To further illustrate the efficacy of MLSGM on the performance of average AUC scores of 6 types of novel abnormalities, we carry out the performance trends of average AUC scores w.r.t the factor of shot number. As can be also observed in Fig. 2, the average AUC score goes up with increasing shot number both for baseline and MLSGM, which indicates that the much more training examples make the better overall performance. According to statistics, MLSGM gains about 3.5 points (MLSGM 0.8273 vs. baseline 0.7924) than baseline w.r.t average AUC score when shot number is 100, whereas MLSGM achieves over 7.6 points (MLSGM 0.7397 vs. baseline 0.6635) above baseline when shot number is only 10. It suggests that MLSGM shows stronger robustness compared to baseline especially when the sample size is much smaller.

---

[1] https://github.com/yaoli/chest_xray_14.

Table 2. AUC scores performance on 6 types of novel abnormalities with different shot number settings w.r.t baseline and MLSGM.

Abnormalities	Approach	n = 10	20	50	100
PTB	baseline	0.6904	0.8052	0.8724	0.8907
	MLSGM	**0.8452**	**0.8411**	**0.9081**	**0.9175**
SCC	baseline	0.7784	0.7868	0.8264	0.8115
	MLSGM	**0.8391**	**0.832**	**0.8776**	**0.8782**
DTS	baseline	0.7119	0.7938	0.8267	0.8183
	MLSGM	**0.7646**	**0.8371**	**0.8767**	**0.8919**
PI	baseline	0.4854	0.5501	0.6667	0.7004
	MLSGM	**0.6494**	**0.6785**	**0.6782**	**0.7272**
ILD	baseline	0.7583	0.724	0.8168	0.845
	MLSGM	**0.7672**	**0.7788**	**0.8288**	**0.8565**
CAL	baseline	0.5567	0.5601	0.5607	0.6885
	MLSGM	**0.5726**	**0.5978**	**0.6421**	**0.6925**

We also compare the AUC scores of the 14 abnormalities from our model with the AUC scores from the methods developed by Wang et al. [1], Yao et al. [12] and CheXNet [3] on the public ChestX-ray14 dataset. It can be found in Table 3 that our models achieve better performances on 10 abnormalities, where baseline and MLSGM hold 3 entries and 7 entries respectively. On the other hand, we also summarize the performance of our incremental model on the 14 types of abnormalities in MyX-ray dataset. As can be observed in Table 3, MLSGM outperforms baseline on 12 types of abnormalities except for Cardiomegaly and Edema. It might again suggest that MLSGM shows stronger robustness and generalized ability than baseline on small dataset. In addition, the performance of

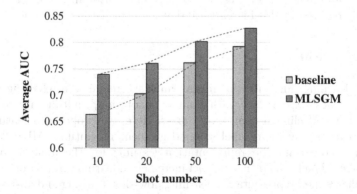

Fig. 2. The comparison of baseline and MLSGM w.r.t average AUC scores of 6 types of novel abnormalities with different low-shot number settings.

Table 3. AUC scores on 14 types of basic abnormalities in ChestX-ray14 and MyX-ray dataset.

Abnormalities	ChestX-ray14					MyX-ray	
	Wang [1]	Yao [12]	CheXNet [3]	baseline	MLSGM	baseline	MLSGM
Atelectasis	0.7158	0.772	0.8209	0.8279	**0.8289**	0.8025	**0.8721**
Cardiomegaly	0.8065	0.904	0.9048	0.9060	**0.9079**	**0.9184**	0.9159
Effusion	0.7843	0.859	0.8831	**0.8891**	0.8887	0.8750	**0.8944**
Infiltration	0.6089	0.695	**0.7204**	0.7203	0.7172	0.6122	**0.6526**
Mass	0.7057	0.792	0.8618	0.8724	**0.8773**	0.7697	**0.8465**
Nodule	0.6706	0.717	0.7766	0.7868	**0.7918**	0.6083	**0.6485**
Pneumonia	0.6326	0.713	**0.7632**	0.7575	0.7599	0.5170	**0.5709**
Pneumothorax	0.8055	0.841	0.8932	**0.9058**	0.9007	0.8612	**0.8757**
Consolidation	0.7078	0.788	0.7939	0.8274	**0.8294**	0.6161	**0.7386**
Edema	0.8345	0.882	0.8932	0.8975	**0.9033**	0.8794	0.8692
Emphysema	0.8149	0.829	**0.9260**	0.9248	0.9225	0.7395	**0.7856**
Fibrosis	0.7688	0.767	0.8044	0.8228	**0.8246**	0.6054	**0.6227**
PT	0.7082	0.765	**0.8138**	0.8099	0.8061	0.7016	**0.7302**
Hernia	0.7667	0.914	0.9387	**0.9680**	0.9651	0.5878	**0.6810**

most basic abnormalities except for Atelectasis, Cardiomegaly and Effusion in MyX-ray are no better than that in ChestX-ray14, and this might indicate that the generalization ability of across dataset in clinic is still required for further improvement.

The time of using ResNet-50 with ChestX-ray14 dataset is about 13 h for training 90 epochs with 4 NVIDIA Titan X GPUs. However, the training of new multi-label classifier for 6 types of novel abnormalities in MyX-ray dataset only took about 11–12 s for 30 epochs. It suggests that the low-shot multi-label scheme has access to feasibility and practicality in clinic. The AI application for the reading and diagnosis of chest X-ray over-all diseases and abnormalities can be possibly realized in this incremental development with cheap time expense.

5   Conclusion

Low-shot learning for medical image analysis remains a challenging but important task as issues face with limited annotation, multiple abnormalities identification and clinical complexity, exist. Here, we proposed a feature regularization prior, say multi-label squared gradient magnitude (MLSGM) that is implemented within the representation learning for the analysis of chest X-ray images. The low-shot multi-label incremental learning framework consists of three phases, i.e., representation learning, low-shot learning and all-label fine-tuning phases, which demonstrates the feasibility and practicality of thoracic abnormalities of CADx in clinical practice. While satisfactory performance with low-shot multi-label abnormality identification is demonstrated in this study, there is still much space to improve the performance in terms of generalization ability across datasets. The future work will explore on one-shot learning with less examples as well as exploration of more types of abnormalities that can be observed in chest X-ray images.

Acknowledgments. This work was supported by National Natural Science Foundation of China (Nos. 61379040, and 61501305), and Sichuan Provincial Open Foundation of Civil-Military Integration Research Institute (Nos. 2017SCII0220 and 2017SCII0219).

References

1. Wang, X., Peng, Y., Lu, L., Lu, Z., Bagheri, M., Summers, R.M.: ChestX-ray8: hospital-scale chest X-ray database and benchmarks on weakly-supervised classification and localization of common thorax diseases. In: 2017 IEEE Conference on Computer Vision and Pattern Recognition (CVPR), pp. 3462–3471, July 2017

2. Wang, X., Peng, Y., Lu, L., Lu, Z., Summers, R.M.: Tienet: text-image embedding network for common thorax disease classification and reporting in chest x-rays. CoRR, abs/1801.04334 (2018)

3. Rajpurkar, P., et al.: Chexnet: radiologist-level pneumonia detection on chest X-rays with deep learning. CoRR, abs/1711.05225 (2017)

4. Hariharan, B., Girshick, R.: Low-shot visual recognition by shrinking and hallucinating features. In: 2017 IEEE International Conference on Computer Vision (ICCV), pp. 3037–3046, October 2017

5. Cheng, J.Z., et al.: Computer-aided diagnosis with deep learning architecture: applications to breast lesions in us images and pulmonary nodules in CT scans. Sci. Rep. **6**, 24454 (2016)

6. Shen, W., et al.: Multi-crop convolutional neural networks for lung nodule malignancy suspiciousness classification. Pattern Recognit. **61**(61), 663–673 (2017)

7. Shen, W., Zhou, M., Yang, F., Yang, C., Tian, J.: Multi-scale convolutional neural networks for lung nodule classification. Inf. Process. Med. Imaging **24**, 588–599 (2015)

8. Suk, H.I., Lee, S.W., Shen, D.: Latent feature representation with stacked autoencoder for AD/MCI diagnosis. Brain Struct. Funct. **220**(2), 841–859 (2015)

9. Bian, C., Lee, R., Chou, Y.-H., Cheng, J.-Z.: Boundary regularized convolutional neural network for layer parsing of breast anatomy in automated whole breast ultrasound. In: Descoteaux, M., Maier-Hein, L., Franz, A., Jannin, P., Collins, D.L., Duchesne, S. (eds.) MICCAI 2017. LNCS, vol. 10435, pp. 259–266. Springer, Cham (2017). https://doi.org/10.1007/978-3-319-66179-7_30

10. Hao, C., et al.: Ultrasound standard plane detection using a composite neural network framework. IEEE Trans. Cybern. **47**(6), 1576–1586 (2017)

11. Wu, L., Cheng, J.Z., Li, S., Lei, B., Wang, T., Dong, N.: FUIQA: fetal ultrasound image quality assessment with deep convolutional networks. IEEE Trans. Cybern. **47**(5), 1336–1349 (2017)

12. Yao, L., Poblenz, E., Dagunts, D., Covington, B., Bernard, D., Lyman, K.: Learning to diagnose from scratch by exploiting dependencies among labels. CoRR, abs/1710.10501 (2017)

13. Li, Z., et al.: Thoracic disease identification and localization with limited supervision. CoRR, abs/1711.06373 (2017)

14. Dave, M., Tapiawala, S., Meng, J.E., Venkatesan, R.: A novel progressive multi-label classifier for class-incremental data. In: IEEE International Conference on Systems, Man, and Cybernetics, pp. 003589–003593 (2017)
15. He, K., Zhang, X., Ren, S., Sun, J.: Deep residual learning for image recognition. In: 2016 IEEE Conference on Computer Vision and Pattern Recognition (CVPR), pp. 770–778, June 2016
16. Deng, J., Socher, R., Fei-Fei, L., Dong, W., Li, K., Li, L.-J.: ImageNet: a large-scale hierarchical image database. In: 2009 IEEE Conference on Computer Vision and Pattern Recognition(CVPR), pp. 248–255, June 2009
17. Russakovsky, O., et al.: ImageNet large scale visual recognition challenge. Int. J. Comput. Vis. **115**(3), 211–252 (2015)

Continuous Convolutional Neural Network with 3D Input for EEG-Based Emotion Recognition

Yilong Yang, Qingfeng Wu[✉], Yazhen Fu, and Xiaowei Chen

School of Software, Xiamen University, Xiamen, China
{yilongyang, fuyazhen, wdenxw}@stu.xmu.edu.cn,
qfwu@xmu.edu.cn

Abstract. Automatic emotion recognition based on EEG is an important issue in Brain-Computer Interface (BCI) applications. In this paper, baseline signals were taken into account to improve recognition accuracy. Multi-Layer Perceptron (MLP), Decision Tree (DT) and our proposed approach were adopted to verify the effectiveness of baseline signals on classification results. Besides, a 3D representation of EEG segment was proposed to combine features of signals from different frequency bands while preserving spatial information among channels. The continuous convolutional neural network takes the constructed 3D EEG cube as input and makes prediction. Extensive experiments on public DEAP dataset indicate that the proposed method is well suited for emotion recognition tasks after considering the baseline signals. Our comparative experiments also confirmed that higher frequency bands of EEG signals can better characterize emotional states, and that the combination of features of multiple bands can complement each other and further improve the recognition accuracy.

Keywords: EEG · Emotion recognition · CNN · Brain-Computer interface

1 Introduction

Emotion plays an important role in human daily life. Due to the recent interest shown by the research community in establishing emotional interactions between humans and computers, the identification of the emotional state of the former is in need [1]. To improve the satisfaction level and the reliability of the human agents who interact or collaborate with machines and robots, intelligent human-machine (HM) systems with the capability of accurately understanding human communications are inevitably required.

Human emotions can be detected by facial expressions [2], speech [3], eye blinking [4] and physiological signals. However, the first three approaches are susceptible to subjective influences of participants, that is, the participants can deliberately disguise their emotions. While the physiological signals such as electroencephalograms (EEG), electrooculography (EOG), blood volume pressure (BVP) are produced spontaneously by human body. Consequently, the physiological signal is more objective and reliable in capturing human real emotional states. Of all of these physiological signals, the EEG signal comes directly from human brain, that is to say, changes in EEG signals can

© Springer Nature Switzerland AG 2018
L. Cheng et al. (Eds.): ICONIP 2018, LNCS 11307, pp. 433–443, 2018.
https://doi.org/10.1007/978-3-030-04239-4_39

directly reflect changes in human emotional states. Furthermore, comparing to functional Magnetic Resonance Imaging (fMRI) and Positron Emission Tomography (PET), the use of EEG is inexpensive, making it a preferred method in studying the brain's response to emotional stimuli.

Due to the advantages of EEG mentioned above, many researchers have focused on emotion recognition from EEG signals. Zheng and Lu used deep neural networks to investigated critical frequency bands for EEG-based emotion recognition and concluded that beta and gamma bands were more suitable for the task [5]. One of the recent works also confirmed that high-frequency bands were more able to distinguish emotional states [6]. Tang et al. used Bimodal Deep Denoising Auto Encoder and Bimodal-LSTM to classify emotion states and achieved mean accuracy of 83.25% on DEAP dataset [7]. Li *et al.* proposed a pre-processing method that transformed the multi-channels EEG data into 2D frame representation and integrated CNN and RNN to recognition emotion states in the trial-level [8]. Li *et al.* extracted Power Spectral Density (PSD) from different EEG channels and mapped it to two-dimensional plane to construct the EEG Multidimensional Feature Image (EEG MFI) and combined CNN and LSTM to deal with the EEG MFI sequences to recognize human emotional states [9].

Most of the studies ignored the importance of the baseline signals (EEG signals recorded under no stimulus); hence, in this paper, we took into account the role of baseline signals and found it can help to improve recognition accuracy significantly. Besides, we proposed a 3D input form for EEG segment and fed it into a continuous convolutional neural network to recognize emotions. The advantage of the 3D input is preserving the spatial information among electrodes while integrating multi frequency bands.

This paper is structured as follows. In Sect. 2, we introduce the DEAP dataset which is widely used in EEG-based emotion recognition field. A detailed description of proposed method is given in Sect. 3. In Sect. 4, we present results that we achieved. Finally, in Sect. 5, we summarize our contributions.

2   DEAP Dataset

The DEAP dataset was first introduced in [10], which was widely used by various EEG-based emotion recognition researches; so we use this dataset to validate our approach. In the dataset, EEG signals of 32 participants were recorded when they were watching 40 1-minute long music videos. Each video was presented to a subject and then she/he was asked to fill a self-assessment for her/his valence and arousal. Valence and Arousal scales from 1 to 9 (1 represents sad/calm and 9 represents happy/excited). During the signal acquisition stage, EEG was recorded at a sampling rate of 512 Hz. The creator of the DEAP dataset also provided a pre-processed version of the dataset so researchers could quickly verify their proposed emotion recognition methods. In the pre-processed dataset, the EEG signals were down-sampled to 128 Hz, only the signal in the 4–45 Hz frequency bandwidth is preserved, and the EOG is also removed. The data collected in each trial was segmented into a 60-second experimental signals (recorded while watching video) and a 3-second pre-trial baseline signals (relax state). The EEG data for each participant includes two arrays: data and labels. Table 1 shows a summary of the DEAP dataset (pre-processed version).

Table 1. DEAP dataset description (pre-processed version)

Overall					
Subjects	Videos	EEG Channels	Sampling rate	Rating scale	Rating values
32	40	32	128 Hz	Arousal Valence	Continuous scale of 1–9

EEG data format for each subject			
Array	Array Shape		Array Content
Data	40 × 32 × 8064 (384 base + 7680 trial)		video/trial × channels × data
Labels	40 × 2		video/trial × label(valence, arousal)

3 Method

3.1 Frequency Pattern Decomposition and Feature Extraction

The original EEG signals can be divided into several different frequency patterns based on intra-band correlation with a distinct behavioral state [11–13]. According to the summary made by Zhang et al. [14], the EEG frequency patterns and the corresponding characters are listed in Table 2. As shown in Table 2, the awareness degree increases with the increase of band frequency. We supposed that emotions are produced while human are in high awareness. This statement is also consistent with our intuitive understanding, if a person is in an unconscious state, he/she is less likely to produce a particular mood. In order to verify our point of view and investigate the effect of different frequency patterns on emotion, we used a Butterworth filter to decompose the original signals into 4 frequency bands (θ, α, β, γ). After decomposition, EEG data of a participant is converted from 40 × 8064 × 32 (video × sample × channel) to 40 × 8064 × 4 × 32 (video × sample × band × channel). This process is depicted in Fig. 2 (Step 1).

Table 2. EEG patterns and corresponding characters

Patterns	Frequency	Brain State	Awareness
(δ) Delta	0.5–4 Hz	Deep sleep pattern	Lower
(θ) Theta	4–8 Hz	Light sleep pattern	Low
(α) Alpha	8–12 Hz	Closing the eyes, relax state	Medium
(β) Beta	12–30 Hz	Active thinking, focus, high alert, anxious	High
(γ) Gamma	30–100 Hz	During cross-modal sensory processing	Higher

Since we focus on identifying emotional states on segmented level, EEG signals are cut into n segments with length l. Hence, the purpose of our work is classifying these segments into correct labels. After segmentation, EEG data is transformed into 40 × n × l × 4 × 32 (video × segment × length × band × channel).

Shi et al. proposed Differential entropy (DE) for EEG-based vigilance estimation and used it to measure the complexity of EEG signals [15]. We adopted DE to characterize EEG signals as it has been proven to be suitable for emotion decoding in previous studies [5, 16]. Its calculation formula is defined as:

$$h(X) = \int_X f(X) log(f(x)) dx \tag{1}$$

Where X is a random variable, $f(x)$ is the probability density function of X. For the series X obeying the Gauss distribution $N\ (\mu, \delta^2)$, its differential entropy can be expressed as:

$$h(X) = \int_{-\infty}^{\infty} \frac{1}{\sqrt{2\pi\delta^2}} e^{\frac{(x-\mu)^2}{2\delta^2}} log\left(\frac{1}{\sqrt{2\pi\delta^2}} e^{-\frac{(x-\mu)^2}{2\delta^2}}\right) dx$$
$$= \frac{1}{2} log(2\pi e\delta^2) \tag{2}$$

Shi et al. have proved that for a specific frequency band i, the differential entropy can be defined as

$$h_i(X) = \frac{1}{2} log(2\pi e\delta_i^2) \tag{3}$$

Where h_i and δ_i^2 denoted the differential entropy of the corresponding EEG signal in frequency band i and the signal variance, respectively. For each EEG segment, differential entropy features were extracted. So the features of a specific frequency band i in segment can be represented by a 1-dimensional vector $v^i \in R^{32}$. This process is illustrated in Fig. 2 (Step 2). The vector is normalized using Z-score normalization. EEG data is further transformed into $40 \times n \times 4 \times 32$ (video $\times$ segment $\times$ band $\times$ channel).

3.2   Taking Baseline Signals into Account

Emotion is a complex state of mind, the slight differences of environment and other external factors have a certain impact on people's emotions. In DEAP dataset, 3-seconds baseline signals generated by subject under none stimulus are collected prior to each experimental data acquisition process. In order to investigate the impact of features of baseline signals on the final classification results, we cut the 3-seconds baseline signals into 3 1-second segments and use the method described earlier to transform each of them into 4 DE feature vector $base\_v \in \mathbb{R}^{32}$. Then the mean DE feature value of these 3 EEG cubes are computed to represent the DE feature of baseline signals. Finally, the DE deviation between experimental EEG (under stimulus) and baseline EEG (under none stimulus) is calculated to represent the emotional state feature of this segment. This step can be expressed as:

$$final\_v_j^i = exper\_v_j^i - \frac{\sum_{k=1}^{3} base\_v_k^i}{3} \left\{final\_v_j^i, exper\_v_j^i, base\_v_k^i\right\} \in R^{32} \tag{4}$$

Where $exper\_v_i^j$ denotes the DE feature vector for frequency band i on segment j. $base\_v_k^i$ is the DE feature vector for frequency band i on baseline signals segment k. $final\_v_j^i$ is the final emotional state feature vector for frequency band i on segment j.

3.3 3D Input Construction

The EEG based BCI system uses a wearable headset with multiple electrodes to capture EEG signals. The International 10–20 System is an internationally recognized method of describing and applying the location of scalp electrodes and the underlying area of the cerebral cortex. The left corner of Fig. 1 is the plane view of the international 10-20 system, where the EEG electrodes filled with yellow are the test points used in DEAP dataset. In EEG electrode map, each electrode is physically neighboring multiple electrodes which records the EEG signals in a certain area of the brain. For the purpose of preserving spatial information among multiple adjacent channels, we transform 1D DE feature vector v^{32} to 2D plane ($h \times w$), according to the electrode distribution map. Where h and w is the maximum number of the vertical and horizontal used electrodes. With DEAP dataset, $h = w = 9$. Hence, the corresponding 2D plane of feature vector v_i^{32} in frequency i is denoted as $p^i \in R^{h \times w}$. Zeroes are used to fill the DEs from channels that are unused in DEAP dataset. This process is depicted in Figs. 1 and 2 (Step 3).

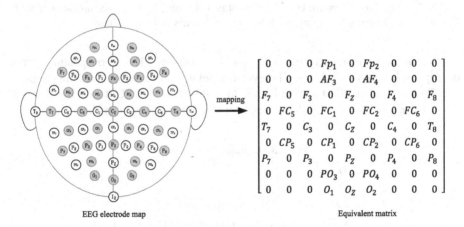

EEG electrode map   Equivalent matrix

Fig. 1. Constructing 2D plane.

So far, we have obtained 4 2D planes for each EEG segment. Our next step is stacking these planes into 3D EEG cubes and use it as inputs to the CNN. This process is depicted in Fig. 2 (Step 4). The 3D organization is inspired by the field of computer vision. On color image classification task, image is represented by Three Primary Colors (Red, Green and Blue). RGB color channel is adopted to organize an image. Specifically, a value of 0 to 255 is used to indicate the intensity of the color in each color channel. We use the representation of color image for reference and adopt similar

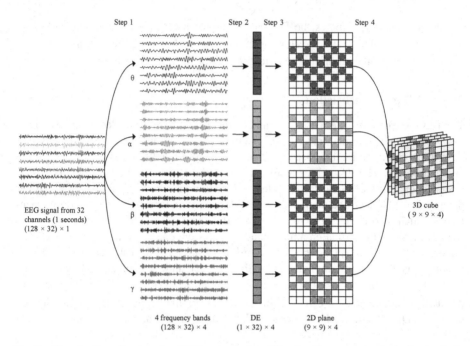

Fig. 2. The pipeline of converting EEG segment into 3D cube. Due to space limitation, only 8 channels were drawn in the figure, but there are 32 channels in DEAP dataset.

representation to construct 3D input for DE features. The corresponding term description is listed in Table 3. The overall pipeline of converting streaming signals into 3D EEG cube is illustrated in Fig. 3. After this process, EEG data is transformed into $40 \times n \times 4 \times 9 \times 9$ (video $\times$ segment $\times$ band $\times$ height $\times$ width).

Table 3. Corresponding terms

		Domain
		Computer Vision	EEG
Term	Color Image	EEG cube
	Color channel (R, G, B)	Frequency band (θ, α, β, γ)
	Color intensity	DE feature value

3.4 Continuous Convolutional Neural Network

CNN has powerful capability in extracting features from image. According to our previous analogy, the constructed 3D EEG cube could be considered as color image, which allows us to make full use of CNN as a powerful tool to extract representative features from input. In this paper, as shown in Fig. 3, we used a continuous convolutional neural network with four convolutional layers to extract features from the input

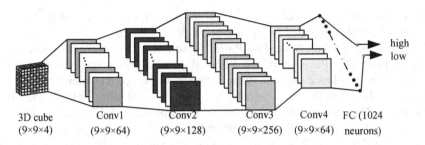

Fig. 3. Continuous convolutional neural network.

cube, a fully connected layer with dropout operation was added for feature fusion, softmax layer was used for final classification. By "continuous" is meant that there is no pooling layer between two adjacent convolutional layers. Although in the field of computer vision the convolutional layer is often followed by a pool layer, we believe that this is not necessary in our model. Since the main function of the pooling layer is to reduce the data dimension at the cost of some information loss, the size of the 3D EEG cube is far smaller than that of the computer vision field. For this reason, pooling layer is discarded in our model. Besides, in each convolutional layer, zero-padding is applied to prevent information from missing at the edge of cube.

More specifically, in the first 3 convolutional layers, the kernel size is set to 4×4 and the stride is set to 1. After convolution operation, the *RELU* activation function was added to endow the model with nonlinear feature transformation capability. We started the first convolutional layer with 64 feature maps and double the feature maps in the following 2 convolutional layers; hence, there were 128 and 256 feature maps in the second and the third layer. To fuse different feature maps and reduce computational cost, a 1 by 1 convolutional layer with 64 feature maps was added. After these 4 continuous convolutional layers, a fully connected layer was added to map the 64 9×9 feature maps into a final feature vector $f \in R^{1024}$. Then the following softmax layer receives f to predict human emotional state.

4 Experiments

4.1 Experiment Setup

In this paper, we focus on segmented-level based emotion recognition; hence, as we claimed in Sect. 3, the EEG data (under stimulus) of a participant was cut into $(40 \times n)$ segments and each of them with length l. Previous researches have shown that time window with size 1 s is suitable for emotion recognition [9, 17]. So in our experiments, the length l was set to 128, which means there are 2400 segments on each participant. We apply the proposed method to recognize valence and arousal from DEAP dataset. More specifically, we choose 5 as threshold to divide the labels into two binary classification problems according to the rated levels (1–9) of arousal and valence, i.e., High/Low valence, High/Low arousal. We perform 10-fold cross validation on each subject's data and the mean of them was denoted as the result of the subject, the result of approach are then reported as an average of all 32 subjects.

We implemented the CNN and MLP with Tensorflow[1] framework and trained it on a NVIDIA Titan XP GPU. The truncated normal distribution function was used to initialize the weight of kernels and the Adam optimizer was adopted to minimize the cross-entropy loss function, the initial learning rate was 10^{-4} and 10^{-2} for CNN and MLP, respectively. The keep probability of dropout operation was 0.5. L2 regularization was added to avoid overfitting and improve generalization capability, the penalty strength of L2 was 0.5 and 0.05 for CNN and MLP respectively. There are 3 hidden layers in MLP, the number of neurons in the first hidden layer is half of that in the input layer, and the next two layers are halved layer by layer. DT was implemented with scikit-learn[2].

4.2 Effects of Baseline Signals, EEG Pattern Decomposition and 3D Input

To examine the impacts of baseline signals on the final classification results, we designed two cases, ran experiments on 3 approaches and compared their results. Case1 represents a situation where DE features of baseline signals are discarded while Case2 represents a situation where DE features of baseline signals are used. To examine the effect of different frequency patterns and their combination on emotion recognition, we designed different experiments for each single band and different combinations of frequency bands. The result is shown in Tables 4 and 5. To show the superiority of 3D input format, we also implement the other two baseline models which use 1-dimensional input to classify emotion. The comparison result is shown in Fig. 4.

Table 4. Classification result on arousal class

	Case	θ	α	β	γ	θ+α	θ+β	θ+γ	α+β
DT	1	58.00	58.11	60.97	64.10	58.15	61.17	63.92	61.02
	2	**67.88**	**67.53**	**68.65**	**70.12**	**70.14**	**71.32**	**72.12**	**71.05**
MLP	1	66.38	64.22	64.27	64.91	67.60	67.86	67.69	67.77
	2	**70.68**	**67.70**	**69.43**	**68.37**	**80.88**	**83.47**	**80.99**	**80.07**
Ours	1	62.99	63.63	67.37	70.29	64.13	67.32	69.53	67.09
	2	**78.09**	**80.81**	**79.97**	**80.98**	**85.30**	**85.43**	**85.99**	**86.75**
	Case	α+γ	β+γ	θ+α+β	θ+α+γ	θ+β+γ	α+β+γ	θ+α+β+γ	
DT	1	63.60	64.11	61.02	63.66	63.70	63.64	63.86	
	2	**72.11**	**71.45**	**71.93**	**72.92**	**72.94**	**73.03**	**73.91**	
MLP	1	65.46	65.57	68.02	68.29	68.64	66.62	69.51	
	2	**80.21**	**81.47**	**86.79**	**85.86**	**86.85**	**86.13**	**88.68**	
Ours	1	69.34	71.02	67.12	69.05	70.02	70.28	69.55	
	2	**87.55**	**86.63**	**88.44**	**89.12**	**88.55**	**89.76**	**90.24**	

---

[1] https://www.tensorflow.org/.

[2] http://scikit-learn.org/stable/.

Table 5. Classification result on valence class

	Case	θ	α	β	γ	θ+α	θ+β	θ+γ	α+β
DT	1	54.84	55.58	60.05	63.51	55.67	59.89	62.67	59.61
	2	**64.14**	**64.73**	**66.50**	**68.25**	**67.03**	**68.35**	**70.27**	**68.73**
MLP	1	63.61	58.00	58.52	61.51	65.17	65.05	65.57	59.77
	2	**68.70**	**61.97**	**65.90**	**64.67**	**80.06**	**82.53**	**80.17**	**78.85**
Ours	1	58.90	60.15	64.84	68.55	60.73	64.80	67.95	65.37
	2	**75.66**	**78.73**	**78.13**	**79.83**	**83.60**	**83.65**	**84.51**	**85.40**
	Case	α+γ	β+γ	θ+α+β	θ+α+γ	θ+β+γ	α+β+γ	θ+α+β+γ	
DT	1	62.69	62.93	59.42	62.21	62.75	62.72	62.52	
	2	**70.53**	**70.02**	**69.95**	**71.53**	**71.44**	**71.08**	**71.93**	
MLP	1	62.32	62.70	66.43	67.21	66.95	64.14	68.11	
	2	**78.07**	**80.85**	**85.32**	**84.92**	**86.15**	**84.47**	**87.82**	
Ours	1	68.50	69.07	65.19	67.95	68.75	68.82	68.56	
	2	**86.77**	**85.47**	**87.34**	**87.89**	**87.50**	**88.80**	**89.45**	

From Tables 4 and 5, we can see that whether using 1D vector as input or 3D cube as input, the use of DE features of baseline signals can greatly improve the accuracy of emotion recognition. Another finding is that the higher the frequency band is, the more beneficial is the recognition of emotions. Furthermore, the combination of all bands can complement each other and contribute to better results. The comparison result of our model with MLP and DT which was summarized in Fig. 4 shows the benefits of 3D representation and the CNN's powerful ability on feature extraction.

4.3   Performance Comparison Among Relevant Methods

We also compared our result with three other reported approaches using DE feature [7], CNN [8], pattern decomposition [18] to recognize emotional states. Li *et al.* transformed 1D EEG sequence into grid-like frame through wavelet and scalogram and designed a hybrid deep learning model to recognize emotions. Tang *et al.* extracted DE features from EEG signals in four frequency bands and time-domain features from peripheral physiological signals and then used Bimodal-LSTM to fuse features from two modal and mine temporal dependency information. Yin *et al.* proposed a multiple-fusion-layer based ensemble classifier of stacked AutoEncoder to recognize emotional states, and also estimated the accuracy by 10-fold cross validation. As shown in Table 6, the proposed continuous convolutional neural network using 3D EEG cube as input surpasses these three approaches on both arousal and valence classification tasks.

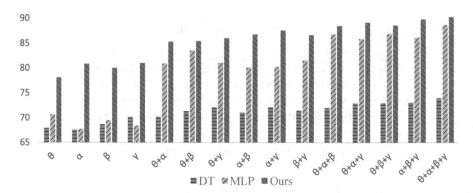

Fig. 4. Result comparison on 1D input and 3D input (arousal class).

Table 6. Average accuracies (%) of different approaches on DEAP dataset

	Li *et al.* [2016]	Tang *et al.* [2017]	Yin *et al.* [2017]	Ours
Arousal	74.12	83.23	84.18	**90.24**
Valence	72.6	83.82	83.04	**89.45**

5   Conclusion

This paper has shown that the DE features of baseline EEG signals can help to improve emotion recognition accuracy significantly. The deviation between the DE feature vectors of the experimental signals and the baseline signals can better characterize emotional states. The proposed 3D EEG representation combines signal features from different frequency bands while maintaining inter-channel spatial information. The experimental results on the continuous convolutional neural network model show that this kind of representation is very effective. Our result also con-firmed that multi-band combination can achieve better results than using single frequency bands. Compared with other relevant methods, the proposed method has achieved the best performance with a mean accuracies of 90.24% and 89.45% for arousal and valence classification tasks on DEAP dataset.

Acknowledgments. This work was supported by the National Key Research and Development Program of China (No. 2017YFC1703303); the Fundamental Research Funds for Central Universities of China (No. 20720180070).

References

1. Alarcao, S.M., Fonseca, M.J.: Emotions recognition using EEG signals: a survey. IEEE Trans. Affect. Comput. (2017)
2. Anderson, K., McOwan, P.W.: A real-time automated system for the recognition of human facial expressions. IEEE Trans. Syst. Man Cybern. Part B Cybern. **36**(1), 96–105 (2006)

3. Petrushin, V.A.: Emotion in speech: recognition and application to call centers. In: Proceedings of Artificial Neural Networks in Engineering, vol. 710 (1999)
4. Soleymani, M., Pantic, M., Pun, T.: Multimodal emotion recognition in response to videos. IEEE Trans. Affect. Comput. 3(2), 211–223 (2012)
5. Zheng, W.L., Lu, B.L.: Investigating critical frequency bands and channels for EEG-based emotion recognition with deep neural networks. IEEE Trans. Auton. Mental Dev. 7(3), 162–175 (2015)
6. Li, J., Zhang, Z., He, H.: Hierarchical convolutional neural networks for EEG-based emotion recognition. Cogn. Comput. 1–13 (2017)
7. Tang, H., Liu, W., Zheng, W.L., Lu, B.L.: Multimodal Emotion Recognition Using Deep Neural Networks. In: Liu, D., Xie, S., Li, Y., Zhao, D., El-Alfy, E.S. (eds.) Neural Information Processing. Lecture Notes in Computer Science, vol. 10637, pp. 811–819. Springer, Cham (2017). https://doi.org/10.1007/978-3-319-70093-9_86
8. Li, X., Song, D., Zhang, P., Yu, G., Hou, Y., Hu, B.: Emotion recognition from multi-channel EEG data through convolutional recurrent neural network. In: BIBM, pp. 352–359 (2016)
9. Li, Y., Huang, J., Zhou, H., Zhong, N.: Human emotion recognition with electroencephalographic multidimensional features by hybrid deep neural networks. Appl. Sci. 7(10), 1060 (2017)
10. Koelstra, S., Muhl, C., Soleymani, M., Lee, J.S., Yazdani, A., Ebrahimi, T., Pun, T., Nijholt, A., Patras, I.: Deap: a database for emotion analysis; using physiological signals. IEEE Trans. Affect. Comput. 3(1), 18–31 (2012)
11. Li, X.: Signal Processing in Neuroscience. Springer, Singapore (2016). https://doi.org/10.1007/978-981-10-1822-0
12. Başar, E., Dumermuth, G.: EEG-Brain dynamics: relation between EEG and brain evoked potentials. Comput. Programs Biomed. 14(2), 227–228 (1980)
13. Steriade, M.: Alertness, Quiet Sleep, Dreaming. Normal and Altered States of Function. Springer, US (1991). https://doi.org/10.1007/978-1-4615-6622-9_8
14. Zhang, X., Yao, L., Kanhere, S.S., Liu, Y., Gu, T., Chen, K.: MindID: Person identification from brain waves through attention-based recurrent neural network (2017). arXiv preprint arXiv:1711.06149
15. Shi, L.C., Jiao, Y.Y., Lu, B.L.: Differential entropy feature for EEG-based vigilance estimation. In: 35th Annual International Conference of the IEEE Engineering in Medicine and Biology Society, pp. 6627–6630 (2013)
16. Duan, R.N., Zhu, J.Y., Lu, B.L.: Differential entropy feature for EEG-based emotion classification. In: 2013 6th International IEEE/EMBS Conference on Neural Engineering, pp. 81–84. (2013)
17. Wang, X.W., Nie, D., Lu, B.L.: Emotional state classification from EEG data using machine learning approach. Neurocomputing 129, 94–106 (2014)
18. Yin, Z., Zhao, M., Wang, Y., Yang, J., Zhang, J.: Recognition of emotions using multimodal physiological signals and an ensemble deep learning model. Comput. Methods Programs Biomed. 140(C), 93–110 (2017)

3D Large Kernel Anisotropic Network for Brain Tumor Segmentation

Dongnan Liu[1]([✉]), Donghao Zhang[1], Yang Song[1], Fan Zhang[2],
Lauren J. O'Donnell[2], and Weidong Cai[1]

[1] School of Information Technologies, University of Sydney,
Sydney, NSW 2006, Australia
dliu5812@uni.sydney.edu.au
[2] Brigham and Women's Hospital, Harvard Medical School, Boston, USA

Abstract. Brain tumor segmentation in magnetic resonance images is a key step for brain cancer diagnosis and clinical treatment. Recently, deep convolutional neural network (DNN) based models have become a popular and effective choice due to their learning capability with a large amount of parameters. However, in traditional 3D DNN models, the valid receptive fields are not large enough for global details from the objective and the large amount of parameters are easy to cause high computational cost and model overfitting. In order to address these problems, we propose a 3D large kernel anisotropic network. In our model, the large kernels in the decoders ensure the valid receptive field is large enough and the anisotropic convolutional blocks in the encoders simulate the traditional isotropic ones with fewer parameters. Our proposed model is evaluated on datasets from the MICCAI BRATS 17 challenge and outperforms several popular 3D DNN architectures.

Keywords: Brain tumor segmentation · Magnetic resonance image
3D deep neural network

1 Introduction

Brain tumor is one of the leading causes of cancer deaths, which is difficult to cure and remains high mortality [23]. Among all varieties of brain tumors diagnosed in adults, gliomas account for about 70% [26]. Thus the detection and segmentation of gliomas is a necessary task for clinical diagnosis of brain cancer. Gliomas are caused by glial cells [8] and can be divided into high and low grade categories. High grade gliomas (HGG) turn out to be more aggressive with a survival time of no more than two years, while the low grade gliomas (LGG) grow slowly and leave a longer life expectancy of several years [19].

Although surgical treatment is the most effective way to directly remove the tumors, radiological treatment is also necessary to slow the growth of the tumors which can not be removed. Nowadays, magnetic resonance imaging (MRI) is one of the most common tools for radiological diagnosis since it is able to image the

© Springer Nature Switzerland AG 2018
L. Cheng et al. (Eds.): ICONIP 2018, LNCS 11307, pp. 444–454, 2018.
https://doi.org/10.1007/978-3-030-04239-4_40

brain structure with detailed information. Various imaging modalities are used to describe the information for different subregions of the tumor, as shown in Fig. 1. For example, T2- and Flair-weighted brain images highlight the tissues with water content, which represent the whole tumor with edema. T1-weighted images highlight the tumor core which contains no water, and contrast enhance T1-weighted (T1ce) images represent the enhanced parts with hyper-intensity in the tumor core [16].

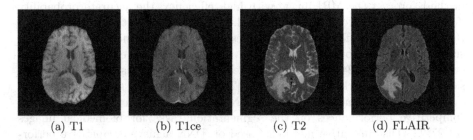

| (a) T1 | (b) T1ce | (c) T2 | (d) FLAIR |

Fig. 1. An example of one slice of a brain MR image in different modalities: T1, T1ce, T2, and FLAIR.

However, brain tumor segmentation is challenging due to the following factors: (1) Brain tumors can appear anywhere in the whole brain with different shapes and sizes across patients. Thus the methods based on processing the shape or size features become less effective. (2) The gradual transition from tumor to edema makes the boundaries between the edema and the tumor core ambiguous and hard to segment. (3) Although T1ce modality is able to highlight the enhanced tumor core, some other parts such as blood vessels and cortical cerebrospinal fluid (CSF) are highlighted as well. In this way, a thresholding method can not be directly applied to segment the enhanced tumor [21]. (4) Due to the pre-processing methods and poor contrast of tumors, the boundaries between tumors and healthy tissues are fuzzy.

In this work, we propose an end-to-end 3D convolutional neural network (CNN) model for the brain tumor segmentation task. The contributions of our work are three-folds. First, we apply anisotropic convolutional blocks to simulate the isotropic blocks with less parameters for memory efficiency. Second, inspired by 2D global convolutional network (GCN) [20], we propose a 3D large convolutional block which is capable of enlarging the receptive field for the feature maps. Third, our proposed model is evaluated on part of the MICCAI brain tumor segmentation (BRATS) challenge 2017 dataset [16] and outperforms some state-of-the-art 3D CNN models.

2   Related Work

Expert annotation for brain tumor segmentation is time-consuming and labor-intensive, thus automatic segmentation algorithms are urgently needed for high efficiency. Brain tumor segmentation approaches can be categorized into generative and discriminative methods. Generative methods apply specific prior knowledge about appearance of the brain tissues, including tumors and healthy parts. Such approaches require little training and largely rely on encoding the prior probability distribution of the spatial relationship between tissues [4,17]. In [22], Prastawa *et al.* propose a typical generative method which aligns the brain tissues to the ICBM brain digital atlas. Then the tumor is detected by comparing its posterior probabilities with that of the healthy tissues. Although generative methods process the MRI image with high efficiency, the accurate probability distribution is hard to encode and model. Discriminative methods directly learn the characteristic differences of the appearance between tumor and healthy tissues. In these methods, first the dense voxel features are extracted from the original images [14,24]. Then the features are fed into a classifier such as decision forests [30], Markov Random Field [24], or clique-based graphic model [14].

Recently, deep neural network (DNN) based architectures have shown competitive performances on the segmentation tasks for biomedical images [13,27,28]. Discriminative methods based on DNN achieve competitive performance compared with other state-of-the-art methods as DNN is able to learn the feature information in accurate details with numerous parameters. In [6], Havaei *et al.* proposes a dual path CNN architecture for analyzing the local and global feature details of the brain tumors. However, the model only applies 2D convolutional blocks, which means that it fails to process the information between different slices. DeepMedic [9] is a dual path architecture which contains 3D convolutional blocks and residual connections. HighRes3DNet [12] is a residual connected DNN model with dilated convolutional blocks which is capable of getting high spatial resolution features. However, in these single connection models, the feature information from low resolution level could easily be lost after passing through a number of convolutional and pooling layers. In DNN, the information from each resolution level represents different features of the original image. Losing the low level feature information results in a lack of detailed information such as curves and edges in the final segmentation. In order to solve this problem, skip connections between the encoders and decoders are applied to 3D DNN such as 3D U-Net [3] and V-Net [18]. In V-Net [18], residual connections are applied on the encoders and decoders to prevent the gradient vanishing during the training process. Even though these methods achieve competitive performance, the complexity of the model tends to be high due to the huge amount of parameters from 3D convolutional kernels, which makes the model difficult to train and easy to overfit. Wang *et al.* [25] decomposes a traditional $3 \times 3 \times 3$ block to one $3 \times 1 \times 1$ block and one $1 \times 3 \times 3$ block in the proposed architecture to process the inter- and intra- slices information respectively for a higher efficiency and lower computational cost. Moreover, 3D anisotropic convolutional blocks also show their effectiveness on the segmentation tasks for other 3D biomedical datasets, such as membrane segmentation for electron microscopy images [11], and liver tumor segmentation for CT scans [15].

3   Methods

In this section, we introduce our proposed 3D large kernel anisotropic network. Our model is composed of residual connection encoders with anisotropic convolutional blocks and decoders with 3D large kernel blocks to enlarge the actual receptive field, as shown in Fig. 2.

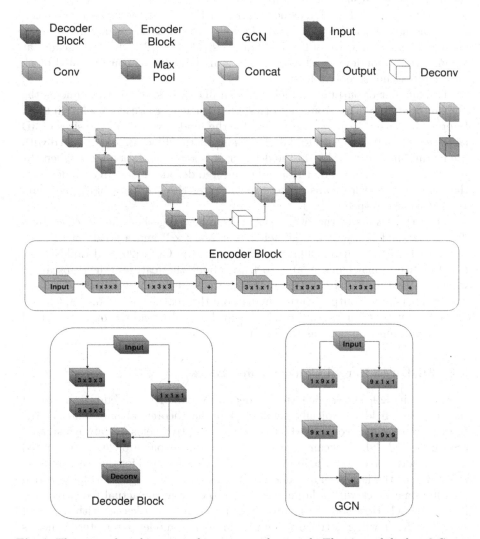

Fig. 2. The network architecture of our proposed network. The sizes of the last 2 Conv blocks are $3 \times 3 \times 3$ and $1 \times 1 \times 1$ respectively. The size and the stride for the Deconv at the highest resolution are $2 \times 2 \times 2$ and $(2, 2, 2)$. We omit the batch normalization and PReLU layers before each convolutional layers in this graph for brevity.

As the pre-activation residual connections proposed in [7] are proved to out-perform the traditional ones, we apply them on encoders. In the first residual connection unit of the first encoder, the output feature channel of the convolutional path is 16. The input image is padded from $H \times W \times D \times 4$ to $H \times W \times D \times 16$. In each encoder, the feature channels are doubled after passing the $3 \times 1 \times 1$ kernel for the next residual unit. For each encoder at different resolution levels, the numbers of output feature maps are 32, 64, 128, 256 and 512, respectively. After passing through each encoder except for the one at the highest resolution level, the output feature maps pass through two ways: downsampling by max pooling layers with a size of $2 \times 2 \times 2$ and a stride of $(2, 2, 2)$ for the encoder at a higher resolution level, and feeding into a 3D large kernel convolutional block for a large actual receptive field.

The number of output channels for each 3D large kernel is the same as the number of feature maps from the corresponding encoder at each resolution level. In order to have a large receptive field for the model, we set all the sizes of 3D large kernels as 9. The output of each 3D large kernel block is concatenated with the feature map upscaled from the decoder at a lower resolution level. Then the result passes through another residual connected decoder block. In each decoder, the number of feature maps reduces to $1/3$ after the $3 \times 3 \times 3$ block path and $1 \times 1 \times 1$ path, respectively.

Then the results of the two paths are summed together and upscaled by a deconvolutional block with a kernel size of $2 \times 2 \times 2$ and a stride of $(2, 2, 2)$. Compared with the upsampling methods applied in GCN [20] and LinkNet [2], the deconvolutional block is able to learn about the features when upscaling, which makes the model consider more information for the final segmentation. The numbers of the output feature maps from the decoders are 256, 128, 64, and 32, respectively. After passing the last convolutional block, the output channel number becomes the same as the input.

3.1   3D Large Kernel Convolutional Block

The work in [29] shows that in the deep CNN model, the empirical size of the receptive field is typically smaller than the theoretical one. Thus in the traditional segmentation models, the actual receptive fields are always smaller than expected, which means the models are only capable of learning limited feature information from the input image. It is thus harmful for the segmentation result due to the lack of global details. To this end, we design a 3D large kernel convolutional block with a large kernel size to enlarge the actual receptive field for the model. However, if we directly apply a large 3D kernel with a size of $K \times K \times K$, it will greatly increase the number of model parameters which is a large computational burden. In order to reduce the computational cost, we apply two convolutional kernels with sizes of $1 \times K \times K$ and $K \times 1 \times 1$ combined in different orders to simulate a convolutional kernel with a size of $K \times K \times K$, as illustrated in Fig. 2.

3.2 Anisotropic Convolutional Block

Due to the large amount of parameters from the 3D DNN model, the isotropic convolutional blocks with a size of $N \times N \times N$ make the computational cost and memory consumption high. Thus we decompose a $N \times N \times N$ kernel into one $1 \times N \times N$ kernel which fuses the features within each slice and one $N \times N \times 1$ kernel which fuses the features between slices. Inspired by the architecture proposed in [15] and [25], we employ the anisotropic convolutional blocks in the encoders. We first apply two $1 \times 3 \times 3$ kernels with residual connection for learning the intra-slice information in sagittal and coronal directions. Then we make the result pass through a $3 \times 1 \times 1$ kernel for inter-slice information processing along the axial direction. In this way, the anisotropic convolutional blocks are able to learn the information of the dataset in all dimensions of the isotropic ones with a higher memory efficiency.

4 Experiments and Results

4.1 Data Description and Implementation Settings

The dataset we used to evaluate our model is from 2017 MICCAI BRATS Challenge [16], which contains 210 HGG cases and 75 LGG cases. For each case, there are four MR images in different modalities: T1, T2, T1ce and FLAIR , with image size $155 \times 240 \times 240$. The organizers have already finished the pre-processing for the dataset, which includes skull-strip and co-registering for the four sequences. All the ground truths are labeled by experts and the results are evaluated by the official evaluation server named CBICA's Image Processing Portal. For each image, the segmentation ground truth contains three classes besides the background, which is shown in Fig. 3: the green part is edema (label 2), the yellow part is enhanced tumor core (label 4), and the red part is tumor core (label 1). In this experiment, we only focus on the segmentation for HGG brain tumor. We select 87 cases from HGG set for training and 40 from the rest of HGG set for testing.

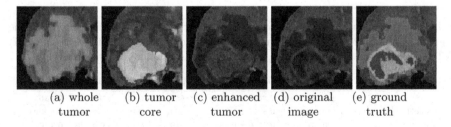

(a) whole   (b) tumor   (c) enhanced   (d) original   (e) ground
tumor   core   tumor   image   truth

Fig. 3. An example of the ground truth label for one HGG brain tumor case. (Color figure online)

We use ADAM [10] as the optimiser with $\beta_1 = 0.9$, $\beta_2 = 0.99$, and $\epsilon = 10^{-8}$. The learning rate for training our model is set as 0.001. For each model in the experiment, the training iteration is 8700 and the cost function is Dice loss using dice score between the predicted results and the ground truth. All of our networks are trained on 2 NVIDIA GTX 1080Ti GPUs implemented in Tensorflow [1] with NiftyNet [5].

4.2   Evaluation Metrics

In order to evaluate the performances of the models, we apply the Dice score and Hausdorff distance mentioned in [16] for the segmentation results of the three tumor regions. For each part of the tumor, we obtain the binary map of the predicted result P and ground truth T all in range $[0, 1]$ and calculate the Dice score:

$$Dice(P, T) = \frac{|P_1 \wedge T_1|}{(|P_1| + |T_1|)/2} \tag{1}$$

where $\wedge$ is the logical AND operation, $|.|$ represents the size of the set and P_1, T_1 mean the numbers of voxels in set P and T having a value of 1 (tumor voxels) respectively. While the Dice score measures the segmentation overlap results, Hausdorff distance is applied to calculate the distance from all the points on the surface of predicted set to those of the ground truth set:

$$Haus(P, T) = max\{ \sup_{p \in \partial P_1} \inf_{t \in \partial T_1} d(p, t), \sup_{t \in \partial T_1} \inf_{p \in \partial P_1} d(t, p)\} \tag{2}$$

where $d(p, t)$ represents the shortest least-squares distance from point p to t. ∂P_1 and ∂T_1 mean the surface of set P_1 and T_1, respectively.

4.3   Results and Comparison

Table 1 and Fig. 4 show the comparison between the result of our proposed method and the state-of-the-art. It can be seen that our proposed method outperforms others in most of the metrics.

Table 1. The experiment results from different models. All the results shown are the average results on the 40 testing images. The unit of Hausdorff distance is mm. ET, WT and TC mean enhanced tumor, whole tumor, and tumor core, respectively.

Metrics	3D U-Net [3]	V-Net [18]	HighRes3DNet [12]	Proposed
Dice-ET	0.7731 ± 0.1673	0.7416 ± 0.1573	0.6593 ± 0.2174	$\mathbf{0.7930 \pm 0.1351}$
Dice-WT	0.8417 ± 0.1188	0.8257 ± 0.1281	0.8032 ± 0.1619	$\mathbf{0.8644 \pm 0.0909}$
Dice-TC	0.8022 ± 0.1764	0.7546 ± 0.1876	0.7205 ± 0.2440	$\mathbf{0.8189 \pm 0.1465}$
Hausdorff95-ET	$\mathbf{10.0988 \pm 20.1343}$	10.2352 ± 18.5055	21.0615 ± 29.4237	10.4811 ± 21.9127
Hausdorff95-WT	15.3867 ± 21.7618	19.6240 ± 24.1664	31.9161 ± 25.6838	$\mathbf{15.0470 \pm 21.8504}$
Hausdorff95-TC	13.687 ± 21.4019	13.4893 ± 18.9466	30.0803 ± 32.3823	$\mathbf{13.1196 \pm 21.9502}$

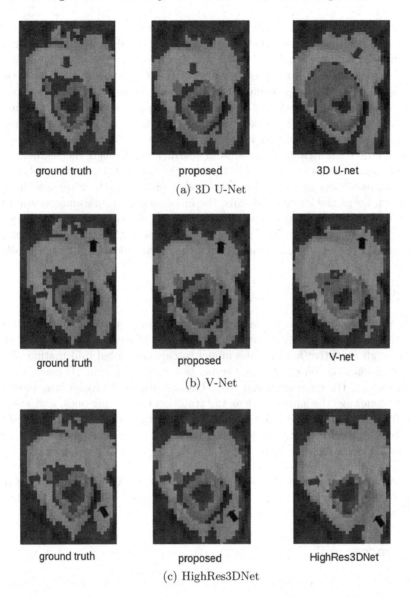

Fig. 4. Visual comparison between our proposed method and the state-of-the-art. (Color figure online)

As shown in Table 1, the standard deviations of the Hausdorff distances are large. It is because there remain large differences of the brain tumor between cases, which makes the difficulty of the brain tumor segmentation for each case vary. In HighRes3DNet [12], although the dialated convolutional blocks produce a large receptive field, the detailed information from low resolution levels would be lost during training due to the lack of skip connections. From the result shown in Fig. 4(c), HighRes3DNet fails to segment part of the details of the tumor core and enhanced tumor, as pointed by the red arrow. As for the whole tumor, part of the background is still misclassified as edema, as pointed by the black arrow. 3D U-Net [3] solves the problem with skip connections and provides a better segmentation result. However, the small actual field produced by the limited size of convolutional blocks makes the model fail to learn the global details of the object. From the result in Fig. 4(a), the ratio of the tumor core to whole tumor is larger than that of the ground truth. Even though the segmentation result of tumor core is correct in a local view, the high ratio from the global view makes the result less accurate. From the result of V-Net [18], although the skip connection keeps the low level details, the large amount of parameters from the isotropic convolutional blocks causes high memory consumption and makes the training hard to converge. In Fig. 4(b), it can be seen that some details at the boundaries of the whole tumor and parts of the enhanced tumor are misclassified.

5   Conclusion

In this work, we propose a 3D large kernel anisotropic network for brain tumor segmentation in MR images. The 3D large kernels after the encoders produce a large enough receptive field for the model to capture global information. Additionally, the simulation of 3D large kernel blocks with several 2D convolutional kernels reduces the computational cost. The anisotropic convolutional blocks in the encoders have the same effect as the traditional isotropic ones, and are proposed for memory efficiency and low risk of overfitting. Evaluated on part of the BRATS 17 dataset, our model is proved to be effective by outperforming some popular 3D DNN architectures.

Acknowledgements. This work was supported in part by Australian Research Council (ARC) grants and National Center for Image-Guided Therapy (NCIGT): P41 EB015898.

References

1. Abadi, M., et al.: Tensorflow: a system for large-scale machine learning. OSDI **16**, 265–283 (2016)
2. Chaurasia, A., Culurciello, E.: Linknet: Exploiting encoder representations for efficient semantic segmentation (2017). arXiv preprint arXiv:1707.03718
3. Çiçek, Ö., Abdulkadir, A., Lienkamp, S.S., Brox, T., Ronneberger, O.: 3D U-Net: Learning dense volumetric segmentation from sparse annotation. In: Ourselin, S., Joskowicz, L., Sabuncu, M.R., Unal, G., Wells, W. (eds.) MICCAI 2016. LNCS, vol. 9901, pp. 424–432. Springer, Cham (2016). https://doi.org/10.1007/978-3-319-46723-8_49
4. Doyle, S., Vasseur, F., Dojat, M., Forbes, F.: Fully automatic brain tumor segmentation from multiple MR sequences using hidden Markov fields and variational EM. In: Procs. NCI-MICCAI BraTS, pp. 18–22 (2013)

5. Gibson, E., et al.: NiftyNet: a deep-learning platform for medical imaging (2017). arXiv preprint arXiv:1709.03485
6. Havaei, M., et al.: Brain tumor segmentation with deep neural networks. Med. Image Anal. **35**, 18–31 (2017)
7. He, K., Zhang, X., Ren, S., Sun, J.: Identity mappings in deep residual networks. In: Leibe, B., Matas, J., Sebe, N., Welling, M. (eds.) ECCV 2016. LNCS, vol. 9908, pp. 630–645. Springer, Cham (2016). https://doi.org/10.1007/978-3-319-46493-0_38
8. Holland, E.C.: Progenitor cells and glioma formation. Curr. Opin. Neurol. **14**(6), 683–688 (2001)
9. Kamnitsas, K., et al.: Deepmedic for brain tumor segmentation. In: Crimi, A., Menze, B., Maier, O., Reyes, M., Winzeck, S., Handels, H. (eds.) BrainLes 2016. LNCS, vol. 10154. Springer, Cham (2016). https://doi.org/10.1007/978-3-319-55524-9_14
10. Kingma, D.P., Ba, J.: Adam: a method for stochastic optimization. In: ICLR (2015)
11. Lee, K., Zung, J., Li, P., Jain, V., Seung, H.S.: Superhuman accuracy on the SNEMI3D connectomics challenge. In: NIPS (2017)
12. Li, W., Wang, G., Fidon, L., Ourselin, S., Cardoso, M.J., Vercauteren, T.: On the compactness, efficiency, and representation of 3d convolutional networks: brain parcellation as a pretext task. In: Niethammer, M., Styner, M., Aylward, S., Zhu, H., Oguz, I., Yap, P.-T., Shen, D. (eds.) IPMI 2017. LNCS, vol. 10265, pp. 348–360. Springer, Cham (2017). https://doi.org/10.1007/978-3-319-59050-9_28
13. Liu, D., et al.: Large kernel refine fusion net for neuron membrane segmentation. In: CVPR Workshops, pp. 2212–2220 (2018)
14. Liu, S., Song, Y., Zhang, F., Feng, D., Fulham, M., Cai, W.: Clique identification and propagation for multimodal brain tumor image segmentation. In: Ascoli, G.A., Hawrylycz, M., Ali, H., Khazanchi, D., Shi, Y. (eds.) BIH 2016. LNCS (LNAI), vol. 9919, pp. 285–294. Springer, Cham (2016). https://doi.org/10.1007/978-3-319-47103-7_28
15. Liu, S., et al.: 3D anisotropic hybrid network: transferring convolutional features from 2D images to 3D anisotropic volumes (2017). arXiv preprint arXiv:1711.08580
16. Menze, B.H., et al.: The multimodal brain tumor image segmentation benchmark (BRATS). IEEE Trans. Med. Imaging **34**(10), 1993–2024 (2015)
17. Menze, Bjoern H., van Leemput, Koen, Lashkari, Danial, Weber, Marc-André, Ayache, Nicholas, Golland, Polina: A generative model for brain tumor segmentation in multi-modal images. In: Jiang, Tianzi, Navab, Nassir, Pluim, Josien P.W., Viergever, Max A. (eds.) MICCAI 2010. LNCS, vol. 6362, pp. 151–159. Springer, Heidelberg (2010). https://doi.org/10.1007/978-3-642-15745-5_19
18. Milletari, F., Navab, N., Ahmadi, S.A.: V-Net: Fully convolutional neural networks for volumetric medical image segmentation. In: Fourth International Conference on 3D Vision (3DV), pp. 565–571. IEEE (2016)
19. Ohgaki, H., Kleihues, P.: Population-based studies on incidence, survival rates, and genetic alterations in astrocytic and oligodendroglial gliomas. J. Neuropathol. Exp. Neurol. **64**(6), 479–489 (2005)
20. Peng, C., Zhang, X., Yu, G., Luo, G., Sun, J.: Large kernel matters-improve semantic segmentation by global convolutional network. In: CVPR, pp. 4353–4361 (2017)
21. Prastawa, M., Bullitt, E., Gerig, G.: Synthetic ground truth for validation of brain tumor MRI segmentation. In: Duncan, J.S., Gerig, G. (eds.) MICCAI 2005. LNCS, vol. 3749, pp. 26–33. Springer, Heidelberg (2005). https://doi.org/10.1007/11566465_4
22. Prastawa, M., Bullitt, E., Ho, S., Gerig, G.: A brain tumor segmentation framework based on outlier detection. Med. Image Anal. **8**(3), 275–283 (2004)

23. Singh, S.K., et al.: Identification of a cancer stem cell in human brain tumors. Cancer Res. **63**(18), 5821–5828 (2003)
24. Subbanna, N.K., Precup, D., Collins, D.L., Arbel, T.: Hierarchical probabilistic gabor and MRF segmentation of brain tumours in MRI volumes. In: Mori, K., Sakuma, I., Sato, Y., Barillot, C., Navab, N. (eds.) MICCAI 2013. LNCS, vol. 8149, pp. 751–758. Springer, Heidelberg (2013). https://doi.org/10.1007/978-3-642-40811-3_94
25. Wang, G., Li, W., Ourselin, S., Vercauteren, T.: Automatic brain tumor segmentation using cascaded anisotropic convolutional neural networks (2017). arXiv preprint arXiv:1709.00382
26. Wen, P.Y., Kesari, S.: Malignant gliomas in adults. N. Engl. J. Med. **359**(5), 492–507 (2008)
27. Zhang, D., et al.: Panoptic segmentation with an end-to-end cell R-CNN for pathology image analysis. In: Frangi, A.F., Schnabel, J.A., Davatzikos, C., Alberola-López, C., Fichtinger, G. (eds.) MICCAI 2018. LNCS, vol. 11071, pp. 237–244. Springer, Cham (2018). https://doi.org/10.1007/978-3-030-00934-2_27
28. Zhang, D., Song, Y., Liu, S., Feng, D., Wang, Y., Cai, W.: Nuclei instance segmentation with dual contour-enhanced adversarial network. In: ISBI, pp. 409–412. IEEE (2018)
29. Zhou, B., Khosla, A., Lapedriza, A., Oliva, A., Torralba, A.: Object detectors emerge in deep scene CNNs. In: ICLR (2015)
30. Zikic, D., et al.: Decision forests for tissue-specific segmentation of high-grade gliomas in multi-channel MR. In: Ayache, N., Delingette, H., Golland, P., Mori, K. (eds.) MICCAI 2012. LNCS, vol. 7512, pp. 369–376. Springer, Heidelberg (2012). https://doi.org/10.1007/978-3-642-33454-2_46

Saliency Supervision: An Intuitive and Effective Approach for Pain Intensity Regression

Conghui Li[1], Zhaocheng Zhu[2], and Yuming Zhao[1(✉)]

[1] School of Electronic Information and Electrical Engineering,
Shanghai Jiao Tong University, Shanghai, China
arola_zym@sjtu.edu.cn
[2] School of Electronics Engineering and Computer Science,
Peking University, Beijing, China

Abstract. Getting pain intensity from face images is an important problem in autonomous nursing systems. However, due to the limitation in data sources and the subjectiveness in pain intensity values, it is hard to adopt modern deep neural networks for this problem without domain-specific auxiliary design. Inspired by human vision priori, we propose a novel approach called saliency supervision, where we directly regularize deep networks to focus on facial area that is discriminative for pain regression. Through alternative training between saliency supervision and global loss, our method can learn sparse and robust features, which is proved helpful for pain intensity regression. We verified saliency supervision with face-verification network backbone [15] on the widely-used *UNBC-McMaster Shoulder-Pain* [10] dataset, and achieved state-of-art performance without bells and whistles. Our saliency supervision is intuitive in spirit, yet effective in performance. We believe such saliency supervision is essential in dealing with ill-posed datasets, and has potential in a wide range of vision tasks.

Keywords: Regression · Saliency supervision · Regularization
Triplet loss · Multi-task training

1 Introduction

Excessive usage of anesthetic will cause bad effects on patinets because the pain intensity of the patient is not well measured by tradition methods like skin conductance algesimeter or heart rate variability, which calls for a reliable approach that reports pain intensity in time. From the perspective of computer vision, the pain intensity task can be viewed as a semantic problem from single images or videos. Our job is aimed at pushing the benchmark of pain intensity to state-of-art level with a new design.

There are a few datasets that have pain intensity labels. Particularly, *UNBC-McMaster Shoulder-Pain* [10] dataset is the only dataset available for per-frame

© Springer Nature Switzerland AG 2018
L. Cheng et al. (Eds.): ICONIP 2018, LNCS 11307, pp. 455–464, 2018.
https://doi.org/10.1007/978-3-030-04239-4_41

visual analysis. It contains only 200 videos of 25 patients who suffer from shoulder pain and repeatedly raise their arms and then put them down (onset-apex-offset). While all frames are labeled with pain intensities, the labels are reported by the patient, which is very subjective. Moreover, in most frames patients got zero pain intensity, making it hard to observe the pattern behind frames. Many methods [1,14] have exploited deep neural networks trained in a data-extensive domain to alleviate the limited training data problem. For example, they finetune a well-trained face verification network with a regularized regression loss for pain intensity regression or valence-arousal estimation. In this case, their initial value is expected to be closer to some optimum for pain intensity regression, due to the similarity between domains. However, we doubt the finetuning procedure in those methods, as pain intensity values is not always a good supervision signal, especially for small dataset like *UNBC-McMaster Shoulder-Pain.*

Many people would give their reasons based on face attributes like eyes or lips when asked to determine individual's pain intensity from his face. Inspired by this fact, we exploit Action Units (AUs) [2], a semantic representation of face attributes, to regularize the training of pain intensity regression. AUs have been utilized for facial expression recognition. For example, FATAUVA-Net [1] train a mid-level network for AU detection using the AU label, and then finetune mid-level network for valence-arousal estimation and facial expression recognition. They use the labeled AU values as additional information and get better result. Pain is also a kind of expression, and here we explicitly supervise the network to focus on the areas of AUs so that the network will pay attention to the areas related to pain which will give a closer initial value to the global optimum than the previous work [14] when finetuning the network for pain intensity regression.

Since deep neural networks work like a black box, we cannot exactly know where the network pay attention. Zeiler proposed a kind of visualizing network method [17] using deconvolution [18] to get the contribution of input to the network's output. As we want to directly supervise the saliency map, which means the network for getting saliency should be able to merge with the origin network and support forward and backward calculation. Deconvolution is a good option for generating saliency map for our task. We add a deconvolution group after the bottleneck layer, making the network an encoder-decoder, note that in this encoder-decoder framework, the framework is different from the task of image generation [11] or image completion [16], the deconvolution group shares the architecture and weights with convolution group, and we hope the decoded saliency map is similar with attention map which means the saliency map observed from the bottleneck layer should follow certain distribution where some areas such as mouth, eyes should be more activated than others.

According to Facial Action Coding System, six common emotions: happiness, sadness, surprise, fear, anger, disgust, contempt are coded by the AU. Since not all AUs are related to a specific expression, we can not directly use the AU value as supervised signal. To avoid the intra-variance AUs that may case wrong supervision, we organize the inputs in the form of triplets [13] to use the relative value of pain intensity. In a triplet, the anchor and positive have the same pain intensity, and the negative has different pain intensity. If an AU is almost the same between anchor and positive and different between anchor and negative, we think such AU is related to pain, otherwise it belongs to irrelevant AUs. We select possible related AUs by comparing the pain intensity and AU labels in every triplet and then build the attention map for the triplet. At the same time, using triplets as input for saliency supervision, we alleviate the imbalance problem of training data through utilizing most image instances whose pain intensity is zero and avoiding overfitting.

As for the encoder-decoder framework, the deconvolution group is used for saliency map generation while not for feature learning, so we freeze the parameter updating in this group. As we need to get the saliency map in time after the convolution group has been updated, we copy the parameter in convolution group to deconvolution group before we calculate the saliency map.

The saliency map is used to supervise the AU local features, and we also take the whole image feature into consideration. The global feature is extracted from the bottleneck layer and supervised by triplet loss, and the margin is determined by the difference of pain intensity between the anchor and negative example. We combine local and global feature through a multi-task learning framework, using alternate training strategy.

Finally we finetune the network from the bottleneck layer supervised by pain intensity with regression loss, and get better result than the previous work [14].

In summary, the contributions of this work include:

1. Use triplet to build the attention map for pain regression task without sampling and datasets balancing, which can make full use of the ill-posed datasets.
2. Supervise the saliency map with attention map so the bottleneck feature is embedded with more information of the interest area related with the pain regression task.
3. Propose a method to train the network as a multi-task problem combining local feature and global feature.

2 Related Works

2.1 Pain Intensity Regression

The work on pain intensity based on computer version has significant improvement due to the deep learning technology and the release of the *UNBC-McMaster Shoulder-Pain*. There are two main streams, video based and image based methods.

Personalized method [7] uses the facial point of face image as input and uses Bi-LSTM [6] to estimate the observed pain intensity (OPI) value. They build individual facial expressiveness score (I-FES) for each person and use Hidden Conditional Random Fields (HCRFs) to merge the sequence result to get personalized visual analog scales (VAS) estimation. Recurrent Convolutional Neural Network Regression (RCR) [21] uses recurrent convolution network leveraging sequence information, and is trained end-to-end yet achieving sub-optimal performance. The other direction is image based method [14] that changes the face verification task to be a regression task, using smooth $L1$ [3] loss and adding center loss [15] to make the result more discrete. FATAUVA-Net presents a deep learning framework for Valence-Arousal (V-A) estimation that employs AUs as mid-level representation where they map the AU labels to the feature maps as a surprised signal, training different branch for different AUs and using the AU detection results for Valence-Arousal (V-A) estimation.

2.2 Regulation Loss

Regulation restricts the size of the parameter space, so the deeper neural network can also have good generalization ability learning from small datasets. $L1$ regulation makes parameters sparsely while $L2$ weight decay restricts the norm of parameters. Recently proposed methods [8,9] focus on small datasets training or semi-supervised learning, which are similar with ours, however, we add the regulation on saliency map. To our best knowledge, no previous work was carried in such direction.

3   Our Method

3.1   Attention Map Generation

Since only some of the AUs are relevant with pain intensity, we organize the inputs images in the form of triplets to use the relative values of pain intensity instead of the absolute values. Building attention map in the form of triplets, we can exclude the irrelevant AUs that wave obviously among same pain intensity or barely change among different pain intensity. Supervised by such attention map, the saliency map areas corresponding to irrelevant AUs will have almost the same brightness, which means the areas corresponding to irrelevant AUs in input image are embedded as little as possible into the bottleneck features, and the areas corresponded with relevant AUs in input images are embedded as much as possible into the final bottleneck features.

In details, we choose the triplet: the anchor and positive have the same pain intensity, and the negative has different. We divide all AUs into set $\mathcal{A}$ and set $\mathcal{B}$. AUs in set $\mathcal{A}$ are relevant AUs whose corresponded areas in saliency map should discriminate in a triplet, and AUs in set $\mathcal{B}$ are irrelevant AUs whose corresponded areas in saliency map should almost be the same in a triplet. AU $\mathbf{k}$ is selected to set $\mathcal{A}$ if it satisfy the following equation, otherwise it was added to set $\mathcal{B}$. Set $\mathcal{A}$ contains relevant AUs that barely change between anchor and positive and wave obviously between anchor and negative.

$$\left| V_a^k - V_p^k \right| < \alpha \quad and \quad \left| V_a^k - V_n^k \right| \geq \alpha \tag{1}$$

V^k is the value of facial AU $\mathbf{k}$. In a triplet, we denote the value of facial AU $\mathbf{k}$ for anchor, positive, negative as $V_{\mathbf{a}}^{\mathbf{k}}$, $V_{\mathbf{p}}^{\mathbf{k}}$, $V_{\mathbf{n}}^{\mathbf{k}}$ respectively. α is the threshold depending on the facial AU value.

Since AUs are corresponding to the semantic pattern of facial attributes, we use areas around each AU to generate the attention map, which is similar to the approach in EAC (enhancing and cropping) Net [5]. Note that EAC Net uses the absolute value of AU to generate the attention map for AU detection task. Different from their approach, we use the relative value of AU in a triplet for pain intensity regression as for not all AUs are related with pain intensity. We use attention map as direct signals for saliency supervision while EAC Net adds attention map to feature map.

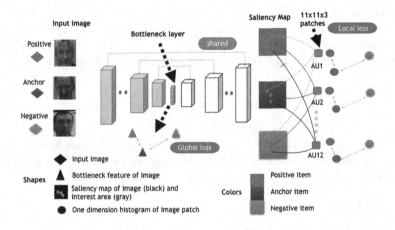

Fig. 1. Our proposed network structure.

3.2 Saliency Map Supervision

Fig. 1 shows our proposed network structure. The saliency map is observed from the bottleneck feature using deconvolution network (white blocks) that has the same architecture and parameters as the convolution network (yellow blocks). For each AU, we find the corresponding patch in anchor, positive and negative item and every patch is 11×11 pixels around the corresponding facial landmarks. The local loss is the sum of triplet loss generated by each AU.

We divide AUs into two set $\mathcal{A}$ and $\mathcal{B}$ in Sect. 3.1. The loss function for two sets is different.

The loss for set $\mathcal{A}$ is:

$$loss_{\mathcal{A}} = \sum_{k \subseteq \mathcal{A}} ([g(P_a^k, P_n^k) - g(P_a^k, P_p^k) + O * W^k]_+) \tag{2}$$

$$W^{\mathbf{k}} = \frac{|V_{\mathbf{a}}^{\mathbf{k}} - V_{\mathbf{n}}^{\mathbf{k}}|}{\sum_{i \subseteq \mathcal{A}} (|V_{\mathbf{a}}^{\mathbf{i}} - V_{\mathbf{n}}^{\mathbf{i}}|)} \tag{3}$$

The loss for set $\mathcal{B}$ is:

$$loss_B = \sum_{k \subseteq \mathcal{B}} ([g(P_a^k, P_n^k) - g(P_a^k, P_p^k) + O * \frac{1}{N}]_-) \tag{4}$$

We denote saliency map patch of AU **k** in anchor, positive, negative as $\mathbf{P_a^k}$, $\mathbf{P_p^k}$, $\mathbf{P_n^k}$ respectively. $\mathbf{g(m, n)}$ is the distance metric which we will discuss in details. **O** is the absolute value between the pain intensity of anchor and negative. $\mathbf{W^k}$ represents the contribution of AU **k** to pain intensity in set $\mathcal{A}$. **N** is the number of AUs in set $\mathcal{B}$, and we average the influence of each AU to pain intensity in set $\mathcal{B}$.

We denote the distance metric by $\mathbf{g(m, n)}$, in practical, we use normalized Earth Mover's Distance [12] between the one dimensional histogram of image patch **m** and **n**, as it is robust to face alignment error and performs better when evaluating the saliency map difference according to the image retrieval method [12].

The total local loss is:

$$loss_L = loss_{\mathcal{A}} + loss_{\mathcal{B}} \tag{5}$$

The supervision signal is saliency map, making the bottleneck feature embedded with similar information of the saliency map. To the best of our knowledge, we first supervise the saliency map for the origin task training.

3.3 Multi-task Training

The local loss we get in Sect. 3.2 focus on the facial AU areas which acts as the regulation loss. We find that if we only use the local loss, the network can not converge, so we combine it with the global feature together through multi-task training strategy and the local loss acts as the regulation loss. The normalized bottleneck feature is taken as the global feature presentation for the whole image, and we train it with triplets loss using Euclidean distance. The single triplet loss is as the following equation:

$$loss_G = [f(P_a, P_n) - f(P_a, P_p) + O * \beta]_+ \tag{6}$$

We denote the bottleneck feature of anchor, positive, negative as $\mathbf{P_a}$, $\mathbf{P_p}$, $\mathbf{P_n}$ respectively. $\mathbf{f(m, n)}$ is the Euclidean distance. **O** is the absolute value between the pain intensity of anchor and negative and we multiply **O** by factor β as the triplet loss margin. During training, we carry hard negative sampling [4] in all triplets in a batch, and we use alternate training strategy to combine the global loss and local loss.

3.4 Pain Regression

After training the network using local feature and global feature, we take the pain regression task. We finetune the network from the bottleneck layer, and as for the pain intensity is set to $[0, 5]$, we use the activation function $\sigma(x) = \frac{5}{1+e^{-x}}$, **x** is the output of bottleneck layer. we choose the smooth $L1$ [3] + $L1$ center loss [15] as our regression loss.

4   Experiments

In this section we will explain our implement details and experiments result.

We choose the related face recognition task and use the state of art pertained model as pretrained model. We use MTCNN [19] to detect the face area, and do face alignment according to the facial landmarks provided by *UNBC-McMaster Shoulder-Pain* dataset. The AU labels are available from *UNBC-McMaster Shoulder-Pain* dataset.

We train the attention network alternatively using basic learning rate 0.001 for global loss, and 0.01 for local loss. When we train the local loss, the deconvolution group is frozen, and share the parameters with the convolution group. During training we first select hard negative examples [4] in a batch, and train the network using $loss_G$, and then we use the same hard negative examples to get the $loss_L$ to train the network again.

When training the pain intensity regression, we set pain intensity value from [0, 15] to [0, 5] as proposed by the previous work [20, 21]. We evaluate the result using 25-cross-valuation. We try to train the network only using the local loss, however the network is not converging, and when the global loss is added, the network performance well which means the local loss acts like regulation loss here.

We compare the difference of directly taking regression for the task, using global feature then finetune the network for regression, combining global feature and local feature then finetune the network for regression. The comparison of saliency map can be found in Fig. 2. The experiment results can be found in Tables 1 and 2.

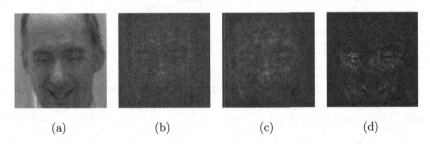

| (a) | (b) | (c) | (d) |

Fig. 2. Saliency maps of different network training approaches. (a) is the input picture, (b) is the saliency map from bottleneck embeddings trained directly by the pain intensity, (c) is the saliency map from bottleneck embeddings trained by the global feature and finetuned by the pain intensity, (d) is the saliency map from bottleneck embedding trained by the combination of global feature and local feature and finetuned by pain intensity.

Brighter areas in the saliency map means that the corresponded areas in input image are more embedded into the bottleneck feature. As we can see from Fig. 2, using global feature and then finetuned by the pain intensity, the model

Table 1. Performance of our proposed methods and related works on the *UNBC-McMaster Shoulder-Pain* dataset for the estimation of pain intensity. Our proposed method 1: Use global feature to train the network and smooth $L1 + L1$ center loss for regression. Our proposed method 2: Use the combination of global feature and local feature to train the network and smooth $L1 + L1$ center loss for regression. MAE is short for mean absolute error deviated from the ground-truth labels over all frames per video. MSE is mean squared error which measures the curve fitting degree. PCC is Pearson correlation coefficient which measures the curve trend similarity (↑ indicates the larger, the better). The best is highlighted in bold.

Methods	MAE↓	MSE↓	PCC↑
our proposed method 1	0.401	0.742	0.643
our proposed method 2	**0.334**	**0.626**	**0.804**
smooth $L1 + L1$ center loss [14]	0.456	0.804	0.651
OSVR-$L1$ [20]	1.025	N/A	0.600
OSVR-$L2$ [20]	0.810	N/A	0.601
RCR [21]	N/A	1.54	0.65

Table 2. Performance of our network when evaluated using the weighted MAE and weighted MSE proposed by [14]. Methods proposed by us are applied with uniform class sampling technique.

Methods	wMAE↓	wMSE↓
our proposed method'1	0.883	1.697
our proposed method 2	**0.727**	**1.566**
smooth $L1 + L1$ center loss + sampling [14]	0.991	1.720
OSVR-$L1$ [20]	1.309	2.758
OSVR-$L2$ [20]	1.299	2.719

focuses some area of face while not accurate enough, when using local feature and global feature together, the model focuses on the eye, mouth areas that related to the pain expression.

5   Summary

In this paper, we propose a novel method for attention based network. We use deconvolution network to get the saliency map of the bottleneck layer, and design the attention map for pain intensity regression task. Through direct regulation on saliency map and multi-task training with global loss, we successfully push forward single image based method by a considerable margin without bells and whistles. We believe our job is great for single image based prediction, and the performance of joint prediction over image sequences should benefit from features extracted by our method, which we leave as a future direction.

References

1. Chang, W., Hsu, S., Chien, J.: Fatauva-net: an integrated deep learning framework for facial attribute recognition, action unit detection, and valence-arousal estimation. In: IEEE Conference on Computer Vision and Pattern Recognition Workshops (CVPRW), pp. 1963–1971 (2017)
2. Ekman, P., Rosenberg, E.: What the face reveals: basic and applied studies of spontaneous expression using the facial action coding system (facs), **68**(1), 83–96 (1997)
3. Girshick, R.B.: Fast R-CNN abs/1504.08083 (2015)
4. Hermans, A., Beyer, L., Leibe, B.: In defense of the triplet loss for person re-identification abs/1703.07737 (2017)
5. Li, W., Abtahi, F., Zhu, Z., Yin, L.: Eac-net: A region-based deep enhancing and cropping approach for facial action unit detection (2017)
6. Ma, X., Hovy, E.H.: End-to-end sequence labeling via bi-directional lstm-cnns-crf (2016)
7. Martinez, D.L., Rudovic, O., Picard, R.: Personalized automatic estimation of self-reported pain intensity from facial expressions. In: IEEE Conference on Computer Vision and Pattern Recognition Workshops, pp. 2318–2327 (2017)
8. Miyato, T., Maeda, S.I., Koyama, M., Ishii, S.: Virtual adversarial training: a regularization method for supervised and semi-supervised learning. IEEE Transactions on Pattern Analysis and Machine Intelligence (2017)
9. Miyato, T., Maeda, S., Koyama, M., Nakae, K., Ishii, S.: Distributional smoothing with virtual adversarial training. Computer Science (2015)
10. Lucey, P., Cohn, J.F., Prkachin, M., Solomon, P.E., Matthews, I.: Painful data: The UNBC-McMaster shoulder pain expression archive database. In: Face and Gesture 2011, pp. 57–64 (2011)
11. Radford, A., Metz, L., Chintala, S.: Unsupervised representation learning with deep convolutional generative adversarial networks (2015)
12. Rubner, Y., Tomasi, C., Guibas, L.J.: The earth mover's distance as a metric for image retrieval. Int. J. Comput. Vis. **40**(2), 99–121 (2000)
13. Schroff, F., Kalenichenko, D., Philbin, J.: Facenet: a unified embedding for face recognition and clustering (2015). CoRR abs/1503.03832
14. Wang, F., et al.: Transferring face verification nets to pain and expression regression (2017). abs/1702.06925
15. Wen, Y., Zhang, K., Li, Z., Qiao, Y.: A discriminative feature learning approach for deep face recognition. In: Leibe, B., Matas, J., Sebe, N., Welling, M. (eds.) ECCV 2016. LNCS, vol. 9911, pp. 499–515. Springer, Cham (2016). https://doi.org/10.1007/978-3-319-46478-7_31
16. Yeh, R.A., Chen, C., Lim, T., Hasegawa-Johnson, M., Do, M.N.: Semantic image inpainting with perceptual and contextual losses (2016). abs/1607.07539
17. Zeiler, M.D., Fergus, R.: Visualizing and understanding convolutional networks. In: Fleet, D., Pajdla, T., Schiele, B., Tuytelaars, T. (eds.) ECCV 2014. LNCS, vol. 8689, pp. 818–833. Springer, Cham (2014). https://doi.org/10.1007/978-3-319-10590-1_53
18. Zeiler, M.D., Krishnan, D., Taylor, G.W., Fergus, R.: Deconvolutional networks. In: Computer Vision and Pattern Recognition, pp. 2528–2535 (2010)
19. Zhang, K., Zhang, Z., Li, Z., Qiao, Y.: Joint face detection and alignment using multi-task cascaded convolutional networks (2016). abs/1604.02878

20. Zhao, R., Gan, Q., Wang, S., Ji, Q.: Facial expression intensity estimation using ordinal information. In: Computer Vision and Pattern Recognition, pp. 3466–3474 (2016)
21. Zhou, J., Hong, X., Su, F., Zhao, G.: Recurrent convolutional neural network regression for continuous pain intensity estimation in video pp. 1535–1543 (2016)

Identification of Causality Among Gene Mutations Through Local Causal Association Rule Discovery

Ruichu Cai[1(✉)], Qiqi Zhen[1], and Zhifeng Hao[2]

[1] School of Computer Science, Guangdong University of Technology,
Guangzhou, China
cairuichu@gmail.com, zqq80ti@outlook.com
[2] School of Mathmatics and Big Data, Foshan University, Foshan, China
zfhao@fosu.edu.cn

Abstract. Detecting the interaction among gene mutations is still an open problem on genetic research. Among various types of interaction, the causality among the gene mutations provides deep insight of the gene mutation and evolution, is the focus of the current research. Different from the global causal network reconstruction method, we propose a local causal discovery method by exploring the causal concept under the association rule discovery framework. Firstly we propose a V-Structure Measure (VSM) to evaluate the causal significance of the local SNPs structures. Secondly, we develop a method called ASymmetric Causal Association Rule Discovery (ASCARD) to mine the reliable causal association rules considering the conflicts among the candidate structures. Finally, the experiments on the synthetic data and WTCCC (Wellcome Trust Case Control Consortium) SNPs dataset shows the effectiveness of the proposed method. Some interesting biological discoveries also show the potential of the real world applications.

Keywords: Gene mutations · Causality · Causal association rule

1 Introduction

Interaction among gene mutations is an important research topic in GWAS, evolution, and so on topics. A lot of important results are reported in this area during the past ten years. For example, Masyukova et al. [1] found the interactions between osm-3 and nphp-4, Wang et al. [2] discoveries genetic interactions in breast cancer.

Association analysis is one of the commonly used tools for interaction discovery. On the high dimensional SNPs data, the association analysis usually returns millions of candidates of the interactions, which are hard to be analyzed by the biologist. On the other hand, some traditional measures of association analysis have poor performance for high-dimensional data, such as FLDA mentioned in [3] that should be modified elaborately in the high-dimensional cases.

© Springer Nature Switzerland AG 2018
L. Cheng et al. (Eds.): ICONIP 2018, LNCS 11307, pp. 465–477, 2018.
https://doi.org/10.1007/978-3-030-04239-4_42

Kevin Beyer [4] shows that as the dimension increases, the distance to the nearest sample is close to that of the farthest one. Moreover, traditional measures fail to process collinearity [5] and random noise [6] problems with high dimensional bioinformatics data. Most methods extract valid features from high-dimensional data through dimension reduction, such as PCA [7], reducing computational complexity as well. However, the principal component is typically hard to create biological explanation model for the discrete bioinformatics data.

Recently, causality among gene mutations becomes increasingly important since it reveals the internal mechanism of gene mutation and evolution, and usually returns more accurate candidates. For example, Takeshi et al. [8] proposed a scan statistical framework to extract causal gene clusters, which showed high accuracy. Catlett et al. [9] presented a method of Reverse Causal Reasoning to interpret the high-throughput data and complemented the analyses using pathway gene sets. As for the interpretability, Martin et al. [10] developed a method of integrated cause-and-effect networks and transcriptomics measurements to quantify and interpret the network's response. Furthermore, causal inference is the useful method to reveal genome evolution [11]. To improve the causality research efficiency, Mihil et al. [12] defined a scheme that had enriched the biomedical domain corpora with causality relationship, which had been verified by forming BioCause. The main challenge of the causal discovery method on the SNPs data is that it is hard to construct a global causal network with millions of variables [14,15]. As shown in Chickering [13], the construction of the Bayesian network is an NP-complete problem even under certain conditions.

Different from the global causal network reconstruction method, we propose a local causal discovery method by exploring the causal concept under the association rule discovery framework. In detail, we develop a method called ASymmetric Causal Association Rule Discovery (ASCARD) to detect causal single nucleotide polymorphisms (SNPs) based on the assumption that local interaction is more evident than the global. Combining association rules mining with V-Structure, ASCARD quantifies the conditional independence by V-Structure Measure (VSM). To deal with the high-dimensional data, a mutual information based measurement is proposed to deduce the local structure and the causal SNPs. To verify ASCARD, we conduct experiments on synthetic data comparing with the association rule mining which uses relative reporting ratio (RR for short) described in [16] and confidence (AR for short) as the proxy respectively and explain the advantage of ASCARD. Finally, we discuss the experimental results biologically on WTCCC (Wellcome Trust Case Control Consortium) SNPs dataset.

2   Method

In this section, the details of the causal rule mining algorithm are provided. In Sect. 2.1, an interestingness measure of the causal rule is proposed by considering the properties of V-Structure. In Sect. 2.2, a mining algorithm is also devised by exploring the monotonic property of the interestingness measure.

2.1   V-Structure Based Causal Rule Interestingness Measure

Causal rules A → C is a special type of association rule, with the requirement that A is the cause of C. To detect the causality among gene mutations,
we develop an ASCARD algorithm to the symmetry problem and to detect
causal SNPs. Considering high-dimension data, ASCARD focuses on local structure detection. Specifically, it measures the conditional independence by mutual
information instead of the traditional confidence between SNPs to infer local
causal structure and causal SNPs with the form of "$SNP_{a1}, SNP_{a2}, ..., SNP_{an} \rightarrow$
$SNP_{c1}, SNP_{c2}, ..., SNP_{cm}$".

Generally, it is difficult to directly recognize the causal-effect pairs among
the SNPs. However, the exiting works on causal discovery have shown that the
V-Structure can be distinguished based on its unique conditional independence
relationship. As shown in Fig. 1, there are four basic structures comprised of three
loci SNPs. As arrow serves as the flow from cause to effect, Fig. 1(a)–(c) are all
subjected to the same conditional independence relationship that variable SNP_1
is conditionally independent of variable SNP_2 given variable SNP_3, so it is hard
to distinguish each of them. However, Fig. 1(d) known as the V-Structure [17]
is the structure with the opposite property that variable SNP_1 is conditionally
dependent on variable SNP_2 given variable SNP_3. With the more stable and
distinguishing statistical properties [18], we can detect the causal SNPs that
conform to V-Structure.

Fig. 1. Four basic structures comprised of three loci SNPs. (a)–(c) are the conditional
independence equivalent, while (d) has the opposite property.

We introduce a concept to evaluate the V-Structure significance of SNPs. Let
causal nodes denote cause and effect nodes denote effect. To simplify the analysis,
we consider the simplest case of the causal node set containing two SNPs pointing to the effect node set with only one SNP, which is shown as Fig. 1(d). We
use $NMI(SNP_1, SNP_2)$ to denote the normalized mutual information between
SNP_1 and SNP_2, and $NMI(SNP_1, SNP_2|SNP_3)$ the conditional mutual information between SNP_1 and SNP_2 given SNP_3. Because SNP_1 and SNP_2 are
independent of each other, the $NMI(SNP_1, SNP_2)$ is expected to be lower. Simi-
larly, it is expected that the $NMI(SNP_1, SNP_2|SNP_3)$ is greater since SNP_1 and
SNP_2 are dependent on each other given SNP_3. In other words, in the case of
Fig. 1(d) $NMI(SNP_1, SNP_2|SNP_3)$ should be greater than $NMI(SNP_1, SNP_2)$.
The V-Structure Measure (VSM) can be derived as follow:

$$VSM(SNP_1, SNP_2 \rightarrow SNP_3) = NMI(SNP_1, SNP_2|SNP_3) - NMI(SNP_1, SNP_2)$$
(1)

It can be considered as the convincing match between the relationship among SNPs and V-Structure if VSM is greater than zero. $\text{NMI}(\text{SNP}_1, \text{SNP}_2)$ is computed as

$$\text{NMI}(\text{SNP}_1, \text{SNP}_2) = 2 * \frac{MI(\text{SNP}_1; \text{SNP}_2)}{H(\text{SNP}_1) * H(\text{SNP}_2)} \tag{2}$$

where $MI(\text{SNP}_1; \text{SNP}_2)$ is the mutual information between SNP_1 and SNP_2, $H(\text{SNP}_1)$ and $H(\text{SNP}_2)$ are the information entropy of SNP_1 and SNP_2 respectively. Denoting the range of SNP_i as S_i, we obtain

$$MI(\text{SNP}_1; \text{SNP}_2) = \sum_{s_1 \in S_1} \sum_{s_2 \in S_2} p(s_1, s_2) \log \frac{p(s_1, s_2)}{p(s_1) * p(s_2)} \tag{3}$$

$$H(\text{SNP}_1) = \sum_{s_1 \in S_1} p(s_1) \log p(s_1) \tag{4}$$

$$H(\text{SNP}_2) = \sum_{s_2 \in S_2} p(s_2) \log p(s_2) \tag{5}$$

Similarly, $\text{NMI}(\text{SNP}_1, \text{SNP}_2 | \text{SNP}_3)$ is computed as

$$\text{NMI}(\text{SNP}_1, \text{SNP}_2 | \text{SNP}_3) = 2 * \frac{MI(\text{SNP}_1; \text{SNP}_2 | \text{SNP}_3)}{H(\text{SNP}_1 | \text{SNP}_3) * H(\text{SNP}_2 | \text{SNP}_3)} \tag{6}$$

where $MI(\text{SNP}_1; \text{SNP}_2 | \text{SNP}_3)$ is the conditional mutual information between SNP_1 and SNP_2 given SNP_3, $H(\text{SNP}_1 | \text{SNP}_3)$ and $H(\text{SNP}_2 | \text{SNP}_3)$ are the conditional information entropy of SNP_1 and SNP_2 respectively given SNP_3, and

$$MI(\text{SNP}_1; \text{SNP}_2) = \sum_{s_3 \in S_3} \sum_{s_1 \in S_1} \sum_{s_2 \in S_2} p(s_1, s_2 | s_3) \log \frac{p(s_1, s_2 | s_3)}{p(s_1 | s_3) * p(s_2 | s_3)} \tag{7}$$

$$H(\text{SNP}_1 | \text{SNP}_3) = \sum_{s_3 \in S_3} \sum_{s_1 \in S_1} p(s_1 | s_3) \log p(s_1 | s_3) \tag{8}$$

$$H(\text{SNP}_2 | \text{SNP}_3) = \sum_{s_3 \in S_3} \sum_{s_2 \in S_2} p(s_2 | s_3) \log p(s_2 | s_3) \tag{9}$$

Now we consider a complex structure whose causal node sets contain three loci SNPs and effect node sets contain two loci SNPs as shown in Fig. 2.

We can divide the structure into two substructures. Each of them has a causal node set with three SNPs pointed to an effect node, as shown in Fig. 3. Each substructure in Fig. 3 can be regarded as the union of many basic V-Structure whose causal node sets contain two SNPs. Figure 3(a) can be regarded as being composed of three V-Structure as shown in Fig. 4. Then we compute the VSM of each V-Structure in Fig. 4 according to the Eq. (1) and select the minimum VSM as the VSM of Fig. 3(a) since VSM of the structure is dependent on the weakness of its substructure. We obtain

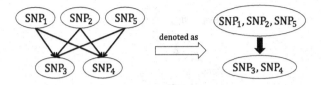

Fig. 2. A structures whose causal node sets contain three SNPs and effect node sets contain two SNPs.

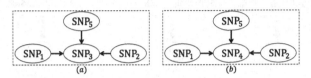

Fig. 3. Two substructures are obtained by divided the structure in Fig. 2.

$$VSM(SNP_1, SNP_2, SNP_5 \rightarrow SNP_3)$$
$$= \min\{VSM(SNP_1, SNP_2 \rightarrow SNP_3),$$
$$VSM(SNP_1, SNP_5 \rightarrow SNP_3),$$
$$VSM(SNP_2, SNP_5 \rightarrow SNP_3)\}$$

$$(10)$$

As for the general structure shown as Fig. 5 whose causal node sets contain n SNPs and effect node sets contain m SNPs can be divided into m substructure. Each VSM can be calculated as the Eq. (11):

$$VSM(SNP_{a1}, SNP_{a2}, ..., SNP_{an} \rightarrow SNP_{c1})$$
$$= \min_{S_1, S_2 \subset \{SNP_{a1}, SNP_{a2}, ..., SNP_{an}\}} VSM(S_1, S_2 \rightarrow SNP_{c1}),$$
$$VSM(SNP_{a1}, SNP_{a2}, ..., SNP_{an} \rightarrow SNP_{c2})$$
$$= \min_{S_1, S_2 \subset \{SNP_{a1}, SNP_{a2}, ..., SNP_{an}\}} VSM(S_1, S_2 \rightarrow SNP_{c2}),$$

$$\vdots$$

$$VSM(SNP_{a1}, SNP_{a2}, ..., SNP_{an} \rightarrow SNP_{cm})$$
$$= \min_{S_1, S_2 \subset \{SNP_{a1}, SNP_{a2}, ..., SNP_{an}\}} VSM(S_1, S_2 \rightarrow SNP_{cm})$$

$$(11)$$

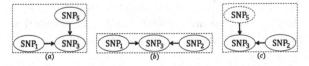

Fig. 4. Fig. 3(a) can be divided into three V-Structures.

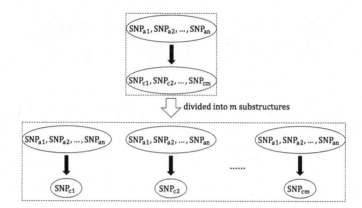

Fig. 5. General structure can be divided into m substructures each of which contains a causal node set with n SNPs pointed to one effect node.

2.2   Causal Rule Discovery

Before given the causal rule discovery algorithm, some properties of VSM are investigated, which are essential to the correctness and efficiency of the algorithm.

Symmetric Property of the Structure. Causal node sets and effect node sets are symmetric if there is without any prior knowledge of SNP-SNP interaction. It means that the structure shown in Fig. 6 can be as meaningful as the one in Fig. 2(b), which can be divided into three substructure shown as Fig. 7.

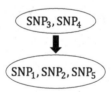

Fig. 6. Without any priori knowledge, the symmetric of Fig. 2 can be meaningful as well.

As some structures in Fig. 7 are conflicted with some in Fig. 3, we compute the VSM of each V-Structure, then select the structure with the maximum VSM as the acceptance. Those who are consistent with the selected are the reliable local causal structure, while the rest would be unacceptable.

Monotonic of VSM. VSM is non-strictly decreasing with the increasing of the number of variables in the causal node set according to the Eq. (10). In other

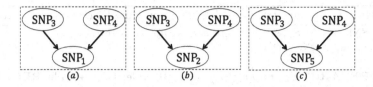

Fig. 7. The structure in Fig. 6 divided into three V-Structures conflicted with Fig. 4.

words, for two causal association rules such as $P_1 \rightarrow C$ and $P_1 \cup S \rightarrow C$, the following

$$\text{VSM}(P_1 \rightarrow C) \geq \text{VSM}(P_1 \cup S \rightarrow C) \tag{12}$$

holds, where S is SNPs set. This provides an efficient way to calculate the VSM of the new rules that are created based on the accepted rules.

Strategy for Update. Based on the accepted causal association rule $P_1 \rightarrow C$, adding SNP to P_1 creates the new causal association rule $P_1 \cup \text{SNP} \rightarrow C$ whose VSM is calculated as follow:

$$\begin{aligned}
&\text{VSM}(P_1 \cup \{\text{SNP}\} \rightarrow C) \\
&= \min\left\{ \text{VSM}(P_1 \rightarrow C), \min_{SNP' \subset P_1} \text{VSM}(\{SNP', \text{SNP}\} \rightarrow C) \right\}
\end{aligned} \tag{13}$$

Reliability of V-Structure. In a V-Structure, causal nodes are conditionally dependent on each other given effect node. If there exists $\text{SNP}_1, \text{SNP}_2 \in P_1$ holding $\text{SNP}_1 \perp \text{SNP}_2|C$, SNP_1, SNP_2 and C do not conform to the topological property of V-Structure, so the rule $P_1 \rightarrow C$ should be rejected.

Now we induce the Algorithms 1 and 2. In algorithm 1, the input contains two SNPs sets S_1 and S_2, the threshold of the support and the VSM, while the output **R** is a set with reliable causal association rules. **R** is empty for initializations. Firstly, we refine causal association rules in two cases of $S_1 \rightarrow S_2$ and $S_2 \rightarrow S_1$ and get two causal association rule sets denoted as R_1 and R_2 respectively. Then we sort R_{12}, the union of R_1 and R_2, in descending order of VSM of the rules in R_{12}. Finally, all the rules in R_{12} are traversed to be added into R if they are not conflicted with each rule $r \in R$.

We give pseudo-code of Algorithm 2 as follow:

The input and the output of Algorithm 2 are similar to Algorithm 1. We make **R** empty for initialization. According to formulation Eq. (11), the Algorithm 2 traverses each SNPs in S_2 that serves as a effect node, and S_1 serves as the causal node set. Based on the updating strategy, we create new causal association rules in line 6, then put those who surpass the threshold of support, VSM and the V-Structure, into **R**.

Algorithm 1. ASymmetric Causal Association Rule Discovery (ASCARD)

Input: SNPs set S_1, S_2, support threshold $supp$, VSM threshold v;
Output: Causal SNPs set R ;
1: $R=\varnothing$;
2: $R_1 = RECAR(S_1, S_2, supp, v)$; // Refining the rule of $S_1 \rightarrow S_2$ by **RECAR**, the Algorithm 2
3: $R_2 = RECAR(S_2, S_1, supp, v)$; // Refining the rule of $S_2 \rightarrow S_1$ by **RECAR**, the Algorithm 2
4: $R_{12} = R_1 \cup R_2$;
5: Sort R_{12} in descending order by VSM;
6: **for** rule $r_{12} \in R_{12}$ **do**
7:     **if** r_{12} is not conflicted with each $r \in R_{12}$ **then**
8:         $R = R \cup \{r_{12}\}$
9:     **end if**
10: **end for**

Algorithm 2. REfine Causal Association Rule (RECAR)

Input: SNPs set S_1, S_2, support threshold $supp$, VSM threshold v;
Output: Causal SNPs set R ;
1: **for** each $SNP_P \in S_1$ and each $SNP_C \in S_2$ **do**
2:     **if** supp($SNP_P \rightarrow SNP_C$)$\geq supp$ **then**
3:         $R = R \cup \{SNP_P \rightarrow SNP_C\}$
4:         **for** $i = 2$ to the max length of the rule in R **do**
5:             **for** rule $P \rightarrow SNP_C \in R$ with i SNPs **do**
6:                 $r = P \cup \{SNP_P\} \rightarrow SNP_C$
7:                 **if** supp(r)$\geq supp$ and VSM(r)$\geq v$ and r passed V-Structure Test **then**
8:                     $R = R \cup \{r\}$
9:                 **end if**
10:             **end for**
11:         **end for**
12:     **end if**
13: **end for**

---

3   Experiments

3.1   Experiment on Synthetic Dataset

We generate 569 sets of synthetic data conforming to the topological property of V-Structure by using R package bnlearn. Then we randomly select 1138 genes of the 22207 genes mentioned above with more than 4 SNPs to constitute 569 pairs of genes. For each gene pair, we replace one SNP data of a gene with effect node data of a set of synthetic data, and two SNPs data of another gene with two causal nodes data of the set of synthetic data. Finally, ASCARD, association rule mining [16] that uses relative reporting ration (RR for short) and the one that uses traditional confidence (AR for short, with a confidence of 0.7), are tested in the experiments.

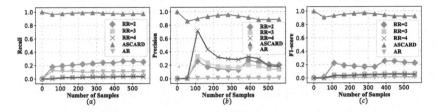

Fig. 8. The performance of ASCARD and association rule mining in synthetic data experiment.

We use recall, precision and F1-score to evaluate the performances of ASCARD and association rule mining, which is shown in Fig. 8.

From Fig. 8, we find that although association rule mining can learn the implicit causal association rules from synthetic data and RR's performance improves by adjusting the threshold of relative reporting ration, ASCARD shows a better performance all the way. The main reason is that the region of SNPs data value is {0,1,2}, while for association rule mining 0 indicates no mutation and both 1 and 2 are regarded as mutation, meaning that the information of the difference between 1 and 2 is not used. ASCARD uses mutual information that contains the information of probability distribution of 0, 1 and 2. In other words, more information is considered in ASCARD, and that is the advantage of ASCARD.

3.2 Experiment on WTCCC SNPs Dataset

We conduct the experiment on a real-world dataset, the WTCCC SNPs dataset. The dataset contains 16179 samples, each of which contains 394747 SNPs. All information about the concerned SNPs is shown through an example on two SNPs, rs3094315 and rs4040617, being shown in Tables 1 and 2. From Table 1, we learn that the major alleles of two SNPs are all A, and the minor alleles are all G. Table 2 shows the part of the dataset, where "rs3094315_G" indicates SNP rs3094315 could mutate into G. Since duplication of the chromosome in a human body, 0 in Table 2 indicates that no chromosome of the sample mutates, so that genotype of the SNPs sites of the sample is AA, 1 indicates that genotype is AG/GA, and 2 indicates GG.

Table 1. Information of SNPs

Chromosome	SNPs	Position	Minor allele	Major allele
1	rs3094315	752566	G	A
1	rs4040617	779322	G	A

Table 2. Samples and Data

FID	IID	rs3094315_G	rs4040617_G
FAM_HT	WTCCC169395	0	0
FAM_HT	WTCCC70517	0	0
FAM_HT	WTCCC69846	1	1
FAM_HT	WTCCC169826	0	0
FAM_HT	WTCCC69962	2	2

Table 3. Some of the experiment results

Cause		Effect		VSM
Gene	SNPs	Gene	SNPs	
PLCG1	rs753381	ATP13A4	rs2280476	0.000240335
	rs4297946			
ABCA9	rs8066118	ABCA10	rs12103556	0.040902519
	rs7215642			
TVP23C-CDRT4	rs2954759	TRIM16	rs2074890	0.038316057
	rs7211982			
	rs2052022			
KRTAP29-1	rs1001191	KRT38	rs897416	0.030948885
	rs758741			
	rs1005196			
ZSCAN32	rs27230	ADCY9	rs2230742	0.013372
	rs27231			
LAMA2	rs9375614	USPL1	rs7984952	0.001677
	rs2306942			
IRAK2	rs696322	MYO16	rs16973313	0.001315
	rs708035			

According to Algorithm 1, the input of ASCARD are two SNPs sets, but here there is only one dataset containing 394747 SNPs. For these 394747 SNPs, they are divided into several SNPs sets for detecting. By NCBI [19] we learn that a gene contains multiple SNPs sites, and 225676 SNPs of the 394747 SNPs are located in 22207 genes regions respectively. Therefore, we remain and divide the 225676 SNPs into 22207 SNPs sets for detecting.

Some results in Table 3 verify recently related work, like PLCG1 → ATP13A4. Recent studies show that the mutation of PLCG2, one of the phospholipase C gamma gene families like PLCG1, has an influence on immunity of microglial cells, which causes the Alzheimer disease (AD) [20]. ATP13A2, similar to ATP13A4, is one of the ATPase gene families, and its mutation causes juvenile Parkinsonism disease (PD). Both are neurodegenerative diseases, and a study on the biochemical association [21] between both diseases is published. Our experiments show that the association between the two diseases can be reflected in the expression of the homologous genes of PLCG2 and ATP13A2. Considering causal association rule PLCG1 → ATP13A4, two SNPs of PLCG1, rs753381 and rs4297946 serve as the causal nodes, while one SNP of ATP13A4, rs2280476, serves as an effect node. Specifically, rs753381 shows the mutation on coding region of gene PLCG1, and rs4297946 implies the 3'-UTR of gene PLCG1 has mutated, which causes rs2280476 and implies the 3'-UTR of gene ATP13A4 has mutated. The mechanism for the rule is that 3'-UTR controls the degradation of mRNA [22] and can even affect gene expression. In conclusion, the experiment results show that the mutation on 3'-UTR and coding region makes gene PLCG1 express abnormally, which causes mutation on 3'-UTR that leads gene ATP13A4 to express abnormally. This is a biochemical reflection of an association between AD and PD.

From the above analysis, it provides a significant research direction for reference that considering the association among gene PLCG1, gene ATP13A4, as well as their homologous genes can promote the development of research on AD and PD.

4 Conclusion

In this work, we develop a method called ASymmetric Causal Association Rule Discovery (ASCARD) to discover the reliable local causal association rules among the gene mutations. The experiments on synthetic data and real-world data verify the effectiveness of ASCARD. The success of ASCARD not only reflects the existing of causality among gene mutations but also shows the locality of the gene mutation interactions. The future work includes accelerating the mining process, filtering out the false positive discoveries using the bioinformatics databases.

Acknowledgments. This study makes use of data generated by the Wellcome Trust Case-Control Consortium. A full list of the investigators who contributed to the generation of the data is available from www.wtccc.org.uk. Funding for the project was provided by the Wellcome Trust under award 076113, 085475 and 090355. This research was also supported by NSFC-Guangdong Joint Found (U1501254), Natural Science Foundation of China (61876043, 61472089), Natural Science Foundation of Guangdong (2014A030306004, 2014A030308008), Science and Technology Planning Project of Guangdong (2015B010108006, 2015B010131015), Guangdong High-level Personnel of Special Support Program (2015TQ01X140), Pearl River S&T Nova Program of Guangzhou (201610010101), and Science and Technology Planning Project of Guangzhou (201604016075).

References

1. Masyukova, S.V., et al.: A screen for modifiers of cilia phenotypes reveals novel MKS alleles and uncovers a specific genetic interaction between osm-3 and nphp-4. PLoS Genet. **12**(2), e1005841 (2016)
2. Wang, W., Xu, Z.Z., Costanzo, M., Boone, C., Lange, C.A., Myers, C.L.: Pathway-based discovery of genetic interactions in breast cancer. PLoS Genet. **13**(9), e1006973 (2017)
3. Tebbens, J.D., Schlesinger, P.: Improving implementation of linear discriminant analysis for the high dimension/small sample size problem. Comput. Stat. Data Anal. **52**(1), 423–437 (2007)
4. Aggarwal, C.C.: Re-designing distance functions and distance-based applications for high dimensional data. ACM Sigmod Rec. **30**(1), 13–18 (2001)
5. Parkhomenko, E., Tritchler, D., Beyene, J.: Sparse canonical correlation analysis with application to genomic data integration. Stat. Appl. Genet. Mol. Biol. **8**(1), 1–34 (2009)
6. Foley, J.W., Katagiri, F.: Unsupervised reduction of random noise in complex data by a row-specific, sorted principal component-guided method. BMC Bioinform. **9**(1), 508 (2008)
7. Liu, L., Zhang, D., Liu, H., Arendt, C.: Robust methods for population stratification in genome wide association studies. BMC Bioinform. **14**(1), 132 (2013)
8. Scherer, S.W., et al.: A scan statistic to extract causal gene clusters from case-control genome-wide rare CNV data. BMC Bioinform. **12**(1), 205 (2011)
9. Catlett, N.L., et al.: Reverse causal reasoning: applying qualitative causal knowledge to the interpretation of high-throughput data. BMC Bioinform. **14**(1), 1–14 (2013)
10. Martin, F., Sewer, A., Talikka, M., Yang, X., Hoeng, J., Peitsch, M.C.: Quantification of biological network perturbations for mechanistic insight and diagnostics using two-layer causal models. BMC Bioinform. **15**(1), 238 (2014)
11. Freudenberg, J., Wang, M., Yang, Y., Li, W.: Partial correlation analysis indicates causal relationships between GC-content, exon density and recombination rate in the human genome. BMC Bioinform. **10**(Suppl 1), S66 (2009)
12. Mihil, C., Ohta, T., Pyysalo, S., Ananiadou, S.: Biocause: Annotating and analysing causality in the biomedical domain. BMC Bioinform. **14**(1), 1–18 (2013)
13. Chickering, D.M.: Learning bayesian networks is NP-complete. In: Fisher, D., Lenz, H.J. (eds.) Learning from Data. Lecture Notes in Statistics, vol. 112, pp. 121–130. Springer, New York (1996). https://doi.org/10.1007/978-1-4612-2404-4_12
14. Cai R., Zhang Z., Hao Z.: SADA: a general framework to support robust causation discovery. In: International Conference on Machine Learning, pp. 208–216 (2013)
15. Cai, R., Zhang, Z., Hao, Z., Winslett, M.: Sophisticated merging over random partitions: a scalable and robust causal discovery approach. IEEE Trans. Neural Netw. Learn. Syst. **29**(8), 3623–3635 (2017)
16. Harpaz, R., Chase, H.S., Friedman, C.: Mining multi-item drug adverse effect associations in spontaneous reporting systems. In: BMC bioinformatics, vol. 11, p. S7. BioMed Central (2010)
17. Pearl, J.: Causality: models, reasoning, and inference. Econ. Theory **19**(675–685), 46 (2003)
18. Cai, R., et al.: Identification of adverse drug-drug interactions through causal association rule discovery from spontaneous adverse event reports. Artif. Intell. Med. **76**, 7–15 (2017)

19. National Center for Biotechnology Information. http://www.ncbi.nlm.nih.gov
20. Sims, R., et al.: Rare coding variants in PLCG2, ABI3, and TREM2 implicate microglial-mediated innate immunity in alzheimer's disease. Nat. Genet. **49**(9), 1373 (2017)
21. Duong, D.M., et al.: Asparagine endopeptidase cleaves α-synuclein and mediates pathologic activities in parkinson's disease. Nat. Struct. Mol. Biol. **24**(8), 632 (2017)
22. de Moor, C.H., Meijer, H., Lissenden, S.: Mechanisms of translational control by the 3' UTR in development and differentiation. Semin. Cell Dev. Biol. **16**(1), 49–58 (2005)

EEG Sparse Representation Based Alertness States Identification Using Gini Index

Muna Tageldin[1], Talal Al-Mashaikki[2], Hamza Bali[3],
and Mostefa Mesbah[1(✉)]

[1] College of Engineering, Department of Electrical and Computer Engineering,
Sultan Qaboos University, Muscat, Oman
mmostefa@gmail.com
[2] College of Science and Engineering,
Hamad Bin Khalifa University, Doha, Qatar
[3] College of Medicine, Department of Physiology,
Sultan Qaboos University, Muscat, Oman

Abstract. Poor alertness experienced by individuals may lead to serious accidents that impact on people's health and safety. To prevent such accidents, an efficient automatic alertness states identification is required. Sparse representation-based classification has recently gained a lot of popularity. A classifier from this class typically comprises three stages: dictionary learning, sparse coding and class assignment. Gini index, a recently proposed method, was shown to possess a number of properties that make it a better sparsity measure than the widely used l_0- and l_1-norms. This paper investigates whether these properties also lead to a better classifier. The proposed classifier, unlike the existing sparsity-based ones, embeds the Gini index in all stages of the classification process. To assess its performance, the new classifier was used to automatically identify three alertness levels, namely awake, drowsy, and sleep using EEG signal. The obtained results show that the new classifier outperforms those based on l_0- and l_1-norms.

Keywords: Gini index · Sparse representation · Alertness classification

1 Introduction

Monitoring vigilance level is expected to be a key factor in preventing many potential accidents related to human activities, such as vehicle driving, air traffic control-ling, and heavy machinery operating [1, 2]. In the medical field, alertness monitoring can be used to monitor the depth of anesthesia during surgery.

Several automated methods for detecting human vigilance levels have been proposed in the literature [1–4]. Most of these methods use electroencephalogram (EEG) as a basis for discriminating among different alertness levels. EEG is considered the "gold standard", as its underlying rhythms (δ (1–3 Hz), θ (4–7 Hz), α (8–12 Hz), β (13–30 Hz)) have been shown to be highly correlated with the different wakefulness and sleep stages [5–7].

© Springer Nature Switzerland AG 2018
L. Cheng et al. (Eds.): ICONIP 2018, LNCS 11307, pp. 478–488, 2018.
https://doi.org/10.1007/978-3-030-04239-4_43

Automatic classification methods comprise three stages: data acquisition, pre-processing (including feature extraction and selection) and decision making. The existing literature on automatic alertness identification focuses primarily on the selection of the classifier, as it is seen as a vital component in the decision process. Several classifiers have been proposed. These include artificial neural network (ANN) [8], support vector machine (SVM) [9], self-organizing map (SOM) [10], clustering-based [11], and fuzzy logic based [12].

Sparse signal representation has recently been proposed as a basis for signal classification. This approach is believed to improve upon the drawbacks of existing classification methods by harnessing the notion of signal sparsity [13]. It has become an important tool in biomedical signal analysis and processing [14]. A Sparsity-based classifier comprises three stages: dictionary learning/selection, sparse coding and class assignment. A number of such methods have been proposed to automatically identify the different levels of human alertness. A brief summary of these methods is shown in Table 1 below.

Table 1. Existing sparse representation based classification methods using EEG [15–17]

	Method 1 [15]	Method 2 [16]	Method 3 [17]
Alertness states	Awake, Drowsy, Sleep	Alert, Sleep	Alert, Drowsy
Dictionary selection	K-SVD (shared among classes)	K-SVD (one per class)	K-SVD (one per class)
Sparse coding	l_1-norm	l_0-norm	l_0-norm
Class assignment	Reconstruction error	Reconstruction error	Reconstruction error

Recently, a new measure of signal sparsity, called Gini index, has been proposed [18]. Gini index was initially used in economics as a measure of the income/wealth distribution of nations' residents [18]. It has later been introduced in signal processing as a measure of signal sparsity. Gini index was shown to possess a number of desirable properties that makes it a better alternative than l_0- and l_1-norms [18]. The authors in [19] used Gini index in the context of signal classification but only at the class assignment stage. Gini index was also used for signal reconstruction in compressive sensing [20]. No attempt, however, was made to investigate the use of Gini index in all stages of sparse representation based classification.

In this paper, we develop a new sparsity-based signal classifier using the Gini index, as a measure of sparsity. To assess its performance, the classifier was used to identify three alertness states (awake, drowsy, and sleep) using features extracted from a single EEG channel. In this study, Gini index was incorporated in all classification stages, namely dictionary learning, sparse coding and class assignment. The classification results were compared to those based on l_0- and l_1-norms.

This paper is organized as follows. Section 2.1 gives a brief description of the EEG data used in the classification. Section 2.2 introduces Gini index as a measure of signal sparsity, while Sect. 2.3 gives an overview of the new sparse representation based classification. Section 3 presents and discusses the experimental results.

2 Sparse Representation-Based Classification

2.1 Sparse Signal Representation

Let $y \in \mathcal{R}^N$ be discrete-time signal. Its sparse representation is defined as a weighted sum of a small number of waveforms d_i (called atoms) that belong to a dictionary D.

$$y = Dx \tag{1}$$

where $x \in \mathcal{R}^N$ (usually $M \ll N$) and $D = [d_1, d_2, d_3, \ldots, d_M] \in \mathcal{R}^{N \times M}$ [16]. The over-complete dictionaries used in the literature are either fixed or learned from the training data. The fixed dictionaries are usually obtained from existing bases, such as discrete Fourier Transform and discrete wavelet transform, or their combinations. The sparse coefficient vector x is found by solving a constrained optimization problem. This last step is known as sparse coding [21]. The learned dictionary approach is usually preferred due to its flexibility and adaptability to specific data/applications [21].

The problem of dictionary learning can be expressed as the following joint optimization problem:

$$\{x_{opt}, D_{opt}\} = argmin \frac{1}{2} \|Y - Dx\|_2^2 + \lambda S(x) \tag{2}$$

where $S(x)$ represents the sparsity measure of x. This joint optimization problem is solved iteratively by first updating the sparse vector while the dictionary atoms $\{d_i\}$ are kept fixed and then updating the dictionary atoms while fixing the sparse vector x. One of the most widely used dictionary learning algorithm is the K-SVD [21, 22].

Sparse coding is the process of finding a sparse vector to represent a signal y using the learned dictionary D. The sparsity of the solution is guaranteed through the introduction of the regularizing sparsity measure S. A good sparsity measure is thought to fulfill six desirable properties discussed in [18]. There are two forms of sparse coding optimization problems, depending on whether the measurement is assumed noise-free or noisy. Due to space limitation, we only discuss the noisy case here.

$$x_{opt} = \max_{x \in R^N} S(x) s.t \|y - Dx\|_2 < \varepsilon \tag{3}$$

There exist several measures of sparsity. Most of these measures are based on vector norms. The most widely used ones are the l_0 (pseudo)-norm and the l_1-norm. The l_0-norm (number of non-zero elements in a vector) is the most intuitive measure of sparsity. In this case, however, the solution of (3) is known to be non-convex and computationally N-P hard. To overcome this problem, greedy algorithms, such as matching pursuit (MP) and its variants [23], were used to find approximate solutions to (3). An alternative solution to the greedy solutions is the convex relaxation. In this approach, l_1-norm is used as a sparsity measure and the resulting convex optimization problem can be efficiently solved using standard algorithms such as, alternating direction method of multi-pliers (ADMM) and proximal point algorithms [23]. Other measures used for sparsity representation are the l_p-norm $(0 < p < 1)$ and the recently proposed Gini index [23]. Among all these measures, only Gini index is known to satisfy the desirable properties discussed in [18, 23] (see also Sect. 2.2).

2.2 Gini Index

Gini index satisfies the following properties: normalization, independence of the size of the vector signal, and independent of the energy in the signal [20]. It has been recently applied to sparse signal representation and signal reconstruction from compressive samples [24, 25].

Assume a vector $s = \{s_1, s_2, \ldots \ldots s_N\}$ organized in a descending order. We defined the Lorenzo curve as:

$$s\left(\frac{i}{N}\right) = \sum_{j=1}^{i} \frac{s_i}{\sum_{k=1}^{N} s_i}, \quad \text{for} \quad i = 1, \ldots, N \tag{4}$$

The Gini index is defined as twice the area between Lorenzo curve and the 45° line (see Fig. 1 below). This is mathematically expressed as:

$$Gini\ index = 1 - 2B(s) \tag{5}$$

where $B(s)$ is the area under Lorenzo curve.

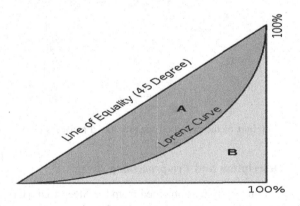

Fig. 1. Lorenzo curve

2.3 Class Assignment

Two approaches are used to assign a signal to one of the classes. The first one is based on the reconstruction error while the other is based on the sparsity measure.

Reconstruction Error (RE): In this approach, the decision is based on the reconstruction error $e_r^j = \left\| y - D^j x^j \right\|_2^2, j = 1, 2, \ldots c$, where c is the class index, and D^j and x^j are the dictionary and the sparsest coefficient vector associated with the *jth* class respectively. A test sample y is said belong to class j (associated with D^j) that achieves the minimum e_r^j [26, 27]. In other words,

$$f = class(y) = \text{argmin}_j \left(\|y - D^j x^j\|_2^2 \right), \quad j = 1, 2, .., c \qquad (6)$$

Sparsity Measure: In this method, the decision to assign a test signal y to a specific class j is based on the sparsity measure $S(x^j)$. In other words, the test signal is assigned to the class whose dictionary achieves the maximum sparsest representation:

$$f = class(y) = \text{argmax}_j \, S(x^j) \quad j = 1, 2, .., c \qquad (7)$$

3   Sparsity-Based Alertness Classification Using Gini Index

A flow chart of the proposed classifier used to classify alertness states is shown in Fig. 2. The proposed classifier used K-SVD for dictionary learning of three dictionaries corresponding to three alertness states. The Gini index was used as a measure of sparsity $(S(x) = Gini(x))$ in all components of the classification process.

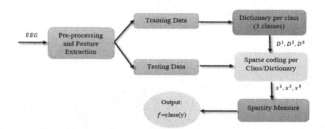

Fig. 2. Flowchart of the proposed sparsity-based alertness classification

3.1   EEG Data Description and Pre-processing

The EEG data used in this study was obtained from the Sleep EDF [Expanded] dataset, part of the public domain Physionet databases [28, 29]. The dataset contains two types of files: SC files for healthy subjects and ST files for subjects who took sleep medication. In this study, we only use the SC records. Each of these records contains two EEG channels (Pz-Oz and Fpz-Cz) sampled at 100 Hz. The record also includes one EOG channel and one EMG channel sampled at 100 Hz and 1 Hz respectively.

As this approach is designed for a single channel, only the channel Pz-Oz was used for classification as it has been shown to give the better results [30–32]. The EEG data was segmented into 30 s epochs and visually annotated using the method developed by Rechtschaffen and Kales (R&K) [33].

Since EEG is a non-stationary signal, Wavelet Packet Transform (WPT) was used to decompose the annotated EEG signal into its sub-bands to provide time-frequency localization. Daubechies 4 wavelet was used in this study as it was previously reported to give the best alertness state classification performance [34].

Feature extraction methods were applied to EEG to highlight key patterns. The selected features used in classification include the four bands' relative powers $(P_\delta/P_{Tot}, P_\theta/P_{Tot}, P_\alpha/P_{Tot}, \text{and } P_\beta/P_{Tot})$, their ratios $(P_{\alpha+\theta}/P_{\alpha+\beta}, P_{\alpha+\theta}/P_\beta, P_\alpha/P_\beta,$ and P_θ/P_β and their sample entropies $(En_\delta, En_\theta, En_\alpha, \text{and } En_\beta)$ [35–38].

3.2 Dictionary Learning

The 12 extracted EEG features for each of the training epochs were utilized for dictionary learning using the K-SVD algorithm (Gini index was utilized in sparse coding). Each class (alertness state) is represented by a separate dictionary $D^j, j = 1, 2, 3$.

3.3 Sparse Coding

To solve the sparse coding optimization problem using Gini index as a sparsity measure, Simultaneous Perturbation Stochastic Approximation (SPSA) was used. Although a number of algorithms can be used, SPSA was chosen for its computation-al efficiency [39].

SPSA algorithm:

Given an objective function $(Gini(x))$ and the constraint $(\|y - Dx\|_2 < \varepsilon)$, the SPSA algorithm operates as follows:

1. *Initialization* $(k = 0)$: In this step, SPSA parameters $(a, c, \alpha, \gamma \in R)$ were initialized (selected by the user).
2. *Parameter update*: The step size a_k and the coefficient c_k at iteration k are computed using following equations:

$$a_k = a/(1 + k + A)^\alpha \tag{8}$$

$$c_k = c/(1 + k)^\gamma \tag{9}$$

3. *Gradient approximation and update*: At the kth iteration, the algorithm simultane-ously perturbs all elements of x_k according to a distributing vector $\Delta_k \in R^N$ whose elements are generated using a Bernoulli distribution [40]. The gradient is approximated by:

$$\hat{g}_k(\hat{x}_k) = \frac{GI(\hat{x}_k + c_k\Delta_k) - GI(\hat{x}_k - c_k\Delta_k)}{2c_k\Delta_k} \tag{10}$$

4. *Projection*: To satisfy the constraint $(\|y - Dx\|_2 < \varepsilon)$, the updated approximate solution vector x_{k+1} is obtained by

$$\hat{x}_{k+1} = P(\hat{x}_k - a_k\hat{g}_k(\hat{x}_k)) \tag{11}$$

where $P = D^{\#}$ (pseudo-inverse) is the projection operator defined for quadratic constraint (noisy).

5. *Stopping criterion*: The algorithm stops when the maximum number of iterations is reached or when $\|\hat{x}_{k+1} - \hat{x}_k\|_2$ is less than a predefined value.

The guideline for selecting SPSA parameters are mentioned in Table 2 [39]. The algorithm's parameters are selected using the training data to achieve the maximum classification accuracy while taking into consideration imposed guide-lines.

Table 2. SPSA parameters and selection guidelines

SPSA parameter	Guidelines
$a \in R$	Larger (algorithm diverges, higher number of iterations needed for reaching minimum)
$c \in R$	Noisy measurements $\rightarrow$ larger c Noiseless measurements $\rightarrow$ smaller c
$\alpha \in R$	$0.6 < \gamma \leq 1$, $\alpha - 2\gamma > 0$, $3\alpha - 2\gamma \geq 0$
$\gamma \in R$	$0.1 < \gamma \leq 0.5$, $\alpha - 2\gamma > 0$, $3\alpha - 2\gamma \geq 0$
$A \in R$	10% or less of maximum number of iterations (preferable $5\% - 10\%$)

In order to assess the performance of the proposed algorithm, we used records from the Sleep EDF database [28, 29]. As drowsy state is a rare event in the database and to prevent the bias caused by class unbalancing, we only selected the (five) subjects who have significant amount of drowsy state epochs. Therefore, 200 epochs (70% training, 30% testing) were randomly selected for each alertness states. The accuracy was selected as a measure of performance:

$$accuracy = \frac{total\ \#\ of\ correctly\ classified\ epochs}{total\ \#\ of\ test\ epochs} \qquad (12)$$

4   Experimental Results and Discussion

Sparse coding using l_0-norm and l_1-norm were solved using orthogonal matching pursuit algorithm (OMP) [41, 42] and spectral projected-gradient (SPG) algorithm [43, 44] respectively. Figure 3 shows the classification performance (accuracy) of different alertness states for five subjects.

We can clearly see that Gini index achieved significantly better results compared to l_0- and l_1-norms. Thus, our results tend to confirm statement "sparsity enhances recognition rate" [13]. The l_0-norm method tends to suffer from inability to classify alertness states (especially drowsy and sleep states). This issue could be partially explained by binary nature of the l_0-norm. This problem can be overcome by using an

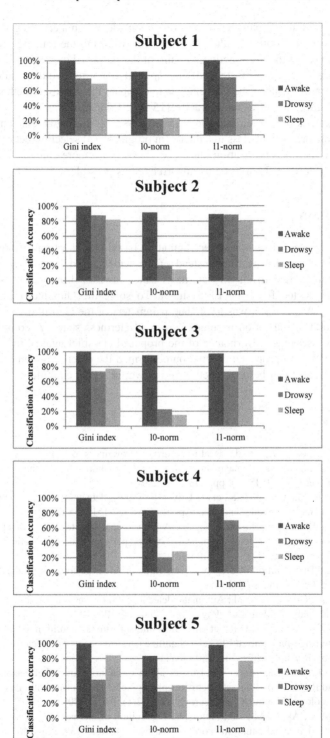

Fig. 3. Comparison between three sparsity measures (l_0-norm, l_1-norm, Gini index) in terms of classification accuracy of alertness states for subjects 1 and 3

adaptive threshold that forces the vector components less than a certain given value to zero. The threshold's optimal value should be found during the training process.

As mentioned in the introduction, the aim of the proposed method is to investigate whether using Gini index, instead of the l_0- and l_1-norms, would result in better classification performance. The next stage is to compare the proposed methods to other alertness classifiers found in the literature [15–17]. Prior to this a issues will worked out: (1) generalization to multi-channels (in progress), (2) finding an efficient approach to deal with the problem of class unbalance (in progress), (3) selecting better features to further reduce the misclassification rates, and (4) finding a more efficient way to select dictionaries by forcing them to be as uncorrelated as possible.

5   Conclusion

This paper proposes a new sparse representation based classification method that uses Gini index as a sparsity measure instead of the widely used l_0- and l_1-norms. This method was then used to classify alertness states (awake, drowsy and sleep) using information extracted from the EEG signal. To solve the Gini index based sparse coding, an SPSA algorithm was used. The parameters of the algorithms were selected to achieve good classification accuracy across all alertness states. A comparison was performed to assess the performance of the proposed classifier against those based on l_0- and l_1-norms. The proposed method outperformed the other two methods and gave an indication that it can be developed into a powerful signal classifier.

References

1. Yan, J.J., Kuo, H.H., et al.: Real-time driver drowsiness detection system based on PERCLOS and grayscale image processing. In: International Symposium on Computer, Consumer and Control (IS3C), pp. 243–246 (2016)
2. Alshaqaqi, B., Baquhaizel, A.S., et al.: Driver drowsiness detection system. In: Workshop on Systems, Signal Processing, and their Applications, pp. 151–155 (2013)
3. Awais, M., Badruddin, N., Drieberg, M.: A Hybrid approach to detect driver drowsiness utilizing physiological signals to improve system performance and wearability. Sensors (Basel) 17(9), 1991 (2017)
4. Nodine, E.: The detection of drowsy drivers through driver performance indicators. Master of Science, Tuffs University (2006)
5. Garcés, C.A., Orosco, L., et al.: Automatic detection of drowsiness in EEG records based on multimodal analysis. Med. Eng. Phys. 36(2), 244–249 (2014)
6. Johnson, R.R., Popovic, D.P., et al.: Drowsiness/alertness algorithm development and validation using synchronized EEG and cognitive performance to individualize a generalized model. Biol. Psychol. 87(2), 241–250 (2011)
7. Cantero, J.L., Atienza, M., et al.: Human alpha oscillations in wakefulness, drowsiness period, and REM sleep: different electroencephalographic phenomena within the alpha band. Neurophysiologie Clinique/Clin. Neurophysiol. 32(1), 54–71 (2002)
8. Kiymik, M.K., Akin, M., et al.: Automatic recognition of alertness level by using wavelet transform and artificial neural network. J. Neurosci. Methods 139, 231–240 (2004)

9. Yu, S., Li, P. et al.: Support vector machine based detection of drowsiness using minimum EEG features. In: SocialCom, pp. 827–835 (2013)
10. Wang, X., Zhang, Y., et al.: Alertness staging based on improved self-organizing map. Trans. Tianjin Univ. **19**(6), 459–462 (2013)
11. Gurudath, N., Riley, H.B.: Drowsy driving detection by EEG analysis using wavelet transform and K-means clustering. Procedia Comput. Sci. **34**, 400–409 (2014)
12. Al-Ani, A., Mesbah, M.: EEG rhythm/channel selection for fuzzy rule-based alertness state characterization. *Neural Comput Appl* (2016)
13. Wright, J., Yang, A.Y., et al.: Robust face recognition via sparse representation. IEEE Trans. Pattern Anal. Mach. Intell. **31**(2), 210–227 (2009)
14. Wen, D., Jia, P., et al.: Review of sparse representation-based classification methods on EEG signal processing for epilepsy detection, brain-computer interface and cognitive impairment. Front. Aging Neurosci. **8**, 172 (2016)
15. Yu, H., Lu, H. et al.: Vigilance detection based on sparse representation of EEG. In: Proceedings of IEEE Engineering in Medicine and Biology Society Conference, pp. 2439–2442 (2010)
16. Zhang, Z., Luo, D., et al.: A Vehicle active safety model: vehicle speed control based on driver vigilance detection using wearable EEG and sparse representation. Sensors **16**(2), 242 (2016)
17. Luo, D.Y., Zhang, Z.T.: A novel vehicle speed control based on driver's vigilance detection using EEG and sparse representation. In: Applied Mechanics and Materials, pp. 607–611 (2014)
18. Hurley, N., Rickard, S.: Comparing measures of sparsity. IEEE Trans. Inf. Theory **55**(10), 4723–4741 (2009)
19. Baali, H., Mesbah, M.: Ventricular ectopic beats classification using sparse representation and Gini index. In: Proceedings of IEEE Engineering in Medicine and Biology Society Conference, pp. 5821–5824 (2015)
20. Zonoobi, D., Kassim, A.A., et al.: Gini index as sparsity measure for signal reconstruction from compressive samples. IEEE J. Sel. Topics Signal Process. **5**(5), 927–932 (2011)
21. Rubinstein, R., Bruckstein, A.M., et al.: Dictionaries for sparse representation modeling. Proc. IEEE **98**(6), 1045–1057 (2010)
22. Aharon, M., Elad, M., et al.: K-SVD: An Algorithm for designing overcomplete dictionaries for sparse representation. IEEE Trans. Signal Process. **5**(11), 4311–4322 (2006)
23. Zhang, Z., Xu, Y., et al.: A Survey of sparse representation: algorithms and applications. IEEE Access **3**, 490–530 (2015)
24. Feng, C., Xiao, L., et al.: Compressive sensing inverse synthetic aperture radar imaging based on Gini index regularization. IJAC **11**(4), 441–448 (2014)
25. Feng, C., Xiao, L., et al.: Parameterized lorenz curve based compressive sensing reconstruction. JDCTA **7**(13), 185–194 (2013)
26. Huang, Z., Liu, Y., et al.: Study on sparse representation based classification for biometric verification, https://arxiv.org/abs/1502.06073 (2015)
27. Gangeh, M.J., Farahat, A.K., et al.: Supervised dictionary learning and sparse representation: a review, https://arxiv.org/abs/1502.05928 (2015)
28. Goldberger, A.L., Amaral, L.A.N., et al.: PhysioBank, PhysioToolkit, and PhysioNet. Circulation **101**(23), E215–E220 (2000)
29. Kemp, B., et al.: Analysis of a sleep-dependent neuronal feedback loop: the slow-wave microcontinuity of the EEG. IEEE Trans. Bio-Med. Eng. **47**(9), 1185–1194 (2000)
30. Berthomier, C., Drouot, X., et al.: Automatic analysis of single-channel sleep EEG: validation in healthy individuals. Sleep **30**(11), 1587–1595 (2007)

31. da Silveira, T.L.T., Kozakevicius, A.J., et al.: Automated drowsiness detection through wavelet packet analysis of a single EEG channel. Expert Sys. Appl. **55**, 559–565 (2016)
32. Pal, N.R., Chuang, C.Y., et al.: EEG-based subject- and session-independent drowsiness detection: an unsupervised approach. EURASIP J. Adv. Signal Process., 519480 (2008)
33. Imtiaz, S.A. at al.: An open-source toolbox for standardized use of PhysioNet Sleep EDF expanded database. In: Proceedings of IEEE Engineering in Medicine and Biology Society Conference, pp. 6014–6017 (2015)
34. Şen, B., Peker, M., Çavuşoğlu, A., Çelebi, F.V.: A comparative study on classification of sleep stage based on EEG signals using feature selection and classification algorithms. J. Med. Syst. **38**(3), 18 (2014)
35. Fell, J., et al.: Discrimination of sleep stages: a comparison between spectral and nonlinear EEG measures. Electroencephalogr. Clin. Neurophysiol. **98**(5), 401–410 (1996)
36. Ji, H., Li, J., Cao, L., Wang, D.: A EEG-based brain computer interface system towards applicable vigilance monitoring. In: Wang, Y., Li, T. (eds.) Foundations of Intelligent Systems. Advances in Intelligent and Soft Computing, vol 122. Springer, Heidelberg (2011). https://doi.org/10.1007/978-3-642-25664-6_87
37. Jap, B., Lal, S., et al.: Using EEG spectral components to assess algorithms for detecting fatigue. Expert Syst. Appl. **36**(2), 2352–2359 (2009)
38. Eoh, H., Chung, M., et al.: Electroencephalographic study of drowsiness in simulated driving with sleep deprivation. Int. J. Ind. Ergon. **35**(4), 307–320 (2005)
39. Spall, J.C.: Simultaneous Perturbation Stochastic Approximation - Introduction to stochastic Search and Optimization. Wiley (2003)
40. Sadegh, P., Spall, J.C.: Optimal random perturbations for stochastic approximation using a simultaneous perturbation gradient approximation. In: Proceedings of the 1997 American Control Conference, pp. 3582–3586 (1997)
41. Cai, T.T., Wang, L.: Orthogonal matching pursuit for sparse signal recovery with noise. IEEE Trans. Inf. Theory **57**(7), 4680–4688 (2011)
42. Shaban, M.: OMP (2015). https://www.mathworks.com/matlabcentral/fileexchange/50584-orthognal-matching-pursuit-algorithm-omp
43. van den Berg, E., Friedlander, M.: Probing the pareto frontier for basis pursuit solutions. SIAM J. Sci. Comput. **31**(2), 890–912 (2008)
44. Berg, E.v.d., Friedlander, M.P.: SPGL1: a solver for large-scale sparse reconstruction (2007). https://www.cs.ubc.ca/~mpf/spgl1/

Attention-Based Network for Cross-View Gait Recognition

Yuanyuan Huang, Jianfu Zhang, Haohua Zhao, and Liqing Zhang[✉]

Key Laboratory of Shanghai Education Commission for Intelligent Interaction
and Cognitive Engineering, Department of Computer Science and Engineering,
Shanghai Jiao Tong University, Shanghai, China
{523541979,c.sis,haoh.zhao,lqzhang}@sjtu.edu.cn

Abstract. Existing gait recognition approaches based on CNN (Convolutional Neural Network) extract features from different human parts indiscriminately, without consideration of spatial heterogeneity. This may cause a loss of discriminative information for gait recognition, since different human parts vary in shape, movement constraints and so on. In this work, we devise an attention-based embedding network to address this problem. The attention module incorporated in our network assigns different saliency weights to different parts in feature maps at pixel level. The embedding network strives to embed gait features into low-dimensional latent space such that similarities can be simply measured by Euclidian distance. To achieve this goal, a combination of contrastive loss and triplet loss is utilized for training. Experiments demonstrate that our proposed network prevails over the state-of-the-art works on both OULP and MVLP dataset under cross-view conditions. Notably, we achieve 6.4% rank-1 recognition accuracy improvement under 90° angular difference on MVLP and 3.6% under 30° angular difference on OULP.

Keywords: Gait recognition · Attention mechanism
Embedding learning

1 Introduction

Human identification based on biometrics, such as face, voice and fingerprints, has been widely used in various applications like security and authorization systems. Gait recognition is one of the biometric identification technologies which exerts walking styles of human-beings to verify identities. Compared to other biometrics, the merits of gait based recognition mainly lie in the following three aspects: (i) Gait data is collected by cameras from a distance, and this means that identification process is contactless and can be carried out without the awareness of the subject. (ii) The walking posture instead of appearance is utilized for identification which promises a compatible performance even when the source data is in low resolution. (iii) Walking postures are much more difficult to imitate than static information like face and fingerprint images.

© Springer Nature Switzerland AG 2018
L. Cheng et al. (Eds.): ICONIP 2018, LNCS 11307, pp. 489–498, 2018.
https://doi.org/10.1007/978-3-030-04239-4_44

However, variations in views, clothes or carried bags, which can prominently change the appearance of subjects still remain challenging for gait based recognition. To address this problem, generative models [7,8,11] transform variant gait features into invariant ones such that comparison can be performed under the same condition. On the contrary, discriminative models seek for invariant gait features under different conditions and compare these variant gait features directly [6,12].

Recently, approaches based on DNN (Deep Neural Network) have shown promising results for gait identification even with variations. Among generative models, Yu et al. [21,22] devised both Generative Adversarial Network and Autoencoder to transform variant GEIs [13] (Gait Energy Images[1]) into invariant ones which are in side view with normal clothes and no carried bags before matching. With respect to discriminative models, Wu et al. [20] proposed a two-input CNN to measure the similarity between two GEIs in an end-to-end manner. Taking gait recognition for classification problem, Shiraga et al. [17] devised an classification CNN called GEINet. Besides, Zhang et al. [2] made use of a Siamese neural network to obtain gait representation in a 2-dimensional space. The contrastive loss they used drove the intra-subject similarity to be small and inter-subject similarity to be large.

However these aforementioned CNN-based approaches deal with different parts of GEI indiscriminately due to the weight-shared strategy of CNN. This strategy neglects spatial heterogeneity of GEIS and may result in a loss of discriminative information, because different human parts in a GEI vary in shape, movement constrains and so on. To address such a problem, we propose an attention-based embedding network for cross-view gait recognition. The contributions of this work mainly lie in:

(i) We design an attention module to optimize the visual feature extraction by employing different saliency weights for different parts.
(ii) We adopt a deeper structure for embedding and explore two loss functions, contrastive loss and triplet loss, which are frequently utilized for recognition problem.
(iii) Our proposed approach surpasses the state-of-the-art methods on both OULP [4] and MVLP [15] dataset when the difference of view angles is large. It demonstrates that our approach can manage cross-view gait recognition problem.

2 Proposed Method

2.1 Network Structure

Our network aims to embed each input GEI into a low dimensional hidden space such that similarities between GEIs can be directly measured by Euclidean

---

[1] A GEI is obtained by averaging a sequence of aligned silhouettes. An example is shown in Fig. 1).

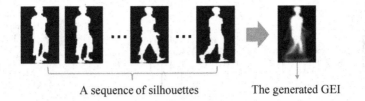

A sequence of silhouettes   The generated GEI

Fig. 1. A sequence of aligned silhouettes extracted from source videos and the GEI generated by these silhouettes. The GEI is calculated by averaging these silhouettes. GEI can represent spatial-temporal information in a single 2D image, hence it is quite popular among CNN-based approaches.

distances in the embedding space. Once the embeddings are calculated, the recognition process only involves seeking the nearest neighbor. That is to say, embeddings can be saved as feature vectors in advance such that only Euclidean distances are calculated during recognition, which will largely improve the computation efficiency.

Figure 2 illustrates the overall architecture of our proposed network. Basically, our embedding network consists three parts: an attention-based CNN ended with fully-connected (FC) layer, a $L2$ normalization layer and a loss layer. Taking an GEI x as input, the output of the FC layer formulates an embedding representation $E(x)$ in $\mathbb{R}^n$ space and the subsequent $L2$ layer confines the embedding to the unit hypersphere, i.e. $\|(E(x))\|_2 = 1$. Here in our network, we set n to be 256. The whole network is trained in an end-to-end manner for the purpose of reducing the distance between GEIs of the same subjects whereas enlarging the distance between GEIs of different subjects. To achieve this goal, both contrastive loss and triplet loss are employed for their great success in face recognition and person re-identification [16,24]. In Sect. 2.2, we will discuss more details about these two loss functions.

The first part of our network, an attention-based CNN, is constructed with five cascaded triplets of convolution, normalization and pooling layers. Our network employs a deeper structure than previous models like [2,17,20], and less parameters are applied in FC layer.

Moreover, inspired by the success of attention mechanism in person re-identification [10] and image classification [19], a spatial attention module is employed to locate the most discriminative regions of visual features for recognition. Due to the weight sharing of kernels, CNNs in previous works deal with different parts of GEI indiscriminately. This may cause a great loss to discriminative information since different human parts in a GEI vary in shape, movement range, individual variation and so on. Spatial attention module incorporated in our network strive to deal with these variations by applying a saliency weight map. We will describe the design of our spatial attention module in Sect. 2.3.

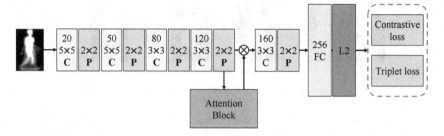

Fig. 2. The structure of the proposed attention-based embedding network for gait recognition. Yellow blocks are convolution layers, while blue ones are pooling layers. Batch Normalization and ReLU activation layers applied after each convolution layers are abbreviated. The 256-dimensional embeddings are generated at FC layer (green) and normalized by L2 normalization layer (orange). (Color figure online)

2.2   Contrastive Loss and Triplet Loss

Given a GEI x as input, our network embeds it as $E(x)$ in a low-dim hidden space. For computation efficiency, distance between samples is measured by the squared Euclidean distance. That is

$$D(x_1, x_2) = \|E(x_1) - E(x_2)\|_2^2 , \tag{1}$$

where x_1 and x_2 are samples from input space. Thus it is desired that $D(x_1, x_2)$ is as small as possible if x_1 and x_2 belong to the same subject, otherwise $D(x_1, x_2)$ is as large as possible. Both contrastive loss function and triplet loss function can help to achieve this goal.

Contrastive loss is frequently used in Siamese networks [2,23]. The motivation of contrastive loss is to cluster samples from the same subjects and to keep clusters away from each other. Thus pairs of samples are employed for training. Specifically, give a pair of GEIs, x_i and x_i', a binary label y denotes whether they are from the same subject, namely $y = 1$ if they are positive pair and $y = 0$ if they are negative pair. Therefore, the loss function can be written as

$$L_{contrastive} = \frac{1}{2M} \sum_{i=1}^{M} [y\|x_i - x_i'\|_2^2 + (1 - y)max(0, \alpha - \|x_i - x_i'\|_2^2)], \tag{2}$$

where α is the expected lower bound of distance between negative pairs and m is the number of pairs extracted from the training set.

Triplet loss was proposed by Schroff et al. [16] for face recognition. Different from contrastive loss, triplet loss is to maximize the margins between samples from different subjects. Consider a triplet of GEIs: x_i, x_i^p(positive) and x_i^n(negative), where x_i and x_i^p belong to the same subject while x_i^n belong to another one. Hence it is required that

$$\|x_i - x_i^p\|_2^2 + \beta < \|x_i - x_i^n\|_2^2, \tag{3}$$

where β is a margin demanded between positive and negative pairs. Based on this inequality, the loss function of triplet loss is defined as

$$L_{triplet} = \frac{1}{2N} \sum_{i=1}^{N} max(0, \|x_i - x_i^p\|_2^2 - \|x_i - x_i^n\|_2^2 + \beta), \qquad (4)$$

where N is the quantity of triplets used for training.

2.3 Attention Module

We now describe the attention module in detail. Similar to Spatial Transform Module [5], our attention module is a single branch connected to the trunk network, as shown in Fig. 3. It generates a saliency weight mask at a pixel level.

The attention module, inspired by the design of Harmonious Attention Module in [10], can be divided into three parts. First of all, the input feature map $M \in \mathbb{R}^{H \times W \times C}$, with height H, width W and C channels is connected to a cross-channel average pooling layer. Specifically, the output of the cross-channel average pooling can be written as

$$P(M) = \frac{1}{C} \sum_{i=1}^{C} M_i, \qquad (5)$$

where M_i is the i-th channel of feature map M. The channel number is reduced to 1 after this operation. Afterwards, a small network is designed to generate saliency weights for each pixel in the mask, which contains two convolution layers and one sigmoid activation layer. At last, the single channel mask is replicated in order to formulate a spatial mask $S \in \mathbb{R}^{H \times W \times C}$.

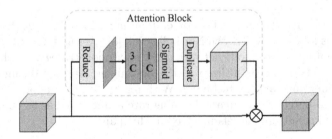

Fig. 3. The structure of the attention module. Blocks with yellow background are convolution layers whose kernel sizes are specified in the picture. Batch Normalization and ReLU activation layers after each convolution layer are abbreviated. (Color figure online)

3 Experiment

In this section, we evaluate the performance of our proposed method on two largest datasets for cross-view gait recognition, OU-ISIR Multi-View Large Population Dataset (MVLP) [15] and OU-ISIR Large Population Dataset (OULP) [4]. We first explore the loss layer and demonstrate the effectiveness of attention module on MVLP. And then our proposed model is compared with the state-of-the-art methods on both MVLP and OULP.

3.1 Datasets and Evaluation

OU-MVLP. Multi-View Large Population Dataset (OU-MVLP), the largest gait data-base, consists of 10307 subjects from 14 view angles, ranging 0–90, 180–270. For each subject and view, two sequences are provided. When evaluating, we divide the total 10307 subjects into two disjoint groups: training group with 5153 subjects and testing group with 5154 subjects.

OULP. The OU-ISIR gait database, Large Population (OULP) is the world's second largest gait dataset which is widely used in previous gait recognition methods. The first version of the dataset called OULP-C1V1 has a collection of 4007 subjects with four different view angles (55, 65, 75, 85) and a wide range of ages. There are 2 sequences for each subject under each view angle. To follow the protocol used in [14], here we only employ a subset of OULP-C1V1 from 1912 subjects. For evaluation, five times of two-fold cross-validation are conducted for each experiment, namely 956 subjects for training and other 956 subjects for testing.

3.2 Setup

For computational convenience, the size of input is set to be 128×96. Each convolution layer is followed by a Batch Normalization [3] layer and a ReLU activation layer. All pooling layers in our network exert max pool strategy. Stochastic gradient descent (SGD) [1] is utilized as the optimizing strategy during training, and the momentums are set to be 0.9. With regard to other hyper parameters, we choose min-batch of size 16 and learning rate is initialized as 0.0008. Softmax loss is applied to all the models involved in this paper.

3.3 Ablation Study

Contrastive Loss vs. Triplet Loss. First of all, we discuss the performance of contrastive loss and triplet loss based on the truck of our network, i.e. no attention module is used. These two loss functions are combined via weight w_c and w_t correspondingly. Then our loss function can be written as

$$L = w_c L_{contrastive} + w_t L_{triplet}. \tag{6}$$

Table 1. Rankd-1 accuracy (%) for cross-view recognition when assembling contrastive loss and triplet loss with different weights.

w_c	w_t	View angle difference			
		0°	30°	60°	90°
1.0	0.0	89.2	54.0	30.6	18.9
0.0	1.0	89.1	54.7	30.6	17.7
1.0	1.0	89.0	56.1	34.2	20.9
1.0	2.0	**89.2**	**57.1**	**35.2**	**22.0**

Table 2. Rank-1 recognition accuracy (%) of our proposed method on MVLP, comparing with the state-of-the-art methods.

Methods	View angle difference			
	0°	30°	60°	90°
Base	**89.2**	57.1	35.2	22.0
Base+Attention	89.1	**57.2**	**36.1**	**23.7**

We explore different values of w_c and w_t and the detailed evaluation on MVLP dataset is shown in Table 1. Obviously, it gives the optimal performance to set $w_c = 1.0$ and $w_t = 2.0$.

Effect of Attention Module. Now we demonstrate the effectiveness of the attention module, using MVLP dataset. The trunk network is trained with and without the attention module (base and base+Attention). Table 2 shows the rank-1 recognition accuracy of these two approaches. It is suggested that the attention module does make an improvement for the cross-view conditions.

3.4   Performance on MVLP

In this section, we evaluate the proposed method on MVLP, comparing with the state-of-the-art methods, including GEINet [17], local @ Bottom (LB) and Mid-level @ Top (MT) proposed by Wu et al. [20] and 3in+2dff proposed by Takemura et al. [18].

Our proposed method, shown in Table 3, achieves 6.4% improvement under 90° angular difference and 3% improvement on average, comparing with the ensemble of 3in and 2dff network (3in+2dff). Only considering single CNNs, our method improves the average accuracy by 6.5%. Note that, with similar accuracy under 0° angular difference, our proposed method obtain over 12.6% improvement to recognition under 90° angular difference, comparing with LB, MB and 2dff. This suggests that our proposed method is able to cope with the cross-view condition even when the difference of angles between gallery view and probe view is relatively large.

Table 3. Comparision with the state-of-the-art methods on MVLP.

Methods	View angle difference				Mean
	0°	30°	60°	90°	
GEINet [17]	85.7	40.3	13.8	5.4	40.7
LB(Wu) [20]	**89.9**	42.2	15.2	4.5	42.6
MB(Wu) [20]	89.3	49.0	20.9	8.2	46.9
3in [18]	85.7	47.8	26.3	15.9	47.9
2dff [18]	89.1	40.8	17.6	7.8	42.9
3in+2dff [18]	89.5	55.0	30.0	17.3	52.7
Ours	89.1	**57.2**	**36.1**	**23.7**	**55.7**

Table 4. Comparison with the state-of-the-art methods on OULP.

Methods	View angle difference				Mean
	0°	10°	20°	30°	
GEINet [17]	94.9	93.9	90.5	88.6	91.6
DeepGait+NN [9]	94	71.6	21.8	2.9	56.2
DeepGait+JB [9]	**97.8**	**97.0**	94.6	89	95.6
Ours	97.6	**97.0**	**95.4**	**92.6**	**96.2**

3.5   Performance on OULP

We also compare with the state-of-the-art methods on OULP dataset, i.e. GEINet [17] and DeepGait [9]. Table 4 shows that our proposed method can generalize well on OULP. Using the same nearest neighbour (NN) classification strategy, our proposed method outperforms both GEINet and DeepGait. Moreover, Joint Bayesian (JB) is utilized in place of NN strategy in [9], and DeepGait+JB shows a great improvement of recognition accuracy. Even compared with DeepGait+JB, our proposed method still achieves 3.6% improvement under 30° angular difference. Hence it demonstrates that our proposed method is more capable of handling cross-view recognition than previous works.

4   Conclusion

In this paper, we propose an attention-based embedding network for cross-view gait recognition. This network embeds gait features into a low-dimensional latent space where similarities are directly measured by Euclidian distance. Once embeddings are calculated, the recognition problem becomes a process of seeking the nearest neighbor in embedding space. Hence it will largely improve computation efficiency to save embeddings as feature vector before recognition. Meanwhile, we explore two loss functions, contrastive loss and triplet loss, which can be adopted to train our network in an end-to-end manner. Experiments are conducted to show that such an embedding network is quite suitable for gait recognition problem.

In addition, an attention module is designed to generate spatial saliency weights for optimization. This mechanism allows us to deal with different human parts in GEIs with disparity. The holistic network, incorporating the attention module, surpasses the state-of-the-art methods on both MVLP and OULP dataset under cross-view conditions, especially when the difference between view angles is large. Specifically, we achieve 6.4% rank-1 recognition accuracy improvement under 90° angular difference on MVLP and 3.6% under 30° angular difference on OULP.

Acknowledgement. The work was supported by the Key Basic Research Program of Shanghai Municipality, China (15JC1400103, 16JC1402800) and the National Basic Research Program of China (Grant No. 2015CB856004).

References

1. Bottou, L., Bousquet, O.: The tradeoffs of large scale learning. In: Advances in Neural Information Processing Systems, pp. 161–168 (2008)
2. Chopra, S., Hadsell, R., LeCun, Y.: Learning a similarity metric discriminatively, with application to face verification. In: IEEE Computer Society Conference on Computer Vision and Pattern Recognition, CVPR 2005, vol. 1, pp. 539–546. IEEE (2005)
3. Ioffe, S., Szegedy, C.: Batch normalization: accelerating deep network training by reducing internal covariate shift. arXiv preprint arXiv:1502.03167 (2015)
4. Iwama, H., Okumura, M., Makihara, Y., Yagi, Y.: The ou-isir gait database comprising the large population dataset and performance evaluation of gait recognition. IEEE Trans. Inf. Forensics Secur. **7**(5), 1511–1521 (2012)
5. Jaderberg, M., Simonyan, K., Zisserman, A., et al.: Spatial transformer networks. In: Advances in Neural Information Processing Systems, pp. 2017–2025 (2015)
6. Jean, F., Bergevin, R., Albu, A.B.: Computing and evaluating view-normalized body part trajectories. Image Vis. Comput. **27**(9), 1272–1284 (2009)
7. Kale, A., Chowdhury, A.R., Chellappa, R.: Towards a view invariant gait recognition algorithm. In: Proceedings of the IEEE Conference on Advanced Video and Signal Based Surveillance, 2003, pp. 143–150. IEEE (2003)
8. Kusakunniran, W., Wu, Q., Zhang, J., Li, H.: Support vector regression for multi-view gait recognition based on local motion feature selection. In: 2010 IEEE Conference on Computer Vision and Pattern Recognition (CVPR), pp. 974–981. IEEE (2010)
9. Li, C., Min, X., Sun, S., Lin, W., Tang, Z.: DeepGait: a learning deep convolutional representation for view-invariant gait recognition using joint Bayesian. Appl. Sci. **7**(3), 210 (2017)
10. Li, W., Zhu, X., Gong, S.: Harmonious attention network for person re-identification. arXiv preprint arXiv:1802.08122 (2018)
11. Makihara, Y., Sagawa, R., Mukaigawa, Y., Echigo, T., Yagi, Y.: Gait recognition using a view transformation model in the frequency domain. In: Leonardis, A., Bischof, H., Pinz, A. (eds.) ECCV 2006. LNCS, vol. 3953, pp. 151–163. Springer, Heidelberg (2006). https://doi.org/10.1007/11744078_12

12. Makihara, Y., Suzuki, A., Muramatsu, D., Li, X., Yagi, Y.: Joint intensity and spatial metric learning for robust gait recognition. In: Proceedings of 30th IEEE Conference on Computer Vision and Pattern Recognition (CVPR 2017), pp. 5705–5715 (2017)
13. Man, J., Bhanu, B.: Individual recognition using gait energy image. IEEE Trans. Pattern Anal. Mach. Intell. **28**(2), 316–322 (2006)
14. Muramatsu, D., Makihara, Y., Yagi, Y.: Cross-view gait recognition by fusion of multiple transformation consistency measures. IET Biom. **4**(2), 62–73 (2015)
15. Takemura, N., Makihara, Y., Muramatsu, D., Echigo, T., Yagi, Y.: Multi-view large population gait dataset and its performance evaluation for cross-view gait recognition. IPSJ Trans. Comput. Vis. Appl. **10**(4), 1–14 (2018)
16. Schroff, F., Kalenichenko, D., Philbin, J.: FaceNet: a unified embedding for face recognition and clustering. In: Proceedings of the IEEE conference on computer vision and pattern recognition, pp. 815–823 (2015)
17. Shiraga, K., Makihara, Y., Muramatsu, D., Echigo, T., Yagi, Y.: Geinet: view-invariant gait recognition using a convolutional neural network. In: 2016 International Conference on Biometrics (ICB), pp. 1–8. IEEE (2016)
18. Takemura, N., Makihara, Y., Muramatsu, D., Echigo, T., Yagi, Y.: On input/output architectures for convolutional neural network-based cross-view gait recognition. IEEE Trans. Circuits Syst. Video Technol. (2017)
19. Wang, F., et al.: Residual attention network for image classification. arXiv preprint arXiv:1704.06904 (2017)
20. Wu, Z., Huang, Y., Wang, L., Wang, X., Tan, T.: A comprehensive study on cross-view gait based human identification with deep CNNs. IEEE Trans. Circuits Syst. Video Technol. **39**(2), 209–226 (2017)
21. Yu, S., Chen, H., Reyes, E.B.G., Norman, P.: Gaitgan: invariant gait feature extraction using generative adversarial networks. In: Proceedings of the IEEE Conference on Computer Vision and Pattern Recognition Workshops, pp. 30–37 (2017)
22. Yu, S., Chen, H., Wang, Q., Shen, L., Huang, Y.: Invariant feature extraction for gait recognition using only one uniform model. Neurocomputing **239**, 81–93 (2017)
23. Zhang, C., Liu, W., Ma, H., Fu, H.: Siamese neural network based gait recognition for human identification. In: 2016 IEEE International Conference on Acoustics, Speech and Signal Processing (ICASSP), pp. 2832–2836. IEEE (2016)
24. Zhang, J., Wang, N., Zhang, L.: Multi-shot pedestrian re-identification via sequential decision making. In: The IEEE Conference on Computer Vision and Pattern Recognition (CVPR), June 2018

Experimental Validation of Minimum-Jerk Principle in Physical Human-Robot Interaction

Chen Wang[1,2], Liang Peng[1], Zeng-Guang Hou[1,2,3(✉)], Lincong Luo[1,2], Sheng Chen[1,2], and Weiqun Wang[1]

[1] State Key Laboratory of Management and Control for Complex Systems, Institute of Automation, Chinese Academy of Sciences, Beijing 100190, China
{wangchen2016,liang.peng,zengguang.hou,luolincong2014,chensheng2016, weiqun.wang}@ia.ac.cn
[2] University of Chinese Academy of Sciences, Beijing 100049, China
[3] CAS Center for Excellence in Brain Science and Intelligence Technology, Beijing 100190, China

Abstract. Human motor control is a complex process, and undergoes changes due to the environmental interactions in physical human-robot interaction (pHRI). This pilot study aims to explore whether human motion under robotic constraints still complies with the same principles as in unconstrained situations, and how humans adapt to non-biological patterns of robot movements. Two typical modes in applications of pHRI (e.g., robot-assisted rehabilitation) are tested in this study. In human-dominant mode, by building spring-damper force fields using a planar rehabilitation robot, we demonstrated that participants' actual motion in reaching movements complied well with the standard minimum-jerk trajectory. However, when the virtual impedance between human force and virtual display was different from the human-robot physical impedance, the actual motion was also in a straight line but had a skewed bell-shaped velocity profile. In robot-dominant mode, by instructing participants to move along with the robot following biological or non-biological velocity patterns, we illustrated that humans were better adapted to biological velocity patterns. In conclusion, minimum-jerk trajectory is a human preferred pattern in motor control, no matter under robotic force or motion constraints. Meanwhile, both visual feedback and haptic feedback are critical in human-robot cooperation and have effects on actual human motor control. The results of our experiments provide the background for modeling of human motion, prediction of human motion and trajectory planning in robot-assisted rehabilitation.

Keywords: Physical human-robot interaction · Motor control Minimum-jerk model · Robot-assisted rehabilitation

This work was supported in part by National Natural Science Foundation of China (Grant #61603386, U1613228, 61720106012, 61533016, 61421004) and Beijing Natural Science Foundation (Grant L172050).

L. Cheng et al. (Eds.): ICONIP 2018, LNCS 11307, pp. 499–509, 2018.
https://doi.org/10.1007/978-3-030-04239-4_45

1   Introduction

Human movement is considered as the control result of central nervous system (CNS), peripheral nervous system (PNS) and musculoskeletal system. When humans perform free point-to-point reaching tasks without strict constraints on accuracy, arm movements of different humans tend to be similar. In order to analyze the invariant law in reaching movements, numerous observations have led to that the point-to-point movements in unconstrained situations comply with the "Minimum-X" models [e.g., minimum-jerk model [1,2], minimum-torque-change model [3], minimum-variance model [4], and minimum-work model [5]]. The computational models of voluntary movements are generally viewed as the result of motor control which involves motor planning and motor execution, and connects three levels of motor system—motor behavior, limb mechanics and neural control [6].

The minimum-jerk principle is the most prominent among these invariant laws, and can predict the qualitative and quantitative features of the arm movements between two targets. Several studies have implemented minimum-jerk principle in human-robot co-manipulation tasks. Maeda et al. proposed estimating human motion using the minimum-jerk model for smooth cooperation. They used nonlinear least-squares method to identify parameters of the model in real-time and estimated position of the human hand was used to determine the desired position of the robot [7]. Furthermore, Corteville et al. demonstrated that humans were able to move along with the minimum-jerk speed profile, a rectangular and a triangular speed profile, but only the minimum-jerk speed profile felt comfortable and natural [8].

Specifically, in physical human-robot interaction (pHRI), the minimum-jerk principle integrated with external measurements of kinematics and kinetics can be used in prediction of human motion and trajectory planning for robot. In the applications like robot-assisted rehabilitation, much of the prior literature defined the desired trajectories based on minimum-jerk model for upper-limb rehabilitation robots. For example, in the robotic assisted rehabilitation following neurological injury, state-of-the-art control strategies direct patients to follow a fixed or adaptive minimum-jerk trajectory, with aims to force patients to learn optimally smooth (minimum jerk) movements [9].

However, there is little concern about whether the minimum-jerk principle in free motion still hold in pHRI. Reed et al. demonstrated that humans behaved differently in human-robot and human-human physical interaction [10]. Further, since humans are physically coupled to the robotic device in pHRI, the dynamics of musculoskeletal system can be constrained by direct physical contact, and visual feedbacks can be augmented or distorted by virtual reality technology [11,12], which lead to that human motor planning and motor execution are different from those in unconstrained situations. This raises the question whether the arm motion determined by human motor control under robotic constraints comply with the minimum-jerk model.

In present work, we focus on the law of motor control when humans and robots work together to accomplish some task, and investigate two kinds of arbitration in shared control. One is human-dominant mode, where impedance control of an upper-limb rehabilitation robot is implemented to build a spring-damper force field, and emphasizes the influence of force constraints in pHRI; the other is robot-dominant mode, where the robot alone leads the movement and the human has to move along with biological or non-biological velocity patterns, which aims to test the effect of motion constraints in pHRI. These two modes are representative of active training and passive training respectively in robot-assisted rehabilitation [13]. Meanwhile, we examine the condition that the virtual impedance between human force and virtual display is different from the physical human-robot impedance, considering that both visual feedback and kinaesthetic haptic feedback are critical in human motor control.

The remaining parts of this paper are organized as follows: Sect. 2 introduces the minimum-jerk model, and describes our experimental set-up including two sets of experiments. The experimental results are presented and discussed in Sect. 3. Finally, the conclusions of our research and remarks for future work are presented in Sect. 4.

2   Method

2.1   Minimum-Jerk Model

For unconstrained point-to-point movements, several authors have investigated that natural hand movements tend to be smooth and graceful. Flash and Hogan postulated that maximizes the smoothness of the movement is a criterion to which the motor control system abides in point-to-point movements [1]. This criterion complies with the minimization of a cost function C, expressed in terms of the mean square of the jerk (derivative of the acceleration) [14]:

$$C = \frac{1}{2} \int_{t_1}^{t_2} \left[\left(\frac{d^3 x}{dt^3} \right)^2 + \left(\frac{d^3 y}{dt^3} \right)^2 \right] dt \tag{1}$$

where (x, y) is the Cartesian coordinates of the hand position.

In essence, minimum-jerk model assumes that among all possible trajectories, the motor control system selects one specific trajectory that satisfies the minimum condition. The actual trajectory generated between two points in two dimensional space can be simplified to:

$$\frac{x(t) - x_i}{x_d - x_i} = \frac{y(t) - y_i}{y_d - y_i} = 10(\frac{t}{t_d})^3 - 15(\frac{t}{t_d})^4 + 6(\frac{t}{t_d})^5 \tag{2}$$

where (x_i, y_i) and (x_d, y_d) are respectively the coordinates of starting point and ending point, t_d is the total movement time. The trajectory of 5-order polynomial form corresponds to a straight line connecting two points with bell-shaped tangential velocity profiles.

2.2 Experimental Set-Up

(1) Apparatus and Behavior Task: We have developed an upper-limb reha-bilitation robot, called CASIA-ARM, which can provide compliant force feedback and virtual environment training [15]. In order to explore the characteristics of human motion in pHRI, we use CASIA-ARM as the experimental platform (see Fig. 1). The design of parallelogram structure and capstan-cable transmission system has small reflected inertia and friction at the end-effector, which is ben-eficial to precise force control and compliant interaction between the robot and patients [16]. In order to measure the force applied by the participant's hand at the end-effector, a 6-axis force sensor is mounted at the tip of the robot arm.

In all experiments, participants seated with their right arm coupled with the robot at the end-effector, and made 2-DOF point-to-point movements in the horizontal plane. The virtual reality display is shown in Fig. 1, where two targets (point A and point B) were displayed as 1-cm-diameter black solid circles on the screen, and corresponded to $(-0.14\,\text{m}, 0.4\,\text{m})$ and $(0.14\,\text{m}, 0.4\,\text{m})$ in workspace, respectively. For repeated goal-directed reaching movements, A and B took turns to become a 2-cm-diameter red solid circle to indicate the end target. Partic-ipants were asked to manipulate the robot handle to control a blue cursor to make reaching movements between two targets via a virtual reality display, and the robot force outputs were controlled differently according to the experiment modes.

(2) Subjects: To examine the characteristics of human motor control in pHRI, 5 right-handed healthy subjects participated in Experiment 1, and 5 others par-ticipated in Experiment 2. They were aged between 23 and 32, 3 females and 7 males, with heights ranging from 162 to 187 cm. All participants were neurolog-ically healthy and had normal or corrected vision.

 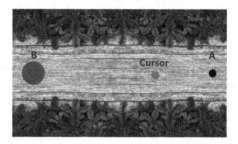

Fig. 1. Experimental Set-up. Left: Demonstration of experimental task using CASIA-ARM, where the participant was directed to reach the red target in the virtual envi-ronment. Right: The virtual reality display of reaching task. (Color figure online)

(3) Design of Experiment 1 (Human-Dominant Mode): For pHRI applications like robot-assisted rehabilitation, the robot emphasizes on compliant interaction with humans. In this study, impedance control was implemented, and a widely used spring-damper force field was built by the robot in Exp. 1A and Exp. 1B. Specifically, Exp. 1A employed standard impedance control to simulate a normal force field, while Exp. 1B was a control experiment to test the effect of kinaesthetic haptic feedback in minimum-jerk trajectory formation.

5 subjects participated in Exp. 1A and Exp. 1B. Each participant performed a total of 100 trails per experiment, and a 1 s break separated trails. The break could assure that initial and final velocity and acceleration were zero in each trial. The interaction force, human hand position and velocity were sampled at 100 Hz.

Experiment 1A: Standard Impedance Condition. In this condition, the robot force outputs were regulated with the human hand position $x(t)$ and velocity $\dot{x}(t)$ according to a predefined impedance relationship as below:

$$\begin{cases} F_x(t) = -k_x(x(t) + 0.14) - b_x\dot{x}(t) \\ \dot{x}(t) = (x(t) - x(t-1))/T_s \end{cases} \tag{3}$$

where $F_x(t)$ denotes the robot force along X axis, b_x and k_x denote the damping and stiffness coefficient respectively, and the equilibrium position of the "spring" is at point A (x = -0.14 m) as in Fig. 1.

In this experiment, the damping and stiffness coefficients can be achieved to make all participants feel comfortable, therefore $b_x = 5\,\text{N} \cdot \text{s/m}$ and $k_x = 50\,\text{N/m}$. Generally speaking, participants received assistance when they made reaching movements from B to A, and were resisted by the robot when moving from A to B.

Besides, the target motion displayed on screen was consistent with participants' hand motion, and the human-robot interaction impedance equaled to the dynamic relationship between human effort and the displayed target motion.

Experiment 1B: Virtual Impedance Condition. In this condition, the robot was locked and kept still, and the force sensor measured the participant's force at the handle. Participants applied endpoint force to control the cursor move from a start target to an end target, and the interaction force was transformed into the cursor movements, and the relationship can be expressed as:

$$\begin{cases} \dot{x}(t) = -[F_x(t) + k_x(x(t) + 0.14)]/b_x \\ x(t+1) = x(t) + \dot{x}(t)T_s \end{cases} \tag{4}$$

where $F_x(t)$ denotes the interaction force along the X axis, $x(t)$ is the X coordinate of the displayed cursor, different from that of the human hand which kept still during the experiment, b_x and k_x denote the damping and stiffness coefficient respectively, which were set the same values as in Exp. 1A.

Importantly, the human hand remained stationary, and the imposed force controlled the displayed motion as manipulating a virtual spring-damper system. In other words, the physical human-robot impedance was infinity, while the virtual impedance between human effort and displayed motion remained the same as in standard impedance condition, thereby participants performed reaching movements under biased force feedback.

(4) Design of Experiment 2 (Robot-Dominant Mode): In human-robot cooperation, higher interaction forces indicated more effort from the participant and a less intuitive interaction [17]. According to this criterion, we designed Experiment 2 to examine whether biological vs. non-biological trajectories affected the efficiency of human-robot interaction, and how human motor control adapt to non-biological velocity patterns. The interaction was tested with three different velocity profiles in robot-dominant interaction: a minimum-jerk profile, a trapezoidal and a constant velocity profile. Therefore, there are three different velocity patterns: biological (minimum-jerk velocity), weakly non-biological (trapezoidal velocity) and strongly non-biological (constant velocity) patterns.

For each velocity profile, the reference position trajectory was developed in the workspace, and then the spatial coordinates of the trajectory were transformed to joint angles by inverse kinematics. Finally, two PID joint angle controllers were conducted to achieve the passive tracking task.

In robot-dominant mode, 5 subjects participated in tracking movements, and each participant performed 100 trails of the reaching movements between point A and point B per pattern (see Fig. 1). The robot was programmed to wait 2 s when reached the goal target, so that participants would not be able to have trial-to-trial adaptation. Importantly, the robot took the leader role during the reaching movements with either biological or non-biological velocity patterns. As in Experiment 1, the interaction force, robot position and velocity were recorded at a sampling frequency of 100 Hz.

3 Results

3.1 Experiment 1

In order to acquire stable motion states and exclude the familiarization process, the last 50 successive trials were selected in each condition. Figure 2 displays the spatial trajectory, displacement and velocity of hand movements for each participant in standard impedance condition, together with the standard minimum-jerk trajectory. Figure 3 shows the corresponding results in virtual impedance condition. Despite variations in magnitude, the displacement and velocity profiles were consistent across participants.

As shown in Fig. 2, participants' hand movements were smooth with typical bell-shaped velocity profiles either reaching A from B or reaching B from A. Further, the average displacement and velocity profiles were evidently coincident with the standard minimum-jerk profiles in the standard impedance condition. In contrast, the velocity profiles in Fig. 3 were skewed bell-shaped and the fluctuation of curves were higher. Therefore, in virtual impedance condition, the divergences between actual profiles and minimum-jerk profiles were obvious both in displacement and velocity.

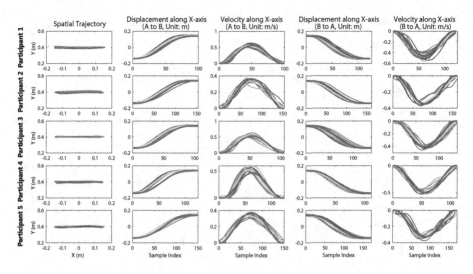

Fig. 2. Spatial trajectories, displacements and velocities of participants' hand during the point-to-point movements in Exp. 1A. First column: trajectories of the hand movements in Cartesian space. Second and third columns: time-series displacements and velocities along the X axis in reaching movements from A to B. The last two columns: corresponding displacements and velocities in opposite direction movements (B to A). Colored lines indicate individual trials, and thick yellow lines indicate the average motion profiles of individual participant across all 50 trials. Thick red lines indicate the minimum-jerk profiles as comparison standards. (Color figure online)

Table 1. RMS errors on all participants for Exp. 1A and Exp. 1B

	Displacement	Displacement	Velocity
	X-axis (m)	Y-axis (m)	X-axis (m/s)
Standard condition (A to B)	0.0240	0.0023	0.0402
Standard condition (B to A)	0.0138	0.0036	0.0486
Virtual condition (A to B)	0.0333	0.0044	0.0846
Virtual condition (B to A)	0.0227	0.0037	0.1143

Table 1 shows the RMS errors of the actual displacement and velocity as comparison between actual hand motion and minimum-jerk generated trajectory. More specifically, in reaching movements from A to B, the deviation errors of displacement and velocity along the X axis in standard condition were 28% and 52% smaller than those in virtual condition, and the deviation error of displacement along the Y axis proved that trajectories were almost straight in both conditions. The RMS errors in opposite direction movements accorded with the same quantitative relationship.

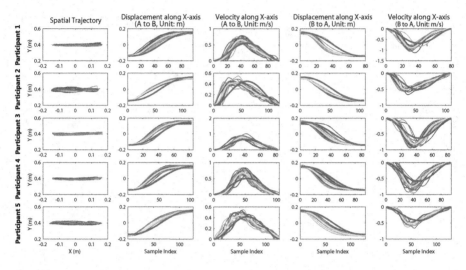

Fig. 3. Spatial trajectories, displacements and velocities of participants' hand during the point-to-point movements in Exp. 1B. The robot kept still, and participants received only visual feedback. First column: trajectories of the hand movements in Cartesian space. Second and third columns: displacements and velocities along the X axis in reaching movements from A to B. Fourth and fifth column: displacements and velocities along the X axis in reaching movements from B to A. Each colored line indicates an individual trail. Each thick yellow line indicates an individual participant average across all 50 trials. Thick red lines indicate the minimum-jerk profiles used as baselines. (Color figure online)

Thus, the results of Experiment 1 can be concluded that participants were able to accomplish reaching movements with minimum-jerk model under the force constraints of two different impedance conditions. In standard impedance condition, the participants' hand motions complied well with the minimum-jerk principle. However, in virtual impedance condition, the displayed motions controlled by human force were also in a straight line but had a skewed bell-shaped velocity curve. The comparisons between two conditions reveal that both visual and kinaesthetic haptic feedbacks are necessary for human motor system to generate a minimum-jerk trajectory as in unconstrained situations.

3.2   Experiment 2

As the tracking movements described in Sect. 2.2 were only in X direction, we evaluated the efficiency of human-robot interaction through the interaction force along the X axis. The last 50 stable trials were selected in each pattern in order to exclude the familiarizing phase. Figure 4 displays the force applied by participants on the robot for the three different task patterns. Comparison of the three red lines showed that the force profiles were smoothest and least variable in biological pattern, and indicated that participants were better adapted

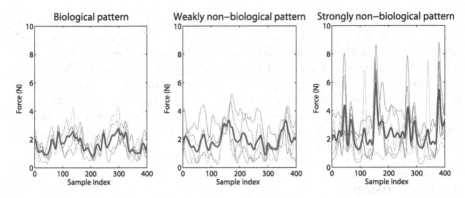

Fig. 4. Interaction force applied by participants along the X axis for biological, weakly non-biological and strongly non-biological patterns. In each pattern, the thin lines indicate individual participant averages across all 50 trials, the thick red line indicates the average force of all participants across all trials. (Color figure online)

to the minimum-jerk profiles offered by the robot. In non-biological patterns, lower forces were achieved in weakly non-biological (trapezoidal velocity) pattern rather than strongly non-biological (constant velocity) pattern, which can be attributed to that the trapezoidal velocity profile had the similar acceleration and deceleration phases as the minimum-jerk profile.

For the statistical analysis of Experiment 2, the RMS force of each participant in an trail is defined as:

$$F_{rms} = \sqrt{\frac{1}{N} \sum_{k=1}^{N} f_k^2} \tag{5}$$

where N is the size of sample set, and f_k represents the measured force of kth sample index. As can be seen in Table 2, the mean value of F_{rms} in biological pattern was 14% smaller than that in weakly non-biological pattern, and 36% smaller than that in strongly non-biological pattern. This supports that the force magnitude was smallest when participants followed a minimum-jerk trajectory. Then we conducted repeated measures analysis of variance (ANOVA) on F_{rms}. Results showed a significant effect of velocity pattern for interaction force (F(2, 11) = 41.67, p = 0.0001), and the ANOVA also detected the difference between participants (F(4, 9) = 12.57, p = 0.0016) due to the different baseline levels of force.

Table 2. Average force and standard deviation of the RMS force for three patterns.

Pattern	Biological	Weakly non-biological	Strongly non-biological
Mean	1.83	2.13	2.88
Std	0.15	0.74	0.80

Therefore, in robot-dominant mode, better performance were achieved when participants moved along with the robot that followed a natural minimum-jerk profile, and the sudden change of acceleration in trapezoidal and constant velocity profiles resulted in higher and fast changing interaction force. The experiments also showed that the going-along motion with biological velocity pattern felt more human-like and natural, as human motor system can supply "minimal-effort" to accomplish the reaching movements following the robot. It can be concluded that the biological velocity profile is the best choice for humans to adapt in pHRI with robotic motion constraints.

4 Conclusion

Human invariant laws like minimum-jerk model are promising in human motion prediction, especially to be used in applications like robot-assisted rehabilitation. In order to answer the question whether the arm motion determined by human motor control still complies with the minimum-jerk principle when humans physically interact with robots, we tested two interaction modes (human-dominant mode vs. robot-dominant mode) in this pilot study.

Experiments with an upper-limb rehabilitation robot named CASIA-ARM demonstrated that humans were able to perform reaching movements with the minimum-jerk principle under force constraints, and both visual feedback and kinaesthetic haptic feedbacks were critical for humans to generate a minimum-jerk trajectory as in unconstrained situations. Our study also provided experimental evidence that human motor control were better adapted to biological velocity patterns and smaller interaction force is achieved under motion constraints.

In summary, the minimum-jerk principle is a human preferred pattern in motor control for goal-directed reaching movements, no matter under robotic force constraints or motion constraints. This conclusion suggests that the invariant laws in free motion can give important insight for physical human-robot interaction. In the future, as the laws of motion in free motion have positive reference value to human movement under robotic constraints, we plan to combine such features into modeling of human motion, intention detection and trajectory planning in robot-assisted rehabilitation.

References

1. Flash, T., Hogan, N.: The coordination of arm movements: an experimentally confirmed mathematical model. J. Neurosci. 5(7), 1688–1703 (1985)
2. Wada, Y., Kaneko, Y., Nakano, E., Osu, R., Kawato, M.: Quantitative examinations for multi joint arm trajectory planning–using a robust calculation algorithm of the minimum commanded torque change trajectory. Neural Netw. 14(4–5), 381–393 (2001)
3. Uno, Y., Kawato, M., Suzuki, R.: Formation and control of optimal trajectory in human multijoint arm movement. Biol. Cybern. 61(2), 89–101 (1989)

4. Harris, C.M., Wolpert, D.M.: Signal-dependent noise determines motor planning. Nature **394**(6695), 780 (1998)
5. Soechting, J.F., Buneo, C.A., Herrmann, U., Flanders, M.: Moving effortlessly in three dimensions: does donders' law apply to arm movement? J. Neurosci. **15**(9), 6271–6280 (1995)
6. Scott, S.H.: Optimal feedback control and the neural basis of volitional motor control. Nat. Rev. Neurosci. **5**(7), 532 (2004)
7. Maeda, Y., Hara, T., Arai, T.: Human-robot cooperative manipulation with motion estimation. In: Proceedings of IEEE/RSJ International Conference on Intelligent Robots and Systems (IROS), vol. 4, pp. 2240–2245 (2001)
8. Corteville, B., Aertbeliën, E., Bruyninckx, H., De Schutter, J., Van Brussel, H.: Human-inspired robot assistant for fast point-to-point movements. In: Proceedings of IEEE International Conference on Robotics and Automation, pp. 3639–3644 (2007)
9. Pehlivan, A.U., Sergi, F., O'Malley, M.K.: A subject-adaptive controller for wrist robotic rehabilitation. IEEE/ASME Trans. Mechatron. **20**(3), 1338–1350 (2015)
10. Reed, K.B., Peshkin, M.A.: Physical collaboration of human-human and human-robot teams. IEEE Trans. Haptics **1**(2), 108–120 (2008)
11. Robertson, J.V.G., Roby-Brami, A.: Augmented feedback, virtual reality and robotics for designing new rehabilitation methods. In: Didier, J.-P., Bigand, E. (eds.) Rethinking physical and rehabilitation medicine. Collection de L'Académie Européenne de Médecine de Réadaptation, pp. 223–245. Springer, Paris (2010). https://doi.org/10.1007/978-2-8178-0034-9_12
12. Matsuoka, Y., Brewer, B.R., Klatzky, R.L.: Using visual feedback distortion to alter coordinated pinching patterns for robotic rehabilitation. J. Neuroeng. Rehabil. **4**(1), 17 (2007)
13. Peng, L., Hou, Z.-G., Kasabov, N., Peng, L., Hu, J., Wang, W.: Implementation of active training for an upper-limb rehabilitation robot based on impedance control. In: Proceedings of 27th Chinese Control and Decision Conference (CCDC), pp. 5453–5458 (2015)
14. Hogan, N.: An organizing principle for a class of voluntary movements. J. Neurosci. **4**(11), 2745–2754 (1984)
15. Peng, L., Hou, Z.-G., Peng, L., Wang, W.: Design of CASIA-ARM: a novel rehabilitation robot for upper limbs. In: Proceedings of IEEE/RSJ International Conference on Intelligent Robots and Systems (IROS), pp. 5611–5616 (2015)
16. Peng, L., Hou, Z.-G., Peng, L., Luo, L., Wang, W.: Robot assisted rehabilitation of the arm after stroke: prototype design and clinical evaluation. Sci. China Inf. Sci. **60**(7), 073201 (2017)
17. Maurice, P., Huber, M.E., Hogan, N., Sternad, D.: Velocity-curvature patterns limit human-robot physical interaction. IEEE Robot. Autom. Lett. **3**(1), 249–256 (2018)

Residual Semantic Segmentation of the Prostate from Magnetic Resonance Images

Md Sazzad Hossain$^{(\boxtimes)}$, Andrew P. Paplinski, and John M. Betts

Faculty of Information Technology, Monash University, Melbourne, Australia
{sazzad.hossain,andrew.paplinski,
john.betts}@monash.edu

Abstract. The diagnosis and treatment of prostate cancer requires the accurate segmentation of the prostate in Magnetic Resonance Images (MRI). Manual segmentation is currently the most accurate method of performing this task. However, this requires specialist knowledge, and is time consuming. To overcome these limitations, we demonstrate an automatic segmentation of the prostate region in MRI images using a VGG19-based fully convolutional neural network. This new network, VGG19RSeg, identifies a region of interest in the image using semantic segmentation, that is, a pixel-wise classification of the content of the input image. Although several studies have applied fully convolutional neural networks to medical image segmentation tasks, our study introduces two new forms of residual connections (remote and neighbouring) which increases the accuracy of segmentation over the basic architecture. Our results, using this new architecture, show that the proposed VGG19RSeg can achieve a mean Dice Similarity Coefficient of 94.57%, making it more accurate than comparable methods reported in the literature.

Keywords: Prostate · MRI images · Semantic segmentation
Deep convolutional neural networks

1 Introduction

Prostate cancer is one of the most common and lethal cancers among men, with over 221,000 cases diagnosed in 2015 in the United States [1]. In Australia, approximately 17,000 males were diagnosed with prostate cancer, and it was the cause of death for approximately 3,500 men in 2017 [2]. Medical imaging is used in many cases to aid the diagnosis and treatment of prostate cancer. MRI is used for the detection of tumours and for treatment planning whereas Trans-Rectal Ultrasound (TRUS) is often used in a clinical setting, to guide needle placement for biopsies and treatments such as brachytherapy [3]. A current goal in medical imaging is to fuse images from these two modalities (MRI and TRUS) by the creation of a 3D model of the patient's prostate. Automatic image segmentation (the identification of organ boundaries) is a necessary task in achieving this goal.

© Springer Nature Switzerland AG 2018
L. Cheng et al. (Eds.): ICONIP 2018, LNCS 11307, pp. 510–521, 2018.
https://doi.org/10.1007/978-3-030-04239-4_46

Despite significant progress in prostate segmentation, fully automatic segmentation still remains challenging due to the deformability of the organ and the typically low resolution of MRI and TRUS images. Therefore, in many situations, segmentation is usually manually performed, which requires skilled clinicians, and is time-consuming. Because of the low accuracy of automatic segmentation, much recent work on MRI-TRUS registration still relies on manual segmentation by experts, for example [4, 5]. To address this challenge, we present a novel method of fully automatic segmentation of the MRI images of prostate using deep convolutional neural networks with residual connections.

Convolutional Neural Networks (CNNs) were designed and implemented primarily to classify objects in images [6]. These networks typically have a structure consisting of kernels or filters, activation functions and subsampling arranged in 16–200 layers. This enables a very detailed feature representation of an image to be made, in order to identify given objects. This gives rise to the terms 'convolutional' and 'deep' in describing these networks. Besides classification, CNNs have been adopted in segmentation tasks as well. Many recent medical imaging studies have used CNNs for automatic segmentation of regions of interest.

Non-CNN based medical image segmentation procedures have previously mainly been based on feature engineering. For example, Liao et al. [7] proposed an automatic feature extraction procedure using representation learning and a deep learning framework to identify the most significant parts of the extracted features. Toth et al. [8] proposed an advanced feature selection algorithm by modifying the typical Active Appearance Model (AAM) utilizing level-set representation. Yan et al. [9] presented a partial contour based segmentation method utilizing an a priori shape. These methods rely greatly on identifying suitable feature information. Another popular segmentation method is based on the concept of the nearest neighbours as in [10]. The CNN based procedures here seems to be more straightforward and more efficient, because CNNs combine both feature extraction and classification.

Feature extraction by deep learning is more effective in the sense that the algorithm 'learns' how to extract the features by utilizing real-world samples. As a consequence, many researchers are now investigating how CNNs may be used for segmentation tasks on medical images. Pereira et al. [11] used a CNN to segment a brain tumour. Havaei et al. [12] segmented brain tumours in MRI slices using two concatenated CNNs where the output probabilities from one CNN was fed to another CNN. Jia et al. implemented a pre-trained CNN, 'VGG-19' [13], to identify the prostate boundary in MRI images. Instead of segmenting at once, they first performed a coarse segmentation using atlas registration to obtain a rough boundary around the prostate, after which the VGG-19 CNN was applied to refine the segmentation. Zhang et al. segmented infant brain tissue by applying CNNs on multi-modal MRI images of brain. Other researchers, [14–16], have achieved segmentation by dividing the images into subregions known as 'patches' and by training a convolutional neural network on those patches. However, according to [17], training with whole image is more efficient and effective than patch-wise training.

More recently, researchers have developed methods to implement deep neural networks for segmentation of images automatically into particular classes using pixel-wise classification of the whole image known as semantic segmentation [17, 18]. For example, when provided an image of a cat and a dog together, semantic segmentation will identify the regions containing cats or dogs in that image. Using this method, Tian et al. [19]

segmented the prostate region from MRI images. Ahmad et al. [20] segmented human thigh quadriceps and Tran [21] presented an automated cardiac segmentation procedure using CNN based semantic segmentation. However, the network models used in these studies for semantic segmentation do not possess neighbouring residual connections, which can notably improve accuracy. The reason is that residual connection bypasses information, i.e. input to a deeper layer rather than just to the adjacent layer. He et al. [22] proved that residual connections are important to effectively train deep neural network models, because residual mapping is easier to optimize than unreferenced mapping of a stacked layer series. In another study [23], performance of several CNN were examined with and without residual connections and it was found that residual connections accelerate training speed and improve accuracy notably.

Therefore, this study uses residual connection on a VGG19 based fully convolutional network (FCN) to deliver more efficient network for automatic identification and localization of the prostate gland. Although remote residual connection between the convolution and the deconvolution part of an FCN has been implemented previously such as U-net [24] and V-net [25], residual connections between neighbouring layers in combination with residual connections between convolution and deconvolution layers in an FCN is still unreported to this date.

To extend the work by previous researchers we introduce a novel semantic segmentation of MRI prostate image slices using an adaptation of the popular deep neural network model, VGG19 [26]. Our new model, VGG19RSeg, modifies the original network by adding residual/skip connections between neighbouring and distant layers, thereby creating a semantic segmentation structure as explained in Sect. 2.3. Hence, this study first introduces an FCN with residual connections between stacked convolution layers and inter-convolution-deconvolution layers.

The following section describes details of the methodology applied in the research reported in this paper, including details of our VGG19RSeg network starting with its predecessor, the VGG19 network. Section 3 then presents computational experiments and results.

2 Methodology

2.1 Convolutional Neural Networks

A significant number of papers and tutorials describe convolutional neural networks in a great detail. To establish our notation, we say that each convolutional layer performs the following feedforward operations:

$$U = convT(X, W) + b; Z = \max(U, 0) \tag{1}$$

where X is a $M \times N \times D_x$ input tensor (a colour image, in particular), W is a $P \times Q \times D$ filter/kernel, $convT(\cdot)$ denotes a tensor, i.e., multidimensional convolution that generates a tensor U typically of dimensions $M \times N \times D_w$. The bias b has a matching dimension. The output of the convolutional layer, Z, represents the output of the ReLU activation function applied on U. The output Z becomes the input tensor for the next layer. It is typically referred to as a feature map. Convolutional layers are typically separated by a subsampling operation that reduces the dimension of the feature maps. The most common method for subsampling operation is 'maxpooling', which replaces a square subregion of a feature map of size $c \times c$ with a single value being equal to the largest value in the subregion. A composition of the convolutional layers is followed by one to three fully connected layers as in "traditional" neural networks. For completeness of the description, we say that during the learning procedure all the weight filters W are modified using a stochastic gradient error-backpropagation algorithm that minimizes the specific loss function L.

2.2 The VGG19 Model

In 2014, Simonyan and Zisserman introduced the very deep network model, VGG16 [26]. This network was composed of 3 sequential convolutional layers, each having a max pooling layer to reduce the volume size. Followed by these convolutional layers, there are two fully connected layers containing 4096 neurons each. The final layer – a softmax layer, follows the fully connected layer, and delivers the ultimate probabilistic output of classification. In the same study, they modified the network by adding 3 extra weight layers thus creating a VGG19 network which outperformed VGG16. Although other deep learning models such as AlexNet, ResNet and InceptionNet have been developed [23], VGG models have proven to be more accurate image classifiers due to their simpler yet very deep network architecture compared to those other models [26]. Furthermore, VGG models appear to have a greater accuracy for semantic segmentation tasks [17, 18].

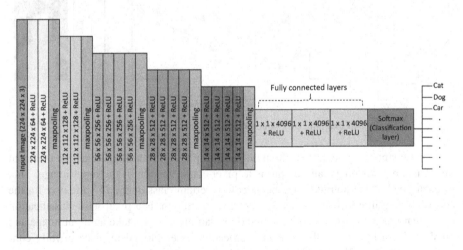

Fig. 1. VGG19 deep neural network model.

As Fig. 1 shows, the VGG19 structure beginning with an RGB image of the size $224 \times 224 \times 3$. There are total 16 convolution layers, where most of them are connected consecutively having 5 maxpooling stages/layers. Each convolution layer output goes through a non-linear ReLU activation function. The whole convolution part of the network can be divided into 5 subregions, where each subregion is followed by a maxpooling layer to reduce the learnable parameters. The first two subregions consist of 2 consecutive convolution layers and the remaining 3 subregions comprise 4 consecutive convolution layers each. Convolution layers in same subregion each produce the same number of feature maps D_w. In the first region layers $D_w = 64$, in the second region $D_w = 128$, in the third region layers $D_w = 256$, and in the fourth and fifth regions $D_w = 512$. The product of the final maxpooling layer is flattened and passes through 3 fully connected layers comprising 4096 neurons each. The output of the final fully connected layer is then processed through classification layer which produces a probabilistic output classifying the input image, for example, as 97% cat, 2% dog, 1% car etc.

2.3   Semantic Segmentation

The structure of the semantic segmentation network consists of two asymmetric parts identified as a convolutional and the de-convolutional parts, respectively, as shown in Fig. 2. Deep neural network models have been designed to classify visual objects by giving a probability of the object belonging to each of the classes for which the neural network has been designed to classify. For example, a picture of a dog may yield a 90% probability associated with "dog", and smaller probabilities associated with the other outputs in order that the total equals to 1.

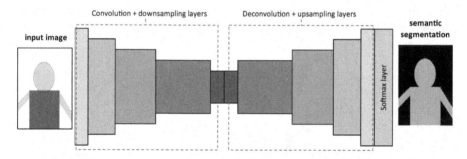

Fig. 2. An example of a structure of a semantic segmentation network.

For semantic segmentation, the input and output are both images, where the input image is a regular image and output is a 'pixel-class' version of the input image. That is, each pixel of the output image belongs to a certain class defined by the user for the corresponding input image. Thus, semantic segmentation is a pixel-wise classification of a given image. Since convolution and downsampling layers take an input image and break it down hierarchically, a reverse procedure is employed in place of fully connected layer (Fig. 2), by which deconvolution, upsampling and finally a softmax classifier, will produce a pixel-classified, i.e. semantically segmented image instead of just one object label. For instance, Fig. 2 shows an image of a human body that is applied to a trained convolutional neural network to identify the regions of the human in the image by a distinguishable colormap. Here, the grey region indicates the human and the black region is the background.

2.4 The Structure of the Residual VGGRSeg Network for Semantic Segmentation of Prostate

Because VGG19 was designed for object classification in images, adaptation for pixel classification requires that its fully-connected layers and the softmax layer are replaced with deconvolution and upsampling layers whose architecture is like a reflection of the convolution and downsampling layers shown in Fig. 2.

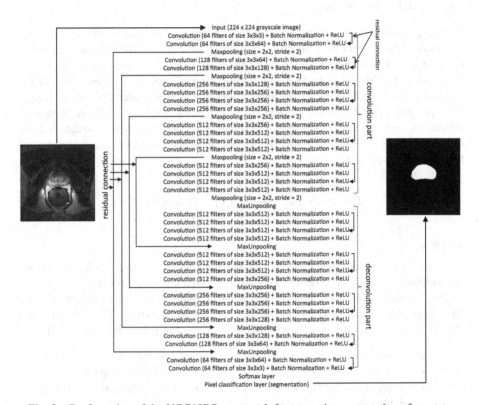

Fig. 3. Configuration of the VGG19RSeg network for semantic segmentation of prostate.

Each slice within the 3D MRI image stack was processed individually as a 2D monochromatic image. Therefore, the initial number of channels of VGG19 was changed from 3 to 1. Since VGG19 was designed to take input images of 224 × 224 pixels, we also resampled each MRI slice to this size to maintain consistency. The primary part of the proposed network is the residual connections between layers as shown in Fig. 3. Firstly, 'Maxpooling' layers from the convolution part and their corresponding mirror reflection layers in the deconvolution part, i.e. 'MaxUnpooling' were residually connected. Networks with such distantly connected layers are also known as Directed Acyclic Graph (DAG) networks. As shown by previous researchers [27, 28], a layer in a DAG network can connect, i.e. receive or provide inputs to a remote layer which makes the network more complex and more effective for classification. Secondly, during convolution, at each stack of convolution layers before 'maxpooling', the first and the last convolution layers were residually connected. Such connections between neighbouring stacked layers were inspired from ResNet structure [22], which won first place in ILSVRC 2015 image classification competition. Therefore, due to residual connections and VGG19 architecture based FCN, the proposed network has been named as VGG19RSeg.

2.5   Training Procedure

To train the segmentation network, the ground truth segmentation of input images was provided as the output of the network for supervised learning. The dataset used in this research was PROMISE12 [29] grand challenge datasets for prostate segmentation from MRI. MRI images from 49 patients, consisting of total 1377 slices, were used in this study. 90% of slices were used for training and the remainder for testing. Images were pre-processed by taking the square root of each pixel value. These were then normalized to be between 0 and 255. As explained in [19], because the prostate is much smaller than the surrounding region in each image, a class-weighted cross entropy loss function was used as Eq. 2.

$$L = -\frac{1}{n}\sum_{i=1}^{N} w_i^{cl}[\hat{P}_i log P_i + (1 - \hat{P}_i)\log(1 - P_i)] \tag{2}$$

where, $w_i^{cl} = \frac{1}{pixels\,of\,class\,x_i}$

Here, L is the loss function, $\hat{P}_i$ and P_i represent the ground truth version and probabilistic version of a pixel i belonging to a given class, respectively.

The network was built and trained in MATLAB. It was executed on Monash University High Performance Computing (HPC) grid, Massive3. Training time was slightly over 5 h. The machine configuration and training parameters are as follows:

Machine configuration:
- No. of processors – 18
- Memory – 120GB
- GPU – Nvidia Tesla K80

Training parameters:
- Initial learning rate – 0.005
- L2 regularization – 0.0005
- Momentum – 0.9
- Mini batch size – 4
- No. of epochs – 50
- Iterations per epoch – 309

3 Results

3.1 Visual Inspection

As we performed segmentation slice-by-slice, some qualitative results, i.e., visualization of deep learning based semantic segmentations on different MRI slices have been given compared, in addition, with the ground truth segmentation in Fig. 4. In this figure, the green and blue overlays on the MRI slices indicate the ground truth segmentations and the proposed deep learning model based segmentations respectively. Visual representation shows that proposed model was able to perform segmentation almost as accurate as the ground truth version.

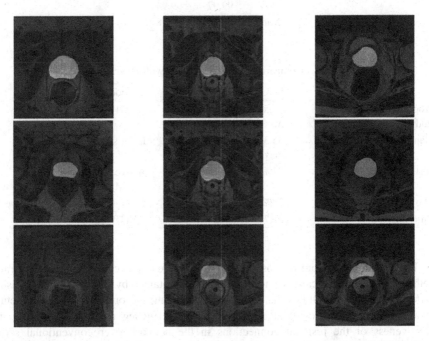

Fig. 4. Prostate segmentation by proposed model (blue) vs. ground truth (green) (Color figure online)

3.2 Quantitative Comparisons

To quantify the accuracy, we follow two common metrics: Dice Similarity Coefficient (DSC) [30] and Intersection-over-Union (IU) as given in Eqs. (3) and (4).

$$DSC = 2\frac{|X \cap Y|}{|X| + |Y|} \tag{3}$$

$$IU = \frac{|X \cap Y|}{|X| + |Y| - |X \cap Y|} \tag{4}$$

where X and Y are two different regions, i.e. sets of pixels in the image which is 'prostate region' and 'background' in our case. The modulus sign '| |' defines the cardinal of the corresponding sets.

Table 1 shows the accuracy measures of the proposed method, which signifies achievement of a high accuracy in the proposed segmentation task. Table 2 compares our result with some other studies.

Table 1. Segmentation accuracy using VGG19RSeg.

	DSC (%)	IU (%)
Average	94.57	91.48
Max	99.97	99.94
Min	80.88	80.18

Table 2. Comparison of VGG19RSeg with other studies.

	DSC (%)	Method
Proposed method of this study	94.57% avg., 99.97% max	VGG19RSeg
Tian et al. [19]	85.3% average, 91.5% max	Long's FCN [17]
Ghasab et al. [31]	87% average, 94% max	Active Appearance Model (AAM)
Cho et al. [32]	78% average	CNN + topological derivative
Other method	87.7% average	VGG19-equipped FCN

Table 2 shows that our proposed method has increased accuracy over other similar methods. The next best case is a non-CNN method obtained by Ghasab et al. [31] using AAM. The table also shows that using a FCN having the original VGG19 structure obtains a lower accuracy than the proposed VGG19RSeg method, illustrating the effectiveness of the residual connections in the network over conventional series connections.

A convex hull-based 3D model of the prostate was created for two different patients from MRI slices segmented using the proposed new method. This model is presented alongside a 3D model created from ground truth segmentations and shown in Fig. 5. Visual inspection shows that the using proposed convolutional neural network leads to the creation an almost identical 3D model to that created using human segmentation.

Ground truth   Proposed method

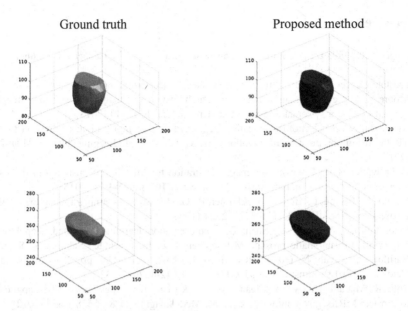

Fig. 5. Ground truth vs. the proposed method for creating a 3D model of the prostate.

4   Conclusion

This study has introduced a highly accurate automatic method for segmenting the prostate in MRI images. Our model has taken a popular deep neural network model, VGG19, which was originally designed for classification, and modified it to build a fully convolutional network with residual connections namely 'VGG19RSeg' for semantic image segmentation. Our results using this proposed deep learning method obtained a mean IU accuracy of 91.48% and a 94.57% DSC accuracy. This is greater than the accuracy of any other comparable method we are aware of. The accuracy of the proposed model is also greater than that of a typical VGG19 based fully convolutional network. Therefore, this study demonstrates the contribution of residual connections in FCN to obtain a greater accuracy in semantic segmentation. Future research will focus on achieving even higher accuracy by using effective pre-processing techniques, as well as adapting this method to other imaging modalities for the purpose of automatically creating and registering 3D models of the patient prostate in real time.

Acknowledgements. Datasets used in this study were part of the PROMISE12 grand challenge for prostate segmentation data sets. The authors wish to thank the Monash University Massive-HPC facility for the provision of high performance computing resources.

References

1. Siegel, R.L., Miller, K.D., Jemal, A.: Cancer statistics, 2016. CA Cancer J. Clin. **66**(1), 7–30 (2016)
2. Prostate Cancer Statistics. https://prostate-cancer.canceraustralia.gov.au/statistics
3. Moore, C.M., et al.: Image-guided prostate biopsy using magnetic resonance imaging-derived targets: a systematic review. Eur. Urol. **63**(1), 125–140 (2013)
4. Khallaghi, S., et al.: Biomechanically constrained surface registration: application to MR-TRUS fusion for prostate interventions. IEEE Trans. Med. Imaging **34**(11), 2404–2414 (2015)
5. Fedorov, A., et al.: Open-source image registration for MRI–TRUS fusion-guided prostate interventions. Int. J. Comput. Assist. Radiol. Surg. **10**(6), 925–934 (2015)
6. LeCun, Y., Bottou, L., Bengio, Y., Haffner, P.: Gradient-based learning applied to document recognition. Proc. IEEE **86**(11), 2278–2324 (1998)
7. Liao, S., Gao, Y., Oto, A., Shen, D.: Representation learning: a unified deep learning framework for automatic prostate MR segmentation. In: Mori, K., Sakuma, I., Sato, Y., Barillot, C., Navab, N. (eds.) MICCAI 2013. LNCS, vol. 8150, pp. 254–261. Springer, Heidelberg (2013). https://doi.org/10.1007/978-3-642-40763-5_32
8. Toth, R., Madabhushi, A.: Multifeature landmark-free active appearance models: application to prostate MRI segmentation. IEEE Trans. Med. Imaging **31**(8), 1638–1650 (2012)
9. Yan, P., Xu, S., Turkbey, B., Kruecker, J.: Discrete deformable model guided by partial active shape model for TRUS image segmentation. IEEE Trans. Biomed. Eng. **57**(5), 1158–1166 (2010)
10. Abdullah, S., Tischer, P., Wijewickrema, S., Paplinski, A.: Parameter-free hierarchical image segmentation. In: Visual Communications and Image Processing (VCIP), pp. 1–4. IEEE, St. Petersburg (2017)
11. Pereira, S., Pinto, A., Alves, V., Silva, C.A.: Brain Tumor segmentation using convolutional neural networks in MRI images. IEEE Trans. Med. Imaging **35**(5), 1240–1251 (2016)
12. Havaei, M., et al.: Brain tumor segmentation with deep neural networks. Med. Image Anal. **35**, 18–31 (2017)
13. Jia, H., Xia, Y., Song, Y., Cai, W., Fulham, M., Feng, D.D.: Atlas registration and ensemble deep convolutional neural network-based prostate segmentation using magnetic resonance imaging. Neurocomputing **275**, 1358–1369 (2018)
14. Tajbakhsh, N., et al.: Convolutional neural networks for medical image analysis: full training or fine tuning? IEEE Trans. Med. Imaging **35**(5), 1299–1312 (2016)
15. Zhang, W., et al.: Deep convolutional neural networks for multi-modality isointense infant brain image segmentation. Neuroimage **108**, 214–224 (2015)
16. Milletari, F., et al.: Hough-CNN: deep learning for segmentation of deep brain regions in MRI and ultrasound. Comput. Vis. Image Underst. **164**, 92–102 (2017)
17. Long, J., Shelhamer, E., Darrell, T.: Fully convolutional networks for semantic segmentation. In: Proceedings of the IEEE Conference on Computer Vision and Pattern Recognition, pp. 3431–3440. IEEE, Boston (2015)
18. Noh, H., Hong, S., Han, B.: Learning deconvolution network for semantic segmentation. In: Proceedings of the IEEE International Conference on Computer Vision, pp. 1520–1528. IEEE, Santiago (2015)
19. Tian, Z., Liu, L., Fei, B.: Deep convolutional neural network for prostate MR segmentation. In: Medical Imaging 2017: Image-Guided Procedures, Robotic Interventions, and Modeling: International Society for Optics and Photonics, vol. 10135, p. 101351L (2017)

20. Ahmad, E., Goyal, M., McPhee, J.S., Degens, H., Yap, M.H.: Semantic Segmentation of Human Thigh Quadriceps Muscle in Magnetic Resonance Images. arXiv preprint arXiv: 1801.00415 (2018)
21. Tran, P.V.: A Fully Convolutional Neural Network for Cardiac Segmentation in Short-Axis MRI. arXiv preprint arXiv:1604.00494 (2016)
22. He, K., Zhang, X., Ren, S., Sun, J.: Deep residual learning for image recognition. In: Proceedings of the IEEE Conference on Computer Vision and Pattern Recognition, pp. 770–778 (2016)
23. Canziani, A., Paszke, A., Culurciello, E.: An Analysis of Deep Neural Network Models for Practical Applications. arXiv preprint arXiv:1605.07678 (2016)
24. Norman, B., Pedoia, V., Majumdar, S.: Use of 2D U-Net convolutional neural networks for automated cartilage and meniscus segmentation of knee MR imaging data to determine relaxometry and morphometry. Radiology, 172322 (2018)
25. Milletari, F., Navab, N., Ahmadi, S.-A.: V-Net: fully convolutional neural networks for volumetric medical image segmentation. In: 2016 Fourth International Conference on 3D Vision (3DV), pp. 565–571. IEEE, Stanford (2016)
26. Simonyan, K., Zisserman, A.: Very Deep Convolutional Networks for Large-Scale Image Recognition. arXiv preprint arXiv:1409.1556 (2014)
27. Yang, S., Ramanan, D.: Multi-Scale Recognition with DAG-CNNs. In: 2015 IEEE International Conference on Computer Vision (ICCV), pp. 1215–1223. IEEE, Santiago (2015)
28. Shuai, B., Zuo, Z., Wang, B., Wang, G.: Dag-recurrent neural networks for scene labeling. In: Proceedings of the IEEE Conference on Computer Vision and Pattern Recognition (2016)
29. MICCAI Grand Challenge: Prostate MR Image Segmentation 2012. https://promise12. grand-challenge.org/
30. Zou, K.H., et al.: Statistical validation of image segmentation quality based on a spatial overlap index1: scientific reports. Acad. Radiol. **11**(2), 178–189 (2004)
31. Ghasab, M.A.J., Paplinski, A.P., Betts, J.M., Reynolds, H.M., Haworth, A.: Automatic 3D modelling for prostate cancer brachytherapy. In: 2017 IEEE International Conference on Image Processing (ICIP), pp. 4452–4456. IEEE, Beijing (2017)
32. Cho, C., Lee, Y.H., Lee, S.: Prostate detection and segmentation based on convolutional neural network and topological derivative. In: 2017 IEEE International Conference on Image Processing (ICIP), pp. 4452–4456. IEEE, Beijing (2017)

The Relationship Between the Movement Difficulty and Brain Activity Before Arm Movements

Tomoki Semoto[✉], Isao Nambu, and Yasuhiro Wada

Graduate School of Engineering, Nagaoka University of Technology,
1603-1 Kamitomioka, Nagaoka, Niigata 940-2188, Japan
semoto@stn.nagaokaut.ac.jp

Abstract. A brain-computer interface (BCI) is a technology that can control external devices using brain activity. It is expected that the flexibility and safety of a BCI will be improved if movement-related information can be extracted from brain activity before executing the movement. In this study, we examined whether movement difficulty levels can be decoded from electroencephalogram (EEG) data. We conducted an experiment where in five participants performed arm reaching movements with three different levels of difficulty, brain activity was measured before these movements. To classify the levels of difficulty, we extracted event-related spectrum perturbation (ERSP) data and performed classification using a relevance vector machine (RVM). Single-trial classification using ERSP data could not obtain high classification accuracy. However, classification accuracies using averaged-trial ERSP data were 66.0% on average (53.9%, 82.3%, 79.6%, 53.1% and 61.1% for each participant). These results show that information related to movement difficulty might be decoded from brain activity before movement, although it is necessary to improve the performance at the single-trial level in future work.

Keywords: EEG · BCI · Relevance Vector Machine (RVM)
Bhattacharyya distance

1 Introduction

Many studies on the relationship between human brain function and human mobility have been conducted. Brain-machine interfaces (BMI) and Brain-computer interface (BCIs), which control external devices (e.g. computers or robots) using brain activity as an input signal, are receiving special attention. The operation of BCI systems can be performed by decoding brain activity data related to different types of movements. In general, BCI studies use brain activity during actual movements or motor imagery of several different body parts (such as arms and legs). However, even if identical movements are performed, the characteristics of those movements are often different. One of these characteristics is movement difficulty. In general, humans feel movement difficulty differently depending on different target locations, movement time, movement speed, movement complexity, etc.

© Springer Nature Switzerland AG 2018
L. Cheng et al. (Eds.): ICONIP 2018, LNCS 11307, pp. 522–529, 2018.
https://doi.org/10.1007/978-3-030-04239-4_47

Fitts's law concerns the relationship between movement difficulty, movement time [1], and the difference between muscle contraction, movement difficulty [2], and brain activity owing to overall differences in movement [3, 4]. Thus, it is evident that movement difficulty affects the performance of human movements. In cases where brain activities related to movement difficulty have emerged before the actual movements, we can expect difficulties decoding the data and incorporating it into BCIs. For example, decoded movement difficulty can be used to create rehabilitation systems that select appropriate loads or schedules. Therefore, in this study, we sought to extract brain activity related to movement difficulty when the participant prepares an arm reaching task, and to test whether different difficulty levels can be classified according to brain activity. An electroencephalogram (EEG) is used to measure brain activity due to its low cost and other participants' limitations (such as physical restrictions).

2   Method

2.1   Experiment and Preprocessing Procedure

Participant

Five healthy right-handed males (aged 22–23 years) participated in this experiment, which was conducted with the approval of the ethics committee of the Nagaoka University of Technology and in accordance with the Declaration of Helsinki. In addition, the details of this experiment were clearly explained to the participants; all participants provided informed consent.

Experiment Procedure

The participants sat on a chair facing a monitor, which displayed their hand position. Target circles were displayed in five directions at intervals of 45° and at a distance of 20 cm from the initial hand position (Fig. 1). The target circle had three radii: 30 mm, 20 mm, and 5 mm (Fig. 2) for different difficulty levels. Both position and radius were displayed at random. The target circle was displayed 1 s before the start of movement cue (Go cue) and participants performed arm movements from the initial position to a target circle within 1.5 s. In this study, the allowable arm movement time was set to 0.4–0.7 s to determine the movement difficulty level within the target radius. This experiment was run for each target size until approximately 100 successful trials were performed.

Recordings

An EEG measurement system (Biosemi ActiveTwo; Biosemi, Amsterdam, The Netherlands) was used to measure brain activity. Sixty-four electrodes were arranged based on the international 10/10 method; one electrode was placed under the right eye to measure electrooculography. In addition, the trajectories of the participants' elbows and shoulders were measured to confirm motor performance. The sampling frequency of the EEG was 1024 Hz, and the movement trajectory was 200 Hz (Fig. 3).

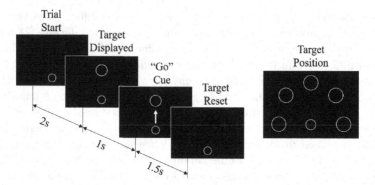

Fig. 1. Timing of movement task and target position.

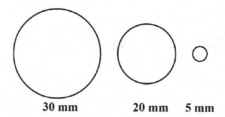

30 mm   20 mm   5 mm

Fig. 2. Target radius for each difficulty level.

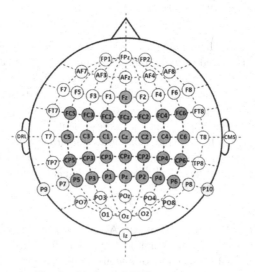

Fig. 3. 64-channel EEG arrangement.

Data Preprocessing

The correct movement trials were extracted based on measured trajectory data. The following conditions were used for extraction:

- The hand position was located within the specified range at the start and end of the movement.
- The hand speed during movements had no multiple peaks.
- The actual movement time was within the average time ± 2 standard deviations (SD).

Baseline corrections were performed on the extracted data to ensure that the average of the baseline (500–0 ms before target was displayed) became 0 using the average of the baseline. Furthermore, a band-pass filter (frequency band 1–50 Hz) was applied and the data were downsampled to 256 Hz. Following this, the blinking artifacts were removed from the brain signals using the electrooculography measured from under the right eye using independent component analysis (ICA) of EEGLAB [5]. In this experiment, 29 channels around the motor cortex (Fz, FC5, FC3, FC1, FCz, FC2, FC4, FC6, C5, C3, C1, Cz, C2, C4, C6, CP5, CP3, CP1, CPz, CP2, CP4, CP6, P5, P3, P1, Pz, P2, P4, P6) of the arranged EEG electrodes were used for classification [6].

Event-related desynchronization (ERDS) was verified in the frequency domain of the brain activity before actual motor function or during motor imagery [7] as follows:

$$ERS(f, t) = |F(f, t)|^2 \tag{1}$$

where $F(f, t)$ is the power spectrum calculated by the short-time Fourier transform at frequency f and time t. Thus, it may be possible to visually show changes in brain activity using the calculation of event-related spectrum perturbation (ERSP), which was calculated using time-frequency analysis in EEGLAB after ICA. The frequency power change (from the baseline) was chosen as the feature value for the following classification analysis. The frequency power during the baseline is:

$$\mu_B = \frac{1}{m} \sum_{t \in Base} |F(f, t)|^2 \tag{2}$$

where $Base$ is the time interval of the baseline and m is number of time samples at the baseline. The ERSP is defined as:

$$ERSP(f, t) = 10 \log_{10} \left(\frac{ERS(f, t) - \mu_B}{\mu_B} \right). \tag{3}$$

Bhattacharyya Distance

The Bhattacharyya distance is a method for selecting high-separability feature values for each class [6, 8]. Using this distance calculation, cost can be reduced by decreasing the amount of data. A larger Bhattacharyya distance calculated between feature values is considered to be more effective for classification. Thus, we calculated the Bhattacharyya distance between each target size for the same channel condition, frequency band (δ:1–4 [Hz], θ:4–8 [Hz], α:8–14 [Hz], β:14–20 [Hz], γ_low:20–30 [Hz], γ_high:30–50 [Hz]), and time (0–500 [ms], 500–1000 [ms]: target display time is 0 [ms]). We used 10 conditions (combinations of time, frequency band, and channel) with the largest Bhattacharyya distances for calculation ERSP and classification.

Classification Method

The same number of training and test data were created for 4-fold cross-validation. This was accomplished through random averaging using the feature value of 70% of training and test data in each class with bootstrapping. In order to classify multi-classes with the created data set, we used a relevance vector machine (RVM) [9], which is a type of machine learning that uses a kernel method. RVM is a widely used method in regression, classification and pattern recognition, etc. In comparison with a support vector machine (SVM), an RVM has the advantage of determining weight parameters using the automatic determination of hyperparameters. Thus, consistent extension to multi-class classification is possible as the output is denoted by membership probability to each class. Therefore, it is possible to create a multi-class classifier that is more consistent than binary classifiers (such as SVMs) by combining the one-vs-rest classification results of the repeated number of classes using the softmax function. Thus, we tested whether information of movement difficulty is included in brain activity before an actual movement in single-trial by measuring classification accuracy. In addition, the classification accuracy of averaged test data was also evaluated.

3 Results

3.1 Bhattacharyya Distance

Figure 4 shows a mapping of the Bhattacharyya distance for each participant.

Legends I-V show the results for each participant.

The large Bhattacharyya distance was confirmed in the left motor cortex in 500–1000 ms for all participants; however, it could not be confirmed as the common trend across all participants in the frequency band.

RVM Classification Results

The classification results of each target size (three movement difficulty levels) for both the averaged and single-trial test data of each participant using RVM are shown in Fig. 5.

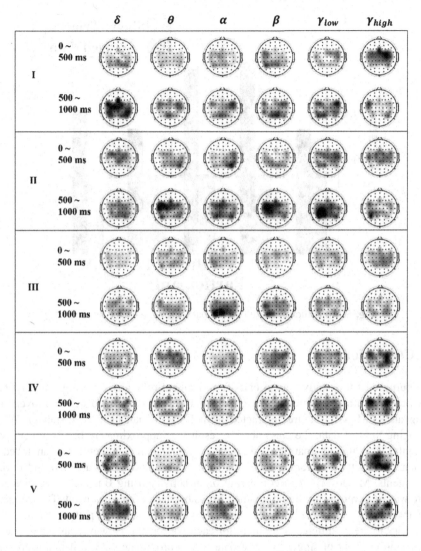

Fig. 4. Bhattacharyya distance map for all participants.

When averaging the test data, the classification accuracies of the single-trial data were approximately at the chance level (33%), which suggests that it is difficult to classify difficulty levels. However, the results for the averaged trial data were 53–82% (53.9%, 82.3%, 79.6%, 53.1% and 61.1% for I, II, III, IV and V, respectively); the mean value of all participants was 66.0%.

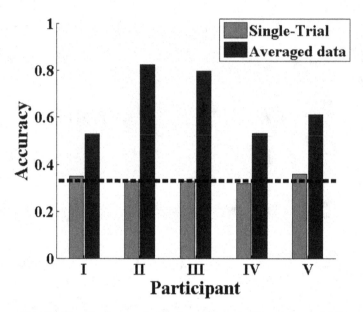

Fig. 5. Classification result for all participants. (Chance level: 0.33)

4 Conclusion

The purpose of this study was to classify movement difficulty by using brain activity that precedes arm movements. We classified EEG data collected before movements using an RVM with features extracted from ERSP (Bhattacharyya distance). When averaging the test data, the classification result was approximately 66.0% for all participants. This suggests that detailed information on planned movements is included in human brain activity before movements (other than muscle contractions) and during movements. Moreover, when comparing brain activity using Bhattacharyya distance, there was no common trend in frequency bands between each participant. This suggests that there may be large individual differences in brain activity patterns. Thus, it is necessary to create different classifiers for each participant. However, as the averaged classification result of single-trial test data was approximately 33%, it is not possible for movement difficulty information to be classifiable in single-trial EEG. We believe that this owes to noise contaminating EEG signals. Hence, it is necessary to produce improvements in classification accuracy by using different feature value selection methods to remove noise. In future work, it will be necessary to discuss trends in brain activity observed in measurements acquired from greater numbers of participants and to improve classification accuracy of movement difficulty in single-trial data. In addition, it is necessary to confirm the existence of similar brain activity when participants perform more complex movements.

Acknowledgments. This research was partially supported by the JSPS KAKENHI (15K12597 and 18K19807), the Tateisi Science and Technology Foundation, and the KDDI Foundation.

References

1. Fitts, P.M.: The information capacity of the human motor system in controlling the amplitude of movement. J. Exp. Psychol. **121**(3), 262–269 (1992)
2. Gribble, P.L., Mullin, L.I., Cothros, N., Mattar, A.: Role of cocontraction in arm movement accuracy. J. Neurophysiol. **89**, 2396–2405 (2003)
3. Kourtis, D., Sebanz, N., Knoblich, N.: EEG correlates of Fitts's law during preparation for action. Psychol. Res. **76**(4), 514–524 (2012)
4. Gundel, A., Wilson, G.F.: Topographical changes in the ongoing EEG related to the difficulty of mental tasks. Brain Topogr. **5**(1), 17–25 (1992)
5. Delorme, A., Makeig, A.: EEGLAB: an open source toolbox for analysis of single-trial EEG dynamics including independent component analysis. J. Neurosci. Methods **134**(1), 9–21 (2004)
6. Morash, V., Bai, O., Furlani, S., Lin, P., Hallett, M.: Classifying EEG signals preceding right hand, left hand, tongue, and right foot movements and motor imageries. Clin. Neurophysiol. **119**(11), 2570–2578 (2008)
7. Chen, X., Bin, G., Daly, I., Gao, X.: Event-related desynchronization (ERD) in the alpha band during a hand mental rotation task. Neurosci. Lett. **541**, 238–242 (2013)
8. Bai, O., Lin, P., Vorbach, P., Li, J., Furlani, S., Hallett, M.: Exploration of computational methods for classification of movement intention during human voluntary movement from single trial EEG. Clin. Neurophysiol. **118**(12), 2637–2655 (2007)
9. Tipping, M.E.: Sparse Bayesian learning and the relevance vector machine. J. Mach. Learn. Res. **1**, 211–244 (2001)

A Deep Learning Assisted Gene Expression Programming Framework for Symbolic Regression Problems

Jinghui Zhong[✉], Yusen Lin, Chengyu Lu, and Zhixing Huang

Guangdong Provincial Key Lab of Computational Intelligence and Cyberspace
Information, School of Computer Science and Engineering,
South China University of Technology, Guangzhou, China
jinghuizhong@gmail.com

Abstract. Genetic programming is a powerful evolutionary algorithm that solves user-defined tasks through the evolution of computer programs. Selecting a proper set of function primitives is a fundamental and challenging operation in applying GP to real applications. Traditional manual design methods require a lot of domain knowledge and are not effective and convenient enough. To address this issue, this paper proposed an automatic function primitive identification mechanism. The key idea is to train a deep convolutional neural network to predict the probability of the existence of a function primitive in the target solution. During the evolution of GP, function primitives with higher probabilities are more likely to be selected to construct solutions. The proposed method is tested on nine benchmark problems and the experimental results have demonstrated the efficacy of the proposed method.

Keywords: Genetic Programming (GP) · Deep Learning (DL)
Convolutional Neural Network (CNN)

1 Introduction

Genetic Programming (GP) is a population-based meta-heuristic search algorithm that solves user-defined tasks through the evolution of computer programs [4,6,9]. Due to its good effectiveness in real applications, GP has attracted increasing attention recently. Various enhanced GP variants have been proposed, such as Cartesian Genetic Programming (CGP) [13], Semantic Genetic Programming (SGP) [3,7,14], Grammatical evolution (GE) [15], Gene Expression Programming (GEP) [5], Linear Genetic Programming (LGP) [2]. So far, GP and its variants have been applied to a range of real applications, including time series prediction, classification, combinatorial optimization, knowledge discovery and data mining [1,19–22].

One fundamental operation in applying GP to real applications is to select a proper set of function primitives to efficiently construct solutions (e.g., a mathematic formula). The function primitives (e.g., numerical functions such as power

© Springer Nature Switzerland AG 2018
L. Cheng et al. (Eds.): ICONIP 2018, LNCS 11307, pp. 530–541, 2018.
https://doi.org/10.1007/978-3-030-04239-4_48

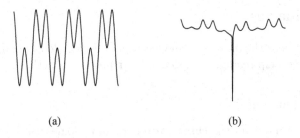

(a)   (b)

Fig. 1. Examples of time series generated by different functions.

function and logarithmic function) are used to link terminals (e.g., variables x and y) to generate new outputs. Traditionally, function primitives are selected manually by experts in an ad-hoc manner, relying on experts' domain knowledge and experience.

For example, the time series in Fig. 1(a) is very likely to involve trigonometric functions (e.g., sin) because of the periodicity reflected in the curve. However, for a more complicated case as shown in Fig. 1(b), it will become more difficult for experts to guess the functions in the target formula. To ensure high quality solutions can be found in cases similar to Fig. 1(b), the function set should be set large enough. However, this will significantly increase the search space, slow down the search efficiency and make GP get trapped into local optima easily. How to select proper function set to balance the effectiveness and efficiency of GP in real applications is still a challenging problem remained to be explored in the GP community. In this scenario, this paper proposes an enhanced GP (named DL-GEP), which utilizes a deep learning based method to select important functions for efficient solution construction. Deep Learning is a hot machine learning method which shows impressive performance in many applications such as image processing and pattern recognition [8,11]. However, little work has been reported in the literature on applying deep learning to improve GP. This paper makes an early attempt to fill in this gap.

Specifically, in the proposed DL-GEP, a function predictor is built based on a deep convolutional neural network. The goal of the function predictor is to predict the probability of the existence of a function primitive in the target formula. A multi-label classification problem is modelled and a large set of artificial functions are generated to train the deep convolutional neural networks, so that it can automatically identify important function primitives in the original function set. By assigning higher selecting probability to those primitives which are more important, the search efficiency of GP can be improved. Furthermore, a new mechanism is proposed to adaptively adjust the selection probabilities of function primitives based on the promising individuals in the current population. The proposed DL-GEP is tested on nine benchmark problems. The experiment results have demonstrated that the DL-GEP can offer very promising performance.

2   Preliminaries

This section briefly introduces some background knowledge and reviews the related works to help readers comprehend our proposed method.

2.1   Deep Learning

Deep Learning refers to the machine learning methodologies based on deep neural networks (DNNs) and recurrent neural networks, among others [11]. Generally, a neural network has the following components: input layer, hidden layer and output layer. The input layer receives raw input data and the output layer outputs the probability of different labels. A DNN is a neural network containing more than one hidden layer. In recent years, the DNN has been developing rapidly and applied to a wide range of applications such as classification and image processing [8,11].

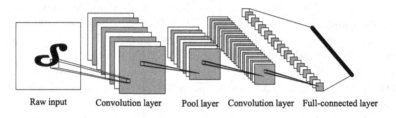

Raw input   Convolution layer   Pool layer   Convolution layer   Full-connected layer

Fig. 2. The schematic of CNN

The Convolutional Neural Network (CNN) is one of the most successful DNNs which is composed of convolution layer, pool layer and fully-connected layer, as shown in Fig. 2. The key idea of CNN is substituting the full connected topology between layers into a convolution layer to reduce the number of weights. The convolution layer maps a group of inputs from last layer into a single output and delivers this output to the activation function. Though the convolution layer filters the input into an output with smaller dimensions, the output still has relatively large dimensions which may make the trained model over-fitting easily. Thus, the pooling layer is introduced to cut down the dimension. The most common pooling layers are maximum pooling and average pooling. The maximum pooling takes the maximum output in a small group of neuron as the single output while the average pooling takes the average of outputs from the group of neurons. The introduction of pool layer guarantees the next layer to be insensitive to the small fluctuation from the previous layers. The full-connected layer is adopted to generate the final outputs. One promising characteristic of CNN is the sharing of weights, which can reduce the number of parameters, the complexity of the network, and can accelerate the training efficiency.

2.2   Deep Learning Meets GP

In the literature, various Evolutionary Algorithms (EAs) such as GP have been proposed to optimize the parameters and structures of Artificial Neural Networks (ANNs) [18]. For example, Maryam proposed to use Cartesian Genetic Programming (CGP) to evolve the structure of ANN [12]. Andrew developed a recurrent CGP (RCGP) to describe the topology of recurrent networks, which has been shown to be very effective in optimizing the structure of the recurrent ANN [17]. Recently, Lamos-Sweeney et al. [10] have proposed a GA-based method to optimize the structure of a DNN, using a similar idea of the Restricted Boltzmann Machine (RBM). The model uses GA to optimize the weights of the layer so that the outputs of each layer are as many as possible to be similar with the inputs. Although various efforts have been made in applying EAs to improve deep learning, little work has been reported on applying deep learning to improve EAs. This paper makes an early attempt to utilize deep learning to automatically identify important function primitives for GP, which can improve the search efficacy of GP.

3   Proposed Method

3.1   The General Framework

Figure 3 illustrates the general structure of the proposed framework, which consists of two modules: the CNN training module and the GP training module. The CNN training module is used to train a CNN to predict the weights of functions in the original function set. Functions with larger weight are more likely to be a component in the target solution. To train the CNN, a set of artificial functions with random function primitives is generated as training data. The training functions should be many enough to achieve high prediction accuracy. Once the well-tuned CNN model is obtained, it then can be used to predict the weights of function primitives for new regression problems.

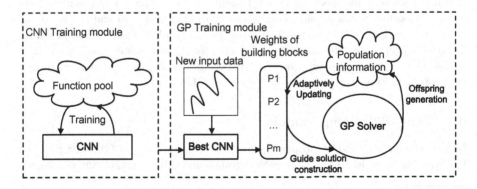

Fig. 3. The general structure of the proposed framework.

Meanwhile, the GP training model is used to learn an optimal formula to fit the given data. In the GP training module, a GP solver selects function primitives in the given function set to construct solutions based on the weights predicted by the well-tuned CNN. Functions with larger weights are more likely to be selected to construct solutions. Further, to improve the search efficiency, the weight vectors are updated adaptively by considering the frequencies of the function primitives in the current population. Suppose $\mathbf{P}_g$ is the current weight vector, and $\mathbf{Q}_g$ is the frequency of function primitives in the survival individuals, the weight vector is updated by

$$\mathbf{P}_{g+1} = \lambda \mathbf{P}_g + (1 - \lambda)\mathbf{Q}_g \tag{1}$$

where λ is the update rate. Usually, $\lambda \in [0, 1]$ is set to a relatively large value (e.g., $\lambda = 0.9$).

3.2   The CNN Training Module

The CNN Training Module is used to generate a weight vector to describe the importance of function primitives. Specifically, the inputs of the CNN training module are images of different categories. The images are the curves of function generated by randomly linking elementary functions such as sin, cos, exp and log. In this study, a GP variant is adopted to generate a large number of random solutions. Each solution represents a random function, and we can generate an image for the function by plotting the curve of the function. The labels of each random solution can be set automatically based on the valid elements in the solution. Noting that each image can have multiple labels if it contains multiple elementary function primitives. For example, a function $exp(sin(x))$ has two labels: one for exp and one for sin. Based on the above steps, we can convert the weight prediction problem to a multi-label classification problem.

Inception-v3 [16] is an excellent CNN-based tool to solve muti-label classification problem. Inception-v3 is applied to solve a larger variety of computer vision tasks and achieves great success by utilizing deeper and wider networks. It can recognize multi-label images and output each probability of the given categories. Besides, it has already been trained and published in the Internet[1]. In this study, we choose Inception-v3 to solve the multi-label classification problem. To apply Inception-v3 to the multi-label classification problem, the only task is to set up the training images and the responding labels as input for training a new top layer that can recognize other classes of images.

3.3   The GP Training Module

In this study, the SL-GP proposed in [22] is adopted as the GP solver in the GP training module. In SL-GEP, each chromosome is fixed-length and contains a

---

[1] The Inception-v3 can be downloaded from http://download.tensorflow.org/models/image/imagenet/inception-2015-12-05.tgz.

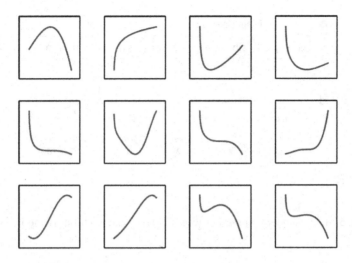

Fig. 4. Example images for training the CNN.

main program part and an automatically defined function (ADF) part. The main program part generates the final output, while the ADF part contains a number of subfunctions which are used as building blocks in the main program part. The main program and ADF parts can be respectively decoded to expression trees by using the breadth-first traversal scheme [6]. Suppose a typical chromosome is given as:

$$[G, exp, G, sin, y, x, x, y, y, +, sin, -, b, b, a, a] \tag{2}$$

where G is an ADF (i.e., sub-function), x and y are terminals, and a and b are the input arguments of ADFs. Suppose that the length of the main program and the ADF parts are 9 and 7 respectively. As illustrated in Fig. 5, the given chromosome can be decoded as:

$$
\begin{aligned}
\Gamma &= G(exp(sin(x)), G(y, x)) \\
&= sin(sin(b) + (b - a) + sin(b) + (b - a) - exp(sin(x)))
\end{aligned}
\tag{3}
$$

Based on the above chromosome representation, the proposed DL-GEP performs the following steps to find the optimal solutions.

Step 1 – Initialization. In this step, a random population of chromosomes are generated. Each chromosome is a vector of symbols, i.e.,

$$X_i = [x_{i,1}, x_{i,2}, \ldots, x_{i,n}] \tag{4}$$

where n is the length of chromosomes. To set the value of $x_{i,j}$, a primitive type is randomly selected among {function, sub-function, terminal, input argument}. Then a random value of the selected feasible type is assigned to $x_{i,j}$. If the

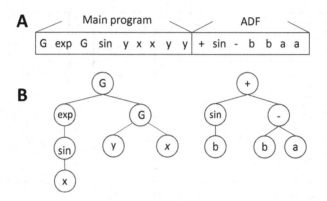

Fig. 5. The decoded expression trees of the example chromosome in (3).

primitive type is function, then the value is selected using a route-wheel selection strategy based on the weights **P** generated by the best CNN. Functions with larger weights have larger selection probabilities.

Step 2 – Reproduction. In this step, the selection probabilities of functions are updated at first using Eq. (1). Then, the genetic operators proposed in SL-GEP [22] (i.e., the mutation, crossover and selection) are performed iteratively to update the population. Specifically, In the mutation, the extended "DE\current-to-best\1" mutation operation is performed on each target vector to generate a mutant vector:

$$Y_i = X_i + F \cdot (X_{best} - X_i) + \beta \cdot (X_{r1} - X_{r2}). \tag{5}$$

where X_{best} is the best-so-far solution and X_{r1} and X_{r2} are two random individuals. In [22], the numerical operators in Eq. (5) (i.e., $\cdot$, $+$ and $-$) are redefined to evolve solutions in a discrete search space. As shown in Fig. 5, each solution is represented by tree structures that comprise of two types of nodes: function and terminal. A function node has one or multiple children (e.g., $+$ and sin), while a terminal node is a leaf node without any child (e.g., variables and constants). Based on the redefined operators, the result of Eq. (5) is equivalent to mutating element of Y_i with a probability vector. The reader is referred to [22] for the implementation details of the mutation operation of SL-GEP.

After the mutation operation, the crossover generates a trial vector for each target vector:

$$u_{i,j} = \begin{cases} y_{i,j}, \text{if} \quad rand(0,1) < CR \quad \text{or} \quad j = k \\ x_{i,j}, \text{otherwise} \end{cases} \tag{6}$$

where U_i is the trial vector, CR is the crossover rate, k is a random integer within $[1, n]$, and $x_{i,j}$, $y_{i,j}$ and $u_{i,j}$ are the jth variables of X_i, Y_i and U_i respectively.

The selection performed following the crossover, the selection operator is performed to select the better one among each pair of target and trial vectors:

$$X_i = \begin{cases} U_i, \text{if} & f(U_i) \le f(X_i) \\ X_i, \text{otherwise} \end{cases} \tag{7}$$

where $f(X)$ is the fitness evaluation function.

The reproduction step is performed repetitively until finding the optimal solution or the termination condition is met (e.g., the maximum number of generations is reached).

4   Experiment Studies

4.1   Results of CNN Training Module

In this subsection, we evaluate the performance of the CNN training module. First of all, we create 100,000 training images by generating 100,000 random functions (i.e., chromosomes), which are composed of seven elementary functions: $\{+, -, *, sin, cos, exp, log\}$. We also select 20,000 testing images from the training images. We focused on predicting the weights of the last four functions (i.e., sin, cos, exp, log). That is, each image is assigned a weight vector with four dimension, i.e., $P = [p_1, p_2, p_3, p_4]$. If log is in a function, p_4 of the corresponding image will set to 1. Otherwise, p_4 is set to 0. Since the Inception-v3 has been trained except the top layer, the only task is to place the function images and the corresponding labels. This experiment iterates for 10,000 times. The final training accuracy is 87.0% and the final test accuracy is 88.1%. The testing results indicate that the trained CNN is able to offer promising prediction accuracy.

4.2   Results of GP Training Module

In this subsection, the GP training module is applied to solve nine benchmark problems, whose target formula are listed in Table 1. For all problems, the function set of all compared algorithms is set to be $\{+, -, \times, \div, sin, cos, exp, ln(|x|)\}$.

First, we investigate the effectiveness of the CNN training module. Table 2 lists the weights of the functions for each problem. It can be observed that the prediction results are correct to some extend. Taking F_5 for example, the CNN classifies F_5 to two classes, sin and cos and successfully removes two useless functions, exp and log. Similar results can be found on other problems such as F_2, F_3, F_7, and F_8. However, the CNN classifies F_9 into wrong classes. This may be because the given regression data is not sufficient enough to capture the landscape feature of F_9.

Next, we compare the proposed DL-GEP with SL-GEP to demonstrate its effectiveness. The common settings of DL-GEP and SL-GEP are set the same as [22]. The DL-GEP contains one additional parameter λ, which is set to 0.95. As in [22], the 10-fold cross validation approach is adopted to test the effectiveness of the compared algorithms. Each algorithm will terminate when the number

Table 1. Nine symbolic regression benchmarks for testing.

P	Objective function	Data set
F_1	$exp(sin(x))$	$U[-1, 1, 200]$
F_2	$cos(x)ln(x)$	$U[-1, 1, 200]$
F_3	$sin(x) + ln(x) - x$	$U[-1, 1, 200]$
F_4	$exp(x)sin(x^2)$	$U[-1, 1, 200]$
F_5	$cos(x^2)sin(x)$	$U[-1, 1, 200]$
F_6	$cos(x) + cos(x^2)$	$U[-1, 1, 200]$
F_7	$sin(x^2)cos(x) - 1$	$U[-1, 1, 200]$
F_8	$sin(x) + sin(x + x^2)$	$U[-1, 1, 200]$
F_9	$ln(x + 1) + ln(x^2 + 1)$	$U[0, 2, 200]$

$U[a, b, c]$ represents c uniform random samples from a to b.

Table 2. Predicted weights of the elementary functions on the nine problems.

P	sin	cos	exp	log
F_1	0.84	0.72	0.67	0.14
F_2	0.71	0.77	0.46	0.74
F_3	0.72	0.67	0.41	0.79
F_4	0.86	0.82	0.53	0.31
F_5	0.90	0.94	0.48	0.12
F_6	0.50	0.84	0.42	0.24
F_7	0.81	0.72	0.40	0.23
F_8	0.94	0.87	0.58	0.22
F_9	0.86	0.87	0.36	0.14

of evaluations reaches a maximum of 1,000,000. When an algorithm converges to a solution with fitting error smaller than 10^{-4}, a successful search convergence or perfect hit is achieved. In the empirical studies, three performance metrics adopted in [22] are used for comparison analysis. The first is the RMSE which is the average testing accuracy of the 100 independent runs based on the 10-fold cross validation. The second is the success rate of achieving perfect hits (denoted as Suc), which is calculated by:

$$Suc = \frac{C_s}{C} \cdot 100\% \tag{8}$$

where C is the number of independent runs and C_s is the number of successful runs achieving a perfect hit. The third metric is the number of fitness evaluations required to achieve a perfect hit (denoted as Run Time, RT). This metric measures the convergence speed of an algorithm. For each problem, a Wilcoxon signed-rank test detects significant differences on the RMSE and the RT.

Table 3. Comparsion results of SL-GEP and DL-GEP.

Problem	SL-GEP			DL-GEP		
	Suc	RMSE	RT	Suc	RMSE	RT
F_1	100	0 $\approx$	549 $-$	100	0	383
F_2	1	0.1240 $+$	992,197 $-$	5	0.1257	977,152
F_3	40	0.0263 $-$	784,942 $-$	53	0.0177	704,517
F_4	100	0 $\approx$	12,479 $\approx$	100	0	10,526
F_5	100	0 $\approx$	33,444 $-$	100	0	15,065
F_6	100	0 $\approx$	10,816 $-$	100	0	4,530
F_7	87	0.0004 $-$	443,884 $-$	95	1.76E-04	338,833
F_8	100	0 $\approx$	14,582 $+$	100	0	17,066
F_9	22	0.0074 $+$	855,736 $+$	21	0.0076	944,694

Symbol $-$, $\approx$, $+$ represents SL-GEP is relatively significantly worse than, similar to and better than DL-GEP according to Wilcoxon signed-rank test at $\alpha = 0.05$.

Table 3 shows the comparison results of DL-GEP and SL-GEP on the nine problems. It can be observed that DL-GEP reported higher Suc on F_2, F_3 and F_7 than SL-GEP. As for RT, DL-GEP performs better than SL-GEP on six out of the nine problems,i.e., F_1, F_2, F_3, F_5, F_6 and F_7. The above results indicate that by using the selection probabilities predicted by the CNN training model, the DL-GEP can converge faster than SL-GEP. To further validate this, we take F_7 for example and plot the evolution curve of the best fitness offered by DL-GEP and SL-GEP on this problem. The evolution curve in Fig. 6 shows that the DL-GEP can find better solution at the beginning, and DL-GEP can converges faster

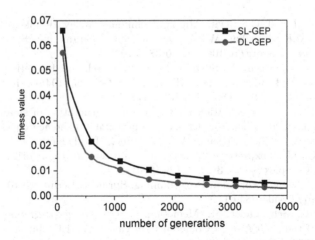

Fig. 6. Evolution of the best fitness values derived from SL-GEP and DL-GEP on F_7

during the evolution process. The above results demonstrate that the proposed deep learning based mechanism is effective to improve the search efficiency and accuracy.

5   Conclusions

In this paper, we proposed a deep learning assisted GP framework for symbolic regression. The key idea is to train a deep convolutional neural network to predict the weights of function primitives for solution construction. The predicted weights are then used to guide the solution construction of GP during the evolution process. The experiments on nine benchmark problems show that the proposed framework is effective to improve the search efficiency.

In this study, the CNN training module are trained using images of 2D function curves. Thus, the proposed framework can only work on one-dimensional symbolic regression problem (e.g., time series prediction problem). As for future work, we plan to extend the proposed method for complicated symbolic regression problems with more input variables. The second future research work is to apply the proposed framework to solve real problems. In addition, developing new machine learning techniques to assigned weights to both terminals and functions is another promising research direction.

Acknowledgments. This work was supported by the National Natural Science Foundation of China (Grant No. 61602181), the Fundamental Research Funds for the Central Universities (Grant No. 2017ZD053), the Program for Guangdong Introducing Innovative and Enterpreneurial Teams (Grant No. 2017ZT07X183), and the Guangzhou Science and Technology Plan Project (Grant No. 201804010245).

References

1. Berry, M.J., Linoff, G.S.: Data Mining Techniques. Wiley, Hoboken (2009)
2. Brameier, M.F., Banzhaf, W.: Linear Genetic Programming. Springer, Boston (2007). https://doi.org/10.1007/978-0-387-31030-5
3. Castelli, M., Vanneschi, L., Silva, S.: Semantic search-based genetic programming and the effect of intron deletion. IEEE Trans. Cybern. **404**(1), 103–113 (2014). https://doi.org/10.1109/TSMCC.2013.2247754
4. Cramer, N.L.: A representation for the adaptive generation of simple sequential programs. In: Proceedings of the 1st International Conference on Genetic Algorithms, pp. 183–187. L. Erlbaum Associates Inc., Hillsdale (1985)
5. Ferreira, C.: Gene expression programming: a new adaptive algorithm for solving problems. Complex Syst. **13**(2), 8–129 (2001)
6. Ferreira, C.: Gene Expression Programming. Springer, Berlin (2006). https://doi.org/10.1007/3-540-32849-1
7. Ffrancon, R., Schoenauer, M.: Memetic semantic genetic programming. In: Proceedings of the 2015 Annual Conference on Genetic and Evolutionary Computation, pp. 1023–1030. ACM (2015)

8. He, K., Zhang, X., Ren, S., Sun, J.: Deep residual learning for image recognition. In: Proceedings of the IEEE Conference on Computer Vision and Pattern Recognition, pp. 770–778 (2016)

9. Koza, J.R.: Genetic Programming: On the Programming of Computers by Means of Natural Selection, vol. 1. MIT press, Cambridge (1992)

10. Lamos-Sweeney, J.D.: Deep learning using genetic algorithms. Dissertations and Theses - Gradworks (2012)

11. LeCun, Y., Bengio, Y., Hinton, G.: Deep learning. Nature **521**(7553), 436 (2015)

12. Mahsal Khan, M., Khan, G.M., Miller, J.: Evolution of optimal ANNs for nonlinear control problems using cartesian genetic programming, vol. 1, pp. 339–346 (2010)

13. Miller, J.F., Thomson, P.: Cartesian genetic programming. In: Poli, R., Banzhaf, W., Langdon, W.B., Miller, J., Nordin, P., Fogarty, T.C. (eds.) EuroGP 2000. LNCS, vol. 1802, pp. 121–132. Springer, Heidelberg (2000). https://doi.org/10.1007/978-3-540-46239-2_9

14. Moraglio, A., Krawiec, K., Johnson, C.G.: Geometric semantic genetic programming. In: Coello, C.A.C., Cutello, V., Deb, K., Forrest, S., Nicosia, G., Pavone, M. (eds.) PPSN 2012. LNCS, vol. 7491, pp. 21–31. Springer, Heidelberg (2012). https://doi.org/10.1007/978-3-642-32937-1_3

15. O'Neill, M., Ryan, C.: Grammatical evolution. IEEE Trans. Evol. Comput. **5**(4), 349–358 (2001)

16. Szegedy, C., Vanhoucke, V., Ioffe, S., Shlens, J., Wojna, Z.: Rethinking the inception architecture for computer vision. In: Proceedings of the IEEE Conference on Computer Vision and Pattern Recognition, pp. 2818–2826 (2016)

17. Turner, A.J., Miller, J.F.: Recurrent cartesian genetic programming of artificial neural networks. Genet. Program. Evolvable Mach. **18**(2), 185–212 (2017)

18. Yao, X.: Evolving artificial neural networks. Proc. IEEE **87**(9), 1423–1447 (1999)

19. Zhong, J., Feng, L., Cai, W., Ong, Y.: Multifactorial genetic programming for symbolic regression problems. IEEE Trans. Syst. Man Cybern. Syst. (2018, in press). https://doi.org/10.1109/TSMC.2018.2853719

20. Zhong, J., Cai, W., Lees, M., Luo, L.: Automatic model construction for the behavior of human crowds. Appl. Soft Comput. **56**, 368–378 (2017)

21. Zhong, J., Feng, L., Ong, Y.S.: Gene expression programming: a survey. IEEE Comput. Intell. Mag. **12**(3), 54–72 (2017)

22. Zhong, J., Ong, Y.S., Cai, W.: Self-learning gene expression programming. IEEE Trans. Evol. Comput. **20**(1), 65–80 (2016)

Automated Tongue Segmentation in Chinese Medicine Based on Deep Learning

Yushan Xue[1], Xiaoqiang Li[1,2(✉)], Pin Wu[1], Jide Li[1], Lu Wang[1], and Weiqin Tong[1,2]

[1] School of Computer Engineering and Science,
Shanghai University, Shanghai, China
{xqli,wupin}@shu.edu.cn
[2] Shanghai Institute for Advanced Communication and Data Science,
Shanghai University, Shanghai, China

Abstract. During the process of tongue diagnosis automation, the segmentation of tongue body from the original image is an important part. Tongue body is a target with irregular edge, and the color of local area is similar to that of lips, so it is difficult to achieve accurate segmentation with traditional segmentation methods. In this paper, we present a tongue image segmentation method based on deep learning. We first apply semantic segmentation to tongue segmentation that use fully convolutional networks to automatically segment tongue body based on preserving tongue shape semantic information. Moreover, we try to combine it with traditional algorithm to optimize the results. The experimental results show that the semantic segmentation method based on neural network is superior to the traditional algorithm in accuracy and efficiency. In addition, comparing with traditional algorithms, the method does not require manual label, which greatly reduces the workload.

Keywords: TCM tongue diagnosis · Convolution neural network
Semantic segmentation

1 Introduction

In traditional Chinese medicine diagnosis and treatment, observation is one of the important methods. As the only one organ that can stick out of the body, tongue is a key to observing in traditional Chinese medicine diagnosis [1]. Doctors can judge the patient's physical condition through some features on the surface of tongue. Traditional tongue diagnosis requires face-to-face communication and diagnosis, and the diagnosis process relies entirely on the experience of doctors [2]. In addition, it needs a doctor to observe each patient's tongue in person and obtain some features of tongue, which will inevitably cause some wrong judgments because of fatigue.

With the application of computers, it is common that some medical diagnosis can be done by computers. In earlier studies, doctors analyzed patient's tongue images by a computer [3]. In recent years, an automatic tongue diagnosis and analysis system has been developed gradually [4, 5]. Tongue image is analyzed by the computer and the analysis results are fed back to doctors. When collecting tongue images, images will

L. Cheng et al. (Eds.): ICONIP 2018, LNCS 11307, pp. 542–553, 2018.
https://doi.org/10.1007/978-3-030-04239-4_49

contain face, lips and teeth. In order to guarantee accuracy of analysis results and eliminate interference of irrelevant factors, automated analysis system will remove complex background information and segment tongue body from original image. It ensures that tongue images analyzed by the system have a pure color background (usually black or white) except for the part of tongue body.

Because tongue body is a kind of target with irregular edge, and the color of local area is similar to that of lips, so it is difficult to achieve accurate segmentation with traditional segmentation methods. In this paper, we try to use fully convolutional networks to automatically segment tongue body based on preserving tongue shape semantic information. And, it can even separate areas where the lips and tongue are similar in color, but the traditional algorithm couldn't do that.

The rest of the paper is organized as follows. Section 2 introduces tongue image segmentation methods, matting methods and semantic segmentation methods. Section 3 introduces semantic segmentation based on deep learning and its application to tongue image segmentation. In Sect. 4, we use semantic segmentation method to complete tongue image segmentation experiment and Sect. 5 concludes the paper.

2   Related Work

In image processing, there are a lot of mature segmentation algorithms. Some of them can successfully apply to tongue image segmentation. In this part, we introduce some tongue image segmentation algorithms, matting methods and semantic image segmentation methods.

2.1   Tongue Segmentation Methods

Ning [6] presents a region merging-based automatic tongue segmentation method. The basic steps of the method are as follows. First, gradient vector flow is modified as a scalar diffusion equation to diffuse the tongue image while preserving the edge structures of tongue body. Then the diffused tongue image is segmented into many small regions by using the watershed algorithm. Third, the maximal similarity-based region merging is used to extract the tongue body area under the control of tongue marker. Finally, the snake algorithm is used to refine the region merging result by setting the extracted tongue contour as the initial curve.

[7] proposes a fully automated active contour initial method that utilizes prior knowledge of the tongue shape and its location in tongue images. Then colorspace information is introduced to control curve evolution. Combining the geometrical Snake model with the parameterized GVFSnack model, a novel approach for automatic tongue segmentation: C^2G^2FSnake (color control-geometric & gradient flow Snack) is proposed. This method increases the curve velocity but decreases the complexity.

In [8], Pang proposes an original technique that is based on a combination of a bi-elliptical deformable template(BEDT) and an active contour model, namely the bi-elliptical deformable contour(BEDC). The BEDT captures gross shape features by using the steepest decent method on its energy function in the parameter space. The BEDC is derived from the BEDT by substituting template forces for classical internal forces, and can deform to fit local details. Their algorithm features fully automatic interpretation of tongue images and a consistent combination of global and local controls via the template force.

2.2 Image Matting

The goal of image matting is to estimate fractional opacity of foreground layer from an image and recover the colors of the foreground/background layers respectively. Typically, given an image I, assumes that I is formed through linearly blending foreground image F and background image B with coefficients α:

$$I = \alpha F + B(1 - \alpha) \tag{1}$$

α ranges between 0 and 1. Image matting refers to the process of image foreground and background segmentation in form of calculating opacity mask α.

Learning based matting (LBM) [9] is a precise and easy-to-use algorithm for solving matting problems. It can handle difficult problems such as translucent foreground and complicated border which make it perfect candidate for current widely used grab cut serious. LBM algorithm consists of two stages: learning stage and solving stage. The first stage solves a serious of small dense matrix equation. The second stage constructs a huge sparse matrix equation and then solves it. However, both of the two steps require heavy computation which makes LBM unsuitable for real-time interactive matting scene.

Trimap skeleton based (TSB) algorithm proposed in [10] is based on LBM and uses a scheme to accelerate LBM and implement it on modern GPU in parallel, which involves learning stage and solving stage. Firstly, the author presents GPU-based method to accelerate the pixel-wise learning stage. Then, trimap skeleton based algorithm is proposed to divide the image into blocks and process blocks in parallel to speed up the solving stage.

2.3 Semantic Segmentation

Yu [11] tried to use dilated convolution for image semantic segmentation. The advantage of dilated convolution is that it increases the receptive field without pooling, which not only can increase accuracy by retaining internal data structure, but also reduce some weights and computational capacity.

Many image semantic segmentation methods use Conditional random field (CRF) [12] as post-processing operations to optimize the semantic segmentation results. It takes the relationship between pixels into consider in classification process. CRF takes each pixel in image as node and the relationship between pixels as edge, which constitutes a condition random field. For each pixel i, there is a category label x_i and a corresponding observation value y_i. We can deduce x_i by y_i.

Deeplab (v1&v2) [13, 14] makes a comprehensive use of Deep Convolution Neural Network (DCNN), atrous convolution (also known as dilated convolution) and CRF. It makes three contributions. First, it uses atrous convolution. Atrous convolution explicitly controls the resolution to improve segmentation results (as Fig. 1). Second, atrous spatial pyramid pooling (ASPP), where four parallel atrous convolution with different atrous rates are applied on top of the feature map, can take both large objects and small objects in images into account. Third, it uses CRF as a separate post-processing stage to improve segmentation results.

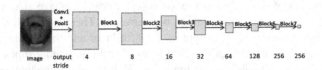

(a) Without atrous convolution

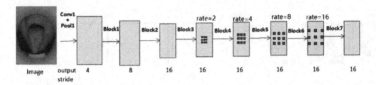

(b) With atrous convolution. Atrous convolution with rate>1 is applied after block3 when output_stride=16

Fig. 1. Cascaded modules with and without atrous convolution.

3 Our Method

Recently, a series of semantic segmentation methods based on Convolution Neural Network (CNN), represented by Fully Convolutional Networks (FCN) [15], have been proposed one after another, repeatedly refreshing image semantic segmentation accuracy. In this paper, we try to use convolutional neural networks to segment tongue body from original tongue image automatically.

For tongue image, segmenting tongue body from original image can be regarded as marking all pixels belonging to tongue in image as tongue and all other pixels as background. Therefore, tongue image can be automatically segmented by image semantic segmentation method.

In training stage of neural network, we input original tongue images and artificially marked mask images into network for training, and then get the network model of tongue image segmentation. In a new tongue image segmentation, we input image directly and use the generated model to segment tongue image. There is no need for manual intervention during the whole process, and the error caused by human factors can be avoided, while the workload is reduced.

3.1 FCN

FCN [15] is the pioneering work of deep learning in image semantic segmentation, whose idea that performs a pixel-level end-to-end semantic segmentation directly is very intuitive. It is based on mainstream CNN model.

All layers of FCN are convolution layers. After multiple convolution and pooling, image is getting smaller and smaller and the resolution is getting lower and lower. In order to classify image on pixel-level, FCN adopts 32 times upsampling to the output image after the fifth convolution layer to restore image to the size of original image. Upsampling is achieved by deconvolution. However, the result obtained by deconvoluting the output of the fifth convolution layer to original image size, which is not accurate enough and some details cannot be recovered. There is a method to improve the segmentation precision. The output of the fifth layer and the fourth layer is concatenated with the feature map of the third layer after 4 times and 2 times upsampling respectively, and then 8 times upsampling is carried out to obtain the final segmentation result, and this structure is called FCN-8s (Fig. 2).

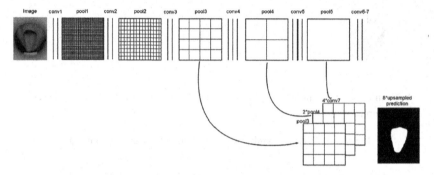

Fig. 2. The network structure of FCN-8s that we used.

3.2 Deeplabv3

The main problems of FCN-8s caused by pooling are the loss of internal data structures and spatial hierarchical information. Pooling increases the receptive field but reduces the resolution in image segmentation. Although image size is restored by upsampling, it is still difficult to eliminate the effects of pooling, because upsampling does not completely recover the lost information.

In Deeplabv3 [16], authors revisit atrous convolution [14] for image semantic segmentation. The advantage of atrous convolution is that it increases the receptive field without pooling, which not only can increase accuracy by retaining internal data structure, but also reduce some weights and computational capacity. In order to handle the problem of segmenting objects at multiple scales, they design modules which employ atrous convolution in parallel (as Fig. 3) to capture multi-scale context by adopting multiple atrous rates. Furthermore, they propose to augment their previously proposed ASPP model [14] with image-level features encoding global context and further boost performance (as Fig. 3). They also propose to improve ASPP by adding one 1x1 convolution layer and three 3x3 atrous convolutions with different rates. And batch normalization is also added after each ASPP to improve segmentation results. Meanwhile, multi-grid atrous convolution is used in both cascade and parallel structures. They adopt different atrous rates within block4 to blocks7. The performance of the model of Deeplabv3 is better than that of FCN-8s.

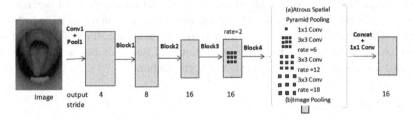

Fig. 3. Parallel modules with atrous convolution (ASPP), augmented with image-level features.

3.3 Deeplabv3-LBM

Shen [17] proposed an automatic image matting method, which combine deep learning method and a traditional method. The method does not need user interaction, which was however essential in most approaches just like we used in LBM described in 2.2. In the paper, Convolution Neural Network (CNN) was used to gain trimaps of images and segmented images based on trimaps by Closed-Form matting. It considers not only image semantic segmentation but also pixel-level image matte optimization. In this paper, we tried to use Deeplabv3 to segment tongue images based on trained model, which could avoid user interaction. From the experiment about LBM, we found that fine trimap can obtain good segmentation. Inspired by the idea proposed in [17], we tried to use LBM to optimize the segmentation result of tongue image based on Deeplabv3. In our study, we used the trained model proposed in 3.2 to get tongue segmentations and generate trimaps. And then use LBM to get the final segmentation of tongue images by inputting original tongue image and trimaps created by Deeplabv3 (Fig. 4).

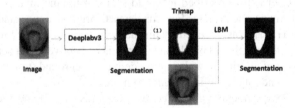

Fig. 4. The structure of Deeplabv3-LBM. (1) Manually changes the thickness of the contour line to 10 pixels.

4   Experiment

In this section, we use LBM, FCN-8s, Deeplabv3 and Deeplabv3-LBM to segment tongue images. And we compared the four methods from two aspects of precision and speed respectively. In our experiment, we use Mean Intersection over Union (mIoU) to evaluate the experimental results, which is the standard measure of semantic segmentation. It calculates the ratio of intersection and union of two sets. In semantic segmentation, these two sets are mask and segmentation.

4.1   Dataset

The dataset we used consists of tongue images and the mask corresponding to each image. The tongue images were collected from hospitals and communities in multiple batches, and the collection equipment was also customized. The original image size is 367 * 489. Each mask corresponding to each image is based on original image segmentation. First, the tongue image was manually segmented (as Fig. 5(b)), and then the segmented tongue image was used to generate a single channel image with only two pixels values of 0 and 255.

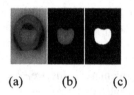

(a)   (b)   (c)

Fig. 5. (a) Original image. (b) Tongue body is manually segmented from (a). (c) Mask corresponding to (a) is generated from (b).

4.2   Experiment Based on LBM

In this part, we used LBM described in 2.3 to segment tongue images. The input of LBM is original image and trimap. For our dataset, original image and trimap were shown in Fig. 6. Trimap was obtained by manually changing contour line thickness of mask (Fig. 5(c)) to 20 pixels, 50 pixels and 70 pixels respectively and setting the color of contour line to grey. In trimap, white area is definite foreground (tongue body), black is definite background, and grey is the uncertain area.

We conducted experiments with different trimaps (as Fig. 7) and found that good segmentation results can be obtained only when the trimap is very fine, which means that it requires a lot of labor. And if the trimap is not subtle enough, the running time of the algorithm will increase. In the experiment, we found that LBM couldn't performer better on the image which contain teeth or the tongue was similar in color to the lip.

Finally, we tested the performance of LBM with 111 original tongue images and corresponding 20-pixels-trimaps. The mIOU was calculated with 111 original tongue images and segmentation by LBM. The result was 89.3%.

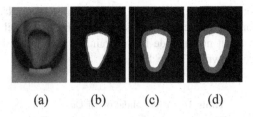

(a)   (b)   (c)   (d)

Fig. 6. (a) Original image. (b) 20-pixels-trimap. (c) 50-pixels-trimap. (d) 70-pixels-trimap.

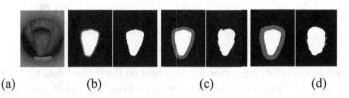

(a)   (b)   (c)   (d)

Fig. 7. (a) Original image. (b) 20-pixels-trimap and segmentation. (c) 50-pixels-trimap and segmentation. (d) 70-pixels-trimap and segmentation.

4.3   Experiment Based on FCN-8s

We finetuned the model weights of the Imagenet-pretrained [18] VGG-16 to adapt them to the tongue image segmentation. We replaced the 1000-way Imagenet classifier in the last layer with a classifier having as many targets as the number of classes of our task (two classes, tongue body and background).

We used 441 tongue images and corresponding masks as training set to train FCN-8s. During the training stage, we input original images and corresponding mask with a single channel of 0 and 255 pixels (as Fig. 5(c)). In testing stage, we evaluated the network model with 111 tongue images and corresponding masks. The evaluation criteria is mIoU and the result was 90.48%. Figure 8 are examples segmented by FCN-8s.

Fig. 8. Original image and segmentation by FCN-8s.

4.4   Experiment Based on Deeplabv3

We adopt the ImageNet-pretrained [18] ResNet [19] to the tongue segmentation by applying atrous convolution to extract dense features. Our implementation was built on TensorFlow [20]. We used the parallel structure of Deeplabv3 (as Fig. 3) and set up several groups of experiments using different numbers of data: 496 images in training set and 56 images in verification set (ratio: 9:1); 441 images in training set and 111 images in verification set (ratio: 8:2); 386 images in training set and 166 images in verification set (ratio: 7:3). Training set was used to train network by inputting original images and corresponding one-channel mask with 0 and 255 pixel values. Validation set consist of images that not include in the training set. The evaluation standard is mIoU and results are shown in Table 1.

Table 1. Cross validation of Deeplabv3

Number of training set	Number of validation set	mIoU
496	56	93.2%
441	111	93.7%
386	166	92%

As shown in the table, the performance is best on the second data set (441 images in training set and 111 images in verification set) when use the trained Deeplabv3 to segment tongue images (Fig. 9).

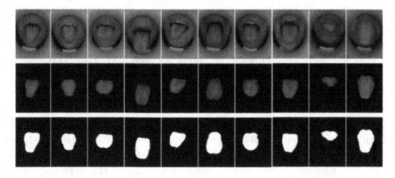

Fig. 9. From top to bottom are input tongue images, manually segmentations and Deeplabv3 segmentations respectively.

4.5 Experiment Based on Deeplabv3-LBM

We manually changed contour line thickness of 111 segmentations that obtained by Deeplabv3 to 10 pixels and set the color of contour line to grey to obtain trimaps. And then, we use LBM described in 2.3 to segment tongue images. The input of Deeplabv3-LBM is original image and 10-pixels-trimap. We tested the performance of Deeplabv3-LBM by mIoU. The result was 86.2%.

4.6 Comparison

As shown in Fig. 10, Deeplabv3 performs well where tongue and lip colors are similar, and traditional algorithms do poorly. Because Deeplabv3 segment tongue images based on semantic information, it could learn more knowledge about difference in shape and texture between lips and tongue.

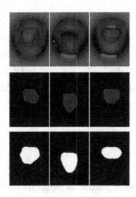

Fig. 10. The first line is original tongue image, the second line is bad segmentation by LBM and the third line is fine segmentation by Deeplabv3.

Table 2 shows the mIoU comparison of FCN-8s, Deeplabv3, LBM and Deeplabv3-LBM on the same data set (441 images in training set and 111 images in verification set). As can be seen from Table 2, the segmentation precision of convolution neural network is higher than that of traditional algorithm, and Deeplabv3 has the highest segmentation precision, far exceeding that of LBM.

Table 2. Comparison of mIoU of FCN-8s, Deeplabv3, LBM and Deeplabv3-LBM

Method	mIoU
FCN-8s	90.48%
Deeplabv3	93.7%
LBM (20px)	89.3%
Deeplabv3-LBM (10px)	86.2%

Because Deeplabv3 makes a comprehensive use of Deep Convolution Neural Network (DCNN) and atrous convolution, which minimizes the information loss in the process of image sampling, Deeplabv3 performs better than FCN-8s. Deeplabv3-LBM is based on LBM, the accuracy of Deeplabv3-LBM is lower than that of LBM. Because the trimap obtained by the network is not as accurate as the trimap obtained by the manual or automated tools.

Then, we tested the speed of FCN-8s, Deeplabv3, LBM and Deeplabv3-LBM to segment tongue images, as shown in Table 3.

Where, LBM use 20-pixels-trimap and Deeplabv3-LBM use 10-pixels-trimap. The accuracy of the trimap determines the speed of the algorithm.

As can be seen from the table above, the segmentation speed of convolution neural network is obviously much faster than traditional algorithm.

Table 3. Speed comparison between FCN-8s, Deeplabv3, LBM and Deeplabv3-LBM

Method	Time (s)
FCN-8s	11.02
Deeplabv3	12.73
LBM (20px)	278.61
Deeplabv3-LBM (10px)	33.31

5   Conclusion

Traditional sematic segmentation algorithms need to manually provide trimap for each image. The result of algorithms relies heavily on human-machine interaction. Moreover, the segmentation precision of traditional algorithm depends on the precision of trimap and segmentation precision has an upper limit with trimap of fixed precision.

However, CNN is more capable of learning semantic information of image than hand-craft features. In this paper, we train network to learn features from tongue images, so that the segmentation of tongue image can be completed automatically instead of relying on manual work. It greatly reduces workload. And semantic information provides more knowledge for tongue segmentation to make it work well. Experiments show that the segmentation precision of CNN is higher than that of traditional algorithm, and it has the obvious advantage in speed. In addition, the segmentation precision of CNN depends on the structure of network, which can be improved continuously with more powerful network in the future work.

Acknowledgement. This work was supported by the Shanghai Innovation Action Plan Project under Grant 16511101200.

References

1. Pang, B., Zhang, D., Wang, K.: Tongue image analysis for appendicitis diagnosis. Inf. Sci. **175**(3), 160–176 (2005)
2. Maciocia, G.: Tongue Diagnosis in Chinese Medicine, vol. 16. Eastland Press, Seattle (1995)
3. Zhang, H.Z., Wang, K.Q., Zhang, D., Pang, B., Huang, B.: Computer aided tongue diagnosis system. In: 27th Annual International Conference of the Engineering in Medicine and Biology Society, IEEE-EMBS 2005, pp. 6754–6757. IEEE (2006)
4. Chiu, C.C.: A novel approach based on computerized image analysis for traditional Chinese medical diagnosis of the tongue. Comput. Methods Programs Biomed. **61**(2), 77–89 (2000)
5. Pang, B., Zhang, D.D., Li, N., Wang, K.: Computerized tongue diagnosis based on Bayesian networks. IEEE Trans. Biomed. Eng. **51**(10), 1083 (2004)
6. Ning, J., Zhang, D., Wu, C., Yue, F.: Automatic tongue image segmentation based on gradient vector flow and region merging. Neural Comput. Appl. **21**(8), 1819–1826 (2012)
7. Shi, M.J., Li, G.Z., Li, F.F.: C^2G^2FSnake: automatic tongue image segmentation utilizing prior knowledge. Sci. China Inf. Sci. **56**(9), 1–14 (2013)

8. Pang, B., Zhang, D., Wang, K.: The bi-elliptical deformable contour and its application to automated tongue segmentation in Chinese medicine. IEEE Trans. Med. Imaging **24**(8), 946–956 (2005)
9. Zheng, Y., Kambhamettu, C.: Learning based digital matting. In: IEEE 12th International Conference on Computer Vision, pp. 889–896. IEEE (2009)
10. Li, X., Cui, Q.: Parallel accelerated matting method based on local learning. In: Lai, S.-H., Lepetit, V., Nishino, K., Sato, Y. (eds.) ACCV 2016. LNCS, vol. 10111, pp. 152–162. Springer, Cham (2017). https://doi.org/10.1007/978-3-319-54181-5_10
11. Yu, F., Koltun, V.: Multi-scale context aggregation by dilated convolutions. arXiv preprint arXiv:1511.07122 (2015)
12. Zheng, S., Jayasumana, S., Romera-Paredes, B., et al.: Conditional random fields as recurrent neural networks. In: Proceedings of the IEEE International Conference on Computer Vision, pp. 1529–1537 (2015)
13. Chen, L.C., Papandreou, G., Kokkinos, I., Murphy, K., Yuille, A.L.: Semantic image segmentation with deep convolutional nets and fully connected CRFs. In: Computer Science, vol. 4, pp. 357–361 (2014)
14. Chen, L.C., Papandreou, G., Kokkinos, I., Murphy, K., Yuille, A.L.: Deeplab: semantic image segmentation with deep convolutional nets, atrous convolution, and fully connected CRFs. IEEE Trans. Pattern Anal. Mach. Intell. **40**(4), 834–848 (2018)
15. Long, J., Shelhamer, E., Darrell, T.: Fully convolutional networks for semantic segmentation. In: Proceedings of the IEEE Conference on Computer Vision and Pattern Recognition, pp. 3431–3440 (2015)
16. Chen, L.C., Papandreou, G., Schroff, F., Adam, H.: Rethinking atrous convolution for semantic image segmentation. arXiv preprint arXiv:1706.05587 (2017)
17. Shen, X., Tao, X., Gao, H., Zhou, C., Jia, J.: Deep automatic portrait matting. In: Leibe, B., Matas, J., Sebe, N., Welling, M. (eds.) ECCV 2016. LNCS, vol. 9905, pp. 92–107. Springer, Cham (2016). https://doi.org/10.1007/978-3-319-46448-0_6
18. Russakovsky, O., Deng, J., Su, H., et al.: Imagenet large scale visual recognition challenge. Int. J. Comput. Vis. **115**(3), 211–252 (2015)
19. He, K., Zhang, X., Ren, S., Sun, J.: Spatial pyramid pooling in deep convolutional networks for visual recognition. In: Fleet, D., Pajdla, T., Schiele, B., Tuytelaars, T. (eds.) ECCV 2014. LNCS, vol. 8691, pp. 346–361. Springer, Cham (2014). https://doi.org/10.1007/978-3-319-10578-9_23
20. Abadi, M., et al.: Tensorflow: a system for large-scale machine learning. In: 12th USENIX Symposium on Operating Systems Design and Implementation, pp. 265–283 (2016)

Deep Feature Learning and Visualization for EEG Recording Using Autoencoders

Yue Yao[✉], Jo Plested, and Tom Gedeon

Research School of Computer Science,
The Australian National University, Canberra, Australia
u6014942@anu.edu.au

Abstract. In this era of deep learning and big data, the transformation of biomedical big data into recognizable patterns is an important research focus and a great challenge in bioinformatics. An important form of biomedical data is electroencephalography (EEG) signals, which are generally strongly affected by noise and there exists notable individual, environmental and device differences. In this paper, we focus on learning discriminative features from short time EEG signals. Inspired by traditional image compression techniques to learn a robust representation of an image, we introduce and compare two strategies for learning features from EEG using two specifically designed autoencoders. Channel-wise autoencoders focus on features in each channel, while Image-wise autoencoders instead learn features from the whole trial. Our results on a UCI EEG dataset show that using both Channel-wise and Image-wise autoencoders achieve good performance for a classification problem with state of art accuracy in both within-subject and cross-subject tests. A further experiment using shared weights shows that the shared weights technique only slightly influenced learning but it reduced training time significantly.

Keywords: Deep learning · Neural network · Autoencoders
Convolutional neural network · EEG · Brain-Computer Interface

1 Introduction

As an important part of Brain-Computer Interfaces (BCIs), EEG has found a variety of interesting and useful applications for users and has become increasingly important in various areas. Especially for the medical field, diagnosis of epilepsy for example, EEG has shown success [1, 2]. Gathered from the scalp, the EEG is a signal containing information about the electrical activity of the brain. Electrodes placed on the scalp are used to detect electrical information from the brain under the scalp, bone and other tissues. Since it is an overall measurement of human brain electrical activity, it may contain a wealth of information. This is the reason why EEG can be applied to diverse areas like personal recognition, disease identification [1], sleep stage classification [3], visual image generation using brain waves [4], and so on.

On the other hand, using EEG signals faces many difficulties. First, being full of information also here means full of noise and interference, making it very hard to extract reliable features. Further, depending on the collection device, EEG will have a

© Springer Nature Switzerland AG 2018
L. Cheng et al. (Eds.): ICONIP 2018, LNCS 11307, pp. 554–566, 2018.
https://doi.org/10.1007/978-3-030-04239-4_50

different format, hence it becomes difficult to construct standard algorithms to extract features from EEG. Third, EEG signals have large individual differences, making it hard for cross-subject tests [5] to achieve high accuracy. These three difficulties make EEG feature engineering still a work in progress.

For feature learning tasks in bioinformatics, a wide range of traditional machine learning algorithms have been applied and achieved success. In some areas, such as for bio-signals like EEG, many well-known algorithms have been applied like support vector machine, random forests, Bayesian networks, and hidden Markov models [6]. The good performance of conventional machine learning algorithms relies heavily on features extracted [7]. Traditional learned features are not always as good as we want since they are not always robust and not designed to counter noise. For this reason, we need algorithms that can learn features from big data automatically.

Deep learning, a neural network based technique, is a rising subfield of machine learning. It has achieved great success in computer vision (CV), natural language processing (NLP) and many other areas in recent years. But unlike CV and NLP which have many successful algorithms and datasets using deep learning, bio-information areas have no widely accepted learning algorithms or even a well-known and popular dataset like imagenet [8] in CV. Human brain waves have commonalities and differences, and it is these commonalities and differences which are exactly the properties that we want. We believe that only when all these properties are better understood can we make it possible to design a robust and recognized deep learning method in this area – even deep learning approaches need some understanding of the structure of the data to extract features well. So that is the reason why both learning and visualization are introduced in this paper.

To address these difficulties, deep learning approaches are utilized in this paper to achieve both learning and visualization. Two autoencoder-based techniques are used for feature learning and dimensionality reduction for short time EEG signals. They are referred to as Channel-wise autoencoders and Image-wise autoencoders. Channel-wise autoencoders are inspired by one-dimensional convolutions. For EEG data, the number of channels is often significantly less than the timescale length, forming an unbalanced matrix input. In such settings, for applying convolutional neural networks based techniques, it is usual to perform one-dimensional convolution for feature extraction [6, 9]. Thus, as a first step in our work, we design a group of channel-wise autoencoders which only focus on features from a single channel using simple fully connected layers. The Image-wise autoencoders are designed based on Fast Fourier Transform (FFT) and CNN. Using FFT, we can obtain the three EEG frequency bands, then we use these frequency bands to achieve an RGB-color visualization (an image) [10]. Then, a CNN based autoencoder is designed to extract features from these color images.

2 Related Work

Convolutional neural networks (CNNs) are feature extraction networks proposed by Lecun [11], based on the structure of the mammalian visual cortex – thus providing structural information about the data via the network topology. The difference between convolution neural networks and the traditional neural networks is the convolution layer. We consider the convolution layers as feature extractors. Then, the fully connected layer serves as a 'classifier' trying to find decision boundaries between each class. From another point of view, the role of the fully connected layer is similar to the kernel method, warping the high-level feature space to make each class approximately linearly separable.

Much CNN based research has been applied to EEG. Depending on the type of the kernel, CNN based work can be divided into normal CNN as well as frequency-based CNN. Normal CNN takes the raw EEG as the input while frequency-based CNN extracts frequency features from raw EEG. Examples of normal CNN approaches include Deep4net [12] and EEGNet [5]. The SyncNet [13] is the latest example of a frequency-based CNN for EEG. An interesting commonality is that one-dimensional convolutions are often applied among convolution procedures [6, 9].

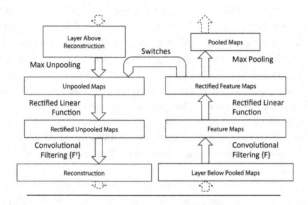

Fig. 1. The structure of DNN [14]

Deconvolution neural network (DNN) was first used for visualization in CV by Zeiler and Fergus [14]. A high-level feature map with many high dimensional features is difficult to interpret intuitively. A DNN projects the response value of the specified convolution layer into the input pixel space by reversing the CNN, thus revealing the contribution made by each pixel of the input image to the response value, thus creating a more comprehensible feature map visualization. These operations are shown in Fig. 1. The right side is the process of forward propagation, while the left side is the process of mapping the response value back to the input pixel space. By using a DNN, an autoencoder with CNN as encoder and DNN as decoder can be easily implemented.

Autoencoder is a sort of compression algorithm, or dimension reduction algorithm, which has similar properties to Principal Components Analysis (PCA). But compared with PCA, the autoencoder has no linear constraints. The autoencoder structure has been widely used for image compression, for example [15], which inspired us to try an autoencoder based learning algorithm. From Fig. 2, an autoencoder can be divided into two parts, an encoder and a decoder. The number of nodes in the hidden layer is

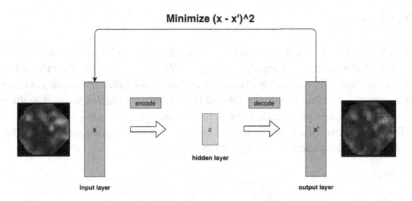

Fig. 2. Structure of autoencoder

generally less than the nodes in the input layer and the output layer. That is, the original input is compressed to a smaller feature vector. In Eq. 1 below, ϕ and ψ stand for encoder and decoder, respectively, and L means squared loss. The objective of the autoencoder is to minimize the difference between the input and the generated output. A CNN based autoencoder [16] uses convolution operations as the encoder and deconvolution operations for the decoder, making it better for operating on image data.

$$\phi, \psi = \operatorname{argmin}_{\phi,\psi} L(X, (\phi \circ \psi)X). \tag{1}$$

A major purpose of our work is to solve the difficulties we identified and advance the state of art in feature abstraction of signal analysis for short time EEG biosignals. Autoencoders are a mature method to extract robust features. Prior to our work, a number of autoencoder related methods have been applied to EEG signals. Stober [17] use convolutional autoencoders with custom constraints to learn features and improve generalization across subjects and trials. It achieved commendable results but it uses CNN directly on the time domain features from EEG signals but not frequency domain features like our methods. But Stober's work inspired us that it could be a general conclusion that the autoencoder based structure can increase the cross-subject accuracy, forming our basic inspiration to try autoencoder based structures.

The most similar work to our model is by Tabar and Halici [18]. They used EEG motor imagery signals and a combined CNN and fully connected stacked autoencoders (SAE) to find discriminative features. They used Short-time Fourier transform (STFT) to build an EEG motor imagery (MI) which is unlike our 3-D electrode location mapping as in our work (described in Sect. 3). Also, their autoencoder design is quite different from ours since they used a CNN followed by an 8-layer SAE. Nevertheless, they have demonstrated that autoencoders can help to learn robust features from EEG signals.

3   Methodology

The block diagram shows the general procedure for Channel-wise Autoencoders and Image-wise Autoencoders, as depicted in Fig. 3. We first pre-process the raw EEG data into a useable form. Then feature extraction and dimensionality reduction are done by using autoencoders. Finally, fully connected (FC) layers are utilized to do classification and evaluation. In our procedure, we extract features prior to applying the classification. To achieve this, two kinds of autoencoders are used to enhance features. That is, Channel-wise autoencoders and Image-wise autoencoders.

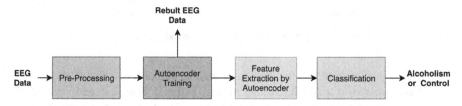

Fig. 3. Structure of general procedure for learning discriminative features based on autoencoders

3.1   Dataset

The dataset we use is from UCI, the EEG dataset from Neurodynamics Laboratory at the State University of New York. It has a total of 122 subjects with 77 diagnosed with alcoholism and 45 control subjects [13, 19]. Each subject has 120 separate trials. If a subject is labeled with alcoholism, all 120 trails belonging to that subject will be labeled as alcoholism. The stimuli they use are several pictures selected from the Snodgrass and Vanderwart picture set. It is a sort time EEG where one trial of EEG signal is of one second length and is sampled at 256 Hz with 64 electrodes. Models are first evaluated using data within subjects, which is randomly split as 7:1:2 for training, validation and testing for one person [5]. We further test them using data across subjects [5], using the same setting as Li [13]. The classification task is to recover whether the subject has been diagnosed with alcoholism or is a control subject. Also, we note that this is not a balanced dataset. It is a two-task classification but alcoholism trials account for more than 70% of the data.

The usual challenges of handling EEG make it more difficult to apply deep learning methods compared with computer vision data or natural language processing data. The UCI EEG dataset is not an exception.

First, a label is usually applied to one trial. But as one trial contains 64 channels and 256 time series data, making it become a 64 × 256 large matrix. In other words, a single EEG trial has 64 × 256 attributes, difficult for a neural network to find meaningful features if treated as 16,384 independent inputs.

Second, EEG is a kind of time-series data but it lacks recognizable patterns in single time slices (1/sampling rate) compared with natural language processing, since each word in NLP often has a specific meaning.

Third, as previous work has shown, if we consider an EEG signal as a picture and directly use a convolution neural network for raw EEG data, there is always a serious problem to determine the size of the kernels to use at each stage [6, 18]. That is, because the original features could be distributed with different time differences in a single trial depending on the scenario (different classification task for example). Furthermore, due to the huge personal differences in EEG data, the cross-subject test result is often far from satisfactory. To address these difficulties, we used two kinds of autoencoders as described in the following.

3.2 Channel-Wise Autoencoders

The key idea of applying a channel-wise autoencoder is to separate the feature extraction procedure into two parts. The channel-wise autoencoder only focuses on features in one channel while the final fully connected layer will combine features across channels to make a final prediction. This is very much like using 1-D convolutions to get channel-wise features followed by a fully connected layer to make a prediction [9].

As shown in Fig. 4. An EEG trial with the 64×256 dimensions will be separated into 64 1×256 signals, then each signal will be the input for one autoencoder only trained on that channel. These 64 autoencoders are just 2 layers of a fully connected neural network with 16 hidden units in the middle. The input of autoencoders will be normalized to $[-1, 1]$ and we use a tanh activation function for the output layer to match the output to $[-1, 1]$ as well. The shared weight technique derived from image compression [15] is also used for signal compression, which takes the transpose of encoder weights for the decoder weights.

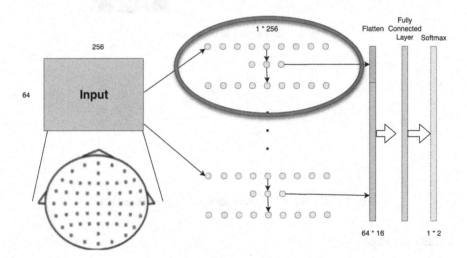

Fig. 4. Structure of channel-wise autoencoders

3.3   Image-Wise Autoencoders

The image-wise autoencoders take the images as input while using a CNN to extract features. The whole procedure is shown in Fig. 5, below is some further explanation.

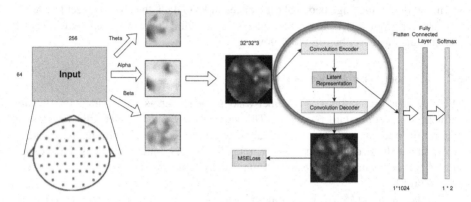

Fig. 5. Structure of image-wise autoencoder

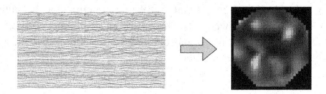

Fig. 6. EEG signal to image example

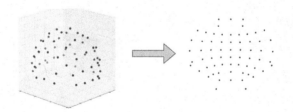

Fig. 7. Transform 3-D coordinate to 2-D coordinate [10]

A. EEG to Image

The method is derived from Bashivan's work [10]. As shown in Fig. 6, it is a method that combines the time-series information and spatial channel locations information over the scalp in a trial of EEG signals. An FFT is performed on the time series to estimate the power spectrum of the signal for each trial (64 × 256). Then, three frequency bands of theta (4–7 Hz), alpha (8–13 Hz), and beta (13–30 Hz) are extracted, and the sum of squared absolute values in these frequency bands are used, forming a 64 × 3 map. To form an RGB EEG image, the theta frequency will be the red channel, alpha the green channel and beta the blue channel. For each frequency band (64 × 1), shown in Fig. 7, Azimuthal Equidistant Projection (AEP) also known as Polar Projection is used to map the three-dimensional 64 channel position into two-dimensional positions on a flat surface. That is, all EEG electrodes positions are mapped into a consistent 2-D space because the original EEG electrodes are distributed over the scalp in a three-dimensional fashion. In this way, each 64 × 1 frequency band can be mapped to a 32 × 32 mesh, forming 32 × 32 × 3 data. The CloughTocher scheme is used for estimating the values in-between the electrodes over the 32 × 32 mesh. Thus, a trial of 64 × 256 EEG signals is transformed to 32 × 32 × 3 color pictures.

B. Autoencoder design

The design of this CNN based autoencoder is inspired by the CNN for CIFAR-10 [20]. The CIFAR-10 dataset consists of 60,000 32 × 32 color images in 10 classes, with 6,000 images per class, with the same input dimension as our generated EEG pictures. Our encoder and decoder are described in Table 1. Our shared weight CNN Autoencoder is in the same structure as the normal CNN autoencoder but the weight of the three deconvolution layers is fixed and derived from the encoder's convolution layer. The Rectified Linear Unit (ReLU) is used for activation layers to speed up the training process while dropout is performed after every activation layer to make the model more robust, since it forces all the layers before the dropout to extract redundant representations. Adam optimizer is used with 1e–4 learning rate and the batch size is set to 64. Xavier normal initialization is used for convolution kernels.

Table 1. The detailed encoder and decoder structure

Encoder	Decoder
Input 32 × 32 × 3 Color Image	Input 16 × 8 × 8 Matrix
3 × 3 conv, 2 × 2 max-pooling	3 × 3 deconv, 2 × 2 max-un-pooling
ReLU, 0.25 dropout	ReLU, 0.25 dropout
3 × 3 conv, 2 × 2 max-pooling	3 × 3 deconv, 2 × 2 max-un-pooling
ReLU, 0.25 dropout	ReLU, 0.25 dropout
3 × 3 conv, ReLU	3 × 3 deconv

3.4 Classification Task

The features extracted from channel-wise autoencoders and image-wise autoencoders will be flattened into a long vector, composed of 16 hidden unit representations ×64 autoencoders in the channel-wise case or 16 × 8 × 8 matrix in the image-wise case. Then we use a feedforward network with three hidden layers. During training of these three fully connected layers using 4e–5 learning rate, the encoder of both the channel-wise and image-wise autoencoders will also be fine-tuned by the classification loss using a much smaller learning rate (1e–7).

4   Results and Discussions

Our experiment was to compare the classification accuracy using normal channel-wise autoencoders, shared weight channel-wise autoencoders, normal image-wise autoencoders and shared weight image-wise autoencoders. The code was written in python and pytorch. All experiments were done on an i5-7500 CPU, Nvidia GTX1050Ti, 8g RAM and Windows environment. Below is the classification result for the different classification task.

Table 2. Comparison between two image-wise autoencoders

Method	Within accuracy	Final loss	Training time (100 epoch)
Shared weight Image-wise Autoencoders	0.897	0.00026	**132.99 s**
Normal Image-wise Autoencoders	**0.917**	**0.00019**	150.68 s

Table 3. Comparison between image-wise autoencoders and common CNN

Method	Within accuracy	Cross accuracy
Normal Image-wise Autoencoders	**0.917**	**0.756**
Image-wise CNN	0.915	0.712

Table 4. Classification accuracy – within-subject tests

Method	Accuracy
Normal Channel-wise Autoencoders	0.864
Shared weight Channel-wise Autoencoders	0.858
Normal Image-wise Autoencoders	**0.917**
Shared weight Image-wise Autoencoders	0.897
EEGNet (Lawhern et al. 2016)	0.878
SyncNet (Li et al. 2017)	**0.923**
DE (Zheng and Lu 2015)	0.821
PSD (Zheng and Lu 2015)	0.816
rEED (O'Reilly et al. 2012)	0.702

Table 5. Classification accuracy – cross-subject tests

Method	Accuracy
Normal Channel-wise Autoencoders	0.731
Shared weight Channel-wise Autoencoders	0.713
Normal Image-wise Autoencoders	**0.756**
Shared weight Image-wise Autoencoders	0.740
EEGNet (Lawhern et al. 2016)	0.672
SyncNet (Li et al. 2017)	**0.723**
DE (Zheng and Lu 2015)	0.622
PSD (Zheng and Lu 2015)	0.605
rEED (O'Reilly et al. 2012)	0.614

The accuracy of prediction on the UCI EEG dataset, from a variety of methods, is given in Tables 4 and 5. The accuracy of other methods is listed from Li's paper [13].

We first check cross-subject result, which is the test format we are more likely to meet in real life for classifying disease, and the part we are focusing on to improve through autoencoders. As shown in Table 5, all of our autoencoders except the shared weight channel-wise autoencoders achieve state-of-art cross-subject test accuracy. We believe this is because autoencoders can encourage feature extraction without over-fitting, and will prevent the model from performing badly on new data. In other words, this prevents our model from learning the disease condition by merely remembering the personal identity, and instead makes our model focus on the common features for alcoholism. This could explain why autoencoders based methods perform best in the cross-subject test. Further evidence is shown in Table 3 in order to show the performance of using autoencoder, we construct an Image-wise CNN which has the same structure of the encoder of Image-wise autoencoder with a three-layer FC as the classifier. The result shows that though it can achieve similar within subject accuracy as Image-wise autoencoder but it performs badly in the cross-subject test. That is, an autoencoder structure helps to improve the ability to extract robust features.

From the result above in the within-subject test, we can see that the accuracy of our autoencoder based method is better than most of the past methods except the SyncNet [13] published last year.

Apart from these, there are also some general conclusions. Image-wise autoencoders perform better than channel-wise autoencoders while the normal autoencoders perform slightly better than shared weight autoencoders. From Table 2, we can see the normal image-wise autoencoder has better within subject accuracy and lower final test loss than the shared weight image-wise autoencoders. Also from Fig. 8, the picture generated by normal Image-wise autoencoder is slightly clearer and more similar to the original image. On the other hand, shared weight image-wise autoencoders have lower training time. This is an advantage of the shared weight technique because it cuts half of the parameters. From these results above, we can see that the image-wise autoencoders find the best discriminative features among different methods. We believe this is because frequency-based feature learning methods can obtain more discriminative information – both our image-wise autoencoders and the SyncNet approach are frequency based and they achieved the best performance.

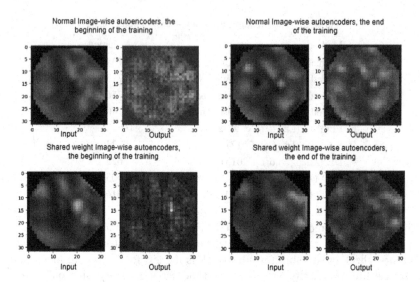

Fig. 8. Image-wise autoencoders' performance

5 Conclusion

Feature extraction for EEG data is very challenging because EEG signals contain a lot of noise, use different collection standards and huge personal differences exist. This paper introduced two kinds of autoencoders: Image-wise autoencoders and Channel-wise autoencoders. These two types of autoencoders were tested for feature extraction, and both achieve state of the art accuracy in cross-subject tests and comparable accuracy in the (less important in classifying disease) within-subject tests. The experiment results demonstrate that the autoencoder based feature learning is discriminative and robust for new data. Also, we found that a shared weight technique can noticeably reduce the training time with only a small discriminative information loss.

6 Limitation and Future Work

Many further experiments should be done. First of all, the UCI dataset also contains other labels that can be classified. We should further test our extracted features with those labels to ensure our extracted features are discriminative for multiple task classifications. Furthermore, since we are using the EEG2image technique, our method should be a general framework without further fine-tuning for other datasets as long as the 3-D electrode location information is provided. Then, other popular datasets like DEAP should be tested in the same setting as the UCI dataset we used. Then we can turn our attention to more frequency based methods since both Image-wise autoencoders and SyncNet are frequency based. Finally, we may try LSTM based work – there also exist many RNN feature extractors for EEG data. If we have realized all these feature extraction methods, we would do more visualization procedures. Unlike the computer vision area, the features of EEG signals are not obvious, so visualization will be a good choice for understanding EEG features. Our ultimate goal is to get a deeper understanding of EEG features and make it possible to design stronger feature extractors and classifiers.

Currently, there is a very limited work for applying deep learning work for bio-signals. Since many successful examples exist in CV and NLP areas, it will be very worthwhile to try them in bioinformatics areas.

References

1. Truong, N.D., Nguyen, A.D., Kuhlmann, L., Bonyadi, M.R., Yang, J., Kavehei, O.: A Generalised Seizure Prediction with Convolutional Neural Networks for Intracranial and Scalp Electroencephalogram Data Analysis (2017). arXiv preprint: arXiv:1707.01976

2. Thodoroff, P., Pineau, J., Lim, A.: Learning robust features using deep learning for automatic seizure detection. In: Machine Learning for Healthcare Conference, pp. 178–190 (2016)

3. Ebrahimi, F., Mikaeili, M., Estrada, E., Nazeran, H.: Automatic sleep stage classification based on EEG signals by using neural networks and wavelet packet coefficients. In: 30th Annual International Conference of the IEEE Engineering in Medicine and Biology Society, EMBS 2008, pp. 1151–1154. IEEE (2008)

4. Palazzo, S., Spampinato, C., Kavasidis, I., Giordano, D., Shah, M.: Generative adversarial networks conditioned by brain signals. In: Proceedings of the IEEE Conference on Computer Vision and Pattern Recognition, pp. 3410–3418 (2017)

5. Lawhern, V.J., Solon, A.J., Waytowich, N.R., Gordon, S.M., Hung, C.P., Lance, B.J.: Eegnet: A compact convolutional network for eeg-based brain-computer interfaces (2016). arXiv preprint: arXiv:1611.08024

6. Min, S., Lee, B., Yoon, S.: Deep learning in bioinformatics. Brief. Bioinform. **18**, 851–869 (2017)

7. Goodfellow, I., Bengio, Y., Courville, A., Bengio, Y.: Deep Learning. MIT Press, Cambridge (2016)

8. Krizhevsky, A., Sutskever, I., Hinton, G.E.: Imagenet classification with deep convolutional neural networks. In: Advances in Neural Information Processing Systems, pp. 1097–1105 (2012)

9. Hajinoroozi, M., Mao, Z., Jung, T.-P., Lin, C.-T., Huang, Y.: EEG-based prediction of driver's cognitive performance by deep convolutional neural network. Signal Process. Image Commun. **47**, 549–555 (2016)

10. Bashivan, P., Rish, I., Yeasin, M., Codella, N.: Learning representations from EEG with deep recurrent-convolutional neural networks (2015). arXiv preprint: arXiv:1511.06448

11. LeCun, Y., Bengio, Y.: Convolutional networks for images, speech, and time series. In: The Handbook of Brain Theory and Neural Networks, p. 3361 (1995)

12. Schirrmeister, R.T., et al.: Deep learning with convolutional neural networks for EEG decoding and visualization. Hum. Brain Mapp. **38**, 5391–5420 (2017)

13. Li, Y., Dzirasa, K., Carin, L., Carlson, D.E.: Targeting EEG/LFP synchrony with neural nets. In: Advances in Neural Information Processing Systems, pp. 4623–4633 (2017)

14. Zeiler, M.D., Fergus, R.: Visualizing and understanding convolutional networks. In: Fleet, D., Pajdla, T., Schiele, B., Tuytelaars, T. (eds.) ECCV 2014, Part I. LNCS, vol. 8689, pp. 818–833. Springer, Cham (2014). https://doi.org/10.1007/978-3-319-10590-1_53

15. Gedeon, T., Catalan, J., Jin, J.: Image compression using shared weights and bidirectional networks. In: Proceedings 2nd International ICSC Symposium on Soft Computing (SOCO 1997), pp. 374–381 (1997)
16. Masci, J., Meier, U., Cireşan, D., Schmidhuber, J.: Stacked convolutional auto-encoders for hierarchical feature extraction. In: Honkela, T., Duch, W., Girolami, M., Kaski, S. (eds.) ICANN 2011, Part I. LNCS, vol. 6791, pp. 52–59. Springer, Heidelberg (2011). https://doi.org/10.1007/978-3-642-21735-7_7
17. Stober, S., Sternin, A., Owen, A.M., Grahn, J.A.: Deep feature learning for EEG recordings (2015). arXiv preprint: arXiv:1511.04306
18. Tabar, Y.R., Halici, U.: A novel deep learning approach for classification of EEG motor imagery signals. J. Neural Eng. **14**, 016003 (2016)
19. Sykacek, P., Roberts, S.J.: Adaptive classification by variational Kalman filtering. In: Advances in Neural Information Processing Systems, pp. 753–760 (2003)
20. Krizhevsky, A., Hinton, G.: Learning multiple layers of features from tiny images (2009)

A Feature Filter for EEG
Using Cycle-GAN Structure

Yue Yao[⊠], Jo Plested, and Tom Gedeon

Research School of Computer Science,
The Australian National University, Canberra, Australia
u6014942@anu.edu.au

Abstract. The brain-computer interface (BCI) has become one of the most important biomedical research fields and has created many useful applications. As an important component of BCI, electroencephalography (EEG) is in general sensitive to noise and rich in all kinds of information from our brain. In this paper, we introduce a new strategy to filter out unwanted features from EEG signals using GAN-based autoencoders. Filtering out signals relating to one property of the EEG signal while retaining another is similar to the way we can listen to just one voice during a party. This approach has many potential applications including in privacy and security. We use the UCI EEG dataset on alcoholism for our experiments. Our experiment results show that our novel GAN based structure can filter out alcoholism information for 66% of EEG signals with an average of only 6.2% accuracy lost.

Keywords: Deep learning · EEG · Brain-Computer interface
Image translation · Generative adversarial nets

1 Introduction

Being an essential input signal of a Brain-Computer Interface (BCI), EEG has been harnessed in a variety of interesting and useful applications for users and has changed our life in various areas. The EEG is defined as the overall measurement of human brain electrical activity using electrodes placed on the scalp. Since it is an overall measurement, this makes EEG applicable to diverse areas like personal recognition [1], disease identification [2, 3], sleep stage classification [4], even to rebuild the picture from a person's eyes [5], and so on.

Taking personal recognition as an example, compared with fingerprint or face recognition, EEG has more advantages in identifying different people because it has a higher safety factor. For instance, if one person's fingerprint is stolen or one person's face is reconstructed by others, it is basically an irreparable problem because both fingerprint and face model are irrevocable without expensive and painful plastic surgery. But for EEG data, if it is hacked by others, users can still reset a new EEG pattern because the EEG recognizer can identify a person by both personal details and personal brain action [1].

© Springer Nature Switzerland AG 2018
L. Cheng et al. (Eds.): ICONIP 2018, LNCS 11307, pp. 567–576, 2018.
https://doi.org/10.1007/978-3-030-04239-4_51

But being full of information also means full of personal privacy issues for personal identification. For example, if we would like to use EEG for a personal recognition task for a bank, the only information we would like to upload is personal identity-related information and not share the full EEG with the bank, as it may contain information related to a disease or condition we may have. But since there currently does not exist a suitable information filtering algorithm, both the bank and hackers will also be able to get our other information like disease information, emotion information and so on.

But to be able to filter out unwanted features faces many difficulties. First, the property we mentioned of EEG being full of information also generally means full of noise and interference, making it hard to filter out unwanted features exactly. Second, filtering out features is not as easy as cutting off a bounding box in computer vision (CV), it is more like a transformation from the whole since EEG is not interpretable for all its features. Third, filtering out unwanted features also means we need to retain normal EEG trial properties, and we have to make sure our desired features are maintained during the operation.

As a result, we consider deep learning methods, which have achieved success in many areas like CV and natural language processing (NLP). In practice, we do not use the idea of subtracting features to filter out properties as such properties are not well-defined. Instead, we choose to generate a new EEG trial without the unwanted features but maintaining the desired features of the original EEG trial signal. So as the result, a generative adversarial network (GAN) based technique is utilized to create such an EEG signal. In this paper, we introduce GAN-based autoencoders, which is as an extension of our previous work [6]. As mentioned earlier, the feature filter of EEG is more like a style transformation. So we are inspired by the idea of Image-to-Image translation [7] introduced in the computer vision area. This approach is designed to map one image distribution to another image distribution in order to achieve a style transformation. In our paper, such a translation mechanism is used for feature filtering.

2   Related Work

EEG2Image is a work designed to transfer EEG signals to images which is derived from Bashivan's work [8]. Shown in Fig. 1, each trial of EEG is transformed to a colored image using both the time-series information and electrode location information. The transformation procedure is as follows. First, for a single trial of EEG signal, Fast Fourier Transform (FFT) is performed to extract three frequency bands, theta (4–7 Hz), alpha (8–13 Hz), and beta (13–30 Hz). Then, calculate the sum of squared absolute values for

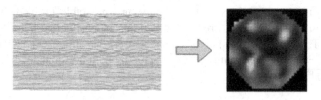

Fig. 1. EEG signal to image example [6]

each frequency band, thereby giving each electrode three scalar values to describe it. Next, using Polar Projection to project 3-D electrode position to 2-D position to create 2-D position sets in a 2-D map with three values to describe it. Then, with CloughTocher scheme to interpolate values between positions, we can produce consistent 2-D color images which reproducibly represent an EEG trial as a full color image.

Generative adversarial networks (GANs) are systems of two neural networks contesting with each other in a minimax game framework [9]. The GAN approach has achieved great success in the image generation area [10–12]. GANs include two main parts, namely a generator and a discriminator. The generator is mainly used to learn the distribution of the real image and produce images in order to fool the discriminator, while the discriminator needs to accept real images while rejecting generated images. Throughout this process, the generator strives to make the generated image more realistic, while the discriminator strives to identify the real image. The key part of GAN is the adversarial loss. For the image generation task, the adversarial loss is very powerful for images in one domain transformed to the other domain since this domain cannot be discriminated by simple rules.

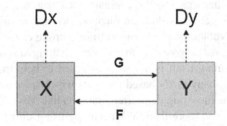

Fig. 2. CycleGAN structure

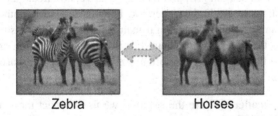

Zebra   Horses

Fig. 3. Image translation example [13]

Image-to-image translation is a kind of system that can learn the mapping between an input image distribution and an output image distribution using two separate image domains [7]. Shown in Fig. 2, given a source distribution X, we are aiming to use a generative model G to map our source distribution X to target distribution Y. An example is shown in Fig. 3, though it is not perfect, the translation system has successfully transformed between the most important features between zebra and horses like the hide color. In this translation system, we do not explicitly tell the neural network to change some features. Instead, we have the prior knowledge of two separate image distributions. As a result, it is possible for us to extract the stylistic differences between two image distributions and then directly translate them from one domain to the other domain.

Cycle-Consistent Adversarial Networks (CycleGAN) is a well-known image-to-image translation for unpaired images [13]. It overcomes the difficulty of getting paired images, and forms an autoencoder-like structure to achieve image translation. In Fig. 2, G is such a generator that generates a domain Y image from domain X, while F is the generator that generates a domain X image from the domain Y. Dx and Dy are two discriminators that are used to justify whether the coming image really belongs to domain X or domain Y, respectively. The training procedure can be separated into two symmetric parts. One is $X \rightarrow G(X) \rightarrow F(G(X))$. In this autoencoder-like loop, the training loss comes from two parts, the first is the discriminator loss which comes from Dy to judge whether G(X) is really from domain Y and the second is the reconstruction loss to judge whether F(G(X)) is the same as X or not. The other loop $Y \rightarrow F(Y) \rightarrow G(F(Y))$ is the same in principle.

But all these GAN methods are based on two hypotheses. One is that it is possible to build a strong classifier that can discriminate such features, and the second is the availability of a reliable generator that can filter out original features and rebuild target features. For the first hypothesis, if we cannot train a strong classifier in normal labeled training, it will be almost impossible for us get a strong discriminator in training, because adversarial training itself is not well designed to help train the discriminator. That is not an issue for many GAN based methods which have achieved great success in the CV area, since the most popular current datasets like MNIST [14] and CIFAR-10 [15] have already achieved more than 90% accuracy using different CNNs to serve as accurate discriminators. In contrast to CV, since the NLP area does not have a universally recognized text classification method for grammar checking, current GAN methods for NLP, like Seqgan [16] and its improved version Leakgan [17] do not have a strong discriminator to guide the generator. For our second hypothesis, we have to have a strong generator which can rebuild features. But building a strong generator is strongly related to the given type of data. For the image translation area, convolution and deconvolution-based methods are often used. The U-net [18] based method is the current state of the art [7].

Image-wise autoencoders [6] are the solution we use to meet the two hypotheses of building a GAN for EEG. An image-wise autoencoder is used to extract discriminative and robust features from EEG images. During the autoencoder training, it can reduce reconstruction loss to a very low level for the test set, making it possible to become a generator for the GAN structure. Furthermore, when we connect the features to a fully connected layer to work as a classifier, it achieves convincing results with more than 90% accuracy in the within-subject test [19], showing it has the ability to be a strong discriminator.

3   Methodology

3.1   UCI EEG Dataset

The dataset we use is from UCI; it is a multi-label dataset. This EEG dataset was created by the Neurodynamics Laboratory at the State University of New York. It has a total of 122 subjects with 77 diagnosed with alcoholism and 45 control subjects. Each subject has 120 separate trials [20]. If a subject is labeled with alcoholism, all 120 trials belonging to that subject will be labeled as alcoholism. The stimulus they use are several pictures selected from the Snodgrass and Vanderwart picture set. As a result, for each trial of EEG signal, there are both alcoholism and stimulus information labels.

3.2   Gan-Based Autoencoder

The Gan-based Autoencoder is mainly used for data filtering and the latent representation of this autoencoder is the filtered result we want. Our Gan-based Autoencoder is the same structure as the CycleGAN structure [13]. We call it a GAN-based Autoencoder mainly because it is principally still in a data->latent representation->original data structure and uses reconstruction loss. So in this autoencoder design, we take this latent representation as our filter result. As introduced before, the training procure can be split into two separate training loops, and each loop has two separate losses. The detailed loss definitions are as follows.

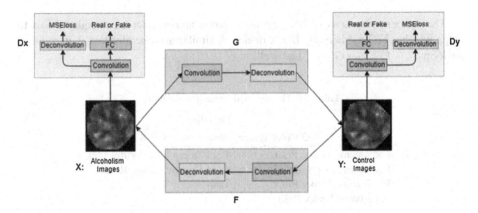

Fig. 4. Structure of GAN-based autoencoder

A. Adversarial Loss:

The adversarial loss is mainly designed to judge whether the incoming image really belongs to a certain distribution. So take loop $X \rightarrow G(X) \rightarrow F(G(X))$ for example, it is designed to map distribution X to distribution Y using generator G. The adversarial loss for this loop is defined as

$$
\begin{aligned}
L_{GAN}(G, D_Y, X, Y) =& E_{y \sim pdata(y)}[\log D_Y(y)] \\
&+ E_{x \sim pdata(x)}[\log (1 - D_Y(G(x)))]
\end{aligned}
\tag{1}
$$

This is a common GAN loss, where G(x) is trying to fool the discriminator D_Y to make the generated image become more similar to image distribution Y. A similar adversarial loss is introduced for loop $Y \rightarrow F(Y) \rightarrow G(F(Y))$.

B. Autoencoder Loss:

The autoencoder loss (reconstruction loss) is mainly used as a regularization term to make sure the generated image is not from random selection, because the target distribution could have multiple choices. The autoencoder loss will help the generator to choose a target image which also maintains some feature(s) from the original image in order to help reduce the reconstruction loss. Also, take loop $X \rightarrow G(X) \rightarrow F(G(X))$ for example, It is defined as:

$$
L_{AL}(G, F) = E_{x \sim pdata(x)}[||F(G(x)) - x||_1]
\tag{2}
$$

The Autoencoder Loss is the same as common autoencoder mean squared loss to judge whether F(G(x)) is really like x or not. A similar autoencoder loss is introduced for loop $Y \rightarrow F(Y) \rightarrow G(F(Y))$ as well.

Table 1. The detailed generator structure

Encoder	Decoder
Input 32 × 32 × 3 Color Image	Input 128 × 8 × 8 Matrix
4 × 4 conv, Leaky ReLU, 4 ×4 conv, Leaky ReLU, 3 × 3 conv, Leaky ReLU, 3 × 3 conv, Leaky ReLU	4 ×4 Deconv, Leaky ReLU, 4 ×4 Deconv, Leaky ReLU, Tanh

Shown in Fig. 4, we use EEG images with alcoholism condition and then map them to an EEG image with the control condition. By doing this transformation, we aim to eliminate alcoholism information from an EEG image while still maintaining its stimulus information. Inspired by the Image-wise autoencoder, shown in Table 1, our modified version of Image-wise autoencoder is now working as our generator, and the combination of Image-wise autoencoder and one fully connected layer works as our discriminator. Adam optimizer is used with 0.0002 learning rate.

3.3   Evaluation Method

The evaluation method for GAN is a difficult problem which needs to take many factors into account [21]. For a long time after the original GAN paper was published, the generated results from GANs still needed to be judged by manual selection in the CV area. But after the critical work from Google brain, the Fréchet Inception Distance (FID) and F1 scores [21] were introduced to judge the generation quality of a GAN. Both the FID and F1 score require a strong pretrained classifier in CV, making it impossible to directly use in the bio-signal area.

Thus, we learn from the idea of using FID and Inception Score (IS) but simply use the idea of training an additional classifier to judge the classification accuracy changes. The classifier we take is still the Image-wise autoencoder with fully connected layer (FC) which is trained separately from adversarial training. In this work, we are trying to filter out alcoholism information while keeping stimulus information. So, the desired best result should be that we get a large alcoholism accuracy reduction while keeping reasonable stimulus accuracy (low stimulus accuracy reduction) through the GAN based autoencoder.

4   Results and Discussion

The picture generated by our GAN-based autoencoder is shown in Fig. 5. These are six generation examples randomly selected from all generation pairs. We can see our GAN works and makes some slight modification to the images. The fact that we cannot see interpretable features from these transformations, means the generated results from our GAN cannot be manually checked. So we turn instead to digital indicators. Here, we only evaluate whether our generated image is really removing features we do not want using the normal Image-wise autoencoder with a classification net [6]. From Fig. 6, we can see that 96.1% of the original images are correctly classified as alcoholism, which is a good result on this dataset and shows our underlying approach works. After our GAN-based autoencoder has processed these images, only 29.8% of the images are classified as alcoholism. That is 2/3 (66.3%) of images have their alcoholism information filtered out. At the same time, only 6.2% accuracy has been lost for stimulus accuracy, and its accuracy still remains well above chance, which is 20% in this case as

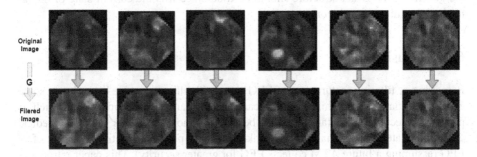

Fig. 5. GAN-based autoencoders output

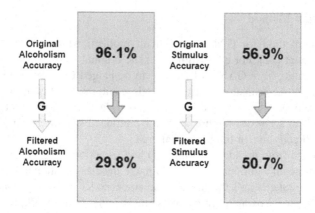

Fig. 6. GAN-based autoencoders performance

there are 5 stimulus conditions. Also, from the figure above, it seems that our GAN-based autoencoder does not change our EEG images much by eye, but it has already filtered out one feature of the original EEG image. That is a very interesting result and further study is needed to determine whether we can remove all alcoholism features while retaining stimulus features without loss of accuracy. As a summary, it turns out that our GAN-based autoencoder can filter out alcoholism information to some extent.

5 Limitation and Future Work

The first limitation is in using accuracy only as performance evaluation. This is an issue because there could be various ways to reduce accuracy like adding random noise or adversarial attacks [22]. One potential solution to this is to check whether such methods can achieve the same performance as GAN-based autoencoder. The second limitation is that there is still a 6.2% accuracy drop in stimulus classification. One possible solution is to try to add a stimulus discriminator to provide a penalty for stimulus information loss. But since the stimulus classifier is currently far from a strong classifier. Our 56.9% is reasonable where chance is 20%, but cannot really be called 'strong'. Thus, the result of adding a stimulus discriminator is not predictable. The third point is future work for the generator, the U-net structure should be tried since it is the current state of the art method for image translation.

6 Conclusion

Removing or filtering features out of EEG signals is difficult, but we have shown some excellent initial results. This approach can lead to many useful applications, such as privacy protection. An example could be where a hospital stores only the medical condition related EEG signal, but the bank stores only personal identification part of an EEG (assuming a future ATM collects EEG for greater security). This paper introduces

GAN-based autoencoders, which transfer the feature filtering task to an image translation task. The experiment results show that our GAN-based autoencoder can filter out a large proportion of unwanted features while mostly keeping desired features, as evaluated by using accuracy drops. Limited by time, the potential of these models is not fully revealed, with further adjustment and fine-tuning, the performance could be increased.

References

1. Kumari, P., Vaish, A.: Brainwave based user identification system: a pilot study in robotics environment. Robot. Auton. Syst. **65**, 15–23 (2015)
2. Truong, N.D., Nguyen, A.D., Kuhlmann, L., Bonyadi, M.R., Yang, J., Kavehei, O.: A Generalised Seizure Prediction with Convolutional Neural Networks for Intracranial and Scalp Electroencephalogram Data Analysis. arXiv preprint arXiv:1707.01976 (2017)
3. Thodoroff, P., Pineau, J., Lim, A.: Learning robust features using deep learning for automatic seizure detection. In: Machine Learning for Healthcare Conference, pp. 178–190 (2016)
4. Ebrahimi, F., Mikaeili, M., Estrada, E., Nazeran, H.: Automatic sleep stage classification based on EEG signals by using neural networks and wavelet packet coefficients. In: 30th Annual International Conference of the IEEE Engineering in Medicine and Biology Society, 2008. EMBS 2008, pp. 1151–1154. IEEE (2008)
5. Palazzo, S., Spampinato, C., Kavasidis, I., Giordano, D., Shah, M.: Generative adversarial networks conditioned by brain signals. In: Proceedings of the IEEE Conference on Computer Vision and Pattern Recognition, pp. 3410–3418 (2017)
6. Yao, Y., Plested, J., Gedeon, T.: Deep Feature Learning and Visualization for EEG Recording Using Autoencoders. Submited to International Conference on Neural Information Processing (ICONIP) 2018 12 (2018)
7. Isola, P., Zhu, J.-Y., Zhou, T., Efros, A.A.: Image-to-image translation with conditional adversarial networks. arXiv preprint arXiv:1611.07004 (2017)
8. Bashivan, P., Rish, I., Yeasin, M., Codella, N.: Learning representations from EEG with deep recurrent-convolutional neural networks. arXiv preprint arXiv:1511.06448 (2015)
9. Goodfellow, I., et al.: Generative adversarial nets. In: Advances in Neural Information Processing Systems, pp. 2672–2680 (2014)
10. Radford, A., Metz, L., Chintala, S.: Unsupervised representation learning with deep convolutional generative adversarial networks. arXiv preprint arXiv:1511.06434 (2015)
11. Liu, M.-Y., Breuel, T., Kautz, J.: Unsupervised image-to-image translation networks. In: Advances in Neural Information Processing Systems, pp. 700–708 (2017)
12. Arjovsky, M., Chintala, S., Bottou, L.: Wasserstein gan. arXiv preprint arXiv:1701.07875 (2017)
13. Zhu, J.-Y., Park, T., Isola, P., Efros, A.A.: Unpaired image-to-image translation using cycle-consistent adversarial networks. arXiv preprint arXiv:1703.10593 (2017)
14. LeCun, Y., Bottou, L., Bengio, Y., Haffner, P.: Gradient-based learning applied to document recognition. Proc. IEEE **86**(11), 2278–2324 (1998). Wiley-IEEE Press, Indianapolis, Indiana
15. Krizhevsky, A., Hinton, G.: Learning multiple layers of features from tiny images (2009)
16. Yu, L., Zhang, W., Wang, J., Yu, Y.: SeqGAN: Sequence Generative Adversarial Nets with Policy Gradient. arXiv preprint arXiv:1609.05473 (2016)
17. Guo, J., Lu, S., Cai, H., Zhang, W., Yu, Y., Wang, J.: Long Text Generation via Adversarial Training with Leaked Information. arXiv preprint arXiv:1709.08624 (2017)

18. Ronneberger, O., Fischer, P., Brox, T.: U-Net: convolutional networks for biomedical image segmentation. In: Navab, N., Hornegger, J., Wells, W.M., Frangi, A.F. (eds.) MICCAI 2015. LNCS, vol. 9351, pp. 234–241. Springer, Cham (2015). https://doi.org/10.1007/978-3-319-24574-4_28

19. Lawhern, V.J., Solon, A.J., Waytowich, N.R., Gordon, S.M., Hung, C.P., Lance, B.J.: Eegnet: a compact convolutional network for eeg-based brain-computer interfaces. arXiv preprint arXiv:1611.08024 (2016)

20. Li, Y., Dzirasa, K., Carin, L., Carlson, D.E.: Targeting EEG/LFP synchrony with neural nets. In: Advances in Neural Information Processing Systems, pp. 4623–4633 (2017)

21. Lucic, M., Kurach, K., Michalski, M., Gelly, S., Bousquet, O.: Are gans created equal? a large-scale study. arXiv preprint arXiv:1711.10337 (2017)

22. Goodfellow, I.J., Shlens, J., Szegedy, C.: Explaining and harnessing adversarial examples. arXiv preprint arXiv:1412.6572 (2014)

Influence of Difference of Spatial Information Obtained from a Moving Virtual Sound Presentation on Auditory BCI

Yuki Onodera[✉], Isao Nambu, and Yasuhiro Wada

Graduate School of Engineering, Nagaoka University of Technology,
1603-1 Kamitomioka, Nagaoka, Niigata 940-2188, Japan
onodera@stn.nagaokaut.ac.jp

Abstract. Brain-computer interface (BCI) technology can control external devices by using human brain activity. In a previous study, a compact BCI system for the estimation of the intended direction was realized by using virtual sound. Some data were averaged to improve the identification rate. Herein, we expected to obtain a higher identification rate with a small amount of averaged data by improving the localization accuracy of users. In this study, to investigate the effect of the difference of auditory stimulation methods (static sound and moving sound) on brain activity, we performed an experiment with six directions of auditory stimulation using each method and measured the brain activity of subjects. We used a variant of regularized Fisher's discriminant analysis to classify brain waves. As a result of comparing the identification rates obtained through each method, individual differences were observed in the effect on the identification rate, and an improvement was observed for moving sound as compared with static sound.

Keywords: Auditory BCI · EEG · P300 · Virtual sound

1 Introduction

Brain-computer interface (BCI) technology, which is used to control external devices such as computers by measuring human brain activity, is attracting research attention. BCI does not require voluntary movements of muscles. Therefore, it is expected to be used for people with serious movement disorders such as amyotrophic lateral sclerosis [1]. In the studies on BCI, event-related potentials (ERPs), which are measured using electroencephalography (EEG) in relation to some events, are commonly used. A typical example is P300 obtained in the oddball paradigm. P300 is a positive potential evoked approximately 300 ms after the presentation of various types of stimuli such as visual and auditory stimuli. While two or more different stimuli are randomly presented, it occurs when attention is paid to one of them (e.g., attending to the low-frequency stimulus from several different frequencies). In the BCI using ERP, visual stimulus presentation is a popular method for the estimation and input of characters or the recognition of visual image (present or absent) [2]. Discrimination of tone is also common in auditory stimulation [3].

© Springer Nature Switzerland AG 2018
L. Cheng et al. (Eds.): ICONIP 2018, LNCS 11307, pp. 577–584, 2018.
https://doi.org/10.1007/978-3-030-04239-4_52

In this study, we focused on auditory BCI using spatial information such as sound source direction. In such a BCI, it is necessary to prepare for many loudspeakers presenting an auditory stimulus and a wide space sufficient to place them. A compact BCI system has been proposed by using a virtual sound source generated by out-of-head sound localization technology [4]. Virtual sounds are generated by using individual sound localization transfer functions (SLTFs). The user localizes the sound image at the same position as the sound source existing in space by listening through earphones. Therefore, this system can present auditory stimulus to a subject from any position without placing a real sound source (i.e., loudspeakers).

As described before, we suggested the possibility of auditory BCI, which estimates the intended direction from the measured EEG (P300) while listening to virtual sounds. Moreover, to improve the performance of the auditory BCI, we increased the number of estimated directions and shortened the trial interval [5]. However, in the practical use of BCI, sufficient identification rate (or accuracy) is required. Usually, the identification rate is improved by averaging data. However, to obtain a sufficient identification rate, it is necessary to increase the number of averaging data sets; hence, long measuring time is required.

In this study, we aimed at obtaining a high identification rate with a small amount of averaging data. One of the means of improving the identification rate is to improve the quality of ERP evoked by auditory stimuli. When the subject listens to the virtual sound, the quality of evoked ERP will also be improved if the perception of direction of sounds becomes easier for the subject. In addition, differences in the method of presentation of stimuli may affect ERPs. Although there are individual differences in localization accuracy when listening to virtual sounds, it has been reported that the localization accuracy is improved by presenting moving sound rather than static sound [6]. Thus, moving virtual sound might be useful for changing ERP patterns and thus improving BCI performance.

In this study, to examine the effect of the differences in the presentation method of auditory stimuli on ERP, static sounds and moving sounds were prepared, and an experiment involving the task of direction discrimination through out-of-head sound localization was carried out for five subjects. During the experiment, we measured the EEG signal of the subjects. Subsequently, the identification rate for each presentation method was calculated from the EEG signal. Finally, we examined the effect of different presentation methods (static vs. moving sound) on the EEG and identification rate.

2 Method

2.1 Subjects

This study was approved by the ethics board of the Nagaoka University of Technology according to the Declaration of Helsinki. Five healthy people (5 males, ages 21–23) participated in this study. All the participants were informed about the experiment and they provided signed consent forms.

2.2   Auditory Stimuli (Virtual Sound)

The out-of-head sound localization can be realized by equalizing the stimulation in both ear canals generated by sound stimuli from a loudspeaker and earphones. To realize this technique, it is necessary to measure the transfer function having speaker characteristics (loudspeaker transfer function: LSTF), transfer characteristics from a real sound source to the subject's ear canal (spatial sound transfer function: SSTF), and transfer characteristics from the earphones to the ear canal (ear canal transfer function: ECTF) (refer Fig. 1). Thus, the SLTF is generated as follows:

$$SLTF(f) = \frac{SSTF(f)}{LSTF(f) \cdot ECTF(f)} \qquad (1)$$

The measurement was carried out by placing a microphone in both ear canals of the listener. Loudspeakers were positioned at intervals of 15° at the radius of 1.5 m around the subject. The height from the floor to the center of the speaker was 1.2 m. When measuring the SSTF, to acquire measurements at intervals of 5° in the horizontal plane, the subject was rotated at intervals of 5°.

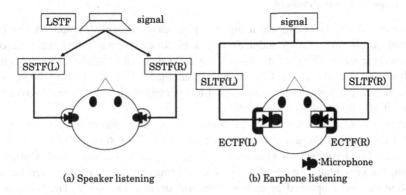

(a) Speaker listening   (b) Earphone listening

Fig. 1. Principle of out-of head sound localization.

The static auditory stimuli were generated by convoluting the measured impulse response and white noise. The length of these stimuli was 500 ms. In this study, the sound sources were allocated in six directions (30°, 90°, 150°, −30°, −90°, and −150°) with the front direction of the subject as 0° (Fig. 2(A)). To generate the moving auditory stimuli, a transfer function with a finer angle is required. Transfer functions of intervals 1° were generated from the transfer functions of intervals 5° via linear interpolation. The stimuli moving in one direction were reproduced by sequentially switching the static stimuli every 1°. With regard to the moving stimuli, 22.2 ms of the auditory stimulus was presented from one angle. Subsequently, adjacent sound sources, which overlapped by 3 ms, were presented. Subsequently, 22.2 ms of the sound was presented from the next angle. The length of the moving stimuli was 500 ms and the stimuli moved at 20° (velocity: 40°/s) during the presentation. The presentation range of the moving stimuli was set to ± 10° with respect to the six directions of the static stimuli. In this study, two kinds of moving stimuli, those in clockwise rotation (cr) and counter-clockwise rotation (cc) around the subject, were prepared (Fig. 2(B)).

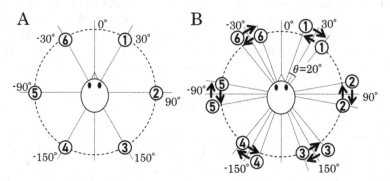

Fig. 2. Presentation position of auditory stimulus. (a) Static stimuli. (b) Moving stimuli.

2.3   Experimental Protocol

We measured the EEG signals using a digital electroencephalograph (ActiveTwo, Biosemi, Amsterdam, Netherlands). Sixty-four electrodes were attached to the subject's scalp in accordance with the international 10–20 system, and reference electrodes were attached to each earlobe. The EEG data were sampled at 256 Hz. During the EEG the measurement, the subjects were instructed to maintain their eyes closed to avoid the effects of visual stimuli and eye blinking.

Each trial was undertaken for 1100 ms with auditory stimuli of 500 ms and an interval time of 600 ms, and each session consisted of 78 ± 6 trials (13 ± 1 trials per direction). The frequency of presentation from each direction was equal. During the session, the presentation method of the stimuli remained constant. However, the presentation method employed in each session was decided randomly. The subjects did not know the details of the sounds presented during the session. In addition, the subjects were instructed to pay attention to one direction (target direction) and to silently count the number of times they localized the sound from the target direction during the session. The target direction was 30°, 90°, 150°, −30°, −90°, or −150°. Therefore, the subjects recognized one of the six directions as the target direction and the remaining five directions as non-target directions. We carried out four sessions for each direction as the target direction. Thirty-six sessions were carried out per day (each of the six directions was set twice as the target direction, and this process was applied to three kinds of presentation methods), and 72 sessions were carried out in two days.

2.4   Identification Method

First, to remove artifacts and noise not related to brain activity, the raw EEG data measured at $F_s = 256$ Hz were filtered using a Butterworth band-pass filter (forth-order, 0.1 to 8 Hz). Second, by setting the stimuli presentation onset as 0 ms in every trial, the data from -100 to 1100 ms were extracted as a single trial. The average over -100 to 0 ms in each trial was set as the baseline and was subtracted from each measurement data set (0 to 1100 ms). The measurement data were downsampled every 10 samples. The total number of measurement data point was 1440, and they were equally divided into training data and test data.

We used a variant of regularized Fisher's discriminant analysis (FDA) proposed by Gonzalez et al. [7] as the classification algorithm. In this method, the regularization parameter for FDA is determined using particle swarm optimization (PSO) with the measurement data as the feature value. Although this method can also select EEG channels to be used as feature values, in this experiment, all the channels were used for the measurement. The FDA projects D-dimensional inputs y onto a scalar space. Therefore, y is expressed as in Eq. (2), and weight w that maximize the distance between the two classes after projection was decided.

$$y = w^T \mathbf{x} \tag{2}$$

The evaluation function is expressed by Eq. (3), where $(m_{target} - m_{non\_target})$ is the difference between the means of each class, and the numerator is the inner product of w and $(m_{target} - m_{non\_target})$. Further, S_w is the in-class covariance matrix and λ is the regularization parameter of FDA determined using PSO.

$$J(w) = \frac{\langle w, m_{target} - m_{non_{target}} \rangle^2}{w^T S_w w + \lambda \|w\|^2} \tag{3}$$

The test data were evaluated using the presumed decision surface. When inputting the test data, the class of the input data is determined by judging which of the target and non-target directions each training data set is close to. Here, the score is defined as the Mahalanobis distance between the input data and the decision surface. The score is 0 at the decision surface and it is judged that the probability of the target is larger as the score is larger and the probability of the non-target is larger if the score is smaller.

With regard to the procedure for the estimation of direction, in this study, one direction out of the six directions in which auditory stimulation was presented was used as the target. Therefore, when determining the identification rate, one data set is obtained from the presentation in each direction (six trials) and the case where the target can be correctly estimated in a data set is considered as the correct answer (six-class classification). This identification was carried out for all data sets, and the correct rate was considered as the identification rate. The direction in which the score value is the maximum in the data set was estimated as the target direction. Regarding the identification rate with respect to the number of trials averaged, averaging was performed on the score of a single trial in each direction, and the direction with the maximum averaging score was estimated as the target direction.

3 Results

Figure 3 shows the average identification rates of the five subjects. Figure 4 respectively shows the identification rates of the each Sub.B and Sub.D. According to Fig. 3 when averaging 10 trials, the identification rates were approximately 62.2% for static sound, 51.0% for cr, and 59.5% for cc conditions. Therefore, it is conceivable that the difference in the direction of moving also affects the EEG. Regarding the influence of the difference in the presentation method on the identification rate, we confirmed that there are individual differences. For Sub.B, when averaging 10 trials, the identification rate was 49.2% for static, 44.2% for cr, and 65.8% for cc. Thus, the identification rate was improved for cc while it was decreased for cr (Fig. 4(a)). For Sub.D, when averaging 10 trials, the identification rate converged to about 98.3% regardless of the presentation method. However, when averaging 3 trials, the identification rate for the moving sound was improved by about 6.7% over static (a similar trend was also confirmed in Sub.C). For other subjects, the trend deteriorated by moving sound could also be confirmed.

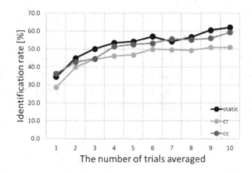

Fig. 3. Results of the identification rate with the grand average across five subjects.

For Sub.B and Sub.D, the average ERP data measured at Fz, Cz, and Pz are shown in Fig. 5. The black lines represent the ERP at the target presentation and the gray lines represent the ERP at the non-target presentation. For ERP in each presentation method, the static stimuli are denoted as "static," the stimuli moving in clockwise rotation are denoted "cr," and the stimuli moving in counter-clockwise rotation are denoted "cc." For the amplitude of ERP in the range 300–500 ms after auditory stimuli presentation, differences between the target and non-target are observed at Fz in cr in Sub.B, and Pz in each presentation method in Sub.D. In terms of the influence of the presentation method on the ERP, differences were observed overall after 500 ms. In contrast, differences were observed in part immediately after the stimulus presentation.

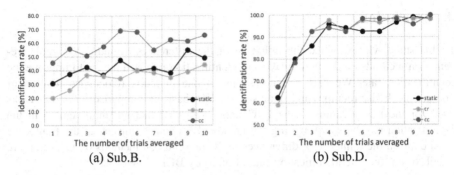

(a) Sub.B.   (b) Sub.D.

Fig. 4. The result of the identification rate of one subject.

In this experiment, the static stimuli were reproduced from 0 ms in the present direction, whereas the moving sound was presented from the position shifted by 10° from the desired position. Therefore, it was conceivable that there was a time lag in the recognition of listening from the attention direction of the subject. However, from the result of Fig. 5, no difference was observed much immediately after the presentation of stimuli (300–500 ms). Accordingly, the subject judged whether the presented sound is target or non-target immediately after the presentation of stimuli. Moreover, there was ambiguity in the recognition of localization direction.

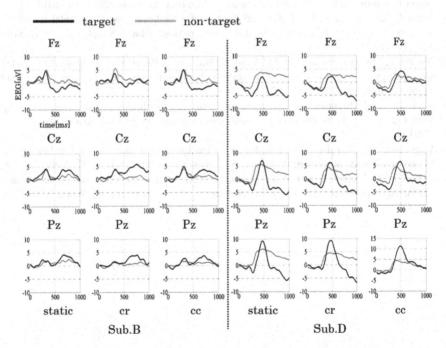

Fig. 5. Comparison of measured EEG owing to the differences in the presentation methods of four subjects.

4 Conclusion

In this study, we examined the influence of the differences in auditory stimuli presentation methods on the brain wave and identification rate as a means to improve the identification rate. We used moving sound in one direction (clockwise rotation or counter-clockwise rotation) and static sound as the basis of comparison. The FDA was used for the identification of the EEG data. We confirmed an improvement in the identification rate by using moving virtual sounds in some subjects. In the future, we must confirm the influence of differences in the presentation methods on EEG in more detail to improve the identification rate of auditory BCI.

Acknowledgments. This work was partly supported by Nagai N-S Promotion Foundation for Science of Perception, Nagaoka University of Technology Presidential Research Grant, and Japan Society for the Science Promotion Kakenhi Grant Number 16K00182.

References

1. Wolpaw, J.R., Birbaumer, N., McFarland, D.J., Pfurtscheller, G., Vaughan, T.M.: Brain-computer interfaces for communication and control. Clin. Neuro-Physiol. **113**(6), 767–791 (2002)
2. Stewart, A.X., Nuthmann, A., Sanguinetti, G.: Single-trial classification of EEG in a visual object task using ICA and machine learning. J. Neurosci. Methods **228**, 1–14 (2014)
3. Farwell, L.A., Donchin, E.: Talking off the top of your head: toward a mental prosthesis utilizing event-related brain potentials. Electroencephalogr. Clin. Neurophysiol. **70**(6), 510–523 (1988)
4. Nambu, I., Ebisawa, M., Kogure, M., Yano, S., Hokari, H., Wada, Y.: Estimating the intended sound direction of the user: toward an auditory brain-computer interface using out-of-head sound localization. PLoS ONE **8**(2), e57174 (2013)
5. Sugi, M., et al.: Improving the performance of an auditory brain-computer interface using virtual sound sources by shortening stimulus onset asynchrony. Front. Neuroscience. **12**, 108 (2018)
6. Wightman, F.L., Kistler, D.J.: Resolution of front-back ambiguity in spatial hearing by listener and source movement. J. Acoust. Soc. America. **105**(5), 2841–2853 (1999)
7. Gonzalez, A., Nambu, I., Hokari, H., Wada, Y.: EEG channel selection using particle swarm optimization for the classification of auditory event-related potentials. Sci. World J. **2014**, 11 (2014)

Association Study of Alzheimer's Disease with Tree-Guided Sparse Canonical Correlation Analysis

Shangchen Zhou[1,2], Shuai Yuan[1], Zhizhuo Zhang[3], and Zenglin Xu[1(✉)]

[1] SMILE Lab, School of Computer Science and Engineering, University of Electronic Science and Technology of China, Chengdu 611731, Sichuan, China
shangchenzhou@gmail.com, shuaiyuan0209@gmail.com, zenglin@gmail.com
[2] School of Computer Science and Technology, Harbin Institute of Technology, Harbin 150001, Heilongjiang, China
[3] Computer Science and Artificial Intelligence Laboratory, Massachusetts Institute of Technology, Cambridge, MA 02139, USA
zhizhuo@mit.edu

Abstract. We consider the problem of finding the sparse associations between two sources of data, for example the sparse association between genetic variations (e.g., single nucleotide polymorphisms, SNPs) and phenotypical features (e.g., magnetic resonance imaging, MRI) in the study of Alzheimer's disease (AD). Despite the success of Canonical Correlation Analysis (CCA) based its sparse variants in a number of applications, they usually neglect the underlying natural tree structures SNPs and MRI data. Specifically, the whole candidate set, genes, SNPs of gene form a path of tree structure in SNPs data, and the whole image, regions of image, features of region form a path of tree structure in the MRI data. In order to model the tree structure of features in both sources of data, in this paper, we propose a Tree-guided Sparse Canonical Correlation Analysis (TSCCA). The proposed model equips CCA with special mixed-norm regularization terms in order to model the underlying multilevel tree structures among both the inputs and outputs. To solve the resulted complicated optimization problem, we introduce an efficient iterative algorithm for TSCCA by rewriting tree-structured regularization into the common form of overlapping group lasso. To evaluate the proposed model, we have designed the simulation study and real world study respectively on Alzheimer's disease. Experimental results on the simulation study have shown that the proposed method outperforms CCA with Lasso and group Lasso. The real world study on Alzheimer's disease has shown that our model can find biologically meaningful associations between SNPs and MRI features.

Keywords: Tree-guided Sparse · Canonical correlation analysis
Association study · Alzheimer's disease

© Springer Nature Switzerland AG 2018
L. Cheng et al. (Eds.): ICONIP 2018, LNCS 11307, pp. 585–597, 2018.
https://doi.org/10.1007/978-3-030-04239-4_53

1   Introduction

Many real world problems in machine learning and science discovery amount to finding a sparse and consistent mapping between one source of high dimensional features to another source of output signals. For example, in social media study, one wants to find the association between the objects in images and the labels raised by users [5]. In bioinformatics, one wants to find the association between a selected set of single nucleotide polymorphism and the output genetic expressions, which is known as expression quantitative trait loci (eQTL) mapping [4]. In the study of many complex diseases such as Alzheimer's disease, identifying associations between genetic variations and intermediate phenotypes is crucial [6,7]. In other words, a key step in this task is to discover cross linkages between genetic risk factors based on genomic data——such as SNPs, and indicative intermediate phenotypes——such as brain regions abnormalities measured by MRI. This result can help us find out a subset of SNPs which may have functional consequences potentially on brain region structures.

Many approaches have been proposed to solve this problem including canonical correlation analysis [8] and its variants with various sparse priors [2,3,6,7,18]. For example, Parkhomenko et al. [18] applied sparse CCA (SCCA) to find relationships between genetic loci and gene expression levels in Utah families; Witten and Tibshirani [3] used SCCA to reveal associations between gene expression and DNA copy variation; and Chen et al. [2] used structured CCA for pathway selection. In general, by assuming the independence of the output variables, sparse priors such as Lasso [21], elastic net [36], group Lasso or equivalently multiple kernel learning [11,24–28,30–32], overlapping group Lasso [1,9,10], and Bayesian automatic relevance determination [15,17] can be applied, where the non-zero coefficients can be interpreted as the markers truly associated with the output variables. Despite the success in some applications, these sparse priors suffer from various problems, for example, the features selected by Lasso may randomly distribute throughout the whole feature set, but neglect structured information in features. Group Lasso aiming modeling the group structure of features either does not consider overlaps among groups or cause an imbalance among different features due to over penalization on large groups.

However, the structure information underneath data in real world problems can be more complex. For Example, in study of person re-identification, Zhou et al. [35] divide the person image into three parts (head, torso and legs), which can be represented by a tree-structured feature. And in the study of Alzheimer's disease, both the SNPs and the MRI features naturally form a tree structure (as shown in Fig. 1): the whole set of corresponding genes related to the AD study forms the root node whose children are the genes and grandchildren are the SNPs; the whole MRI image forms the root of a tree with different regions (e.g. Left Hippocampus) of the image being children and each feature (e.g., volume of Left Hippocampus) being the grandchildren. In order to model the tree structure of features in both sources of data, in this paper, we propose a Tree-guided Sparse Canonical Component Analysis (TSCCA). The proposed model equips CCA with special mixed-norm regularization terms in order to model the underlying multilevel tree (i.e., hierarchical) structures among both the inputs and outputs motivated from literature on tree-structured regularization [13,14]. To solve the resulted complicated optimization problem, we introduce an efficient iterative algorithm for TSCCA by rewriting tree-structured regularization into the common form of overlapping group lasso. To evaluate the proposed model, we have designed a simulation study and a real world study on Alzheimer's disease. Experimental results on the simulation study have shown that the proposed method outperforms CCA with Lasso and CCA with group Lasso. The real world study on Alzheimer's disease has shown that our model can discover meaningful correlations between SNPs and MRI features.

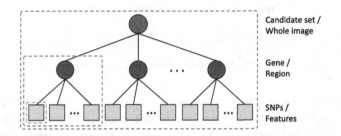

Fig. 1. The illustration the tree structures of SNPs and MRI data. The whole feature set of genes and MRI features are considered as foot nodes of the two trees, respectively; the internal nodes are groups of features, which are grouped based on corresponding genes and regions, respectively; The individual features of SNPs and MRI are considered as leaf nodes.

2   Methods

Before presenting the model, we first introduce some basic notations. In our paper, there are two sources of data (e.g., SNPs and MRI), denoted by $X = [x_1, \ldots, x_n]$ and $Y = [y_1, \ldots, y_n]$, respectively. Here both X and Y contain n subjects, and described by p-dimensional and q-dimensional feature vectors, respectively.

2.1   Model

Sparse Canonical Correlation Analysis. Given the two data sources X and Y, the variance matrices of data X, Y are denoted by Σ_{XX} and Σ_{YY}, respectively. And the covariance matrix between X and Y is denoted by Σ_{XY} or Σ_{YX}. The CCA method aims to find two projection vectors α and β, i.e., the linear combinations of two data variables, which maximize the correlation between these two linear combinations (i.e., $\alpha^\top X$ and $\beta^\top Y$), as shown in the Eq. (1):

$$\max_{\alpha,\beta} \ \alpha^\top \Sigma_{XY}\beta \tag{1}$$

$$\text{s. t. } \alpha^\top \Sigma_{XY}\alpha = 1, \ \beta^\top \Sigma_{XY}\beta = 1.$$

In practical applications, the amount of samples may be much smaller than the high dimensions of data (i.e., $p, q \gg n$). To address this issue, sparse methods such as Lasso [21], elastic net [36], and group Lasso [32], can be imposed on the projection vectors as the penalty function, as shown in the Eq. (2).

$$\min_{\alpha,\beta} \ -\alpha^\top \Sigma_{XY}\beta + \lambda_1\Psi(\alpha) + \lambda_2\Phi(\beta) \tag{2}$$

$$\text{s. t. } \alpha^\top \Sigma_{XX}\alpha \le 1, \ \beta^\top \Sigma_{YY}\beta \le 1,$$

where $\Psi(\cdot)$ and $\Phi(\cdot)$ denote the penalty function on α and β, respectively.

Tree-Structure Modelling. As shown in Fig. 1, the features of SNPs and MRI data form a tree structure, respectively. In order to select important features in the hierarchical tree structures, we introduce the tree-structured lasso into the CCA model. In tree structure, we observe that the overlapping between nodes may exist: any child node overlaps with its parent nodes, actually, the child node's feature set belongs to that of its parent node; the nodes from the same depth in the tree do not overlap.

Let T be an index tree with a depth d, and let $T_i = \{G_1^i, G_2^i, \ldots, G_{n_i}^i\}$ denote the set of nodes corresponding to the depth i, where $n_0 = 1$, $G_1^0 = \{1, 2, \ldots, p\}$ and $n_i \ge 1$, $i = 1, 2, \ldots, d$. Then this set of nodes satisfies the following conditions: (1) the nodes from the same tree depth have non-overlapping features, i.e., $G_j^i \cap G_k^i = \emptyset$, $\forall_i \in \{1, 2, \ldots, d\}$, $j \ne k, 1 \le j, k \le n_i$; and (2) If $G_{j_0}^{i-1}$ is the parent node of a non-root node G_j^i, then $G_j^i \subseteq G_{j_0}^{i-1}$.

Given the tree structure, the tree-structured regularization can be defined as:

$$\Phi(x) = \sum_{i=0}^{d}\sum_{j=1}^{n_i} \omega_j^i \|x_{G_j^i}\|_2 + \frac{h}{2}x^\top x, \tag{3}$$

where $x \in \mathbb{R}^p$, $\omega_j^i \ge 0$ $(i = 0, 1, \ldots, d, j = 1, 2, \ldots, n_i)$ denotes the pre-defined weight for the node G_j^i, $\|\cdot\|$ denotes the Euclidean norm, h is a tuning parameter, and $x_{G_j^i}$ is a vector composed of the feature of x with indices in G_j^i. To address the problem of collinearity, the ridge penalty $\frac{h}{2}x^\top x$ is introduced in Eq. (3). Tree-guided group Lasso can be seen as a special case of the overlapping group Lasso [13], which models the overlapping groups through a tree structure.

Each group of coefficients $x_{G_j^i}$ in Eq. (3) is weighted with ω_j^i. For convenience of subsequent discussion, we denote $\omega_v = \omega_j^i$, where v is the jth node of the depth i. Existing overlapping group Lasso methods may bring an imbalanced penalty among different features. To address this downside, we follow a weight-balanced scheme. We define ω_v in terms of two quantities g_v and s_v, where $g_v + s_v = 1$. The s_v is the weight for selecting the variables according to each of the children node v separately, and g_v is the weight to select these nodes jointly. Now, for a given tree T, we apply the Eq. (4) operation recursively, from the root node towards the leaf nodes.

$$\sum_{i=0}^{d_2} \sum_{j}^{m_i} \omega_j^i \|\beta_{G_j^i}\|_2 = W(v_{root}), \tag{4}$$

where

$$W(v) = \begin{cases} s_v \cdot \sum_{c \in \text{Children}(v)} |W(c)| + g_v \cdot \|\beta_{G_v}\|_2, & (a) \\ |\beta_{G_v}|, & (b) \end{cases}$$

In the above equation, (a) represents that v is an internal node, and (b) represents that v represents the leaf node.

It is shown the relationship in Eq. (5) holds between ω_v's and (s_v, g_v)'s [13].

$$\omega_v = \begin{cases} g_v \prod_{m \in \text{Ancestors}(v)} s_m, \\ \prod_{m \in \text{Ancestors}(v)} s_m, \end{cases} \tag{5}$$

Actually, the above weighting scheme extends the elastic-net-like penalty hierarchically, which result in a balance of our model. It is easy to verify that the following proposition holds [13]:

Proposition 1. *For feature k, the sum of the weights ω_v for the nodes $v \in T$ whose feature group G_v contains the feature k equals one, i.e., as following holds:*

$$\sum_{v:k \in G_v} \omega_v = \prod_{m \in \text{Ancestors}(v_{leaf})} s_m$$

$$+ \sum_{l \in Ancestors(v_{leaf})} g_l \prod_{m \in \text{Ancestors}(l)} s_m = 1.$$

Tree-Guided Sparse CCA. In this study, since the features can be naturally be represented using certain tree structures as shown in Fig. 1, we incorporate the tree-guided sparse penalty on the both source of features into the design of CCA models. This results in a tree-guided sparse CCA (TSCCA), as shown in the following equation:

$$\min_{\alpha,\beta} \ -\alpha^\top \Sigma_{XY} \beta + \lambda_1 \Phi_1(\alpha) + \lambda_2 \Phi_2(\beta) \tag{6}$$

$$\text{s. t. } \alpha^\top \Sigma_{XY} \alpha \leq 1, \ \beta^\top \Sigma_{XY} \beta \leq 1,$$

where $\Phi_1(\alpha) = \sum_{i=0}^{d_1} \sum_{j}^{n_i} \psi_j^i \|\alpha_{G_j^i}\|_2 + \frac{h_1}{2} \alpha^\top \alpha$ and $\Phi_2(\beta) = \sum_{i=0}^{d_2} \sum_{j}^{m_i} \omega_j^i \|\beta_{G_j^i}\|_2 + \frac{h_2}{2} \beta^\top \beta$.

2.2 Optimization Algorithm

The main difficulty in solving Eq. (6) grows from the tree-structured regularization. Motivated by [2], we transform the tree-structured regularization into the common form of the overlapping group lasso. Let the nodes of the index tree be numbered from left to right and from top to bottom (e.g. $G_1^0 = 1, G_1^1 = 2, ..., G_j^i = k, ..., G_{n_d}^d = l$). Then the tree-structured regularization can be rewritten as:

$$\Phi(x) = \sum_{k=1}^{l} \omega_k \|x_k\|_2 + \frac{h}{2} x^\top x. \tag{7}$$

For illustrating easily, we assume that $\Phi_1(\alpha) = \|\alpha\|_1$. Thus, we get the following equation:

$$\min_{\alpha, \beta} -\alpha^\top \Sigma_{XY} \beta + \lambda \sum_{k=1}^{l} \omega_k \|\beta_k\|_2 + \frac{h}{2} \beta^\top \beta \tag{8}$$

$$\text{s. t. } \alpha^\top \Sigma_{XY} \alpha \leq 1, \; \beta^\top \Sigma_{XY} \beta \leq 1, \; \|\alpha\|_1 \leq 1.$$

Let the β domain be denoted as $Q_1 = \{\beta | \|\beta\|_2 \leq 1\}$, $v = \frac{1}{\tau} Y^\top X \alpha$ and $\gamma = \frac{\lambda}{\tau}$. Then the optimization of Eq. (6) respecting β can be written as:

$$\min_{\beta \in Q_1} f(\beta) \equiv l(\beta) + \Phi(\beta), \tag{9}$$

where $l(\beta) = \frac{1}{2} \|\beta - v\|_2$ is the Euclidean distance loss function and $\Phi(\beta)$ is the rewritten tree-structured regularization. According to the Theorems 1–3 in [2] and the Fenchel duality theorem, the optimization problems in the form like Eq. (9) can be solved by the Excessive Gap Method. The details of proof and calculation process can be found in [2].

3 Results

3.1 Simulation Study

At first, we design a simulation study to examine TSCCA to estimate the accuracy on finding cross linkages between the two source of data, to investigate whether TSCCA can improve the detection power compared to the other sparse model.

Simulation Data. For generating the ground truth, we simulate two data sets X and Y, consisting of $p = 400$ and $q = 300$ variables (features), which the index tree structures in α and β are given as a priori. Data X and Y are divided into $G_X = 20$ and $G_Y = 10$ groups respectively as internal nodes. For simplicity, the feature size of internal nodes are the same in both X and Y. We set the depth to be 3, i.e., $d = 3$. Each individual feature in X and Y is a leaf node, and the whole feature set in both X and Y is considered as the foot node. We set the sample size n as 200. To append the potential correlation between variables in X and variables in Y, a latent variable is set $\Upsilon = \{\gamma_i | i = 1, \ldots, n\}$ with a Gaussian distribution $N(0, \sigma_\gamma^2)$ and normalize all the variables to the unit length, which have the similar effect on the correlated variables from two source of data. We generate X and Y with each sample $x_i \sim N(\gamma_i \alpha, \sigma_e^2 \Sigma_{XX})$, and $y_i \sim N(\gamma_i \beta, \sigma_e^2 \Sigma_{YY})$, where $\alpha = [\alpha_1, \ldots, \alpha_p]$, $\beta = [\beta_1, \ldots, \beta_q]$ are the projection vectors of X and Y, respectively, where $\alpha_j \neq 0$, $\beta_k \neq 0$, if x_j, y_k are the correlated variables, σ_e^2 is the variance of noise variable, and Σ_{XX} and Σ_{YY} are the variance-covariance matrices of X, Y data which are used to simulate the tree-structured group effect within each dataset. We set $\Sigma_{ij} = \rho^{|i-j|}$, while variables i and j are correlated variables, i.e., they are included in the same group, where ρ is preferred to be 0.5 referring to [16, 32].

For performance evaluation on the sparsity, we adopt the F1-measure to evaluating different models. We compare the proposed model with current state-of-the-art sparse association study methods, including CCA with lasso sparse penalty (denoted by CCA-Lasso) and CCA with Group Lasso penalty (denoted by CCA-Group Lasso).

In the following, we first compare the recovering results of correlated variables from two source of date among different sparse CCA methods.

Figures 2 and 3 show the results of recovered projection vectors α and β by CCA-Lasso, CCA-Group Lasso and TSCCA methods, respectively. It shows that TSCCA can estimate the α and β more accurate than the other two sparse models. The CCA-Lasso misses out some true variables, and CCA-Group Lasso selects more noise variables. The TSCCA not only can recover the tree structure between two source of date accurately but also can distinguish noise variables.

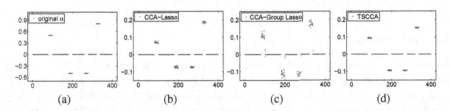

$$
\begin{array}{cccc}
\text{(a)} & \text{(b)} & \text{(c)} & \text{(d)}
\end{array}
$$

Fig. 2. Performance comparison in recovering α of three sparse CCA models. (a) Ground Truth α; (b–d) α recovered by CCA-Lasso, CCA-Group Lasso and TSCCA.

Secondly, we discuss the how the sample size of data effects the recovery performance. We increase the sample size n from 50 to 250, with step of 50, comparing the performance of these sparse models. Figure 4(a) presents the F1 score with different sample size. It shows that, with the sample size increasing, TSCCA keeps the best F1-measure among these methods.

Finally, we discuss how noise in data affects the model performance. To compare these models under different noise levels, we fixed other conditions but adjust the standard deviation σ_e, starting from 0.1 to 1 with step 0.1, manipulating the relation coefficient between two datasets. From Fig. 4(b), we can see that TSCCA is more robust than other methods whenever the noise level increase.

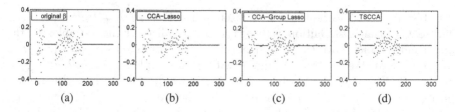

(a)   (b)   (c)   (d)

Fig. 3. Performance comparison in recovering β of three sparse CCA models. (a) Ground Truth β; (b–d) β recovered by CCA-Lasso, CCA- Group Lasso and TSCCA.

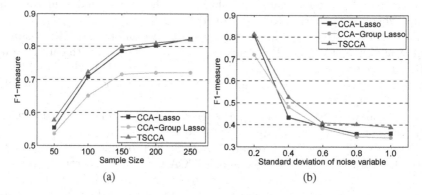

(a)   (b)

Fig. 4. A comparison of three SCCA models in different situations. (a) F1-measures for three methods when varying sample sizes from 50 to 250. (b) F1-measures for three methods when varying the standard deviation of noises.

3.2   AD Study

We conduct association analysis study based on a dataset called Alzheimer's Disease Neuroimaging Initiative (ADNI). The ADNI study is a longitudinal multi-site observational study of the elderly individuals with different cognition conditions (i.e., normal cognition, mild cognitive impairment (MCI) or AD). We applied our model TSCCA to discover the correlation between genetic variations and brain regions atrophy measured by MRI.

The ADNI dataset is available on adni.loni.ucla.edu. After removing samples with missing values, it contains 618 samples (182 normal, 302 MCI and 134 AD) and each sample includes 924 SNPs (selected as the top SNPs with high discriminative power to separate normal subjects from MCI and AD) and 328 MRI features (measuring the brain atrophies in different regions about surface area, volume or cortical thickness with FreeSurfer software). Based on the feature belongs to different brain regions, we divided the 328 MRI features into 118 groups, which are considered as internal nodes. We first obtain the gene that each SNP belongs to by analyzing the SNPs information downloaded from www. ncbi.nlm.nih.gov/projects/SNP/dbSNP.cgi?list=rslist, then we divided the 937 SNPs into 529 groups based on the SNP belongs to different genes, which are also considered as internal nodes. Each individual feature of SNPs and MRI is considered as leaf node, and the whole feature set of both SNPs and MRI is considered as foot node, hence SNPs and MRI data naturally be represented using two certain tree structures with depth three.

We apply our proposed method TSCCA on association discovery of Alzheimer's Disease using SNPs and MRI features. Lastly, the heatmap Fig. 5 shows the biclustering of the SNPs-MRI cross linkages, witch reveals meaningful correlations between genetic variations and brain atrophy. For example, the top ranked 19 SNPs are included in a few genes, such as PVRL2 (rs8105340), TOMM40 (rs2075650), HK2 (rs3771773), MAGI2 (rs508990), of which some have been studied more carefully in AD (www.alzgene.org); In selected 22 MRI features, Fusiform, Middle Temporal, and Hippocampus play a important role in formating long-term memory.

4   Discussion

Since the dimensions of SNPs and MRI features are far larger than sample size in ADNI dataset, the standard CCA models have the problem of overfitting, which cannot be applied directly. To handle this problem, many methods impose an $L1-$norm on α and β as a penalty function to shrink the coefficients of the irrelevant variables toward zero [19, 22, 23]. Corresponding models are referred to sparse CCA. However, sparse learning with $L1-$norm is limited which neglects the latent rich structural information among data. Actually, data structure as prior information is crucial to improve model performance and interpret ability, especially when leaning from high-dimensional data.

Recently, To take advantage of the structure prior knowledge of data, various extensions of $L1-$norm penalty have been proposed, such as the elastic net [36], the Group Lasso penalty [32], overlapping Group Lasso [2], as well as the mixed-norm tree-structured penalty [12–14]. However, the tree-structured penalty has not been incorporated into the CCA framework for AD study. The main challenge arises from the computational side. For solving the group-sparsity regularized optimization problem, [14] showed the Moreau-Yosida regularization associated with the tree structured group Lasso admits an analytical solution, and design efficient algorithm for solving the problem for smooth convex loss functions; and Kim [13] adopted the smoothing proximal gradient approach to solve the problem.

For associations discovery study, [29, 33, 34] presented Bayesian multiview learning models for joint associations discovery and disease prediction in the AD study. Different from the proposed unsupervised association study, those methods also require the given disease status as an input. It is also important to note that there are some approaches aim to make diagnosis based on MRI imaging data [20], while the focus of this paper is to find the associations.

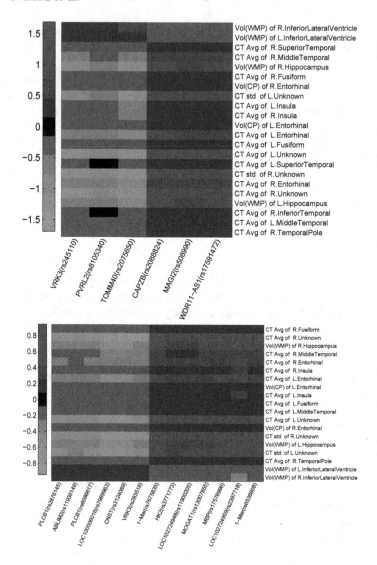

Fig. 5. The heatmap estimated by TSCCA, which shows the associations between SNPs and MRI features. Where, the SNP names are given at the bottom and the MRI features are listed on the right.

5   Conclusions

In this paper, we presented, TSCCA, a tree-guided sparse CCA model for associations discovery in the Alzheimer's disease study. We also develop an efficient optimization algorithm for the proposed model. Our experimental results on the simulation study have demonstrated that the proposed method can discover more rich structure information. Experimental results on the real world AD study has shown that the proposed method can output biologically meaningful results and we plan to apply our model to a wide range of applications in bioinformatics and computer vision involving the tree structure.

Acknowledgements. Datasets used in this paper are obtained from the Alzheimer's Disease Neuroimaging Initiative (ADNI) database (ADNI official website: adni.loni.ucla.edu). The investigators who contributed to the design and implementation of ADNI and/or collected data can be found on ADNI official website.

This work was in part supported by grants of NSF China (No. 61572111), a 985 Project of UESTC (No. A1098531023601041) and a Fundamental Research Project of China Central Universities (No. ZYGX2016Z003).

References

1. Chen, J., Bushman, F.D., Lewis, J.D., Wu, G.D., Li, H.: Structure-constrained sparse canonical correlation analysis with an application to microbiome data analysis. Biostatistics **14**(2), 244–258 (2013)
2. Chen, X., Liu, H., Carbonell, J.G.: Structured sparse canonical correlation analysis. In: International Conference on Artificial Intelligence and Statistics, pp. 199–207 (2012)
3. Daniela, M., Tibshirani, R.: Extensions of sparse canonical correlation analysis, with applications to genomic data. Stat. Appl. Genet. Mol. Biol. **383**(1), 1–27 (2009)
4. Du, L., et al.: Pattern discovery in brain imaging genetics via scca modeling with a generic non-convex penalty. Sci. Rep. **7**(1), 14052 (2017)
5. Eisenschtat, A., Wolf, L.: Linking image and text with 2-way nets. In: The IEEE Conference on Computer Vision and Pattern Recognition (CVPR) (2017)
6. Hao, X., et al.: Mining outcome-relevant brain imaging genetic associations via three-way sparse canonical correlation analysis in alzheimer's disease. Sci. Rep. **7**, 44272 (2017)
7. Hao, X., et al.: Identification of associations between genotypes and longitudinal phenotypes via temporally-constrained group sparse canonical correlation analysis. Bioinformatics **33**(14), i341–i349 (2017)
8. Hotelling, H.: Relations between two sets of variates. Biometrika **28**, 321–377 (1936)
9. Jacob, L., Obozinski, G., Vert, J.P.: Group lasso with overlap and graph lasso. In: Proceedings of the 26th Annual international Conference on Machine Learning, pp. 433–440. ACM (2009)
10. Jenatton, R., Audibert, J.Y., Bach, F.: Structured variable selection with sparsity-inducing norms. J. Mach. Learn. Res. **12**, 2777–2824 (2011)
11. Kang, Z., Lu, X., Yi, J., Xu, Z.: Self-weighted multiple kernel learning for graph-based clustering and semi-supervised classification. In: Proceedings of the Twenty-Seventh International Joint Conference on Artificial Intelligence (IJCAI), pp. 2312–2318 (2018)
12. Kim, S., Xing, E.P.: Tree-guided group lasso for multi-task regression with structured sparsity. In: Proceedings of the 27th International Conference on Machine Learning (ICML 2010), pp. 543–550 (2010)
13. Kim, S., Xing, E.P., et al.: Tree-guided group lasso for multi-response regression with structured sparsity, with an application to eqtl mapping. Ann. Appl. Stat. **6**(3), 1095–1117 (2012)

14. Liu, J., Ye, J.: Moreau-yosida regularization for grouped tree structure learning. In: Advances in Neural Information Processing Systems, pp. 1459–1467 (2010)
15. MacKay, D.J.: Bayesian interpolation. Neural Comput. **4**(3), 415–447 (1991)
16. Meier, L., Van De Geer, S., Buhlmann, P.: The group lasso for logistic regression. J. R. Stat. Soc. Ser. B (Stat. Methodol.) **70**(1), 53–71 (2008)
17. Neal, R.M.: Bayesian Learning for Neural Networks, vol. 118, p. 118. Springer Science & Business Media, New York (1996)
18. Parkhomenko, E., Tritchler, D., Beyene, J.: Genome-wide sparse canonical correlation of gene expression with genotypes. BMC Proc. **1**(Suppl. 1), S119 (2007)
19. Parkhomenko, E., Tritchler, D., Beyene, J.: Sparse canonical correlation analysis with application to genomic data integration. Stat. Appl. Genet. Mol. Biol. **8**(1), 1–34 (2009)
20. Que, X., Ren, Y., Zhou, J., Xu, Z.: Regularized multi-source matrix factorization for diagnosis of Alzheimer's disease. In: Neural Information Processing - 24th International Conference, ICONIP, pp. 463–473 (2017)
21. Tibshirani, R.: Regression shrinkage and selection via the lasso. J. R. Stat. Soc. Ser. B (Methodol.) **58**, 267–288 (1994)
22. Witten, D.M., Tibshirani, R., Hastie, T.: A penalized matrix decomposition, with applications to sparse principal components and canonical correlation analysis. Biostatistics p. kxp008 (2009)
23. Witten, D.M., Tibshirani, R.J.: Extensions of sparse canonical correlation analysis with applications to genomic data. Stat. Appl. Genet. Mol. Biol. **8**(1), 1–27 (2009)
24. Xu, Z., Jin, R., King, I., Lyu, M.R.: An extended level method for efficient multiple kernel learning. In: Advances in Neural Information Processing Systems 21, Proceedings of the Twenty-Second Annual Conference on Neural Information Processing Systems (NIPS), pp. 1825–1832 (2008)
25. Xu, Z., Jin, R., Yang, H., King, I., Lyu, M.R.: Simple and efficient multiple kernel learning by group lasso. In: Proceedings of the 27th International Conference on Machine Learning (ICML 2010), pp. 1175–1182 (2010)
26. Xu, Z., Jin, R., Ye, J., Lyu, M.R., King, I.: Non-monotonic feature selection. In: Proceedings of the 26th Annual International Conference on Machine Learning (ICML), pp. 1145–1152 (2009)
27. Xu, Z., Jin, R., Zhu, S., Lyu, M.R., King, I.: Smooth optimization for effective multiple kernel learning. In: Proceedings of the Twenty-Fourth AAAI Conference on Artificial Intelligence (2010)
28. Xu, Z., King, I., Lyu, M.R., Jin, R.: Discriminative semi-supervised feature selection via manifold regularization. IEEE Trans. Neural Networks **21**(7), 1033–1047 (2010)
29. Xu, Z., Zhe, S., Qi, Y., Yu, P.: Association discovery and diagnosis of alzheimer's disease with bayesian multiview learning. J. Artif. Intell. Res. **56**, 247–268 (2016)
30. Yang, H., Xu, Z., King, I., Lyu, M.R.: Online learning for group lasso. In: Proceedings of the 27th International Conference on Machine Learning (ICML-10), pp. 1191–1198 (2010)
31. Yang, H., Xu, Z., Lyu, M.R., King, I.: Budget constrained non-monotonic feature selection. Neural Networks **71**, 214–224 (2015)
32. Yuan, M., Lin, Y.: Model selection and estimation in regression with grouped variables. J. R. Stat. Soc. Ser. B (Stat. Methodol.) **68**(1), 49–67 (2006)
33. Zhe, S., Xu, Z., Qi, Y., Yu, P.: Sparse bayesian multiview learning for simultaneous association discovery and diagnosis of alzheimer's disease. In: Proceedings of the Twenty-Ninth AAAI Conference on Artificial Intelligence, pp. 1966–1972 (2015)

34. Zhe, S., Xu, Z., Qi, Y., Yu, P., et al.: Joint association discovery and diagnosis of alzheimer's disease by supervised heterogeneous multiview learning. In: Pacific Symposium on Biocomputing, vol. 19. World Scientific (2014)
35. Zhou, S., Yao, H., Yu, W., Wang, Y.: Tree-guided group sparse based representation for person re-identification. In: Proceedings of the International Conference on Internet Multimedia Computing and Service, pp. 14–17. ACM (2016)
36. Zou, H., Hastie, T.: Regularization and variable selection via the elastic net. J. R. Stat. Soc. Ser. B (Stat. Methodol.) 67(2), 301–320 (2005)

Relevance of Frequency of Heart-Rate Peaks as Indicator of 'Biological' *Stress* Level

Meena Santhanagopalan[✉], Madhu Chetty, Cameron Foale, Sunil Aryal, and Britt Klein

Federation University, Ballarat, Australia
{msanthanagopalan,madhu.chetty,c.foale,sunil.aryal,
b.klein}@federation.edu.au
http://www.federation.edu.au

Abstract. The *biopsychosocial* (BPS) model proposes that health is best understood as a combination of bio-physiological, psychological and social determinants, and thus advocates for a far more comprehensive investigation of the relationships between 'mind-body' health. For this holistic analysis, we need a suitable measure to indicate participants' 'biological' *stress*. With the advent of wearable sensor devices, health monitoring is becoming easier. In this study, we focus on bio-physiological indicators of *stress*, from wearable devices using the heart-rate data. The analysis of such heart-rate data presents a set of practical challenges. We review various measures currently in use for *stress* measurement and their relevance and significance with the wearables' heart-rate data. In this paper, we propose to use the novel 'peak heart-rate count' metric to quantify level of 'biological' *stress*. Real life biometric data obtained from digital health intervention program was considered for the study. Our study indicates the significance of using frequency of 'peak heart-rate count' as a 'biological' *stress* measure.

Keywords: *Biopsychosocial* model · Bio-physiological data
Wearable · Heart-rate · *'Heart-rate peak count'* · 'Biological' *stress*
Biometric · Big data

1 Introduction

The *biopsychosocial* approach to health-care big data, systematically considers biological, psychological, and social factors and their complex interactions in understanding health, illness and health care delivery [2,10]. The *biopsychosocial* model of health [3]; comprising of biological factors (e.g. genetic, biochemical, physiological), psychological factors (e.g. mood, personality, behavior, neurocognitive) and social factors (e.g. cultural, familial, socioeconomic), represents the underlying trans-disciplinary nature of *biopsychosocial* dataset.

© Springer Nature Switzerland AG 2018
L. Cheng et al. (Eds.): ICONIP 2018, LNCS 11307, pp. 598–609, 2018.
https://doi.org/10.1007/978-3-030-04239-4_54

As a key part of our broader effort, to build a holistic *biopsychosocial* health metric, we studied the bio-physiological factors revealed by the biometric data. In this study, we aimed to measure the bio-physiological indicator of *stress*, using heart-rate data. In order to do so, it is necessary to identify and utilize an effective measure of heart-rate data, that can capture the magnitude of *stress*. We propose to use 'peak heart-rate count', as the measure of heart-rate data to indicate *stress*.

Stress [14], can be defined as a 'response to change in order to maintain the state of stability or homology that the body has maintained against the stimulus to break the mental and physical balance and the stability of the body'. *stress* can be used to indicate different things, depending on context in which it is examined. In the context of this paper, we measured 'biological' *stress*, experienced by the participant. In this study, such 'biological' *stress* refers to the ability of the participant to cope with the frequency of high heart-rate values marked by *peak(s)*.

In a clinical setting, heart-rate variability (HRV) data collected from electrocardiogram (ECG) records, have been a very popular measure to indicate *stress* [1,6,9,11,16]. The gold standard for clinical experiments has been to use HRV, and the best practice is to use ECG machines to capture HRV. Some prior studies [1,11] show the effectiveness of using *peak(s)* in ECG, as a measure to indicate *stress*. However, the use of ECG limits the analysis to controlled clinical settings. It cannot be easily used in uncontrolled 'real-world' settings.

With the digital revolution in wearable technology, users can now monitor their biometrics daily, in a 'real-world' uncontrolled setting. Such wearable technologies generate biometric big data. Wearables such as *heart-rate monitors*, offer the potential to leverage the strengths of the *biopsychosocial* model [3] of health. Prior work, reported so far, has focused predominantly on well-controlled clinical environments. Reviewing available literature [5,8,13], it is noticed that there is work that examines the heart-rate variability data, collected from wearable devices, to indicate *stress*. However, these studies have been done in an experimental setting and are validated using self-reported *stress* levels. The study done by Hao et al. [5] presents the practical challenges associated with measuring *stress* in real-life situations. To the best of our knowledge, there has been no study that has previously examined the heart-rate data from wearable devices in a 'real-world' setting to indicate 'biological' *stress*.

In this paper, we propose to use the novel 'peak heart-rate count' metric to quantify level of 'biological' *stress*, in order to compare the *stress* level between the experimental and control groups as well as between the mindfulness and physical activity sub-groups; within the experimental group. For each of the sample participant, 'heart-rate peak count' is computed for each day. The daily 'heart-rate peak count' are aggregated and the median 'heart-rate peak count' is computed for each participant in the two groups during the trial period. The spread of 'heart-rate peak count' is compared between the participants in the experimental and control group; as well as between the mindfulness and physical activity subgroups; within the experimental group. This was done, using measures such as range and IQR (Inter Quartile Range). Analyzing this kind of 'real-world' heart-rate data from wearable sensor devices, presents a practical set of limitations and challenges. The paper also includes an evaluation of the significance and challenges of using heart-rate data, obtained in 'real-world' setting from wearable devices; as a measure of 'biological' *stress*. The study in this paper proposes a novel 'peak heart-rate count', as a metric to quantify the level of 'biological' *stress*. The metric is applied to investigate real life *biopsychosocial* dataset for experimental studies.

The rest of the paper is organized as follows. Section 2 provides the relevant background information on the biopsychosocial dataset used for this study. Section 3 presents the relevance of using frequency of heart-rate *peak* as a 'biological' *stress* measure. Section 4 provides the details of the proposed method for measuring heart-rate data. In Sect. 5.2, we present quantitative analysis of biometric data set, using 'heart-rate peak count', as a measure of 'biological' *stress*. Section 5.3 concludes the paper.

2   *Biopsychosocial* Data

The study presented in this paper, is part of a larger research project that evaluates the *biopsychosocial* factors using 'real-world' *biopsychosocial* dataset, to determine the well-being of a subject. In this section, we present relevant background information on the *'digital health'* program. We describe the nature of heart-rate data, collected from the wearable device worn by the participants, and its relevance to develop a new metric called 'heart-rate peak count' to indicate level of 'biological' *stress*.

The *biopsychosocial* model is a broad view that attributes disease outcomes to the intricate, variable interaction of bio-physiological, psychological and social factors. The underlying *biopsychosocial* dataset used for this study is illustrated in Fig. 1. The physiological factors are represented by the biometric data collected from BASIS smart-watch (Basis Health Tracker for Fitness, Sleep & stress (2014 Ed.)); wearable device.

In 2015, the Faculty of Health (FOH), Federation University rolled out a *'digital health'* intervention program, inviting local Australians volunteers, aged between 19 to 59 to participate in the program. The goal of the *'digital health'* program was to investigate if the participants in the experimental group benefited from the interventions conducted. The *'digital health'* program enabled the collection of the unique *biopsychosocial* data. The biometric data used for the present study, is a subset of the *biopsychosocial* dataset. This study focuses on the heart-rate data collected as part of biometrics, using the *heart-rate monitor* device, that was worn by the participants. The participants were divided into two groups:

1. Experimental group - participants in this group, were randomly allocated to one of the two active *'digital health'* intervention conditions; i.e. (physical activity or mindfulness), during the trial period.

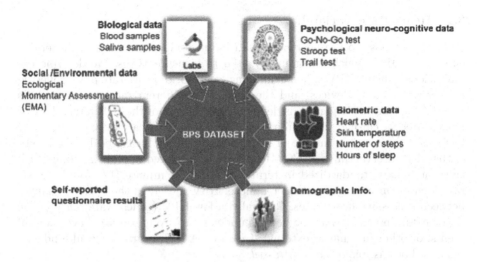

Fig. 1. *Biopsychosocial* dataset (BPS dataset)

2. Wait-list control group - participants in this group, did not participate in any intervention during the trial period of first 8 weeks.

The participants in both the experimental and the control group were asked to wear a BASIS smart-watch during the course of the trial, The biometrics such as heart-rate, skin temperature, GSR and number of steps were collected. The wearable was equipped with an optical heart-rate sensor, that continually tracked beats per second throughout the day, irrespective of what activity the participant was engaged in; i.e. *Inactive, Light activity, Moderate activity, Off* or *Sleep*. The activities and activity levels were set-up on the participants' wearable devices aligned to the experiment requirements. The control group participants however did not receive access to the *'digital health'* intervention programs (i.e. access to mindfulness or physical activity exercises), during the trial. The duration of the trial was for 8 weeks. After the post assessment, the control group had a delayed start in week 9 and could then pick one of the two interventions. There was a 1 month follow up which the control group participants did not go through. A sample of 20 participants' heart-rate data was used for this study; 10 from experimental group and 10 from the control group. The samples from the experimental group included 5 samples from the mindfulness subgroup and 5 samples from the physical activity subgroup.

3   Heart-Rate Count as 'Biological' *Stress* Measure

In this section, we describe the limitations of using HRV derived from wearable 'heart-rate' data; a measure to indicate 'biological' *stress*. We also describe the different biometric data that was collected by the BASIS smart-watch wearable device.

3.1   Heart-Rate Variability

While we propose to use heart-rate data in our studies, we present a brief discussion on HRV, which is commonly used to measure *stress*. We describe the limitations of using HRV, when examining 'real-world' heart-rate data, collected by wearable sensor devices, and the need for a different measure, to address this. HRV, a.k.a heart period, is defined as the length of the R-R interval. HRV is measured in cardiac time units (beat-by-beat) [12]. Heart-rate, on the other hand, is defined as the inverse of heart period usually measured by the length of the 'R-R interval' in the ECG. Heart-rate is measured in real time units such as seconds and standardized in terms of beats per minute [12]. While heart-rate focuses on average beats per minute, HRV measures the specific changes between successive heart-rates. Generally, a low HRV (or less variability in the heartbeats), indicates that the body is under *stress* from exercise, psychological events, or other internal or external stress-ors. A high heart-rate could indicate that the body is 'biologically' *stressed*.

In the context of 'real-world' biometric data collected from wearable devices, we observe the following limitations in the application of HRV to determine the level of 'biological' *stress*.

1. Although HRV can be extracted from time series heart-rate data, conditions such as missing lengths of biometric data, when the wearable was switched off or removed from hand, diminished the quality of data. The heart- rate data collected for the study was from a 'real-world' setting. The participation in the *'digital health'* intervention program was voluntary. This implied that participants were not obligated to hold on to the wearable all the time, during the experiment. This could have compromised the quality of HRV data if extracted from existing heart-rate data.
2. Imputing missing heart-rate data was not considered favorable. The biometric heart-rate data was collected by the wearable device in 'real-time', when the participants were engaged in different kinds of activities. Imputing this kind of 'real-time' biometric data is complex and difficult to achieve.

Owing to the above stated limitations, heart-rate was the choice of measure instead of HRV.

3.2   Wearable Biometric Data

Over the last few years, the world has seen wearable technology cover a sizable segment of the technology industry [4]. Smart watches are wrist worn devices that has 'real-time' fitness and activity tracking capabilities. Wearables accessories such as wrist bands and smart watches are in unique position to evolve and carry forward the more integrative, holistic, *biopsychosocial* model. The data collected from wearable, has the potential to indicate various biometric measures to the user, such that they can appropriately take actions to regulate them, steering the person towards elevated wellbeing.

The biometric data for the present study was generated from the BASIS smart-watch wearable, that was worn by the participants during the course of the *'digital health'* intervention program. The technology in this wearable device includes wrist sensors that allows for collection of biometric data from human vital signs and activities. The biometrics monitored by the BASIS watch include heart-rate, skin temperature, galvanic skin response (GSR) and other derived statistics such as number of hours of sleep. The wearable activity sensors measured attributes of gross user activity, different from narrowly focused vital signs sensors such as either accelerator magnitude (for the sensor) or acceleration (for number of steps).

4   The Proposed Method

In this paper, as stated earlier, we propose to explore the 'heart-rate peak count' as a metric on heart-rate data to measure 'biological' *stress* levels. This section provides the details of the proposed method to define and measure heart-rate peaks.

4.1   Defining Peak

We consider two design parameters to define *peak(s)* in the heart-rate time series data. The first parameter is the *interval* of *peak*. In our definition, this has been set to a 5 min limit. By setting the *peak interval* to 5 min, we treat the recurring *peak(s)* within a 5 min time span, as one single *peak* for that time span. Considering to set such a time *interval*, allows us to eliminate instances of counting separate recurring *peak(s)*, that may arise when a participant removes their wearable device and puts them back within a short span of 5 min. The second parameter used, is *maximum threshold* to define a *peak*.

A *heart-rate data point* is defined as a *peak* if the magnitude of the data point is >*maximum threshold* and the *interval* of identified *peak* is >5 min. We have following two approaches for defining *maximum threshold*.

1. Sum of Mean and SD (MSD) - *maximum threshold* is the sum of mean daily heart-rate and standard deviation computed on this mean.
2. Sum of Median and IQR (MedIQR) - *Maximum threshold* is the *sum* of median heart-rate for the day and IQR of heart-rate data for the day.

MSD approach is not suitable since the computation of mean and standard deviation values assumes that the distribution of heart-rate time series data is normal, whereas in reality the data distribution is skewed and not continuous in nature. Hence, We propose to use MedIQR approach which is simple yet robust to the out-lier data.

4.2   Reference for Measuring Peak

Resting heart-rate is defined as the number of heart contractions per minute while at rest [15]. In our work, the resting heart-rate is used to define *peak(s)* in the heart-rate distribution. The heart-rate data for the 20 sample participants is filtered for the activities *Sleep* and *Inactive* in order to compute the resting heart-rate.

Studies indicate that measurements of resting heart-rate taken in a sitting position tends to be 1–2 beats per minute, higher than in supine position [15]. Physical exercise or any external intervention such as intake of some kinds of medication to lower blood pressure, can affect the resting heart-rate [7]. As part of data setup for this study, the task was to investigate if the wearable reported heart-rate for activities *Sleep* and *Inactive*, truly described the heart-rate when the participant was at rest. If they did, this would allow us to use the resting heart-rate as the mean heart-rate.

The accelerator magnitude is a good reference measure to validate the resting heart-rate and was used to check this correlation of data collected by the wearable device during the activities *Sleep* and *Inactive*. A sample of 5 participants were chosen randomly for whom heart-rate values and accelerator magnitude values were available for different activity conditions.

5   Analysis

In this section, we present the analysis of the *biopsychosocial* data set using 'heart-rate peak count' as the measure of 'biological' *stress*. The participation to the *'digital health* program was voluntary and so the participants were not obligated to keep the wearables on throughout the experiment. The heart-rate time series data collected was not continuous in nature. The interruptions in the heart-rate data ranged from a few seconds to sometimes days, when no biometric heart-rate data was collected. This affected the quality of heart-rate data collected and reduced the quantity of good quality samples that could be used for study. The nature of the data distribution and the low quality of data, limited statistical tests such as ANOVA (Analysis of variance) to be applied on the data set. The IQR describes the H spread (mid-spread) and is the measure of the statistical dispersion between the 75th percentile (also denoted as quartile 3 or Q3) and 25th percentile (also denoted as quartile 1 or Q1). It is a measure of variability based on dividing the range of heart-rate data points into quartiles.

5.1   Variations in Accelerator Magnitude

Analyzing the heart-rate and accelerator magnitude data revealed the following.

1. Maximum value of accelerator magnitude dropped when the activity was *Sleep*.
2. Average values for accelerator magnitude dropped during activities *Sleep* or *Inactive* compared to conditions *Light activity* or *Moderate activity*.

Based on the above results, we conclude that heart-rate values during activity conditions *Sleep* and *Inactive* describe heart-rate during rest and can be used as mean heart-rate to define *peak(s)*.

5.2 'Peak Heart-Rate Count' Metric

The distribution of the heart-rate data used for the study was not normal. The distribution of heart-rate data for a sample participant is illustrated in Fig. 2.

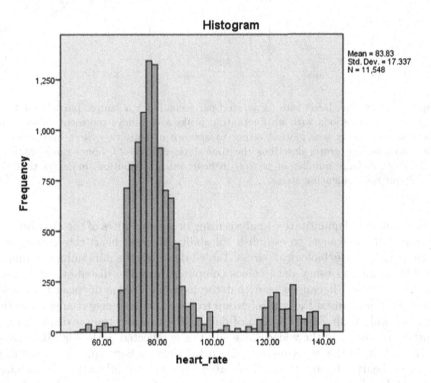

Fig. 2. Distribution of heart-rate data for sample participant on a random day, during conditions *Sleep* and *Inactive*. X-axis: heart- rate and Y-axis: frequency of the heart-rate.

In this study peak heart-rate is defined, as the data point, which is greater than the *maximum_threshold* value. The *maximum_threshold* value is the sum of median heart-rate and IQR, for the participant for the given day. The magnitude of stress was set to 5 min. The interruptions in the heart-rate time series data and the poor quality of 'real-world' heart-rate data limited the use of statistical tests, to compare the levels of 'biological' *stress* between the experimental and the control groups and between the mindfulness and physical activity subgroups (Fig. 3).

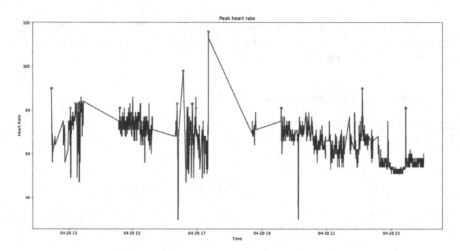

Fig. 3. Plot of peak heart-rate (generated per second) for a sample participant for a single day. The periods with no fluctuating peaks and valleys represent timestamps where no heart-rate was plotted owing to absence of activity - *Sleep* and *Activity*. The x-axis on the graph describes the time date: 2015-04-20, hours range 13:00:00 to 23:00:00. A large number of *peak(s)* in heart-rate distribution, indicates that the participant is experiencing stress.

We conducted quantitative analysis using other attributes of the distribution, i.e. the IQR and range, to establish suitability of 'peak heart-rate count' as a measure to indicate 'biological' *stress*. Table 1 describes the minimum, maximum, Q1, Q3, mean and range data values computed from the distribution of 'peak heart-rate count'. It can be seen that the maximum value of 'peak heart-rate count' (49) has caused the control group to have a wider range compared to the experimental group. The H spread of data in the control group is dispersed less consistently owing to the wider range and this is reflected in the high IQR value of 17.25. Within the experimental-physical activity sub-group, we can see that the 'peak heart-rate count' are distributed more consistently within a narrower range of 4 and IQR of 2.5. The experimental-mindfulness subgroup has a larger range of 12 and IQR of 6.5 which reflects the wider spread between the data points. The box and whiskers plot depicts the shape of the distribution, median and its variability. The 'peak heart-rate count' for the experimental-physical activity sub-group participants lies within a smaller range between 24 and 28 whereas the 'peak heart-rate count' for the experimental- mindfulness sub-group participants lies within a wider range of 16 to 28. The IQR is depicted by the shaded area (see Fig. 4), which shows relatively consistent distribution of 'peak heart-rate count' for the experimental-physical activity sub-group as against the experimental-mindfulness sub-group.

Table 1. Summary of computed Range and IQR values

Variables	Mindfulness group	Physical activity group	Control group
Minimum	16	24	14.5
Q1	20.5	25	19.375
Median	26	27	26.75
Q3	27	27.5	36.625
Maximum	28	28	49
Mean	23.5	26.3	28.35
IQR	6.5	2.5	17.25
Range	12	4	34.5

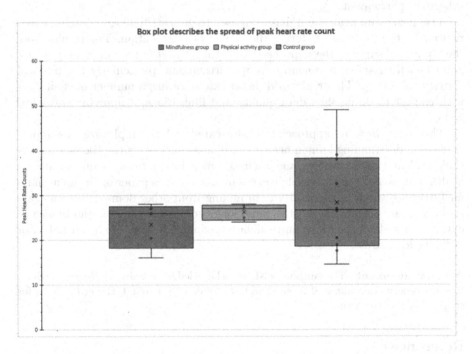

Fig. 4. Box plot describes spread of median 'peak heart-rate count' within IQR for different groups. Whiskers describe spread of *out-lier* median 'peak heart-rate count' values.

5.3 Discussion and Conclusion

The proposed new metric, 'heart-rate peak count' and the associated new MedIQR method is seen to be useful in analyzing the distribution of the heart-rate peaks, using range and IQR values. Its application on 'real-world' heart-rate data collected from wearable devices presents a number of challenges with reference to diminished data quality and limits the application of statistical tests. The observations made from the real-world biometric data are limited by the measurement errors.

Based on the readings of the IQR measure and range, the participants in the experimental-physical activity sub-group were observed to have 'peak heart-rate count' within a smaller range of 4 with recording a maximum 'peak heart-rate count' of 28 and a minimum 'peak heart-rate count' of 24. On the other hand, the participants in the experimental-mindfulness activity sub-group were observed to have a 'peak heart-rate count' within a wider range of 12 with recording a maximum 'peak heart-rate count' of 28 and a minimum 'peak heart-rate count' of 16. The IQR and range of the control group participants were larger than the experimental group which indicates that the *'digital health'* intervention has a higher probability of positive impact on the experimental-physical activity sub-group participants.

The 'heart-rate peak count' has been used to indicate 'biological' stress, experienced by the participant during the course of intervention. The results from the study can support the hypothesis that the *'digital health'* intervention program benefited the experimental group participants, particularly the physical activity sub-group. The 'real-world' heart-rate data has a number of challenges with respect to diminished data quality and limitations of applying statistical tests.

The *biopsychosocial* approach to health-care big data, emphasizes the importance of understanding human health and illness considering the Biological, psychological and social factors and their complex interactions in understanding health, illness and health care delivery. The work in this paper opens up avenues for further research in the reality data mining from other domains such as psychological and social data on the same longitudinal time line as the biometric data, to provide correlations and validate a composite measure as an indicator of 'biological' *stress*.

Acknowledgement. The authors wish to acknowledge, Faculty of Health, Federation University Australia and Australian Government Research Training Program for supporting this research.

References

1. Ahmed, M.U., Begum, S., Islam, M.S.: Heart Rate and Inter-Beat Interval Computation to Diagnose Stress Using ECG Sensor Signal. Report 1929, Mälardalen University, Västerås, Sweden (2010)
2. Engel, G.L.: The clinical application of the biopsychosocial model. Am. J. Psychiatry **137**(5), 535–544 (1980)
3. Borrell-Carrió, F., Suchman, A.L., Epstein, R.M.: The Biopsychosocial model 25 years later: principles, practice, and scientific inquiry. Ann. Fam. Med. **2**(6), 576–582 (2004)
4. Forbes (2018). https://www.forbes.com/sites/paullamkin/2016/02/17/wearable-tech-market-to-be-worth-34-billion-by-2020

5. Hao, T., et al.: Towards Precision Stress Management: Design and Evaluation of a Practical Wearable Sensing System for Monitoring Everyday Stress (2017). https://doi.org/10.2196/iproc.8441
6. Hye-Geum, K., Eun-Jin, C., Dai-Seg, B., Young, H.L., Bon-Hoon, K.: Stress and heart rate variability: a meta-analysis and review of the literature. Psychiatry Investig. 15(3), 235–345 (2018). 2018 Korean Neuropsychiatric Association
7. Kang, S.J., Kim, E., Ko, K.J.: Effects of aerobic exercise on the resting heart rate, physical fitness, and arterial stiffness of female patients with metabolic syndrome. J. Phys. Ther. Sci. 28(6), 1764–1768 (2016)
8. Kroll, R.R., Boyd, J., Maslove, D.M.: Accuracy of a wrist-worn wearable device for monitoring heart rates in hospital inpatients: a prospective observational study (2016). https://doi.org/10.2196/jmir.6025
9. Nikolopoulos, S., Alexandridi, A., Nikolakeas, S., Manis, G.: Experimental analysis of heart rate variability of long-recording electrocardiograms in normal subjects and patients with coronary artery disease and normal left ventricular function. J. Biomed. Inform. 36(3), 202–217 (2003)
10. Priest, J.B., Roberson, P.N., Woods, S.B.: In our lives and under our skin an investigation of specific psycho-biological mediators linking family relationships and health using the biobehavioural family model. Family Process (2018). https://doi.org/10.1111/famp.12357
11. Rebergen, D., Nagaraj, S., Rosenthal, E., Bianchi, M., van Putten, M., Westover, M.: Adarri: a novel method to detect spurious r-peaks in the electrocardiogram for heart rate variability analysis in the intensive care unit. J. Clin. Monit. Comput. 32(1), 53–61 (2018). http://www.es.mdh.se/publications/1929-
12. Richards, J.E.: The statistical analysis of heart rate a review. Psychophysiology 17(2), 153–166 (1980)
13. Sano, A., Taylor, S.M., McHill, A.W., Barger, L.K., Klerman, E., Picard, R.: Identifying Objective Physiological Markers and Modifiable Behaviors for Self-Reported Stress and Mental Health Status Using Wearable Sensors and Mobile Phones: Observational Study (2018). https://doi.org/10.2196/jmir.9410
14. Selye, H.: The Stress of Life. McGraw-Hill, New York (1956)
15. Sharashova, E.: Decline in Resting Heart Rate, its Association with Other Variables, and its Role in Cardiovascular Disease. Thesis, pp. 1–81 (2016)
16. Williams, D., Cash, C., Rankin, C., Bernardi, A., Koenig, J., Thayer, J.: Resting heart rate variability predicts self-reported difficulties in emotion regulation: a focus on different facets of emotion regulation. Front. Psychol. 6 (2018)

Development of a Real-Time Motor-Imagery-Based EEG Brain-Machine Interface

Gal Gorjup[1]([⊠]), Rok Vrabič[1], Stoyan Petrov Stoyanov[2],
Morten Østergaard Andersen[3], and Poramate Manoonpong[4,5]([⊠])

[1] Faculty of Mechanical Engineering, University of Ljubljana, Ljubljana, Slovenia
gorjup.gal@gmail.com, rok.vrabic@fs.uni-lj.si
[2] Berger Neurorobotics, Odense, Denmark
stoyan@bergerneurorobotics.com
[3] SDU Biotechnology, The University of Southern Denmark, Odense, Denmark
moan@kbm.sdu.dk
[4] The Maersk Mc-Kinney Møller Institute, The University of Southern Denmark,
Odense, Denmark
poma@mmmi.sdu.dk
[5] IST, Vidyasirimedhi Institute of Science and Technology, Rayong, Thailand

Abstract. EEG-based brain-machine interfaces offer an alternative means of interaction with the environment relying solely on interpreting brain activity. They can not only significantly improve the life quality of people with neuromuscular disabilities, but also present a wide range of opportunities for industrial and commercial applications. This work focuses on the development of a real-time brain-machine interface based on processing and classification of motor imagery EEG signals. The goal was to develop a fast and reliable system that can function in everyday noisy environments. To achieve this, various filtering, feature extraction, and classification methods were tested on three data sets, two of which were recorded in a noisy public setting. Results suggested that the tested linear classifier, paired with band power features, offers higher robustness and similar prediction accuracy, compared to a non-linear classifier based on recurrent neural networks. The final configuration was also successfully tested on a real-time system.

Keywords: Electroencephalography · Brain-machine interface
Brain-computer interface · Motor imagery · Digital filtering
Feature extraction · Classification

1 Introduction

Electroencephalography (EEG) is a non-invasive method for recording electrical activity of the brain, where the data is collected by means of electrodes positioned on the scalp. The obtained voltage signal is very weak and requires amplification

© Springer Nature Switzerland AG 2018
L. Cheng et al. (Eds.): ICONIP 2018, LNCS 11307, pp. 610–622, 2018.
https://doi.org/10.1007/978-3-030-04239-4_55

before digital conversion and storage. Further steps generally consist of noise reduction procedures, followed by methods for interpreting the obtained data which are governed by the target application (Fig. 1).

In the case of EEG-based brain-machine interfaces (BMI), the goal is to decode the user's intent from the recorded signals. This allows the user to interact with his/her environment without using the default neuromuscular pathways. Such interfaces are mainly used to help people whose medical conditions hinder their muscle manipulation abilities, but alternative uses (e.g. mental state monitoring) are also being actively developed.

The structure of a generic EEG-based brain-machine interface can be seen in Fig. 1. The brain signals are captured and digitised using EEG recording equipment. The general approach in signal processing is to first filter out the noise and artefacts, then extract relevant features, and finally feed them to a trained classifier that produces a prediction of the user's intent. This classification output can then be used to control an arbitrary system that may also send feedback to the user. Several EEG-based brain-machine interfaces have already been developed [1,2], exploiting different control mechanisms.

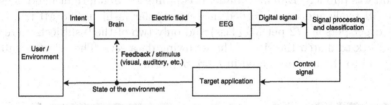

Fig. 1. Structure of a generic EEG-based brain-machine interface.

An EEG-based BMI can exploit different signal features that can be manipulated and controlled by the user through a set of mental strategies. This work focuses on frequency features of the signals (brain rhythms), which are associated with activity levels within different areas of the brain cortex [3]. The rhythms can be detected in localised frequency bands over different regions of the scalp. The BMI developed in this work uses rhythms detected over the motor cortex. The rhythms exhibit changes in amplitude related to real or imagined movement of various body parts [4,5]. Through the mental strategy of motor imagery (MI), these amplitude changes can be voluntarily triggered by imagining limb movement, most commonly of the hands and feet. Several interfaces based on this principle have already been developed [6–8].

However, most motor imagery based brain-machine interfaces rely on laboratory conditions and professional recording equipment with wet electrodes to produce results with high reliability and prediction accuracy [7,9,10]. This work attempts to place the interface in a real world setting by testing a wide array of signal processing and classification methods not only on data sets recorded in laboratory conditions, but also on sets that contain notable amounts of noise. In addition, it explores the usage of novel classifiers based on recurrent neural networks that can process sequential temporal features. The best performing methods are in the end also tested in a noisy environment using compact, low cost hardware with dry electrodes where the goal is to determine a possibility of using the simple and reliable real-time interface for everyday use.

This article first presents the used data sets and software framework in Sects. 2 and 3, respectively. Sections 4 and 5 introduce the implemented filtering, feature extraction and classification methods used for processing the motor imagery EEG signals. The implemented methods are then tested and evaluated in Sect. 6 with respect to the research goals. Section 6.3 presents real-time operation of the system with dry electrodes using previously found optimal processing methods and Sect. 7 provides conclusion of this work.

2 Data Set Descriptions

For offline testing and method selection, three motor imagery data sets were used.

The first used data set was recorded for the purposes of a BCI competition [11], and was publicly available online. It contains left hand, right hand, legs and tongue MI trials and was recorded in a monopolar fashion. 288 MI trials were recorded per person (72 per MI class) and only two of the 9 subjects were used in this work to match the size of the following data sets. The selected subjects correspond to the two presented in more detail in [12].

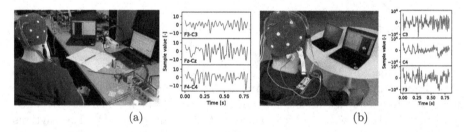

Fig. 2. Recording setup and sample recorded signals for (a) bipolar wet data and (b) monopolar dry data.

The second set was recorded for the purposes of this work and contains only left and right hand MI classes. It was recorded in a bipolar fashion, using wet electrode pairs F7-T3, F3-C3, Fz-Cz, F4-C4, and F8-T4, with respect to the international 10–20 system [13] (Fig. 2a). The signals were captured using an MLAEC1 EEG Electro-Cap System, an ML138 Octal Bioamp and a PowerLab 16/35 data acquisition system (all from AD Instruments). The set consists of two subjects and 96 trials per subject (48 per MI class).

The third set was also recorded within the frame of this work and again contains left and right hand MI classes. It was recorded in a monopolar fashion using dry electrodes at locations C3, C4, F3 and F4 (according to the international 10–20 system) with ear reference (Fig. 2b). The set consists of 192 trials per subject, 96 per MI group, recorded on two subjects. Hardware details are listed in Sect. 6.3.

The second and third data set were recorded in the entrance hall to an engineering research building - an environment containing electromagnetic noise, as well as potential distractions for the subjects in the form of random by-passers and their conversations.

In all of the recordings, the MI trials followed the timing scheme presented in Fig. 3. Each trial started with a fixation cross appearing on the screen, which prompted the subject to focus. After 2 s, a cue indicating which class of motor imagery the subject should perform appeared on screen. The subjects were asked to perform motor imagery until the fixation cross disappeared at $t = 6$ s. After a short break, the next trial initiated. No feedback was provided to the subjects while recording.

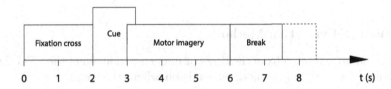

Fig. 3. Reference data trial timing scheme. Image adapted from [11].

3   Software Framework

In this work, a software framework was developed to support implementations of different processing methods. It was developed as a set of class hierarchies in the Python language. It was chosen because it is user friendly and well supported in terms of libraries that provide high level data structures and tools for signal processing and machine learning. Speed was not a primary objective as the brain rhythm changes are relatively slow and highly optimised processing algorithms would not significantly improve the overall response time.

4   Signal Processing

A traditional approach to EEG signal processing was taken, where filtering, feature extraction, and classification methods were implemented and evaluated. The methods were developed to support multivariate temporal data as all of the data sets contained recordings from multiple channels.

4.1   Signal Filtering

The family of digital temporal filters was considered for the purpose of reducing the amount of noise in the signals, as well as for isolating the frequency bands of interest from the recordings.

Even though finite impulse response (FIR) filters have been suggested for scientific EEG signal processing purposes [14], infinite impulse response (IIR) filters were found to be more appropriate in this setting. They offer significantly lower delay times than FIR filters, which greatly influences the responsiveness of the system. Although FIR filters are always stable, IIR instability issues can be easily avoided with proper design [14,15]. In addition, the filtered signals were not recorded for the purpose of detailed inspection but were instead passed to various machine learning techniques. The results were therefore not significantly affected by minor signal distortions, as long as the same filter was applied to all of the processed data. With respect to the findings, a 5th order band-pass Butterworth IIR filter was chosen to be the default filter for isolating the frequency bands of interest and removing the undesired noise components from the signals.

4.2   Feature Extraction Methods

Features can be extracted from the time- or frequency-domain of the EEG signal. The choice of extraction methods depends on where the sought information is encoded. In this case, the implemented features were focused on the frequency domain. Four methods were investigated: Band power, Autoregressive, Hjorth parameters, and FFT features.

(1) Band power features rely on isolating a frequency band of interest from the signal using a band-pass filter and then squaring and averaging the result to obtain a temporal power signal in that band. These features were already extensively used in EEG experiments [4,16], giving solid results. They are computationally inexpensive and offer a very high temporal resolution which is equal to the signal's sampling rate. In this case, the features were constructed using a set of 5th order band-pass Butterworth IIR filters in frequency bands 8–12 Hz, 12–16 Hz and 16–28 Hz, where the first and last band correspond to mu and beta brain rhythm frequencies that hold information about imagined movement.

(2) Autoregressive (AR) modelling is a parametric spectral method, meaning that it encodes spectral information into a selected number of parameters [15]. Those parameters are therefore very appropriate for distinguishing between signals that differ in the frequency spectrum, such as motor imagery recordings. The method implementation computes estimates of the AR coefficients by solving the Yule-Walker equations directly [15], using biased estimates for the autocorrelation sequence. The optimal number of AR parameters per EEG channel was here found to be 7, according to the Akaike Information Theoretic Criterion [15] and result inspection.

(3) Hjorth parameters [17] can easily be calculated in the time domain, but also contain information from the frequency domain, making them suitable for this application. For each of the EEG channels, three parameters were computed.

(4) The Fast Fourier Transform (FFT) was the basis for the last set of features. Segments of separate EEG channels were transformed and then averaged in frequency bands of interest or output as a whole sequence.

5   Classification Methods

Two classification methods were implemented in the work and their training was window based: the MI trials (refer to Fig. 3) were split into windows of length 1 s, with an 80% overlap. The windows whose centres fell between 4.5 and 5.5 s in the trial scheme were isolated and used for classifier training, after extracting the relevant features. Trial splitting, as well as feature extraction and training was performed simultaneously on all the relevant channels.

5.1   Linear Classifier

The first classification method was based on the linear discriminant analysis (LDA) classifier [18] and also served as a benchmark for system validation, comparing it to results obtained in [12]. The one-versus-rest classification scheme was used, where a separate LDA classifier was trained for each of the MI groups.

5.2   Neural Network Based Nonlinear Classifier

The second method was based on recurrent neural networks, chosen by their ability to process sequences and produce non-linear decision boundaries. The core of the networks were Long Short-Term Memory (LSTM) units, which were designed to solve the vanishing gradient problems in sequence modelling through their gated structure. The architecture of an example LSTM-based classifier can be seen in Fig. 4. The figure presents an unfolded architecture, showing how the network processes the sequences through time. On the lowest level, vectors from the input x are passed to the first LSTM layer L_1, which passes its current results to the second LSTM layer L_2 (each circle in the figure represents an array of units). In the next time step, the first layer receives a new vector from the input sequence and its own recurrent connection (same holds for the second layer). When the whole sequence has passed through the L_1 and L_2 layers, a fully connected layer f gathers the results from separate LSTM units in L_2 into the appropriate number of MI classes y. The network was trained through gradient-based optimisation of the chosen categorical cross-entropy loss function [19]. The optimisation algorithm used was Adam [20], which is an adaptive learning rate algorithm with incorporated momentum. The network was trained in minibatches of 50 trials for 200 epochs. To prevent overfitting, the weight decay

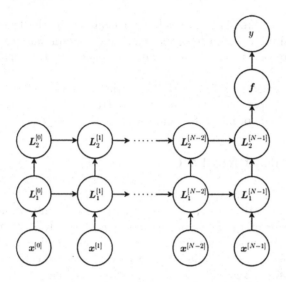

Fig. 4. Unfolded neural classifier with two LSTM layers.

regularisation strategy [19] was applied to weights in the input and recurrent connections of the LSTM layers, as well as on the final fully connected layer.

In the design, several network hyperparameters were left to be optimised: number of LSTM layers, number of units in each layer, initial learning rate for the Adam algorithm and three weight decay parameters. These parameters were found through a massive random search approach that required extensive parallelisation over roughly 500 CPU units in a cluster computing network (SLING [21]). The search was executed in two stages, where the first found coarse parameter values and the second refined them. For every data set presented in Sect. 2 and every feature group described in Sect. 4, roughly 2500 networks with random hyperparameter values (in defined limits) were trained and cross-validated (5-fold in the first stage, 10-fold in the second) to find relevant prediction accuracy. The prediction accuracy was used as a basis for selecting the optimal hyperparameter values. The overall network performances were compared and optimal hyperparameter values were found to be: 1 LSTM layer with 50 units, Adam learning rate in interval $[10^{-3.9}, 10^{-3.3}]$, LSTM input and recurrent connection weight decay in interval $[10^{-5}, 10^{-2}]$ and the final fully connected layer weight decay of $10^{-3.5}$.

6   Experimental Results

6.1   Feature Extractor Selection

In the first stage, all of the implemented feature extraction methods were tested with both classifiers. For the neural classifier, the tests were implicitly performed during the hyperparameter optimisation process and it was found that the feature group of band power sequences in ranges of 8–12 Hz, 12–16 Hz, and 16–28 Hz gave the best performance.

For the linear classifier, feature test results for subjects 1 and 2 from the publicly available data set can be seen in Fig. 5. The figure presents the 5 tested feature groups (band power, autoregressive, Hjorth parameters, FFT features and their combination), with respect to their average prediction accuracy and standard deviation obtained through 10-fold cross validation with the linear classifier. In this case, the band power features (ranges of 8–12 Hz, 12–16 Hz and 16–28 Hz) gave highest prediction accuracy with small variance.

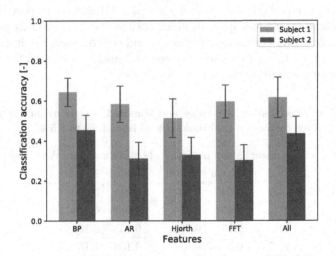

Fig. 5. Linear classifier performance with respect to different feature extraction methods for the publicly available data set. The methods include band power (BP), autoregressive (AR), Hjorth parameters (Hjorth), FFT features (FFT) and their combination (All).

The results showed that the band power features offer the best prediction accuracy over all data sets and both classifiers. Considering also their robustness and low computational complexity, they were chosen as the default feature extraction method for a real-time motor-imagery-based EEG brain-machine interface.

6.2   Classifier Selection

In the second stage, the implemented classifiers, paired with the selected band power features from the previous section, were compared with the goal of determining which is the most appropriate for the available types of data and training paradigm. The classifiers were trained and tested on the 6 subjects from the three available data sets described in Sect. 2. The MI trials were split into windows of length 1 s, with an 80% overlap. Windows with centres between 4.5 and 5.5 s in the trial scheme were used for training the classifiers. Unbiased temporal prediction accuracies were obtained through 10-fold cross validation. Results can be seen in Table 1, which contains mean values for all the data sets. Note that the Reference data set consists of four classes, while the others consist of two.

Fig. 6 presents results for selected subjects in more detail by means of temporal accuracy plots. The plots represent averaged classification accuracy through the trials (refer to trial scheme in Fig. 3) and the shaded blue area indicates its standard deviation. The vertical red and cyan lines indicate the cues that prompted the user to start and stop imagined movement. Before the MI start, classification accuracy moves around 25% and 50% for the subjects. These two percentages correspond to random guessing between 4 and 2 MI classes, respectively. After the cue, the average classification accuracy rises as the subjects start performing the motor imagery and drops again after the end cue. An ideal result would show a sharp rise in accuracy after the start cue and maintain it until the trial end, with minimum standard deviation.

Table 1. Mean classification accuracies and standard deviations for the linear and neural classification method in trial time interval from 4.5 s to 5.5 s.

Data set	Subject	Mean accuracy [ ]		Mean standard deviation [ ]	
		Linear	Neural	Linear	Neural
Reference	1	0.643	0.667	0.071	0.070
	2	0.454	0.414	0.074	0.110
Bipolar	1	0.651	0.655	0.138	0.163
	2	0.536	0.534	0.156	0.159
Dry	1	0.462	0.520	0.099	0.077
	2	0.717	0.692	0.091	0.140

Inspecting results presented in Fig. 6, as well as those from the other subjects in Table 1, it was found that the linear classifier is a more appropriate choice for the interface. It offers the same or better performance than the neural classifier at significantly lower complexity. It is also robust, deterministic and allows faster training through its closed-form optimisation solution.

The neural classifier, on the other hand, has a longer training time due to its nonlinear nature and the chosen iterative optimisation methods. The classifier's performance also substantially depends on hyperparameter values which take quite some time and effort to determine for a specific problem. However, the neural classifier may have the potential to outperform the linear if a different training approach was taken and more data was available.

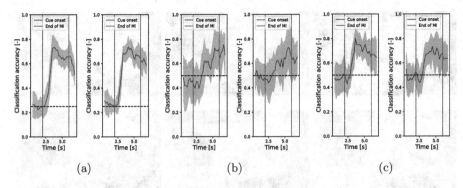

Fig. 6. Comparison of linear and neural classifier performance using the band power feature for (a) subject 1 from the publicly available data set; (b) subject 1 from the recorded bipolar data set with wet electrodes and (c) subject 2 from the recorded monopolar data set with dry electrodes. In (a), (b) and (c), temporal accuracy plots are shown for the linear (left) and neural (right) classifier.

6.3   Real-Time Motor Imagery Based Control

The experiment was performed by one untrained, able-bodied male subject aged 24, who has previously participated as subject 2 in the last recorded data set. The test was based on two class motor imagery, where the subject imagined movement of his left and right hands. The EEG signal was captured with dry reusable electrodes from Florida Research Instruments, mounted at locations C3 and C4 (according to the international 10–20 system) in monopolar montage with reference on the left and ground on the right mastoid. The signals were amplified and digitised with a battery powered prototype board developed by Berger Neurorobotics [22], which was streaming the data to a dedicated computer through a Bluetooth connection. The board had a fixed amplifier sensitivity of 100 μV and a sampling frequency of 240 Hz. The setup can be seen in Fig. 2b.

Taking into account the findings in Sect. 6, the linear one-versus-rest LDA classifier was chosen to be used in the tests, along with the band power features in ranges of 8–12 Hz, 12–16 Hz and 16–28 Hz. The classifier was trained on data from subject 2 in the dry data set, using the training paradigm described in Sect. 6.2 and channels C3 and C4.

The trained classifier, along with required supporting signal processing methods and framework functionality, was then ported to the dedicated computer receiving the live data stream. The received data was accumulated and passed to the classifier in segments of length 1 s with an 80% overlap, mirroring its window-based training paradigm. With this setting, the classifier could produce a new label every 0.2 s. To increase system stability, the classifier was expanded with functionality to refrain from predicting a class label if its confidence was too low.

In the experiment, live visual feedback in the form of a white square target on a black background was available to the subject. A valid label produced by the

Fig. 7. Snapshots of the real-time experiment with their respective timings. Top row shows right hand MI and bottom row shows left hand MI. The vertical red line indicates the initial square position in each row. The video of the experiments can be seen at www.manoonpong.com/SMC2018/video.mp4 (Color figure online)

classifier caused the target to move a short distance in the left or right direction, depending on the predicted class. If no valid label was produced, the target stayed still. The results of the real-time experiment are shown as snapshots in Fig. 7.

The implemented system was responsive and exhibited no noticeable lag between subject mental activity and target movement. The developed framework could process the provided input data stream in an efficient and timely manner, appropriate for the implemented real time application.

During testing, it was evident that the target movement was influenced by the subject's mental activity and was not merely a product of random noise. In the best cases, the subject could steadily move the target towards the left and right edges of the feedback screen through continuous focus on the appropriate MI class. However, the subject's ability to reliably control the target deteriorated significantly with increasing fatigue. As the subject lost focus, the target exhibited quivering behaviour due to random misclassification.

In the best cases, the subject was able to steadily control movement of the target through motor imagery, which proves that the system could in principle already be used in real life applications. It would, however, still need a considerable amount of tuning before any such usage as its performance seemed quite unstable in a number of trials.

The task at hand was physically and mentally tiring for the subject because a high level of focus and minimal body movement were crucial for a successful trial. As the subject grew increasingly fatigued with time, he lost control over the target. The cause of this might be in different brain activity patterns that appeared when the subject was tired and could not be interpreted by the system. The performance issue could therefore be partially resolved through subject training and collecting additional data to train the classifier, as well as through determining the subject's fatigue level and adjusting the processing accordingly.

Future work will be focused on introducing adaptive methods with online adjustment of the control and feature classification parameters [23].

7 Conclusion

This paper presents the development of a real-time motor-imagery-based EEG brain-machine interface. Several filtering, feature extraction, and classification methods were investigated, tested and compared with three data sets obtained through different recording methods, with varying amounts of environmental noise. The experiments show that band power features based on IIR filters, paired with the one-versus-all LDA classifier give the best classification accuracies while maintaining a comparably high level of robustness. The alternative, neural network based method was shown to be less robust and harder to tune. One of the reasons for this is the relatively low amount of training data, which limited the generalisation capabilities of the network. In addition, the trial splitting and training scheme disconnected and shuffled adjacent trial sections, which could have provided additional information to the neural classifier.

The best performing signal processing and classification methods were tested in real-time, using wireless hardware with dry electrodes in a noisy environment. The results were promising, showing that the interface performs well even with an untrained subject, who was able to control marker movement in a two-class MI scheme interpreted by the one-versus-all LDA classifier. As was shown in the offline analysis, its robustness facilitated good performance in the noisy environment.

Acknowledgement. This work is supported by Centre for BioRobotics (CBR) at University of Southern Denmark (SDU, Denmark) and Horizon 2020 Framework Programme (FETPROACT-01-2016–FET Proactive: emerging themes and communities) under grant agreement no. 732266 (Plan4Act).

References

1. McFarland, D., Wolpaw, J.: EEG-based brain-computer interfaces. Current Opinion Biomed. Eng. **4**, 194–200 (2017)
2. Wolpaw, J.R., Birbaumer, N., McFarland, D.J., Pfurtscheller, G., Vaughan, T.M.: Brain-computer interfaces for communication and control. Clin. Neurophysiol. Official J. Int. Fed. Clin. Neurophysiol. **113**(6), 767–791 (2002)
3. Schomer, D.L., Lopes da Silva, F.H.: Electroencephalography: basic principles, clinical applications, and related fields. Lippincott Williams & Wilkins, Philadelphia (2011)
4. Pfurtscheller, G., Lopes Da Silva, F.H.: Event-related EEG/MEG synchronization and desynchronization: Basic principles. Clin. Neurophysiol. **110**(11), 1842–1857 (1999)

5. Pfurtscheller, G., Neuper, C.: Dynamics of sensorimotor oscillations in a motor task. In: Graimann, B., Pfurtscheller, G., Allison, B. (eds.) Brain-Computer Interfaces. The Frontiers Collection. Springer, Heidelberg (2009). https://doi.org/10.1007/978-3-642-02091-9_3

6. Blankertz, B., et al.: The Berlin Brain-Computer Interface presents the novel mental typewriter Hex-o-Spell. In: 3rd International BCI Workshop and Training Course, pp. 2–3. Graz (2006)

7. Pfurtscheller, G., Brunner, C., Leeb, R., Scherer, R.: The graz brain-computer interface. In: Graimann, B., Pfurtscheller, G., Allison, B. (eds.) Brain-Computer Interfaces. The Frontiers Collection. Springer, Heidelberg (2009). https://doi.org/10.1007/978-3-642-02091-9_5

8. Wolpaw, J.R., McFarland, D.J., Vaughan, T.M.: Brain-computer interface research at the Wadsworth Center. IEEE Trans. Rehabil. Eng. 8(2), 222–226 (2000)

9. Prakaksita, N., Kuo, C.Y., Kuo, C.H.: Development of a motor imagery based brain-computer interface for humanoid robot control applications. In: 2016 IEEE International Conference on Industrial Technology (ICIT), pp. 1607–1613. IEEE, Taipei (2016)

10. Tibor Schirrmeister, R., et al.: Deep learning with convolutional neural networks for EEG decoding and visualization. Hum. Brain Mapp. 38, 5391–5420 (2017)

11. Brunner, C., Leeb, R., Müller-Putz, G., Schlögl, A., Pfurtscheller, G.: BCI Competition 2008 - Graz data set A. Graz (2008)

12. Naeem, M., Brunner, C., Leeb, R., Graimann, B., Pfurtscheller, G.: Seperability of four-class motor imagery data using independent components analysis. J. Neural Eng. 3, 208–216 (2006)

13. Klem, G.H., Lüders, H.O., Jasper, H.H., Elger, C.: The ten-twenty electrode system of the International Federation. Electroencephalogr. Clin. Neurophysiol. Suppl. 52, 2–5 (1999)

14. Widmann, A., Schröger, E., Maess, B.: Digital filter design for electrophysiological data - a practical approach. J. Neurosci. Methods 250, 34–46 (2015)

15. Alessio, S.M.: Digital Signal Processing and Spectral Analysis for Scientists. Springer International Publishing, Switzerland (2016)

16. Pfurtscheller, G., et al.: Current trends in graz brain-computer interface (BCI) research. IEEE Trans. Biomed. Eng. 8(2), 216–219 (2000)

17. Hjorth, B.: EEG analysis based on time domain properties. Electroencephalogr. Clin. Neurophysiol. 29(3), 306–310 (1981)

18. Hastie, T., Tibshirani, R., Friedman, J.: The Elements of Statistical Learning. Springer, New York (2009)

19. Goodfellow, I., Bengio, Y., Courville, A.: Deep Learning. MIT Press (2011)

20. Kingma, D.P., Ba, J.: Adam: A method for stochastic optimization. In: 3rd International Conference for Learning Representations. San Diego (2015)

21. Slovenian Initiative for National Grid (SLING). http://www.sling.si

22. Berger Neurorobotics. http://bergerneurorobotics.com/

23. Sun, S., Zhou, J.: A review of adaptive feature extraction and classification methods for EEG-based brain-computer interfaces. In: Proceedings of the International Joint Conference on Neural Networks, pp. 1746–1753. IEEE, Beijing (2014)

Robust Eye Center Localization Based on an Improved SVR Method

Zhiyong Wang[1], Haibin Cai[2], and Honghai Liu[1(✉)]

[1] State Key Laboratory of Mechanical System and Vibration,
Shanghai Jiao Tong University, Shanghai, China
{yzwang_sjtu,honghai.liu}@sjtu.edu.cn
[2] Group of Intelligent System and Biomedical Robotics, School of Creative
Technologies, University of Portsmouth, Portsmouth, UK
jlcaihaibin@gmail.com
http://bbl.sjtu.edu.cn/

Abstract. Eye center localization is an important technique in gaze estimation, human computer interaction, virtual reality, etc., which attracts a lot of attention. Although a great deal of progress has been achieved over the past few years, the accuracy declines dramatically due to the low input image resolution, poor lighting conditions, side face, and eyes status such as closed or covered. To handle this issue, this paper proposes an improved support vector regression (SVR) method to detect the eye center based on the facial feature localization. Several image processing techniques were tried to improve the accuracy, and results showed that the SVR combining a Gaussian filter could get a better accuracy.

Keywords: Eye center localization · SVR · Gaussian filter

1 Introduction

Generally, the role of gaze in human communication is well known to be crucial due to its ability to convey intentions and describe attention. It gets many research focuses in many fields such as human computer interaction, virtual reality, etc. Accurate eye pupil center localization is the basis of the gaze estimation. Although promising progress has been achieved over the past few years, the accuracy of eye center localization declines dramatically due to the low resolution, poor lighting conditions of input image, side face, and complex eyes status such as closed or covered. Nowadays, face detection and facial feature points localization have been mature techniques. It is reasonable and implementable to extract and analyze the eye region separately, which makes the eye pupil center localization more accurate and faster.

Support vector regression (SVR), proposed by Drucker *et al.* [1] in 1997, was given a comprehensive introduction by Smola and Scholkopf [2]. Its basic principle is described as follows. For sample $(\mathbf{x}, \mathbf{y})$, the traditional regression model is calculating the loss between the model output $\mathbf{f}(\mathbf{x})$ and actual output $\mathbf{y}$.

© Springer Nature Switzerland AG 2018
L. Cheng et al. (Eds.): ICONIP 2018, LNCS 11307, pp. 623–634, 2018.
https://doi.org/10.1007/978-3-030-04239-4_56

If and only if the output $\mathbf{f}(\mathbf{x})$ and actual output $\mathbf{y}$ are identical, the loss is equals to zero. But in SVR, it is assumed that the deviation between $\mathbf{f}(\mathbf{x})$ and $\mathbf{y}$ is acceptable within the threshold of ε. That is, the loss is calculated unless the absolute error between $\mathbf{f}(\mathbf{x})$ and $\mathbf{y}$ is greater than ε.

This paper gave tentative work to improve the accuracy and robustness of the existing eye center localization methods. An improved SVR method combining with Gaussian filter and eye state judgment strategy was proposed in this paper. And the results showed that eye centers could be located accurately using our proposed method in several public data base. The basic principles of SVR and linear regression (LR) are introduced as follows.

SVR has been successfully applied in the detection of facial feature points [3], and plenty of source codes have been exposed on the OpenFace [4]. It is compute-efficient and robust against image resolution variation, illumination change, and face angle conversion.

LR is a statistical analysis method that uses regression analysis in mathematical statistics to determine the quantitative relationship between two or more variables. It is a commonly used regression method to eliminate linear error and improve the accuracy.

2   Related Works

The methods that can detect the eye center are generally divided into two categories: (1) model based method, and (2) appearance based method. In this part, some previous work are reviewed.

2.1   Model Based Methods

The model based methods, also known as feature based methods, detect the eye center using some priori knowledge of the eyes such as the dramatic change in the brightness at the edge of the iris and the geometric shape of iris. Daugman *et al.* [5] proposed an Integro-Differential Operator (IDO) method which utilized the drastic intensity change between the iris and sclera. Experimental results achieved good performance for high resolution eye images. In 2015, Cai *et al.* modified the IDO by optimizing the number of kernels and got a desirable accuracy in localizing the eye center in lower resolution image [6]. On the other hand, Hough transform, a classical method to find the circle, was also applied to find eye center as the visual part of the iris is around a circular arc [7]. Timm *et al.* [8] proposed a simple but efficient eye center localization method by calculating the means of gradient for each points in the image. Besides, the color and radial symmetry of the eyes were used to locate the eye center [9,10]. Most of the these model-based methods performed well in dealing with the high resolution images, but the accuracy will drop when the image quality decreases such as in low resolution or poor illumination and even worse when the eyes are closed.

2.2 Appearance Based Methods

The appearance based methods focus on the eye holistic appearance and surrounding structures, which needs a lot of data to train models to detect the eye center. Thus, many machine learning algorithms have been applied to locate the eye center. Markuvs et al. [11] trained an ensemble of randomized regression trees for the eye center localization. Bayesian models was employed in [12] to locate the eye center. Considering the state of the eyes, Gou et al. [13] proposed a cascaded regression framework to detect the eye center. Although these methods need a large amount of training data for model training, a good result can be achieved even the images are in low resolution.

3 Method

This paper proposed a novel method which combined classical image processing method and machine learning algorithms to detect the eye pupil center. Two datasets, LFPW [14] and HELEN [15], were used to train the model. This model is based on the characteristics of the intensity distribution around the eyes. In order to enhance the original image quality, some strategies were used to image preprocessing methods such as Gaussian filter, histogram equalization, gradient features, etc. before training the support vector regression model and linear regression model. The specific method is shown in Fig. 1.

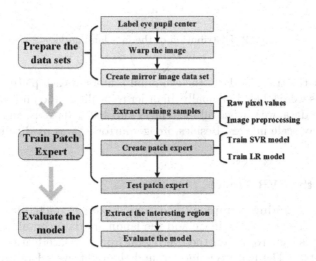

Fig. 1. The framework of the improved SVR eye center detect method.

3.1   Data Preprocessing

To train the eye localization model, we used two large image datasets LFPW and HELEN which were used to train the models detecting facial landmarks. The LFPW training set contains 811 PNG images and test set has 224 PNG images. There are 2000 JPG images and 330 images correspondingly in the HELEN dataset. These two datasets are regarded as the most comprehensive datasets because they collect samples of different skin color, race, age, and situation in various resolutions. Besides, the original images, the 68 landmarks of face are also provided in the datasets. The distribution of landmarks are showed in Fig. 2.

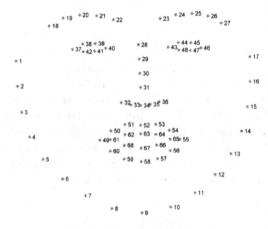

Fig. 2. Distribution of the face landmarks.

To collect the training data, the eye pupil centers of each picture in the two large datasets were labeled manually. In order to facilitate the unified processing of subsequent data, we warped all the images into 400*400 pixels, and transferred them into gray scale images. Besides, image mirroring was used to increase the training data.

3.2   Train the SVR Model

Extract the Training Samples. The samples were extracted from the input images. Both the gray scale image and the landmark data containing 68 points were read. It is easy to obtain the eye region from the face feature points distribution matrix. The feature points around the right eye sclera are related to point 37, 38, 39, 40, 41, 42 (37 as the outer corner feature point, 40 as to the inner eye corner, 38 and 39 for the upper eyelid edge points, 41 and 42 of lower eyelid edge points, the specific distribution is shown in Fig. 3.

The hexagonal region with isoclines is our interesting region. Likewise, point 43, 44, 45, 46, 47, 48, are the related points around the sclera of the left eye. Following we will take the right eye as an example to explain this method.

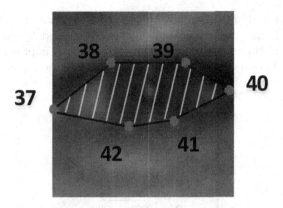

Fig. 3. Distribution of feature points around the corner of the eye.

The distance between the eye corners can be calculated with some horizontal coordinates of point 37 and point 40. For the convenience of processing, we extract a square area with a center that is the horizontal midpoint of point 37 and point 40. The width of the square equals the distance of eye corners. This eye area is usually 27*27 pixels. In order to better extract the characteristics of the iris of human eyes in different states, bilinear interpolation was used for the interest area, which was enlarged to 51*51 pixels. The enlarged iris radius is about 10 to 12 pixels. To make the template contain the iris completely, 25*25 pixels is taken as the template size, that is, the samples dimension is 625. In this way, the area of interest of 51*51 pixels can be sampled according to the template size. Total 27*27 samples, also calling patches, can be extracted. As for the labels, if we take this problem as a classification problem, the true eye pupil center will be the only one positive sample. In this case, the training will fail. Thus the Gaussian distribution was used to generate the labels.

The center of the Gaussian distribution, $(0, 0)$, is the actual center of the iris localization coordinates, and the variance equals 1. The responses of the Gaussian distribution were set to be the labels of patches correspondingly. The patch center is farther away from the center of the iris, the corresponding response is close to 0. Inversely, the patch center is closer to the iris center, the label will be closer to 1. Thus a binary classification problem was turned into a fuzzy problem. In this way, the problem of sample imbalance between positive and negative samples can be solved.

Because the actual coordinates of the six feature points around the eye have been obtained, the iris center can only be located within the hexagon. The hexagon taking the six feature points as the vertex were used to narrow the sampling range. Only the labels of which coordinates were within the hexagon were retained. Meanwhile, all the outer ones were set to 0. Thus, each patch got a label, and the patches with a label of 0 will be discarded to further improve the pertinence of the sample. After completing this step, we got sample data for training the model.

There were two kinds of input feature used to train the model: the raw pixel values and the gradient intensity values calculated by convolution with Soble operator. In order to simplify the calculation and improve the speed, the absolute value addition strategy is used to approximate the gradient. The specific calculation method is showed as followings:

$$\begin{cases} Gx = \begin{bmatrix} -1 & 0 & +1 \\ -2 & 0 & +2 \\ -1 & 0 & +1 \end{bmatrix} * A \\ Gy = \begin{bmatrix} -1 & -2 & -1 \\ 0 & 0 & 0 \\ +1 & +2 & +1 \end{bmatrix} * A \cdot \\ G = \sqrt{G_x^2 + G_y^2} \\ G \approx |G_x + G_y| \end{cases} \quad (1)$$

Meanwhile, this paper tried two classical image processing methods, Gaussian blur and histogram balance, to improve the accuracy. Gaussian filter is used to smooth the interest area, eliminate Gaussian noise and ignore details. Histogram equalization is beneficial to improve the contrast of the images, especially for the images with a concentrated pixel distribution. There are four manners tried totally as shown in Fig. 4. Correspondingly, the processed the input eye regions are showed as Fig. 5.

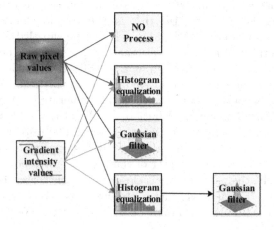

Fig. 4. The four manners used in the image preprocessing.

Create the Patch Expert. The sample data was divided into three parts 80% of them was used to train SVR model, and the inputs were the patches and labels. Thus, a 625 dimension SVR model, ω, could be obtained. In order to further improve the accuracy of the model, 10% of the sample data was used

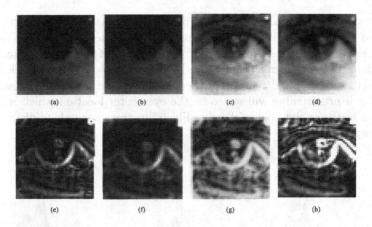

Fig. 5. Examples after the different process. Where (a)~(d) is the original images, processed by Gaussian filter, histogram equalization, and both of these two processing;(e)~(h) is the corresponding gradient images.

to train the linear fitting model, of which the input is trained SVR model ω, patches, and labels. The output is parameter a and b. The specific calculation method is as follows:

$$\begin{cases} y = ax + b \\ x = \omega \cdot patch \;. \\ y = label \end{cases} \qquad (2)$$

Combining the SVR and LR, a 627 dimensional patch expert was trained. The remaining 10% of the sample dataset was used to calculate the error and correlation of the training model.

$$\text{response} = -\frac{1}{e^{a \cdot \omega x + b}}. \qquad (3)$$

where, x is the pixels of the patch in 625 dimension. The response is belong to 0 to 1. And the RMSE (root mean squared error) and correlation were calculated.

3.3   Evaluate the SVR Model

To evaluate the model gathered from the last step, two pubic available dataset including GI4E [16], and BioID [17], were used to detect the eye pupil center.

Datasets. The BioID database has been regarded to be a challenging dataset for its images have a relative low resolution of 384*288 pixels. There are 1521 grayscale images which have various eye conditions such as poor illumination, wearing glasses, closed eyes, etc. The GI4E database contains 1236 images of 103 subjects with 12 gaze directions. The resolution of the images are 800*600 pixels, so it is considered to be a normal web camera setup.

Detect the Eye Pupil Center. The square areas of interest should be extracted and scaled up to 51*51 pixels. All of the response of the 729 test patches can be calculated through the patch expert. If the biggest response is larger than 0.15, the localization with the biggest response will be selected to be the detected eye pupil center. On the contrary, if the biggest response is less than 0.15, a preset value will set to be the eye center location which is related to the six feature points. The following Fig. 6 illustrate a couple of distributions of the responses.

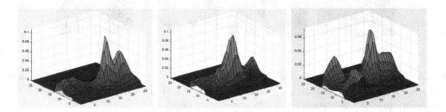

Fig. 6. The distributions of responses for some test images.

4   Experimental Results

4.1   Accuracy Measurement

The method calculating the accuracy of detect eye pupil center was first introduced by Jesorsky *et al.* [17]. Considering the large difference between two eyes and distances of two pupil centers (among different individuals), the normalized error was regarded as the evaluation criterion, which can be computed as:

$$\varepsilon = \frac{\max(d_r, d_l)}{d}. \tag{4}$$

where d_l and d_r are the Euclidean distances between the detected pupil center and actual pupil center for the left eye and right eye, respectively. The labeled eye centers are regarded as the ground truth. d is the actual Euclidean distance between two eye pupil centers, correspondingly.

4.2   Results

It is acknowledged that the Gaussian filter can eliminate Gaussian noise, and histogram equalization can improve image contrast. In this paper, we tried these two methods to improve the quality of the input images. Besides, the gradient features are commonly used to find the object boundary, which may be a reasonable method to detect the boundary of the pupil due to the obvious difference between inside and outside in the pupil region. Table 1 shows the test results of the database Gi4e. It can be discovered that no matter which image processing method was used, the results of raw pixels is better than the

Table 1. Comparison of maximum normalized error in the GI4E database

Patch type	Process	e < 0.05	e < 0.1	e < 0.25
Raw pixel	none	96.32%	99.65%	100.00%
Gradient	gradient	83.79%	98.86%	100.00%
Raw pixel	Gaussian filter	96.84%	99.74%	100.00%
Gradient	Gaussian filter	82.47%	99.04%	100.00%
Raw pixel	Histogram	95.71%	99.21%	100.00%
Gradient	Histogram	80.98%	99.47%	100.00%
Raw pixel	Histogram + Gaussian filter	96.67%	99.65%	100.00%
Gradient	Histogram + Gaussian filter	84.14%	99.74%	100.00%

gradient. Meanwhile, only the method using the Gaussian filter alone gathered a better accuracy. Although the histogram equalization can improve the image contrast, the result reveals that the histogram equalization does not make any improvement and even makes result worse. The results show that which method is able to detect the eye pupil center accurately if the image resolution is not too low. And the raw pixel values with the Gaussian filter performs best, similar conclusion can be achieved as well in the BioID dataset. The accuracy to detect eye center is as high as 96.84%, which is enough for the applications such as gaze estimation, and eye tracking. The Fig. 7 shows some results of the eye localization.

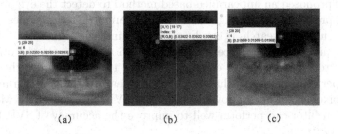

Fig. 7. The results of the eye localization

Table 2 presents the comparison of the proposed method with some state-of-the-art methods using the same database BioID and the same accuracy evaluation criterion. Although the accuracy based on the proposed method was not the highest when the error was less than 0.05, the result was better than the state-of-the-art methods as error is less than 0.1.

To verify the robustness of the proposed algorithm, we also tested the dataset LFPW. Table 3 shows the results for the dateset LFPE. Although the results are not as high as the method mentioned in the previous related works, it is robust to locate the eye center in the rough condition with an acceptable accuracy.

Table 2. Comparison of maximum normalized error in the LFPW database

Patch type	e < 0.05	e < 0.1	e < 0.25
Campadelli2009 [18]	80.70%	93.20%	99.30%
Asadifard2010 [19]	47.00%	86.00%	96.00%
Timm2011 [8]	82.5%	93.40%	98.00%
Valenti2012 [20]	86.10%	91.70%	97.90%
Leo2014 [21]	80.70%	87.30%	94.00%
Anjith2016 [22]	85.00%	94.30%	–
Poulopoulos2017 [10]	87.10%	98.00%	100.00%
Proposed	**82.15%**	**98.70%**	**100.00%**

Table 3. Comparison of maximum normalized error in the LFPW database

Process	e < 0.05	e < 0.1	e < 0.25
None	81.42%	97.34%	100.00%
Gauss filter	84.07%	98.23%	100.00%
Histogram	83.19%	97.34%	100.00%
Histogram + Gauss filter	83.78%	98.23%	100.00%

5   Conclusion

This paper proposed an appearance-based method to detect the eye pupil center with an improved SVR. Though this method cannot get the highest accuracy among the current research, it can be robust against the variations of face state and image quality with an acceptable estimation accuracy. On the other hand, Gaussian filter and histogram equalization were tried to improve the accuracy in this paper. Results show that the gradient features and histogram process for every pixel make no improvement for the SVR-based method. Meanwhile, the Gaussian filter can perform well to improve the accuracy of SVR-based eye center localization.

Acknowledgments. This work is supported by the National Natural Science Foundation of China (No. 61733011, 51575338).

References

1. Drucker, H., Burges, C.J.C., Kaufman, L.: Support vector regression machines. In: Advances in Neural Information Processing Systems, vol. 28, no. 7, pp. 779–784 (1996)
2. Smola, A.J., Schölkopf, B.: A tutorial on support vector regression. Stat. Comput. **14**(3), 199–222 (2004)

3. Baltrusaitis, T., Robinson, P., Morency, L.P.: Constrained local neural fields for robust facial landmark detection in the wild. In: IEEE International Conference on Computer Vision Workshops, pp. 354–361. IEEE Computer Society (2013)
4. Baltrusaitis, T., Robinson, P., Morency, L.P.: OpenFace: an open source facial behavior analysis toolkit. In: IEEE Winter Conference on Applications of Computer Vision, pp. 1–10. IEEE (2016)
5. Daugman, J.G.: High confidence visual recognition of persons by a test of statistical independence. IEEE Computer Society (1993)
6. Cai, H., Liu, B., Zhang, J., Chen, S., Liu, H.: Visual focus of attention estimation using eye center localization. IEEE Syst. J. **11**(3), 1–6 (2017)
7. Huan, N.V., Kim, H.: A novel circle detection method for iris segmentation. In: Congress on Image and Signal Processing, pp. 620–624. IEEE Computer Society (2008)
8. Timm, F., Barth, E.: Accurate eye centre localisation by means of gradients. In: Visapp 2011 - Proceedings of the Sixth International Conference on Computer Vision Theory and Applications, Vilamoura, Algarve, Portugal, 5–7 March, pp. 125–130. DBLP (2011)
9. Skodras, E., Fakotakis, N.: Precise localization of eye centers in low resolution color images. Butterworth-Heinemann (2015)
10. Poulopoulos, N., Psarakis, E.Z.: A new high precision eye center localization technique. In: IEEE International Conference on Image Processing, pp. 2806–2810. IEEE (2017)
11. Markuš, N., Frljak, M., Pandžić, I.S., Ahlberg, J., Forchheimer, R.: Eye pupil localization with an ensemble of randomized trees. Pattern Recognit. **47**(2), 578–587 (2014)
12. Everingham, M., Zisserman, A.: Regression and classification approaches to eye localization in face images. In: International Conference on Automatic Face and Gesture Recognition, pp. 441–448. IEEE Computer Society (2006)
13. Gou, C., Wu, Y., Wang, K., Wang, K., Wang, F.Y., Ji, Q.: A joint cascaded framework for simultaneous eye detection and eye state estimation. Pattern Recognit. **67**(1), 23–31 (2017)
14. Belhumeur, P.N., Jacobs, D.W., Kriegman, D.J., Kumar, N.: Localizing parts of faces using a consensus of exemplars. IEEE Trans. Pattern Anal. Mach. Intell. **35**(12), 2930–2940 (2013)
15. Le, V., Brandt, J., Lin, Z., Bourdev, L., Huang, T.S.: Interactive facial feature localization. In: Fitzgibbon, A., Lazebnik, S., Perona, P., Sato, Y., Schmid, C. (eds.) ECCV 2012. LNCS, vol. 7574, pp. 679–692. Springer, Heidelberg (2012). https://doi.org/10.1007/978-3-642-33712-3_49
16. Villanueva, A., Ponz, V., Sesma-Sanchez, L., Ariz, M., Porta, S., Cabeza, R.: Hybrid method based on topography for robust detection of iris center and eye corners. ACM Trans. Multimed. Comput. Commun. Appl. **9**(4), 1–20 (2013)
17. Jesorsky, O., Kirchberg, K.J., Frischholz, R.W.: Robust face detection using the hausdorff distance. In: Bigun, J., Smeraldi, F. (eds.) AVBPA 2001. LNCS, vol. 2091, pp. 90–95. Springer, Heidelberg (2001). https://doi.org/10.1007/3-540-45344-X_14
18. Campadelli, P., Lanzarotti, R., Lipori, G.: Precise eye localization through a general-to-specific model definition. In: British Machine Vision Conference, Edinburgh, UK, September, pp. 187–196. DBLP (2006)
19. Asadifard, M., Shanbezadeh, J.: Automatic adaptive center of pupil detection using face detection and CDF analysis. Lecture Notes in Engineering & Computer Science, vol. 2180, no. 1, pp. 13–14 (2010)

20. Valenti, R., Gevers, T.: Accurate eye center location through invariant isocentric patterns. IEEE Trans. Pattern Anal. Mach. Intell. **34**(9), 1785–1798 (2012)
21. Leo, M., Cazzato, D., De, M.T., Distante, C.: Unsupervised eye pupil localization through differential geometry and local self-similarity matching. PLoS ONE **9**(8), e102829 (2014)
22. George, A., Routray, A.: Fast and accurate algorithm for eye localisation for gaze tracking in low-resolution images. IET Comput. Vis. **10**(7), 660–669 (2017)

Hardware

Hopfield Neural Network with Double-Layer Amorphous Metal-Oxide Semiconductor Thin-Film Devices as Crosspoint-Type Synapse Elements and Working Confirmation of Letter Recognition

Mutsumi Kimura[1,2,3]($\boxtimes$), Kenta Umeda[4], Keisuke Ikushima[1],
Toshimasa Hori[1], Ryo Tanaka[1], Tokiyoshi Matsuda[3],
Tomoya Kameda[5], and Yasuhiko Nakashima[2]

[1] Department of Electronics and Informatics,
Ryukoku University, Seta, Otsu 520-2194, Japan
mutsu@rins.ryukoku.ac.jp,
{t17m009, t130185, t18m013}@mail.ryukoku.ac.jp
[2] Graduate School of Science and Technology, Nara Institute of Science
and Technology, Takayama, Ikoma 630-0192, Japan
nakashim@is.naist.jp
[3] High-Tech Research Center, Innovative Materials and Processing Research
Center, Seta, Otsu 520-2194, Japan
toki@rins.ryukoku.ac.jp
[4] Graduate School of Materials Science, Nara Institute of Science
and Technology, Takayama, Ikoma 630-0192, Japan
umeda.kenta.ua9@ms.naist.jp
[5] Graduate School of Information Science, Nara Institute of Science
and Technology, Takayama, Ikoma 630-0192, Japan
kameda.tomoya.kg0@is.naist.jp

Abstract. Artificial intelligences are essential concepts in smart societies, and neural networks are typical schemes that imitate human brains. However, the neural networks are conventionally realized using complicated software and high-performance hardware, and the machine size and power consumption are huge. On the other hand, neuromorphic systems are composed solely of optimized hardware, and the machine size and power consumption can be reduced. Therefore, we are investigating neuromorphic systems especially with amorphous metal-oxide semiconductor (AOS) thin-film devices. In this study, we have developed a Hopfield neural network with double-layer AOS thin-film devices as crosspoint-type synapse elements. Here, we propose modified Hebbian learning done locally without extra control circuits, where the conductance deterioration of the crosspoint-type synapse elements can be employed as synaptic plasticity. In order to validate the fundamental operation of the neuromorphic system, first, double-layer AOS thin-film devices as crosspoint-type synapse elements are actually fabricated, and it is found that the electric current continuously decreases along the bias time. Next, a Hopfield neural network is really assembled using a field-programmable gate array (FPGA) chip and the

© Springer Nature Switzerland AG 2018
L. Cheng et al. (Eds.): ICONIP 2018, LNCS 11307, pp. 637–646, 2018.
https://doi.org/10.1007/978-3-030-04239-4_57

double-layer AOS thin-film devices, and it is confirmed that a necessary function of the letter recognition is obtained after learning process. Once the fundamental operations are confirmed, more advanced functions will be obtained by scaling up the devices and circuits. Therefore, it is expected the neuromorphic systems can be three-dimensional (3D) large-scale integration (LSI) chip, the machine size can be compact, power consumption can be low, and various functions of human brains will be obtained. What has been developed in this study will be the sole solution to realize them.

Keywords: Hopfield neural network
Double-layer amorphous metal-oxide semiconductor (AOS) thin-film device
Crosspoint-type synapse elements · Letter recognition · Artificial intelligence
Neural network · Modified hebbian learning

1 Introduction

Artificial intelligences [1, 2] are essential concepts for diverse applications in smart societies, for example, letter recognition, image recognition, information retrieval, information provision, language translation, caption composition, expert system, autonomous driving, artificial brain, etc. Neural networks [3, 4] are typical schemes as artificial intelligences that imitate operation principle of human brains. However, the neural networks are conventionally realized using long and complicated software and high-performance Neumann-architecture computer hardware, which is not optimized for neural networks, and the machine size is unbelievably bulky and power consumption is incredibly huge. For example, one of the most famous cognitive systems in the industry field [5] occupies the size of several refrigerators and consumes the power of approximately a hundred kW. On the other hand, neuromorphic systems are composed solely of optimized hardware, and the machine size and power consumption can be reduced. However, a neuromorphic system [6] is currently a hybrid system, namely, multiple neuron elements are virtually emulated using time sharing of one neuron element, analogue values are stored using multiple bits of digital memories, learning function is not implemented in itself, etc., so the abovementioned problems are only partially resolved.

Therefore, we are investigating neuromorphic systems especially with amorphous metal-oxide semiconductor (AOS) thin-film devices [7, 8]. In this study, we have developed a Hopfield neural network with double-layer AOS thin-film devices as crosspoint-type synapse elements. Hopfield neural networks [9, 10] are neural networks where input terminals of all neuron elements are connected to output terminals of all neuron elements through synapse elements. Here, we propose modified Hebbian learning [11, 12] as a learning rule done locally without extra control circuits, where the conductance deterioration of the crosspoint-type synapse elements can be ingeniously employed as strength plasticity of synaptic connections. Although some prior articles were published on crosspoint-type synapse elements [13–17], AOS thin-film devices are used especially in this study.

In order to validate the fundamental operation of the neuromorphic system, first, double-layer AOS thin-film devices as crosspoint-type synapse elements will be actually fabricated, and it will be found whether the electric current continuously decreases along the bias time. Next, a Hopfield neural network will be really assembled using a field-programmable gate array (FPGA) chip and the double-layer AOS thin-film devices, and it will be confirmed whether a necessary function of the letter recognition is obtained after learning process. Once the fundamental operations are confirmed, more advanced and useful functions will be obtained by scaling up the devices and circuits in the far future obeying the academic history of neural networks. Therefore, because the AOS thin-film devices [18–20] have potential possibility that they can be layered using printing process with low cost, it is expected the neuromorphic systems can be three-dimensional (3D) large-scale integration (LSI) chip in the near future, the machine size can be excellently compact, and power consumption can be remarkably low even in comparison with other neuromorphic systems [21]. Moreover, because human brains are substantially materialized using the neuromorphic systems, various inherent functions will be also obtained, for example, self-organization, self-teaching, parallel distributed computing, fault tolerance, damage robustness, etc. What has been developed in this study will be the sole solution to realize them.

2 Double-Layer Amorphous Metal-Oxide Semiconductor Thin-Film Device

The device structure of the double-layer AOS thin-film device as the crosspoint-type synapse element is shown in Fig. 1. First, a quartz glass substrate is used, whose thickness is 1 mm and size is 3×3 cm. Next, a bottom Ti thin-film electrode is deposited using vacuum evaporation through a metal mask, whose line and space widths are 150 and 150 μm. Sequentially, an amorphous Ga-Sn-O (α-GTO) thin-film layer is deposited using radio-frequency (RF) magnetron sputtering with a ceramic target of In:Ga:Zn = 1:1:1 and sputtering gas of Ar of 20 sccm for two minutes, and another α-GTO thin-film layer is deposited with sputtering gas of O_2 of 20 sccm for four minutes, whose thicknesses are both several ten nm. The reason why the two oxygen (O)-poor and O-rich AOS thin-film layers are used is explained later. Next, another top Ti thin-film electrode is again deposited. Finally, the double-layer AOS thin-film device as the crosspoint-type synapse element is completed, where the two AOS thin-film layers are sandwiched between the top and bottom Ti thin-film electrodes.

The conductance deterioration of the double-layer AOS thin-film device as the crosspoint-type synapse element is shown in Fig. 2. First, the bias voltage is applied between the top and bottom Ti electrodes, whose voltage is 3.3 V. Next, the electric current is measured through the double-layer AOS thin-film device. It is found that the electric current continuously decreases along the bias time. The reason why the conductance deterioration is different between the cases when a positive voltage is applied to the top Ti electrode and when a negative voltage is applied is explained later. It is suggested that the conductance deterioration of the double-layer AOS thin-film devices can be ingeniously employed as strength plasticity of synaptic connections under the modified Hebbian learning.

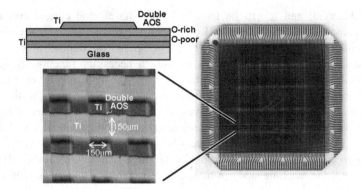

Fig. 1. Device structure of the double-layer AOS thin-film device as the crosspoint-type synapse element.

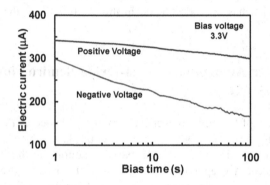

Fig. 2. Conductance deterioration of the double-layer AOS thin-film device as the crosspoint-type synapse element.

The deterioration mechanism of the double-layer AOS thin-film device as the crosspoint-type synapse element is shown in Fig. 3. When a positive voltage is applied to the top Ti electrode, oxygen ions (O^{2-}) are attracted in the O-rich layer, and no outstanding phenomenon occurs. When a negative voltage is applied, O^{2-} are repulsed toward the O-poor layer, and O^{2-} and oxygen vacancy (Vo) are recombined and disappear. Because it is known that the electric current is due to the Vo, the conductance deterioration occurs.

3 Hopfield Neural Network

The system architecture of the Hopfield neural network is shown in Fig. 4. The Hopfield neural network is really assembled using neuron elements composed in an FPGA chip and crosspoint-type synapse elements composed by the double-layer AOS thin-film devices. Hopfield neural networks are neural networks where input terminals of all neuron elements are connected to output terminals of all neuron elements through

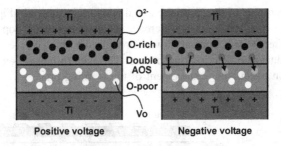

Positive voltage   **Negative voltage**

Fig. 3. Device structure of the double-layer AOS thin-film device as the crosspoint-type synapse element.

synapse elements. First, during the learning process, a signal pattern is inputted to the neuron elements, and the same signal pattern is outputted and applied to the top Ti electrodes of the double-layer AOS thin-film devices and simultaneously bottom Ti electrodes through the switch parts. Next, during the recognizing process, a signal pattern is inputted, the same signal pattern is applied to the top Ti electrode, and the signal pattern is immediately released. Some analog signals are transmitted from the bottom Ti electrodes and inputted to the neuron elements, and some signal pattern is outputted after dynamic behavior of the Hopfield neural network.

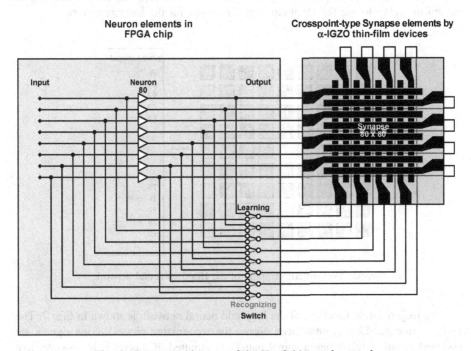

Fig. 4. System architecture of the Hopfield neural network.

The learning process of the Hopfield neural network is shown in Fig. 5. During the learning process, when the voltage difference exists between the top and bottom Ti electrodes and the electric current flows through the double-layer AOS thin-film devices, the conductance deterioration occurs. The pattern generation of the conductance deterioration means the achievement of the learning process.

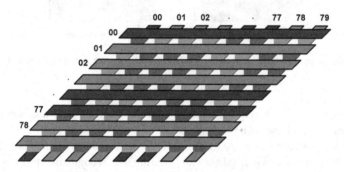

Fig. 5. Learning process of the Hopfield neural network.

The pixel pattern mapping of the Hopfield neural network is shown in Fig. 6. A two-dimensional pixel pattern of 9×9 is transformed to a one-dimensional signal pattern of eighty to use the Hopfield neural network for the letter recognition.

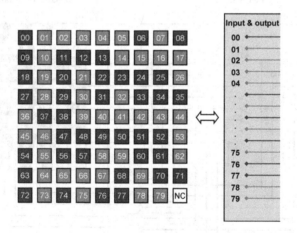

Fig. 6. Pixel pattern mapping of the Hopfield neural network.

The majority-rule handling of the Hopfield neural network is shown in Fig. 7. The majority-rule handling is introduced during the recognizing process. Nine signals are assigned to one pixel. When a signal pattern is inputted, if a pixel is in on-state, five signals among the corresponding nine signals are randomly set to on-state, whereas if a pixel is in off-state, five signals are set to off-state, and the nine signals are inputted to the neuron elements. On the other hand, when a signal pattern is outputted, if all the corresponding signals are in on-state, the pixel is recognized as on-state, whereas otherwise the pixel is recognized as off-state. As aforementioned, when a positive voltage is applied to the top Ti electrode, the conductance deterioration is a little smaller, whereas when a negative voltage is applied, the conductance deterioration is a little larger. This means that the positive voltage is more easily transmitted through the double-layer AOS thin-film devices. When a signal pattern is outputted, the signals are apt to be in on-state. Therefore the abovementioned rule is effective when a signal pattern is outputted.

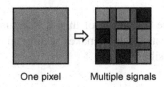

One pixel   Multiple signals

Fig. 7. Majority-rule handling of the Hopfield neural network.

4 Letter Recognition

An experimental example of the letter recognition is shown in Fig. 8. First, during the learning process, alphabet letters of "**T**" and "**L**" are learned. Initially, a pixel pattern of "**T**" is transformed to a signal pattern, and the signal pattern is inputted to the neuron elements for one second. The same signal pattern is outputted and applied to the top Ti electrodes of the double-layer AOS thin-film devices as the crosspoint-type synapse elements and simultaneously bottom Ti electrodes, and the conductance deterioration occurs. Successively, a signal pattern of "**L**" is inputted. Then the learning processes are repeated several hundred times. Next, during the recognizing process, alphabet letters of "**T**" and "**L**" are reproduced. Initially, a one-pixel flipped pattern, namely, slightly distorted pixel pattern, of "**T**" is transformed to a signal pattern with the majority-rule handling, namely, five signals are set to on-state when the pixel is in on-state, and the signal pattern is inputted. The same signal pattern is applied to the top electrodes, and the signal pattern is immediately released. Subsequently, some revised signal pattern is outputted from the neuron elements and transformed to some revised pixel pattern, namely, the pixel is set to on-state when all the signals are in on-state. Then the recognizing processes are repeated for different flipped pixel patterns. Successively, one-pixel flipped patterns of "**L**" are inputted, and some patterns are outputted. Finally, it is confirmed that the revised pixel pattern is the same as the pixel patterns of "**T**" and "**L**".

Experimental results of the letter recognition is shown in Fig. 9. It is confirmed that the revised pixel patterns are the same as the pixel patterns of "**T**" and "**L**" for all the one-pixel flipped patterns. Because simple pattern matching is available once the pixel patterns are reproduced, it can be concluded that a necessary function of the letter recognition is confirmed. Although a two-dimensional pixel pattern of 9 × 9 is used in

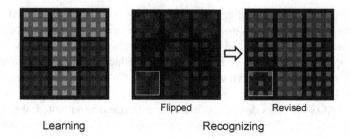

Fig. 8. Experimental example of the letter recognition

this presentation, because the scalability of this architecture is excellent, namely, the same device structure is repeated, which makes scaling up easy, some pixel pattern of higher resolution is also possible.

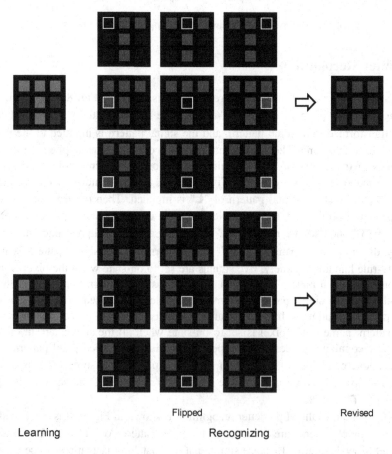

Fig. 9. Experimental results of the letter recognition.

5   Conclusion

We have developed a Hopfield neural network with double-layer AOS thin-film devices as crosspoint-type synapse elements. Here, we proposed modified Hebbian learning done locally without extra control circuits, where the conductance deterioration of the crosspoint-type synapse elements can be employed as synaptic plasticity. In order to validate the fundamental operation of the neuromorphic system, first, double-layer AOS thin-film devices as crosspoint-type synapse elements were actually fabricated, and it was found that the electric current continuously decreases along the bias time. Next, a Hopfield neural network was really assembled using a FPGA chip and the double-layer AOS thin-film devices, and it was confirmed that a necessary function of the letter recognition is obtained after learning process. Once the fundamental operations are confirmed, more advanced functions will be obtained by scaling up the devices and circuits. Therefore, it is expected the neuromorphic systems can be 3D LSI chip, the machine size can be compact, power consumption can be low, and various functions of human brains will be obtained. What has been developed in this study will be the sole solution to realize them. Although it was only confirmed that a necessary function of the letter recognition is obtained, it is hoped that diverse applications are realized using our study, because the system concept is general and variable in principle.

Acknowledgments. This work is partially supported by KAKENHI (C) 16K06733, Yazaki Memorial Foundation for Science and Technology, Support Center for Advanced Telecommunications Technology Research, Research Grants in the Natural Sciences from the Mitsubishi Foundation, the Telecommunications Advancement Foundation, RIEC Nation-wide Cooperative Research Projects, collaborative research with ROHM Semiconductor, collaborative research with KOA Corporation, and Innovative Materials and Processing Research Center.

References

1. McCarthy, J., Minsky, M.L., Rochester, N., Shannon, C.E.: A proposal for the dartmouth summer research project on artificial intelligence. In: Dartmouth Conference (1956)
2. Russell, S., Norvig, P.: Artificial Intelligence: A Modern Approach. Pearson Education, Prentice Hall (2009)
3. McCulloch, W.S., Pitts, W.: A logical calculus of the ideas immanent in nervous activity. Bull. Math. Biophys. **5**, 115–133 (1943)
4. Wasserman, P.D.: Neural Computing: Theory and Practice. Coriolis Group, Scottsdale (1989)
5. Ferrucci, D., et al.: Building watson: an overview of the deep QA project. AI Mag. **31**(3), 59–79 (2010)
6. Merolla, P.A., et al.: A million spiking-neuron integrated circuit with a scalable communication network and interface. Science **345**, 668–673 (2014)
7. Kimura, M., Koga, Y., Nakanishi, H., Matsuda, T., Kameda, T., Nakashima, Y.: In-Ga-Zn-O thin-film devices as synapse elements in a neural network. IEEE J. Electron Devices Soc. **6**, 100–105 (2017)

8. Kameda, T., Kimura, M., Nakashima, Y.: Neuromorphic hardware using simplified elements and thin-film semiconductor devices as synapse elements - simulation of hopfield and cellular neural network. In: Liu, D., Xie, S., Li, Y., Zhao, D., El-Alfy, E.S. (eds.) Neural Information Processing. ICONIP 2017. Lecture Notes in Computer Science, vol. 10639. Springer, Cham (2017). https://doi.org/10.1007/978-3-319-70136-3_81

9. Hopfield, J.J., Tank, D.W.: Neural computation of decisions in optimization problems. Biol. Cybern. **52**, 141–152 (1985)

10. Hopfield, J.J., Tank, D.W.: Computing with neural circuits: a model. Science **233**, 625–633 (1986)

11. Kimura, M., Morita, R., Sugisaki, S., Matsuda, T., Kameda, T., Nakashima, Y.: Cellular neural network formed by simplified processing elements composed of thin-film transistors. Neurocomputing **248**, 112–119 (2017)

12. Kimura, M., Nakamura, N., Yokoyama, T., Matsuda, T., Kameda, T., Nakashima, Y.: Simplification of processing elements in cellular neural networks. In: Hirose, A., Ozawa, S., Doya, K., Ikeda, K., Lee, M., Liu, D. (eds.) ICONIP 2016. LNCS, vol. 9948, pp. 309–317. Springer, Cham (2016). https://doi.org/10.1007/978-3-319-46672-9_35

13. Chen, Y., et al.: Nanoscale molecular-switch crossbar circuits. Nanotechnol. **14**, 462–468 (2003)

14. Jo, S.H., Chang, T., Ebong, I., Bhadviya, B.B., Mazumder, P., Lu, W.: Nanoscale memristor device as synapse in neuromorphic systems. Nano Lett. **10**, 1297–1301 (2010)

15. Merolla, P., Arthur, J., Akopyan, F., Imam, N., Manohar, R., Modha, D.S.: A digital neurosynaptic core using embedded crossbar memory with 45 pJ per Spike in 45 nm. In: 2011 IEEE Custom Integrated Circuits Conference (CICC), pp. 1–4 (2011)

16. Alibart, F., Zamanidoost, E., Strukov, D.B.: Pattern classification by memristive crossbar circuits using ex situ and in situ training. Nat. Commun. **4**, 2072 (2013)

17. Hu, M., et al.: Dot-product engine for neuromorphic computing: programming 1T1 M crossbar to accelerate matrix-vector multiplication. In: The 53rd Annual Design Automation Conference (DAC 2016) (2016)

18. Matsuda, T., Umeda, K., Kato, Y., Nishimoto, D., Furuta, M., Kimura, M.: Rare-Metal-Free High-Performance Ga-Sn-O Thin Film Transistor. Sci. Rep., srep 44326 (2017)

19. Nomura, K., et al.: Three-dimensionally stacked flexible integrated circuit: amorphous oxide/polymer hybrid complementary inverter using n-type a-In-Ga-Zn-O and p-type poly-(9,9-dioctylfluorene-co-bithiophene) thin-film transistors. Appl. Phys. Lett. **96**, 263509 (2010)

20. Okamoto, R., Fukushima, H., Kimura, M., Matsuda, T.: Characteristic evaluation of Ga-Sn-O films deposited using mist chemical vapor deposition. In: The 2017 International Meeting for Future of Electron Devices, Kansai (IMFEDK 2017), pp. 74–75 (2017)

21. Prezioso, M., Merrikh-Bayat, F., Hoskins, B.D., Adam, G.C., Likharev, K.K., Strukov, D.B.: Training and operation of an integrated neuromorphic network based on metal-oxide memristors. Nature **521**, 61–64 (2015)

FPGA Based Hardware Implementation of Simple Dynamic Binary Neural Networks

Shunsuke Aoki, Seitaro Koyama, and Toshimichi Saito[✉]

Hosei University, Koganei, Tokyo 184-8584, Japan
tsaito@hosei.ac.jp

Abstract. This paper studies hardware implementation of a simple dynamic binary neural network that can generate various periodic orbits. The network is characterized by local binary connection and signum activation function. First, using a simple feature quantity, stability of a target periodic orbit is considered. Second, using a FPGA board, a test circuit is implemented. The signum activation function is realized by a majority decision circuit and the binary connection is realized by switches and inverters. The circuit operation is confirmed experimentally.

Keywords: Dynamic binary neural networks · Stability · FPGA

1 Introduction

This paper studies hardware implementation of a 3-1 dynamic binary neural networks (3-1 DBNN) where each neuron transforms 3 binary inputs to 1 binary output via signum activation function [1]. The 3-1 DBNN is characterized by local binary connection and is described by a difference equation of binary state variables. Depending on the connection/threshold parameters, the 3-1 DBNN can generate various binary periodic orbits (BPOs) some of which are applicable to control signal of switching circuits [2,3]. The local binary connection is suitable for hardware implementation and low-power consumption as compared with real connection of recurrent neural networks and full binary connection of the DBNN [4,5]. The 3-1 DBNN is related to various systems including logical circuits [6–8], associative memories [9,10], and cellular automata [11,12]. The hardware of 3-1 DBNNs is important to observe nonlinear dynamics and to consider engineering applications.

First, we consider stability of BPOs that is basic to design the hardware. For simplicity, we focus on one target binary periodic orbit (TBPO) that is related to a control signal of a switching power converter [3]. Presenting a simple feature quantity, we define the direct stability of a TBPO such that closest neighbors fall directly into the TBPO. Using the feature quantity, we have confirmed that sparsity of the 3-1 DBNN is valid to reinforce direct stability of the TBPO.

© Springer Nature Switzerland AG 2018
L. Cheng et al. (Eds.): ICONIP 2018, LNCS 11307, pp. 647–655, 2018.
https://doi.org/10.1007/978-3-030-04239-4_58

Second, we design a test circuit of the 3-1 DBNN where the activation function is realized by a majority decision circuit and the binary connection is realized by switches and inverters. Using the Verilog, the test circuit is implemented on a FPGA board. Using the test circuit, the TBPO generation is confirmed experimentally. Note that this is the first paper presenting the FPGA based implementation of the 3-1 DBNN and direct stability of the TBPO.

2   3-1 Dynamic Binary Neural Networks

In this section, as preparations, we introduce the DBNN and the 3-1 DBNN presented in [1]. Applying delayed feedback to a two-layer network, the DBNN is constructed. The dynamics is described by

$$x_i^{t+1} = F\left(\sum_{j=1}^{N} w_{ij} x_j^t - T_i\right), \ i = 1 \sim N, \ \ F(x) = \begin{cases} +1 \text{ if } x \geq 0 \\ -1 \text{ if } x < 0 \end{cases} \quad (1)$$

The signum activation function F outputs $+1$ or -1 and $x_i^t \in \{-1, +1\} \equiv \boldsymbol{B}$ is the i-th binary state at discrete time t. Let $\boldsymbol{x}^t \equiv (x_1^t, \cdots, x_N^t)$. As an initial state vector $\boldsymbol{x}^1$ is applied, the DBNN generates a sequence of binary vectors $\{\boldsymbol{x}^t\}$. The connection parameters are ternary $w_{ij} \in \{-1, 0, 1\}$ and $\boldsymbol{W} = (w_{ij})$ is referred to as the connection matrix. The threshold parameters are integer.

For convenience in hardware implementation, we have simplified the DBNN into the 3-1 DBNN of local binary connection [1]:

$$\begin{aligned} &x_i^{t+1} = F\left(w_{ii_a} x_{i_a}^t + w_{ii_b} x_{i_b}^t + w_{ii_c} x_{i_c}^t - T_i\right), i = 1 \sim N \\ &x_{i_a} \in \{x_1, \cdots, x_N\}, x_{i_b} \in \{x_1, \cdots, x_N\}, x_{i_c} \in \{x_1, \cdots, x_N\}, \\ &w_{ii_a} \in \{-1, +1\}, w_{ii_b} \in \{-1, +1\}, w_{ii_c} \in \{-1, +1\}, \\ &x_{i_a} \neq x_{i_b} \neq x_{i_c}, T_i \in \{-4, -2, 0, +2, +4\}. \end{aligned} \quad (2)$$

The network configuration is shown in Fig. 1. The local binary connection is suitable for hardware implementation and low-power consumption as compared with real connection of recurrent neural networks and ternary connection of DBNNs. Equation (2) is abbreviated by $\boldsymbol{x}^{t+1} = \boldsymbol{F}(\boldsymbol{W}\boldsymbol{x}^t - \boldsymbol{T})$. Let $X \equiv w_{ii_a} x_{i_a}^t + w_{ii_b} x_{i_b}^t + w_{ii_c} x_{i_c}^t$. As shown in Fig. 1(b), X can be either of 4 values $\{-3, -1, +1, +3\}$, hence the threshold T_i can be restricted in either of four values $\{-4, -2, 0, +2, +4\}$. The i-th neuron selects three binary inputs $(x_{i_a}^t, x_{i_b}^t, x_{i_c}^t)$ out of N binary states $(x_1^t, \cdots, x_N^t)$ at time t. After that the neuron outputs the i-th binary state x_i^{t+1} at time $t+1$. Repeating in this manner, the 3-1 DBNN can output various binary sequences.

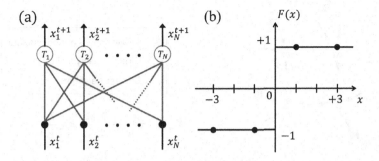

Fig. 1. The 3-1 DBNN. (a) Network configuration. Red branch: $w_{ij} = +1$. Blue branch: $w_{ij} = -1$. Green circle: T_i. (b) Signum activation function. (Color figure online)

3   Target Binary Periodic Orbit and Direct Stability

We consider stability of a target binary periodic orbit (TBPO) from the 3-1 DBNN. First, the TBPO with period p is defined by

$$z^1, \cdots, z^p, \cdots, \quad \begin{cases} z^t = z^s \text{ for } |t-s| = np \\ z^t \neq z^s \text{ for } |t-s| \neq np \end{cases} \tag{3}$$
$$z^t \equiv (z_1^t, \cdots, z_N^t)^\top \in B^N$$

where n denotes positive integers. In a TBPO $(z^1, \cdots, z^p)$, each element z^t is referred to as a target binary periodic point (TBPP). In order to consider local stability of a TBPO, we give several definitions. A binary vector ξ is said to be a closest neighbor of a TBPO if ξ is not a TBPP and differs Hamming distance 1 from either TBPP: $HD(\xi, z^s) = 1$ for some s. A binary vector q is said to be a direct repair point if q is a closest neighbor and falls directly into the TBPO as shown in Fig. 2: $F(q) = z^s$ for some s. The feature quantity is defined by

$$\alpha_d = \frac{\#\text{DRPs}}{\#\text{CNBs}}, \ 0 \le \alpha \le 1$$
$$\#\text{DRPs: The number of direct repair points} \tag{4}$$
$$\#\text{CNBs: the number of closest neighbors}$$

The direct stability of a TBPO is said to become stronger if α_d increases. In order to characterize sparsity of the connection matrix, we define the sparsity rate

$$\text{SR} = \frac{\text{The number of zeros in} W}{N^2 - N}, \ 0 \le \text{SR} \le 1 \tag{5}$$

where each input is assumed to be connected to at least one neuron. The sparsity of the connection matrix is said to increase if SR increases. For example, sparsity of the 3-1 DBNN is characterized by $\text{SR} = N(N-3)/(N^2 - N)$.

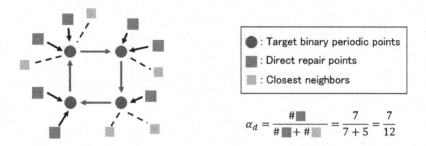

$$\alpha_d = \frac{\#\blacksquare}{\#\blacksquare + \#\square} = \frac{7}{7+5} = \frac{7}{12}$$

Fig. 2. Local stability of a TBPO. in this figure $\alpha_d = 7/12$.

This paper considers the following TBPO with period 8. This TBPO is related to a control signal of a switching power converter [2].

$$
z^1 = \begin{pmatrix} +1 \\ +1 \\ -1 \\ -1 \\ -1 \\ -1 \\ +1 \\ +1 \end{pmatrix}
z^2 = \begin{pmatrix} +1 \\ +1 \\ +1 \\ -1 \\ -1 \\ -1 \\ -1 \\ +1 \end{pmatrix}
z^3 = \begin{pmatrix} +1 \\ +1 \\ +1 \\ +1 \\ -1 \\ -1 \\ -1 \\ -1 \end{pmatrix}
z^4 = \begin{pmatrix} -1 \\ +1 \\ +1 \\ +1 \\ +1 \\ -1 \\ -1 \\ -1 \end{pmatrix}
$$

$$
z^5 = \begin{pmatrix} -1 \\ -1 \\ +1 \\ +1 \\ +1 \\ +1 \\ -1 \\ -1 \end{pmatrix}
z^6 = \begin{pmatrix} -1 \\ -1 \\ -1 \\ +1 \\ +1 \\ +1 \\ +1 \\ -1 \end{pmatrix}
z^7 = \begin{pmatrix} -1 \\ -1 \\ -1 \\ -1 \\ +1 \\ +1 \\ +1 \\ +1 \end{pmatrix}
z^8 = \begin{pmatrix} +1 \\ -1 \\ -1 \\ -1 \\ -1 \\ +1 \\ +1 \\ +1 \end{pmatrix}
\tag{6}
$$

The following 3-1 DBNN can generate the TBPO:

$$
z^{i+1} = W_1 z^i
$$
$$
i = 1 \sim 8, z^9 \equiv z^1
$$

$$
\alpha_d = 48/48
$$
$$
SR = 40/56
$$

$$
W_1 = \begin{pmatrix}
+1 & 0 & 0 & -1 & 0 & 0 & 0 & +1 \\
+1 & +1 & 0 & 0 & -1 & 0 & 0 & 0 \\
0 & +1 & +1 & 0 & 0 & -1 & 0 & 0 \\
0 & 0 & +1 & +1 & 0 & 0 & -1 & 0 \\
0 & 0 & 0 & +1 & +1 & 0 & 0 & -1 \\
-1 & 0 & 0 & 0 & +1 & +1 & 0 & 0 \\
0 & -1 & 0 & 0 & 0 & +1 & +1 & 0 \\
0 & 0 & -1 & 0 & 0 & 0 & +1 & +1
\end{pmatrix}
\quad
T = \begin{pmatrix} 0 \\ 0 \\ 0 \\ 0 \\ 0 \\ 0 \\ 0 \\ 0 \end{pmatrix}
\tag{7}
$$

We can get Eq. (7) by trial and error. Note that direct stability of the TBPO is the strongest. For simplicity, we assume $T_i = 0$ for $i = 1 \sim N$ hereafter. In order to consider relation between the connection sparsity and direct stability of the

TBPO, we show two examples of DBNNs. The first example is characterized by the most sparse connection:

$$z^{i+1} = W_0 z^i$$
$$i = 1 \sim 8, z^9 \equiv z^1$$

$$\alpha_d = 0/48$$
$$SR = 56/56$$

$$W_0 = \begin{pmatrix} 0 & 0 & 0 & 0 & 0 & 0 & 0 & +1 \\ +1 & 0 & 0 & 0 & 0 & 0 & 0 & 0 \\ 0 & +1 & 0 & 0 & 0 & 0 & 0 & 0 \\ 0 & 0 & +1 & 0 & 0 & 0 & 0 & 0 \\ 0 & 0 & 0 & +1 & 0 & 0 & 0 & 0 \\ 0 & 0 & 0 & 0 & +1 & 0 & 0 & 0 \\ 0 & 0 & 0 & 0 & 0 & +1 & 0 & 0 \\ 0 & 0 & 0 & 0 & 0 & 0 & +1 & 0 \end{pmatrix} \quad (8)$$

This DBNN can generate the TBPO and but the direct stability is the weakest. This DBNN is equivalent to the shift register [8] that is one of the most basic sequential circuits. The second example is characterized by full binary connection:

$$z^{i+1} = W_2 z^i$$
$$i = 1 \sim 8, z^9 \equiv z^1$$

$$\alpha_d = 4/48$$
$$SR = 0/56$$

$$W_2 = \begin{pmatrix} +1 & +1 & +1 & -1 & +1 & +1 & +1 & +1 \\ +1 & +1 & -1 & -1 & -1 & -1 & -1 & +1 \\ +1 & +1 & +1 & -1 & +1 & -1 & -1 & +1 \\ +1 & +1 & +1 & +1 & +1 & +1 & -1 & +1 \\ -1 & -1 & +1 & +1 & +1 & -1 & -1 & -1 \\ -1 & +1 & -1 & +1 & +1 & +1 & -1 & +1 \\ -1 & -1 & -1 & +1 & -1 & +1 & +1 & -1 \\ +1 & -1 & -1 & -1 & +1 & +1 & +1 & +1 \end{pmatrix} \quad (9)$$

This DBNN can generate the TBPO and the direct stability is weak.

Adding one branch successively to the most sparse DBNN in Eq. (8), the connection grows and SR decreases. In the connection growing process, one branch is added to either of all possible positions and one DBNN with the largest α_d is preserved for further connection growing. The result is shown in Fig. 3. As SR

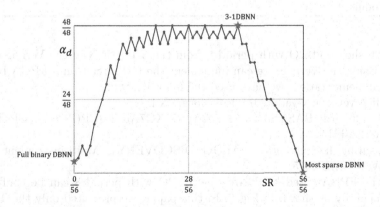

Fig. 3. Direct stability (α_d) of the TBPO for sparsity rate SR. α_d is the maximum value for each SR.

decreases, α_d increases and the direct stability of the TBPO can be reinforced. As SR decreases further, the direct stability is weakened eventually. We can see that appropriate sparsity is necessary to realize strong direct stability. According to Fig. 3, the most sparse of the strongest stability (SR = 40/56 of the 3-1 DBNN) is appropriate.

4   FPGA Based Hardware Implementation

We present a FPGA based hardware of the 3-1 DBNN in Eq. (7). The local binary connection (SR=40/56) is suitable to reduce power consumption. First, we have designed basic circuit as shown in Figs. 4 and 5 where the signum activation function of three inputs is realized by the simplest majority decision circuit. The positive/negative connection is realized by switches and inverters. Adjusting the switches, the circuit can realize all the connection patterns of the 3-1 binary neurons. Algorithm 1 shows outline of the Verilog source code for the 3-1 binary neurons. Using this circuit and D-Flip-Flop(D-FF), the 3-1 DBNN can be realized.

Algorithm 1. 3-1 binary neuron

---
module 3-1BinaryNeuron(input x_a^t, input x_b^t, input x_c^t, output x_i^{t+1})
// And getes and OR gates
wire AND1;
wire AND2;
wire AND3;
wire OR;
// 3-1 Binary Neuron
assign AND1 = $x_a^t \wedge \neg x_b^t$;
assign AND2 = $x_a^t \wedge x_c^t$;
assign AND3 = $x_b^t \wedge x_c^t$;
assign OR = AND1∨AND2;
assign x_i^{t+1} = OR ∨ AND3;
endmodule

Fig. 6 shows TBPO with period 8 from the 3-1 DBNN in the Verilog simulation. Using the Verilog, we can implement the test circuit on a FPGA board. In the implementation, we have used the following tools:

Verilog version: vivado 2017.4 platform (Xilinx).

FPGA board: BASYS3 (Xilinx Artix-7 XC7A35T-ICPG236C), Clock frequency: 12.2[kHz].

Measuring instrument: ANALOG DISCOVERY2, Multi-instrument software: Waveforms 2015

Using the FPBA-based hardware, the TBPO with period 8 can be confirmed experimentally as shown in Fig. 7. In this paper, we presented only the TBPO with period 8. But the 3-1 DBNN is a simple circuit so if N increases, we can realize in row power consumption.

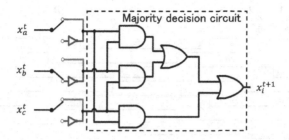

Fig. 4. Majority decision circuit of 3-1 binary neuron.

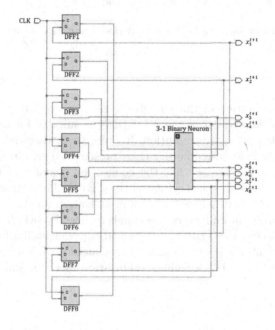

Fig. 5. Basic circuit design of the 3-1 DBNN.

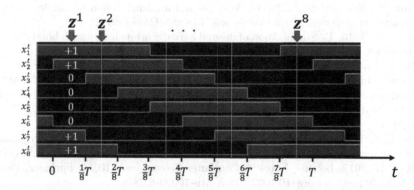

Fig. 6. TBPO from the 3-1 DBNN in verilog simulation.

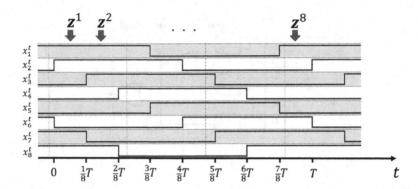

Fig. 7. Measured waveform of TBPO with period 8 in a FPGA board ($T = 655\mu s$).

5 Conclusions

Hardware implementation of the 3-1 DBNN have been discussed in this paper. The local binary connection of the 3-1 DBNN is suitable for hardware implementation with low power consumption. In order to characterize direct stability of a TBPO, a simple feature quantity α_d is presented. Performing basic numerical experiment, it is clarified that the connection sparsity of the 3-1 DBNN is suitable to reinforce direct stability of the TBPO. The test circuit is implemented in a FPGA board where the signum activation function is realized by a majority decision circuit and the connection is realized by switches and inverters. Using the test circuit, the TBPO is confirmed experimentally. Future problems include development of a learning/stabilization method of desired TBPOs, hardware implementation of the learning/stabilization method, and its engineering applications.

References

1. Aoki, S., Koyama, S., Saito, T.: Analysis and implementation of simple dynamic binary neural networks. In: Proceedings of the IJCNN (2018)
2. Sato, R., Saito, T.: Stabilization of desired periodic orbits in dynamic binary neural networks. Neurocomputing **248**, 19–27 (2017)
3. Bose, B.K.: Neural network applications in power electronics and motor drives - an introduction and perspective. IEEE Trans. Ind. Electron. **54**(1), 14–33 (2007)
4. Koyama, S., Aoki, S., Saito, T.: Simple feature quantities for analysis of periodic orbits in dynamic binary neural networks. IEICE Trans. Fund. **E101−A**, 727–730 (2017)
5. Aoki, S., Saito, T.: Stability of periodic orbits and fault tolerance in dynamic binary neural networks. In: Liu, D., Xie, S., Li, Y., Zhao, D., El-Alfy, E.S. (eds.) ICONIP 2017. Lecture Notes in Computer Science, vol. 10636. Springer, Cham (2017). https://doi.org/10.1007/978-3-319-70090-8_78
6. Gray, D.L., Michel, A.N.: A training algorithm for binary feed forward neural networks. IEEE Trans. Neural Netw. **3**(2), 176–194 (1992)

7. Chen, F., Chen, G., He, Q., He, G., Xu, X.: Universal perceptron and DNA-like learning algorithm for binary neural networks: non-LSBF implementation. IEEE Trans. Neural Netw. **20**(8), 1293–1301 (2009)
8. Saravanan, S., Lavanya, M., Vijay Sai, R., Kumar, R.: Design and analysis of linear feedback shift register based on various tap connection. Procedia Eng. **38**, 640–646 (2012)
9. Araki, K., Saito, T.: An associative memory including time-variant self-feedback. Neural Netw. **7**(8), 1267–1271 (1994)
10. Jiang, X., Gripon, V., Berrou, C., Rabbat, M.: Storing sequences in binary tournament-based neural networks. IEEE Trans. Neural Netw. **27**(5), 913–925 (2016)
11. Chua, L.O.: A Nonlinear Dynamics Perspective of Wolfram's New Kind of Science, vol. I and II. World Scientific, Singapore (2005)
12. Wada, W., Kuroiwa, J., Nara, S.: Completely reproducible description of digital sound data with cellular automata. Phys. Lett. A **306**, 110–115 (2002)

Fast Depthwise Separable Convolution
for Embedded Systems

Byeongheon Yoo, Yongjun Choi, and Heeyoul Choi[✉]

School of Computer Science and Electrical Engineering, Handong Global University,
Pohang 37554, South Korea
byeongheonyu@gmail.com, drgnjoon@gmail.com, heeyoul@gmail.com

Abstract. Convolutional neural networks (CNNs) have achieved outstanding performance in many applications. However, as the total number of layers has increased and the model structure has become compound, the computational cost comes into question. The large models
cannot operate in embedded or mobile environments where hardware
resources are quite limited. To overcome these problems, there have been
several attempts like reducing the depth of networks, pruning, quantization or low rank approximation. Depthwise separable convolution (DSC)
was proposed to reduce computation especially in convolutional layers
by separating one convolution into a spatial convolution and a pointwise convolution. In this paper, we apply DSC to the YOLO network for
object detection and propose a faster version of DSC, *FastDSC* by replacing the pointwise convolution with general matrix multiplication. Experiments on the NVIDIA Jetson TX2 board show that FastDSC speeds up
DSC for object detection.

Keywords: Network optimization · Depthwise separable convolution
Pointwise convolution · General matrix multiplication

1 Introduction

The advances of deep learning has dramatically improved the performance
of pattern recognition especially in image recognition and speech recognition
[3,7,12,13]. When a task is well-defined with enough data, deep learning can
outperform human level intelligence in the specific task. To achieve better performance, deep neural networks increase the number of nodes and layers and
make the architecture more compound. However, deep networks with such a
large and complex model are computationally too expensive to run the models
on limited computing resources like embedded systems, usually without GPUs.
Even when enough computing resources are provided, running deep networks
spends a lot of electrical power which prevents deep learning models from working on battery based mobile devices.

There have been several attempts to obtain smaller networks which have less
computational cost to solve the same task. Knowledge distillation trains smaller

© Springer Nature Switzerland AG 2018
L. Cheng et al. (Eds.): ICONIP 2018, LNCS 11307, pp. 656–665, 2018.
https://doi.org/10.1007/978-3-030-04239-4_59

(shallower or thinner) networks whose performance is compatible to deep networks with much less computation cost. Once a deep model is trained for a task, the trained model can teach new smaller networks for the same task [8,17]. Instead of pre-training a deep model, pruning methods can eliminate unnecessary weights whose values are close to zero during the training process. Also, quantization of the weights can reduce the computation cost by representing the parameters with less bits (even with just 1 bit) [2]. Low rank approximation is another approach to reduce the number of parameters and computations based on matrix factorization [11].

In addition to the approaches above, depthwise separable convolution (DSC) was proposed especially to reduce the computation cost in convolutional neural networks (CNNs), as in MobileNet and ShuffleNet [9,19]. While conventional convolution layers apply 3-dimensional (3D) kernels (or filters) to input (or feature maps), DSC layers perform a spatial convolution with 2D kernels and then a 1×1 (or pointwise) convolution with 1D kernels sequentially. DSC was successfully applied to CNN architectures [1].

In this paper, after testing DSC with the MNIST classification as in MobileNet [9], we apply DSC to the YOLO network to speed up running CNNs in limited computing resources. That is, we replace each convolution layer with DSC (one spatial convolution and one pointwise convolution). With DSC, the model size and computations are significantly reduced, which is more preferable for embedded systems, but it comes with a little loss of performance. Then, we propose a fast version of depthwise separable convolution, *FastDSC*, by replacing every pointwise convolution with general matrix multiplication (GEMM). To apply GEMM for the pointwise convolution, reshape operators should be applied accordingly.

In experiments, we compare the three models, (conventional CNN, DSC, and FastDSC) on image classification and object detection tasks. For image recognition on MNIST dataset [14], the model consists of two convolution layers, one maxpooling and a fully connected layer. For object detection on VOC dataset [4], we use the YOLO model [16] to evaluate performance and execution time. Experiments are conducted on two different systems: (1) Linux server with Intel Xeon Processor E5-2620 and GTX1080, and (2) NVIDIA Jetson TX2 with Dual Denver 2 + Quad ARM A57 and 256-core Pascal GPU. On both systems, we conduct the same experiments with CPU and with GPU.

The paper is organized as follows. Section 2 briefly reviews the features of the convolution neural networks and depthwise separable convolution layers. Section 3 describes the proposed FastDSC with the process of applying GEMM to the pointwise convolution. Section 4 presents experiment results of different models. Conclusion and future works follow in Section 5.

2   Background

2.1   Convolutional Neural Networks

Convolutional neural networks (CNNs) are the winning models in many pattern recognition competitions [12,18]. CNNs can be trained purely in the supervised fashion without pretraining. CNNs are based on neuroscientific ideas like the Hubel-Wiesel model, which has simple and complex cells [5]. Each node in a higher layer is connected to a small region of the lower layer, which implements local receptive fields. Also, CNNs share the weights in the same feature map. The reason that CNNs can train the networks without pretraining is that the local connection prevents the error information in backpropagation from vanishing, and weight sharing integrates the scattered error information.

When an image is processed with a fully connected layer, the 3D image must be stretched into one dimension, which loses the spatial information of the image. CNNs keep the spatial information of images by applying convolution of the same kernels over the image. CNNs consist of convolution and pooling layers to extract data regularities from images, and followed by an activation function for which ReLU is usually considered [15]. Also, batch normalization may be included depending on whether batch mode [10] is used or not. After extracting features, fully connected layers at the end of the model are responsible for classification of the image. Since CNNs share the parameters as filters, the number of parameters to be learned is relatively small compared to the corresponding fully connected layers. CNNs can use enough filters to extract regularities from images for classification.

2.2   Depthwise Separable Convolution

Although CNNs have relatively a small number of parameters, the size of each kernel is still large, since the size is proportional to the number of input channels, which could easily reach several hundreds. To reduce the computation cost in CNNs, depthwise separable convolution (DSC) was proposed and successfully applied [1].

DSC is a combination of 2D spatial convolution and 1×1 (or pointwise) convolution as shown in Fig. 1. In other words, while the conventional convolution is performed on all channels of input corresponding to the filter size, DSC decomposes the standard convolution into two layers: spatial convolution (or filtering stage) and pointwise convolution (or combining stage). These two convolutions are applied sequentially.

A spatial convolution applies a 2D filter to each input channel. When a single filter is applied to each channel, the amount of computation is significantly reduced compared to the standard convolution, but only the input channel corresponding to each filter is calculated. Thus, a pointwise convolution is used to combine the results of spatial convolution to create a new feature. In total with the spatial and the pointwise convolution layers, the computation cost is still very low compared to the standard convolution. For example, when a 3×3 filter

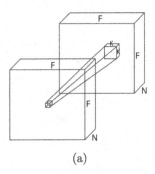 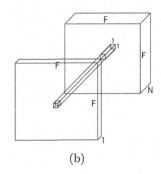

(a)   (b)

Fig. 1. Depthwise separable convolution is combination of (a) spatial convolution, applying a 2D kernel on each channel and (b) pointwise convolution, applying a 1D kernel on each pixel across all channels.

is applied to 32 input channels to obtain 64 output channels as in the MNIST classification model, the amount of computation with DSC decreases by around 8 times compared to the standard convolution algorithm.

When a powerful GPU is available and the model is quite small as in the MNIST classification, DSC is hard to achieve a lot of speedup compared to the standard convolution even with much less computations. It is because DSC has sequential runs of the spatial and the pointwise convolutions, while standard convolution can run one 3D convolution at the same time in parallel on a GPU.

3   Fast Depthwise Separable Convolution

While DSC reduces the number of computations and parameters, the pointwise convolution step requires most of the parameters and computations in the model. According to [9], the pointwise convolution occupies 95% of Multi-Adds and 75% of parameters.

To make the pointwise convolution step in DSC more efficient, we apply general matrix multiplication (GEMM) to the pointwise convolution, which leads to *FastDSC*. FastDSC can be processed as follows. When the spatial convolution is performed in a depthwise separable convolution, the output is 3D (width, height, and input channel). We reshape the result of the spatial convolution into a 2D matrix (width × height, and input channel). At this step, the cells corresponding to one of the patches constitute one row as shown in Fig. 2(a). The pointwise filters are also reconstructed as a single matrix (input channel, and output channel). Thus, in this matrix, each column is a pointwise filter that we originally intended to apply as shown in Fig. 2(b). Applying GEMM to the two previous matrices produces a 2D matrix (width × height, and output channel). As a final step, we reshape the result of the matrix multiplication to the original image shape (width, height, and output channel) as presented in Fig. 2(c). This step produces exactly the same result as the pointwise convolution, but with a more efficient implementation.

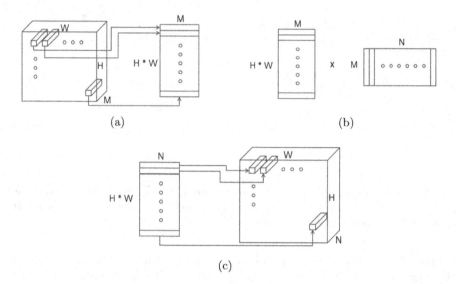

Fig. 2. The *FastDSC* process for the pointwise convolution. (a) Reshape the output of spatial convolution into a matrix, (b) multiply the reshaped image and filters, and (c) reshape the result of GEMM into the original image data shape.

4   Experiments

We evaluate three models (conventional convolution, DSC, and FastDSC) on four environments (CPU or GPU on Linux server or NVIDIA Jetson TX2). The Linux server has an Intel Xeon Processor E5-2620 and a GTX1080 GPU, and NVIDIA Jetson TX2 has a Dual Denver 2 + Quad ARM A57 and a 256-core Pascal GPU.

For image classification and object recognition, we use MNIST [14], and the VOC dataset [4], respectively. The MNIST dataset includes hand written digit images (60K train images and 10 K test images) of 10 classes (0–9 digits), and the VOC dataset includes 2007 and 2012 train set, and 2007 test set of 20 objects (16,551 train images and 4,952 test images).

4.1   Image Classification

For the MNIST classification, the baseline model consists of 2 standard convolution layers, a max pooling layer, and a fully connected layer, as presented in Table 1. In the table, each row includes the type of layer and the stride when the type is convolution. The filter shape is presented by (filter width × filter height × input channel) × output channel, or simply by (filter width × filter height) for max pooling. The input size to the layer is (width × height × input channel).

Given the baseline model, we applied DSC and FastDSC as shown in Figs. 1 and 2, and the result model architectures are presented in Tables 2 and 3, respectively. We tested the models in both the CPU and the GPU environments, where

Table 1. Baseline model architecture with standard convolution for the image classification on the MNIST dataset

Type/Stride	Filter shape	Input size
Convolution/1	$(3 \times 3 \times 1) \times 32$	$28 \times 28 \times 1$
Max pooling	2×2	$28 \times 28 \times 32$
Convolution/2	$(3 \times 3 \times 32) \times 64$	$14 \times 14 \times 32$
Fully connected	$(7 \times 7 \times 64) \times 10$	$7 \times 7 \times 64$

Table 2. DSC model architecture for the image classification on MNIST

Type/Stride	Filter shape	Input size
Convolution/1	$(3 \times 3 \times 1) \times 32$	$28 \times 28 \times 1$
Max pooling	2×2	$28 \times 28 \times 32$
Spatial conv/2	$(3 \times 3 \times 32) \times 32$	$14 \times 14 \times 32$
Pointwise conv/1	$(1 \times 1 \times 32) \times 64$	$7 \times 7 \times 32$
Fully connected	$(7 \times 7 \times 64) \times 10$	$7 \times 7 \times 64$

Table 3. FastDSC model architecture for the image classification on MNIST

Type/Stride	Filter shape	Input size
Convolution/1	$(3 \times 3 \times 1) \times 32$	$28 \times 28 \times 1$
Max pooling	2×2	$28 \times 28 \times 32$
Spatial conv/2	$(3 \times 3 \times 32) \times 32$	$14 \times 14 \times 32$
Reshape & GEMM	$(32, 64)$	$7 \times 7 \times 32$
Fully connected	$(7 \times 7 \times 64) \times 10$	$7 \times 7 \times 64$

CPU has less computation power than GPU for inference in neural networks. Tables 4 and 5 show the experiment results on Linux server and NVIDIA Jetson TX2, respectively.

The two modified models (i.e., DSC and FastDSC) are faster than the baseline model in terms of execution time on both CPU and GPU. Especially, with CPU on the TX2 board where the computing power is less than Linux server, DSC or FastDSC are much faster than the baseline model. The computational complexity and parameters for the convolution are also much smaller than those of the standard convolution model. The test speed of FastDSC is almost the same as DSC in both environments. The reason for this result is that the model itself is simply composed of just two convolution layers, and only one pointwise convolution layer is replaced by GEMM.

In terms of accuracy, there is no big difference in the three models because the MNIST classification is an easy task which can be solved with very simple models. Also, there is no accuracy difference between Linux and TX2, because

Table 4. Comparison of three models on a Linux server. The variances are presented in the parentheses. The test times are the total time in seconds, taken for the 10,000 test images on CPU and GPU, respectively.

Model	Accuracy(%)	CPU(sec)	GPU(sec)	Parameters
Baseline	98.77(0.01)	2.88 (0.09)	0.67 (0.001)	50186
DSC	98.76(0.01)	2.22 (0.02)	0.58 (0.001)	34122
FastDSC	98.78(0.01)	2.27 (0.005)	0.58 (0.6e−04)	34122

Table 5. Comparison of each model on the TX2 board.

Model	Accuracy(%)	CPU(sec)	GPU(sec)	Parameters
Baseline	98.77(0.01)	89.3 (0.057)	2.64 (0.001)	50186
DSC	98.76(0.01)	12.83 (0.003)	2.40 (0.003)	34122
FastDSC	98.78(0.01)	13.25 (0.006)	2.45 (0.002)	34122

for classification we use the same weights trained on the Linux server, since we are interested in test time to run the models with test data on different environments.

4.2   Object Detection

To see the difference of the models in a big task, we compare them with the YOLO network on the VOC data sets. We modified the structure of the original YOLO version2 network, by applying DSC and FastDSC to every standard convolution layer, which lead to YOLO-DSC and YOLO-FastDSC, respectively.

Table 6 shows the composition of the YOLO-DSC network architecture. Like DSC in MobileNet, a convolution layer includes a batch normalization and activation function except for the last convolution layer. A negative number in Route layers indicates the previous layer number that is object of concatenation. For example, Route(−1, −4) means that it concatenates the layers above one and 4 lines. YOLO-FastDSC replaces all the pointwise convolutions in YOLO-DSC with matrix multiplication (GEMM) as in Fig. 2.

Tables 7 and 8 show mAP (mean average precision) and test time of the three models (original YOLO version2, YOLO-DSC, and YOLO-FastDSC) for object detection on the VOC data set. We did not use pretrained weights in training, unlike the existing YOLO manual, because we wanted to see the performance difference of the models. Thus, we trained each model by using the Xavier weight initialization [6] from scratch. When the YOLO and YOLO-DSC models are compared, the mAP of the YOLO-DSC is lower than that of YOLO, but the execution time is much faster in YOLO-DSC as shown in Tables 7 and 8. In addition, the model file size of YOLO-DSC and YOLO-FastDSC is 28MB which is around just 14% of the original YOLO version2 model size, 194 MB.

Table 6. YOLO-DSC architecture. A DSC layer consists of a spatial convolution and a pointwise convolution.

Type	Filters	Size	Stride	Output
DSC	3/32	3 × 3/1 × 1	1	416 × 416 × 3
Max pooling		2 × 2	2	208 × 208 × 32
DSC	32/64	3 × 3/1 × 1	1	208 × 208 × 32
Max pooling		2 × 2	2	104 × 104 × 64
DSC	64/128	3 × 3/1 × 1	1	104 × 104 × 64
Convolution	64	1 × 1	1	104 × 104 × 64
DSC	64/128	3 × 3/1 × 1	1	104 × 104 × 64
Max pooling		2 × 2	2	52 × 52 × 128
DSC	128/256	3 × 3/1 × 1	1	52 × 52 × 128
Convolution	128	1 × 1	1	52 × 52 × 128
DSC	128/256	3 × 3/1 × 1	1	52 × 52 × 128
Max pooling		2 × 2	2	26 × 26 × 256
DSC	256/512	3 × 3/1 × 1	1	26 × 26 × 256
Convolution	256	1 × 1	1	26 × 26 × 256
DSC	256/512	3 × 3/1 × 1	1	26 × 26 × 256
Convolution	256	1 × 1	1	26 × 26 × 256
DSC	256/512	3 × 3/1 × 1	1	26 × 26 × 256
Max pooling		2 × 2	2	13 × 13 × 512
DSC	512/1024	3 × 3/1 × 1	1	13 × 13 × 512
Convolution	512	1 × 1	1	13 × 13 × 512
DSC	512/1024	3 × 3/1 × 1	1	13 × 13 × 512
Convolution	512	1 × 1	1	13 × 13 × 512
DSC	512/1024	3 × 3/1 × 1	1	13 × 13 × 1024
DSC	1024/1024	3 × 3/1 × 1	1	13 × 13 × 1024
DSC	1024/1024	3 × 3/1 × 1	1	13 × 13 × 1024
Route(−9)				26 × 26 × 512
Convolution	64	1 × 1	1	26 × 26 × 64
Reorg			2	13 × 13 × 256
Route(−1, −4)				13 × 13 × 1280
DSC	1280/1024	3 × 3/1 × 1	1	13 × 13 × 1024
Convolution	125	1 × 1	1	13 × 13 × 125

In the YOLO-DSC and YOLO-FastDSC models, basically they are the same model with different computation orders, the accuracies must be the same. At first, we applied matrix multiplication as the order of Fig. 2(b). To speed up in YOLO-FastDSC, we changed the order of matrix multiplication from the order

Table 7. Model comparison in performance and test time per image on Linux. The variances are presented in the parentheses.

Model	mAP	CPU(sec)	GPU(sec)
YOLO [16]	0.5901	0.9058 (0.0054)	0.0091 (0.0001)
YOLO-DSC	0.5115	0.337 (0.003)	0.0060 (0.00001)
YOLO-FastDSC	0.5115	0.25 (0.003)	0.0073 (0.0001)

Table 8. Model comparison in performance and test time per image on NVIDIA Tx2

Model	mAP	CPU(sec)	GPU(sec)
YOLO [16]	0.5901	132.58 (0.552)	0.103 (0.1e−03)
YOLO-DSC	0.5115	8.91 (0.05)	0.038 (0.1e−04)
YOLO-FastDSC	0.5115	8.69 (0.02)	0.039 (0.3e−04)

of (input × filter) to (filter × input). By using this trick, YOLO-FastDSC runs faster than YOLO-DSC on both Linux server and TX2 board with CPU where the number of computing units is very small. That is, YOLO-FastDSC can run faster than YOLO-DSC, when the computation power is limited. Note that the execution time on the GPU was measured from the beginning to the end of the work on all kernels.

5   Conclusion and Future Work

In this paper, we proposed FastDSC to pointwise convolution in DSC. After testing DSC and FastDSC on the MNIST classification, we applied them to the YOLO version2 model. The number of parameters and the computing time have decreased significantly in object detection. Also, the small size of model files is attractive in embedded systems with a limited memory space. Furthermore, FastDSC which replaces a pointwise convolution with a matrix multiplication provides speedup especially in harsh environment where powerful computing resources like GPU are not available. As future work, we can apply FastDSC based CNNs to embedded chips with a poor computing power in restricted systems.

Acknowledgement. This work was partly supported by Institute for Information & communications Technology Promotion(IITP) grant funded by the Korea government(MSIT) (No. 2018-0-00749,Development of virtual network management technology based on artificial intelligence) and the National Program for Excellence in Software funded by the Ministry of Science, ICT and Future Planning, Republic of Korea (2017-0-00130).

References

1. Chollet, F.: Xception: deep learning with depthwise separable convolutions. CoRR abs/1610.02357 (2016). http://arxiv.org/abs/1610.02357
2. Courbariaux, M., Bengio, Y., David, J.: Binaryconnect: training deep neural networks with binary weights during propagations. CoRR abs/1511.00363 (2015). http://arxiv.org/abs/1511.00363
3. Dahl, G.E., Sainath, T.N., Hinton, G.E.: Improving deep neural networks for LVCSR using rectified linear units and dropout (2013)
4. Everingham, M., Van Gool, L., Williams, C.K.I., Winn, J., Zisserman, A.: The pascal visual object classes (VOC) challenge. Int. J. Comput. Vis. **88**(2), 303–338 (2010)
5. Fukushima, K.: Neocognitron: a self-organizing neural network for a mechanism of pattern recognition unaffected by shift in position. Biol. Cybern. **36**(4), 193–202 (1980)
6. Glorot, X., Bengio, Y.: Understanding the difficulty of training deep feedforward neural networks. In: Proceedings of the International Conference on Artificial Intelligence and Statistics (AISTATS 2010). Society for Artificial Intelligence and Statistics (2010)
7. Hinton, G., et al.: Deep neural networks for acoustic modeling in speech recognition. IEEE Signal Process. Mag. **29**, 82–97 (2012)
8. Hinton, G., Vinyals, O., Dean, J.: Distilling the knowledge in a neural network. arXiv preprint arXiv:1503.02531v1 (2015)
9. Howard, A.G., et al,: Mobilenets: efficient convolutional neural networks for mobile vision applications. CoRR abs/1704.04861 (2017). http://arxiv.org/abs/1704.04861
10. Ioffe, S., Szegedy, C.: Batch normalization: accelerating deep network training by reducing internal covariate shift. CoRR abs/1502.03167 (2015). http://arxiv.org/abs/1502.03167
11. Kim, Y., Park, E., Yoo, S., Choi, T., Yang, L., Shin, D.: Compression of deep convolutional neural networks for fast and low power mobile applications. CoRR abs/1511.06530 (2015). http://arxiv.org/abs/1511.06530
12. Krizhevsky, A., Sutskever, I., Hinton, G.E.: ImageNet classification with deep convolutional neural networks. In: Advances in Neural Information Processing Systems, pp. 1097–1105 (2012)
13. LeCun, Y., Bengio, Y., Hinton, G.E.: Deep learning. Nature **521**(7553), 436–444 (2015)
14. LeCun, Y., Cortes, C.: MNIST handwritten digit database (2010). http://yann.lecun.com/exdb/mnist/
15. Nair, V., Hinton, G.E.: Rectified linear units improve restricted Boltzmann machines (2010)
16. Redmon, J., Farhadi, A.: YOLO9000: better, faster, stronger. CoRR abs/1612.08242 (2016). http://arxiv.org/abs/1612.08242
17. Romero, A., Ballas, N., Kahou, S.E., Chassang, A., Gatta, C., Bengio, Y.: Fitnets: hints for thin deep nets. CoRR abs/1412.6550 (2014). http://arxiv.org/abs/1412.6550
18. Schmidhuber, J.: Deep learning in neural networks: an overview. Neural Netw. **61**, 85–117 (2015)
19. Zhang, X., Zhou, X., Lin, M., Sun, J.: ShuffleNet: an extremely efficient convolutional neural network for mobile devices. CoRR abs/1707.01083 (2017). http://arxiv.org/abs/1707.01083

An Analog Circuit Design
for k-Winners-Take-All Operations

Xiaoyang Liu[1($\boxtimes$)] and Jun Wang[2,3($\boxtimes$)]

[1] School of Automation, Huazhong University of Science and Technology,
Wuhan, China
xyliu6@hust.edu.cn
[2] Department of Computer Science, City University of Hong Kong,
Kowloon Tong, Hong Kong
[3] Shenzhen Research Institute, City University of Hong Kong, Shenzhen, China
jwang.cs@cityu.edu.hk

Abstract. This paper designs an analog circuit for k-winners-take-all (kWTA) operations. The circuit is stable and finite-time convergent. The stable state of the circuit is equivalent to the optimal solution of the kWTA. Simulation results via SPICE substantiate the efficiency of the design.

Keywords: k-winners-take-all · Analog circuit design
Recurrent neural network

1 Introduction

Winner-take-all (WTA) problem is to select the largest element from a collection of inputs. k-winners-take-all (kWTA) problem is an extension of WTA problem. It is to select k largest elements from n inputs ($1 \leq k \leq n$) [11]. WTA and kWTA operations have many applications in various fields, such as associative memories [5,14], sorting [6,13], image processing [2], etc.

There are many WTA and kWTA models [1,4,8,9,18,21,22]. A simple kWTA model with only one state variable is proposed in [4]. A fast-converge kWTA model with a hard-limiting step activation function is proposed in [8]. The kWTA model proposed in [21] adopts both single state variable and hard-limiting step activation function.

Various WTA and kWTA circuits are proposed [3,7,12,13,15–17,19,20]. They have various precision, accuracy and speed. This paper presents an analog kWTA circuit based on the kWTA model in [21] with theoretically guaranteed

Jun Wang—This work was supported in part by the Research Grants Council of the Hong Kong Special Administrative Region of China, under Grants 14207614 and 11208517, and in part by the National Natural Science Foundation of China under grant 61673330.

© Springer Nature Switzerland AG 2018
L. Cheng et al. (Eds.): ICONIP 2018, LNCS 11307, pp. 666–675, 2018.
https://doi.org/10.1007/978-3-030-04239-4_60

global stability, high precision and fast convergence speed. Compared with existing kWTA circuits, the proposed circuit is theoretically guaranteed stable and the operation time can be easily controlled. The rest of the paper is organized as follows. In Sect. 2, preliminaries of kWTA problem, the kWTA model used in this paper, and some basic operation circuits are introduced. Section 3 describes and analyzes the presented kWTA circuit. Section 4 illustrates the simulation results. Finally, the conclusion is given in Sect. 5.

2   Preliminaries

2.1   The k-Winners-Take-All Model

The kWTA problem can be formulated as the function [10]

$$x_i = f(u_i) = \begin{cases} 1, & \text{if } u_i \in \{k \text{ largest elements of } \boldsymbol{u}\} \\ 0, & \text{otherwise} \end{cases} \tag{1}$$

where $x_i \in \boldsymbol{x} = (x_1, x_2, \ldots, x_n)^{\mathrm{T}}$ is the output vector, and $u_i \in \boldsymbol{u} = (u_1, u_2, \ldots, u_n)^{\mathrm{T}}$ is the input vector.

It is shown in [21] that the kWTA problem can be formulated as the following model assuming that the solution is unique

$$\epsilon \frac{dy}{dt} = \sum_{i=1}^{n} x_i - k, \tag{2}$$

$$x_i = g_\infty(u_i - y), \quad i = 1, 2, \ldots, n; \tag{3}$$

where ϵ is a time constant, $y \in \mathbb{R}$ is the state variable, and $g_\infty(\cdot)$ is the Heaviside step activation function which is defined as

$$g_\infty(\rho) = \begin{cases} 0, & \rho < 0 \\ 1, & \rho \geq 0. \end{cases} \tag{4}$$

This model has one neuron and $2n$ connections, where n is the input number. It has advantages of being global stable, finite-time convergent, and having high-resolution. $\boldsymbol{x}$ corresponding to the equilibrium of the state equation (2) is a solution to the kWTA problem (1).

2.2   Basic Circuit Elements

Analog Adder. An analog adder circuit composed of four resistors R_1–R_4 and one operational amplifier (OA) A is show in Fig. 1. v_1–v_3 are input voltages, and v_{out} is the output voltage. v_+ and v_- are positive and negative input voltages of the operational amplifier, respectively.

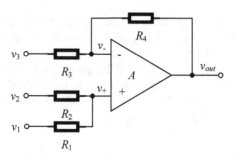

Fig. 1. An analog adder

According to Kirchhoff's current law (KCL) and properties of OA

$$\frac{v_3 - v_-}{R_3} = \frac{v_- - v_{out}}{R_4},$$

$$v_{out} = (1 + \frac{R_4}{R_3})v_- - \frac{R_4 v_3}{R_3}, \tag{5}$$

and

$$\frac{v_1 - v_+}{R_1} = \frac{v_+ - v_2}{R_2},$$

$$v_+ = \frac{R_2 v_1 + R_1 v_2}{R_1 + R_2}. \tag{6}$$

According to properties of OA

$$v_+ = v_-. \tag{7}$$

Combining (5), (6), and (7), the output voltage is expressed as

$$v_{out} = \frac{(R_3 + R_4)(R_2 v_1 + R_1 v_2)}{R_3(R_1 + R_2)} - \frac{R_4 v_3}{R_3}. \tag{8}$$

If $R_1 = R_2$ and $R_3 = R_4$, then

$$v_{out} = v_1 + v_2 - v_3. \tag{9}$$

The output of the circumstance of any number of inputs can be computed accordingly.

Analog Integrator. An analog integrator circuit composed of a capacitor, resistors, and one OA is shown in Fig. 2.
According to properties of OA

$$v_+ = v_- = 0, \tag{10}$$

$$i_c = \frac{v_{in} - v_-}{R} = \frac{v_{in}}{R}. \tag{11}$$

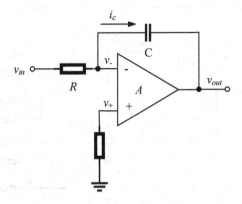

Fig. 2. An analog integrator

Assume that the initial voltage of the capacitor C is 0, then the voltage of C is

$$v_c = \frac{1}{C} \int i_c \mathrm{d}t$$

$$= \frac{1}{RC} \int v_{in} \mathrm{d}t. \tag{12}$$

Therefore the output voltage of the integrator is

$$v_{out} = -v_c = -\frac{1}{RC} \int v_{in} \mathrm{d}t. \tag{13}$$

3 Circuit Design

The circuit is designed with available electronic devices, so it can be easily implemented with a practical circuit. The devices models will be detailed in Sect. 4.

3.1 Design

The presented kWTA circuit is shown in Fig. 3. It is composed of one capacitor, resistors, transistors, and $n+2$ operational amplifiers. $\boldsymbol{u} = (u_1, u_2, \ldots u_n)$ is the input voltage vector, $\boldsymbol{x} = (x_1, x_2, \ldots x_n)$ is the output voltage vector, k is an input voltage whose voltage amplitude represents the number of winners under the dimension of volt, and y is the state variable. There are totally n neurons and $2n$ connections.

From the circuit diagram, when $R_1 = R_2 = \cdots = R_{n+1}$ and $R'_1 = R'_2 = \cdots = R'_n = R_a$, following equations can be got

$$v_x = k - (x_1 + x_2 + \ldots + x_n), \tag{14}$$

$$y = -\frac{1}{RC} \int v_x \mathrm{d}t. \tag{15}$$

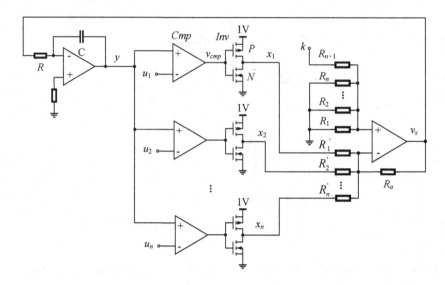

Fig. 3. The kWTA circuit

Substituting (14) into (15)

$$y = \frac{1}{RC} \int (\sum_{i=1}^{n} x_i - k) dt. \tag{16}$$

That is

$$RC\frac{dy}{dt} = \sum_{i=1}^{n} x_i - k, \tag{17}$$

which is in accordance with (2) with $RC = \epsilon$. Because the stable state of x_i ($i = 1, 2, \ldots, n$) is $0\,\mathrm{V}$ or $1\,\mathrm{V}$, so the number of winners is equal to the voltage sum of the these winners under the dimension of volt. When the circuit reaches steady state, (17) is equal to 0 which means that $\sum_{i=1}^{n} x_i = k$, and therefore k represents the winner number.

A comparator Cmp and an inverter Inv execute the comparison operation and the Heaviside step activation function together. Take the first input u_1 as an example: When $u_1 > y$, $v_{cmp} = V_{--}$, otherwise $v_{cmp} = V_{++}$, where $V_{++} > 0$ and $V_{--} < 0$ are the high and low power voltages of the operational amplifier in the comparator, respectively. When $v_{cmp} = V_{--}$, $v_{cmp} < 0\,\mathrm{V}$ and $1 - v_{cmp}$ is bigger that the opening threshold voltage v_{th} of P, so the P-channel transistor P is open and $x_1 = 1\,\mathrm{V}$, otherwise the N-channel transistor N is open and $x_1 = 0\,\mathrm{V}$. That is

$$x_i = g_\infty(u_i - y) = \begin{cases} 0, & u_i < y \\ 1, & u_i > y, \end{cases} \quad i = 1, 2, \ldots, n. \tag{18}$$

Equation (18) is in accordance with (3). Considering the previous conclusion that (17) is equal to (2), the presented circuit can execute kWTA operations.

3.2 Specifications Analysis

Range of Inputs. Denote power voltages of operational amplifiers in the integrator, comparators, and the adder by V_{I++} and V_{I--}, V_{c++} and V_{c--}, and V_{a++} and V_{a--}, where $V_{x++} > 0$ and $V_{x--} < 0$, the letter x stands for I, c, or a. It is known that input voltages of an operational amplifier should not exceed power voltages, therefor any u_i and y must be smaller than V_{c++} and larger than V_{c--}. On the other hand, y is produced by an integrator, that is, $|y| < V_{I++}$ and $|y| < |V_{I--}|$, therefore $|y| < V_{c++}$ and $|y| < |V_{c--}|$ are always satisfied if $V_{I++} >= V_{c++}$ and $V_{I--} <= V_{c--}$. Another condition should be satisfied is that $k < V_{a++}$ and $\sum_{i=1}^{n} x_i \le n < V_{a++}$. Therefor, assume that operation speeds of every components are coordinated, the circuit is feasible if $V_{c++} > \max(\boldsymbol{u})$ and $V_{c--} < \min(\boldsymbol{u})$, $V_{a++} > k$, and $V_{a++} > n$.

The Initial Value of the State Variable. The initial value y_0 of the state variable y effects the convergence time, [21] gives initialization approaches of y under different distribution of inputs. In the presented circuit, y_0 is represented by the reverse initial voltage of the capacitor C, and it is set to 0.

Co-operation Speeds of Electronic Devices. The operation speeds of every components of the circuits should keep pace with each other. The integration speed of the integrator, which plays an essential role in the operation speed of the kWTA circuit, is controlled by the value of RC. By adjusting the value of R and C, the operation speed of the circuit is easily controlled. If RC is very small, high-speed operational amplifiers are needed to constitute comparators and the adder.

4 Simulation Results

High-speed OAs ML339 and LM324 are selected for comparators and the adder, and the integrator, respectively, in SPICE simulations.

4.1 Constant Inputs

Consider ten inputs with voltages increase from 0.01 V to 0.1 V with interval 0.01 V, and $k = 5$. Figure 4(a) shows inputs, Fig. 4(b) shows outputs of the designed kWTA circuit, and Fig. 4(c) shows the curve of the state variable y. Stable x_i's with 1 V are winners, and those with 0 V are losers. At first $y = 0$, so all x_i's are initialized with 1 V. After about 1.4×10^{-5} s, x_1 to x_5 which are corresponding to five smallest inputs u_1 to u_5 go down to 0 V, remaining x_6 to

x_{10} being 1 V, and all x_i's are stable after that. Simulation results show that the circuit successfully determines five largest inputs which are close to each other in amplitude in about 1.4×10^{-5} s, proving that the circuit is high-speed and high-precision. y is stable at about 0.0592 V, which is between 0.05 V and 0.06 V, conforming that $\bar{y}$ is a threshold.

4.2   Time-Vary Inputs

Consider an input vector of sinusoidal signals $u_p(t) = sin[40\pi(5t + (p-1)\pi)](p = 1, 2, \ldots, 10)$ and $k = 5$. Figure 5 plots simulation results. Figure 5(a) shows input signals, and for better view, ten output curves are depicted in five subfigures Fig. 5(b)–(f). There are always five x_i's with 1 V, which represent five winners, corresponding to five largest u_i's at that time. The circuit successfully selects five winners from time-vary inputs instantly.

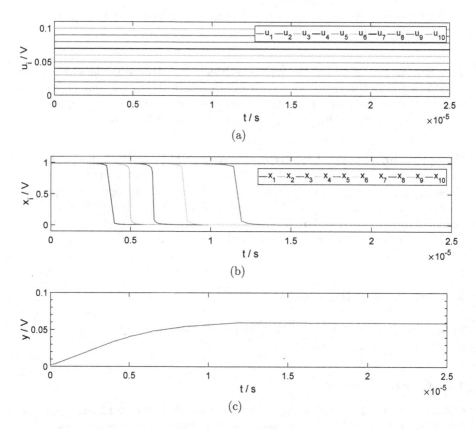

Fig. 4. Simulation results for ten constant inputs with $k = 5$. (a) Inputs. (b) Outputs of the kWTA circuit. (c) State variable.

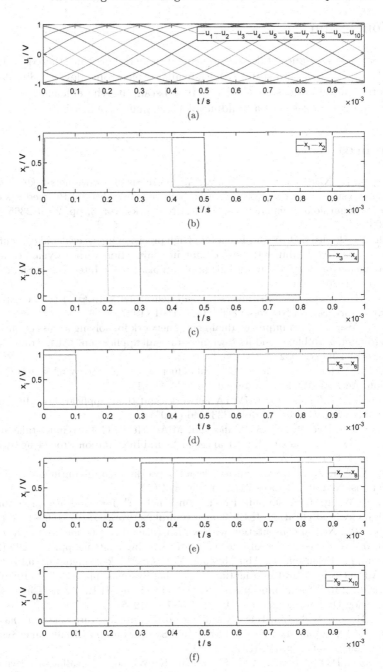

Fig. 5. Simulation results for ten sinusoidal inputs with $k = 5$. (a) Inputs. (b)–(e) Outputs of the kWTA circuit.

5 Conclusion

In this paper, a k-winners-take-all circuit is designed. It is constructed with high speed, high precision, and stability with available electronic devices. Simulations on constant inputs and sinusoidal inputs substantiate the efficiency of it. The kWTA circuit may serve as a building block in many applications.

References

1. Ferreira, L., Kaszkurewicz, E., Bhaya, A.: Synthesis of a k-winners-take-all neural network using linear programming with bounded variables. In: Proceedings of the International Joint Conference on Neural Networks, vol. 3, pp. 2360–2365. IEEE (2003)
2. Fish, A., Akselrod, D., Yadid-Pecht, O.: High precision image centroid computation via an adaptive k-winner-take-all circuit in conjunction with a dynamic element matching algorithm for star tracking applications. Analog Integr. Circ. Sig. Process **39**(3), 251–266 (2004)
3. Hsu, T.C., Wang, S.D.: k-winners-take-all neural net with $\Theta(1)$ time complexity. IEEE Trans. Neural Networks **8**(6), 1557–1561 (1997)
4. Hu, X., Wang, J.: An improved dual neural network for solving a class of quadratic programming problems and its k-winners-take-all application. IEEE Trans. Neural Networks **19**(12), 2022–2031 (2008)
5. Krogh, A., Hertz, J., Palmer, R.G.: Introduction to the Theory of Neural Computation. Addison-Wesley, Reedwood City CA (1991)
6. Kwon, T.M., Zervakis, M.: KWTA networks and their applications. Multidimension. Syst. Signal Process. **6**(4), 333–346 (1995)
7. Lazzaro, J., Ryckebusch, S., Mahowald, M.A., Mead, C.A.: Winner-take-all networks of O(n) complexity. In: Advances in Neural Information Processing Systems, pp. 703–711 (1989)
8. Liu, Q., Wang, J.: Two k-winners-take-all networks with discontinuous activation functions. Neural Networks **21**(2–3), 406–413 (2008)
9. Liu, S., Wang, J.: A simplified dual neural network for quadratic programming with its kwta application. IEEE Trans. Neural Networks **17**(6), 1500–1510 (2006)
10. Maass, W.: Neural computation with winner-take-all as the only nonlinear operation. In: Advances in Neural Information Processing Systems, pp. 293–299 (2000)
11. Majani, E., Erlanson, R., Abu-Mostafa, Y.S.: On the k-winners-take-all network. In: Advances in Neural Information Processing Systems, pp. 634–642 (1989)
12. Opris, I.E.: Rail-to-rail multiple-input min/max circuit. IEEE Trans. Circuits Syst. II: Analog Digital Sig. Process. **45**(1), 137–140 (1998)
13. Ou, S.H., Lin, C.S., Liu, B.D.: A scalable sorting architecture based on maskable WTA/MAX circuit. In: IEEE International Symposium on Circuits and Systems, ISCAS 2002, vol. 4, p. IV. IEEE (2002)
14. Pouliquen, P.O., Andreou, A.G., Strohbehn, K.: Winner-takes-all associative memory: a hamming distance vector quantizer. Analog Integr. Circ. Sig. Process **13**(1–2), 211–222 (1997)
15. Sekerkiran, B., Cilingiroglu, U.: A CMOS k-winners-take-all circuit with O(N) complexity. IEEE Trans. Circuits Syst. II: Analog Digital Sig. Process. **46**(1), 1–5 (1999)

16. Serrano-Gotarredona, T., Linares-Barranco, B., Andreou, A.G.: A high-precision current-mode WTA-MAX circuit with multi-chip capability. In: Adaptive Resonance Theory Microchips, pp. 89–115. Springer, Boston (1998). https://doi.org/10.1007/978-1-4419-8710-5_4

17. Sum, J.P.F., Leung, C.S., Tam, P.K.S., Young, G.H., Kan, W.K., Chan, L.W.: Analysis for a class of winner-take-all model. IEEE Trans. Neural Networks 10(1), 64–71 (1999)

18. Sum, J.P.-F., Leung, C.-S., Ho, K.: Analysis on Wang's kWTA with stochastic output nodes. In: Lu, B.-L., Zhang, L., Kwok, J. (eds.) ICONIP 2011. LNCS, vol. 7064, pp. 268–275. Springer, Heidelberg (2011). https://doi.org/10.1007/978-3-642-24965-5_29

19. Urahama, K., Nagao, T.: K-winners-take-all circuit with O(N) complexity. IEEE Trans. Neural Networks 6(3), 776–778 (1995)

20. Wang, J.: Analogue winner-take-all neural networks for determining maximum and minimum signals. Int. J. Electron. 77(3), 355–367 (1994)

21. Wang, J.: Analysis and design of a k-winners-take-all model with a single state variable and the heaviside step activation function. IEEE Trans. Neural Networks 21(9), 1496–1506 (2010)

22. Wolfe, W.J., Mathis, D., Anderson, C., Rothman, J., Gottler, M., Brady, G., Walker, R., Duane, G., Alaghband, G.: K-winner networks. IEEE Trans. Neural Networks 2(2), 310–315 (1991)

NVM Weight Variation Impact on Analog Spiking Neural Network Chip

Akiyo Nomura[1]([✉]), Megumi Ito[1], Atsuya Okazaki[1], Masatoshi Ishii[1],
Sangbum Kim[2], Junka Okazawa[1], Kohji Hosokawa[1], and Wilfried Haensch[2]

[1] IBM Research - Tokyo, Tokyo, Japan
{ainomura,megumii,a2ya,ishiim,khosoka,junka}@jp.ibm.com
[2] IBM Research, Yorktown Heights, NY, USA
{SangBum.Kim,whaensch}@us.ibm.com

Abstract. In extremely energy-efficient neuromorphic computing using analog non-volatile memory (NVM) devices, device variability arises due to process variation and electro/thermo-dynamics of NVM devices, such as phase change memory and resistive-RAM. Thus, for realizing NVM-based neuromorphic computing, it is important to quantitatively analyze the impact of synaptic device variability on neural network training accuracy and assess requirements for NVM devices. We investigated the analysis using simulations focusing on a spiking neural network (SNN)-based restricted Boltzmann machine (RBM). MNIST dataset simulation results revealed that more than 500 steps of conductance achieve comparable performance to the previous study of software-based simulation on SNN-based RBM. We also observed that at least a less than 10% of variation in conductance update for each synaptic device is required for achieving comparable performance to the result with no variation. These results provide baselines for designing and optimizing the characteristics of NVM devices.

Keywords: Spiking neural network · Neuromorphic computing
Restricted Boltzmann machine · Spike-timing-dependent plasticity
Phase change memory · Resistive-RAM

1 Introduction

Neuromorphic computing using non-volatile memory (NVM) has been attracting great interests as a way of achieving very low power consumption. NVM arrays are used in order to represent the matrices of synaptic weights in artificial neural networks (ANNs) and spiking neural networks (SNNs), where the conductance of an NVM device corresponds to the synaptic weight. This makes it possible not only to realize power-efficient multiply-accumulate operation performed by NVM arrays but also to overcome the von-Neumann bottleneck thanks to the computation using NVM arrays as synaptic devices [1]. In particular, an NVM-array-based SNN chip has the potential to play an important role in applications

© Springer Nature Switzerland AG 2018
L. Cheng et al. (Eds.): ICONIP 2018, LNCS 11307, pp. 676–685, 2018.
https://doi.org/10.1007/978-3-030-04239-4_61

for mobile and Internet of Things (IoT) devices [2,3] because SNN can process sparse data at the edge with ultra-low power consumption [4,5].

NVM devices are expected to be utilized in neuromorphic computing, but there are challenges to use a NVM device as a synaptic device. When analog synaptic weights are desired, the major challenges are constraints due to the characteristics of the NVM device, such as limitations in the number of steps in the conductance range and device variability. Device variability is categorized into two types: variation for each synaptic device (device-to-device variation) and variation with synaptic weight modulation (cycle-to-cycle variation). Cycle-to-cycle variation arises from physical mechanisms; synaptic weight modulation for NVM devices, such as phase change memory (PCM) or resistive-RAM (ReRAM), is accompanied by atomic-level change based on electro/thermo-dynamics [6–9]. Some studies have reported the impact of the characteristics of NVM devices, such as NVM variation on ANNs [10] and the impact of ReRAM variation on the restricted Boltzmann machine (RBM) [11].

In this paper, we report on the impact of the characteristics of NVM devices on SNN-based RBM using two models: (1) synaptic weight is represented by conductance of one NVM device (bipolar model), and (2) synaptic weight is represented by conductance of two NVM devices (G_p and G_m model). Our simulator covered the NVM characteristics, such as the limitation in the number of steps of weight, variation of the maximum and minimum of weight, and device-to-device variation and cycle-to-cycle variation of weight with update events. MNIST dataset simulation results revealed that more than 500 steps of conductance achieves comparable performance to the previous study of software-based simulation on SNN-based RBM for both the bipolar model and G_p and G_m model. We also observed that at least a less than 10% device-to-device variation is required. Our quantitative results can be guidelines for improving the properties of NVM devices and systems using NVM devices.

2   Simulation Method

2.1   Restricted Boltzmann Machine for Spiking Neural Network Chip

We implemented an event-based C++ simulator for a restricted Boltzmann machine (RBM) using a contrastive divergence (CD) algorithm with a spike-timing-dependent plasticity (STDP) update rule based on Neftci's model [12]. The update value Δw_{ij} is represented by the difference between the data phase and model phase as described below [12]

$$\Delta w_{ij} \propto \langle v_i h_j \rangle_\mathrm{data} - \langle v_i h_j \rangle_\mathrm{model} \tag{1}$$

where v_i and h_j are the activities of visible neurons and hidden neurons, respectively. Synaptic weight increases in the data phase, and the weight decreases in the model phase until learning converges; $\Delta w_{ij} \propto 0$. Figure 1(a) shows a modified STDP learning window for facilitating hardware implementation.

The update rule was expressed using positive weight update in the data phase and negative weight update in the model phase. All the results were evaluated by the network as shown in Fig. 1(b). In the input, we added 40 label neurons, assigning 4 neurons for each label. For quantitative evaluation, the MNIST hand-written dataset was used, which consists of 28×28-pixel handwritten images [13]. Each black and white image was converted into the input spikes on the basis of a Poisson distribution. We use a 60,000-image MNIST training database for learning and 10,000 images for inference.

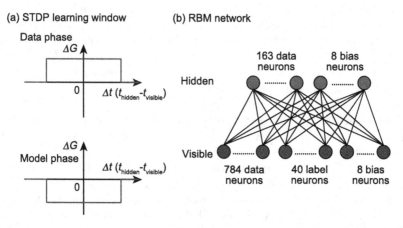

Fig. 1. (a) Schematic illustration of STDP learning window. Horizontal axis Δt shows the delay from spike time at the hidden side (t_{hidden}) to visible side (t_{visible}). Vertical axis shows the update value of weight. STDP update rule was expressed using positive weight update in the data phase and negative weight update in the model phase. (b) Schematic illustration of RBM network. Network had 784 visible neurons, 163 hidden neurons, 40 label neurons, and 8 bias neurons on visible and hidden layers.

2.2   NVM Characteristic Model

We investigated the impact of the characteristics of NVM devices on the MNIST benchmark by using the bipolar model and G_{p} and G_{m} model. Comparing two models indicates the difference in the impact of variability between using one NVM device and two NVM devices. Figure 2 shows the weight update for the bipolar model (a) and G_{p} and G_{m} model (b). For the bipolar model, ΔG represents the weight update value for each STDP update event, which corresponds to normalized conductance of an NVM device; the conductance change value is divided by the total conductance. For the G_{p} and G_{m} model, the weight update values for each STDP update event for G_{p} and G_{m} are represented as ΔG_{p} and ΔG_{m}, respectively. $\Delta G_{\text{p}} - \Delta G_{\text{m}}$ corresponds to the total weight update value for each update event. $G_{\text{p}} - G_{\text{m}}$ corresponds to the total weight in the model. The characteristics covered in this paper are the limitation in the number of steps of weight and device variability. The impact of the number of steps of weight was

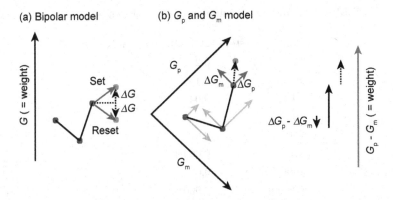

Fig. 2. (a) Schematic illustration of weight update in bipolar model. ΔG represents weight update value for each update. (b) Schematic illustration of weight update in the G_p and G_m model. ΔG_p and ΔG_m represent weight update value for each update for G_p and G_m, respectively. $G_p - G_m$ is the total weight using two devices. $\Delta G_p - \Delta G_m$ is the total weight update value for each update using two devices.

investigated by changing ΔG while fixing the weight range $G_{max} - G_{min}$. This simulator implemented three variation models: (1) variation of the maximum of weight G_{max} and minimum of weight G_{min}, and (2) device-to-device variation and (3) cycle-to-cycle variation of ΔG using a normal distribution. By changing the standard deviation σ, we evaluated the quantitative impacts.

3   Simulation Results

3.1   Number of Steps of Weight

We investigated the impact of the number of steps of weight on simulated error by changing ΔG while fixing the weight range $G_{max} - G_{min}$ as 2 for the bipolar model since $\Delta G/(G_{max} - G_{min})$ corresponds to the number of steps of weight. For the G_p and G_m model, $\Delta G_p(= -\Delta G_m)$ were changed while fixing the weight range just like the bipolar model. We investigated the difference between using one NVM device and using two NVM devices, where each NVM device has the same characteristic for both models. Table 1 shows the parameters used for evaluating the impact of the number of steps of weight. Figure 3(a) shows the dependence of the number of steps on the average error at 15–20 epochs for the bipolar model. At 500 steps, there is a clear cliff in the error rate. This also applies to the G_p and G_m model as shown in Fig. 3(b). These results show that fewer than 500 steps cannot improve MNIST score. However, the results with more than 500 steps achieved comparable performance to the previous study of software-based simulation on SNN-based RBM [12].

Table 1. Parameters for number of steps of weight variation.

Number of steps	$\Delta G,\ \Delta G_{\mathrm{p}}(=-\Delta G_{\mathrm{m}})$
10	0.2
20	0.1
50	0.04
100	0.02
200	0.01
500	0.004
1000	0.002
10,000	0.0002

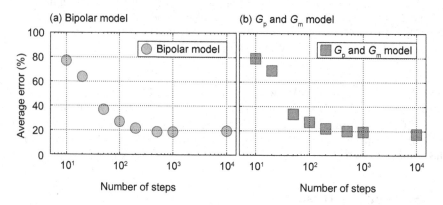

Fig. 3. (a) Dependence of number of steps on average error for bipolar model. Average error was calculated from the average of the simulated error values at the 15–20 epochs. (b) Dependence of number of steps on average error for G_{p} and G_{m} model. Average error was calculated from the average of the simulated error values at the 15–20 epochs the same as for the bipolar model.

3.2   Weight Variation of NVM

This section covers the device variability of an NVM device, specifically, the variation of the maximum weight G_{max} and minimum weight G_{min} and the device-to-device variation and cycle-to-cycle variation of ΔG using the variation model mentioned in Sect. 2.2. The number of steps of weight was fixed to 1000 steps using the parameters in Table 1.

First, we evaluated the variation of G_{max} and G_{min}. The value of the variation for both the G_{max} and G_{min} of each device was set using a normal distribution $3\sigma = \frac{G_{\mathrm{max}}-G_{\mathrm{min}}}{2}$ so that 3σ did not exceed the weight range. For the bipolar model, the change ratio from the baseline (error rate with $\sigma = 0$) due to the variation of the maximum weight G_{max} and minimum weight G_{min} was 2.5%. For the G_{p} and G_{m} model, the change ratio due to the variation was 2.1%.

Even with the worst case of the variation ($3\sigma = \frac{G_{\max} - G_{\min}}{2}$), the change ratio in both models was less than 5%. These results show that it has a high tolerance to the variation of $G_{\max}$ and $G_{\min}$ for both the bipolar and G_{p} and G_{m} models since the average number of steps of each synapse was 1000, which achieves comparable performance to the previous study of software-based simulation on SNN-based RBM.

Second, Fig. 4(a) shows the results of the cycle-to-cycle variation of ΔG for the bipolar model. The results are for the training epoch dependence on simulated error when $\sigma/\mu = 0, 1, 3, 5$, and 7, where μ is the average of ΔG in a uniform distribution. Figure 4(b) shows the results of the cycle-to-cycle variation of ΔG_{p} and ΔG_{m} for the G_{p} and G_{m} model. The error rate changed by 10% or less even when σ was 7 times larger than the μ of ΔG. These results show acceptable performances comparable to the results with $\sigma = 0$ in both the bipolar and G_{p} and G_{m} models. Cycle-to-cycle variation is inevitable due to physical phenomena when using NVM devices, such as PCM and ReRAM, due to the electro/thermo-dynamics. These findings show that this variation does not have a serious impact on MNIST score with 1000 steps in the weight range (conductance range).

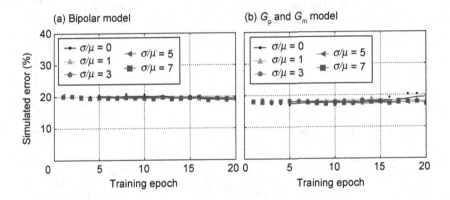

Fig. 4. Training epoch dependence on simulated error for the (a) bipolar model and (b) G_{p} and G_{m} model. Each line shows the moving average of past 5 epochs. In (a), black line shows the result when $\sigma = 0$ (no variation for ΔG). Solid red lines show the results with $\sigma/\mu = 0, 1, 3, 5$, and 7. In (b), black line shows the result when $\sigma = 0$ (no variation for ΔG_{p} and ΔG_{m}). Solid blue lines show the results with $\sigma/\mu = 0, 1, 3, 5$, and 7. (Color figure online)

Finally, Fig. 5 shows the results of the device-to-device variation of ΔG for the (a) bipolar model and (b) G_{p} and G_{m} model. These figures represent that even $\sigma/\mu = 0.2$ leads to more than 30% reduction of accuracy compared to the result with $\sigma = 0$ (no variation). From the results in Fig. 5(a–b), we calculated the change ratio from the baseline ($\sigma = 0$). Figure 5(c) shows the impact of the device-to-device variation on change ratio by changing σ/μ, where μ is the average of ΔG in a normal distribution. The average error rate was calculated using the simulated error values at 15–20 epochs. These results were obtained when $\sigma/\mu = 0, 0.05, 0.1, 0.2$, and 0.3. The change ratio for the G_p and G_m model is better than that for the bipolar model. One reason for this is the additivity of the normal distribution by using two NVM devices in the model. For the G_p and G_m model, using two NVM devices, where $\Delta G_p = (-\Delta G_m) = 0.002$ and σ of both ΔG_p and ΔG_m is 6×10^{-4}, is equivalent to using one NVM device, where $\Delta G = 0.004$ and σ of ΔG is $\sqrt{2} \times 6 \times 10^{-4}$. Therefore, σ/μ in the G_p

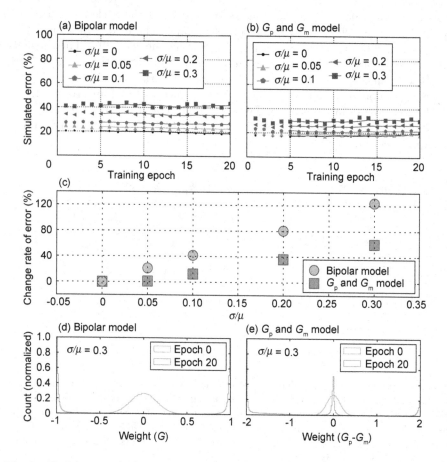

Fig. 5. Training epoch dependence on simulated error for the (a) bipolar model and (b) G_p and G_m model. Each line shows the moving average of past 5 epochs. (c) σ/μ dependence on change ratio from the error rate with $\sigma = 0$. The average error rate was calculated from the data at the 15–20 epochs for bipolar model (red circles) and G_p and G_m model (blue squares). σ and μ are dispersion and average of device-to-device variation of ΔG in a normal distribution. A histogram of weight G at epoch 0 and epoch 20, where $\sigma/\mu = 0.3$ for the (d) bipolar model and (e) G_p and G_m model. (Color figure online)

and G_m model is less than that in the bipolar model. Figure 5(d) and (e) show a histogram of weight G for the bipolar model and weight $G_p - G_m$ for the G_p and G_m model, respectively. The deterioration in error rate stems from the weights of some synapses that were stacked in 0, G_{max}, and G_{min} for both models. To investigate this further, we looked into the impact of device-to-device variation of the set and reset operation on error rate for the bipolar model. Figure 6(a) shows the device-to-device variation of ΔG, where the set and reset operation have different variations. In contrast, Fig. 6(b) shows device-to-device variation of ΔG, where the set and reset operation have the same variation. Surprisingly, in the case in Fig. 6(b), device-to-device variation has much less influence on the error rate than in the case in Fig. 6(a). These results indicate that device-to-device variation of set and reset operation leads to stacked weight in 0, G_{max}, and G_{min}. The results of device-to-device variation show that careful designing is required to use NVM devices that have variations.

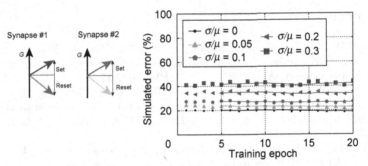

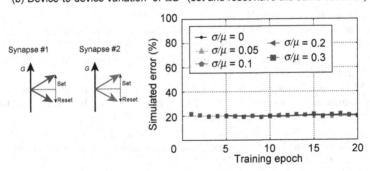

Fig. 6. (a) Device-to-device variation of ΔG dependence on error rate in the bipolar model. ΔG of each synapse has device-to-device variation, and the set and reset have different variations. Each line shows the moving average of past 5 epochs. (b) Device-to-device variation of ΔG dependence on error rate in the bipolar model. ΔG of each synapse has device-to-device variation, but the set and reset of each synapse have the same variation. Each line shows the moving average of past 5 epochs.

4   Conclusion

We investigated the impact of NVM characteristics on an SNN-based RBM simulator. NVM characteristics covered in this paper are the limitation in the number of steps of weight, variation of G_{max} and G_{min}, and device-to-device variation and cycle-to-cycle variation of ΔG using the bipolar model and G_{p} and G_{m} model. Our results revealed that more than 500 steps of conductance achieve comparable performance to software-based simulation on SNN-based RBM. The results for variation of G_{max} and G_{min} and cycle-to-cycle variation of ΔG show that this system has high tolerance to these variations. We also observed that at least less than 10% device-to-device variation of ΔG is required since device-to-device variation of the set and reset operation leads to deterioration in error rate. The change ratio from the baseline ($\sigma = 0$) was better for the G_{p} and G_{m} model than for the bipolar model thanks to additivity of the normal distribution of variability. Our quantitative results can be guidelines for improving the properties of an NVM device and system using an NVM device.

References

1. Burr, G.W., Shelby, R.M., Sebastian, A., Kim, S., Kim, S., Sidler, S., Virwani, K., Ishii, M., Narayanan, P., Fumarola, A., et al.: Neuromorphic computing using non-volatile memory. Adv. Phys.: X **2**(1), 89–124 (2017)
2. Kim, S., et al.: NVM neuromorphic core with 64k-cell (256-by-256) phase change memory synaptic array with on-chip neuron circuits for continuous in-situ learning. In: 2015 IEEE International Electron Devices Meeting, pp. 17.1.1–17.1.4. IEEE (2015)
3. Ito, M., et al.: Lightweight Refresh Method for PCM-based Neuromorphic Circuits. In: 2018 IEEE 18th International Conference on Nanotechnology (in press)
4. Merolla, P.A., Arthur, J.V., Alvarez-Icaza, R., Cassidy, A.S., Sawada, J., Akopyan, F., Jackson, B.L., Imam, N., Guo, C., Nakamura, Y., et al.: A million spiking-neuron integrated circuit with a scalable communication network and interface. Science **345**(6197), 668–673 (2014)
5. Davies, M., et al.: Loihi: a neuromorphic manycore processor with on-chip learning. IEEE Micro **38**(1), 82–99 (2018)
6. Suri, M., et al.: Phase change memory as synapse for ultra-dense neuromorphic systems: application to complex visual pattern extraction. In: 2011 IEEE International Electron Devices Meeting, pp. 4.4.1–4.4.4. IEEE (2011)
7. Kuzum, D., Jeyasingh, R.G.D., Lee, B., Wong, H.S.P.: Nanoelectronic programmable synapses based on phase change materials for brain-inspired computing. Nano Lett. **12**(5), 2179–2186 (2012)
8. Burr, G.W., et al.: Experimental demonstration and tolerancing of a large-scale neural network (165 000 synapses) using phase-change memory as the synaptic weight element. IEEE Trans. Electron Devices **62**(11), 3498–3507 (2015)
9. Ielmini, D.: Modeling the universal set/reset characteristics of bipolar RRAM by field- and temperature-driven filament growth. IEEE Trans. Electron Devices **58**(12), 4309–4317 (2011)
10. Gokmen, T., Vlasov, Y.: Acceleration of deep neural network training with resistive cross-point devices: design considerations. Front. Neurosci. **10**, 333 (2016)

11. Eryilmaz, S.B., Kuzum, D., Yu, S., Wong, H.S.P.: Device and system level design considerations for analog-non-volatile-memory based neuromorphic architectures. In: 2015 IEEE International Electron Devices Meeting, pp. 4.1.1–4.1.4. IEEE (2015)

12. Neftci, E., Das, S., Pedroni, B., Kreutz-Delgado, K., Cauwenberghs, G.: Event-driven contrastive divergence for spiking neuromorphic systems. Front. Neurosci. **7**, 272 (2014)

13. LeCun, Y., Bottou, L., Bengio, Y., Haffner, P.: Gradient-based learning applied to document recognition. Proc. IEEE **86**(11), 2278–2324 (1998)

Author Index

Agarwal, Saurabh 209
Al-Mashaikki, Talal 478
Andersen, Morten Østergaard 610
Aoki, Shunsuke 647
Aryal, Sunil 598
Aziz, Fayeem 74

Bali, Hamza 478
Bauer-Wersing, Ute 143, 279
Betts, John M. 510
Biddulph, Alexander 120

Cai, Haibin 623
Cai, Ruichu 465
Cai, Weidong 444
Cao, Mingna 349
Cao, Yang 97
Chalup, Stephan K. 74, 120
Chan, C. K. 246
Chen, Bo 420
Chen, Caiping 189
Chen, Mingjian 358
Chen, Sheng 235, 499
Chen, Xiaowei 433
Chen, Yiko 246
Cheng, Jie-Zhi 420
Cheng, Lin 168
Cheng, Long 26
Cheng, Qingrong 62
Chetty, Madhu 598
Choi, Heeyoul 656
Choi, Yongjun 656
Chung, I-Fang 369

Dai, Yiming 349
Dheenadayalan, Kumar 222
Ding, Bo 40
Ding, Xinghao 410
Du, Baolin 410
Duan, Zhiping 62

Fan, Yibo 52
Foale, Cameron 598

Franzius, Mathias 143
Fu, Yazhen 433

Gao, Xiaoyang 269
Gaurav, Ashish Kumar 209
Gedeon, Tom 554, 567
Geng, Mingyang 40
Gorjup, Gal 610
Gu, Xiaodong 62
Guan, Zhuoqun 26
Guo, Shin-Ning 369

Ha, Mingming 258
Haensch, Wilfried 676
Hao, Zhifeng 465
Haris, Muhammad 143
Hasler, Stephan 279
He, Gang 52
Hori, Toshimasa 637
Hosokawa, Kohji 676
Hossain, Md Sazzad 510
Hou, Zeng-Guang 235, 499
Hou, Zengguang 316
Houliston, Trent 120
Huang, Jianping 26
Huang, Jiyue 398
Huang, Jun 420
Huang, Kai 26
Huang, Man-Ling 369
Huang, Tingwen 189
Huang, Wanrong 303
Huang, Yuanyuan 489
Huang, Yue 410
Huang, Zhixing 530

Ikushima, Keisuke 637
Ishii, Masatoshi 676
Ito, Megumi 676

Jian, Zhiyong 26
Jin, Yulan 387
Juang, Chia-Feng 369
Jung, Seul 339

Kameda, Tomoya   637
Kang, Yu   97
Kasabov, Nikola   358
Kim, Sangbum   676
Kimura, Mutsumi   637
Klein, Britt   598
Koyama, Seitaro   647
Kreger, Jennifer   279

Lei, Kai   398
Li, Changlong   420
Li, Chuandong   189
Li, Conghui   455
Li, Fu   108
Li, Guangbin   327
Li, Haojie   3
Li, Hongfei   189
Li, Jide   542
Li, Ruoxiang   108
Li, Tieshan   269
Li, Xiaoqiang   542
Li, Yueheng   15
Liang, Xu   316
Liao, Xiaofeng   132
Liao, Zehua   180
Lin, Cheng-Jian   369
Lin, Hsueh-Yi   369
Lin, Yusen   530
Lin, Zhipeng   303
Liu, Derong   15, 258
Liu, Dongnan   444
Liu, Honghai   623
liu, Linlin   26
Liu, Xiaoyang   666
Liu, Yuqi   3
Liu, Zhiqin   420
Lo, Ying-Chih   369
Lu, Changsheng   358
Lu, Chengyu   530
Luo, Biao   15
Luo, Lincong   235, 499
Luo, Yanhong   327

Manoonpong, Poramate   610
Mansouri, Deloula   378
Matsuda, Tokiyoshi   637
Mendes, Alexandre   120
Mesbah, Mostefa   478
Mu, Nankun   132
Muralidhara, V. N.   222

Nakashima, Yasuhiko   637
Nambu, Isao   522, 577
Nirala, Mehul Kumar   209
Nomura, Akiyo   676

O'Donnell, Lauren J.   444
Okazaki, Atsuya   676
Okazawa, Junka   676
Onodera, Yuki   577
Ouyang, Deqiang   189

Paplinski, Andrew P.   510
Peng, Liang   235, 316, 499
Peng, Zhouhua   3
Plested, Jo   554, 567

Qi, Qi   410
Qiao, Tong   291

Ren, Shixin   316

Saito, Toshimichi   647
Sang, Chen   387
Santhanagopalan, Meena   598
Semoto, Tomoki   522
Shan, Qihe   269
Shen, Ying   398
Shi, Bo   349
Shi, Dianxi   108
Si, Shangchun   398
Sinha, Sayan   209
Song, Yang   444
Srinivasaraghavan, Gopalakrishnan   222
Stoyanov, Stoyan Petrov   610
Sui, Xin   168

Tageldin, Muna   478
Tanaka, Ryo   637
Tang, Yuhua   303
Tong, Weiqin   542
Tu, Enmei   358

Umeda, Kenta   637

Vrabič, Rok   610

Wada, Yasuhiro   522, 577
Wan, Tao   387
Wang, Aijuan   132

Wang, Chao   420
Wang, Chen   235, 499
Wang, Dan   3
Wang, Ding   258
Wang, Haotian   303
Wang, Huaimin   40
Wang, Jiasen   155
Wang, Jiaxing   316
Wang, Jun   155, 666
Wang, Lu   542
Wang, Peijun   86
Wang, Qingfeng   420
Wang, Weiqun   235, 316, 499
Wang, Xuefeng   97
Wang, Zhiyong   623
Wei, Qinglai   180, 258
Welsh, James S.   74
Wen, Guanghui   86
Wen, Mei-Chin   369
Wong, Aaron S. W.   74
Wong, K. Y. Michael   246
Wu, Pin   542
Wu, Qingfeng   433
Wu, Ting   291
Wu, Wenjie   3
Wu, Yiming   291

Xu, Jian   291
Xu, Jiu   52
Xu, Ming   291
Xu, Shusong   387
Xu, Xianyun   168
Xu, Yancai   258
Xu, Zenglin   585
Xu, Zihao   316
Xue, Shan   15
Xue, Yushan   542

Yan, Haofeng   291
Yan, Min   246
Yang, Dongsheng   327
Yang, Jie   358

Yang, Shaowu   108
Yang, Wenjing   303
Yang, Yilong   433
Yang, Yongqing   168
Yang, Zhuo   52
Yang, Zhuoyue   108
Yao, Yue   554, 567
Yoo, Byeongheon   656
Yu, Liang   52
Yu, Wenwu   86
Yu, Wenxin   52
Yu, Xinghuo   86
Yuan, Shuai   585
Yuan, Xiaohui   378

Zhang, Donghao   444
Zhang, Fan   444
Zhang, Jianfu   489
Zhang, Lei   40
Zhang, Li   349
Zhang, Liqing   489
Zhang, Qichao   200
Zhang, Sai   349
Zhang, Yongjun   108
Zhang, Zhiqiang   52
Zhang, Zhizhuo   585
Zhao, Bo   258
Zhao, Dongbin   200
Zhao, Haohua   489
Zhao, Yuming   455
Zhen, Qiqi   465
Zheng, Hao   358
Zhong, Jinghui   530
Zhou, Shangchen   585
Zhou, Xing   40
Zhou, Xuehai   420
Zhou, Ying   420
Zhu, Rongrong   97
Zhu, Yanmei   349
Zhu, Zhaocheng   455
Zhuang, Hang   420
Zhuang, Mingyong   410

Printed in the United States
By Bookmasters